Cytochromes P450

Role in the Metabolism and Toxicity of Drugs and other Xenobiotics

Issues in Toxicology

Series Editors

Professor Diana Anderson, *University of Bradford, UK*
Dr Michael D Waters, *Integrated Laboratory Systems, Inc, N. Carolina, USA*
Dr Timothy C Marrs, *Edentox Associates, Kent, UK*

This Series is devoted to coverage of modern toxicology and assessment of risk and is responding to the resurgence in interest in these areas of scientific investigation.

Ideal as a reference and guide to investigations in the biomedical, biochemical and pharmaceutical sciences at the graduate and post graduate level.

Titles in the Series:

Cytochromes P450: Role in the Metabolism and Toxicity of Drugs and other Xenobiotics
Edited by Costas Ioannides, *University of Surrey, Guildford, UK*

Hair in Toxicology: An Important Bio-Monitor
Edited by Desmond John Tobin, *University of Bradford, Bradford, UK*

Male-mediated Developmental Toxicity
Edited by Diana Anderson and Martin H Brinkworth, *University of Bradford, Bradford, UK*

Visit our website at www.rsc.org/issuesintoxicology

For further information please contact:
Sales and Customer Care, Royal Society of Chemistry, Thomas Graham House, Science Park, Milton Road, Cambridge, CB4 0WF, UK
Telephone +44 (0)1223 432360, Fax: +44 (0)1223 426017, Email: sales@rsc.org

Cytochromes P450

Role in the Metabolism and Toxicity of Drugs and other Xenobiotics

Edited by

Costas Ioannides
Molecular Toxicology Group, School of Biomedical and Molecular Sciences, University of Surrey, Guildford, UK

RSC Publishing

ISBN: 978-0-85404-274-6

A catalogue record for this book is available from the British Library

Published by The Royal Society of Chemistry,
Thomas Graham House, Science Park, Milton Road,
Cambridge CB4 0WF, UK

Registered Charity Number 207890

For further information see our web site at www.rsc.org

Preface

From uniqueness to multiplicity

As an introduction to the fifteen chapters of this new survey on the latest developments in the field of cytochrome P450, a short glimpse on the course of events during the last fifty years may be appropriate.

During half a century, cytochrome P450 in its original *uniqueness* as an optically "wrong" cytochrome has attracted many investigators, who have contributed to the unveiling of a bewildering *multiplicity* of biologically important functions of the by now very large superfamily of cytochrome P450 enzymes. The various enzymes and the specificities of the multitude of the iron porphyrins are reflected in this book.

As a principal biological function of cytochrome P450 is the oxidative metabolism of drugs and other xenobiotic substances, a backward look into the "Detoxication Mechanisms" (2nd edition in 1959) by R. T. Williams (often adoringly nicknamed as the 'Beilstein-Williams') is worthwhile. Page 4 contains the following statement: 'The mechanism of these 'oxidation' reactions, however, is unknown except in a few cases'. Thereafter, however, with its discovery in 1958 and with the advent of more refined spectroscopic methodologies, through the double wavelength spectrophotometry, the mysterious enzyme system began to reveal its secrets in a swift stream of investigative successes. They could superficially and very incompletely be summarised in the following terms:

- Binding of substrates as photometrically characterised by different binding spectra of either type I or type II.
- Binding of oxygen as well as of NADPH and the creation of the oxidative reaction cycle.
- The induction of the enzyme protein by phenobarbital or 3-methylcholanthrene (cytochrome P448).
- The description of oxidative metabolism pathways began after the discovery of cytochrome P450, e.g. the O-demethylation of model substrates

Issues in Toxicology
Cytochromes P450: Role in the Metabolism and Toxicity of Drugs and other Xenobiotics
Edited by Costas Ioannides

such as nitroanisole, which by optically measuring the formation of 'yellow' nitrophenol allowed for the characterisation of the enzyme kinetics even in turbid microsomal suspensions.[1] Ring hydroxylation and the structural preconditions for specific points of attack on polycyclic aromatic hydrocarbons (bay region), N-oxidation *etc.* were most important further subjects of investigational interest.

- Attempts at the elucidation of the reaction mechanism by the use of inhibitors e.g. carbon monoxide (the action spectrum), SKF 525-A and metyrapone revealed more of the nature of cytochrome P450. Inhibition kinetics suggested inhibitor binding at a location different from the active centre itself. This has led to the concept of physical changes in the protein configuration during the reaction cycle by invoking a'hinge' mechanism as modelled in a 'mouse trap'.[2]
- Understanding the function of the haem iron was achieved by determining its spin states.
- Finally, after numerous, long and difficult attempts, the purification of cytochrome P450 was achieved, and opened the way to the amino acid sequence and further realisation of its diversity. Observation of species differences pointed to genetic and evolutionary variations in drug metabolism. The role of cytochrome P450 in the oxidative steroid metabolism became essential in explaining sex differences.

In order to gain the ultimate insight into a natural phenomenon, the two generally possible pathways for a scientific approach have been used (sometimes certainly inadvertently) also by the investigators of cytochrome P450 functions, namely a hypothesis-driven experimentation as well as the random explorative method. Both processes are intimately intertwined, and have proven to be successful. The shrewd and extensive utilisation of the molecular biology methodology very rapidly led to a vast body of enzymes calling for a classification of the plethora of different cytochromes P450, the superfamily, into families and subfamilies.[3] This is aptly exemplified by the ten chapters in Part B of this book, dealing with ten subfamilies and two families of cytochrome P450.

A short internet search has revealed a very intense as well as an almost uniform distribution of interest into the cytochrome families discussed here: the subfamilies of CYP1 and CYP2 have shown hits between 45 000 and 14 000 times, with the exception of CYP1B and CYP2F, which showed only 2 000 and 800 hits, respectively. The CYP3 and CYP4 families generated about 28 000 hits. The interpretation of these figures will be left to the readers of this monograph.

Part A of this monograph introduces the reader to the current knowledge of the evolutionary development of cytochrome P450 structure and function. Furthermore, it deals with the role of this enzyme in the formation of reactive

[1]K. J. Netter, *Naunyn-Schmiedeberg's Arch exp Pathol Pharmakol.*, 1960, **238,** 292.
[2]K. J. Netter, G.F. Kahl, M.P. Magnussen, *Naunyn-Schmiedebergs Arch Pharmacol.*, 1969, **265**, 205.
[3]D. W. Nebert, M. Adesnik, M.J. Coon, R.W. Estabrook, F.J.Gonzalez, F.P. Guengerich, I.C. Gunsalus, E.F. Johnson, B. Kemper, W. Levin, I.R. Phillips, R. Sato, and M.R. Waterman, *DNA*, 1987, **6**, 1.

intermediates. Figuratively speaking, Part C offers an insight into another aspect of cytochrome P450 research, namely its regulation through receptor-mediated stimuli – as opposed to enzyme induction or inhibition. Equally important are the recently recognised variations in the pharmacokinetics of drugs emanating from the ingestion of phytochemicals, which can act as inducers and as inhibitors of the drug metabolising enzymes. In an attempt to 'translate' the current data on this one of several drug metabolising systems into clinical application, the closing chapter highlights the role of cytochromes P450 in the treatment of neoplastic growth.

This most recent member of the large 'family' of previous monographs devoted to cytochromes P450 and drug metabolism, indirectly could provide a basis for a more rational approach in the discovery and development of new drugs and therapeutic procedures. The knowledge on this major drug metabolising system assembled in this book may possibly be of help in better understanding why certain drugs in humans do not fulfill the expectations resulting from successful preclinical experiments. Recently this has led to costly disappointments in establishing new effective tumour therapies in humans, because obviously human and experimental mammalian organisms in themselves are far more complex and complicated interacting systems than e.g. cultured cells *etc.* A promising drug candidate acted on the ErbB epidermal growth factor receptor on the outside cell wall, which then is believed to be internalised in an inactive state and, therefore, to downregulate the replication pathway; but obviously frequently occurring mutant receptors had a low internalisation rate, and hence the drug could not be effective. This experience has refined drug design and development by introducing another biological variant into the already multifaceted interplay between organisms and drugs, which requires novel approaches to correctly predict drug actions from preclincical testing regimes. Consequently, these have to be adapted to ever more diversified biological conditions. Understanding and coping with the multitude of interrelations has been provisionally labeled as 'Systems Biology'. Whether it will survive as a fact-filled new era in drug discovery and regulation remains to be seen. But 'Systems Biology' begins to become a new concept-or just a *terminus technicus*-describing a more subtle testing or precalculating of desired effects by taking into account as many biological features as possible.[4]

It is to be hoped that the combined knowledge and expertise of the authors of this monograph might be able to contribute to an eventual success of Systems Biology in developing new biologically – and therapeutically-active xenobiotics.

K. J. Netter
Department of Pharmacology & Toxicology
University of Marburg
Marburg
Germany

[4]B. Borell, The Scientist, August 2007, **21**, No. *8*, 37.

Contents

Part A

Issues in Toxicology
Cytochromes P450: Role in the Metabolism and Toxicity of Drugs and other Xenobiotics
Edited by Costas Ioannides

Part A

CHAPTER 1

Cytochrome P450 Structure and Function: An Evolutionary Perspective

DAVID F.V. LEWIS[a] AND YUKO ITO[b]

[a] School of Biomedical and Molecular Sciences, University of Surrey, Guildford, Surrey GU2 7XH, UK; [b] Department of Bioscience and Bioinformatics, Kyusku Institute of Technology, 680-4 Kawazu, Iizuka-City, Fukuoka, 820-8502, Japan

Table of Contents

Issues in Toxicology
Cytochromes P450: Role in the Metabolism and Toxicity of Drugs and other Xenobiotics
Edited by Costas Ioannides

1.1 Introduction

The cytochromes P450 constitute a superfamily of haem-thiolate enzymes which are ubiquitous in nature.[1,2] Figure 1.1 shows the many fields in which P450s play important roles, thus highlighting their relevance to several branches of biological science. During the course of evolution, the P450 structure developed to bind the following entities: oxygen, carbon-based substrates, a haem group and redox partners, such as an iron-sulphur redoxin, an NADPH-dependent FAD- and FMN-containing flavoprotein reductase and cytochrome b_5.[1–6] In eukaryotic P450 systems, a membrane phospholipid bilayer such as that present in the smooth endoplasmic reticulum is also able to bind, as summarised in Tables 1.1 and 1.2. Mitochondrial P450s, such as CYP11 in the adrenal cortex, retain some of the prokaryotic P450 characteristics in possessing an iron-sulphur redoxin (specifically adrenodoxin) as a redox partner rather than utilising an NADPH-dependent flavoprotein reductase.[7–9] This finding suggests that the mitochondria and, indeed, other cell organelles may have had bacterial origins, as has been reported previously from RNA comparisons.[10,11] Over 5000 P450 sequences have been reported (Osamu Gotoh, personal communication) and it appears that these enzymes are present in all five biological kingdoms. Consequently, P450s will have undergone modifications during the course of evolution in order to adapt to the changes in environmental and cellular conditions.

P450 structures are known from X-ray crystallographic determinations[12–32] and a list of these is provided in Table 1.3; surprisingly, there are relatively few differences across species ranging from bacteria to mammalia. Such differences include orientations of the B′ and F helices, together with the extent of polypeptide in the loop regions between certain helical motifs, such as those of the

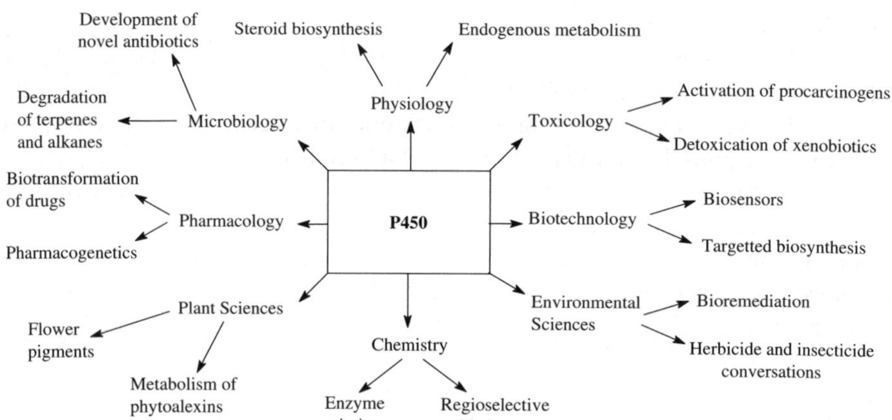

Figure 1.1 The fields of P450 research showing the various inter-relationships between them. A number of the above areas are inter-related and the scheme is not intended to be comprehensive but to provide a flavour of the breadth of the P450 field.

Table 1.1 Classification of P450-containing systems from various sources.[9,42]

Components in electron transfer pathway	Type of system	Environment
NADH → FAD (ferredoxin reductase) → FeS (ferrodoxin) → P450 (haemoprotein)	Bacterial	Cytosolic
NADH → FAD (adrenodoxin reductase) → FeS (adrenodoxin) → P450 (haemoprotein)	Mitrochondrial	Membrane-bound
NADPH → FAD → FMN (flavoprotein reductase) → P450 (haemoprotein)	Microsomal	Membrane-bound
NADPH → FAD → FMN (flavoprotein reductase) → P450 (haemoprotein)	*Bacillus megaterium*-3	Cytosolic

Notes: 1. In the *Bacillus megaterium*-3 system, the flavoprotein and haemoprotein domains are fused into a single polypeptide. 2. The FAD and FMN cofactors are generally bound into a single flavoprotein although exceptions exist where they can form separate proteins. 3. Systems involving fused redoxin and FMN domains are also known to exist, in addition to those where the redoxin and haemoprotein domains are fused. 4. CYP55 is an example of a P450 where there are in fact no redox partners involved in the catalytic system, which effects the reduction of nitric oxide to nitrous oxide.

Table 1.2 Regions of the P450 structure associated with binding various system components.[1]

Component	*Regions of the enzyme structure involved in binding component*
Haem	I and L helices, two conserved basic residues and conserved tryptophan
Dioxygen	I helix 'kink' close to conserved distal threonine
Redox Partner	Conserved basic residues close to the proximal haem face
Substrate	Substrate recognition site (SRS) regions involved B′, F and I helices together with $\beta_1(4)$ and $\beta_4(2)$ turn regions
Phospholipid	N-terminal peptide comprising the first 20–40 residues

Notes: 1. The invariant cysteine which ligates the haem iron is also involved with haem binding and oxygen activation. 2. Conserved acidic residue preceding the conserved distal threonine is involved in the catalytic cycle charge relay system. 3. A conserved tryptophan residue may also be associated with redox partner and electron transfer to the haem, together with a conserved phenylalanine residue proximal to the haem face. 4. A number of ion-pairs in the active site may be associated with proton transfer to the haem-bound oxygen, and a conserved acidic residue in the L helix could effect water ingress to the active site. 5. A number of interhelical loop regions may also be involved in the binding of P450 to the phospholipid membrane in the microsomal system.

F-G loop and the H-I loop. Also, there is an additional N-terminal peptide of some 40 residues which serves as a membrane anchor in eukaryotic P450s, and this is absent in prokaryotic sequences. Furthermore, it is not only in the primary sequence but also in the spatial orientation of secondary structural elements with respect to the haem moiety that P450s can vary. This may explain why homology models do not always correctly reproduce the active site topographies encountered in the actual crystal structures, although a relatively high homology (ie. over 40%) should provide a fairly good match between homology models and crystal structures. Indeed, for enzymes of the CYP2C family, it is found that models of CYP2C8 and CYP2C9, based on the CYP2C5 template, do indeed match closely with the actual crystal structures,[33,34] and the relationship between degree of fit and sequence homology is explored further in a following section.

The tertiary structure of P450 possesses certain well-conserved ion-pairs (such as the ExxR motif) and other features (like the polyproline motif near the N-terminus) which determine its overall folding pattern, together with various hydrogen bonded, π-π stacking and hydrophobic contacts between amino acid residues, some of which are associated with haem and substrate binding. However, certain conserved basic residues on the surface of the enzyme tend to be utilised for the binding of redox partners. These charged surface residues are largely conserved across the superfamily for redox partner binding, and this can be illustrated by the example of putidaredoxin (an iron-sulphur redoxin) binding to $P450_{cam}$,[35] and also that of adrenodoxin binding to its reductase[36] where it is apparent that the redoxin fits closely within a depression on the reductase surface. The structure of P450 resembles a triangular prism in overall shape with the haem moiety located approximately at the centre of one triangular face. The haem group lies in a depression, which is ideally suited, in terms of shape and complementary surface residues, for the binding of redox partners

Table 1.3 Unique cytochrome P450 crystal structures (from Protein Data Bank, 2006).

CYP	*Species*	*Resolution (Å)*	*Substrate*	*Protein data-bank code*
101A1	*Pseudomonas putida*	1.62	camphor[a]	2cpp
102A1	*Bacillus megaterium*	1.65	N-palmitoylglycine	1jpz
107A1	*Saccharopolyspora eyrthraea*	2.10	6-deoxyery-thronolide B[a]	1oxa
108A1	*Pseudomonas sp.*	2.30	none present	1cpt
119A1	*Sulfolobus solfataricus*	1.50	none present[a]	1io7
119A2	*Sulfolobus tokodaii*	3.00	none present	1ue8
121A1	*Mycobacterium tuberculosis*	1.06	none present[a]	1n4o
152A1	*Bacillus subtilis*	2.10	myristic acid	1izo
154A1	*Streptomyces coelicolor*	1.85	none present	1odo
154C1	*Streptomyces coelicolor*	1.92	none present	1gwi
158A2	*Streptomyces coelicolor*	1.50	none present	1slf
165A3	*Amycolatopsis orientalis*	1.70	none present	1lfk
167A1	*Sorangium cellulosum*	1.93	epithilone B	1q5d
175A1	*Thermus thermophilus*	1.80	none present	1n97
51A1	*Mycobacterum tuberculosis*	1.55	estriol[a]	1x8v
55A1	*Fusarium oxysporum*	1.00	nitric oxide	1jfb
2A6	human	1.90	coumarin[a]	1z10
2B4	rabbit	1.60	none present[a]	1po5
2C5	rabbit	2.10	diclofenac[a]	1nr6
2C8	human	2.70	none present	1pq2
2C9	human	2.00	flurbiprofen	1r9o
2D6	human	3.00	none present	2f9q
3A4	human	2.05	none present[a]	1tqn

[a]Another crystal structure has been reported for this enzyme which contains a bound inhibitor.

such as redoxins, cytochrome b_5 and reductase.[35] However, there are certain subtle distinctions between the entire binding regions of these various redox partners and the corresponding P450 enzymes[8] which may explain why certain redox partners tend to bind at particular surface locations on the P450 enzymes.

1.2 Evolutionary Aspects

It is generally thought that life emerged around 3.5 billion years ago, which is about a billion years after the Earth's formation (~4.55 billion years ago) although it is possible that the development of life may have occurred even earlier.[37] P450s are ubiquitous within living systems and the enzymes have been found in all biological kingdoms, but some Archaea (early prokaryotes) such as *Escherichia coli*, do not appear to contain any P450 enzymes whatsoever. However, two P450s have been isolated and crystallised from the thermophilic bacterial species *Sulfolobus solfataricus* and *Thermus thermophilus*.[35] These organisms are apparently able to exist in the hot sulphurous conditions encountered in the thermal vents of undersea volcanic fissures, which are thought to represent the likely environment for early life formation around 3.5 billion years ago. Interestingly, the presence of iron and sulphur in such oceanic vents may have assisted in the generation of proteins requiring such elements including, for example, cytochrome P450 and its iron-sulphur ferredoxin redox partner.[2] Figure 1.2 represents an abbreviated phylogenetic tree for certain P450s, and shows how this accords with the general development of terrestrial biota.

As the oxygen levels in the atmosphere began to increase around 2.5 billion years ago, protective systems would have developed to ensure species survival in a more aerobic environment, and eurakyotes are thought to have emerged about 2.1 billion years ago.[38–40] This would have afforded some protection from the deleterious effects of free oxygen, and it is also possible that an early role of P450 at this time may have been in the detoxification of O_2 itself.[41] In microbial species, P450s are involved in the biosynthesis of antibiotics, certain toxins and for the generation of secondary metabolites.[42] The full characterisation of such activities is currently being investigated, particularly with respect to the development of novel therapeutic agents *via* protein engineering of bacterial P450 enzymes, such as those from various *Streptomyces* species.[42]

With the development of metazoa,[43] the endogenous roles of P450 changed to steroid biosynthesis, together with that of fatty acid and prostanoid/eicosanoid metabolism.[44] When animal species started to colonise land areas in the Devonian period[45] about 400 million years ago, plants developed toxins to deter animal predators, and many of these toxic compounds are known to be synthesised in part *via* P450-mediated pathways.[44] However, it is thought that animal species began to develop new P450 enzymes specifically for the detoxification of these harmful plant products, and a phylogenetic analysis of P450s across the superfamily appears to support this viewpoint.[46–52] Insect species also developed P450s with detoxifying roles. For example, the black swallowtail butterfly, *Papilio polyxenes*, possesses a specific P450 (CYP6B1) for metabolising the plant toxin methoxsalen (xanthotoxin), which itself may have been biosynthesised *via* the mediation of P450 enzymes.[53] Furthermore, another insect-plant co-evolutionary role of P450 emerged when flowering plants appeared about 125 million years ago, as various P450s are known to be responsible for the biosynthesis of flower pigments such as the anthocyanins.[54] It is thought, therefore, that the exogenous roles of P450s may have developed over a geological timescale *via* a co-evolutionary process, commonly termed plant-animal 'warfare' where animals developed certain

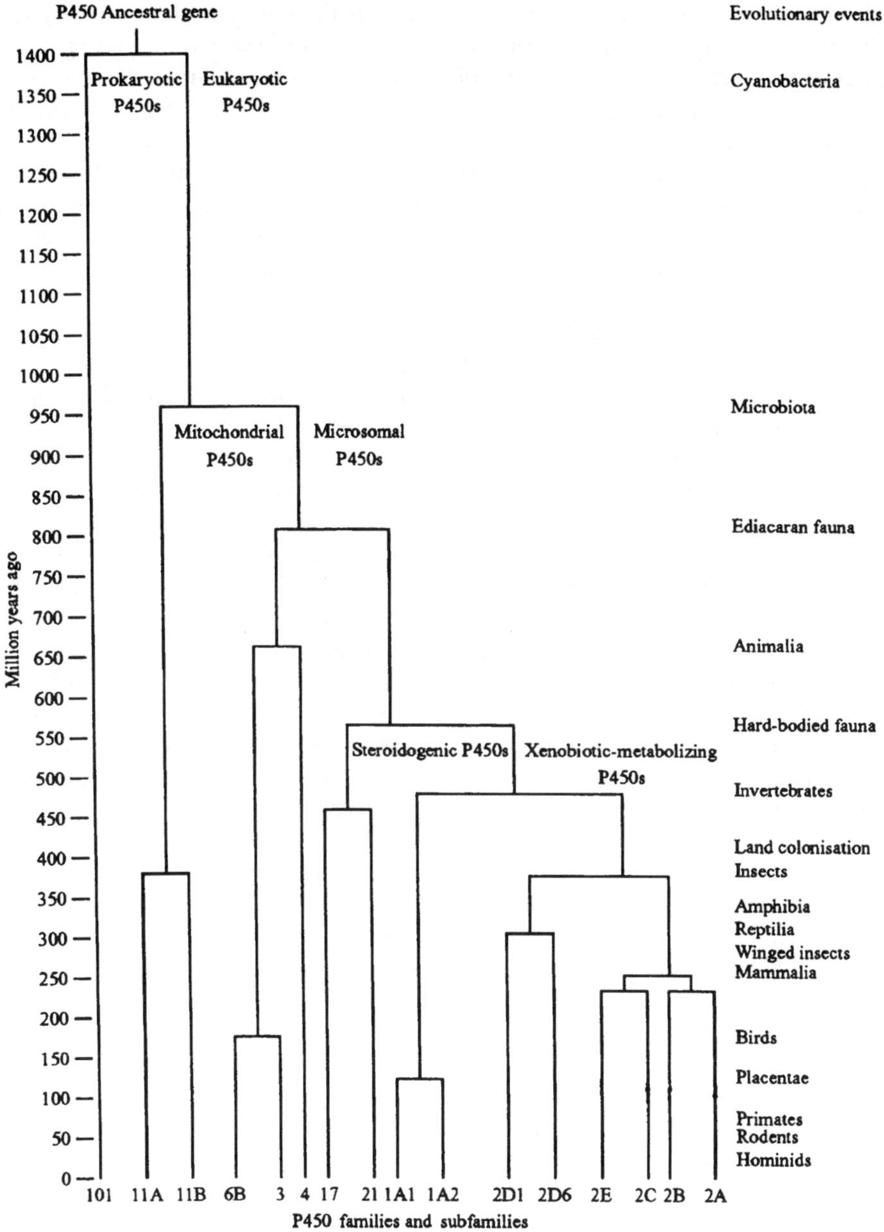

Figure 1.2 An abbreviated version of the P450 phylogenetic tree showing the parallel development of certain species of biota.

xenobiotic-metabolising P450s to detoxify the deleterious effects of plant toxins which had been biosynthesised to deter animal predators.

The huge diversity of P450 functionality[55] may, therefore, have arisen from the increase in atmospheric oxygen which was harnessed for oxidative metabolism and biosynthesis. For P450 utilises the inherent chemical power of the dioxygen molecule and controls its activation *via* two consecutive reductions to peroxide, which probably represents the precursor for the active oxygen species that is inserted into carbon-based substrates.[56] The development of P450's changing roles mirrors the evolution of terrestrial biota as indicated in Figure 1.2, and it should also be recognised that periodic mass extinctions[57] have played their part in the rise of the mammalia and, eventually, to mankind itself. For example, the elaboration of the CYP2 family occurred after the major global extinction event at the end of the Permian period, approximately 250 million years ago.

As mentioned previously (*vide supra*), P450s are not found in *E. coli* and other primitive anerobic bacteria (ie. the Archebacteria) although they are present in certain thermophiles.[58] These may provide a possible clue to the origins of P450 in abyssal thermal vents where there would have been the relatively high concentrations of iron and sulphur required for the formation for haem-thiolate proteins, as well as the iron-sulphur redoxins which constitute the earliest form of P450 redox partner. The relatively high percentage and clustering of aromatic amino acid residues present in these thermophilic bacterial P450s (from *Sulfolobus solfataricus* and *Thermus thermophilus*) are thought to provide a likely explanation for the thermal and high-pressure stability of these enzymes,[35] thus enabling them to tolerate such extreme environments as may have occurred extensively in early terrestrial prehistory. The substrates for these P450s remain to be determined but one can realistically assume that oxygen levels would have been relatively low during this stage of biological development, possibly at a fraction of 1%. When atmospheric oxygen levels eventually started to increase, perhaps P450 enzymes had some role in detoxifying oxygen before the development of other biological defense systems for dealing with reactive oxygen species (ROS).[41] In evolutionary terms, it would appear that dioxygen was not in the abundance it is today for most of the development of biological systems.[59] It is interesting to note that the only fungal P450 to have had its crystal structure determined is, in fact, a nitric oxide reductase (CYP55) from *Fusarium oxysporum*, which utilises only nitric oxide in a coupling reaction, without the use of oxygen, to form nitrous oxide. However, the vast majority of P450 reactions (of which there are over 50 different types) involve splitting of the dioxygen molecule and subsequent monooxygenation of substrates[5,60,61] while the unusual P450 reactions include: reduction, desaturation, oxidative ester cleavage, ring expansion, ring formation, aldehyde scission, dehydration, isomerisation, *ipso* attack of aromatic rings, one-electron oxidation, coupling reactions, rearrangement of fatty acid and prostaglandin hydroperoxides, and oxidative deamination. The more common P450-catalysed reactions are: aliphatic and aromatic hydroxylation, N-hydroxylation, N- and S-oxidation, O-, S- and N-dealkylation, aromatisation, dehalogenation, dehydro-halogenation, epoxidation, deformylation and the reduction of nitro compounds, N-oxides, quinones, epoxides, azo compounds and certain halogen compounds. It is in the mammalian P450s where a wide diversity of reactions are characterised and, to some extent, this is due to the relatively large number of P450s in such species, together with their various endogenous and exogenous roles.

In the human P450 complement, there is the well-documented situation known as genetic polymorphism, where genetic defects in individual P450s (mainly CYP2D6 and CYP2C19) can give rise to significant changes in metabolic capacity towards certain substrates, including a number of drugs in clinical use.[62–73] Consequently, there is considerable current interest in the screening of novel compounds that are destined for human exposure, such that adverse drug reactions can be avoided.

1.3 Binding Functions of P450

The apoprotein in P450 performs a number of specific functions by virtue of its primary, secondary and tertiary structure; some of these will be described in the remaining sections of this chapter, and Table 1.2 provides a summary. It is evident that the unusual functionalities of P450 enzymes evolved during the course of biological development from prokaryotic to eukaryotic organisms. Being haemoproteins, P450s contain a haem prosthetic grouping which, in common with most haemoproteins, is able to bind dioxygen although, in haem-thiolate enzymes like P450, oxygen becomes activated *via* the unique properties of the thiolate ligand.[74] In addition to the iron-sulphur covalent linkage, the haem moiety is bound by two conserved basic amino acid residues which form ion-pairs with the two haem propionate sidechains. A generally well-conserved tryptophan residue in the C helix, normally encountered in most mammalian P450s, is able to form a hydrogen bond with one of the haem propionate head-groups.[1,2] Furthermore, the relatively planar haem group is bound in an essentially hydrophobic cleft within the P450 protein, formed by an intersection of the I and L helices, where a number of complementary residues enter into hydrophobic contacts with the haem structure. Also, a generally well-conserved serine residue in the C helix, located fairly close to the haem group, is thought to represent a phosphorylation site for initiating haem degradation (reviewed in reference 1).

The Fe^{2+}/Fe^{3+} redox potential in P450 is modulated by substrate binding such that reduction of the enzyme by its redox partner is facilitated.[75,76] In an ideal system like that of $P450_{cam}$ (CYP101), the binding of camphor makes the Fe^{2+}/Fe^{3+} redox potential become less negative, such that $P450_{cam}$ lies on a potential gradient between those of NADH, putidaredoxin and dioxygen, as shown in Figure 1.3. This mechanism prevents reduction of $P450_{cam}$ in the absence of the camphor substrate and there is a degree of similarity to that encountered in other bacterial systems, together with most mammalian P450s, although the situation is not so clear-cut in the microsomal system.[77] The binding of redox partners shows both similarities and differences between various P450s, depending on the type of system concerned.[35] In $P450_{cam}$ (CYP101), a group of basic residues surrounding the proximal haem face appear to form electrostatic interactions with a complementary cluster of acidic

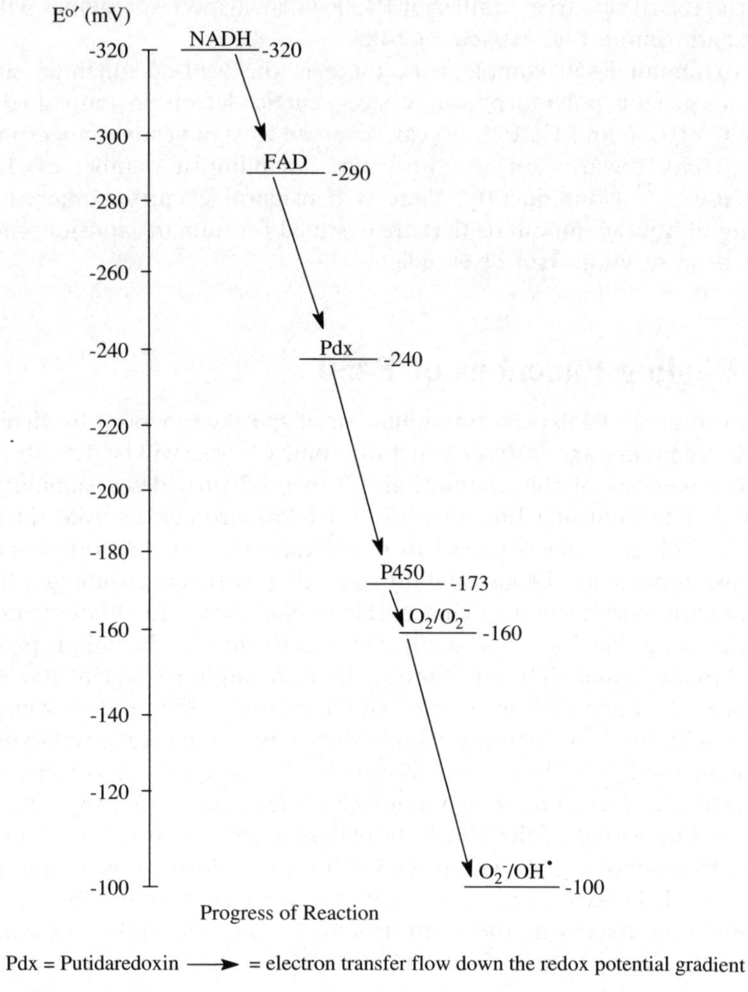

Figure 1.3 Redox potentials of various components in the $P450_{cam}$ system, including redox partners and oxygen species.

sidechains in the iron-sulphur protein, putidaredoxin, which enables electron transfer from the Fe_2S_2 centre to the haem *via* the mediation of a C-terminal tryptophan residue in putidaredoxin.[78] A well-conserved proximal phenylalanine in P450 is thought to facilitate electron transport to the haem group, possibly by aromatic π-π stacking interactions, and Phe350 in $P450_{cam}$ is an example of this type of grouping.[35] This residue is present in both prokaryotic and eukaryotic P450s, whereas it is possible that a conserved tryptophan (*vide supra*) which is present in many, but not all, microsomal P450s could have undergone transference from the ferredoxin redox partner during the course of evolution. However, there is also evidence to suggest that this tryptophan plays a role in haem binding rather than solely acting as a conduit for electrons during the reduction process.[36]

Interestingly, cytochrome b_5 can also reduce $P450_{cam}$ (CYP101), and the way in which the b_5 redox partner is likely to interact with CYP101 probably resembles that of the putidaredoxin interaction and this finding is, therefore, suggestive of another evolutionary linkage between prokarytic and eukaryotic P450 systems. Apparently, there is a coupling between the iron redox and spin-state equilibria in certain P450 systems[75,76] which can, in certain instances, also exhibit correlations with both substrate binding affinity and metabolic rate of the P450-mediated reactions.[79,80] This would indicate that the P450 system has developed over an evolutionary timescale with inbuilt mechanisms for coupling its various biophysical chemistry equilibria to the optimised metabolism of substrates, depending on the cellular environment.

Several basic residues on the surface of P450 are thought to be associated with the interaction of its reductase redox partner.[8] Apart from those already discussed, which are situated in the proximal haem region, there is also a conserved surface lysine (situated between the K and K′ helices) that is present in both microsomal and some bacterial P450s (often forming a PKG motif) and which may serve as a point for ion-paired contact with flavoprotein reductase in the microsomal P450s and, possibly, for ferredoxin reductase in certain bacterial P450s.[7,8] Similarly, adrenodoxin is able to bind to its reductase *via* charge-pair interactions,[36] and there are certain similarities between the putidaredoxin binding interaction with $P450_{cam}$ and that of the FMN and haem domains in $P450_{BM3}$,[35] thus indicating that P450 redox partner interactions have adapted during the process of evolution by subtle 'fine-tuning' of key amino acids involved in these contact points.

One of the major differences between bacterial and microsomal P450s lies in the fact that the former are cytosolic whereas the latter are membrane-bound. There is an N-terminal 'anchor' peptide of between 20–40 residues in length, composed of essentially hydrophobic amino acids, which is thought to span the phospholipid bilayer in the microsomal P450 systems, and sequence comparisons indicate that this feature is absent in bacterial forms of the enzyme. A number of surface 'loop' regions, that are present in microsomal P450s as opposed to those from bacterial species, may also be associated with membrane phospholipid interactions.[35] The binding of phospholipid is, therefore, achieved primarily by the additional stretch of N-terminal peptide relative to bacterial sequences, and this probably forms an essentially helical conformation which is thus able to span the entire thickness of a phospholipid bilayer.[35] It is not clear to what extent the rest of the P450 is embedded in the membrane, although it is apparent that both lateral and transverse movement of the enzyme is possible, together with rotation of the P450 about an axis perpendicular to the plane of the membrane itself. Furthermore, there is evidence for either octameric or hexameric clusters of P450 subunits which may be orientated around a central reductase.[81] The so-called 'flip-flop' or transverse motion of P450 within the microsomal membrane could be related to the conformational and/or flotation changes which may accompany substrate binding, thus leading to a preferred orientation of the enzyme for reduction by the flavoprotein reductase. It is thought that between 35Å and 45Å of the P450 structure is located above the microsomal membrane[35,82] which implies that roughly 20% of the enzyme is embedded within the phospholipid itself. Presumably, this situation would favour the binding of essentially lipophilic substrates *via* passage through the membrane bilayer, and it is clear that lipophilicity plays a role in substrate binding to microsomal P450s.[83] The overall structure of a P450 such as CYP101 resembles a triangular prism with a side length of about 60Å[15] and microsomal P450s are probably of a similar size. Consequently, one can imagine that around 10Å of the P450 structure may be embedded within the smooth endoplasmic reticular membrane, and this would be consistent with the experimental findings of Schwarz and colleagues.[81,84] Figure 1.4 compares the crystal structures of two bacterial P450s with one fungal and one mammalian form of the enzyme, thus showing the overall similarity between their tertiary folds.

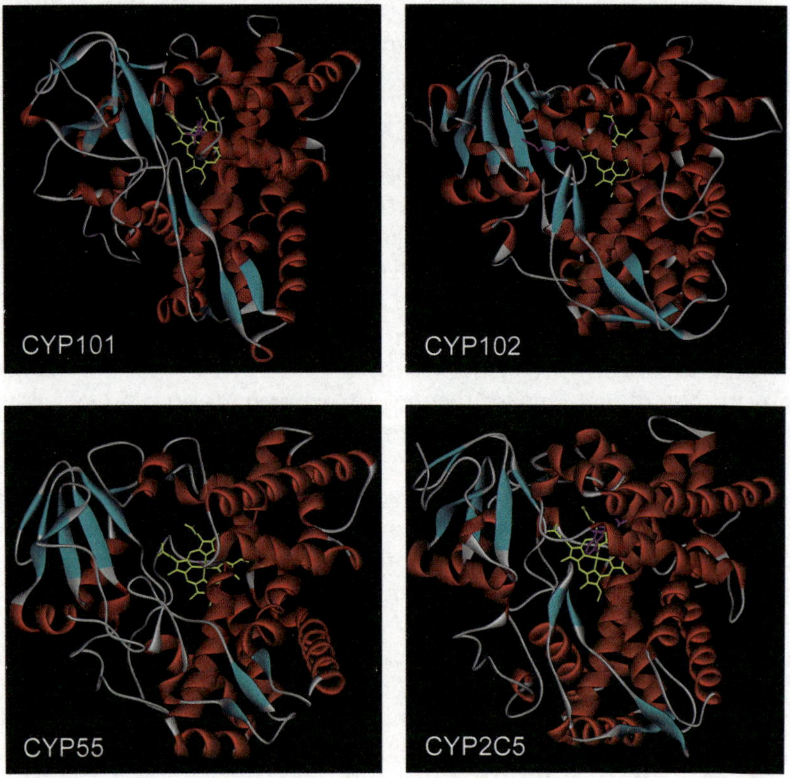

Figure 1.4 Comparison between P450 crystal structures for CYP101, CYP102, CYP55 and CYP2C5, showing α-helical and β-sheet regions in red and blue, respectively. The haem groups are shown in yellow, whereas the substrates are magenta. The protein data base codes for these structures are given in Table 1.10.

Oxygen binding at the haem iron is effected *via* initial ingress at a 'kink' in the I helix, following an initial reduction of the substrate-bound complex, due to the relatively strong affinity of high-spin iron(II) for the electron-deficient dioxygen molecule, which is presumably in the triplet ground state. The distal haem face is ligand-free, following desolvation of the essentially hydrophobic active site region by ingress of the substrate molecule, and oxygen is known to bind to this position.[19] The Fe(II)/Fe(III) redox potential of P450 is altered by the binding of substrate, becoming less negative, which thus facilitates reduction by its redox partner. Upon oxygen ligation to the reduced P450, it is thought that an electron is transferred from iron(II) to the dioxygen molecule which, presumably, then forms the superoxide anion, O_2^-. The O_2/O_2^- redox potential is $-160\,mV$ under physiological conditions and, therefore, electron transfer from the reduced P450 is thermodynamically favourable when slightly more negative than the oxygen/superoxide couple (Figure 1.3). Residue changes in the oxygen-binding pocket of the distal I helix brings about uncoupling of the distal charge relay and this gives rise to P450s which exhibit unusual reactions, such as those encountered in allene oxide synthase (CYP74), thromboxane synthase (CYP5), prostacyclin synthase (CYP8), erythromycin synthase (CYP107A1) and nitric oxide reductase (CYP55) where the typical pattern of GG x D/ET (where x may be any residue) for the oxygen-binding pocket is fundamentally altered.[85] This is, therefore, a further example by which P450 functionality can be 'fine-tuned' by subtle changes to the apoprotein.

1.4 Substrate Binding and Selectivity

The ability to bind and metabolise a large number of substrates of diverse chemical class is a fundamental feature of P450 activity. However, it should be recognised that this functionality depends very much on the actual enzyme concerned. For example, the steroidogenic P450s which are involved in the biosynthesis of various steroid hormones (namely: CYP11, CYP17, CYP19 and CYP21, as shown in Figure 1.5) are exquisitely selective towards their specific substrates whereas CYP3A4, which constitutes a significant proportion ($\sim$30%) of the human hepatic P450 complement, is able to metabolise structurally diverse xenobiotics of which over 1000 are currently known.[86,87] In addition, there is the possibility of metabolic activation of chemicals in certain instances, depending upon the type of substrate and P450 enzyme involved, especially CYP1 and CYP2E.[88] Endogenous roles of several drug-metabolising P450s have also been identified, such as CYP2C9[89,90] and it has been well established that the induction of many mammalian P450s are, in general, mediated by ligand interaction with various nuclear receptor proteins.[91,92] The endogenous roles of P450, such as steroid biosynthesis, probably developed prior to exogenous functions although, in some cases, these may have occurred concurrently. The 57 human P450s exhibit a mixture of both exogenous and

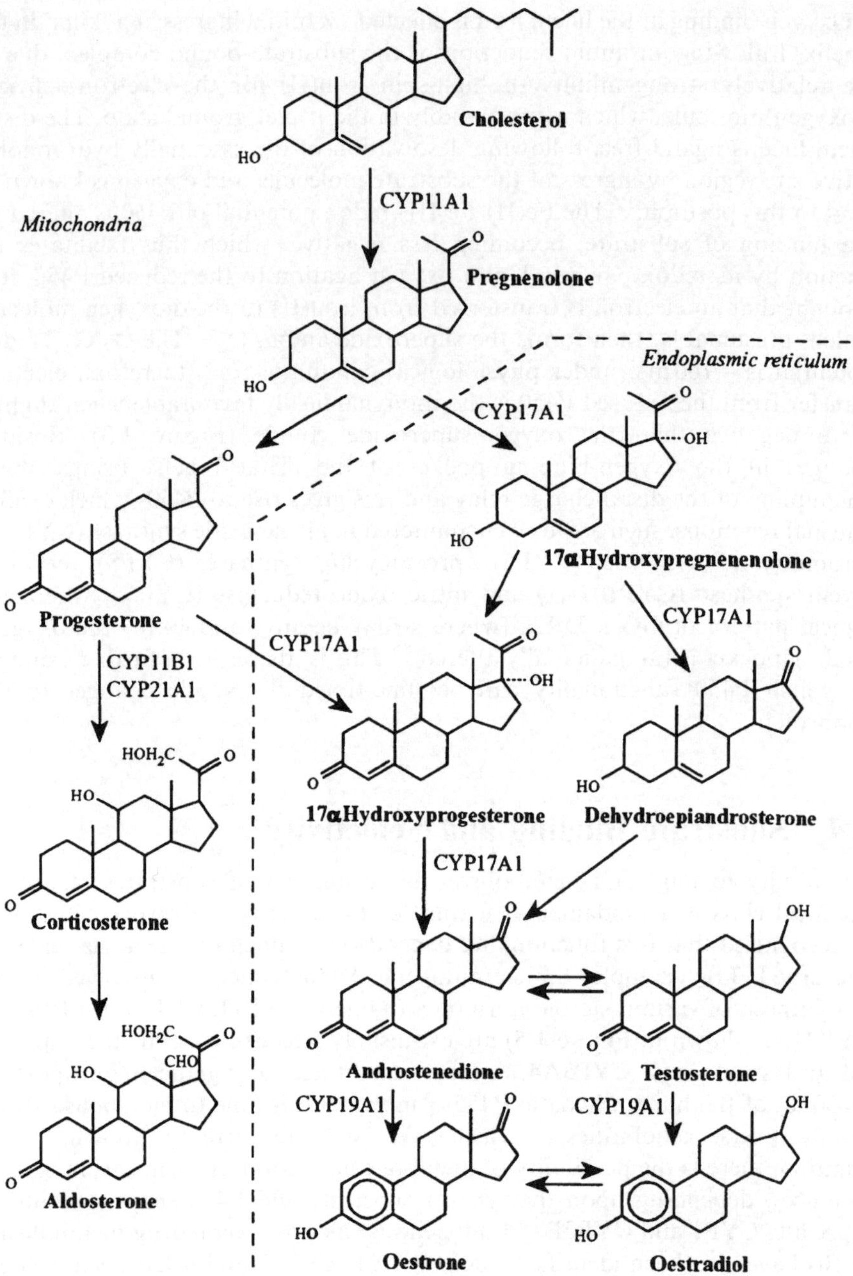

Figure 1.5 Steroid hormone biosynthetic pathways in mammalia showing the distinction between conversions carried out in the mitochondria and endoplasmic reticulum.

Table 1.4 Human cytochrome P450 enzymes (57 in total) and their substrate classes.[150]

Substrates	*P450 Enzymes involved*
Fatty Acids	4A11, 4B1, 4F12, 2J2
Eicosanoids	4F2, 4F3, 4F8, 5A1, 8A1
Vitamins	24A1, 26A1, 26B1, 27B1
Steroids	7A1, 7B1, 8B1, 11A1, 11B1, 11B2, 17A1, 19A1, 21A2, 27A1, 39A1, 46A1, 51A1
Xenobiotics	1A1, 1A2, 1B1, 2A6, 2A13, 2B6, 2C8, 2C9, 2C18, 2C19, 2D6, 2E1, 2F1, 3A4, 3A5, 3A7

Notes: 1. Other known human P450s for which the functionalities remain to be established include: 2A7, 2R1, 2S1, 2U1, 2W1, 3A43, 4A22, 4F11, 4F22, 4V2, 4X1, 4Z1, 20A1, 26C1 and 27C1. 2. Some P450s (eg. 1A2 and 1B1) could be placed in another class, such as steroid metabolism. 3. Over 75% of the human liver P450 complement is comprised of xenobiotic metabolising enzymes.

endogenous functionality, as summarised in Table 1.4, where the different substrate classes are shown.

Essentially, the binding and selectivity of P450 substrates are both linked to the particular characteristics of the enzyme's active site region.[93] It has been generally established that so-called Substrate Recognition Sites are present in all P450s, and these have been extensively mapped out using the techniques of site-directed mutagenesis[94] following multiple sequence alignment studies for selected CYP2 family proteins;[95] Table 1.5 shows an alignment of the six SRSs for the human CYP2 family. Furthermore, molecular modelling of various P450s based on sequence homology with those for which the crystallographic coordinates are available, has proved to be extremely useful for understanding and exploration of the structural determinants governing both selectivity and binding affinity towards their respective substrates and inhibitors.[2,96–104] It is possible to derive certain methodologies for the estimation of binding affinity between P450s and their substrates using a variety of techniques, such as homology modelling (*vide supra*), and from quantitative structure-activity relationships (QSARs). In addition, there are several docking programs available for orientating molecules within enzyme active sites and, as detailed later, we have found that AutoDock version 3.05[105,106] gives satisfactory binding energies for human P450 substrates, together with reproducing the experimentally observed positions of metabolism in the majority of cases studied thus far.[107] Table 1.6 shows that, for CYP2C9 substrates, there is a good correlation between AutoDock-calculated and experimental binding energies. In addition, QSAR analysis of 90 human P450 substrates shows that the combination of substrate log P (octanol/water partition coefficient) number of hydrogen bonds and π-π stacking interactions formed on binding to the enzyme produces very good correlations with binding energy data, where the R values (correlation coefficients) obtained are about 0.98.[100] Neural network analysis provides a useful means of examining the parametric contributions to P450 selectivity based on a number of substrate physicochemical quantitites, and this technique

Table 1.5 Alignment of SRS regions in CYP2 family enzymes showing substrate contact residues and SDMs.

CYP				*SRS1*					*SRS2*				*SRS3*					*SRS4*				*SRS5*			*SRS6*		
2C5	S_{99}	V_{100}	I_{102}	L_{103}	K_{108}	A_{113}	F_{114}	N_{204}	V_{205}	L_{208}	S_{209}	P_{220}	A_{221}	A_{237}	K_{241}	D_{290}	G_{293}	A_{294}	E_{297}	T_{298}	L_{359}	N_{362}	L_{363}	N_{471}	G_{472}	F_{473}	V_{474}
2A6	E_{103}	Q_{104}	T_{106}	F_{107}	K_{112}	$\boldsymbol{V}_{117}$	F_{118}	I_{208}	$\boldsymbol{F}_{209}$	T_{212}	S_{213}	S_{224}	S_{225}	L_{241}	E_{245}	$\boldsymbol{N}_{297}$	$\boldsymbol{I}_{300}$	G_{301}	$\mathbf{E}_{304}$	$\mathbf{T}_{305}$	I_{366}	S_{369}	L_{370}	V_{478}	G_{479}	$\boldsymbol{F}_{480}$	A_{481}
2B6	K_{100}	I_{101}	M_{103}	V_{104}	R_{109}	$\boldsymbol{I}_{114}$	$\boldsymbol{F}_{115}$	T_{205}	$\boldsymbol{F}_{206}$	I_{209}	S_{210}	S_{221}	G_{222}	L_{238}	N_{242}	S_{294}	$\boldsymbol{F}_{297}$	$\boldsymbol{A}_{298}$	$\mathbf{E}_{301}$	$\mathbf{T}_{302}$	$\boldsymbol{L}_{363}$	G_{366}	$\boldsymbol{V}_{367}$	C_{475}	G_{476}	$\boldsymbol{V}_{477}$	$\boldsymbol{G}_{478}$
2C8	$\mathbf{N}_{99}$	S_{100}	$\mathbf{I}_{102}$	S_{103}	$\mathbf{K}_{108}$	$\mathbf{I}_{113}$	$\mathbf{S}_{114}$	N_{204}	F_{205}	L_{208}	N_{219}	P_{220}	L_{221}	V_{237}	$\mathbf{R}_{241}$	$\boldsymbol{D}_{293}$	V_{296}	$\boldsymbol{A}_{297}$	E_{300}	$\mathbf{T}_{301}$	$\boldsymbol{V}_{362}$	G_{365}	$\boldsymbol{V}_{366}$	K_{474}	G_{475}	I_{476}	V_{477}
2C9	$\boldsymbol{I}_{99}$	F_{100}	$\mathbf{L}_{102}$	A_{103}	$\boldsymbol{R}_{108}$	V_{113}	$\boldsymbol{F}_{114}$	N_{204}	I_{205}	L_{208}	S_{209}	$\mathbf{S}_{220}$	$\mathbf{P}_{221}$	V_{237}	$\boldsymbol{K}_{241}$	$\mathbf{D}_{293}$	G_{296}	$\boldsymbol{A}_{297}$	E_{300}	$\mathbf{T}_{301}$	$\boldsymbol{L}_{362}$	S_{365}	$\boldsymbol{L}_{366}$	N_{474}	G_{475}	F_{476}	A_{477}
2C19	$\boldsymbol{H}_{99}$	F_{100}	$\mathbf{L}_{102}$	A_{103}	$\mathbf{R}_{108}$	$\mathbf{V}_{113}$	F_{114}	N_{204}	I_{205}	V_{208}	S_{209}	$\mathbf{P}_{220}$	$\mathbf{T}_{221}$	L_{237}	$\boldsymbol{E}_{241}$	$\mathbf{D}_{293}$	G_{296}	$\boldsymbol{A}_{297}$	E_{300}	$\mathbf{T}_{301}$	$\boldsymbol{I}_{362}$	S_{365}	$\boldsymbol{L}_{366}$	N_{474}	G_{475}	F_{476}	A_{477}
2D6	P_{103}	V_{104}	I_{106}	$\boldsymbol{T}_{107}^{*}$	F_{112}	$\boldsymbol{F}_{120}$	L_{121}	G_{212}^{*}	L_{213}	$\boldsymbol{E}_{216}$	S_{217}	V_{227}	P_{228}	Q_{244}	L_{248}	$\boldsymbol{D}_{301}$	$\boldsymbol{S}_{304}$	A_{305}	V_{308}	T_{309}	V_{370}	G_{373}	$\boldsymbol{V}_{374}^{*}$	$\mathbf{F}_{481}$	A_{482}	$\boldsymbol{F}_{483}$	L_{484}
2E1	G_{101}	L_{103}	A_{105}	F_{106}	R_{110}	I_{115}	F_{116}	N_{206}	F_{207}	L_{210}	S_{211}	P_{222}	S_{223}	V_{239}	K_{243}	D_{295}	F_{298}	A_{299}	E_{302}	$\mathbf{T}_{303}$	V_{364}	N_{367}	L_{368}	I_{476}	G_{477}	F_{478}	G_{479}

Key: Italics = contact residue with substrates in either the crystal structure or model of the enzyme
Bold = sdm residue, including those performed on another enzyme in the same subfamily
* = corresponds to a change encountered in an allelic variant of the enzyme
SDM = site-directed mutagenesis

Notes: 1. CYP2C5, CYP2A6, CYP2C8, CYP2C9 and CYP2D6 have had their crystal structures determined. CYP2B6 and CYP2C19 have been modelled by homology with the CYP2B4 and CYP2C9 crystal structures, respectively. 2. The analogous residues in the CYP2C5 sequence are included for comparison as this enzyme has been used as a crystallographic template for modelling various human P450s from the CYP2 family, including CYP2E1. 3. The protein data bank codes for the relevant crystal structures investigated are as follows: 2C5 (1n6b and 1nr6), 2A6 (1z10), 2C8 (1pq2), 2C9 (1r9o) and 2D6 (2f9q). 4. The CYP2B6 enzyme was modelled from the 1suo (CYP2B4) crystal structure and CYP2C19 was constructed from the 1og5 (CYP2C9) crystallographic coordinates. 5. The CYP2E1 enzyme was produced *via* homology with the CYP2C5 crystal structure (1n6b) where the sequence identitiy is 59%.

Table 1.6 CYP2C9 substrates: comparison between experimental and calculated binding energies (kcal mol^{-1}).

Compound	*Route of metabolism*	K_m (μM)	ΔG^{expt}_{bind}	ΔG^{calc}_{bind}
1. Diclofenac	4′-hydroxylation	13.11 (ave)	−6.9250	−6.94
2. Warfarin (S)	7-hydroxylation	6.0	−7.4070	−7.22
3. Tienilic acid	5-hydroxylation	5.0	−7.5193	−7.57
4. Flurbiprofen (S)	4′-hydroxylation	1.9	−8.1154	−8.22
5. Ibuprofen (S)	3-hydroxylation	21	−6.6353	−7.14
6. Tolbutamide	4-methyl hydroxylation	82	−5.7961	−5.75
7. Phenytoin	4-hydroxylation	23.3 (ave)	−6.5712	−6.47
8. Torasemide	methyl hydroxylation	40	−6.2383	−6.69
9. Celecoxib	methyl hydroxylation	3.5	−7.7391	−8.17
10. Mefenamic acid	3′-methyl hydroxylation	7	−7.3121	−7.35

ave = average of several values available in the literature.
Correlation between experimental and calculated binding energies is as follows:
$\Delta G^{expt}_{bind} = 0.982\ \Delta G^{calc}_{bind}$ $s = 0.243$; $R = 0.948$; $R^2 = 0.90$

has been found to give satisfactory results by correctly discriminating the enzyme involved in 95% of human P450 substrates studied from a dataset of 64 compounds.[108] There is evidence that the six SRS regions correspond with four areas of α-helical and two strands of β sheet structure in P450s, and protein sequence analysis from multiple alignment can indicate the likely amino acid residues for further examination *via* site-directed mutagenesis studies,[94,109–112] such that enzyme selectivity determinants can thus be explored. Consequently, the combination of molecular modelling, site-directed mutagenesis and X-ray crystallography can be employed in the important exercise of understanding selectivity *via* building up substrate templates within the relevant enzymes' active sites, which also tend to show good agreement with experimental findings on the known routes of P450-mediated metabolism.[113,114]

These various findings serve to build up an emerging picture of the P450 active site regions that are responsible for substrate binding and selectivity, particularly with regard to human drug metabolism. The quality of homology models can be evaluated by comparison with the recently determined crystal structures of CYP2C8, CYP2C9, CYP2A6, CYP2D6 and CYP3A4, where the average root-mean-square difference (rmsd) between matched α-carbons on the model and crystal structure displays a well-defined relationship (Figure 1.6) with primary sequence identity relative to the CYP2C5 template, as described in a following section; Table 1.7 provides the relevant data. Although at first sight the relation appears to be approximately linear, careful analysis reveals that it is, in fact, exponential in nature with a function which is similar to that reported by Lesk and Chothia[115] for the core residues of various protein structures, including those of hemoproteins.

Substrate selectivity (Table 1.8) for the human drug-metabolising enzymes is somewhat complex as some of these enzymes exhibit distinct, but overlapping, selectivities. However, one is now in a position to make certain definitive statements regarding this important area, especially due to the current interest

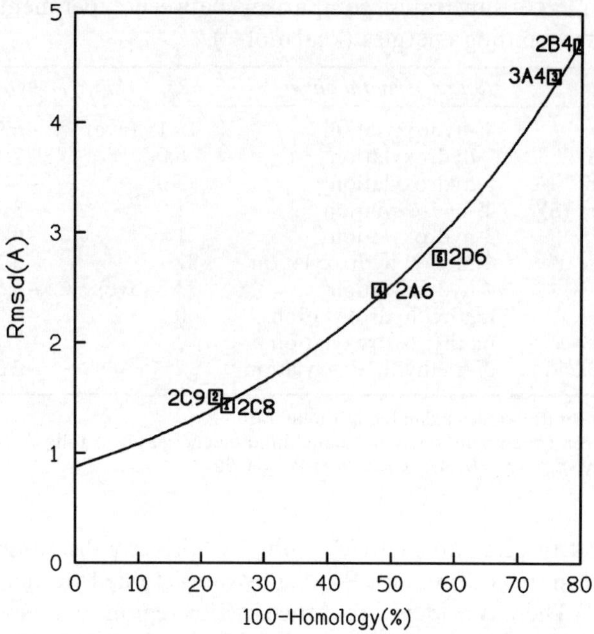

Figure 1.6 A plot of root-mean-square distance (rmsd) between matched α-carbon atoms versus percentage sequence identity for several mammalian P450s of known crystal structure, using the data presented in Table 1.7.

Table 1.7 Percentage sequence identity and Rmsd values for mammalian P450s.

CYP model	*Sequence identity (%)*	*Rmsd (Å)*	*Crystallographic target*	*Crystallographic template*
2A6	51.9481	2.4749	2A6	2C5
2B4	20.2299	4.7034	2B4	102
2C8	75.7576	1.4291	2C8	2C5
2C9	77.7056	1.5044	2C9	2C5
2D6	42.3913	2.7742	2D6	2C5
3A4	24.3478	4.4240	3A4	2C5

Rmsd = average root mean square distance between matched α-carbon atoms in the fitted structures.

Sequence Identity = the percentage primary sequence identity between target and template P450s based on the matched residues involved.

Equations relating Rmsd with Percentage Sequence Identity for the six structures shown above are as follows:-

1. Rmsd = 0.056 (100−% identity) (±0.005) R = 0.983
2. % identity = 98.166−17.135 Rmsd (±1.785) R = 0.979
3. log Rmsd = 0.009 (100−% identity)−0.045 (±0.0004) R = 0.996

Table 1.8 Characteristics of human drug-metabolising P450 substrates.

CYP	*Types of substrates*	*Average log P*	*Typical substrate*	*log P*
1A2	Planar heterocyclic amines and amides	2.01	MeIQ	1.98
2A6	Fairly small-sized molecules	1.44	Losigamone	1.46
2B6	Basic, medium-sized molecules	2.54	Bupropion	2.54
2C8	Acidic, fairly large-sized molecules	3.38	Rosiglitazone	3.20
2C9	Acidic, medium-sized molecules	3.20	Naproxen	3.18
2C19	Medium-sized amines and amides	2.56	Proguanil	2.53
2D6	Basic, medium-sized molecules	3.08	Propranolol	3.09
2E1	Structurally diverse small molecules	2.07	4-Nitrophenol	2.04
3A4	Structurally diverse large molecules	3.10	Nifedipine	3.17

Notes: 1. Molecular size (M_r) distinguishes CYP2E1 (small) and CYP3A4 (large) from other substrate classes (medium). 2. Molecular planarity (area/depth^2) distinguishes CYP1A2 and CYP2E1 (high) from CYP2B6 (low) relative to other substrate classes (medium). 3. Compound acid/base character (pK_a) distinguishes CYP2C9 (acidic) and CYP2D6 (basic) from other substrate classes, such as CYP2A6 and CYP2E1 (neutral). 4. Number of fused aromatic rings distinguishes CYP1A2 substrates from other substrate classes, and this is related to molecular planarity. 5. The log P volume is related to a combination of molecular size and compound polarity, which is a function of the number of hydrogen bond donor and acceptor atoms. The average log P for a relatively large number of substrates ($\sim$16 for each enzyme) is close to that of a typical substrate in each case. P, octanol/water partition coefficient; MeIQ, 2-amino-3,4-dimethylimidazo[4,5-*f*]quinoline.

in high-throughput screening programmes for the development of new chemical entities (NCEs) that are destined for human exposure. In this respect, it is now well-established that drug metabolism represents a pipeline 'bottleneck' for NCE development and many of the challenges to those engaged in this endeavour appear to centre around pharmacokinetics issues related almost entirely to P450-mediated pathways. Table 1.9 provides a listing of selective human P450 substrates for the major drug-metabolising P450s.

The substrate selectivities for the eight human P450s which comprise the major drug-metabolising enzymes appear to be largely determined by compound molecular size (M_r), acid/base characteristics (pK_a) and the ability to form hydrogen bonds with the enzyme, which relates to the number of hydrogen bond acceptors and donors in the molecule (N_{HB}^{A+D}). These three structural/physicochemical parameters are able to provide a 95% concordance with enzyme selectivity for a dataset of 64 compounds using a neural network analysis approach.[108] The reason for this is due to the fact that pK_a is sufficient to discriminate CYP2C9 (acidic) and CYP2D6 (basic) substrates, whereas M_r differentiates substrates of CYP3A4 (large) and CYP2E1 (small) from the

Table 1.9 Selective human CYP substrates.[86,153]

CYP1A2	Tienilic acid	Debrisoquine
Antipyrine	Ibuprofen	Sparteine
Caffeine	Flurbiprofen	
Clozapine	Mefenamic acid	**CYP2E1**
Clomipramine		Chlorzoxazone
Imipramine	**CYP2C19**	Enflurane
Paracetamol	Amitriptyline	Halothane
(acetaminophen)	Citalopram	Paracetamol
Phenacetin	Clomipramine	(acetaminophen)
Propranolol	Diazepam	Salicylic acid
Tacrine	Imipramine	Benzene
Theophylline	Mephobarbital	Ethanol
	Omeprazole	Dimethylnitrosamine
CYP2A6	Proguanil	4-Nitrophenol
Coumarin	Propranolol	Dapsone
Nicotine	S-Mephenytoin	
Cotinine		**CYP3A4**
Fadrozole	**CYP2D6**	Amiodarone
SM-12502	Aprindine	Lidocaine
Losigamone	Encainide	Propafenone
4-Nitroanisole	Mexiletine	Quinidine
2,6-Dichlorobenzonitrile	Propafenone	Ifosfamide
Quinoline	Metoprolol	Tamoxifen
Indole	Propranolol	Toremifene
	Perphenazine	Vinblastine
CYP2B6	Haloperidol	Alprazolam
7-Benzyloxyresorufin	Thioridazine	Diazepam
7-Ethoxy-4-trifluoro-	Zuclopenthixol	Midazolam
methylcoumarin	Codeine	Triazolam
Deprenyl	Dextromethorphan	Diltiazem
Testosterone	Dihydrocodeine	Felopidine
Benzphetamine	Ethylmorphine	Nifedipine
Bupropion	Hydrocodone	Verapamil
PNU 249173	Tramadol	Cortisol
Cinnarizine	Fluoxetine	Ethynyloestradiol
7-Ethoxycoumarin	Paroxetine	Testosterone
Arteether	Amitriptyline	Carbamazepine
	Clomipramine	Clomipramine
CYP2C9	Desipramine	Cyclosporin A
Antipyrine	Imipramine	Erythromycin
Diclofenac	N-Desmethylclomipramine	Imipramine
Phenytoin	Nortriptyline	Omeprazole
S-Warfarin	Trimipramine	Proguanil
Tolbutamide	Maprotiline	Terfenadine
R-Naproxen	MDMA	

remaining compound set. The utilisation of the third descriptor, N_{HB}^{A+D}, is more subtle and facilitates distinction between those P450s which exhibit a preference for polar molecules, such as CYP2A6, for example. Figure 1.7 presents the structures of a number of CYP2C9 substrates orientated such that their sites of

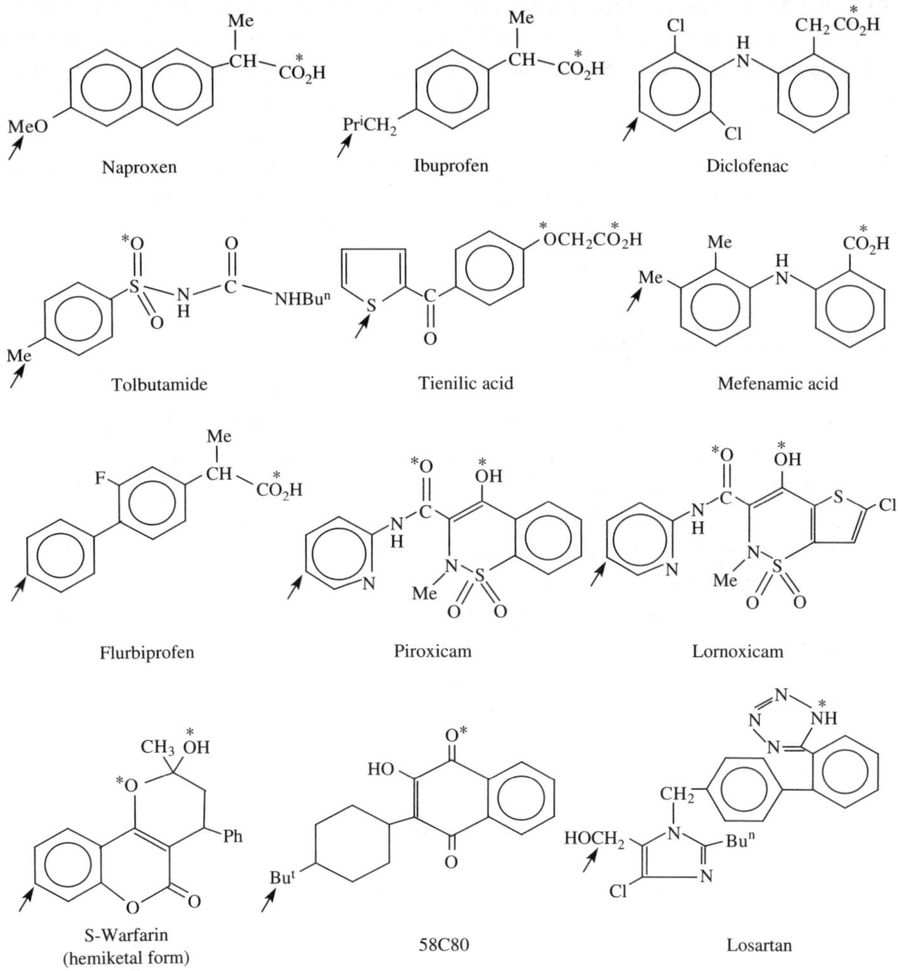

Figure 1.7 A comparison between typical substrates of CYP2C9 showing structural similarities, sites of metabolism (↑) and key interacting groups. Hydrogen bond donor/acceptor atoms are also marked (*). Bu refers to the butyl group, with the n superscript representing normal butyl and t the tertiary butyl form.

metabolism and specific binding groups would reinforce if the molecules were superimposed, and a template of such compounds is able to fit within the CYP2C9 active site such that there are common interactions with key amino acid residues. For more precise definition of selectivity determinants, it is necessary to employ some form of active site modelling or undertake an examination of superimposed substrate templates, and these two procedures can be readily combined by building up matched structures of typical substrates within the binding sites of crystal structures and good quality molecular models of human P450s. The so-called SRS regions[95] of CYP2 family enzymes have remained a target for site-directed mutagenesis experiments on mammalian P450s for some time (reviewed by Domanski and Halpert[94]) and it is possible to show that there is a match-up between SDMs in the SRSs of CYP2 family P450s and known substrate contacts encountered in the crystallographically resolved three-dimensional structures. Table 1.5 presents a summary of this information, from which it can be appreciated that there are key positions for amino acids determining substrate selectivity in the CYP2 family.

1.5 P450 Catalysis

In the P450 catalytic cycle, as shown in Figure 1.8, most intermediary stages have now been fairly well characterised,[19,56,116–122] and crystal structures are now available for the substrate-free, substrate-bound, reduced form, oxygen-bound form and oxene intermediate, albeit only for the $P450_{cam}$ (CYP101) system. CYP101 from the bacterium *Pseudomonas putida* is probably one of the most extensively characterised P450 enzymes and, consequently, this is the one where much biophysical chemistry is known from the employment of a variety of biochemical, physical and theoretical techniques, including, for example, Mössbauer spectroscopy.[78,116,123]

From a consideration of the bond lengths for the haem moiety ligands in the CYP101 crystal structures, it would appear that the dioxygen-bound state

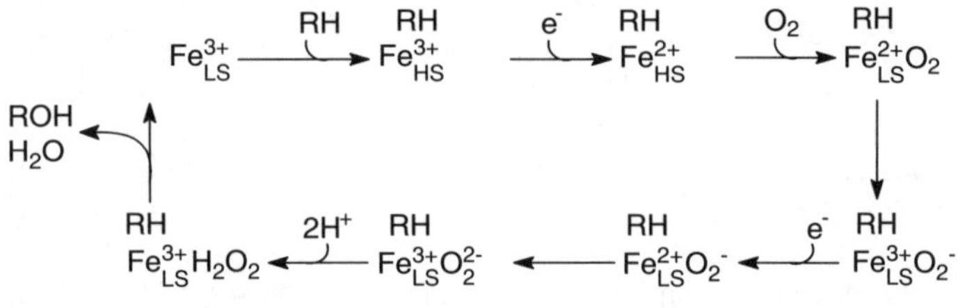

LS = low-spin HS = high spin

Overall reaction: $RH + O_2 \xrightarrow[2e^-]{2H^+} ROH + H_2O$

RH = substrate ROH = oxygenated metabolite

Figure 1.8 The P450 catalytic cycle showing key intermediates at the haem locus and the various changes in the iron redox- and spin-states. The two reducing equivalents ($2H^+$, $2e^-$) are supplied by NADPH and mediated *via* P450 oxidoreductase (flavoprotein) containing FMN and FAD co-factors. There is considerable debate about the precise nature of the active oxygen species which actually performs the final stage of substrate oxygenation. An iron-bound peroxy species and an iron oxene intermediate represent the two main candidates at present, and there is evidence for both of these.

probably comprises superoxide coordinated to iron (III), whereas the iron-oxo intermediate (sometimes referred to as an iron-oxene) may be considered as a single oxygen atom bound to iron(III), although there are several ways of formulating this species,[119] such as $[Fe{=}O]^{3+}$, and some of these can also involve higher oxidation states of iron such as Fe(IV) or even Fe(V) species. However, it is clear that the cysteine ligand present as the thiolate form plays a crucial role in the P450 catalytic cycle, where it is thought to be important for activation of the dioxygen complex following the initial reduction of the substrate-bound enzyme.[74] Standard tabulations of atomic and covalent radii concord well with the observed Fe–S and Fe–O distances encountered in the crystal structures of various P450s, although it is apparent that the Fe–S bond length changes somewhat during the course of the reaction cycle, this being largely dependent upon the electronic state of iron; it may also be affected by the presence of inhibitors which are able to ligate the haem iron, such as carbon monoxide.[124] A summary of the P450 haem geometry is presented in Figure 1.9.

As P450-mediated catalysis can be viewed as the consecutive two-stage reduction of dioxygen to, firstly, superoxide and then peroxide, there is an opportunity for the generation of deleterious reactive oxygen species (ROS) if uncoupling of the cycle occurs such that the substrate is not oxygenated fully. Although most P450 oxygenations can be mechanistically formulated in terms of the electrophilic $[Fe{=}O]^{3+}$ species, there is evidence for iron-peroxo and/or iron hydroperoxo complexes as being the active oxygenating intermediates in certain cases, especially where nucleophilic attack is required to concord with experimental observations.[56] The first reduction is generally regarded as the rate-determining step, although this stage is clearly dependent upon the initial binding of substrate to low-spin iron(III) in the resting state of the enzyme. Evidence is strong for the rapid, high-affinity binding of dioxygen following the first reduction, and it is likely that an equilibrium exists between iron(II)-dioxygen and iron(III)superoxide, with protonation of superoxide representing a pathway favouring the latter state. The second reduction is then thought to occur, and formation of iron(III) as in the previously discussed equilibrium would favour this, with a subsequent equilibrium state emerging between iron(II)superoxide (possibly protonated) and iron(III)peroxide, for which evidence also exists. Probably the iron(III)peroxo species will readily become protonated to form the iron(III)hydroperoxy intermediate, and this latter state has been postulated as possessing both electrophilic and nucleophilic characteristics which would appear to be ideal for many P450 oxygenations.[56] However, the loss of a water molecule following a second protonation step yields the iron(III)oxo species which is currently favoured as the prime candidate for the actual oxygenating species in most P450-mediated reactions and, mechanistically, this would appear to be most likely.[118,119]

In a series of impressive papers, Shaik and colleagues have shown *via ab initio* molecular orbital calculations that the postulated active oxygen intermediate, namely the iron(III)oxo species, is a remarkably stable entity that is also capable of performing an energetically favourable oxygenation of selected substrate molecules (reviewed in Shaik and De Visser[121]). In fact, the work of

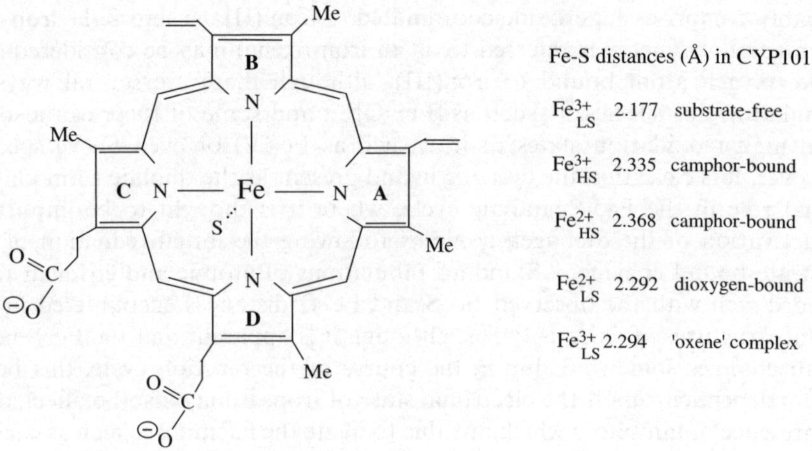

Figure 1.9 Structural geometry of the haem moiety in P450 showing Fe–S distances at key stages in the catalytic cycle, as obtained from crystallographic studies. 1. A, B, C and D designate the nomenclature of the four pyrrole rings as viewed from above the distal haem face. 2. The Fe–S bond length changes during the course of the catalytic cycle, this being dependent upon the redox and spin-states of the iron atom. 3. In general, the change from low-spin (LS) to high-spin (HS) increases the Fe–S distance slightly, and Fe(III) generally shows a shorter Fe–S bond than for Fe(II). 4. The effect of the changes in iron spin-state on the haem geometry usually involves a movement of the iron atom out of the porphyrin ring plane and towards the cysteine sulphur when iron is in the high-spin state due to its increased ionic radius. 5. The Fe–N distances tend to lie close to a value of 1.95–2.05Å with only slight variations between different P450s and within the reaction cycle itself. 6. The Fe–O bond lengths in the dioxygen-bound and 'oxene' complex structures indicate the presence of bound superoxide and a bare oxygen atom, respectively, on the basis of the relevant atomic and covalent radii. Moreover, the O–O bond distance of 1.253Å in the dioxygen complex is consistent with the formation of a superoxide moiety where the O–O bond length is 1.26Å. 7. Haem binding is achieved by an Fe–S covalent bond with the invariant cysteine* sulphur atom, together with two ionic interactions with conserved basic residues for the propionate head groups, and various hydrophobic contacts with complementary residues within the I and L helices which effectively 'sandwich' the haem group.
*The cysteine is present within a well-conserved decapeptide that constitutes the P450 signature motif by which the presence of the enzyme can be identified *via* sequence analysis.

Sligar and colleagues[19] on the crystallographic determination of atomic resolution of various stages in the $P450_{cam}$ (CYP101) catalytic cycle indicates the possible formation of an iron-oxene state, and these coworkers have also reported the isolation of the dioxygen-bound complex of reduced $P450_{cam}$ which may be regarded as the iron(III)superoxide, as discussed previously. However, others have demonstrated the possibility of an iron-peroxo species as being capable of carrying out P450-mediated oxygenations in some cases,[125–127] although the consensus view is that the iron-oxo species is probably the most common active oxygen intermediate in P450 oxygenations, and this is certainly consistent with the likely reaction mechanisms of many P450-mediated conversions.[61,128]

The redox potentials in P450 systems are important to the progress of the catalytic cycle, whereby substrate binding lowers the E° value of the iron(II)/iron(III) couple (becoming less negative, in fact) thus enabling reduction and electron transfer by the redoxin, reductase or cytochrome b_5 redox partner, depending on the type of system under consideration (reviewed in reference 1). In this way P450 becomes optimally positioned for the establishment of a smooth potential gradient from NADPH/NADP (or NADH/NAD) to oxygen/superoxide, irrespective of whether one examines the bacterial, mitochondrial or microsomal P450 systems (see Figure 1.3 for the $P450_{cam}$ redox pathway). Although the FAD/FMN combination in flavoprotein reductase is a better electron transfer mediator than the iron-sulphur cluster arrangement in redoxins, it is found that the bacterial systems are generally somewhat faster in terms of catalytic turnover than are the microsomal systems, and this may be due to the fact that the bacterial P450s are cytosolic whereas the microsomal P450s are membrane bound (reviewed in references 1 and 2). Presumably, the phospholipid in the endoplasmic reticulum system is able to modulate the rate of electron transfer which is likely to be rate-determining. However, the fused P450 system in $P450_{bm3}$ (CYP102) which contains both a flavoprotein and haem domain is significantly faster (turnovers are in the order of $4000\,min^{-1}$) in performing substrate oxygenations than, say, $P450_{cam}$ where the redox partner is the iron-sulphur protein putidaredoxin.[36,129,130] The relatively rigid arrangement of this fused P450 system would ensure a facile electron transfer between the various cofactors involving the FAD→FMN→haem electron transport pathway in CYP102 relative to that encountered in the microsomal system.[36]

The molecular electrostatic potential energy of P450 surface residues may change during the course of the catalytic cycle, especially at those points where substrate binding, redox partner interaction and electron transfers occur. This may well have an important bearing on the mechanism by which the actual reduction process itself is effected, including a rationale for the triggering of redox partner binding at the appropriate stage in the catalytic cycle. Desolvation of the P450 active site by incoming substrate tends to lower the redox potential of P450 (ie. it becomes less negative) which facilitates the first reduction step, as already discussed. Apparently, there is a clear relationship between percentage haem exposure and iron(II)/iron(III) redox potential (E° value) in haemoproteins, including P450, and this also correlates with oxygen affinity in terms of pK_{O_2} values for other series of haemoproteins.[131]

The initial binding of a substrate molecule will desolvate water from the active site region, thus altering the electrical characteristics of the assembly due to the exposure of ion-pairs and other charged residues from the electrically insulating effect of bound water molecules, which essentially act like a dielectric layer within the haem environment. There is a generally conserved acidic residue close to the invariant cysteine haem ligand which appears to control water access to the active site *via* a hydrogen-bonded network, so 'activation' of this residue by loss of water from the distal haem face possibly acts like a 'switch' to turn on the electric field gradient surrounding the proximal haem face, thus attracting either reductase or cytochrome b_5 in the microsomal system for triggering reduction of the iron.[12] It is feasible that the extent of desolvation at the haem locus, which accompanies substrate binding, may be proportional to the substrate-bound redox potential of the P450 and, hence, the rate of the first reduction process itself. This may help to explain the dependance of substrate lipophilicity upon the overall rate of metabolism which is found in certain series of P450-mediated oxygenations. Although there are other factors involved, when comparing the rate-dependancy for compounds in a congeneric series one can frequently find that compound lipophilic character, in the form of either log P or log $D_{7.4}$, shows a reasonable correlation with the metabolic rate constant when expressed logarithmically. However, electronic factors are also likely to play key roles in determining the relative rate of P450-mediated reactions.

The enzyme changes its conformation upon substrate binding, as determined by comparison between bound and free P450s. A structural fit between substrate-bound and substrate-free CYP2C5 indicates which residues and general regions of the structure are mainly affected by substrate binding. Figure 1.10 shows the root-mean-square distance (rmsd) value plotted against residue number, with the various SRS regions[95] indicated for completeness. It is apparent that only a few residues alter their position markedly upon the binding of a typical substrate, although it is likely that the accompanying

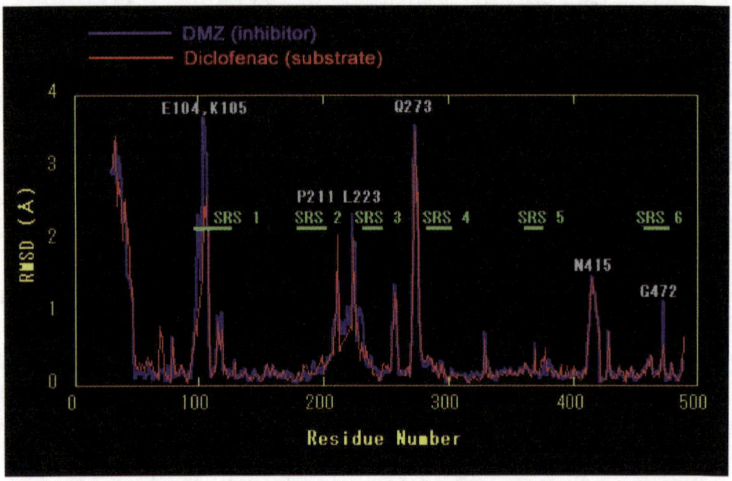

Figure 1.10 A plot of root-mean-square distance (RMSD) (Å) versus residue number for substrate- and inhibitor-bound CYP2C5 relative to that of the substrate-free structure. The location of substrate recognition sites (SRS) regions are shown, together with amino acid residues which correspond to the peak rmsd positions showing movement in substrate binding. DMZ = dimethyl-2-phenyl-2H-pyrazol-3-ylbenzenesulphonamide.

conformational change triggers the binding of the redox partner such that reduction is effected, thus leading to dioxygen activation and insertion of a single oxygen atom into the substrate molecule. Orientation of the substrate within the active site and, specifically, relative to the haem will determine the site of metabolism. In this respect, the spatial disposition of a relatively small number ($\sim$6) of key active site amino acid residues is crucial for determining the ultimate course of metabolism and some of these amino acids can also play a role in defining the substrate selectivity of the enzyme. Thus, although most residues are largely conserved across the superfamily, those within the SRSs may differ markedly between individual P450s where they operate in directing the nature of preferred substrates and, ultimately, determine their sites of oxygenation.

1.6 Structural Modelling of P450s

Three-dimensional structural models of P450s can be obtained *via* X-ray crystallography and these are generally recognised to be of immense value in understanding the way in which P450s operate. Homology models may also be constructed from a suitable crystallographic template, and the likely level of accuracy in such experiments is essentially determined by the level of sequence identity between template and target sequences. Until the advent of the rabbit CYP2C5 crystal structure, the most promising template for mammalian microsomal P450s was the haemoprotein domain of the unique bacterial P450 from *Bacillus megaterium* (CYP102) which shares a degree of similarity with many membrane-based eukaryotic P450s in the possession of an FAD/FMN-containing flavoprotein reductase redox partner (reviewed in reference 1). However, the CYP2C5 crystallographic coordinates have proved to be a significant advance on CYP102 as a structural template for homology modelling of human P450s associated with drug metabolism, especially those of the CYP2 family where the sequence identities are greater than 40%.[33,34] By and large, it is reasonable to assume that a molecular model, built *via* homology with a high-resolution crystal structure, can be regarded as a satisfactory paradigm comparable to a crystallographic determination, provided that the sequence identity is greater than about 30%. For example, a plot of root-mean-square distances (rmsd) versus primary sequence identity (%) for matched α-carbons between several mammalian P450 crystal structures and homology models (as shown in Figure 1.6) provides a clear indication of this situation, where there is a strong logarithmic relationship ($R = 0.99$) between rmsd and sequence homology.

There are now 24 unique P450 crystal structures of which 16 are from bacterial species, one is of a fungal form, and seven are from mammalian species, of which two are rabbit enzymes and five are of human P450s associated with drug metabolism. The overall tertiary structures of these enzymes show a generally well-conserved common core (as shown in Figure 1.4 for selected structures and Table 1.10) which is presumably essential for the basic functionality of P450, irrespective of the species and tissue source. However, there are clearly certain differences between prokaryotic and eukaryotic P450

Table 1.10 Properties of typical P450 crystal structures.

CYP	*Resolution (Å)*	*PDB code*	*Dimensions*	*Secondary structural content*
101	1.63	2cpp	60Å × 30Å	45% α-helix 15% β-sheet
102	2.70	1fag	54Å × 35Å	48% α-helix 12% β-sheet
2C5	2.10	1nr6	60Å × 40Å	47% α-helix 10% β-sheet
55	1.00	1jfb	63Å × 48Å	45% α-helix 10% β-sheet

Rmsd and identity matrix for matched P450s

Identity % Rmsd (Å)	*101*	*102*	*55*	*2C5*
101	–	18.9	30.8	23.5
102	17.2	–	27.6	28.7
55	26.6	36.0	–	18.1
2C5	30.6	33.1	15.9	–

PDB, Protein Data Bank
Rmsd, root-mean-square distance.

structures which reflect the changes in redox partner and cellular location, namely: cytosolic, in the case of bacteria, and membrane-bound for mammalia. For example, there is an additional sequence of about 20–40 residues at the N-terminus of mammalian P450s which is not found in bacterial P450s, and this is thought to represent a membrane-anchor peptide that probably spans the phospholipid biolayer *via* an essentially α-helical structure.[35] In general, the prokaryotic P450s require an iron-sulphur redoxin as a redox partner, whereas most mammalian P450s utilise an FAD and FMN-containing flavoprotein reductase to supply reducing equivalents to P450. Exceptions exist, however, as mitochondrial P450s resemble those from bacterial sources in possessing redoxins as redox partners and, in addition, at least one prokaryotic P450 (ie. CYP102 from *Bacillus megaterium*) is known to employ a reductase domain directly linked to its haemoprotein portion,[36] and this is thought to be a main factor in explaining the exceptionally high turnover number exhibited by this enzyme.

1.6.1 Molecular Modelling and Dynamics of P450s

It is possible to construct three-dimensional molecular models of P450s based on homology with a suitable crystallographic template.[108,132,133] The majority of these unique crystal structures (24 in total) stem primarily from bacterial sources, whereas six are from mammalian species (including human) and one represents a fungal form ($P450_{nor}$). In the latter instance, the P450-catalysed reaction is nitric oxide reduction, which employs no oxygen in the overall process whereby two NO molecules are combined to form nitrous oxide. Over 5000 P450s are currently known from sequence determinations, and it is apparent that considerable variety exists in the P450 complement of different species. For example, there are 18 P450s in *Streptomyces coelicolor* and 33 in

Streptomyces avermitillis, whereas 57 human P450 enzymes are known; the exact functions of some of these still remain to be determined.[134] The well-conserved structural core of P450 ensures that homology modelling is likely to yield useful results, particularly where sequence identity is relatively high (ie. greater than 30%). In fact, a consideration of the overall three-dimensional structures of P450s from bacteria to mammalia (Figure 1.4) shows that the changes observed are not particularly large, as can be appreciated by an observed rmsd of 5.9Å between CYP102 and CYP2C5 from their structural overlay based on the sequence homology of matched residues (Table 1.10).

For human P450s constructed from rabbit CYP2C5, a rabbit form and the first mammalian P450 to have its X-ray structure determined, there is an approximately linear relationship between goodness-of-fit (measured as average rms distances between matched α-carbons in model and crystal structure) and primary sequence identity expressed as a percentage. The relevant data are provided in Table 1.7 and a graph of the corresponding relationship is shown in Figure 1.6. For example, the CYP2C8 and CYP2C9 models exhibit very good agreement with crystallographic coordinates, whereas that of CYP3A4 is relatively poor, and these findings are commensurate with the differences in sequence identity or homology. The details of the procedures utilised and the overall methodology for homology modelling is beyond the scope of this chapter and will only be outlined here, although the reader is referred to a recent publication by de Groot and coworkers[108] for more information. Briefly, the modelling process involves the generation of a satisfactory sequence alignment, followed by residue replacement, deletion, loop insertion, poor residue contact alleviation and then full energy minimisation using a combination of molecular mechanics and molecular dynamics simulations to explore the conformational flexibility of the enzyme at physiologically relevant temperatures.[108] The overall degree of confidence in X-ray crystal structures is shown in Table 1.11, where some of the known P450 structures are listed in certain categories to indicate the likely quality of these determinations.

Table 1.11 Confidence[a] in structural features of proteins determined by X-ray crystallography.[154]

Resolution	*3Å*	*2.5Å*	*2Å*	*1.5Å*
Chain tracing	Fair	Good	Good	Good
Secondary structure	Fair	Good	Good	Good
Sidechain conformations	–	Fair	Good	Good
Orientation of peptide planes	–	Fair	Good	Good
Protein hydrogens visible	–	–	–	Good
P450 crystal structures	2D6	2C9	2C5	101
conforming to each range[b]		2C8	2A6	119
		108	3A4	102

[a]The estimates of confidence in the structural features are approximate and it should be appreciated that they actually depend strongly on the crystallographic data.
[b]Corresponding P450 crystal structures which are representative of each category.

The fundamental questions facing P450 modellers and other researchers in, for example, the pharmaceutical industry are: what gives rise to substrate selectivity, what directs the course of a compound's metabolism and how can this be predicted, together with which methods represent ways of achieving a quantifiable estimate of binding, catalytic rate and overall clearance values for drug metabolism? In recent years, several of these key aspects of P450 structure, functionality and mechanism have been addressed with varying degrees of success. Homology modelling from P450 crystal structures, both bacterial and mammalian, has facilitated significant exploration of the molecular determinants for substrate selectivity, and has also enabled further insights into the recognition process for substrates and their orientation for metabolism by complementary active site amino acid residues. It is clear that certain key active site residues play important roles in the molecular recognition process and, in some cases, modulate the catalytic functionalities of P450 enzymes. Modelling has the powerful quality of conveying a large quantity of molecular structural information simultaneously into the visual field to provide considerable intuitive support for the observer engaged in the discovery process. It is often a challenge, however, to assimilate such information, and to decide on the appropriate procedures for achieving further understanding of the structure-function relationships involved.

The *in silico* techniques involved in constructing three-dimensional models of P450s by homology can be outlined[108] as follows:

1. Alignment of the template and target sequences
2. Replacement of relevant amino acid residues in the crystallographic template
3. Deletion and insertion of residues, where the latter task requires loop searching of the protein databank (PDB)
4. Alleviation of unfavourable steric contacts produced by the residue replacement process, and
5. Energy minimization of the raw structure *via* molecular mechanics and molecular dynamics.

Although there are several variations on this general scheme, it is apparent that such an approach can yield useful material for further study, including the use of site-directed mutagenesis[94] for analysing selected amino acids in order to test out various hypotheses surrounding their role in enzyme activity. Table 1.12 gives a summary of active site and substrate template volumes for a number of human P450s involved in drug metabolism, and there is a good level of concordance between these findings.

It is now possible to make direct comparisons between homology models and actual crystal structures of human P450s, and the results show that the homology modelling process is indeed a valid method for constructing P450s.[34] As might be expected, there is a clear correlation between goodness-of-fit (rmsd) and primary sequence identity when one compares models with the relevant crystallographic coordinates. Table 1.7 shows the results for a number

Table 1.12 Active site and substrate template volumes ($Å^3$) for high-quality P450 structures.

CYP	*Active site*	*Substrate template*	*Minimised energy ($kcal\,mol^{-1}$)*	*Comments*
2A6	340	345	−1419.660	Refined crystal structure
2B6	260 (2B4 value)	564	−1308.117	Model constructed from CYP2B4[a]
2C8	1386	1013	−1204.880	Refined crystal structure
2C9	1137	519	−1285.690	Refined crystal structure
2C19	1110	806	−1479.431	Model constructed from CYP2C9[b]
2D6	540	N/A	−1143.472	Refined crystal structure
3A4	1438	1264	−1431.405	Refined crystal structure
3A5	1438 (3A4 value)	1282	−1387.338	Model constructed from CYP3A4[c]

[a] = 4-Chlorophenylimidazole-bound structure (pdb code: 1suo).
[b] = Warfarin-bound structure (pdb code: 1og5).
[c] = Metyrapone-bound structure (pdb code: 1wof).
Notes: 1. In general, the substrate template size is within the volume limitations of the active site. In the case of CYP2C9, however, there is a very tight overlay of substrates which are highly superimposable, thus leading to a significantly lower volume than that available in the haem environment. 2. The minimised energy values, produced after 200 iterative cycles of molecular mechanics, indicate that low energy stable conformations have been obtained from relaxation of the raw structure.

of such determinations, from which it can be readily appreciated that the logarithm of rmsd is linearly related to sequence identity when this is expressed as 100% identity. The overall relationship bears some degree of similarity with that reported for other protein structures and their sequences,[115,135] and it is thought that most amino acid substitutions tend to preserve the overall protein fold.[136] The individual rmsd placements along the entire protein sequence for CYP2C9 are presented as Figure 1.11, which compares the actual crystallographic data with that of a homology model constructed from the CYP2C5 template. This indicates that there are certain peaks along the sequence where the rmsd value is relatively large, although the general rmsd values are reasonably low. The likely cause of the peaks in rmsd is probably due to the effects of substrate binding on the conformation of the enzyme, and it is interesting to note that the high rmsd values all occur in SRS regions. For the automated docking of substrates or inhibitors within the active sites of refined crystal structures and homology models, we have found that the AutoDock software[105,106] gives satisfactory agreement with experimental binding energies[137] obtained from literature values for K_m, K_S or K_i (Lewis and Ito, unpublished results).

It is important to emphasise that it is usually necessary to repair and refine the raw crystal structures prior to any further modelling studies. This is due to the fact that the original crystal structure requires initial relaxation using molecular mechanics, because various changes will have occurred during the crystal formation process as a result of packing forces. Furthermore, it is

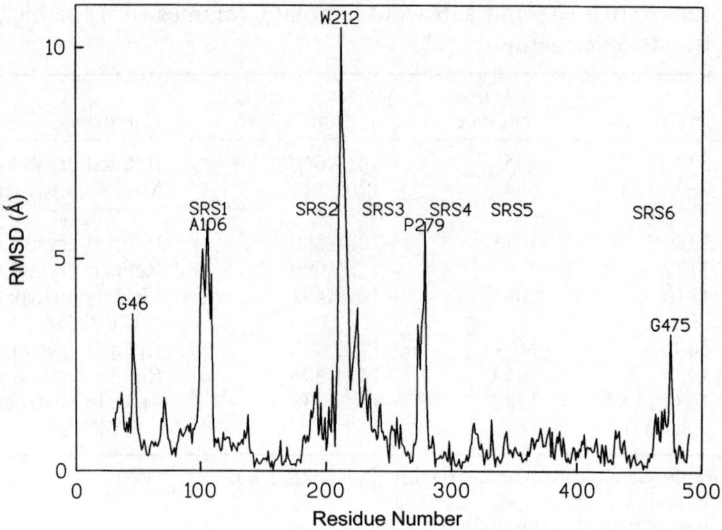

Figure 1.11 Root-mean-square distance (rmsd) profile for the CYP2C9 crystal structure and a homology model of the enzyme based on the CYP2C5 crystallographic template, where rmsd (Å) is plotted against residue number. Rmsd is the distance between matched α-carbons in the superimposed structures. The substrate recognition sites (SRS) regions are shown, together with those residues corresponding to the peak positions on the graph.

frequently necessary to add or remove certain amino acid residues from the structure such that the enzyme conforms to the wild-type, because residues are often changed to aid the crystallisation process. Thermal fluctuation of certain lengths of peptide may have resulted in an inability to resolve the entire structure satisfactorily, and these regions are thus omitted from the crystallographic coordinates. Occasionally, the backbone has been sufficiently resolved but some of the sidechains' conformations could not be determined. Again, these need to be added and then the entire structure energy minimised to provide a refined version. Such good quality models and refined high-resolution crystal structures, where there is a good degree of confidence in the information, represent satisfactory starting points for further study and it is often particularly helpful to have a bound inhibitor or substrate present in the enzyme's active site, as this offers an insight into the likely model of binding for other selective compounds. In the sphere of bacterial P450s,[42] for example, where there is current interest in utilising the modified enzymes in the targetted biosynthesis of antibiotics, it has been possible to employ the same homology modelling procedures outlined above for constructing certain P450s sequenced from *Streptomyces coelicolor*. In particular, it has been shown that one can readily produce, from the CYP154C1 crystallographic template, the 3-D structures of CYP105D5 (present in *S. coelicolor*) and CYP105M1 (present in *S. clavuligerus*) that are involved in the biosynthesis of actinorhodin and clavulanic acid, respectively (Lewis and Avignone-Rossa, unpublished results).

Molecular dynamics (MD) is an important technique for exploring the conformational space of biologically important macromolecules, such as proteins and nucleic acids, at physiologically relevant temperatures. Although computationally intensive, it is possible to utilise Newton's equations of motion to describe the likely behaviour of proteins such as P450 over realistic time-frames (nanoseconds) and in the presence of solvent molecules. MD has been employed to compare the conformational fluctuations in human P450 crystal structures and homology models, to show that there are close analogies between these two forms of 3-D structure in that, essentially, both represent a different starting conformation of the same protein. MD is a useful procedure for testing the validity of newly constructed protein models in order to derive measures of confidence in the energy-minimised structures obtained from molecular mechanics. It has been found that the Amber Force Field is most appropriate for running molecular dynamics simulations on proteins, and this is the one used to investigate the quality of homology models by comparing MD simulations with those carried out on crystal structures. In fact, it is found that the behaviour of both models and X-ray structures of P450s under dynamics is essentially the same, thus providing some degree of confidence in the homology modelling technique.

In addition to MM and MD procedures, there are several types of quantum mechanical (QM) approach available involving molecular orbital (MO) calculations of electronic structure which can be applied to the P450 system. In particular, it has been reported that Density Functional Theory (DFT) represents the most appropriate method for performing MO calculations on systems containing transition elements, and several groups of coworkers have studied the various stages in the P450 catalytic cycle using such procedures.[117,121] Although the calculations are computationally intensive, it has been possible to explain much of the catalysis of this system *via* DFT, albeit only on the haem-thiolate moiety.

1.6.2 Lipophilicity Relationships in P450 Substrate Binding and Selectivity

It has long been known that lipophilic character of a substrate has a bearing on P450 activity.[138–145] This relationship has been quantified for many of the important drug-metabolising enzymes, and it is clear that there is a relatively straightforward relationship between compound lipophilicity, obtained from experimental log P values, and substrate binding affinity expressed in terms of either K_m or K_S values.[83] For example, with substrates of CYP2B6, it is possible to discern a strong linear relationship between log P and -log K_m (Figure 1.12), where conversion to the corresponding ΔG values enables estimation of the average binding energy component which is unrelated to compound lipophilicity. For these data, it appears that the average binding energy, presumably a combination of π-π stacking and hydrogen bond interactions is $-2.4\,\text{kcal mol}^{-1}$, thus indicating that there is a common hydrogen-bonded and π-π stacked

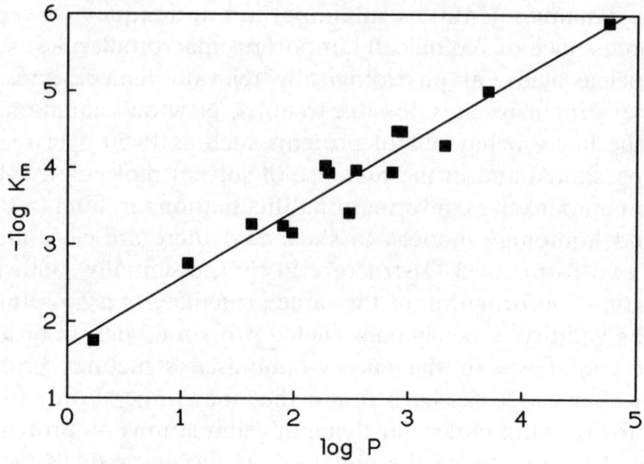

Figure 1.12 Lipophilicity relationship for selected CYP2B6 substrates where $-\log K_m$ is plotted against log P for 16 compounds. K_m is the Michaelis constant for CYP2B6-mediated metabolism and P is the octanol/water partition coefficient.

interaction with substrates in the active site region, and this is confirmed by molecular modelling studies based on the CYP2B4 crystal structure. Similar findings have been observed for other P450s, although there are often outliers and/or secondary parallel lines for different substrates which suggest that additional active site interactions are present, since the difference in intercepts indicates either an extra hydrogen bond ($-2\,\text{kcal mol}^{-1}$) or a π-π stacking ($-1\,\text{kcal mol}^{-1}$) interaction being formed between substrate and the enzyme's active site region. The combination of molecular modelling and site-directed mutagenesis techniques is particularly useful for investigating such possibilities. Consequently, theoretical studies represent important *in silico* probes for pursuing investigations of the structure and function of P450s, especially where they complement experimental findings and thus provide an explanation of the observed facts. However, many potential structural descriptors exist[146,147] and, therefore, extensive QSAR analysis should lead to further insights of P450 structure and function towards specific substrates and inhibitors.

One of the features of the P450 superfamily of enzymes is its wide diversity of substrates, possibly approaching a million separate compounds, and from most chemical classes of organic molecules. Moreover, P450 enzymes display altered substrate selectivities across the superfamily. In some cases, there is exquisite selectivity bordering on specificity towards a single compound, as in the example of the mammalian steroid hormone biosynthetic pathway (Figure 1.5) showing that many of the stages whereby androgens and oestrogens may be synthesised from cholesterol are catalysed by P450s of the CYP11, 17, 19 and 21 families. This highly regulated and well-coordinated pathway for the synthesis of steroids is an example of how tightly engineered enzyme systems have evolved during the early stages of metazoan development for physiologically important hormonal functions. In addition to steroid biosynthesis, endogenous P450 substrates include: fatty acids, eicosanoids, prostanoids and certain vitamins, such as vitamin D_3, together with some of their percursors. The known substrates of human P450 enzymes are presented in Table 1.9, which shows that certain forms are primarily associated with xenobiotic metabolism although some of these are also known to metabolise endogenous compounds such as the steroid hormones (Figure 1.5). It is thought that these may have originally evolved to provide homeostatic mechanisms for maintaining hormone levels at certain stages of the organism's development, such as for hormonal imprinting in the neonate.[148] The substrate selectivity of the xenobiotic-metabolising P450s, however, represents a broad and complex area, especially with respect to species differences, even those between rat, mouse and human, for example.[149]

It is possible to rationalise human P450 substrate selectivity and routes of metabolism based on the evidence from mammalian crystal structures and the three-dimensional models derived from them. The current homology models show very satisfactory correspondences with the recently reported human P450 crystal structures, and it is now possible to derive expressions for the goodness-of-fit in terms of rmsd values (where rmsd is the root mean square distance between matched α-carbons) and percentage sequence identity for the fitted structures, as discussed previously. The results imply that an overall homology of 40% and above between target and template sequences should produce models which can be expected to compare favourably with crystallographic findings.[115,135] The relationship between rmsd and percentage sequence homology can be formulated as follows for six known mammalian P450 crystal structures:

$$\begin{aligned} \text{rmsd} &= 0.056(100 - \%\text{homology}) \\ &\quad\;(\pm 0.005) \end{aligned}$$

where the correlation coefficient of 0.98 indicates the strength of this linear relationship. Careful analysis of the data reveals that the relation is actually exponential and, consequently, one can derive a logarithmic expression which gives slightly better statistical signifcance, with a correlation coefficient of 0.99 for six points, corresponding to CYP2A6, CYP2C8, CYP2C9, CYP2D6, CYP3A4 and CYP2B4 (see Figure 1.6).

1.7 Conclusions

The structure of P450s has remained essentially the same over the time course of biological evolution, with only minor variations to enable membrane binding for accommodating the move from prokaryotes to eukaryotes, and to allow for the occupancy of extreme environments such as those encountered in thermal oceanic vents. The many classes of reaction types, and huge chemical diversity of substrates, have been engineered over the evolutionary timescale by alteration of certain key amino acid residues whilst maintaining a generally conserved tertiary fold. Inherent conformational flexibility in certain P450s has provided the xenobiotic-metabolising forms with the possibility of accepting differently sized substrate molecules (eg. in CYP3A4 and CYP2C8), whereas those P450s with specific endogenous functions exhibit extremely narrow substrate selectivities (eg. CYP11, CYP17, CYP19 and CYP21). Over twenty unique P450 structures have been resolved crystallographically, mostly from bacteria, although several from mammalial species have recently become available, thus reflecting our anthropocentric viewpoint. Examination of further evolutionary relationships would necessitate the X-ray crystal structures of P450s from other animalia such as birds, reptiles, fish and amphibia, although much can be deduced from protein sequence comparisons. P450 has probably been the most extensively studied enzyme system known, as a considerable battery of spectroscopic and other physical techniques have been applied in an effort to understand its structure and function.[1,2] There have also been extensive biochemical and molecular biological studies which, when complemented by robust *in silico* tools such as QM/MM/MD calculations, provide an intriguing insight into the unique biophysical chemistry of this most versatile of Nature's catalysts.

Although the crystal structures of 24 unique P450s are now available out of over 5000 enzymes known from sequencing, it is apparent that, in general, there is a retention of secondary and tertiary structure across species as diverse as bacteria and mammalia. The common structural core of P450 has probably been largely conserved during the course of evolution, although modifications have occurred in order to incorporate the initially cytosolic enzyme in prokaryotes to a membrane-bound one in eukaryotic species. In addition, the variations in substrate selectivity have been 'tailored' over the course of biological species development by key amino acid residue changes in the SRS regions, primarily. Furthermore, redox partners have altered over evolutionary periods, changing from the iron-sulphur redoxins in most bacterial systems and in mitochrondrial P450s to flavoprotein reductase in the microsomal system, together with the unusual *Bacillus megaterium* P450 (CYP102) which comprises fused reductase and haemoprotein domains. There is considerable potential for harnessing the important catalytic properties of P450s, such as CYP102, for biodegradation and other commercial applications[150,151] including the design of industrial catalysts for processes which would otherwise be extremely protracted and difficult. However, the main interest in P450s lies in their major role in the metabolism of drugs and other xenobiotics where safety assessment of novel chemical entities is particularly relevant to the pharmaceutical industry.[152] In conclusion, it should be recognised that much is understood about P450 but there are likely to be further surprises in our quest for further knowledge of this extraordinary enzyme family.

Abbreviations

CYP is the standard abbreviation for cytochrome P450 when referring to the particular gene, enzyme family, subfamily or individual enzyme. However,

P450 (plural: P450s) is employed in reference to cytochromes P450 in general. CYP is italicised with reference to specific P450 genes.

ER	endoplasmic reticulum
QSAR	quantitative structure-activity relationship
bya	billion years ago
mya	million years ago
P	octanol/water partition coefficient
D	distribution coefficient at pH 7.4
SRS	Substrate recognition site
rmsd	root mean square distance
ROS	reactive oxygen species
E^o	redox potential
pK_{O_2}	-log K for oxygen binding where K is the equilibrium constant
Rmsd	root-mean-square distance
NCE	new chemical entity
SDM	site-directed mutagenesis
PDB	protein databank
M_r	relative molecular mass
pK_a	$\log K_a$ where K_a is the acid-base dissociation constant

Acknowledgements

The financial support of GlaxoSmithKline Research & Development Limited, Merck Sharp & Dohme Limited, ExxonMobil Biomedical Sciences Incorporated, British Technology Group Limited and the University of Surrey is gratefully acknowledged by DFVL. Yuko Ito would like to thank the Japanese Foundation for funding a Visiting Scientist Scholarship, and the Daiwa Anglo-Japanese Foundation for a travel grant.

References

1. D.F.V. Lewis, *"Cytochromes P450: Structure, Function and Mechanism"*, Taylor & Francis, London, 1996.
2. D.F.V. Lewis, *"Guide to Cytochromes P450 Structure and Function"*, Taylor & Francis, London, 2001.
3. P.R. Ortiz de Montellano (ed.), *"Cytochrome P450"*, Plenum, New York, 1986.
4. P.R. Ortiz de Montellano (ed.), *"Cytochrome P450"*, 2nd edition, Plenum, New York, 1995.
5. P.R. Ortiz de Montellano (ed.), *"Cytochrome P450"*, 3rd edition, Kluwer/Plenum, New York, 2005.
6. G.G. Gibson and P. Skett, *"Introduction to Drug Metabolism"*, 3rd edition, Nelson Thornes, Cheltenham, 2001.
7. D.F.V. Lewis and P. Hlavica, *Biochim. Biophys. Acta*, 2000, **1460**, 353.

8. P. Hlavica, J. Schulze and D.F.V. Lewis, *J. Inorg. Biochem.*, 2003, **96**, 279.
9. K.N. Degtyarenko and A.I. Archakov, *FEBS Lett.*, 1993, **332**, 1.
10. M.L. Wheelis, P. Kander and C.R. Woese, *Proc. Nat. Acad. Sci. USA*, 1992, **89**, 2930.
11. C.R. Woese, *Microbiol. Rev.*, 1987, **51**, 221.
12. D.C. Haines, D.R. Tomchick, M. Machius and J.A. Peterson, *Biochemistry*, 2001, **40**, 13456.
13. H. Li and T.L. Poulos, *Nature Struct. Biol.*, 1997, **4**, 140.
14. H. Li and T.L. Poulos, *Current Topics in Medicinal Chemistry*, 2004, **4**, 1789.
15. T.L. Poulos in: "Cytochrome P450", P.R. Ortiz de Montellano, ed., Plenum, New York, 1986, 505.
16. T.L. Poulos, B.C. Finzel and A.J. Howard, *J. Molec. Biol.*, 1987, **195**, 687.
17. K.G. Ravichandran, S.S. Boddupalli, C.A. Hasemann, J.A. Peterson and J. Deisenhofer, *Science*, 1993, **261**, 731.
18. P. Rowland, F.E. Blaney, M.G. Smyth, J.J. Jones, V.R. Leydon, A.K. Oxbrow, C.J. Lewis, M.G. Tennant, S. Modi, D.S. Eggleston, R.J. Chenery and A.M. Bridges, *J. Biol. Chem.*, 2006, **281**, 7614.
19. I. Schlichting, J. Berendzen, K. Chu, A.M. Stock, S.A. Maves, D.E. Benson, R.M. Sweet, D. Ringe, G.A. Petsko and S.G. Sligar, *Science*, 2000, **287**, 1615.
20. G.A. Schoch, J.K. Yano, M.R. Wester, K.J. Griffin, C.D. Stout and E.F. Johnson, *J. Biol. Chem.*, 2004, **279**, 9497.
21. E.E. Scott, Y.A. He, M.R. Wester, M.A. White, C.C. Chin, J.R. Halpert, E.F. Johnson and C.D. Stout, *Proc. Nat. Acad. Sci. USA*, 2003, **100**, 13196.
22. E.E. Scott, M.A. White, Y.A. He, E.F. Johnson, C.D. Stout and J.R. Halpert, *J. Biol. Chem.*, 2004, **279**, 27294.
23. H. Shimizu, S.-Y. Park, Y. Shiro and S.-I. Adachi, *Acta Crystallogr.*, 2001, **D58**, 81.
24. M.R. Wester, E.F. Johnson, C. Marques-Soares, P.M. Dansette, D. Mansuy and C.D. Stout, *Biochemistry*, 2003, **42**, 6370.
25. M.R. Wester, E.F. Johnson, C. Marques-Soares, S. Dijols, P.M. Dansette, D. Mansuy and C.D. Stout, *Biochemistry*, 2003, **42**, 9335.
26. M.R. Wester, J.K. Yano, G.A. Schoch, C. Yang, K.J. Griffin, C.D. Stout and E.F. Johnson, *J. Biol. Chem.*, 2004, **279**, 35630.
27. P.A. Williams, J. Cosme, V. Sridhar, E.F. Johnson and D.E. McRee, *Molec. Cell*, 2000, **5**, 121.
28. P.A. Williams, J. Cosme, A. Ward, H.C. Angove, D.M. Vinkovic and H. Jhoti, *Nature*, 2003, **424**, 464.
29. P.A. Williams, J. Cosme, D.M. Vinkovic, A. Ward, H.C. Angove, P.J. Day, C. Vonrhein, I.J. Tickle and H. Jhoti, *Science*, 2004, **305**, 683.
30. J.K. Yano, M.-H. Hsu, K.J. Griffin, C.D. Stout and E.F. Johnson, *Nature Structural and Molecular Biology*, 2005, **12**, 822.

31. J.K. Yano, M.R. Wester, G.A. Schoch, K.J. Griffin, C.D. Stout and E.F. Johnson, *J. Biol. Chem.*, 2004, **279**, 38091.
32. Y. Zhao, M.A. White, B.K. Muralidhara, L. Sun, J.R. Halpert and C.D. Stout, *J. Biol. Chem.*, 2006, **281**, 5973.
33. D.F. V. Lewis, Y. Ito and P.S. Goldfarb, *Expert Opinion in Drug Metabolism and Toxicology*, 2006, **1**, 5.
34. D.F.V. Lewis, Y. Ito and P.S. Goldfarb, *Current Drug Metab.*, 2006, **7**, 589.
35. T.L. Poulos and E.F. Johnson, in: *Cytochrome P450*, ed. P.R. Ortiz de Montellano, Kluwer/Plenum, New York, 2005, p. 87.
36. M.J. I. Paine, N.S. Scrutton, A.W. Munro, A. Gutierrez, G.C.K. Roberts and C.R. Wolf, in: *Cytochrome P450*, ed. P.R. Ortiz de Montellano, Kluwer/Plenum, New York, 2005, p. 115.
37. S.J. Mojzsis, G. Arrhenius, K.D. McKeegan, T.M. Harrison, A.P. Nutman and C.R.L. Friend, *Nature*, 1996, **384**, 55.
38. C. Bryant, *Biologist*, 1993, **40**, 58.
39. D.E. Canfield and A. Teske, *Nature*, 1996, **382**, 127.
40. A.M. Knoll, *Science*, 1992, **256**, 622.
41. R.H. Wickramasinghe and C.A. Villee, *Nature*, 1975, **256**, 509.
42. S.L. Kelly, D.E. Kelly, C.J. Jackson, A.G.S. Warrilow and D.C. Lamb, in: *Cytochrome P450*, ed. P.R. Ortiz de Montellano, Kluwer/Plenum, New York, 2005, p. 585.
43. S.B. Hedges, P.H. Parker, C.G. Sibley and S. Kumar, *Nature*, 1996, **381**, 226.
44. D.F.V. Lewis and G. Sheridan, *The Scientific World*, 2001, **1**, 151.
45. R.F. Doolittle, D.-F. Feng, S. Tsang, G. Cho and E. Little, *Science*, 1996, **271**, 470.
46. F.J. Gonzalez and D.W. Nebert, *Trends in Genetics*, 1990, **6**, 182.
47. D.R. Nelson and H.W. Strobel, *Molecular Biology of Evolution*, 1987, **4**, 572.
48. D.R. Nelson, *Comparative Biochemistry and Physiology*, 1998, **C121**, 15.
49. D.R. Nelson, *Archiv. Biochem. Biophys.*, 1999, **369**, 1.
50. D.R. Nelson, T. Kamataki, D.J. Waxman, F.P. Guengerich, R.W. Estabrook, R. Feyereisen, F.J. Gonzalez, M.J. Coon, I.C. Gunsalus, O. Gotoh, K. Okuda and D.W. Nebert, *DNA and Cell Biology*, 1993, **12**, 1.
51. D.R. Nelson, L. Koymans, T. Kamataki, J.J. Stegeman, R. Feyereisen, D.J. Waxman, M.R. Waterman, O. Gotoh, M.J. Coon, R.W. Estabrook, I.C. Gunsalus and D.W. Nebert, *Pharmacogenetics*, 1996, **6**, 1.
52. D.W. Nebert, D.R. Nelson, M.J. Coon, R.W. Estabrook, R. Feyereisen, Y. Fujii-Kuriyama, F.J. Gonzalez, F.P. Guengerich, I.C. Gunsalus, E.F. Johnson, J.C. Loper, R. Sato, M.R. Waterman and D.J. Waxman, *DNA and Cell Biology*, 1991, **10**, 1.
53. F.P. Guengerich, *American Scientist*, 1993, **81**, 440.
54. K.A. Nielsen and B.L. Møller, in: *Cytochrome P450*, ed. P.R. Ortiz de Montellano, Kluwer/Plenum, New York, 2005, p. 553.

55. D. Mansuy, *Comparative Biochemistry and Physiology*, 1998, **C121**, 5.
56. M. Newcomb and R.E.P. Chandrasena, *Biochem. Biophys. Res. Commun.*, 2005, **338**, 394.
57. D.F.V. Lewis and J.L.C.M. Dorne, *The Scientic World*, 2006, **6**, 718.
58. D.F.V. Lewis and A. Wiseman, *Enzyme and Microbial Technology*, 2005, **36**, 375.
59. P. Cloud, *Paleobiology*, 1976, **2**, 351.
60. F.P. Guengerich, *Chem. Res. Toxicol.*, 2001, **14**, 611.
61. F.P. Guengerich, *Arch. Biochem. Biophys.*, 2003, **409**, 59.
62. A.K. Daly, *Current Topics in Medicinal Chemistry*, 2004, **4**, 1733.
63. W.E. Evans and M.V. Relling, *Science*, 1999, **286**, 487.
64. J.B. Houston, K.E. Kenworthy and A. Galetin, in: *Drug Discovery and Development*, eds. J.S. Lee, R.S. Obach and M.B. Fisher, Marcel Dekker, New York, 2003, p. 211.
65. M. Ingelman-Sundberg, *Drug Metab. Disp.*, 2001, **29**, 570.
66. M. Ingelman-Sundberg, M. Oscarsson and R. A. McLellan, *Trends in Pharmacological Sciences*, 1999, **20**, 342.
67. C. Ioannides (ed.), *"Cytochromes P450: Metabolic and Toxicological Consequences"*, CRC Press, Boca Raton, Florida, 1996.
68. M. Murray, in: *Modern Biopharmaceuticals: Design, Development and Optimization*, ed. J. Knaeblin, Wiley-VCH Verlag, Weinheim, Volume 4, 2005, p. 1603.
69. A. Rannug, A.-K. Alexandrie, I. Persson and M. Ingelman-Sundberg, *Journal of Occupational and Environmental Medicine*, 1995, **37**, 25.
70. A.D. Rodrigues, *Drug Metab. Disp.*, 2005, **33**, 1567.
71. A.D. Rodrigues and T.H. Rushmore, *Current Drug Metab.*, 2002, **3**, 289.
72. T. Shimada, H. Yamazaki, M. Mimura, Y. Inui and F.P. Guengerich, *J. Pharmacol. Exp. Thera.*, 1994, **270**, 414.
73. U. Yasar, A.M. Bennet, E. Eliasson, S. Lundgren, B. Wiman, U. de Faire and A. Rane, *Pharmacogenetics*, 2003, **13**, 715.
74. B.K. Hawkins and J.H. Dawson, *Frontiers in Biotransformation*, 1992, **7**, 216.
75. S.G. Sligar, *Biochemistry*, 1976, **15**, 5399.
76. S.G. Sligar, D.L. Cinti, G.G. Gibson and J.B. Schenkman, *Biochem. Biophys. Res. Commun.*, 1979, **90**, 925.
77. F.P. Guengerich and W.W. Johnson, *Biochemistry*, 1997, **36**, 14741.
78. E.J. Mueller, P.J. Loida and S.G. Sligar, in: *Cytochrome P450*, ed. P.R. Ortiz de Montellano, Plenum, New York, 1995, p. 83.
79. J. Blanck, H. Rein, M. Sommer, O. Ristau, G. Smettan and K. Ruckpaul, *Biochem. Pharmacol.*, 1983, **32**, 1683.
80. J. Blanck, O. Ristau, A.A. Zhukov, A.I. Archakov, H. Rein and K. Ruckpaul, *Xenobiotica*, 1991, **21**, 121.
81. D. Schwarz, *Frontiers in Biotransformation*, 1991, **5**, 94.
82. T.H. Bayburt and S.G. Sligar, *Proc. Nat. Acad. Sci., USA*, 2002, **99**, 6725.

83. D.F.V. Lewis, M.N. Jacobs and M. Dickins, *Drug Discovery Today*, 2004, **9**, 530.
84. D. Schwarz, V. Krüger, A.A. Chernoglov, S.A. Usanov and A. Stier, *Biochem. Biophys. Res. Commun.*, 1993, **195**, 889.
85. D. Mansuy and J.-P. Renaud, in: *Cytochrome P450*, ed. P.R. Ortiz de Montellano, Plenum Press, New York, 1995, p. 537.
86. S. Rendic, *Drug Metab. Rev.*, 2002, **34**, 83.
87. S. Rendic and F.J. DiCarlo, *Drug Metab. Rev.*, 1997, **29**, 413.
88. C. Ioannides and D.F.V. Lewis, *Current Topics in Medicinal Chemistry*, 2004, **4**, 1767.
89. I. Fleming, U.R. Michaelis, D. Bredenkötter, B. Fisslthaler, F. Dehghani, R.P. Brandes and R. Busse, *Cirulation Research*, 2001, **88**, 44.
90. B. Fisslthaler, R. Popp, L. Kiss, M. Potente, D.R. Harder, I. Fleming and R. Busse, *Nature*, 1999, **401**, 493.
91. M. Dickins, *Current Topics in Medicinal Chemistry*, 2004, **4**, 1745.
92. P. Honkakoski and M. Negishi, *Biochem. J.*, 2000, **347**, 321.
93. D.F.V. Lewis, *Biochem. Pharmacol.*, 2000, **60**, 293.
94. T.L. Domanski and J.R. Halpert, *Current Drug Metab.*, 2001, **2**, 117.
95. O. Gotoh, *J. Biol. Chem.*, 1992, **267**, 83.
96. D.F.V. Lewis, *J. Inorg. Biochem.*, 2002, **91**, 502.
97. D.F.V. Lewis, *The Scientific World*, 2003, **3**, 536.
98. D.F.V. Lewis, *Arch. Biochem. Biophys.*, 2003, **409**, 32.
99. D.F.V. Lewis, *Current Med. Chem.*, 2003, **10**, 1955.
100. D.F.V. Lewis, *Current Drug Metab.*, 2003, **4**, 331.
101. D.F.V. Lewis, *Pharmacogenomics*, 2003, **4**, 387.
102. D.F.V. Lewis, *Pharmacogenomics*, 2004, **5**, 305.
103. D.F.V. Lewis, *Toxicol. in vitro*, 2004, **18**, 89.
104. D.F.V. Lewis, *Expert Opinion in Drug Metabolism and Toxicology*, 2005, **1**, 5.
105. G.M. Morris, D.S. Goodsell, R. Huey and A.J. Olsen, *J. Comp.-Aided Molec. Des.*, 1996, **10**, 293.
106. G.M. Morris, D.S. Goodsell, R.S. Halliday, R. Huey, W.E. Hart, R.K. Belew and A.J. Olsen, *J. Comp. Chem.*, 1998, **19**, 1639.
107. Y. Ito, D.F.V. Lewis, P.S. Goldfarb, P.S. and H. Kondo, *J. Mol. Mod. Molec. Graph.*, 2006, in press.
108. M.J. de Groot, D.F.V. Lewis and S. Modi, in: *Comprehensive Medicinal Chemistry II*, Volume 5, eds. D.J. Triggle and J.B. Taylor, Elsevier, Oxford, 2006, p. 809.
109. E.F. Johnson, *Trends in Pharmacological Sciences*, 1992, **13**, 122.
110. E.F. Johnson, *Drug Metab. Disp.*, 2003, **31**, 1532.
111. R.L. P. Lindberg and M. Negishi, *Nature*, 1989, **339**, 632.
112. M. Ridderstrom, I. Zamora, O. Fjellstrom and T.B. Andersson, *J. Med. Chem.*, 2001, **44**, 4072.
113. K.R. Korzekwa and J.P. Jones, *Pharmacogenetics*, 1993, **3**, 1.
114. S. Modi, M.J. Sutcliffe, W.U. Primrose, L.-Y. Lian and G.C.K. Roberts, *Nature Structural Biology*, 1996, **3**, 414.

115. A.M. Lesk and C. Chothia, *Philosophical Transactions of the Royal Society London A*, 1986, **317**, 345.
116. C. Jung, V. Schünemann, F. Lendzian, A.X. Trautwein, J. Contzen, M. Galander, L.H. Böttger, M. Richter and A.-M. Barra, *Biolog. Chemi.*, 2005, **386**, 1043.
117. T. Kamachi and K. Yoshizawa, *J. Amer. Chem. Soc.*, 2003, **125**, 4652.
118. T.M. Makris, R. Davydov, I.G. Denisov, B.M. Hoffman and S.G. Sligar, *Drug Metab. Rev.*, 2002, **34**, 691.
119. T.M. Makris, I. Denisov, I. Schlichting and S.G. Sligar, in: *Cytochrome P450*, ed. P.R. Ortiz de Montellano, Kluwer/Plenum, New York, 2005, p. 149.
120. J.C. Schöneboom, F. Neese and W. Thiel, *J. Amer. Chem. Soc.*, 2005, **127**, 5840.
121. S. Shaik and S.P. De Visser, in: *Cytochrome P450*, ed. P.R. Ortiz de Montellano, Kluwer/Plenum, New York, 2005, p. 45.
122. M.T. Green, J.H. Dawson and H.B. Gray, *Science*, 2004, **304**, 1653.
123. P.G. Debrunner, in: *Applications of Mössbauer Spectroscopy*, Volume 1, ed. R.L. Cohen, Academic Press, New York, 1976, p. 171.
124. D.F.V. Lewis, Y. Ito and P.S. Goldfarb, *Curr. Med. Chem.*, 2006, **13**, 2645.
125. M.J. Coon, A.D.N. Vaz and L.L. Bestervelt, *FASEB J.*, 1996, **10**, 428.
126. M.J. Coon, A.D.N. Vaz, D.F. McGinnity and H.-M. Peng, *Drug Metab. Disp.*, 1998, **26**, 1190.
127. J.M. Pratt, T.I. Ridd and L.J. King, *Chem. Commun.*, 1995, 2297.
128. F.P. Guengerich, in: *Enzyme Systems that Metabolise Drugs and other Xenobiotics*, ed. C. Ioannides, Wiley, Chichester, 2002, p. 33.
129. J.S. Miles, A.W. Munro, B.N. Rospendowski, W.E. Smith, J. McKnight and A.J. Thomson, *Biochem. J.*, 1992, **288**, 503.
130. A.J. Fulco, *Ann. Rev. Pharmacol. Toxicol.*, 1991, **31**, 177.
131. A.W. Addison and S. Burman, *Biochim. Biophys. Acta*, 1985, **828**, 362.
132. M.J. de Groot, S.B. Kirton and M.J. Sutcliffe, *Current Topics in Medicinal Chemistry*, 2004, **4**, 1803.
133. S. Ekins, M.J. de Groot and J.P. Jones, *Drug Metabol. Disp.*, 2001, **29**, 936.
134. F.P. Guengerich, in: *Cytochrome P450*, ed. P.R. Ortiz de Montellano, Kluwer/Plenum, New York, 2005, p. 377.
135. C. Chothia and A.M. Lesk, *EMBO J.*, 1986, **5**, 823.
136. I. Ladunga and R.F. Smith, *Protein Engineering*, 1997, **10**, 187.
137. C. Bauer, A.M. Osman, G. Cercignani, N. Gialluca and M. Paolini, *Biochem. Pharmacol.*, 2001, **61**, 1049.
138. M. Vermeir, N. Boens and K.P.M. Heirweigh, *Biochim. Biophys. Acta*, 1992, **1107**, 93.
139. D.A. Smith, *European Journal of Pharmaceutical Sciences*, 1994, **2**, 69.
140. D.A. Smith, M.J. Ackland and B.C. Jones, *Drug Discovery Today*, 1997, **2**, 406.
141. D.A. Smith, M.J. Ackland and B.C. Jones, *Drug Discovery Today*, 1997, **2**, 479.

142. C. Hansch, *Acc. Chem. Res.*, 1993, **26**, 147.
143. C. Hansch and L. Zhang, *Drug Metabol. Rev.*, 1993, **25**, 1.
144. C. Hansch, S.B. Mekapati, A. Kurup and S.P. Verma, *Drug Metabol. Rev.*, 2004, **36**, 105.
145. H. Gao and C. Hansch, *Drug Metab. Rev.*, 1996, **28**, 513.
146. D.J. Livingstone, *Journal of Chemical Information and Computer Sciences*, 2000, **40**, 195.
147. M. Karelson and V.S. Lobanov, *Chem. Rev.*, 1996, **96**, 1027.
148. D.J. Waxman and R.K.H. Chang, in: *Cytochrome P450*, ed. P.R. Ortiz de Montellano, Plenum, New York, 1995, 391–417.
149. F.P. Guengerich, *Chem.-Biol. Interact.*, 1997, **106**, 161.
150. F.P. Guengerich, *Environmental Health Perspectives*, 1995, **105** (Suppl 5), 25.
151. F.P. Guengerich, *Nature Reviews*, 2002, **1**, 359.
152. G.T. Tucker, J.B. Houston and S.-M. Huang, *Brit. J. Clin. Pharmacol.*, 2001, **52**, 107.
153. K. Brosen, *International Clinical Psychopharmacology*, 1996, **11**(Suppl 1), 23.
154. A.M. Lesk, *"Introduction to Protein Architecture"*, Oxford University Press, Oxford, 2001.

CHAPTER 2

Generation of Reactive Intermediates by Cytochromes P450

HERMANN M. BOLT AND PETER H. ROOS

Institut für Arbeitsphysiologie an der Universität Dortmund, Leibniz Research Centre for Work Environment and Human Factors, Ardeystr. 67, D-44139 Dortmund, Germany

Table of Contents

Issues in Toxicology
Cytochromes P450: Role in the Metabolism and Toxicity of Drugs and other Xenobiotics
Edited by Costas Ioannides

2.1 Introduction

Studies on biological toxication and detoxication processes have provided explanations for the toxicity for many chemicals. In a previous book, the role of metabolism in chemical toxicity has been highlighted.[1] Although many phase I, phase II and phase III pathways may be involved in the generation of toxic metabolites from nontoxic precursors, toxication mechanisms mediated by CYP enzymes play a predominant role. Prominent examples are alkanes, alkenes, arenes, amines, thiones and thioethers.

As a consequence of the diverse profiles of CYP enzyme expression, methods of modelling the nature and quantities of biologically reactive intermediates have received attention. The multiplicity of pathways and the chemistry of CYP reactions involved have been discussed in a previous book devoted to cytochromes P450.[1,2] Based on these principles, the present chapter will concentrate on applications of the general paradigms for biologically reactive CYP metabolites, which link these metabolites with human and/or animal toxicity.

Important chemical classes of toxic intermediates are electrophilic compounds, free radicals, carbenes and nitrenes, and activated oxygen species. Representative CYP-mediated metabolic activation pathways of chemicals to toxic reactive intermediates are summarised in Figure 2.1 and Table 2.1.[1]

Specifically, CYP isoenzymes are often involved in the bioactivation of procarcinogens to ultimate carcinogens. One of the most thoroughly studied examples is the metabolism of the procarcinogen benzo[a]pyrene *via* enantioselective 7,8-epoxidation by CYP1A1 and enantioselective hydrolysis

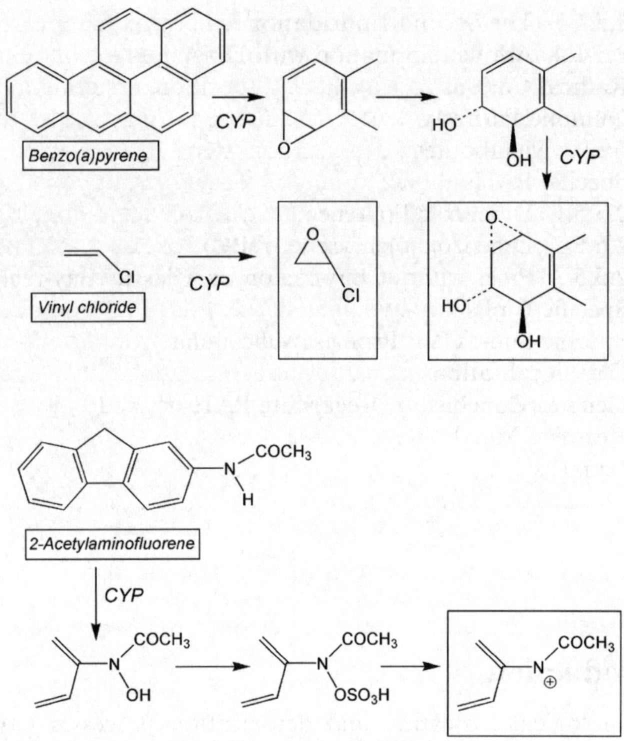

Figure 2.1 Different examples of the CYP-mediated metabolic activation of chemicals to toxic reactive intermediates (after Vermeulen[1]).

Table 2.1 Chemically reactive toxic initiators, types of interactions, and biological targets.

Initiators	*Examples*	*Reaction type*	*Targets*
Electrophiles	Michael acceptors Acyl halides Quinones	Covalent binding	Thiols
Radicals	CCl_4	Covalent binding	Thiols, nucleic acids, (phospho)lipids
	Halothane	H abstraction	Thiols, nucleic acids, (phospho)lipids
	Adriamycin	Reduction	Molecular oxygen
Reactive oxygen species	Superoxide	Reduction	Metal ions
	Peroxide	Oxidation	Thiols
	OH Radical	H abstraction	Thiols, nucleic acids, (phospho)lipids
Heavy metals	Cd, Hg, Pb, Pt, Au, Ag	Complexation	Thiols, nucleic acids
Polycations	Gentamycin Kanamycin	Charge-charge interaction	Phospholipids

Adapted from Vermeulen.[1]

by epoxide hydrolase to two enantiomers of the proximate carcinogen (±)-benzo[a]pyrene-7,8-dihydrodiol.[3] (−)*trans*-Benzo[a]pyrene-7,8-dihydrodiol is further metabolised by CYP1A1 at the 9,10-position, resulting in the ultimate carcinogen (+)*anti*-benzo[a]pyrene-*7R,8S*-diol-epoxide. Besides toxication by 7,8- and 4,5-epoxidation, also detoxication steps by 9,10- and 2,3-epoxide formation, phenol and quinone formation occur. In addition, quinone formation may result in potentially toxic one-electron semiquinone radicals. At the level of the proximate carcinogen, a variety of other CYP-mediated reactions and phase II metabolic reactions can lead to detoxication.[1]

This classical example has highlighted the role of a critical quantitative balance of toxication and detoxication mechanisms.[1] Important research during the last decades has been directed toward elucidating tissue- and cell-specific determinants of this balance, in order to explain organotropisms and cellular events in chemical toxicity.

2.2 Determinants of Biologically Reactive Metabolites

2.2.1 General Principles of Biological Reactivity

Oxidation of aliphatic and aromatic π-electron systems by CYP enzymes is a mechanism that is underlying relevant cases of biological activation.[2] In the 1970s, the example of the 7,8-dihydroxy-9,10-epoxy-7,8,9,10-tetrahydrobenzo[a]pyrenes as the ultimate mutagenic and carcinogenic metabolite of benzo[a]pyrene was a breakthrough case in the application of quantum mechanical calculations to assess the possible roles of potential biologically-reactive entities (diol-epoxides) in genotoxicity and carcinogenesis.[3] These early calculations were based on the pertubational molecular orbital (PMO) procedure of Dewar.[4] This enabled a prediction of relative reactivity, expressed as the ease of carbonium ion formation, for different diol epoxides from different polycyclic aromatic hydrocarbons. The calculated reactivities were taken to predict the respective diol epoxide mutagenic activities. The greater calculated reactivities for diol epoxides in which the oxirane oxygen formed part of a "bay region" provided a decisive theoretical support for the "bay region" theory.[5]

Semiempirical quantum mechanical methods and applications to studies of reactive intermediates of polycyclic aromatic hydrocarbons were subsequently refined.[6,7] It was confirmed that the carcinogenicity of individual polyclyclic aromatic hydrocarbons was correlated with a lack of stability of the respective bay-region epoxide rings.[7] In this context, a pivotal role was assigned to the ease of carbocation formation[6] and to the carbocation delocalisation energies associated with the epoxide ring opening.[8]

Later, a variety of *ab initio* quantum chemical methods were employed in this field, including calculations for the attack of the ultimate carcinogens on DNA constituents. Such elements were integrated into global QSAR (quantitative structure-activity relationship) programmes for prediction of toxicity.[9]

2.2.2 Application to Aliphatic Epoxides

Beginning the early 1980s, attempts were published at modelling quantitatively the properties of biologically reactive aliphatic epoxides, in order to predict their genotoxicity and/or carcinogenicity. In these publications semiempirical quantum chemical methods were applied. Considering the general mechanism of epoxide ring opening (see Scheme 2.1), key issues for modelling were the epoxide C–O bond strength, the expoxide's tendency towards protonation, and the reactivity of the intermediate carbocation.[10–12] In particular, the protonation step has been addressed as a means to characterise potential carcinogens, because epoxide carcinogenicity was found associated with relatively strong negative electrostatic potential near the oxygens, *i.e.* the protonation sites.[13]

Later, *ab initio* molecular orbital methods provided additional descriptors for biologically reactive intermediates,[13,14] but some semiempirical methods,

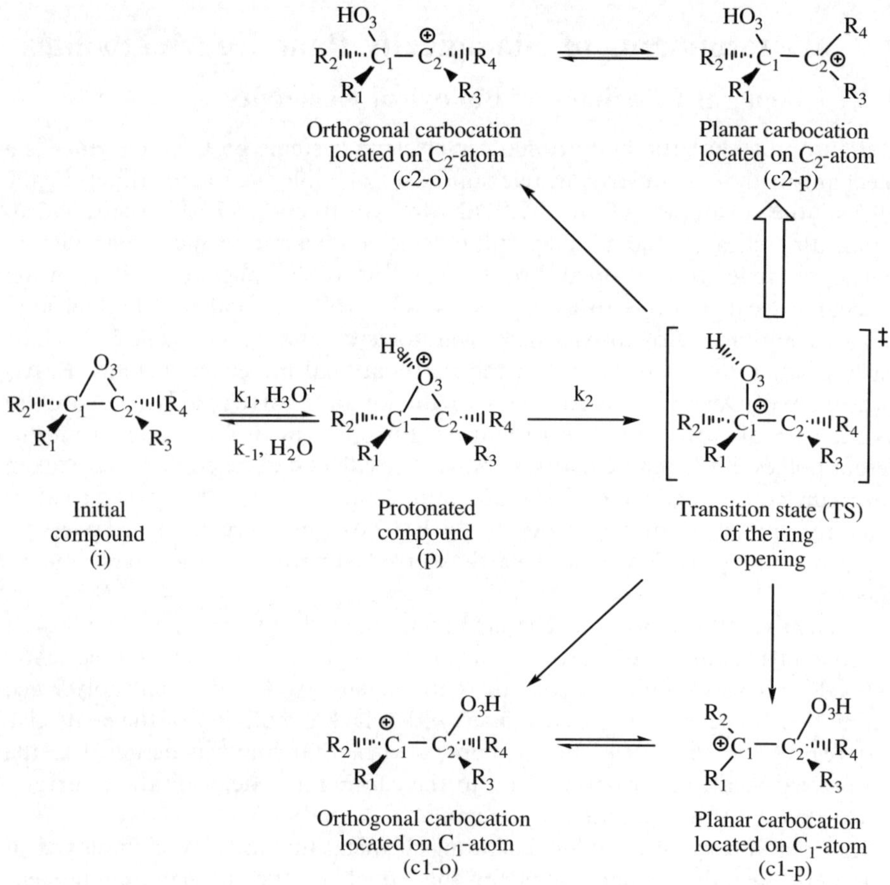

Scheme 2.1 Reaction pathways of oxirane (epoxide) hydrolysis considered. Reprinted from Kirkovski *et al.*,[16] with permission of *Alliance Communications Group, a division of Allen Press, Inc.*

notably AM1, remained an established procedure to calculate descriptors for this purpose.[15] As an example, for methyl-substituted oxiranes (ethylene oxide derivatives) both the AM1 and MNDO methods produced excellent correlations between the logarithm of the experimental rate constant k_{eff} of epoxide hydrolysis with the calculated total energy E_t of ring opening, as shown in Figure 2.2.[16]

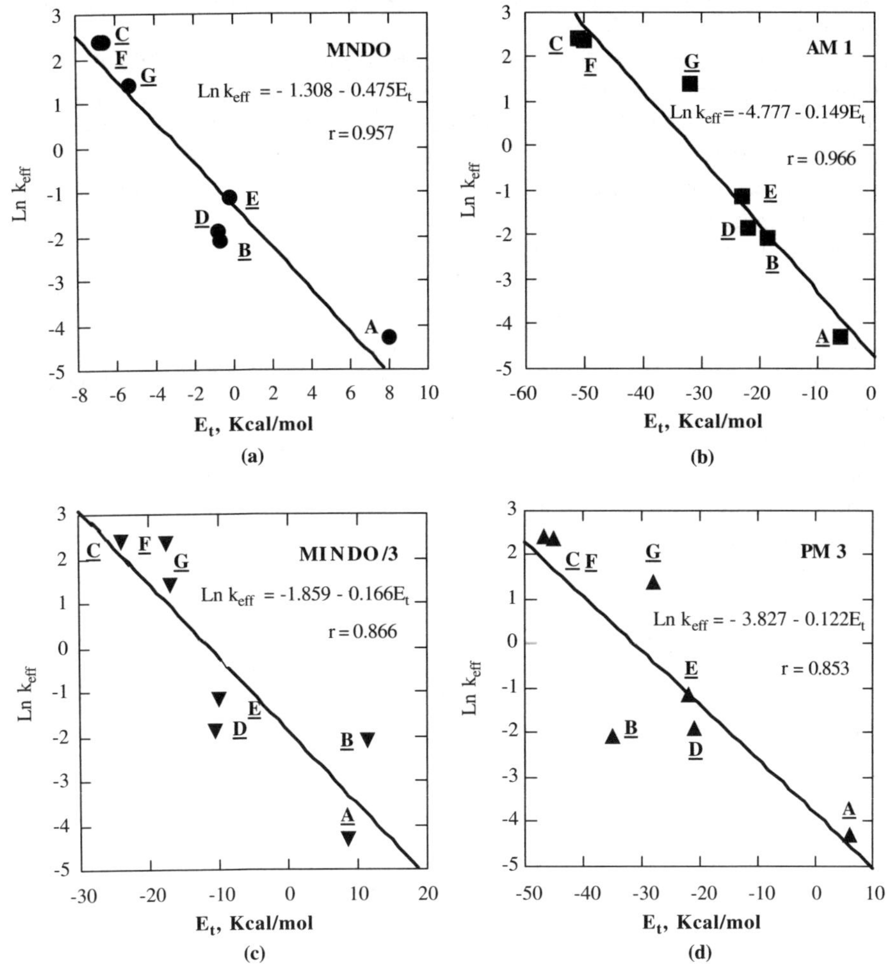

Figure 2.2 Linear correlations of experimentally determined hydrolysis rate constans (Ln k_{eff}, $M^{-1}s^{-1}$) of epoxides (A–G) and the total energies of hydrolysis process (Et, Kcal/mol), calculated by (a) MNDO, (b) AM1, (c) MINDO/3, and (d) PM3. A: oxirane (ethylene oxide); B: 2-methyl-oxirane (proylene oxide); C: 2,2-dimethyl-oxirane; D: *trans*-2,3-dimethyl-oxirane; E: *cis*-2,3-dimethyl-oxirane; F: 2,2,3-trimethyl-oxirane; G: tetramethyl-oxirane. Reprinted from Kirkovski *et al.*,[16] with permission of *Alliance Communications Group, a division of Allen Press, Inc.*

Modelling studies of the biological epoxide intermediates of aliphatic olefins arrived at the conclusion that, with regard to the epoxide reactivity, an intermediate level is required for an "optimum" for carcinogenicity. This paradigm was useful for carcinogenicity rankings and predictions.[12,13]

2.2.3 Structure-Genotoxicity Relationship: Aliphatic Olefins and Their Epoxides

The classical example for this avenue of thinking was published in the 1980s by Jones and Mackrodt.[17,18]

The human carcinogen vinyl chloride is bioactivated by CYP2E1 to the reactive intermediate chloroethylene oxide (chlorooxirane; *Formulae 2.1:D*) that reacts with nucleic acid bases to produce cyclic "etheno" adducts. This is seen as the key to vinyl chloride genotoxicity and carcinogenesis. The characteristic tumour induced by vinyl chloride, in humans as well as in experimental animals, is liver haemangiosarcoma.[19]

Jones and Mackrodt[17] have used the MINDO/3 procedure to describe the stabilities of a number of chlorinated epoxides (*Formulae 2.1: A–E*). These figures were related to bacterial mutagenicity (in *E. coli* K_{12}) of the parent chlorinated olefins in the presence of a metabolising system *in vitro*.[20]

The mutagenicity data for chloroethylenes were plotted against the two-centre bond energies (E_{C-O}) of the corresponding epoxides. The result is shown in Figure 2.3 (top panel). It was concluded that aliphatic epoxides needed an "optimum" stability and reactivity for genotoxicity, as characterised by a window of the 2-centre C–O bond energy of the more reactive C–O bond of the oxirane ring between $-14.5\,\text{eV}$ and $-12.8\,\text{eV}$.[17]

The theory was further refined[18] based on quantitative data for the metabolic rates of halogenated olefins *in vivo* and for the quantities of pre-neoplastic foci induced by these compounds in rat liver.[21] The respective results are shown in the lower panel of Figure 2.3. This example demonstrated that for the modelling of conditions *in vivo*, the consideration of toxicokinetic aspects was absolutely crucial, as the rates of oxidative metabolism of individual halogenated olefins were scattered over 2 orders of magnitude. From the refined model it

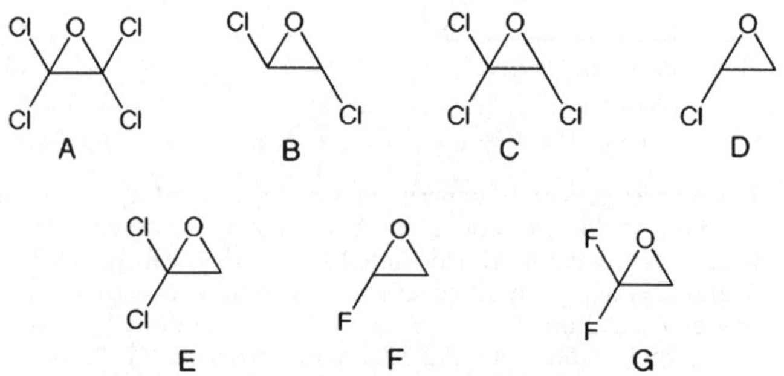

Formulae 2.1 Structures of epoxide metabolites derived from halogenated ethylenes.

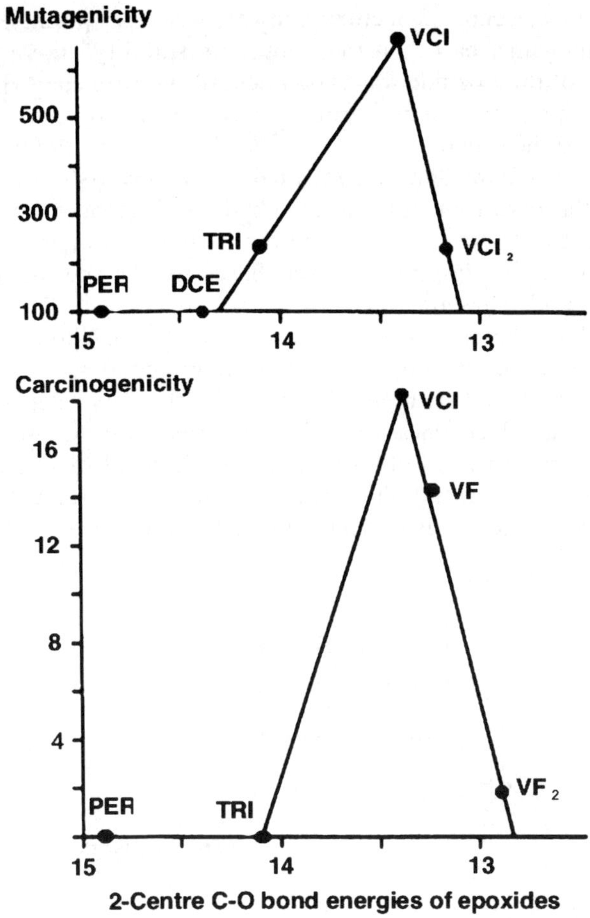

Figure 2.3 The "optimum stability" theory of epoxides. Mutagenicity (Greim *et al.*,[20] top panel) and carcinogenicity (Bolt *et al.*,[21] bottom panel) of the metabolites of halogenated ethylenes are plotted against the 2-centre C–O bond energies [eV] of the corresponding epoxides (Jones and Mackrodt[17,18]). Carcinogenicity and mutagenicity of the epoxides (see Formulae 1: A–G) require an "optimum" between stability and reactivity. According to Bolt,[22] with modifications.

was concluded that a reactivity range between –14.1 and –12.9 eV would be indicative of potentially carcinogenic oxiranes.[18] This range included the epoxides of ethylene, propylene, vinyl fluoride, vinyl chloride, vinyl bromide, 1,3-butadiene, acrylonitrile and acrylamide. All these compounds proved to be carcinogenic, on the basis of their reactive epoxide intermediates.[22] On the contrary, 2,2-dimethyl-oxirane (2-methyl-1,2-epoxypropane), the epoxide of isobutene (2-methylpropene) was predicted not to be sufficiently reactive for clear genotoxicity.[18] Subsequently, it was experimentally confirmed that this epoxide was not mutagenic in the *S. typhimurium* assay.[23]

The epoxide of vinyl acetate, 2-acetoxy-oxirane, was also predicted to be genotoxic and carcinogenic, based on the "optimum stability" theory.[18] The mutagenicity *in vitro* of this epoxide was experimentally confirmed.[24] However, vinyl acetate does not present itself as a systemic carcinogen;[25] the reason is that it is not bioavailable to the general circulation.[26] Both upon oral and inhalation exposure vinyl acetate is immediately converted by carboxylesterases, at the port of entry into the organism, into acetaldehyde and acetic acid.[27] These primary metabolites exert local effects, resulting at high doses in local carcinogenesis within the upper respiratory tract. Based on the specific mode of action, a threshold model has been proposed for this effect.[28] This is summarised in Figure 2.4. The case again demonstrates the key importance of toxicokinetics for the biological action of reactive intermediates.

Taking these elements together, Figure 2.5 compiles the major determinants for the *in vivo* biological effectiveness of reactive intermediates. On the side of toxicokinetics, there are processes of formation and disposal (toxication and detoxication) of the reactive metabolite. The latter needs a certain degree of reactivity to effectively react with critical cellular constituents. At the same

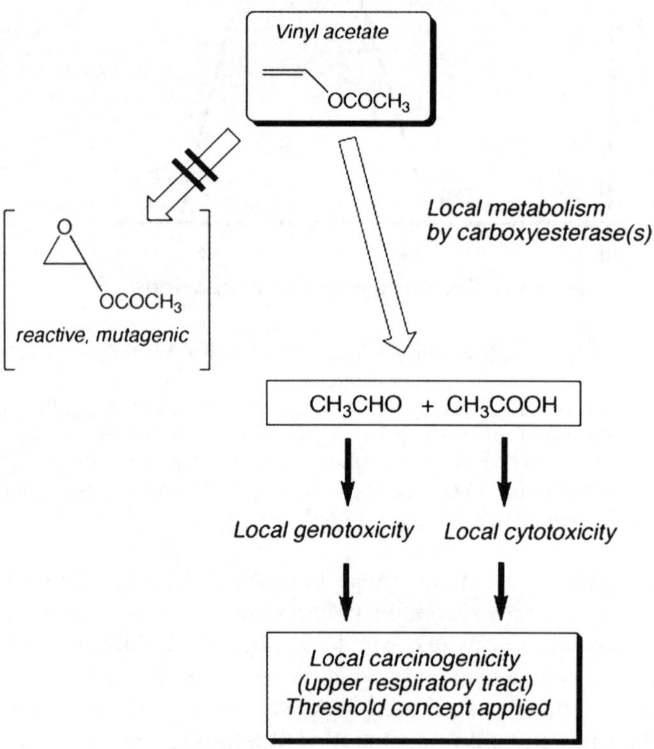

Figure 2.4 Vinyl acetate is not systemically bioavailable, because it is converted at the port of entry into the organism into acetaldehyde and acetic acid. Therefore, the potentially carcinogenic epoxide is not formed.

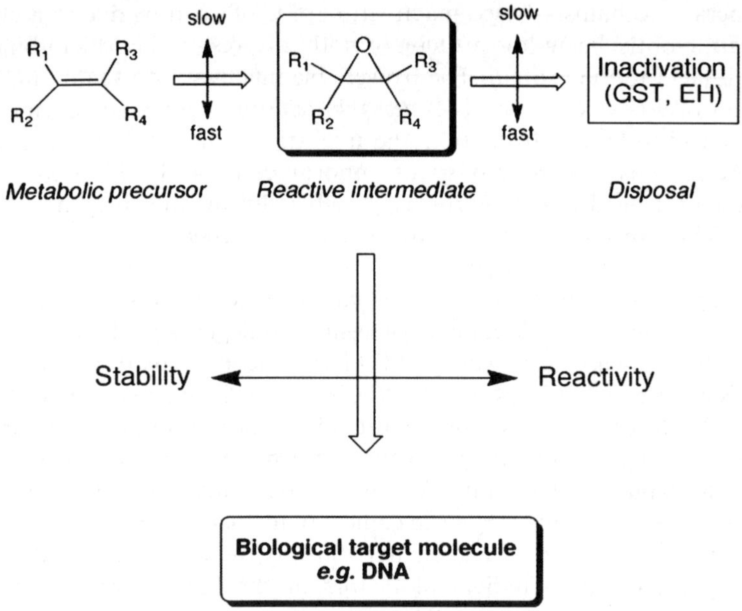

Figure 2.5 Determinants of the biological effectiveness of a reactive intermediate: CYP bioactivation of aliphatic olefins and determinant factors for the epoxide.

time, it must possess sufficient stability to reach the target. The target may be a constituent of the cell where the reactive metabolite is formed, or of a different cell type within the same organ, or it may be located at a different organ.

Examples highlighting these principles will be described here.

2.3 Suicide Substrates for CYP Enzymes

The inhibition of human CYPs is a common mechanism leading to drug-drug interactions.[29] Such inhibition can be reversible or irreversible. Irreversible inhibition usually derives from activation of a drug by CYPs into a reactive metabolite, which tightly binds to the enzyme's active site, leading to a long-lasting inactivation. This process is called "mechanism-based inhibition" or "suicide inhibition". Very classical examples in toxicology are halomethanes such as carbon tetrachloride, chloroform and bromo-dichloromethane.[30]

In the 1980s and early 1990s research into selective inhibitors of CYP enzymes has mainly been driven by mechanistic paradigms, in order to differentiate between properties of individual CYPs or CYP subclasses.[31] Relevant examples are provided by compounds with terminal acetylenic (ethinyl) structures that can produce suicide inhibition without haem destruction.[32] As some pharmacologically important steroids have 17α-ethinyl substituents, this mechanism is important for a variety of drugs. Prominent examples are the contraceptive steroids 17α-ethinyloestradiol[33] and gestodene.[34]

In general, mechanism-based inactivation of CYPs can be due to a chemical modification of the haem, the protein, or both, as a result of covalent binding of modified haem to the protein. The irreversible inactivation usually implies the formation of a covalent bond between the metabolite and the enzyme, which may also lead to hapten formation and may trigger an autoimmune-response. For these reasons, it is now of great importance to study the mode of CYP inhibition of new drugs as early as possible during the drug development process. The current literature covers many aspects of biology and biochemistry, as recently reviewed.[35]

Thus, the aspect of biological reactive intermediates has become important in the design of safer drugs. Automated screening strategies have been proposed to detect mechanism-based inactivators of CYP enzymes particularly relevant to drug metabolism, such as CYP3A4, CYP2C9, CYP2C19, CYP2D6 and CYP1A2.[36]

Since CYP3A4 is the most abundant CYP isoform in human liver, metabolising about 60% of therapeutic drugs, much research has been directed towards mechanism-based inhibition of this particular isoenzyme. However, precise predictions are difficult, since clinical outcomes of CYP3A4 inactivation depend on complex factors related not only to the enzyme, but also to the interacting drug and to individual factors of the patient. Moreover, many CYP3A4 inactivators are also P-glycoprotein substrates/inhibitors, further confounding *in vitro-in vivo* extrapolations. Nevertheless, a large number of clinically important drugs have been identified as mechanism-based CYP3A4 inhibitors. Relevant examples include antibiotics or chemotherapeutics (erythromycin, isoniazid), anticancer drugs (tamoxifen), antidepressants (fluoxetine, midazolam), anti-HIV agents (ritonavir, delavirdine), antihypertensives (dihydralazine, verapamil), steroids and steroid receptor modulators (gestodene, raloxifene), and herbal drug constituents (e.g. bergamottin and glabridin). Compared with reversible inhibition, mechanism-based inhibitors of CYP3A4 appear more prone to cause unfavourable drug-drug interactions, as the inactivated CYP3A4 has to be replaced by newly synthesised protein.[37]

Studies with experimental animals have shown that various CYPs can be inactivated by chemicals. Formation of CYP-Fe^{2+}-metabolite complexes have been investigated in detail to understand both CYP enzyme mechanisms and drug-drug interactions. Drugs containing mono- or dialkylamino groups attached to the molecule *via* aliphatic structures are potential candidates to form CYP-metabolite complexes. In this respect, the effects of several macrolide antibiotics, such as triacetyloleandomycin and erythromycin, on the CYP system have been studied.[38,39] Treatment of rats with triacetyloleandomycin results in an increase in hepatic microsomal CYP3A-enzymes which are, however, enzymically inactive.[39,40] The responsible metabolite traces back to the dimethylamino-group of the molecule which is finally transformed to a nitroso-group by stepwise demethylation and oxidation of the nitrogen. It has been suggested that the nitroso-compound forms a stable complex with the haem iron of CYP3A.[41] CYP3A4 appears to be more susceptible to inactivation by erythromycin than CYP3A5.[42] In a similar way, several tricyclic antidepressants and structurally related compounds may inactivate CYP enzymes. For example, rat CYP2C11 forms a metabolite complex with amitriptyline, nortriptyline, imipramine, desipramine, clomipramine, doxepine and orphenadrine.[43–46]

CYP1A1 is inactivated by an as yet unidentified mechanism by the furanocoumarin coriandrin. The site giving rise to the reactive metabolite is presumably the furano-ring. An epoxide or an α,β-unsaturated aldehyde formed by CYP enzymes have been proposed as the putative intermediate.[47] Inactivation of CYP enzymes by naturally-occuring compounds, such as coumarins, is relevant with respect to CYP-dependent activation of procarcinogens. It has been shown that coriandrin and other furanocoumarins may diminish DNA-adduct formation in mouse epidermis caused by benzo[*a*]pyrene and 7,12-dimethylbenz[*a*]anthracene.[48]

2.4 Formation of Reactive Intermediates: Toxicological Determinants

2.4.1 Bioactivation, Organotropism and Species Differences of *N*-Nitrosamines

N-Nitrosamines are a toxicologically important group of carcinogenic compounds requiring metabolic activation. Their activation includes several steps, as exemplified in Figure 2.6 for *N*-nitroso-dialkylamines.[49,50]

The decisive initial α-C-hydroxylation, mediated by CYP enzymes leads to formation of the *N*-nitrosoalkyl-α-hydroxyalkylamine as the proximal carcinogen. This intermediate is unstable and decomposes with elimination of the oxidised side chain as an aldehyde, leading to the formation of an alkyl-diazohydroxide, diazonium ion and carbonium ion as the ultimate DNA-alkylating species.

In experimental animals, the preferred target organ of symmetrical *N*-nitrosamines (with two identical alkyl groups) is the liver, with little influence of the route of application. Unsymmetrically substituted *N*-nitrosamines produce preferably oesophageal tumours, both after oral and parenteral dosing. By contrast, cyclic *N*-nitrosamines induce a broad spectrum of tumours at a variety of locations. Each individual *N*-nitrosamine displays its characteristic organotropism.[51] This is primarily caused by organ-specific bioactivation processes, for which the specificity and tissue distribution of relevant CYP isoenzymes are decisive determinants.[52]

Examples for such a complex interplay in *N*-nitrosamine bioactivation are *N*-nitrosodibutylamine and *N*-nitroso-methyl-alkylamines with an even number (C_8-C_{12}) of C-atoms in the second side chain. These compounds are oxidised at the ω-position by successive actions of CYP, ADH (alcohol dehydrogenase) and AlDH (aldehyde dehydrogenase), leading to a carboxylic acid side chain. From *N*-nitroso-dibutylamine *N*-nitroso-*N*-butyl-3-carboxypropylamine is formed, and C_8-C_{12} carboxylic acids are degraded by fatty acid β-oxidation to the respective *N*-substituted *N*-nitroso-carboxypropylamine. These proximate intermediates act as transport forms as they are readily excreted in the urine. Within the epithelium of the urinary tract, a final bioactivation takes place to

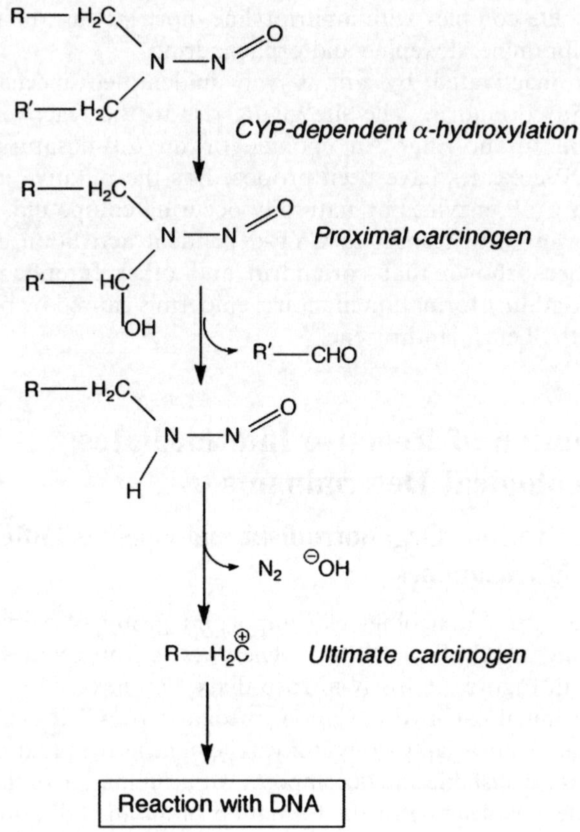

Figure 2.6 General activation pathway for *N*-nitroso-diallkylamines.

the ultimate carcinogen(s). This explains the organotropism of these compounds towards the urinary bladder. CYP2E1 and CYP2A13 have been addressed as specific enzymes in the activation of *N*-nitrosodibutylamine.[51]

The role of individual CYP enzymes has been most extensively studied in the case of tobacco-specific *N*-nitrosamines, and the most prominent example is 4-(methylnitrosamino)-1-(3-pyridyl)-1-butanone ("NNK"). The metabolism and bioactivation pathways of NNK are complex, and major CYP-mediated processes involved are pyridine oxidation, α-methylene oxidation and α-methyl oxidation,[53] as shown in Figure 2.7.

It is thought that the relevant steps of NNK metabolism are catalysed by CYP enzymes, with little involvement of other enzymes.[54] CYP1A2, CYP2A, CYP2B1 and CYP3A appear important for NNK metabolism in rats and mice, where CYP2E1 appears minimally involved.[53] In humans, CYP1A2, CYP2A6 and CYP3A4 are most important in the hepatic activation of NNK, and CYP2D6 has been extensively studied because of its possible involvement in lung cancer. It catalyses the α-hydroxylation of NNK, but with a low activity.[55–57]

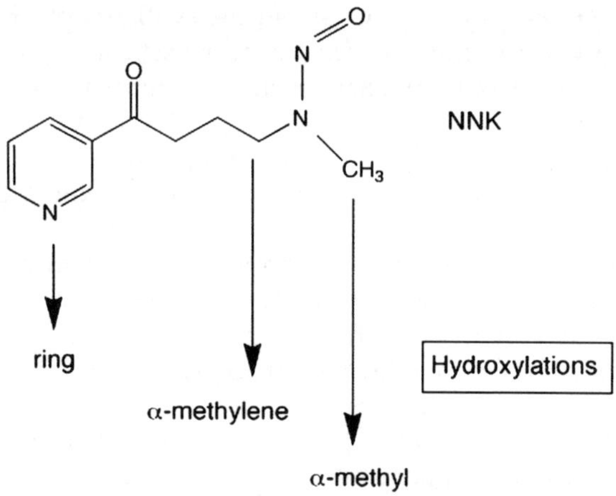

Figure 2.7 Sites of metabolism of the tobacco-specific *N*-nitrosamine, NNK. NNK, 4-(methylnitrosamino)-1-(3-pyridyl)-1-butanone.

Table 2.2 CYP enzymes likely to be involved in the metabolism of the tobacco-specific nitrosamine NNK in rat and human liver and lung (adopted from Hecht,[53] modified).

Hydroxylation at	*Rat liver*	*Rat lung*	*Human liver*	*Human lung*
α-methylene group	**1A1, 2B1, 3A**	**1A, 2B1**, 2A	**2A6, 3A4**, 2D6	
α-methyl group	**1A2, 2A1, 3A**, 2B1	2A	**1A2**	**2A6**
pyridine ring	**2B1**, 1A2, 3A		2C8, 3A4	

Boldtype: strong evidence; normal type: weak evidence.

Again, a consideration of the involvement of different CYP enzymes in the metabolism of NNK (Table 2.2) provides a rationale for the organ and species differences.[53]

Thereby, the *N*-nitrosamines present themselves as a class of very relevant chemical carcinogens for which CYP enzymes are pivotal in determining organotropism and species differences.

2.4.2 Endogenous Compounds: Biological Activation of Oestrogens

The pathogenesis of human mammary cancer is influenced by endogenous and exogenous hormonal factors. An influence of oestrogens on breast cancer development is supported by cell culture studies, animal models, human observational studies, and by the fact that antioestrogens reduce the risk of developing breast cancer.[58] The latter points to a specific role of endogenous oestrogens in breast cancer development.

Aromatase (CYP19) is the key enzyme in the synthesis of phenolic steroids. Aromatase is expressed in human breast cancer tissue at higher levels than in normal breast tissue based on enzyme activity measurements, immunocytochemistry and mRNA quantitation. *In situ* produced oestrogen could probably play a more important role than circulating oestrogens in breast tumour promotion.[59] A classical and most important site for aromatase is the adipose tissue.[60]

In recent years, the use of oestrogen-progestin combinations for hormone replacement therapy (HRT) has been shown to be associated with a significant increase of breast cancer risk.[61] HRT-associated breast cancer risk has been particularly attributed to combined oestrogen-progestin treatments and to long-term use of more than 10 years.

Oestrogenic steroids in HRT preparations are metabolised by drug metabolising enzymes with overlapping substrate specificities, which also metabolise the endogenous oestrogens. The main pathways are shown in Figure 2.8. A most important CYP enzyme for the inactivation of oestrogens is CYP3A4.[62] In the light of known multiple enzyme polymorphisms, a variety of recent publications have reported pharmacogenomic approaches to considering the genetic background influencing the HRT-associated breast cancer risks,[63] as there seems to be a link between HRT use risk and modifying polymorphisms of drug/xenobiotic enzymes. For example, an association of the *CYP1B1*3/*3* genotype (homozygotic change 4326C>G, resulting in 423Val>Leu) of cytochrome CYP1B1 with increased breast cancer risk in Swedish women having used HRT for 4 years and longer has been highlighted.[64] Moreover, it has been suggested that the use of HRT may increase the risk of breast cancer among women with specific combinations of at-risk genotypes of COMT and GST genes,[65] pointing to relevant gene-gene interactions among drug metabolising enzymes.

Although the carcinogenic effects of oestrogens are mainly attributed to hormonal properties, there is considerable interest in oestrogens acting chemically as carcinogens by binding to cellular macromolecules.[66,67] The natural oestrogen, 17β-oestradiol (E_2), is converted in the organism to bioreactive intermediates through the formation of the catechol oestrogens, 2-OH-E_2 and 4-OH-E_2, and the 16α-hydroxylation product, 16α-OH-E_2. Human CYP1B1 is a key enzyme in this metabolism particularly in forming 4-hydroxy estrogens[68] which have been implicated in the carcinogenic process.[69] Among others, CYP1B1 is expressed in endocrine-regulated tissues, such as the mammary gland, uterus, and ovaries.[70]

CYP1B1 exceeds other P450 enzymes in both oestrogen hydroxylation activity and expression levels in breast tissue. To determine whether inherited variants of the polymorphic CYP1B1 differ from wild-type CYP1B1 (*CYP1B1*1*) in oestrogen hydroxylase activity, several recombinant variants of CYP1B1 were expressed: variant 1 (codon 48Arg→Gly), variant 2 (codon 119Ala→Ser), variant 3 (codon 432Val→Leu; *CYP1B1*3*), variant 4 (codon453Asn→Ser), and the combined variant 5 (48Gly, 119Ser, 432Leu, 453Ser; CYP1B1*7). When assessing E_2 hydroxylation activities, the wild-type

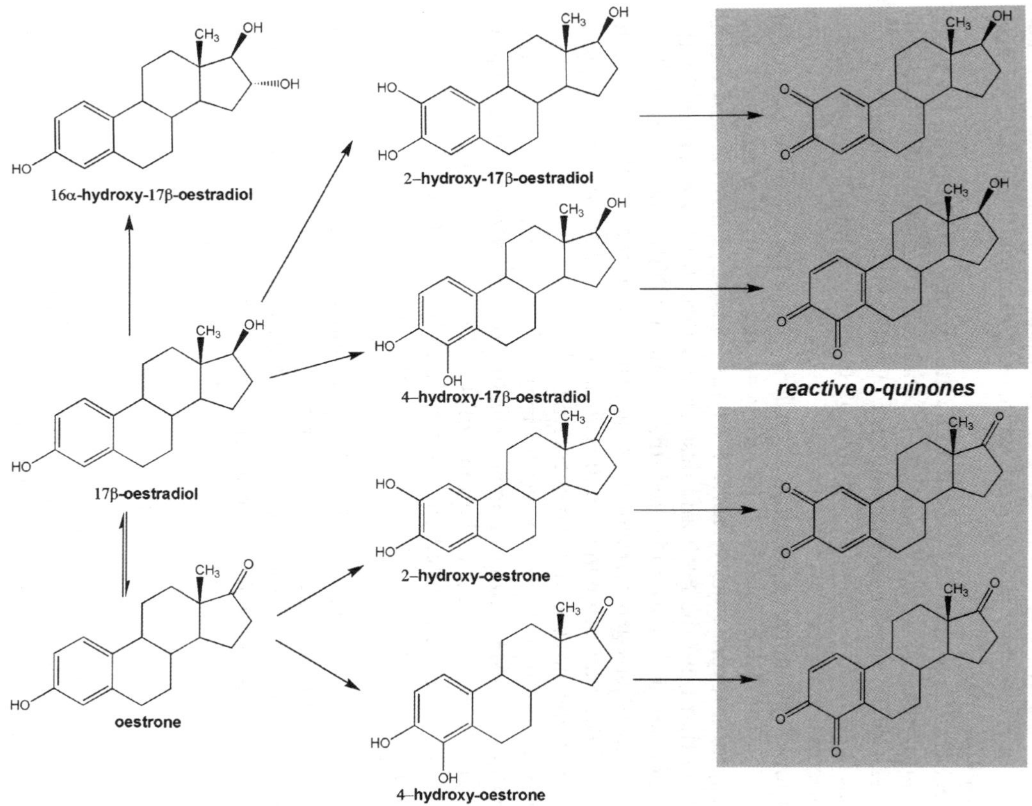

Figure 2.8 Metabolism of 17β-oestradiol: Formation of the catechol oestrogens 2-hydroxy-oestradiol and 4-hydroxy-oestradiol is catalysed by cytochromes CYP1A1 and CYP1B1. The primary catechol metabolites can be subsequently oxidised to reactive o-quinones or O-methylated by catechol O-methyl transferase (COMT). Oestrone is formed from 17β-oestradiol by 17β-hydroxysteroid-dehydrogenase and is subject to further hydroxylation reactions, in much the same way as 17β-oestradiol.

CYP1B1*1 formed 4-OH-E_2 as the major product, followed by 2-OH-E_2 and 16α-OH-E_2. The CYP1B1 variants similarly formed 4-OH-E_2 as the major product but displayed 2.4-fold (variant 4, *CYP1B1*7*) to 3.4-fold (variant 2) higher catalytic efficiency than the wild-type enzyme (*CYP1B1*1*). The variant enzymes also exceeded wild-type CYP1B1 with respect to 2- and 16α-hydroxylation activity. Thus, inherited alterations in CYP1B1 oestrogen hydroxylation activity may be associated with significant changes in oestrogen metabolism and, thereby, may possibly be linked to interindividual differences in breast cancer.[71]

The catechol 2-OH- and 4-OH-oestrogens are further oxidised by CYP enzymes to *o*-quinones (3,5-cyclohexadiene-1,2-diones), as shown with 2-hydroxyoestrone (2-OHE_1) and 4-hydroxyoestrone (4-OHE_1) in rat liver microsomes.[72] These quinones are known to be capable of binding covalently to biological macro-molecules.[73]

With reporter plasmids containing the wild or mutated putative oestrogen responsive element (ERE) on the *CYP1B1* gene and the wild or mutated oestrogen receptor α-expression vectors, luciferase assays using Ishikawa cells demonstrated that both the ERE and oestrogen α-receptor are essential for the transactivation by E_2. This indicated that human CYP1B1 is regulated by oestrogen *via* the oestrogen α receptor. A clinical significance of the oestrogen-regulated *CYP1B1* expression in the homeostasis of oestrogens and in oestrogen-dependent carcinogenesis have been suggested.[70]

The toxic potential of the oestrogen quinones is decreased by conjugation with glutathione (GSH), either nonenzymically or catalysed by glutathione *S*-transferase P1 (GSTP1). Incubations of GSH and recombinant purified CYP1B1 with 2-OHE_2 resulted in conjugation through C-4 and C-1 (*i.e.*, 2-OHE_2-4-SG and 2-OHE_2-1-SG), whereas in the case of 4-OHE_2 yielded only 4-OHE_2-2-SG. The production of GSH-oestrogen conjugates was dependent on the concentrations of GSTP1, but the quantitative conditions allow accumulation of the bioreactive oestrogen quinones.[74]

Besides CYP1B1, also CYP1A1 has been discussed as a phase I enzyme metabolising oestrogens. There is interaction with phase II enzymes, such as GSTP1 and catechol-*O*-methyltransferase (COMT). Using recombinant and purified enzymes, the oxidation of E_2 to 2-OHE_2 and 4-OHE_2 is efficiently catalysed by both CYP1B1 and CYP1A1. COMT generated two products from 2-OHE_2, 2-methoxyoestradiol and 2-hydroxy-3-methoxyoestradiol, but only one product from 4-OHE_2, 4-methoxyoestradiol. Similarly, GSTP1 yielded two conjugates, 2-OHE_2-1-SG and 2-OHE_2-4-SG, from the corresponding quinone 2-hydroxyoestradiol-quinone and one conjugate, 4-OHE_2-2-SG, from 4-hydroxyoestradiol-quinone. A kinetic model for the oxidative metabolism of the parent hormone E_2 revealed differences in rate constants for the C-2 and C-4 metabolites. This finding demonstrated a tightly regulated interaction of phase I and phase II enzymes, in which the latter decreased the concentration of catechol oestrogens and oestrogen quinones, thereby reducing the potential of these oxidative oestrogen metabolites to induce DNA damage.[75]

The ratio of oestrogen 2- to 4-hydroxylation depends on the specific constitutive and modulated CYP-profile in a tissue, and is influenced by allelic variants of CYP1 enzymes. The amount of each metabolite formed is an important determinant of tumour development. While 4-hydroxy oestrogens are predominantly responsible for DNA-adduct formation, the 2-hydroxy derivatives are believed to have antiangiogenic properties and inhibit tumour growth. On this basis, Doherty *et al.*[76] defined low and high risk alleles of CYP1 enzymes. According to their definition *CYP1A1 T6235C* and *CYP1A1* $Ile^{432}Val$ are low risk and *CYP1A1* $Thr^{461}Asn$, *CYP1A2 A734C* and *CYP1B1* $Leu^{432}Val$ are high risk alleles.

Furthermore, both CYP1A1 and CYP1B1 can demethylate 2-$MeOE_2$ and 2-OH-3-$MeOE_2$ to 2-OHE_2, whereas CYP1B1 additionally demethylates 4-$MeOE_2$ to 4-OHE_2. Methoxyoestrogens could probably exert a feedback inhibition on CYP1A1- and CYP1B1-mediated oxidative oestrogen metabolism, thereby reducing the potential for oestrogen-induced DNA damage.[77] The 3,4-quinones of oestradiol and oestrone form adducts with DNA predominantly with guanine and adenine residues. The resulting adducts are unstable and may lead to apurinic sites which can give rise to mutations.[67]

This example of oestrogens shows that important biologically reactive intermediates are also formed by CYP enzymes from endogenous substrates.

2.5 Application to Polycyclic Aromatic Hydrocarbons

Polycyclic aromatic hydrocarbons (PAH) are well-known ubiquitous environmental pollutants, which are generated by the incomplete combustion of organic compounds such as petrol, coal, wood and charbroiled meats. Industrial emissions of PAH are derived for example, from aluminium production plants, petroleum refineries, fossil fuel power plants, paper mills, coal tar and asphalt production plants. As a consequence, particle-associated PAH are found in the air.[78–80] Due to emissions and deposits, soils of industrial sites can be highly contaminated with PAH achieving pollutant concentrations up to several grams of PAH/kg soil.[81] Because of the persistence of PAH, particularly of the higher condensated ones, in soils, considerable PAH concentrations can be found also at abandoned industrial sites. Besides originating from traffic[82] and industrial processes, PAH are also present in tobacco smoke. However, a high percentage of PAH intake by humans is *via* the diet amounting to about 70% for non-smokers.[83] Exposure to PAH is associated with increased health risk and there is no doubt that PAH are initiators of chemical cancerogenesis. Thus, it is of great interest to understand the underlying biochemical and cellular processes initiated by PAH, starting from PAH exposure and uptake and ending up in an adverse effect. Cytochromes P450 play a major role in the mediation of PAH carcinogenicity. It has long been known that PAH require metabolic activation to elicit adverse effects. To understand CYP-mediated PAH toxicity and the potencies of reactive

PAH metabolites, the following aspects appear important and should be considered:

(1) Involvement of several CYP enzymes in the metabolism of PAH
(2) Formation of multiple reactive metabolites by CYP enzymes from parent PAH and from PAH metabolites
(3) Modulation of CYP expression and activity by PAH *via* receptor-mediated processes
(4) Tissue-specific expression profiles of CYP enzymes
(5) Relationship between location of metabolite formation and location of targeted molecules.

These different aspects highlight complex interactions between PAH, the CYP system and the target molecules attacked by reative metabolites, particularly when the diversity of PAH in mixtures and the diversity of CYP enzymes involved are considered. Besides CYP induction by PAH and metabolism of PAH by CYP enzymes, competition between several PAH and CYP inhibition by PAH play a role.[84] As a result, the formation rate for reactive PAH metabolites and the coupling efficiency to subsequent 'neutralising' activities of phase II enzymes in a given tissue can hardly be predicted. Therefore, the emphasis of this review will be on the capacity of cells and tissues to metabolise PAH. Tissue-specific basic CYP profile and CYP inducibility have an impact on the quantitative and qualitative metabolite profiles and are determinants of site-specific susceptibilities to the carcinogenic effects.

Carcinogenic PAH require metabolic activation to elicit adverse effects. The mechanisms involved are well studied and documented. In particular, the initial formation of reactive metabolites by cytochromes P450 is a main contributor to this process. Dependent on size and structure of PAH, different types of metabolites can be formed to varying degrees. The resultant metabolite patterns are also dependent on individual and tissue-specific enzyme profiles, bearing in mind that in most cases several enzymes are involved in the metabolism of a specific PAH, acting in alternative pathways or consecutively.

Metabolism of PAH by cytochromes P450 is complex. Several sites of the molecules are suitable for one or two-electron oxidations. Additionally, these sites include prochiral C-atoms so that reactions lead to stereoisomeric forms. Metabolites resulting from certain pathways are reactive and may form adducts with sensitive biomolecules such as DNA. For PAH, the following pathways leading to reactive metabolites may be distinguished:

- Formation of dihydrodiolepoxides (DDE) requiring two CYP-catalysed steps and epoxide hydrolase
- Formation of phenols *via* radical cations by 1-electron oxidation
- Formation of o-quinones *via* catechols by involvement of aldo-keto reductase (dihydrodioldehydrogenase) coupled with formation of reactive oxygen species

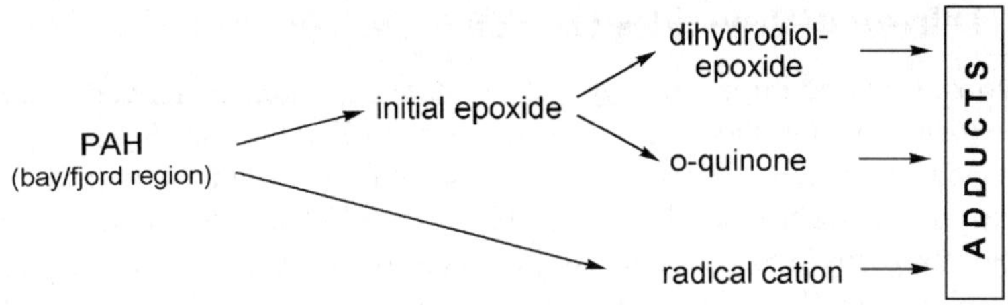

Figure 2.9 Formation of reactive metabolites, capable of forming adducts with DNA, from PAH possessing a bay or fjord region. Intermediate metabolites are omitted to give a generalised overview.

– Formation of reactive sulphates derived from PAH carrying a ring-located secondary benzylic hydroxyl or from hydroxymethyl-PAH.

Figure 2.9 shows a generalised scheme illustrating the formation of reactive metabolites from PAH, excluding the sulphate pathway.

The contribution of the individual pathways in the *in vivo* metabolism of PAHs is largely unclear. It is even a matter of debate whether the radical anion formation has any relevance *in vivo*, considering the rates of formation, and is involved in the carcinogenic process.[85] Analyses of adducts formed *in vivo* reveals that about 95% are obviously derived from DDEs. It has to be noted, however, that adducts which are presumably delineated from PAH radical anions are unstable and lead to depurination, a fact which aggravates an evaluation of their formation rate. On the other hand, there are hints that the DDEs may not be primarily responsible for the carcinogenicity of PAHs, a point which will be discussed later.

For the study of PAH metabolism, carcinogenicity and adduct formation, different experimental approaches have been applied. Simplified *in vitro* systems have clear limitations with respect to an *in vivo* extrapolation. Two examples illustrate these limitations and strengthen our view on the *in vivo* situation. Assay systems which include only single heterologously expressed CYP enzymes and purified target DNA, such as calf thymus DNA, exclude competing reactions performed by several CYPs so that a certain pathway will be preferred. Furthermore, targeted DNA is directly accessible for the reactive metabolites formed in *in vitro*-assays. This is not the case *in vivo*, where reactive metabolites are generated in the cytosol and have to cross the nuclear or mitochondrial membranes before reaching the DNA. Due to their high chemical reactivity, PAH radical anions formed by microsomal CYP enzymes may not get in close contact with nuclear or mitochondrial DNA.

Following identification of the metabolites formed from individual PAH, it has been possible to investigate their mutagenic, carcinogenic and DNA adduct-forming potentials directly with chemically-synthesised compounds. It has been observed that the fjord region DDE of benzo[c]phenanthrene was among the most potent carcinogenic PAH dihydrodiol epoxides.[86,87] However, this metabolite is poorly formed *in vivo*, and the parent compound benzo[*c*]-phenanthrene exhibits only low carcinogenic activity, at least on mouse skin.[87]

2.5.1 Dihydrodiolepoxides (DDE)

DDEs constitute one important type of ultimate carcinogens derived from PAH. They are formed in a multistep process from parent PAH possessing a bay or fjord region which are spatially defined substructures of PAH as indicated in Figure 2.10.

Oxidation of respective PAHs by cytochromes P450 occurs just adjacent to the bay or fjord regions, and the resultant products are therefore called bay- or fjord-region (dihydro)diolepoxides. All enzymic steps catatysed by several cytochromes P450 and by epoxide hydrolase are highly stereoselective so that formation of one specific stereoisomer is clearly favoured. This is a toxicologically important fact because different stereomeric DDEs show different reactivity and carcinogenic potential. The relative degrees of stereoselectivity are summarised in Table 2.3 for benzo[*a*]pyrene as the most thoroughly studied PAH compound.

A reaction scheme for the formation of BaP-dihydrodiol-epoxide dealing only with the stereoselectively favoured products is shown in Figure 2.11.

Dissection of the reaction sequence from parent PAH to their dihydrodiol epoxides with respect to the major enzymes involved in each specific step may be important, but is often not possible based on literature data. In most cases, the final DDEs or their mutagenic potential are assessed delivering an integrating result over the whole process.[89,90] The specific contributions of

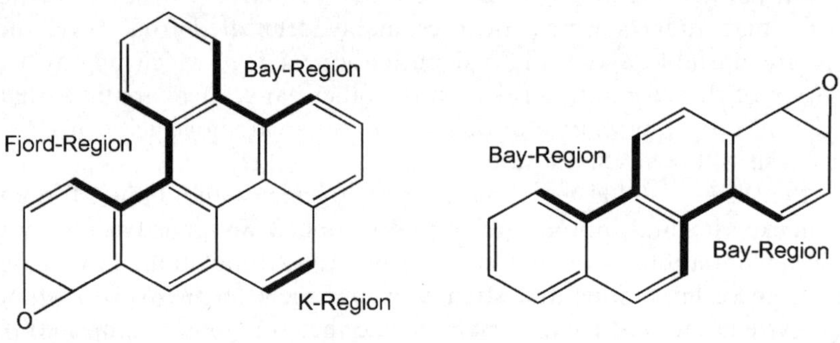

Figure 2.10 Primary arene epoxides adjacent to the fjord region of dibenzo[*a,l*]pyrene and to the bay region of chrysene.

Table 2.3 Stereoselectivity of enzyme-catalysed reactions leading to dihydrodiolepoxides from benzo[a]pyrene (BaP), according to Xue & Warshawsky.[88] Values give the percentage of stereoisomer formed from the precursor to the left. Values in parentheses are related to 100% of the parent compound benzo[a]pyrene. Diol refers to dihydrodiol and diol epoxide to dihydrodiol epoxide.

	%	*Epoxide*	%	*Diol*	%	*Diol epoxide*
BaP	95	7R,8S	100 (95)	7R,8R	86 (82)	7R,8S,9S,10R
					14 (13)	7R,8S,9R,10S
	5	7S,8R	100 (5)	7S,8S	97 (4.9)	7S,8R,9S,10R
					3 (0.2)	7S,8R,9R,10S

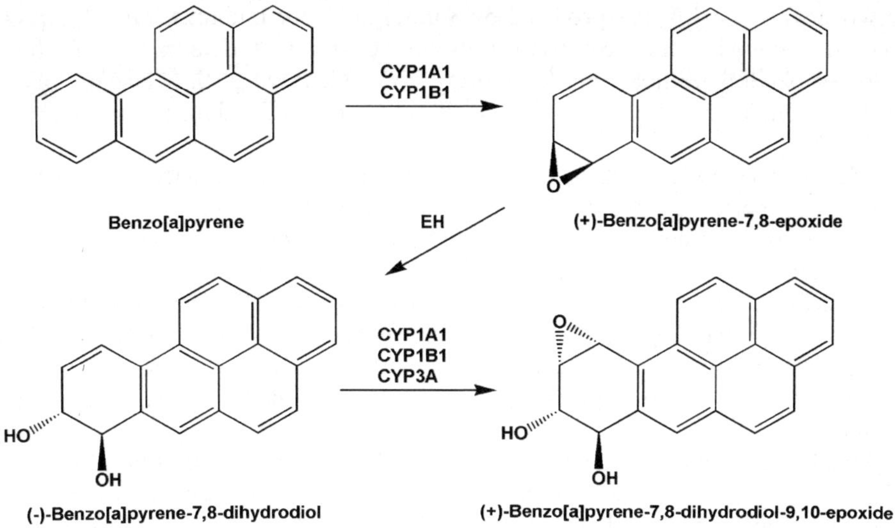

Figure 2.11 Formation of the stereochemically preferred bay region dihydrodiol epoxides from benzo[a]pyrene. Major enzymes involved are indicated. EH, epoxide hydrolase.

CYP-enzymes to a single step can be important in the evaluation of tissue-specific PAH effects based on the underlying CYP profile.

2.5.1.1 Formation of the Primary Epoxide

The primary enzymic attack by cytochromes P450 leads to an arene epoxide (arene oxide) and occurs on the opposite side of a terminal aromatic ring which forms part of the bay-regions, the so-called M-region or non-K-region (Figure 2.11). PAH known to be metabolised in this way are for example benzo[b]fluoranthene, benzo[*c*]phenanthrene, benzo[*a*]anthracene, chrysene, benzo[*g*]chrysene, dibenzo-[*a,l*]pyrene and 7,12-dimethylbenzanthracene. Also from the simplest polycyclic aromatic hydrocarbon possessing a bay region, phenanthrene, a DDE can be formed by cytochromes P450. There are several CYP-enzymes catalysing the initial epoxidation, CYP1A1 being the first one identified. By means of heterologously expressed CYPs, clear assignments of activities to specific CYP enzymes became possible. Shou *et al.*[91] demonstrated that several human CYP-enzymes contribute to the formation of *trans*- benzo[*a*]pyrene-7,8-dihydrodiol. From the set of enzymes tested (excluding CYP1A1 and CYP1B1), CYP2C9, CYP1A2 and CYP2B6 were the most effective with activities of 57.9, 19.7 and 14.8 pmol dihydrodiol min^{-1} $nmol^{-1}$ CYP, respectively. Schwarz *et al.*[92] determined the activities of several heterologously expressed CYP1A1 variants including the 'wild type'. The latter formed the 7,8-dihydrodiol with an activity of about 410 pmol min^{-1} $nmol^{-1}$ CYP, which is 7, 21 or 28 times higher compared with CYP2C9, CYP1A2 and CYP2B6, respectively. Comparable data for CYP1B1, an enzyme discovered in 1994,[93] was provided by Kim *et al.*[94] For human CYP1B1 expressed in baculovirus-infected insect cells an activity was determined of about 170 pmol dihydrodiol min^{-1} $nmol^{-1}$ CYP identifying this enzyme as a potential major contributor for this activity. Comparative data are summarised in Table 2.4.

Methylated PAH, such as 5-methylchrysene and 5,6-dimethylchrysene, in which the methyl groups are located adjacent to the bay region can undergo the same type of epoxidation and subsequent metabolism as typical bay- or fjord-region PAH. With respect to the reactivity of the ultimate carcinogen, there is an influence of the methyl groups. Fjord region DDEs are usually more reactive than bay region DDEs, and the methyl bay region DDEs are just intermediate. Structurally, the additional methyl residue mimics a fjord region, as shown for 5-methylchrysene compared to chrysene and benzo[*g*]chrysene (Figure 2.12).

It is being noted that the rotational flexibility of the C-methyl bond is quite different from that of a ring-integrated carbon-carbon bond. With respect to carcinogenic potential, the compounds follow the order benzo[*g*]chrysene > 5-methylchrysene > chrysene.

An analogous array of compounds can also be created starting with benzo[a]pyrene *via* 11-methyl benzo[*a*]pyrene to dibenzo[*a,l*]pyrene (Figure 2.13). Carcinogenicity of the 11-methyl derivative is higher compared with the non-substituted benzo[*a*]pyrene,[95] and dibenzo[*a,l*]pyrene is the most carcinogenic PAH known so far.[96,97]

Table 2.4 Formation of benzo[a]pyrene-7,8-dihydrodiol by heterologously expressed human cytochromes P450 and epoxide hydrolase. In the case of various CYP1A1 allelic variants were examined. mEH, microsomal epoxide hydrolase. Activity: pmol metabolite/min/nmol P450.

Enzyme	*Activity*	*Remark*	*References*
1A1	570 ± 100	baculovirus-infected insect cells	Kim *et al.*[94]
1A1 + mEH	380 ± 120	baculovirus-infected insect cells	Kim *et al.*[94]
1A1.1	410	baculovirus-infected insect cells	Schwarz *et al.*[92]
1A1.2	180	baculovirus-infected insect cells	Schwarz *et al.*[92]
1A1.4	360	baculovirus-infected insect cells	Schwarz *et al.*[92]
1A2	< 20	baculovirus-infected insect cells	Kim *et al.*[94]
1A2	19.7	vaccinia virus-infected Hep G2-cells	Shou *et al.*[91]
1B1	180 ± 40	baculovirus-infected insect cells	Kim *et al.*[94]
1B1 + mEH	170 ± 60	baculovirus-infected insect cells	Kim *et al.*[94]
1B1	40 ± 20	expressed in yeast	Kim *et al.*[94]
1B1 + mEH	310 ± 80	expressed in yeast	Kim *et al.*[94]
2B6	14.8	vaccinia virus-infected Hep G2-cells	Shou *et al.*[91]
2C8	3.1	vaccinia virus-infected Hep G2-cells	Shou *et al.*[91]
2C9	57.9	vaccinia virus-infected Hep G2-cells	Shou *et al.*[91]
2E1	5.4	vaccinia virus-infected Hep G2-cells	Shou *et al.*[91]
3A4	6.5	vaccinia virus-infected Hep G2-cells	Shou *et al.*[91]

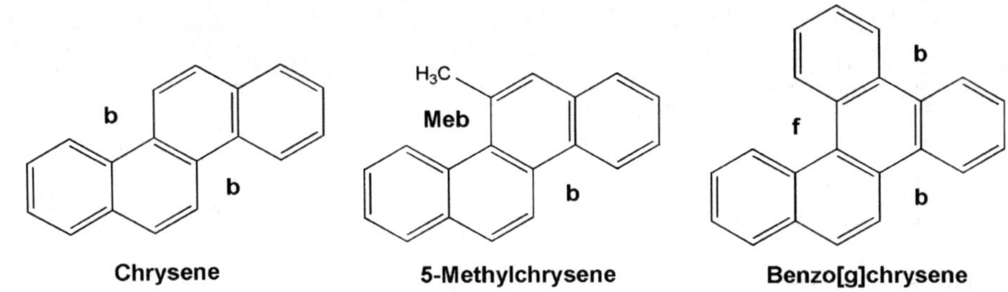

Figure 2.12 Structures of chrysene, 5-methylchrysene and benzo[*g*]chrysene. A fjord-like structure (Meb) is generated by the methyl substituent in 5-methylchrysene. b, bay region; f, fjord region.

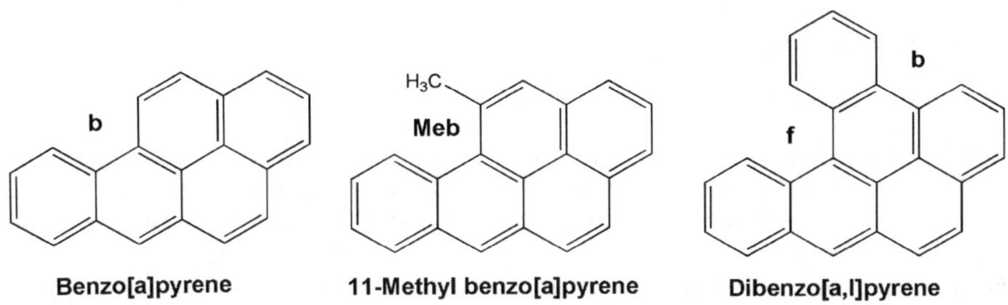

Figure 2.13 Structures of benzo[*a*]pyrene, 11-methyl benzo[*a*]pyrene and dibenzo[*a,l*]pyrene. A fjord-like structure (Meb) is generated by the methyl substituent in 11-methyl benzo[*a*]pyrene. b, bay region; f, fjord region.

2.5.1.2 Hydrolysis of the Primary Epoxide

The second step in DDE formation is the hydrolysis of the PAH epoxide by membrane bound epoxide hydrolase (EH), is a prerequisite for the subsequent CYP-dependent formation of the ultimate carcinogen. The attack by microsomal EH is regio-specific and occurs at the least hindered carbon of the oxirane ring.[98] Furthermore, the reaction is stereoselective so that exclusively the *trans*-isomer of the resulting dihydrodiol is formed.[99] Regardless of absolute configuration, non K-region epoxides are converted to *trans*-dihydrodiols by attack at the allylic carbon.[100] Bay- and fjord-region DDEs are usually no or only poor substrates for EH. Sayer *et al.*[99] have shown, at least for benz[*a*]anthracene, that 'reverse' diolepoxides can be moderate substrates for EH resulting in the formation of tetrahydrotetrols (Figure 2.14). According to Thakker *et al.*[101] the 'reverse' benzo[*a*]pyrene 9,10-diol 7,8-epoxide is only a weak mutagen.

The vicinal diols constitute a branch point in PAH metabolism: they can give rise to quinones by means of aldo keto reductase (AKR) (*vide infra*) or, alternatively, will be oxidised to DDEs by cytochromes P450. Clearly, the extent of involvement of the two pathways depends on the profile of involved enzymes and the enzyme affinities for the different substrates. However, little is known on

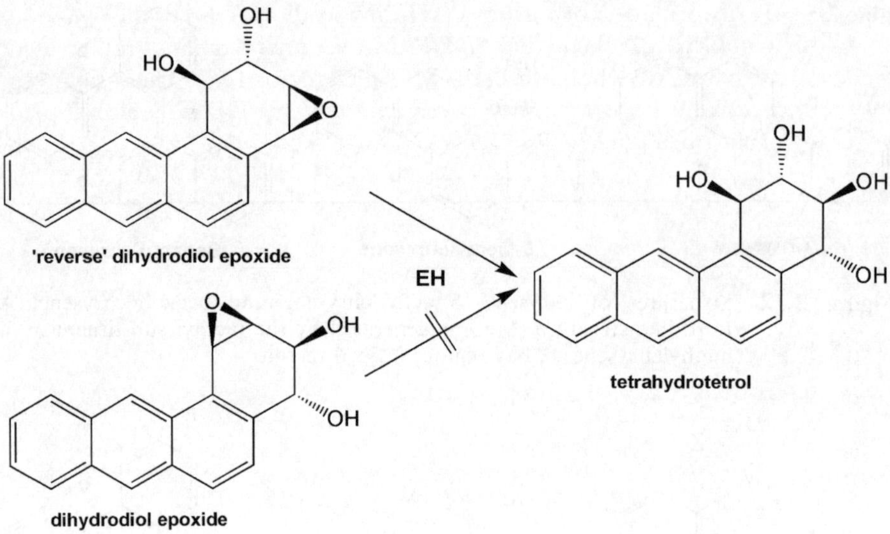

Figure 2.14 Conversion of the 'reverse' bay region 1,2-dihydrodiol 3,4-epoxide of benz[*a*]anthracene to the tetrahydrotetrol by epoxide hydrolase. The 3,4-dihydrodiol 1,2-epoxide is not metabolised by EH. EH, epoxide hydrolase.

the quantitative contributions of the alternative pathways in different tissues characterised by different profiles of xenobiotics metabolising enzymes.

2.5.1.3 The Second Epoxidation

This step leads to the proposed ultimate carcinogens the bay- or fjord-region DDEs. Besides CYP1A1 and CYP1B1, CYP3A4 (previous name: $P450_{NF}$) may be substantially involved in this step, as shown for benzo[*a*]pyrene by Shimada *et al.*[102] It is noted that formation of a DDE from a parent PAH via CYP3A4 requires the initial epoxidation by enzymes other than CYP3A4.

The final epoxidation reaction has been studied with the *trans*-dihydrodiol precursors as substrates. Some data obtained with liver microsomal preparations from 3-methylcholanthrene- and aroclor 1254-treated rats are summarised in Table 2.5. In most cases, the *anti*-dihydrodiol epoxide is formed preferentially. An exception is benzo[*c*]phenanthrene which is converted by liver microsomes of methylcholanthrene-treated rats to about 73% *syn*- and 27% *anti*-stereomer Table 2.5.[103]

2.5.1.4 Adduct Formation with DNA

The finally formed DDEs are reactive electrophiles which can react with DNA as a critical target. Due to their high reactivity, a reaction with weak nucleophiles such as water is also possible. In this case a PAH tetrahydrotetrol is formed. The reactivity and stability of a DDE is dependent on the parent PAH, its configuration and on the chemical environment. It has been shown, for

Table 2.5 Transformation of various PAH-dihydrodiols (DD) to the *anti*- and *syn*-dihydrodiol epoxides (DDE) by liver microsomes of rats treated with 3-methylcholanthrene (3MC) or Aroclor 1254. The ratio given in column 5 is the ratio between *anti*- and *syn*-DDE. Table adopted from Luch *et al.*[104] References: A, Thakker *et al.*[103,105,106]; B, Luch *et al.*[104]

Dihydrodiol substrates	*Inducer*	*% anti-DDE*	*% syn-DDE*	*Ratio*	*Ref.*
Benz[a]anthracene-3, 4-DD	3MC	3	–	100:0	A
Dibenzo[a,l]pyrene-11,12-DD	Aroclor 1254	27.1 ± 3.5	3.2 ± 0.5	90:10	B
Benzo[g]chrysene-*trans*-11,12-DD	Aroclor 1254	3.7 ± 0.4	0.6 ± 0.1	86:14	B
Benzo[c]chrysene-*trans*-9,10-DD	Aroclor 1254	18.4 ± 1.3	8.1 ± 0.3	70:30	B
Benzo[a]pyrene-*trans*-7,8-DD	3MC	56	35	62:38	C
Benzo[c]phenanthrene-*trans*-3,4-DD	3MC	1.08	2.92	27:73	D

example, that the DDE of 7,14-dimethyldibenz[a,j]anthracene is more reactive than that of dimethylbenz[a]anthracene. The half-lives of *anti*- and *syn*-benzo[*a*]pyrene-DDE in culture medium amount to 8 min and 0.5 min, respectively. These are increased in aqueous medium and in the presence of proteins and cell membranes, so that a transport over longer distances can be expected *in vivo*. A further reaction competing with DNA adduct formation is NADPH-mediated reduction of the DDE, resulting in a vicinal triol. A non-ezymic reaction with glutathione is very slow. The anti-carcinogenic effect of some compounds, such as ellagic acid, is due to their reaction with the DDE, thereby partially preventing DNA adduct formation.[107]

Adduct formation of PAH-DDEs with DNA shows some specificities with respect to stereochemistry and purine base selectivity. Under epoxide ring opening, the reactive DDEs bind to the amino groups of adenine (N^6) and guanine-residues (N^2) of DNA *via* their benzylic carbon-atom forming *cis*- or *trans*-adducts. Binding to N^7 of guanine and the exocyclic amino group of cytosine is also possible. In general terms, DDEs of planar PAH, in particular the *RSSR*-isomers, prefer binding to guanine whereas DDEs of non-planar ones such as of benzo[c]phenanthrene prefer binding to adenine.[88] A known exception is the *RSSR*-DDE of 5,6-dimethylchrysene, resulting in about 56% guanine and 44% adenine adducts.[108] The *anti*-dihydrodiolepoxide of 7, 12-dimethylbenz[*a*]anthracene does not show any preference for adenine and guanine and produces equal amounts of *cis*- and *trans*-adducts.[109] In contrast, the *anti*-DDE of 7-methylbenz[*a*]anthracene preferentially reacts with guanine and preferentially produces the *trans*-adduct.[110] For *syn*-5-methylchrysene 1,2-dihydrodiol 3,4-epoxide, *cis*-adducts predominated over *trans*-adducts in

Table 2.6 Base preferences of adduct formation by isomers of several PAH-DDEs with DNA. The results were obtained *in vitro* by incubation of respective dihydrodiol epoxides (0.1 mg/ml) and DNA (1 mg/ml). Gua, guanine; Ade, adenine. References: A, Sayer *et al.*[112]; B, Szeliga and Dipple[113]; C, Chadha *et al.*[114]; D, Cheh *et al.*[115]; E, Szeliga *et al.*[108,116]; F, Agarwal *et al.*[117]

	% total adducts				
	R,S,S,R		*S,R,R,S*		
PAH	*Gua*	*Ade*	*Gua*	*Ade*	*Ref.*
Benzo[a]pyrene	95.0	5.0	85.0	15.0	A
7-Methylbenz[a]anthracene	83.7	16.3	64.3	35.7	B
Dibenz[a,j]anthracene	79.7	20.3	74.7	25.3	C
Benz[a]anthracene	79.0	21.0	62.6	37.5	D
5,6-Dimethylchrysene	55.5	44.1	30.4	69.5	E
Benzo[g]chrysene	32.2	67.8	58.0	41.9	F
Benzo[c]phenanthrene	33.5	66.5	41.7	58.3	G

both DNA and nucleotide reactions.[110] In most cases studied so far, *anti*-DDEs with *RSSR* or *SRRS*-configuration form trans-adducts while *syn*-DDEs with *RSRS*- or *SRSR*-configuration form *cis*-adducts with DNA. An exception is the 7,12-dimethylbenz[*a*]anthracene-*SRRS*-dihydrodiol epoxide which as an *anti*-DDE preferentially forms the *cis*-adduct.[110,111] Selected data on base preferences of PAH adduct formation are summarised in Table 2.6, according to Xue and Warshawsky.[88]

2.5.2 Radical Cations

CYP-dependent one-electron oxidation of PAH can result in the formation of radical cations. These highly reactive metabolites bind to cellular nucleophiles and are carcinogenic. CYP enzymes are not able to convert all PAH species to radical cations. Formation of radical cations is only possible for PAH molecules with an ionisation potential (IP) lower than 7.35 eV.[118] Application of this threshold value, based on IP-data from literature, is not easy because differing IP-values have been published for the various compounds. In Table 2.7, PAHs and their respective ionisation potentials are listed with notes on carcinogenicity and formation of carcinogenic metabolites. In addition LUMO (Lowest Unoccupied Molecular Orbital) energies are listed, since it has been shown that lower LUMO energies are correlated with higher mutagenicity/carcinogenicity at least for different methylated benz[*a*]anthracenes.[119]

Radical cation formation usually occurs preferentially at the carbon position with the highest charge density in the final cation, coinciding with the preferred site for electrophilic subsitution,[123] except when this position is sterically hindered by adjacent (methyl) substituents. In this case, the position with the second highest charge density is attacked. Preferred carbon positions of radical cation formation are for example C6 and to a lesser degree C1 and C3 of

Table 2.7 Ionisation potential, LUMO energy, structural features and relative carcinogenicity (carc.) increasing from + to + + + + + of selected PAH. B, bay region; F, fjord-region; Methyl-B, bay-region with adjacent methyl group (see Figures 2.4 and 2.5). S, bay-region like structure including a ring with 5 carbon atoms. References (ref.): A, Cavalieri and Rogan[118]; B, Hites and Simonsick[120]; C, Bogan and Lamar[121]; D, Harvey.[122]

PAH	*IP (eV)*			*LUMO (eV)*	*Bay/ Fjord- region*	*Carc.*
	Ref. A	*Ref. B*	*Ref. C*	*Ref. B*		*Ref. A*
PAH not forming radical cations						
Phenanthrene	8.19	8.479	8.03	−0.480	B	−
Benzo[c]phenanthrene	7.93	8.292	—	−0.686	F	+
Chrysene	—	8.261	7.81	−0.716	B	±
5-Methylchrysene	7.7	8.244	—	−0.727	Methyl-B	+ + +
Dibenz[a,h]anthracene	7.57	8.148	—	−0.828	2B	+ + +
Benz[a]anthracene	7.54	8.11	7.56	−0.829	B	±
Pyrene	7.50	8.03	7.53	−0.878	−	−
Anthracene	7.43	8.049	7.43	−0.841	−	−
PAH forming radical cations						
7-Methylbenz[a]anthracene	7.37	8.093	—	−0.883	B	+ + +
7,12-Dimethylbenz[a]-anthracene	7.22	—	—		Methyl-B	+ + + + +
Benzo[a]pyrene	7.23	7.830	7.21	−1.095	B	+ + + +
Dibenzo[a,h]pyrene (Dibenzo[b,def]chrysene)	6.97	7.644	—	−1,318	2B	+ + + +
Dibenzo[a,l]pyrene	7.26	7.899	—	−1.101	F + B	+ + + + +
Anthanthrene (Dibenzo[def,mno]chrysene)	6.96	7.621	—	−1.316	−	+
3-Methylcholanthrene	7.12	—	—		B	+ + + +
Radical cation formation not studied						Carc. Ref. D
Dibenzo[c,p]chrysene	—	8.151	—	−0.928	2F + B	?
Indeno[1,2,3-cd]pyrene	—	8.024	—	−1.257	S	+ +
Benzo[ghi]perylene	—	7.943	—	−1.068	B	−

benzo[*a*]pyrene, C10 of dibenzo[*a,l*]pyrene and the 7- and 12-methyl groups of 7,12-dimethylbenz[a]anthracene. Due to their high reactivity, PAH radical cations immediately react with accessible nucleophiles. Thus, direct evidence of their formation is difficult to obtain.[124] Reaction with DNA involves, preferentially the N^7-position of guanine and N^7 or N^3-positions of adenine.[125] The resultant adducts, however, are unstable and lead to subsequent depurination. Consequently, generation of PAH radical cations and subsequent formation of DNA adducts cannot be demonstrated easily. However, purines released from DNA upon PAH exposure have been detected in mouse skin.[126]

Besides "pure" PAH, respective heterocyclic aromatic compounds, such as dibenz[*c,g*]carbazole with an ionisation potential of about 7.3 eV, can also be metabolised *via* the radical cation pathway by cytochromes P450.[127] According

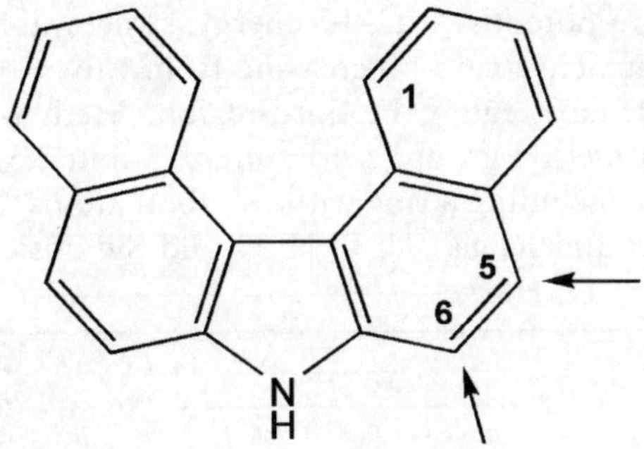

Figure 2.15 Regio-selectivity of radical cation formation for dibenzo[*c*,*g*]carbazole at positions 5 and 6.

to the charge-density rule already discussed (*vide supra*), radical cation formation occurs regio-selectively at carbons 5 or 6 (Figure 2.15).[88]

2.5.3 Quinone Pathway

The quinone pathway deviates from the dihydrodiolepoxide pathway with the primary non-K region dihydrodiol being a common substrate. It is oxidised by aldo-keto reductase (dihydrodiol dehydrogenase) to the respective *o*-quinone. This enzyme does not form quinones from K-region dihydrodiols but requires localisation of the dihydrodiol at a terminal benzo-ring which shares only two carbon atoms with the neighbouring ring.[128] These two pathways compete for the common substrate, the dihydrodiol. Little, however, is known on the contribution of the two pathways in various tissues with different enzyme profiles and different CYP/aldo-keto reductase ratios.

The primary product of the aldo-keto reductase reaction is an α-hydroxy ketone which immediately tautomerises to the more stable aromatic catechol.[88] The next step, leading to the semiquinone, is not so well defined with respect to the components involved. It is known that it is catalysed by peroxidative enzyme activities, resulting in the release of hydrogen peroxide which subsequently gives rise to further reactive oxygen species, such as hydroxy radicals. Oxidation of the semiquinone with molecular oxygen leads the final *o*-quinone and concomitant formation of superoxide radical anions which are known to cause oxidative DNA-damage. The *o*-quinone can be metabolised or reacts in different ways:

(1) NADPH-dependent reduction to the *o*-hydroquinone by NQO1 (DT-diaphorase).
(2) Conjugation with thiols such as glutathione as a detoxification reaction.
(3) As α,β-unsaturated ketones they can form Michael-adducts with DNA at the distant double-bonded carbon.
(4) Finally, the quinones can undergo redox-cycling which can result in continuous release of superoxide anion radicals and subsequent oxidative DNA damage (Figure 2.16).

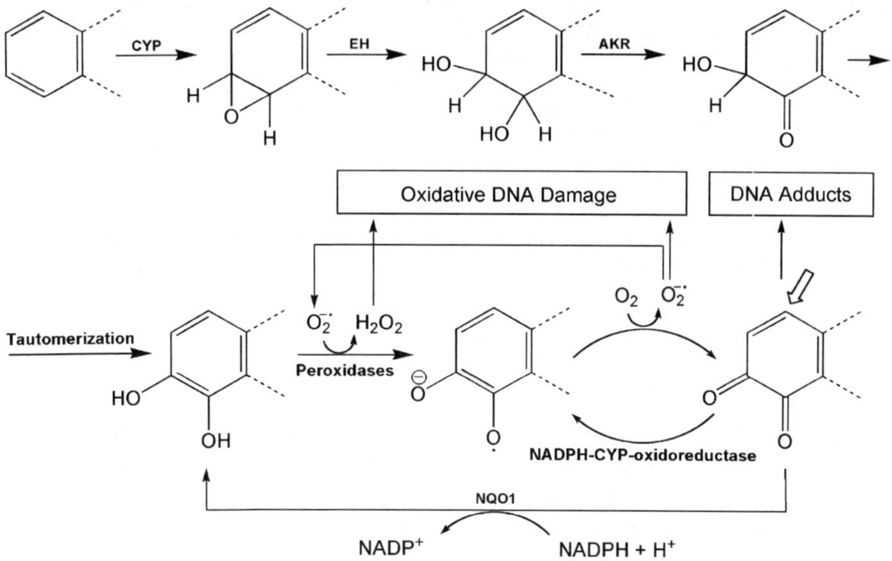

Figure 2.16 Accepted scheme of *o*-quinone formation of a model PAH (only a terminal benzo ring is shown) via the primay epoxide and *trans*-dihydrodiol metabolite, generation of reactive oxygen species (ROS) and redox cycling. Information derived from the following sources was integrated: Palackal *et al.*,[129] Nioi and Hayes,[130] Xue and Warshawsky,[88] and Samuni *et al.*[131] ROS can be formed by peroxidase activity leading to the o-semiquinone. However, it is not clear whether redox-cycling and 'continuous' formation of the superoxide radical anion really occur.[132]

Thus, the reactive metabolites generated *via* this multi-enzyme process, including cytochromes P450, are the *o*-quinone and the reactive oxygen species. Reaction of *o*-quinones as Michael-acceptors with DNA bases results in stable and unstable adducts. The former arise from the reaction with N^6 of adenine and N^2 of guanine,[133,134] and the latter from the reaction with N^7 of guanine, and lead to formation of apurinic sites.[135]

Reactive oxygen species (ROS) are released in the course of quinone formation. However, it is not clear whether redox cycling really occurs. Early studies by Kalyanaraman *et al.*[131] have shown that, in contrast to *p*-quinones, *o*-quinones derived from catechol oestrogens are not substrates for redox cycling.

2.5.4 'Polar Metabolites'

For several PAH there is evidence for the formation of DNA adducts, which are derived from more polar metabolites. These emanate from either triol epoxides or from *bis*-dihydrodiols.[136,137] The triol epoxides are essentially classical DDEs carrying an additional hydroxyl group at a distant site. Examples are

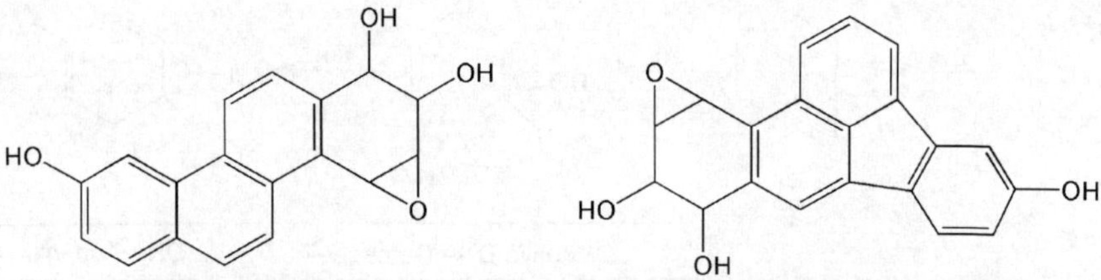

Figure 2.17 Examples of triol epoxides or dihydrodiol epoxide phenols as more polar metabolites.

9-hydroxy-chrysene-1,2-dihydrodiol-3,4-epoxide and 5-hydroxy-benzo[*b*]-fluoranthene-9,10-dihydrodiol-11,12-epoxide (Figure 2.17). The putative final reactive metabolite formed, *via* the *bis*-dihydrodiol pathway, is also an epoxide, as shown for benzo[*a,h*]anthracene in Figure 2.17.

Dibenzo[*a,h*]anthracene (DB[*ah*]A) applied to mouse skin produced one major and three minor adducts, separated by thin layer chromatography. A similar pattern was obtained with DB[*ah*]A-*trans*-3,4-dihydrodiol, the potential precursor of the bay region 3,4-dihydrodiol-1,2-epoxide.[138] Extensive transformation of DB[*ah*]A to its 3,4-dihydrodiol, catalysed by liver enzymes, has been noted.[139] However, initial metabolism studies did not provide an evidence for the formation of the 3,4-dihydrodiol-1,2-epoxide.[140] In fact, the dihydrodiol epoxide pathway is of minor importance for DB[*ah*]A.[141,142] Experiments showed that the major adduct of DB[*ah*]A, described by Carmichael *et al.*,[138] was derived from the *trans,trans*-3,4:10,11-tetrahydrotetrol ("3,4,10,11-bis-diol"). The alternative *bis*-diols *trans,trans*-3,4:12,13-tetrahydrotetrol and *trans,trans*-3,4:8,9-tetrahydrotetrol did not produce any or only a minor adduct spot, respectively. Thus, the major pathway of DNA adduct formation from DB[*ah*]A in mouse skin includes the 3,4-dihydrodiol and the 3,4:10,11-tetrahydrotetrol. As the potential ultimate carcinogen, the reactive metabolite DB[*ah*]A-*trans,trans*-3,4:10,11-tetrahydrotetrol-1,2-epoxide has been proposed.[138,143] Its formation from the 3,4-dihydrodiol-1,2-epoxide by epoxidation in positions 10 and 11 and subsequent hydrolysis by epoxide hydrolase has been excluded.[144] The pathways are shown in Figure 2.18.

Following metabolic activation of dibenzo[*a,l*]pyrene by human recombinant CYP1A1 and CYP1B1, DNA-adducts with higher polarity than those derived from the fjord region dihydrodiolepoxide were found. Based on the low formation rate of the K-region trans-8,9-dihydrodiol, it is assumed that this reaction is not the important initial step in the genotoxic activation of DB[*a,l*]P to the *trans*-8,9:11,12-bis-dihydrodiol or the *trans*-8,9:13,14-bis-dihydrodiol.[145] It has been suggested, however, that the triolepoxide derived from the DB[*a,l*]P-7-phenol, *i.e.* the 7-hydroxy-DB[*a,l*]P-11,12-dihydrodiol-13,14-epoxide, is a good candidate precursor of one of the unidentified polar DNA-adducts.[145]

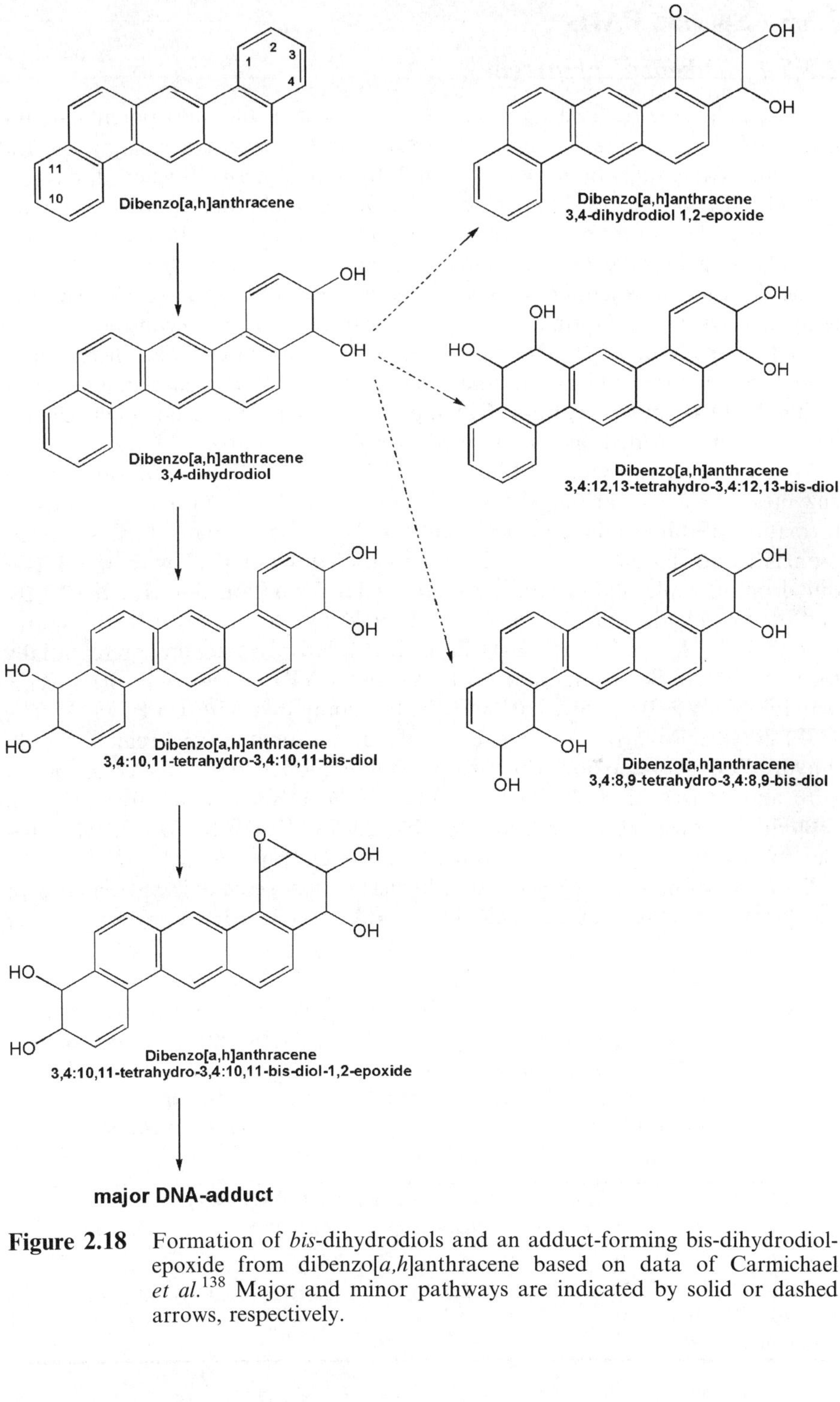

Figure 2.18 Formation of *bis*-dihydrodiols and an adduct-forming bis-dihydrodiol-epoxide from dibenzo[*a,h*]anthracene based on data of Carmichael *et al.*[138] Major and minor pathways are indicated by solid or dashed arrows, respectively.

2.5.5 Specific PAHs

2.5.5.1 Dibenzo[a,l]pyrene

Dibenzo[*a,l*]pyrene (DB[*al*]P) (dibenzo[*def,p*]chrysene) is the most potent carcinogenic PAH known so far[96,97] and has replaced in this respect benzo[*a*]pyrene and 7,12-dimethylbenz[*a*]anthracene,[146,147] and therefore merits further discussion. DB[*al*]P differs from benzo[*a*]pyrene by having an additional terminal aromatic ring so that the former bay-region becomes a fjord-region; furthermore, an additional bay-region is created. The additional ring leads to molecular overcrowding in the fjord-region resulting in considerable distortion of the molecule. The two terminal rings A and F form a distortion angle of 27.6°. Thus, in contrast to many other PAH-molecules, DB[*al*]P is non-planar.[148] As a consequence, there can be spatial restrictions in adduct formation and orientation of the non-planar structure within the DNA-molecule. Possibly, the planar parts are intercalated between the bases and the distorted ones are located over the ribose moiety.[149]

Metabolic activation of DB[*al*]P by CYP1A1, CYP1B1 and other CYP enzymes occurs at various sites of the molecule. CYP1A1-activity leads to formation of dihydrodiols at positions C8/C9, C11/C12 and C13/C14, to the 7-phenol and a quinone.[145,150] Luch *et al.*[145] did not deal with the 13,14-dihydrodiol, while Shou *et al.*[150] found similar formation rates for the 13,14- and the 8,9-dihydrodiol. The latter has been shown to be only a weak tumour initiator.[146] Clearly, the *trans*-DB[*al*]P-11,12-derivative is the preferentially formed dihydrodiol by human CYP1A1 and CYP1B1, whereby the specific activity of CYP1B1 is 5 to 10 times higher compared with CYP1A1.[145] The 11,12-dihydrodiol is the precursor of the ultimate fjord-region 11,12-dihydrodiol-13,14-epoxide which forms stable adducts with DNA. Studies *in vitro* and *in vivo* have shown that most of the DNA adducts formed from DB[*al*]P as substrate are unstable (Table 2.8) and may be formed via one-electron oxidation to radical cations or *via* the DDE pathway.[151–153]

Data shown in Table 2.8 show that highest carcinogenicity is associated with the highest amounts of unstable DNA-adducts formed, as in the case of

Table 2.8 Formation of stable DNA-adducts from dibenzo[a,l]pyrene (DB[a,l]P) and its dihydrodiol (DD) and dihydrodiol epoxide (DDE) metabolites in mouse skin (μmol adduct/mol DNA) compared to the carcinogenic potential of the individual compounds (++++ > +++ > ++ > +). Data adopted from Cavalieri *et al.*[152] and Todorovic *et al.*[153]

	% stable DNA adducts	*DDE-derived DNA-adducts guanine*	*adenine*	*Carcinogenicity*
DB[a,l]P	1	0.10	0.09	++++
DB[a,l]P-11,12-DD	20	2.53	1.88	+++
syn-DB[a,l]P-DDE	68	20.2	6.2	++
anti-DB[a,l]P-DDE	97	25.3	13.9	+

the parent compound DB[*a,l*]P. The primary metabolite, the *trans*-11,12-dihydrodiol, displays decreased carcinogenicity and is associated with increased formation of stable DNA adducts, indicating that carcinogenic metabolites are generated *via* pathways deviating from the 11,12-dihydrodiol 13,14-epoxide pathway. Another metabolic branch point is the *trans*-11,12-dihydrodiol. Apparently, it is only partially transformed into the DDE and gives rise to secondary metabolites that form unstable adducts with DNA. Adducts derived from the DDE-*syn*-stereoisomer show a tendency towards depurination and this may explain its higher carcinogenicity compared with the *anti*-stereoisomer. Such a conclusion is based on the concept that formation of unstable DNA adducts is related to the carcinogenic potential.

It has been reported,[145,150] however, that the high carcinogenicity of DB[*a,l*]P is preferentially mediated by the fjord region (-)-anti-DB[*a,l*]P dihydrodiol epoxide, based on the observation that in mouse skin DNA adducts of this metabolite are found exclusively upon application of the *trans*-DB[*a,l*]P-11,12-dihydrodiol.[154] After oral application of DB[*a,l*]P, specific DNA-adducts derived from the (-)-anti fjord region DDE can be detected in the lungs of mice.[155]

2.5.5.2 *Dibenzo[c,p]chrysene (DBC)*

The specific feature of this hexacyclic PAH is the presence of two fjord-regions and one bay-region (Figure 2.19). Its metabolism and formation of DNA-adducts have been studied by Mahadevan *et al.*[156] The principal observation is that the expected conversion to its different fjord-region DDEs occurs only to a minor extent, at least in human MCF-7 cells. While the potential precursor DBC-11,12-dihydrodiol does not appear to be a suitable substrate for CYP1B1, the DBC-1,2-dihydrodiol can be converted into the respective dihydrodiol epoxide. To emphasise the peculiarities of DBC metabolism, it should be noted that

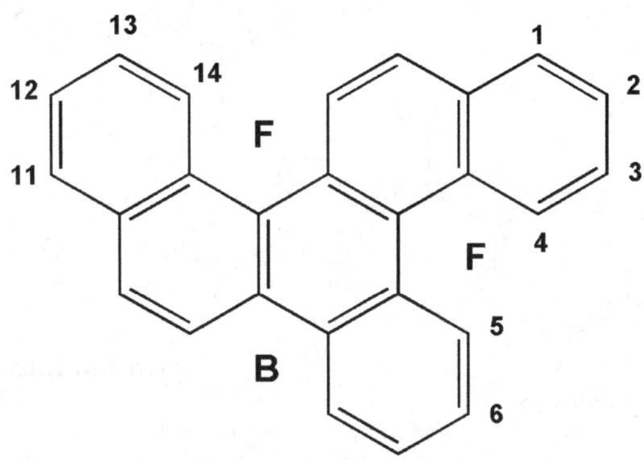

Figure 2.19 Structure of dibenzo[*c,p*]chrysene. Fjord-regions (F) and bay-region (B) are indicated. Numbered carbon atoms indicate regions of possible dihydrodiol epoxide formation.

DDEs are efficiently formed in MCF-7 cells from benzo[*g*]chrysene, and dibenzo-[*a,l*]pyrene.[157,158] Direct application of both DBC dihydrodiol epoxides, the 1,2-dihydrodiol-3,4-epoxide and the 11,12-dihydrodiol-13,14-epoxide, to MCF-7 cells efficiently results in formation of DNA-adducts.[156]

2.5.5.3 PAH without Bay-region or 'Classic' Bay-region

PAH devoid of bay- or fjord-regions may become mutagenic after metabolic activation by liver microsomes. Benzo[*ghi*]perylene shows a bay-region like structure. The adjacent benzo-rings, however, have 3 carbon atoms in common with the complete system and, therefore, cannot be activated to DDEs. The hexacyclic anthanthrene (dibenzo[*def,mno*]chrysene) does not possess any bay- or bay-like region (Figure 2.20).

Benzo[*ghi*]perylene generates at least 17 ethyl acetate-extractable metabolites *in vitro* following incubation with microsomes from Aroclor 1254-treated rats. Primary metabolites are the 7-phenol and the K-region 3,4-epoxide. Both can be further oxidised to quinones, either *via bis*-phenols or the dihydrodiol, respectively. Experimental data obtained by means of expoxide hydrolase inhibition and direct application of individual metabolites suggest that predominantly the 3,4-epoxide gives rise to the mutagenicity of benzo[*ghi*]perylene.[159] The major metabolic pathways are shown in Figure 2.21.

Anthanthrene has a comparatively low ionisation potential of 6.96 eV.[160] Therefore, the 6 (12)-radical cation can be formed as primary metabolite in addition to the 4,5-epoxide, the 1-phenol and the 3-phenol. However, polynuclear quinones derived from the radical cation obviously play only a minor role in anthanthrene-derived mutagenicity. In this respect, the CYP1-dependent formed 3-phenol is the most important metabolite, from

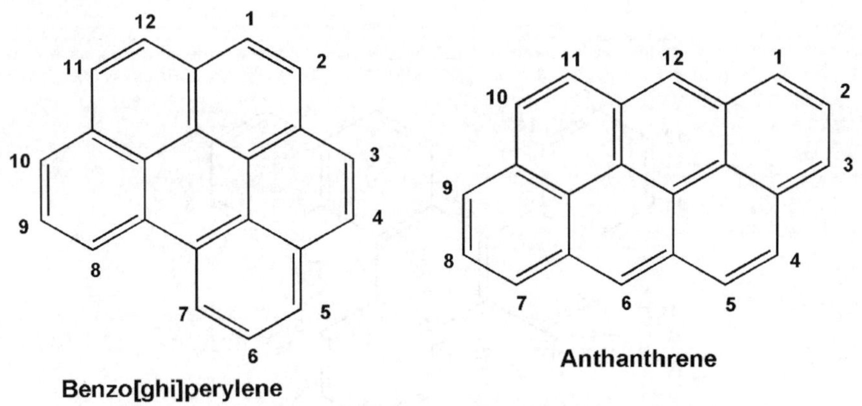

Figure 2.20 Structures of the non-bay region/non-fjord-region PAH molecules benzo[ghi]perylene and anthanthrene. Benzo[*ghi*]perylene contains a type of 'pseudo-bay-region': The adjacent benzo rings share 3 carbon atoms (instead of 2 in a classic bay-region) with further rings in the system.

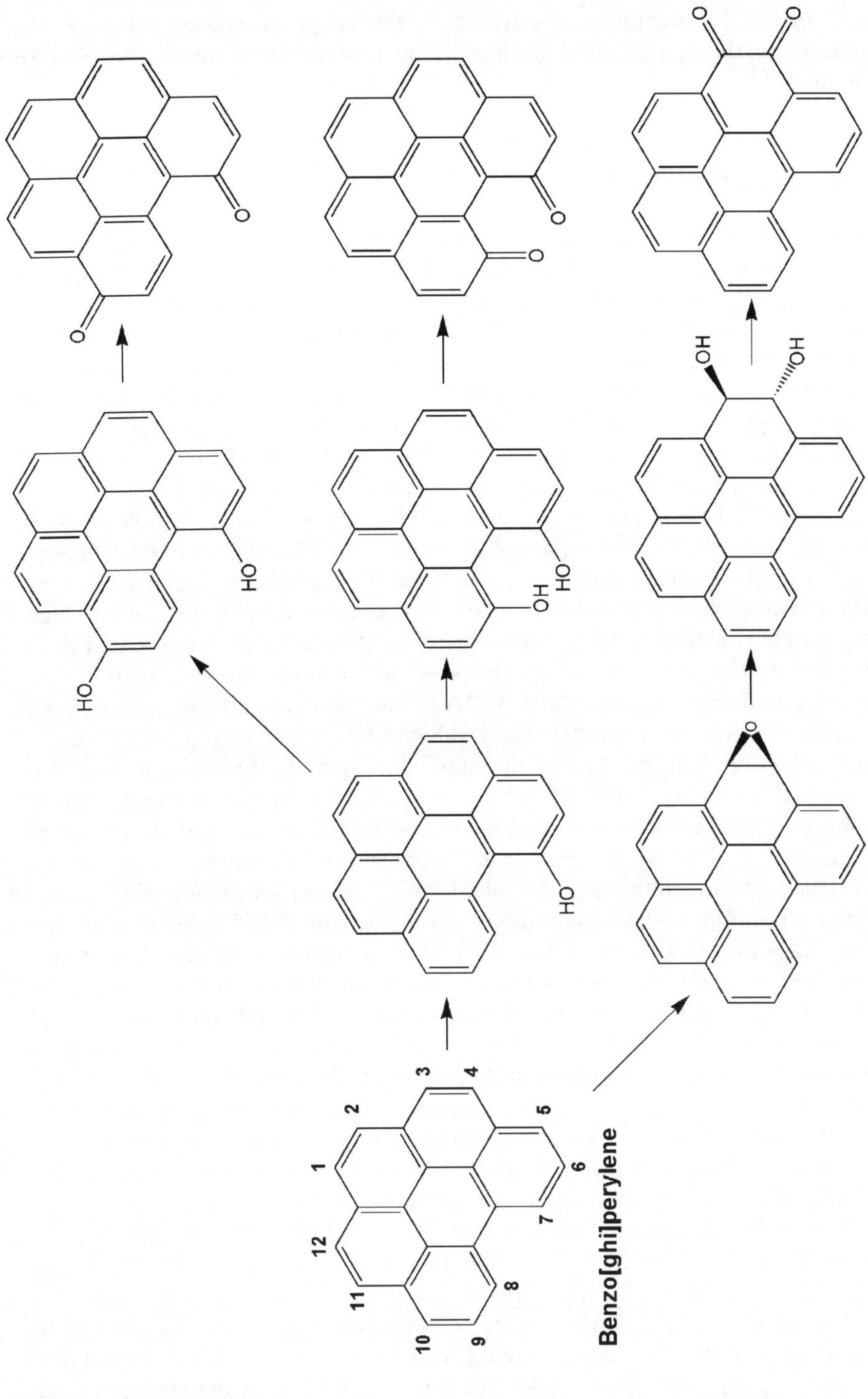

Figure 2.21 Major pathways of benzo[*ghi*]perylene metabolism according to Platt and Grupe.[159]

which further metabolites, such as the 3,6-bis-phenol, the 3,6-quinone and the 4,5-dihydrodiol-9-phenol, are derived. CYP enzymes responsible for the oxidation in the K-region of anthanthrene probably belong to the CYP2B subfamily.[161]

2.5.6 Specific CYPs

Major CYP enzymes involved in the metabolism of PAH are the members of family 1, *i.e.* CYP1A1, CYP1A2 and CYP1B1. There are a number of studies comparing metabolism and mutagenic activation of PAH by these 3 enzymes. Furthermore, allelic variants of CYPs have been investigated to evaluate their impact on PAH activation. In this context, numerous epidemiological studies on CYP1- and, to a lesser extent, CYP3A-polymorphisms and the occurrence of several types of cancer have been carried out. Such data is important for understanding the relationships and the significance of CYPs in PAH-induced cancer development. Reference may be made to some selected articles.[162–166]

By means of heterologously expressed CYP-enzymes, it was possible to comparatively study the relative impact of CYP1A1, CYP1A2 and CYP1B1. Shimada *et al.*[89] examined the potential of these three CYP enzymes to convert several PAH-dihydrodiols into mutagenic metabolites, as assessed by the *umu*-assay. They found striking differences between the different CYP1-enzymes. A number of PAH dihydrodiols, the precursors of the proposed ultimate carcinogenic DDEs, were not converted by CYP1A1 or CYP1A2, but were exclusively activated by CYP1B1. Among them was the 11,12,-dihydrodiol of the very potent procarcinogen dibenzo[*a,l*]pyrene. Under the conditions used by Shimada *et al.*,[89] the dibenzo[*a,l*]pyrene-11,12-dihydrodiol induced the highest mutagenic activity among the 21 PAH-dihydrodiols tested. Further PAH-dihydrodiols converted exclusively by CYP1B1 to mutagens were: benz[*a*]anthracene-1,2-dihydrodiol, 7,12-dimethylbenz[*a*]anthracene-3,4-dihydrodiol, benzo[*c*]phenanthrene-3,4-dihydrodiol, fluoranthene-2,3-dihydrodiol, 5,6-dimethylchrysene-1,2-dihydrodiol and benzo[*g*]chrysene-11,12-dihydrodiol. (+)-Benzo[*a*]pyrene-7,8-dihydrodiol is activated by all three CYP1-enzymes, but with different efficiency, following the order 1B1 > 1A1 > 1A2. Subsequent studies did not confirm differential activities of CYP1B1 and CYP1A1.[90] Anyway, the importance of possible small differences in CYP-specific actvities remains doubtful, as it will be later discussed. The conversion efficiencies for the (−)-benzo[*a*]pyrene-7,8-dihydrodiol are similar for CYP1A1 and CYP1B1.[89] Another substrate for which the enzyme specificities have become clear is 5-methylchrysene-1,2-dihydrodiol; its activation rate by CYP1B1, 1A1 and 1A2 is expressed by the ratios 18:2.4:1.

The profiles of primary metabolites formed from dibenzo[*a,l*]pyrene by human CYP1A1, CYP1A2 and CYP1B1, as well as by rat CYP1A1 and CYP1A2, have been studied with the respective heterologously expressed enzymes in V79 cells.[167] The data are summarised in Table 2.9. In contrast to the corresponding rat enzyme, human CYP1A2 is obviously unable to metabolise dibenzo[*a,l*]pyrene. This is also true for other CYP enzymes studied by the

Table 2.9 Metabolic transformation of dibenzo[a,l]pyrene by heterologously expressed cytochromes P450 in V79-cells. Each CYP-expressing cell line was exposed to 1 μM dibenzo[a,l]pyrene. –, below detection limit. Data from Schober *et al.*[167]

Metabolite formed	*nmol metabolite formed by*				
	human CYP1A1	*human CYP1A2*	*human CYP1B1*	*rat CYP1A1*	*rat CYP1A2*
trans-8,9-Dihydrodiol	0.15	–	–	0.55	0.73
trans-11,12-Dihydrodiol	0.88	–	2.10	0.05	0.11
cis-11,12-Dihydrodiol	0.15	–	0.39	–	–
7-Phenol	0.47	–	0.30	–	–
trans-11,12-Dihydrodiol phenol	0.18	–	0.10	–	–
(*r,t,t,c*)-11,12,13,14-Tetraol	0.17	–	0.16	–	0.14

authors, namely human CYP2A6, CYP2E1, CYP3A4 and rat CYP2B1, CYP2E1. The data show clear species-specific and enzyme-specific differences in the product profile.

Polymorphisms of CYP1B1 within the human population have been investigated in detail. Some of the allelic variants in the coding region of the CYP1B1 gene based on single nuceotide polymorphisms (SNP) lead to a change in amino acids, and can thus have an impact on enzyme parameters such as K_m, v_{max} and substrate specificity. To clarify this point, the mutagenic activation of several PAH-dihydrodiols has been studied in the presence several recombinant allelic CYP1B1 variants. No differences in activity was found for 10 PAH-dihydrodiols between CYP1B1 variants with the following amino acid substitutions: $A^{119}S$, $L^{432}V$ and $A^{119}S+L^{432}V$ represented by the alleles *2, *3/*5 and *6/*7, respectively (http://www.cypalleles.ki.se/).[90] Thus, these results do not deliver a mechanistic explanation for an association of these polymorphisms with the incidence of, for example, head and neck cancer.[168] In a more recent study, some rarer CYP1B1 alleles were included.[169] Among the alleles investigated, some were identified as 'low activity' alleles such as $G^{61}E$, $G^{365}W$, $D^{374}N$, $P^{437}L$, $R^{469}W$ and 1192Δ10. Constructed alleles including all possible combinations of the common SNPs $R^{48}G$, $A^{119}S$, $L^{432}V$ and $N^{453}S$ revealed altogether 16 different CYP1B1 enzymes, which only marginally differed in their enzymic properties, K_m and k_{cat}.[169] In a similar approach, Shimada *et al.*[170] observed slight, but significant, differences in the conversion of benzo[*a*]pyrene to its 7,8-dihydrodiol by CYP1B1 variants differing at amino acid position 432 (L *vs.* V). In conclusion, the common allelic CYP1B1 variants do not lead to relevant differences in CYP1B1 enzymic properties. However, some rare mutations in the CYP1B1 gene may result in lower activity.[169]

Recently, a further CYP enzyme, CYP2W1, has been recognised as being involved in PAH metabolism and able to generate reactive metabolites.[171] Concomitantly, its tumour-specific expression was described,[172] and it was proposed as a drug target in cancer chemotherapy.[173]

2.5.7 Specific Sites: CYP Profiles, Subcellular CYP-localisations

Organs, tissues and cells show quantitatively and qualitatively different CYP profiles and express different capacities to generate reactive metabolites from PAH. In combination with tissue[174] or cell-specific accumulation of PAH,[175] this may determine site-specific susceptibilities to adverse PAH effects. Among the major CYP enzymes involved in PAH metabolism are members of the families CYP1 and CYP3. The former are regulated *via* the Ah-receptor and are modulated by PAH having sufficient affinity for this receptor. Besides the Ah-receptor, other regulatory proteins determine the differential tissue-specific expression of the three mammalian CYP1 enzymes CYP1A1, CYP1A2 and CYP1B1. The enzymes of the family CYP3 are under the control of the pregnane X-receptor and, like CYP1 enzymes, show differential tissue-specific expression.[176–180] While orthology of the three CYP1 enzymes in various species is obvious, the phylogenetic relationships of CYP3 enzymes from different species are not clear, and complicate inter-species comparisons.

In the liver of experimental animals and humans, CYP1A1 is not expressed constitutively, but can be induced, at least in animals, by Ah-receptor agonists such as PAH.[181,182] CYP1B1 is similarly not present in liver. By contrast, about 15% of human liver microsomal cytochrome P450 is made up by CYP1A2 and about 30% by CYP3A.[183] Thus, PAH entering the non-induced liver are received by CYP1A2, which differs in its metabolic activities from CYP1A1 with respect to the regio-specificity and the product profile. A striking substrate example is dibenzo[*a,l*]pyrene which is not at all metabolised by human CYP1A2 but which is, however, a substrate for human CYP1A1 and CYP1B1, as well as rat CYP1A1 and CYP1A2.[167] In contrast to CYP1A2, CYP1B1 is constitutively expressed in various extrahepatic tissues,[176,184–186] and CYP1A1 can be induced in several organs by PAH.[187] Furthermore, CYP1B1 is over-expressed in various tumour cells and has therefore been considered as a tumour marker.[188]

Dibenzo[*a,l*]pyrene (DB[*a,l*]P) is known to induce mammary carcinomas.[96] It is also established that CYP1B1, which is able to metabolise DB[*a,l*]P, is expressed in breast cancer cells.[189–191] In combination with the facts that (1) DB[*a,l*]P does not undergo primary metabolism in the liver due to presence only of the non-metabolising CYP1A2, and that (2) PAH tend to accumulate in adipose tissue,[174] one can formulate the hypothesis that the breast tissue forms a storage site for PAH, so that a continuous release of PAH, such as DB[*a,l*]P, leads to sufficiently high levels of CYP1B1-mediated reactive metabolites to induce adverse genotoxic effects. CYP1B1 is considered to mediate adverse PAH effects also in other tissues, *e.g.* formation of lymphomas by 7,12-dimethylbenz[*a*]anthracene.[192] Toxic effects of benzo[*a*]pyrene and 7,12-dimethylbenz[*a*]anthracene in different lineages of bone marrow cells are also mediated by CYP1B1. The two PAH compounds, however, show different effects on the cells, which have been explained by differential metabolite profiles generated from the parent compounds by CYP1B1.[193]

Considering the reactivity of PAH metabolites and the fact that some of them are (very) short-lived, it is of interest not only to know the tissue of origin where they were generated but also the subcellular localisation. This aspect is rarely considered but appears important with respect to possible interactions of highly reactive metabolites with DNA, either nuclear or mitochondrial. The major CYP enzymes of the family 1 involved in PAH metabolism are predominantly located on the endoplasmic reticulum. It is known, however, that these may also occur in other membrane systems.[194] Their occurrence in the nuclear and mitochondrial membranes may be of particular relevance with respect to DNA adduct formation. The finding that CYP1B1 is present also in the nuclear membrane[185] has largely been ignored. Recently, an enzymically active splice variant of CYP1A1, based on a cryptic intron in exon 2, was found restricted to nuclear and mitochondrial membranes in ovarian cancer cell lines.[195]

2.5.8 General Conclusions Regarding PAH-derived Reactive Metabolites

Although many aspects of PAH-metabolism, formation of reactive PAH metabolites and DNA adducts have been investigated, there are open questions and missing data in the puzzle. Because of the diversity of reactive metabolites formed from even a single PAH, and their differing stabilities, reactivities towards target molecules such as DNA and stabilities of adducts formed, the causal chain from PAH uptake to an adverse (carcinogenic) effect cannot be easily delineated. One must clearly differentiate between correlations and causal chains. It has to be questioned, for example, which particular reactive metabolites are most responsible for carcinogenesis. Traditionally, the vicinal bay- or fjord-region dihydrodiol epoxides are considered to be the ultimate carcinogens. This notion, however, is not entirely supported by data in the literature, some of which may be contradictory. In a recent study, dibenzo[*a,l*]pyrene as the parent compound, the primary (−)-dihydrodiol and the anti- and syn-DDE were applied to mouse skin.[153] Although the DDEs led to higher amounts of stable DNA-adducts than the parent compound and the dihydrodiols, their tumour initiating activity was lower. While 97% stable adducts were formed from *anti*-dibenzo[*a,l*]pyrene-dihydrodiol epoxide, the amounts formed by dibenzo[*a,l*]pyrene-dihydrodiol and the parent compound were only 20% and 1%, respectively. It appears that carcinogenicity is decreasing starting from the parent compound and leading to its DDE. In this case, tumourigenicity is not correlated with formation of stable DNA adducts *via* the fjord-region DDE. It may be hypothesised that unstable depurinating adducts, which can be generated *via* the radical cation pathway of dibenzo[*a,l*]pyrene metabolism, are important.[151,152,160,196] Melendez-Colon *et al.*[197] have argued that formation of stable adducts, rather than of apurinic sites, is responsible for cancer initiation. They showed that the highest amount of stable adducts and the lowest amount of apurinic sites were generated by the potent carcinogen dibenzo[*a,l*]pyrene compared with benz[*a*]pyrene and 7,12-dimethylbenz[*a*]anthracene. In contrast, others found that the overwhelming majority of adducts formed from dibenzo-[*a,l*]pyrene *in vivo* and *in vitro* were unstable depurinating adducts.[152,153] In our opinion, strict evidence that the fjord-region DDEs are predominantly responsible for the carcinogenic effect of PAH has not been presented thus far. On the other hand, PAH radical cations are very reactive with half-lives in the millisecond range so it is unlikely that they survive to react with nuclear DNA when they are formed *in vivo* by ER-localised CYPs.[124] Thus, the subcellular site of metabolite generation is important. In this context, occurrence of CYP1-enzymes in the nuclear and mitochondrial membranes may be quite important.[185,194,195] Despite these points, it is clear that PAH-derived DNA adducts play a significant role in PAH induced carcinogenesis. The observation that formation of PAH-derived adducts leads to an activation of p53, however, without increasing p21-expression is of particular interest.[198] In this way, the cell cycle is not arrested due to non-inhibited activity of CDKs (Cyclin-Dependent Kinases). To better understand the mode of PAH action, the identification of relevant metabolites and the conduct of studies on the tissue-specific distribution of CYP enzymes is still necessary.

2.6 Final Remarks

In the entire field of toxicology, CYP enzymes play a distinct role in the generation of biologically reactive intermediates. Most prominent examples for such processes are olefinic and aromatic systems, including PAH, which were discussed here in detail. Decisive factors that widely determine the nature and organotropism of adverse effects are individual CYP specificities for an array of substrates, parameters of CYP enzyme kinetics, CYP tissue distribution, regulation of individual CYPs, stability and reactivity of the reactive intermediates formed, and modulation of the kinetics of reactive intermediates by phase II and phase III processes. Much progress has been achieved in these fields that now allow modelling and predictions to be made, relevant both in pharmacology and toxicology. Yet, there are many knowledge gaps that need to be addressed in order, to improve our understanding of the relevant processes and the biological effects of individual chemicals of relevance.

References

1. N.P.E. Vermeulen, in *"Cytochromes P450, Role in the Metabolism and Toxicity of Drugs and Other Xenobiotics"*, ed. C. Ioannides, CRC Press, Boca Raton, FL, 1996, 29.
2. F.P. Guengerich, in *"Cytochromes P450, Role in the Metabolism and Toxicity of Drugs and Other Xenobiotics"*, ed. C. Ioannides, CRC Press, Boca Raton, FL, 1996, 55.
3. R.E. Lehr and D.M. Jerina, *Arch. Toxicol.*, 1977, **39**, 1.
4. M.J.S. Dewar, *"The Molecular Orbital Theory of Organic Chemistry". McGraw Hill*, New York NY, 1969.

5. D.M. Jerina and J.W. Daly, in *"Drug Metabolism"*, eds. D.V. Parke and R.L. Smith, Taylor and Francis, London, 1976, 13.
6. M.T. Poulsen and G.H. Loew, *Cancer Biochem. Biophys.*, 1981, **5**, 81.
7. S.M. Adams and L.S. Kaminski, *Mol. Pharmacol.*, 1982, **22**, 459.
8. B.D. Silverman, *Chem. Biol. Interact.*, 1985, **53**, 313.
9. J.J. Kauffman, P.C. Hariharan, W.S. Koski and K. Balasubramaniam, *Prog. Clin. Biol. Res.*, 1985, **172A**, 263.
10. P. Politzer, P. Trefonas 3rd, I.R. Politzer and B. Elfman, *Ann. N. Y. Acad. Sci.*, 1981, **367**, 478.
11. G.H. Loew, E. Kurkjian and M. Rebagliati, *Chem. Biol. Interact.*, 1983, **43**, 33.
12. P. Politzer, in *"Chemical Carcinogens – Activation Mechanisms, Structural and Electronic Factors and Reactivity"*, eds. P. Politzer and F.J. Martin jr, Bioreactive Molecules 5,Elsevier, Amsterdam, 1988, 208.
13. P. Politzer and J.S. Murray, *Mol. Toxicol.*, 1987, **1**, 1.
14. S. Trohalaki and R. Pachter, *SAR QSAR Environ. Res.*, 2003, **14**, 131.
15. S. Trohalaki, E. Gifford and R. Pachter, *Computers and Chemistry*, 2000, **24**, 421.
16. L.I. Kirkovsky, S.A. Lermontov, S.I. Zavorin, I.I. Sukhozhenko, V.I. Zavelsky, R. Thier and H.M. Bolt, *Environ. Toxicol. Chem.*, 1998, **17**, 2141.
17. R.B. Jones and W.C. Mackrodt, *Biochem. Pharmacol.*, 1982, **31**, 3710.
18. R.B. Jones and W.C. Mackrodt, *Biochem. Pharmacol.*, 1983, **32**, 2359.
19. H.M. Bolt, *Crit. Rev. Toxicol.*, 2005, **35**, 307.
20. H. Greim, G. Bonse, Z. Radwan and D. Henschler, *Biochem. Pharmacol.*, 1975, **24**, 2013.
21. H.M. Bolt, R.J. Laib and J.G. Filser, *Biochem. Pharmacol.*, 1982 **31**, 1.
22. H.M. Bolt, in *"Chemical Carcinogens – Activation Mechanisms, Structural and Electronic Factors and Reactivity"*, eds. P. Politzer and F.J. Martin jr, *Bioreactive Molecules* 5, Elsevier, Amsterdam, 1988, 177.
23. U. Jorritsma, M. Cornet, P. van Hummelen, H.M. Bolt, A. Vercruysse, M. Kirsch-Volders and V. Rogiers, *Mutagenesis*, 1995, **10**, 101.
24. P. Simon, B. Epe, P. Mützel, D. Schiffmann, D. Wild, H. Ottenwälder, N. Fedtke, H.M. Bolt and D. Henschler, *J. Biochem. Toxicol.*, 1986, **1**, 43.
25. R.J. Laib and H.M. Bolt, *Carcinogenesis*, 1986, **7**, 841.
26. P. Simon, J.G. Filser and H.M. Bolt, *Arch. Toxicol.*, 1985, **57**, 191.
27. M.S. Bogdanff, R. Sarangapani, D.R. Plowchalk, A. Jarabek and M.E. Andersen, *Toxicol. Sci.*, 1999, **51**, 19.
28. J.G. Hengstler, M.S. Bogdanffy, H.M. Bolt and F. Oesch, *Annu. Rev. Pharmacol. Toxicol.*, 2003, **43**, 485.
29. F.P. Guengerich, *Adv. Pharmacol.*, 1997, **43**, 7.
30. J.W. Allis, B.L. Brown, G. Zhao and R.A. Pegram, *Toxicology*, 2001 **161**, 67.
31. M. Murray and G.F. Reidy, *Pharmacol. Rev.*, 1990, **42**, 85.
32. N.E. Hopkins, M.K. Foroozesh and W.L. Alworth, *Biochem. Pharmacol.*, 1992, **44**, 787.

33. F.P. Guengerich, *Life Sci.*, 1990, **47**, 1981.
34. F.P. Guengerich, *Chem. Res. Toxicol.*, 1990, **3**, 363.
35. E. Fontana, P.M. Dansette and S.M. Poli, *Curr. Drug Metab.*, 2005, **6**, 413.
36. H.K. Lim, N. Duczak jr, L. Brougham, M. Elliot, K. Patel and K. Chan, *Drug Metab. Dispos.*, 2005, **22**, 1211.
37. S. Zhou, E. Chan, L.Y. Lim, U.A. Boelsterli, S.C. Li, J. Wang, Q. Zhang, M. Huang and A. Xu, *Curr. Drug Metab.*, 2004, **5**, 415.
38. G. Danan, V. Descatoire and D. Pessayre, *J. Pharmacol. Exp. Ther.*, 1981, **218**, 509.
39. D. Mansuy, M. Delaforge, E. LeProvost, J.P. Flinois, S. Columelli and P. Beaune, *Biochem. Biophys. Res. Commun.*, 1981, **103**, 1201.
40. P.H. Roos, B. Golub-Ciosk, P. Kallweit, D. Kauczinski and W.G. Hanstein, *Biochem. Pharmacol.*, 1993, **45**, 2239.
41. T. Miura, M. Iwasaki, M. Komori, H. Ohi, M. Kitada, H. Mitsui and T. Kamataki, *J. Antimicrob. Chemother.*, 1989, **24**, 551.
42. D.J. McConn 2nd, Y.S. Lin, K. Allen, K.L. Kunze and K.E. Thummel, *Drug Metab. Dispos.*, 2004, **32**, 1083.
43. M. Murray, *Mol. Pharmacol.*, 1992, **42**, 931.
44. M. Murray and S.L. Field, *Biochem. Pharmacol.*, 1992, **43**, 2065.
45. P.H. Roos and A. Mahnke, *Biochem. Pharmacol.*, 1996, **52**, 73.
46. P.H. Roos, *Xenobiotica*, 1999, **29**, 629.
47. Y. Cai, W. Baer-Dubowska, M.J. Ashwood-Smith, O. Ceska, S. Tachibana and J. DiGiovanni, *Chem. Res. Toxicol.*, 1996, **9**, 729.
48. Y. Cai, H. Kleiner, D. Johnston, A. Dubowski, S. Bostic, W. Ivie and J. DiGiovanni, *Carcinogenesis*, 1997, **18**, 1521.
49. H. Druckrey, *Krebsgeschehen*, 1973, **5**, 73.
50. H. Druckrey, *Xenobiotica*, 1973, **3**, 271.
51. W. Köhl and G. Eisenbrand, in: *"Lehrbuch der Toxikologie"*, H. Marquardt, S. Schäfer, Eds, p. 747, Wissenschaftliche Verlagsgesellschaft, Stuttgart, 2004.
52. Y. Mori, H. Yamazaki, K. Toyoshi, A. Denda and Y. Konishi, *Carcinogenesis*, 1986, **7**, 375.
53. S.S. Hecht, *Chem. Res. Toxicol.*, 1998, **11**, 559.
54. J. Lamoureux and A. Castonguay, *Carcinogenesis*, 1997, **18**, 1979.
55. C.L. Crespi, B.W. Penman, H.V. Gelboin and F.J. Gonzalez, *Carcinogenesis*, 1991, **12**, 1197.
56. T.J. Smith, Z.Y. Guo, F.J. Gonzalez and F.P. Guengerich, *Cancer Res.*, 1992, **52**, 1757.
57. B.W. Penman, J. Reece, T. Smith, C.S. Yang, H.V. Gelboin, F.J. Gonzalez and C.L. Crespi, *Pharmacogenetics*, 1993, **3**, 28.
58. E. Anderson, *Maturitas*, 2004, **48**(Suppl 1), S13.
59. S. Chen, *Front. Biosci.*, 1998, **3**, 922.
60. H.M. Bolt and P. Göbel, *Horm. Metab. Res.*, 1972, **4**, 312.
61. L. Holmberg, H. Anderson and Habits Steering and Data Monitoring Committees, *Lancet*, 2004, **363**, 453.
62. H.M. Bolt, *Drug Metab. Rev.*, 2004, **36**, 497.

63. C. Justenhoven, U. Hamann, B. Pesch, V. Harth, S. Rabstein, C. Baisch, C. Vollmert, T. Illig, Y.D. Ko, T. Brüning and H. Brauch, *Cancer Epidemiol. Biomarkers Prev.*, 2004, **13**, 2059.
64. T. Rylander-Rudqvist, S. Wedren, F. Granat, K. Humphreys, S. Ahlberg, E. Weiderpass, M. Oscarson, M. Ingelman-Sundberg and I. Persson, *Carcinogenesis*, 2003, **24**, 1533.
65. K. Mitrunen, V. Kataja, M. Eskelinen, V.M. Kosma, D. Kang, S. Benhamou, H. Vainio, M. Uusitupa and A. Hirvonen, *Pharmacogenetics*, 2002, **12**, 67.
66. J. Russo, M. Hasan Lareef, G. Balogh, S. Guo and I.H. Russo, *J. Steroid Biochem. Mol. Biol.*, 2003, **87**, 1.
67. E. Cavalieri, D. Chakravarti, J. Guttenplan, E. Hart, J. Ingle, R. Jankowiak, P. Muti, E. Rogan, J. Russo, R. Santen and T. Sutter, *Biochim. Biophys. Acta*, 2006, **1766**, 63.
68. C.L. Hayes, D.C. Spink, B.C. Spink, J.Q. Cao, N.J. Walker and T.R. Sutter, *Proc. Natl. Acad. Sci. USA.*, 1996, **93**, 9776.
69. W. Yue, R.J. Santen, J.P. Wang, Y. Li, M.F. Verderame, W.P. Bocchinfuso, K.S. Korach, P. Devanesan, R. Todorovic, E.G. Rogan and E.L. Cavalieri, *J. Steroid Biochem. Mol. Biol.*, 2003, **86**, 477.
70. Y. Tsuchiya, M. Nakajima, S. Kyo, T. Kanaya, M. Inoue and T. Yokoi, *Cancer Res.*, 2004, **64**, 3119.
71. I.H. Hanna, S. Dawling, N. Roodi, F.P. Guengerich and F.F. Parl, *Cancer Res.*, 2000, **60**, 3440.
72. S.L. Iverson, L. Shen, N. Anlar and J.L. Bolton, *Chem. Res. Toxicol.*, 1996, **9**, 492.
73. H.M. Bolt and H. Kappus, *J. Steroid Biochem.*, 1974, **5**, 179.
74. D.L. Hachey, S. Dawling, N. Roodi and F.F. Parl, *Cancer Res.*, 2003, **63**, 8492.
75. S. Dawling, D.L. Hachey, N. Roodi and F.F. Parl, *Chem. Res. Toxicol.*, 2004, **17**, 1258.
76. J.A. Doherty, N.S. Weiss, R.J. Freeman, D.A. Dightman, P.J. Thornton, J.R. Houck, L.F. Voigt, M.A. Rossing, S.M. Schwartz and C. Chen, *Cancer Epidemiol. Biomarkers Prev.*, 2005, **14**, 357.
77. S. Dawling, N. Roodi and F.F. Parl, *Cancer Res.*, 2003, **63**, 3127.
78. P.B. Farmer, R. Singh, B. Kaur, R.J. Sram, B. Binkova, I. Kalina, T.A. Popov, S. Garte, E. Taioli, A. Gabelova and A. Cebulska-Wasilewska, *Mutat. Res.*, 2003, **544**, 397.
79. L. Pozzoli, S. Gilardoni, M.G. Perrone, G. De Gennaro, M. De Rienzo and D. Vione, *Ann. Chim.*, 2004, **94**, 17.
80. P. Vineis and K. Husgafvel-Pursiainen, *Carcinogenesis*, 2005, **26**, 1846.
81. P.H. Roos, S. Tschirbs, F. Pfeifer, P. Welge, A. Hack, M. Wilhelm and H.M. Bolt, *Toxicology*, 2004, **205**, 181.
82. M. Ruchirawat, D. Settachan, P. Navasumrit, J. Tuntawiroon and H. Autrup, *Toxicol. Lett.*, 2007, **168**, 200.
83. D.H. Phillips, *Mutat Res.*, 1999, **443**, 139.
84. T. Shimada and F.P. Guengerich, *Chem. Res. Toxicol.*, 2006, **19**, 288.

85. T. Shimada, *Drug Metab. Pharmacokinet.*, 2006, **21**, 257.
86. S. Amin, D. Desai and S.S. Hecht, *Carcinogenesis*, 1993, **14**, 2033.
87. W. Levin, R.L. Chang, A.W. Wood, D.R. Thakker, H. Yagi, D.M. Jerina and A.H. Conney, *Cancer Res.*, 1986, **46**, 2257.
88. W. Xue and D. Warshawsky, *Toxicol. Appl. Pharmacol.*, 2005, **206**, 73.
89. T. Shimada, C.L. Hayes, H. Yamazaki, S. Amin, S.S. Hecht, F.P. Guengerich and T.R. Sutter, *Cancer Res.*, 1996, **56**, 2979.
90. T. Shimada, J. Watanabe, K. Kawajiri, T.R. Sutter, F.P. Guengerich, E.M. Gillam and K. Inoue, *Carcinogenesis*, 1999, **20**, 1607.
91. M. Shou, K.R. Korzekwa, K.W. Krausz, C.L. Crespi, F.J. Gonzalez and H.V. Gelboin, *Cancer Lett.*, 1994, **83**, 305.
92. D. Schwarz, P. Kisselev, I. Cascorbi, W.H. Schunck and I. Roots, *Carcinogenesis*, 2001, **22**, 453.
93. T.R. Sutter, Y.M. Tang, C.L. Hayes, Y.Y.P. Wo, E.W. Jabs *et al.*, *J. Biol. Chem.*, 1994, **269**: 13092.
94. J.H. Kim, K.H. Stansbury, N.J. Walker, M.A. Trush, P.T. Strickland and T.R. Sutter, *Carcinogenesis*, 1998, **19**, 1847.
95. K. Prout, A.D. Smith, G.H. Daub, D.E. Zacharias and J.P. Glusker, *Carcinogenesis*, 1992, **13**, 1775.
96. E.L. Cavalieri, E.G. Rogan, S. Higginbotham, P. Cremonesi and S. Salmasi, *J. Cancer Res. Clin. Oncol.*, 1989, **115**, 67.
97. P.D. Devanesan, P. Cremonesi, J.E. Nunnally, E.G. Rogan and E.L. Cavalieri, *Chem. Res. Toxicol.*, 1990, **3**, 580.
98. L.T. Laughlin, H.F. Tzeng and S. Lin, *N. Armstrong, Biochemistry*, 1998, **37**, 2897.
99. J.M. Sayer, H. Yagi, P.J. van Bladeren, W. Levin and D.M. Jerina, *J. Biol. Chem.*, 1985, **260**, 1630.
100. S.K. Yang, *Biochem. Pharmacol.*, 1988, **37**, 61.
101. D.R. Thakker, H. Yagi, R.E. Lehr, W. Levin, M. Buening, A.Y. Lu, R.L. Chang, A.W. Wood, A.H. Conney and D.M. Jerina, *Mol. Pharmacol.*, 1978, **14**, 502.
102. T. Shimada, M.V. Martin, D. Pruess-Schwartz, L.J. Marnett and F.P. Guengerich, *Cancer Res.*, 1989, **49**, 6304.
103. D.R. Thakker, W. Levin, H. Yagi, H.J. Yeh, D.E. Ryan, P.E. Thomas, A.H. Conney and D.M. Jerina, *J. Biol. Chem.*, 1986, **261**, 5404.
104. A. Luch, K.L. Platt and A. Seidel, *Carcinogenesis*, 1998, **19**, 639.
105. D.R. Thakker, W. Levin, H. Yagi, M. Tada, D.E. Ryan, P.E. Thomas, A.H. Conney and D.M. Jerina, *J. Biol. Chem.*, 1982, **257**, 5103.
106. D.R. Thakker, H. Yagi, H. Akagi, M. Koreeda, A.H. Lu, W. Levin, A.W. Wood, A.H. Conney and D.M. Jerina DM, *Chem. Biol. Interact.*, 1977, **16**, 281.
107. A.W. Wood, M.T. Huang, R.L. Chang, H.L. Newmark, R.E. Lehr, H. Yagi, J.M. Sayer, D.M. Jerina and A.H. Conney, *Proc. Natl. Acad. Sci. USA*, 1982, **79**, 5513.

108. J. Szeliga, B.D. Hilton, G.N. Chmurny, J. Krzeminski, S. Amin and A. Dipple, *Chem. Res. Toxicol.*, 1997, **10**, 378.
109. C.A. Bigger, J.T. Sawicki, D.M. Blake, L.G. Raymond and A. Dipple, *Cancer Res.*, 1983, **43**, 5647.
110. K. Peltonen, B.D. Hilton, J. Pataki, H. Lee, R.G. Harvey and A. Dipple, *Chem. Res. Toxicol.*, 1991, **4**, 305.
111. K. Peltonen, S.C. Cheng, B.D. Hilton, H. Lee, C. Cortez, R.G. Harvey and A. Dipple, *J. Org. Chem.*, 1991, **56**, 4181.
112. J.M. Sayer, A. Chadha, S.K. Agarwal, H.J.C. Yeh, H. Yagi and D.M. Jerina, *J. Org. Chem.*, 1991, **56**, 20.
113. J. Szeliga and A. Dipple, *Chem. Res. Toxicol.*, 1998, **11**, 1.
114. A. Chadha, J.M. Sayer, H.J.C. Yeh, H. Yagi, A.M. Cheh, L.K. Pannell and D.M. Jerina, *J. Am. Chem. Soc.*, 1989, **111**, 5456.
115. A.M. Cheh, A. Chadha, J.M. Sayer, H.J.C. Yeh, H. Yagi, L.K. Pannell and D.M. Jerina, *J. Org. Chem.*, 1993, **58**, 4013.
116. J. Szeliga, H. Lee, R.G. Harvey, J.E. Page, H.L. Ross, M.N. Routledge, B.D. Hilton and A. Dipple, *Chem. Res. Toxicol.*, 1994, **7**, 420.
117. S.K. Agarwal, J.M. Sayer, H.J.C. Yeh, L.K. Pannell, B.D. Hilton, M.A. Pigott, A. Dipple, H. Yagi and D.M. Jerina, *J. Am. Chem. Soc.*, 1987, **109**, 2497.
118. E. Cavalieri and E. Rogan, *Environ. Health Perspect.*, 1985, **64**, 69.
119. D.F. Lewis and D.V. Parke, *Mutat Res.*, 1995, **328**, 207.
120. R.A. Hites and W.J. Simonsick, *"Calculated Molecular Properties of Polycyclic Aromatic Hydrocarbons". Elsevier*, Amsterdam, 1987.
121. B.W. Bogan and R.T. Lamar, *Appl. Environ. Microbiol.*, 1995, **61**, 2631.
122. R.G. Harvey, *"Polycyclic Aromatic Hydrocabons: Chemistry and Carcinogenicity". Cambridge University Press*, Cambridge, 1991.
123. P. Cremonesi, E.L. Cavalieri and E.G. Rogan, *J. Org. Chem.*, 1989, **54**, 3561.
124. F.P. Guengerich, *Chem. Res. Toxicol.*, 2001, **14**, 612.
125. E.L. Cavalieri and E.G. Rogan, *Xenobiotica*, 1995, **25**, 677.
126. E.G. Rogan, P.D. Devanesan, N.V. RamaKrishna, S. Higginbotham, N.S. Padmavathi, K. Chapman, E.L. Cavalieri, H. Jeong, R. Jankowiak and G.J. Small, *Chem. Res. Toxicol.*, 1993, **6**, 356.
127. W. Xue, D. Zapien and D. Warshawsky, *Chem. Res. Toxicol.*, 1999, **12**, 1234.
128. N.T. Palackal, M.E. Burczynski, R.G. Harvey and T.M. Penning, *Biochemistry*, 2001, **40**, 10901.
129. N.T. Palackal, M.E. Burczynski, R.G. Harvey and T.M. Penning, *Chem. Biol. Interact.*, 2001, **130–132**, 815.
130. P. Nioi and J.D. Hayes, *Mutat Res.*, 2004, **555**, 149.
131. A.M. Samuni, E.Y. Chuang, M.C. Krishna, W. Stein, W. DeGraff, A. Russo and J.B. Mitchell, *Proc. Natl. Acad. Sci. USA*, 2003, **100**, 5390.
132. B. Kalyanaraman, W. Korytowski, B. Pilas, T. Sarna, E.J. Land and T.G. Truscott, *Arch. Biochem. Biophys.*, 1988, **266**, 277.
133. M. Shou, R.G. Harvey and T.M. Penning, *Carcinogenesis*, 1993, **14**, 475.

134. N. Balu, W.T. Padgett, G.R. Lambert, A.E. Swank, A.M. Richard and S. Nesnow, *Chem. Res. Toxicol.*, 2004, **17**, 827.
135. K.D. McCoull, D. Rindgen, I.A. Blair and T.M. Penning, *Chem. Res. Toxicol.*, 1999, **12**, 237.
136. S. Nesnow, C. Davis, W. Padgett, M. George, G. Lambert, F. Meyers, J. Allison, L. Adams and L.C. King, *Chem. Res. Toxicol.*, 1998, **11**, 1596.
137. W. Baer-Dubowska, *Acta Biochim. Pol.*, 1999, **46**, 263.
138. P.L. Carmichael, K.L. Platt, M.N. She, S. Lecoq, F. Oesch, D.H. Phillips and P.L. Grover, *Cancer Res.*, 1993, **53**, 944.
139. M. Nordqvist, D.R. Thakker, W. Levin, H. Yagi, D.E. Ryan, P.E. Thomas, A.H. Conney and D.M. Jerina, *Mol. Pharmacol.*, 1979, **16**, 643.
140. K.L. Platt and I. Reischmann, *Mol. Pharmacol.*, 1987, **32**, 710.
141. S. Lecoq, M.N. She, A. Hewer, P.L. Grover, K.L. Platt, F. Oesch and D.H. Phillips, *Carcinogenesis*, 1991, **12**, 1079.
142. S. Lecoq, W. Pfau, P.L. Grover and D.H. Phillips, *Chem. Biol. Interact.*, 1992, **85**, 173.
143. K.L. Platt and M. Schollmeier, *Chem. Res. Toxicol.*, 1994, **7**, 89.
144. S. Nesnow, J. Ross, S. Beck, J. Lasley, G. Nelson, G. Lambert, K.L. Platt and S.C. Agarwal, *Carcinogenesis*, 1994, **15**, 2225.
145. A. Luch, W. Schober, V.J. Soballa, G. Raab, H. Greim, J. Jacob, J. Doehmer and A. Seidel, *Chem. Res. Toxicol.*, 1999, **12**, 353.
146. E.L. Cavalieri, S. Higginbotham, N.V. RamaKrishna, P.D. Devanesan, R. Todorovic, E.G. Rogan and S. Salmasi, *Carcinogenesis*, 1991, **12**, 1939.
147. S. Higginbotham, N.V. RamaKrishna, S.L. Johansson, E.G. Rogan and E.L. Cavalieri, *Carcinogenesis*, 1993, **14**, 875.
148. A. Kaufman Katz, H.L. Carrell and J.P. Glusker, *Carcinogenesis*, 1998, **19**, 1641.
149. H.L. Carrell, J.P. Glusker, R.C. Moschel, W.R. Hudgins and A. Dipple, *Cancer Res.*, 1981, **41**, 2230.
150. M. Shou, K.W. Krausz, F.J. Gonzalez and H.V. Gelboin, *Carcinogenesis*, 1996, **17**, 2429.
151. K.M. Li, R. Todorovic, E.G. Rogan, E.L. Cavalieri, F. Ariese, M. Suh, R. Jankowiak and G.J. Small, *Biochemistry*, 1995, **34**, 8043.
152. E.L. Cavalieri, E.G. Rogan, K.M. Li, R. Todorovic, F. Ariese, R. Jankowiak, N. Grubor and G.J. Small, *Chem. Res. Toxicol.*, 2005, **18**, 976.
153. R. Todorovic, P. Devanesan, E. Rogan and E. Cavalieri, *Chem. Res. Toxicol.*, 2005, **18**, 984.
154. S.L. Ralston, H.H.S. Lau, A. Seidel, A. Luch, K. -L. Platt and W.M. Baird, *Polycyclic Aromat Comp.*, 1994, **6**, 199.
155. B. Mahadevan, A. Luch, C.F. Bravo, J. Atkin, L.B. Steppan, C. Pereira, N.I. Kerkvliet and W.M. Baird, *Cancer Lett.*, 2005, **227**, 25.
156. B. Mahadevan, A. Luch, J. Atkin, T. Nguyen, A.K. Sharma, S. Amin and W.M. Baird, *Chem. Biol. Interact.*, 2006, **164**, 181.
157. R. Agarwal, S.L. Coffing, W.M. Baird, A.S. Kiselyov, R.G. Harvey and A. Dipple, *Cancer Res.*, 1997, **57**, 415.

158. S.L. Ralston, S.L. Coffing, A. Seidel, A. Luch, K.L. Platt and W.M. Baird, *Chem. Res. Toxicol.*, 1997, **10**, 687.
159. K.L. Platt and S. Grupe, *Chem. Res. Toxicol.*, 2005, **18**, 700.
160. E.L. Cavalieri and E.G. Rogan, *Free Radic. Res. Commun.*, 1990, **11**, 77.
161. K.L. Platt, C. Degenhardt, S. Grupe, H. Frank and A. Seidel, *Chem. Res. Toxicol.*, 2002, **15**, 332.
162. H. Bartsch, U. Nair, A. Risch, M. Rojas, H. Wikman and K. Alexandrov, *Cancer Epidemiol. Biomarkers Prev.*, 2000, **9**, 3.
163. J. Watanabe, T. Shimada, E.M. Gillam, T. Ikuta, K. Suemasu, Y. Higashi, O. Gotoh and K. Kawajiri, *Pharmacogenetics*, 2000, **10**, 25.
164. J.A.G. Agúndez, *Curr. Drug Metab.*, 2004, **5**, 211.
165. H. Dally, L. Edler, B. Jager, P. Schmezer, B. Spiegelhalder, H. Dienemann, P. Drings, V. Schulz, K. Kayser, H. Bartsch and A. Risch, *Pharmacogenetics*, 2003, **13**, 607.
166. P.H. Roos and H.M. Bolt, *Expert Opin. Drug Metab. Toxicol.*, 2005, **1**, 187.
167. W. Schober, A. Luch, V.J. Soballa, G. Raab, J.J. Stegeman, J. Doehmer, J. Jacob and A. Seidel, *Chem. Biol. Interact.*, 2006, **161**, 37.
168. Y. Ko, J. Abel, V. Harth, P. Bröde, C. Antony, S. Donat, H.P. Fischer, M.E. Ortiz-Pallardo, R. Thier, A. Sachinidis, H. Vetter, H.M. Bolt, C. Herberhold and T. Brüning, *Cancer Res.*, 2001, **61**, 4398.
169. J.S. Mammen, G.S. Pittman, Y. Li, F. Abou-Zahr, B.A. Bejjani, D.A. Bell, P.T. Strickland and T.R. Sutter, *Carcinogenesis*, 2003, **24**, 1247.
170. T. Shimada, J. Watanabe, K. Inoue, F.P. Guengerich and E.M. Gillam, *Xenobiotica*, 2001, **31**, 163.
171. Z.L. Wu, C.D. Sohl, T. Shimada and F.P. Guengerich, *Mol. Pharmacol.*, 2006, **69**, 2007.
172. M. Karlgren, A. Gomez, K. Stark, J. Svard, C. Rodriguez-Antona, E. Oliw, M.L. Bernal, S. Ramon, Y. Cajal, I. Johansson and M. Ingelman-Sundberg, *Biochem. Biophys. Res. Commun.*, 2006, **341**, 451.
173. M. Karlgren and M. Ingelman-Sundberg, *Expert Opin. Ther. Targets*, 2007, **11**, 61.
174. H. Obana, S. Hori, T. Kashimoto and N. Kunita, *Bull. Environ. Contam. Toxicol.*, 1981, **27**, 23.
175. Z. Zhong, M. Goto, T. Hijioka, H. Oide, F.C. Kauffman and R.G. Thurman, *Drug Metab. Dispos.*, 1994, **22**, 680.
176. X. Ding and L.S. Kaminsky, *Annu. Rev. Pharmacol. Toxicol.*, 2003, **43**, 149.
177. D. Choudhary, I. Jansson, J.B. Schenkman, M. Sarfarazi and I. Stoilov, *Arch. Biochem. Biophys.*, 2003, **414**, 91.
178. D. Choudhary, I. Jansson, I. Stoilov, M. Sarfarazi and J.B. Schenkman, *Arch. Biochem. Biophys.*, 2005, **436**, 50.
179. A. Mahnke, P.H. Roos, W.G. Hanstein and G.G. Chabot, *Biochem. Pharmacol.*, 1996, **51**, 1579.
180. A. Mahnke, D. Strotkamp, P.H. Roos, W.G. Hanstein, G.G. Chabot and P. Nef, *Arch. Biochem. Biophys.*, 1997, **337**, 62.

181. P.H. Roos, S. Tschirbs, P. Welge, A. Hack, D. Theegarten, G. Mogilevski and M. Wilhelm, *Arch. Toxicol.*, 2002, **76**, 326.
182. P.H. Roos, *Arch. Toxicol.*, 2002, **76**, 75.
183. T. Shimada, H. Yamazaki, M. Mimura, Y. Inui and F.P. Guengerich, *J. Pharmacol. Exp. Ther.*, 1994, **270**, 414.
184. J. Hakkola, M. Pasanen, O. Pelkonen, J. Hukkanen, S. Evisalmi, S. Anttila and A. Rane, *Carcinogenesis*, 1997, **18**, 391.
185. L. Muskhelishvili, P.A. Thompson, D.F. Kusewitt, C. Wang and F.F. Kadlubar, *J. Histochem. Cytochem.*, 2001, **49**, 229.
186. G.I. Murray, W.T. Melvin, W.F. Greenlee and M.D. Burke, *Annu. Rev. Pharmacol. Toxicol.*, 2001, **41**, 297.
187. P.H. Roos, S. Tschirbs, A. Hack, P. Welge and M. Wilhelm, *Xenobiotica*, 2004, **34**, 781.
188. G.I. Murray, M.C. Taylor, M.C. McFadyen, J.A. McKay, W.F. Greenlee, M.D. Burke and W.T. Melvin, *Cancer Res.*, 1997, **57**, 3026.
189. J.A. McKay, W.T. Melvin, A.K. Ah-See, S.W. Ewen, W.F. Greenlee, C.B. Marcus, M.D. Burke and G.I. Murray, *FEBS Lett.*, 1995, **374**, 270.
190. M.C. McFadyen, S. Breeman, S. Payne, C. Stirk, I.D. Miller, W.T. Melvin and G.I. Murray, *J. Histochem. Cytochem.*, 1999, **47**, 1457.
191. S. Haas, C. Pierl, V. Harth, B. Pesch, S. Rabstein, T. Bruning, Y. Ko, U. Hamann, C. Justenhoven, H. Brauch and H.P. Fischer, *Int. J. Cancer*, 2006, **119**, 1785.
192. J.T. Buters, S. Sakai, T. Richter, T. Pineau, D.L. Alexander, U. Savas, J. Doehmer, J.M. Ward, C.R. Jefcoate and F.J. Gonzalez, *Proc. Natl. Acad. Sci. U S A*, 1999, **96**, 1977.
193. N. Galvan, T.J. Page, C.J. Czuprynski and C.R. Jefcoate, *Toxicol. Appl. Pharmacol.*, 2006, **213**, 105.
194. M. Seliskar and D. Rozman, *Biochim. Biophys. Acta*, 2007, **1770**, 458.
195. Y.K. Leung, K.M. Lau, J. Mobley, Z. Jiang and S.M. Ho, *Cancer Res.*, 2005, **65**, 3726.
196. E.L. Cavalieri and E.G. Rogan, *Pharmacol. Ther.*, 1992, **55**, 183.
197. V.J. Melendez-Colon, A. Luch, A. Seidel and W.M. Baird, *Carcinogenesis*, 1999, **20**, 1885.
198. Q.A. Khan, K.H. Vousden and A. Dipple, *Oncology*, 1999, **57**, 258.

Part B

CHAPTER 3

The CYP1A Subfamily

BHAGAVATULA MOORTHY

Department of Pediatrics, Baylor College of Medicine, Houston, TX 77030, USA

Table of Contents

Issues in Toxicology
Cytochromes P450: Role in the Metabolism and Toxicity of Drugs and other Xenobiotics
Edited by Costas Ioannides

3.1 Introduction

The CYP1A subfamily comprises 2 members, i.e. CYP1A1 and CYP1A2, which are inducible by 2,3,7,8-tetrachlorodibenzo-*p*-dioxin (TCDD), polycyclic aromatic hydrocarbons [PAHs, e.g., 3-methylcholanthrene (MC), benzo[a]-pyrene (BP)], β-naphthoflavone (BNF) and other compounds. The CYP1A1 and 1A2 enzymes play critical roles in the metabolic activation of carcinogenic PAHs and heterocyclic aromatic amines/amides (HAA) to electrophilic reactive intermediates, respectively, leading to a wide range of toxicities and cancer. The induction of CYP1A enzymes is mediated by the aryl hydrocarbon receptor (AHR)-mediated mechanisms. The impact of CYP1A induction on the carcinogenic and toxic potentials of environmental, occupational, dietary, and therapeutic chemicals has been the central focus of human risk evaluation and has broadly influenced the fields of Cancer Research, Toxicology, Pharmacology, Molecular Biology, Pharmacogenomics, Toxicogenomics and Risk Assessment over the past five decades. From the early discovery of CYP1A induction to the establishment of CYP1A enzymes as the principal P450s for bioactivation of PAH and HAA in *in vitro* assays, and to the recent realization of an essential protective role of CYP1A in BP-induced lethality and carcinogenesis with CYP1A knockout mice, the understanding of the interrelation between CYP1A induction and chemical safety has followed a full circle. This unique pathway of CYP1A research underscores the importance of whole animal and human studies in chemical safety evaluation.

3.2 Human CYP1A Enzymes and Their Role in Drug Metabolism

The human CYP1A1 and 1A2 enzymes are ~70% identical in their amino acid sequences. Both have been studied extensively because of their roles in the activation of carcinogens. CYP1A1 is expressed at only very low levels in human liver and is essentially an extrahepatic enzyme.[1–3] Cigarette smoking induces this enzyme in lung, placenta, and lymphocytes.[4] Attempts to correlate the inducibility of the enzyme with cancer have been inconclusive,[5–8] although there seems to be a correlation between cancer and absolute level of the enzyme.[9–10] An association has been observed between an *Msp*I polymorphism and lung cancer incidence in a Japanese study,[11] but not in Caucasians.[12–13] The *Msp*I polymorphism is considered to be linked to a Val/Ile polymorphism in the protein coding region,[14] but linkage disequilibrium has been observed.[13] A distinct polymorphism has been reported in African-Americans,[15] but its relation to cancer is not known. CYP1A1 is mainly involved in PAH metabolism,[16,17] but is also involved in the metabolism of some drugs.[18–20] Recently six drugs that have been approved by the FDA and are marketed therapeutics, namely omeprazole, nimodipine, leflunomide, mexiletine, atorvastatin, flutamide (Figure 3.1), induce CYP1A1 without causing dioxin-like toxic effects, suggesting that induction of CYP1A1 is a non-specific biomarker of direct AHR affinity and activation, and lend further support to the hypothesis that CYP1A1 induction is not synonymous with dioxin-like toxicity.[18] In humans, CYP1A induction shows large interindividual variation and, furthermore, suggests a causative role of CYP1A induction in the occurrence of cancer in certain human populations. Subsequent reports showed a strong correlation between high CYP1A inducibility and cancer of the lung, larynx, and oral cavity that are in direct contact with cigarette smoke.[21] However, it is noteworthy to point out that no association between lung cancer and CYP1A inducibility has been reported in a number of studies.[22]

Human CYP1A2, in contrast, is expressed essentially predominantly in the liver and not in extrahepatic tissues,[19] although there are occasional reports of low extra-hepatic expression.[23–26] Both human *CYP1A1* and *1A2* genes are under the regulation of the *Ah* locus, involving the interaction of Ah receptor (AHR)/AHR nuclear translocator (ARNT) heteromeric complexes with upstream enhancer elements.[27] However, the CYP1A2 gene also contains five putative elements, xenobiotic response elements (XRE) or dioxin response elements (DRE) sites, to explain the tissue-selective induction.[28] One of the non-invasive screens to assess the expression of CYP1A2 in humans is the caffeine N-demethylase activities.[29–30] Levels of CYP1A2 and caffeine N-demethylase activities vary about 40-60-fold among the population.[29–30] There are some papers describing the genetic polymorphism of P4501A2; however, there appear to be no major

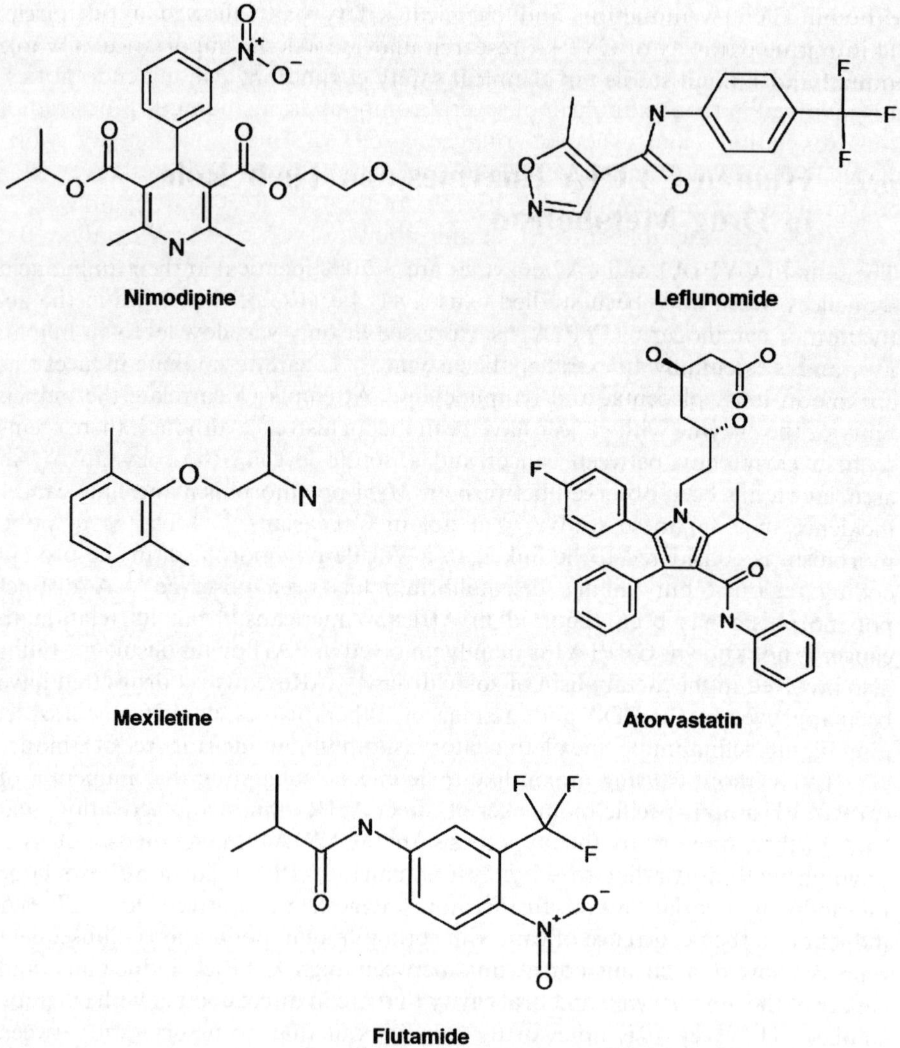

Figure 3.1 Examples of some drugs that induce CYP1A1 in rats.

differences in the coding or upstream regions of the genes of individuals with high and low P450 1A2.[31] Recent resequencing of the human *CYP1A1_CYP1A2* locus (39.6 kb) in five major geographically isolated subgroups led to the identification of 85 single nucleotide polymorphisms (SNPs).[32] However, no SNP or haplotype on the *CYP1A2* gene has yet been identified that can unequivocally be used to predict the metabolic phenotype in any patient.[32]

There are many drugs that are metabolised by CYP1A2, which include acetaminophen (paracetamol),[33] antipyrine,[34,35] bufuralol,[36] ondansetron,[22] phenacetin,[37] tacrine,[38] tamoxifen (N-demethylation),[21] theophylline[32,39] and warfarin.[40] The significance of CYP1A2 in drug oxidation is not great, although dietary alterations leading to theophylline ineffectiveness can now be interpreted in terms of this enzyme.[41] The anti-ulcer drug omeprazole can induce CYP1A2 although not particularly effectively.[42–43] Among endobiotics, it is involved in the metabolism of several compounds, including 17β-oestradiol 2-hydroxylation[44] and uroporphyrinogen. CYP1A2's metabolic activity is primarily hydroxylation and demethylation of compounds through oxidative metabolism. This enzyme accounts for 10–15% of the total P450 activity in liver.[45]

There is considerable interest in the ability of CYP1A2 to catalyse the N-hydroxylation of carcinogenic arylamines and heterocyclic amines.[30] Since they are found in charbroiled meats and in cigarette smoke, the role of this enzyme in metabolizing the heterocyclic amines is of considerable interest.[46] The oxidised products of CYP1A2 can bind to DNA bases, either directly or following conjugation with acyl or sulphate groups.[47] There is some epidiemiological evidence showing that elevated levels of CYP1A2 can be a predisposing factor to colon cancer, although the risk is marginal unless there is a high level of N-acetyltransferase and high consumption of charbroiled meat.[31]

3.3 Ontogenic Expression of CYP1A Enzymes

The absolute contribution that each isoform makes to the total hepatic CYP pool varies in a species-specific manner. Our current knowledge of the relationship among embryological, foetal, and neonatal hepatic metabolism of xenobiotics and their toxicity is poor. Such studies have been limited, due in part, to the low level of total CYP, which is consequently extremely difficult to measure, particularly because subcellular fractions are often excessively contaminated with haem.[48] Recently, techniques such as *in situ* hybridisation, Western blotting, real time RT-PCR and cDNA microarrays have greatly contributed to our understanding of the ontogeny of CYPs. Determination of mRNA levels for both genes has indicated that CYP1A1 is constitutively expressed in 7-day-old embryos, and CYP1A2 occurs near to the time of birth.[49,50] In general, levels of CYP1A1 and 1A2 are low to negligible at most ages, although a weak band for 1A1 can be detected in some 5 day-old rats and a stronger band in adult rats for CYP1A2.[48] In rabbits, CYP1A2 is detected in 6 day-old neonates, in weanlings, and adults; none is detected between ages 9 and 21 days and prenatally.[48] In the case of CYP1A2, it is detected at 6 days and increases with age.[51] Constitutive levels of CYP1A1 are less than 5%, regardless of age. Interestingly, expression of CYP1A1 in the rat is switched off during the perinatal period, with no protein being detected 24 hours prior, and a few days following birth.[48] This phenomenon may be related to its proposed role in maintaining the levels of ligands that modulate cell division, growth and neuroendocrine function.[52]

Constitutive levels of CYP1A2 are low throughout development in the rat and are not detected by specific antibodies to CYP1A2[48] or by monoclonal antibodies that cross-react with CYP1A1 and 1A2.[48] Using highly specific and well-characterised antibodies, it is now apparent that the constitutive expression of CYP1A1 and 1A2 is separated only by a few days, with 1A1 occurring 3-4 days before birth, and 1A2 occurring near birth.[41,51]

PAHs differentially affect the levels of CYP1A1 and 1A2 in both rats[53] and rabbits.[51] 3-Methylcholanthrene (MC) induces 1A1 24 hours pre- and post-partum, indicating transplacental induction. Induction of CYP1A1 occurs 5–6 days after birth, the level of induction being about 30-fold.[51,54] Using anti-sense probes, *in situ* hybridisation studies have shown that in mice, MC induces CYP1A1 mRNA as early as 12 days of gestation,[55] emphasising the potential role of these enzymes in activating teratogenic agents to more reactive species during organogenesis.[56] MC treatment of animals leads to developmentally advanced livers, which not only have specific content of total P450 approaching that of adults, but also, immunohistochemically, resembling livers of animals at least 10–20 days older in their distribution of CYP1A2.[51] These results would have important implications for the toxicology of compounds activated by this enzyme. For example, Juchau *et al.*[57] showed that the incidence of foetal malformations elicited by exposure to 2-acetlyaminofluorene (AAF) were markedly increased following transplacental induction with MC, but not phenobarbitone.

In human liver, the CYP1A1 and 1A2 are low to absent in foetuses.[58] CYP1A1 is almost undetectable in adult liver, even after induction with cigarette smoking.[59] However, CYP1A2 is constitutively expressed in liver and is inducible by smoking.[59–60]

3.4 Role of CYP1A1 Enzymes in Human Carcinogenesis

Human cancer is thought to arise from an interplay of endogenous factors and environmental exposures.[61] Humans and other living organisms are constantly exposed to a large number of potentially genotoxic environmental chemicals, including the ubiquitous PAHs, aromatic amines, allylbenzenes and nitrosamines.[62,63]

3.4.1 Human Exposure to PAHs

PAHs are formed as products of incomplete pyrolysis of organic materials and are present in considerable quantities in fossil fuel, from which they are released by combustion.[64,65] They are also present in cigarette smoke, particulate matter present in the ambient atmosphere, and charcoal broiled meats, etc.[62] Each of the aforementioned sources produces mixtures containing between 100 and 300 different individual hydrocarbons, and the estimated total annual emission for just one of them, benzo[a]pyrene (BP), is 1200 tons.[65] American sources indicate an intake of total PAHs in the order of 6–16 μg/day. In addition, PAHs are occupational carcinogens to which sweepers, tar workers, cotton spinners, fishermen and tool setters are exposed to significant amounts.[61,66] Because only μg amounts are required to initiate tumours in mouse skin, people are placed at increased risk of developing cancer as a result of significant pollution of the environment by PAHs.[67]

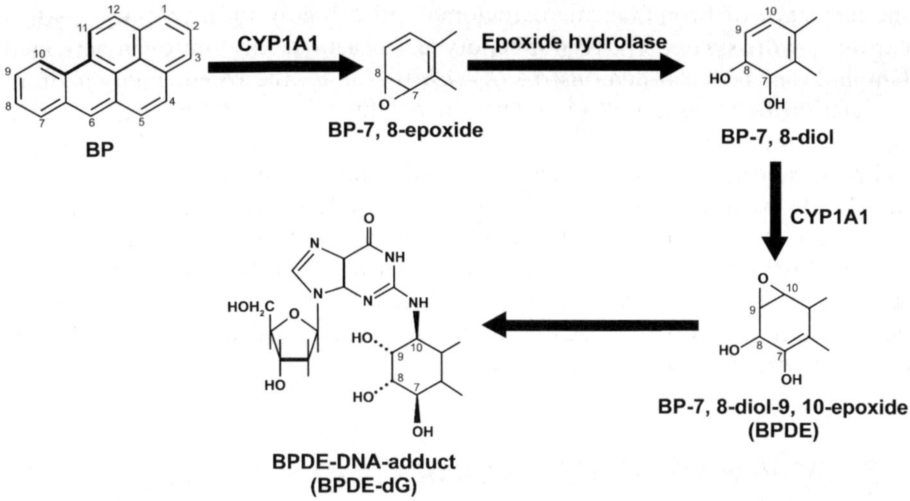

Figure 3.2 Metabolic activation of BP to DNA-binding carcinogenic metabolites. The major metabolic activation pathway of BP entails the CYP1A1-dependent metabolic activation at the 7, 8 bond to yield an epoxide, which then undergoes epoxide hydrolase-mediated hydration to BP-7, 8-diol. BP-7, 8-diol is in turn transformed to BPDE by CYP1A1-mediated epoxidation at the 9,10 position. The major BPDE-DNA adduct results from nucleophilic attack of the exocyclic N^2 position of guanine on the C-10 position of (+)*anti*-BPDE.

3.4.2 Role of CYP1A Enzymes in PAH Metabolism

PAHs, e.g., BP, dibenz[a,c]anthracene (DBA), and MC are carcinogenic to laboratory animals and may be involved in the aetiology of human cancers associated with exposure of PAH mixtures, such as coal tar and cigarette smoke.[68–70] The CYP1A enzymes are primarily involved in metabolic activation of PAHs and other environmental chemicals to reactive intermediates that can form DNA adducts, a critical molecular event in the initiation of carcinogenesis[71–76] (Figure 3.2). We have recently shown that exposure of rats to potent carcinogens, e.g., MC[77] or DBA,[78] leads to multiple DNA adducts, which are persistent for several weeks after withdrawal of the PAHs. In fact, expression of the *CYP1A1* gene has been associated with development of PAH-mediated cancers and other disorders in rodents.[79] The CYP1A enzymes are also inducible by cigarette smoke in humans.[80–88] The bioactivation of PAHs are discussed in more detail in Chapter 2.

3.4.3 Genetic Polymorphisms of CYP1A and Human Cancers

Specific *CYP1A1* genetic polymorphism closely correlates with smoking-induced lung cancer risk in the Japanese,[89] while other CYP1A1 germline polymorphisms, associated with the genetic predisposition for lung cancer, are related to cigarette smoking-associated *p*53 mutations.[90] Correlation between the incidence of bronchogenic carcinoma and a highly inducible aryl hydrocarbon hydroxylase (AHH) phenotype measured in mitogen-activated lymphocytes has been demonstrated.[91] Furthermore, the recent findings of an association between *CYP1A1* restriction polymorphism and breast cancer in African-American women[92,93] suggests a role for CYP1A1 in the development of breast cancer. Other studies have shown that humans having *CYP1A1* (MspI and NcoI) polymophisms to be more susceptible to developing squamous cell carcinoma of the larynx.[94] Data are emerging that genetic differences in *CYP1A1* regulatory factors, such as the AHR and (ARNT) contribute to the varying levels of CYP1A1 expression.[95] Thus, CYP1A1 appears to be a key determinant of individual susceptibility to human cancers associated with certain environmental chemical exposures.[96]

3.5 Molecular Mechanisms of CYP1A Induction

The molecular mechanisms of CYP1A1 induction by TCDD have been extensively studied.[97–99] The AHR plays a pivotal role in the induction of CYP1A1 by TCDD or PAHs, such as MC.[97–99] TCDD or PAHs upon entry into the cell, binds with high affinity to the cytosolic AHR, which then complexes with a second related protein, the ARNT.[96–99] Before binding of the ligand (Figure 3.3), the AHR in the cytosol is bound to heat shock proteins. After ligand binding, the AHR is released from the inactive state, allowing it to translocate to the nucleus, where it binds to the ARNT. This ligand-activated transcription factor (AHR-ARNT heterodimer) interacts with Ah responsive elements (AHREs), which are located in multiple copies of the upstream portion of the *CYP1A1* promoter, resulting in transcriptional

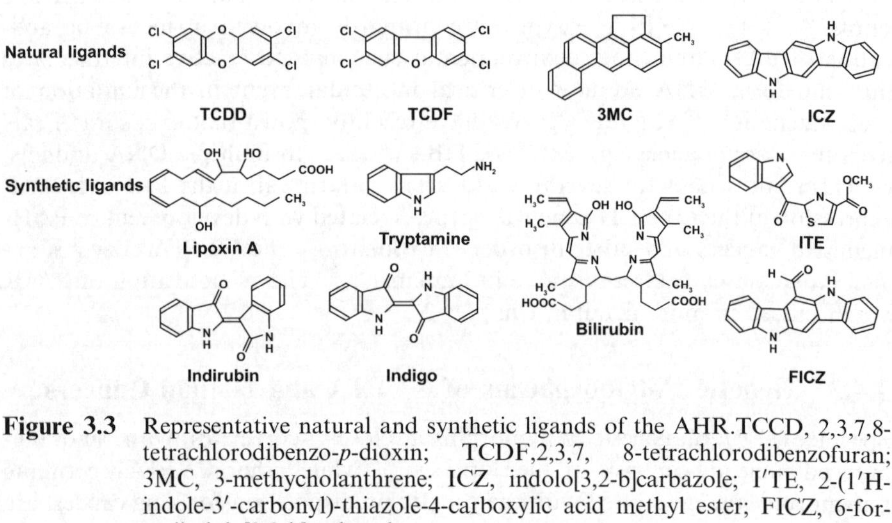

Figure 3.3 Representative natural and synthetic ligands of the AHR.TCCD, 2,3,7,8-tetrachlorodibenzo-*p*-dioxin; TCDF,2,3,7, 8-tetrachlorodibenzofuran; 3MC, 3-methycholanthrene; ICZ, indolo[3,2-b]carbazole; I′TE, 2-(1′H-indole-3′-carbonyl)-thiazole-4-carboxylic acid methyl ester; FICZ, 6-formylindolo[3,2-b]carbazole.

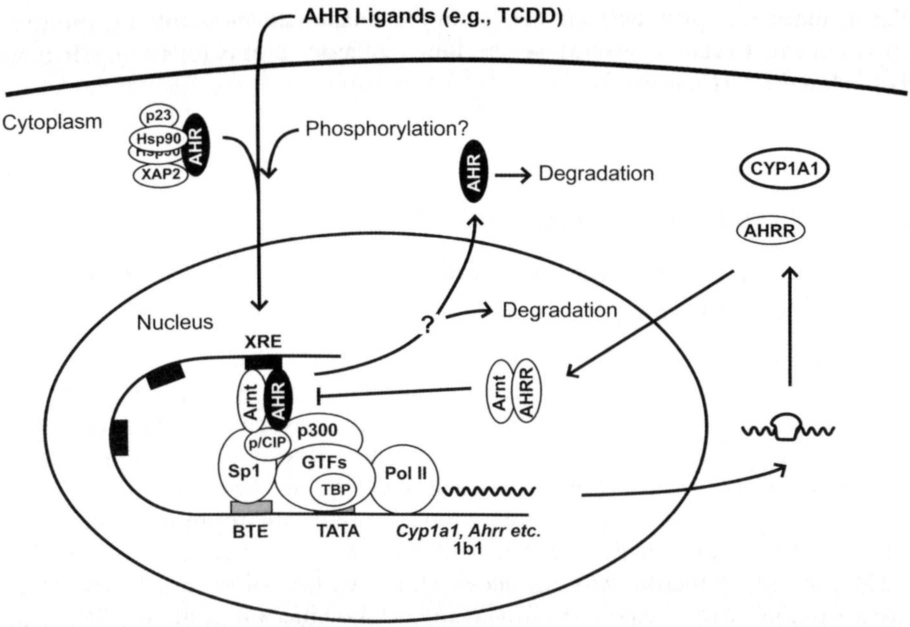

Figure 3.4 Molecular mechanisms of CYP1A induction by the AHR. After ligand binding, the AHR is released from the inactive state, allowing it to translocate to the nucleus, where it binds to the ARNT. This ligand-activated transcription factor (AHR-ARNT heterodimer) interacts with AHREs, which are located in multiple copies of the upstream portion of the *CYP1A1* promoter, resulting in transcriptional activation of many genes on the *Ah* locus, including *CYP1A1*, *CYP1A2*, and other phase II genes. AHRR, Ah receptor repressor; BTE, basal transcriptional element.

activation of many genes on the *Ah* locus, including *CYP1A1*, *CYP1A2,* glutathione S-transferase (GST-α), NAD(P)H quinone reductase (NQO1), UDP glucuronyl transferase (UDPGT) and aldehyde dehydrogenase (ADH).[100,101] This process is affected by numerous coactivators, including nuclear factor 1 (NF1), specificity protein (SP)1 and other transcription factors and an AHR repressor (Figure 3.4). In addition to the AHREs, several negative regulatory elements (NREs) are located on the *CYP1A1* promoter in humans [102,103] as well as in rodents.[103] Using EMSA and DNase I foot printing techniques, Hines *et al.*[104] and Sterling *et al.*[103] have identified 2 nuclear proteins that interact specifically with the NREs, thereby ensuring low CYP1A1 activity in the non-induced state. Entry into the nucleus of the inducer-AHR complex leads to positive regulation *via* binding of the complex to the AHREs on the one hand, and inhibition of negative regulation on the other, eventually leading to *CYP1A1* induction. Furthermore, glucocorticoid-responsive elements within intron 1 of the *CYP1A1* gene and additional elements on ;the upstream portion of the *CYP1A1* gene[105] have been identified.

Thus, induction of *CYP1A1* is a complex phenomenon involving multiple mechanisms. Recently, several natural and synthetic ligands for the AHR have been described (Figure 3.3).

3.6 Persistent Induction of CYP1A1

3.6.1 AHR Contributes to Sustained Induction of CYP1A1 by Persistent Chemicals

Persistent induction of CYP enzymes, as well as of other drug metabolising enzymes, has been reported for carcinogenic chemicals, such as TCDD,[106,107] polychlorinated biphenyls (PCBs),[108] and the pesticide DDT (Dichloro Diphenyl Trichloro ethane).[109] For these agents, persistent induction appears to be the result of the long biological half-life of these chemicals.[106–111] Anderson *et al.* [108] have demonstrated a positive tumour-promoting effect and prolonged induction of P450 enzymes in mice given a single dose of Aroclor 1254, and this induction was correlated with retention of several PCBs in the tissues of mice over the course of the study. TCDD has a half-life of 19 days in rats.[110] Exposure of rat or mouse hepatoma cells to TCDD results in binding of TCDD to the cytosolic AHR, followed by translocation of the TCDD-AHR complex to the nucleus, leading to induction of *CYP1A1* gene expression. This event is followed by drastic decline in receptor activity,[111] the half-life of AHR in Hepa-1 cells being only 8 h.[112] On the other hand, in animals the AHR appears to be more stable, and *in vivo* studies have detected low levels of AHR in the nucleus of hamsters exposed to TCDD after as many as 35 days, and it has been proposed that 5% of AHR can maintain relatively high levels of CYP1A1.[113] In humans, the AHR appears to be less susceptible to TCDD-induced responses.[114] Because TCDD is highly persistent in the cell, and is at least 30,000 times more potent than MC in inducing CYP1A1,[115] it appears that mechanisms of sustained induction of CYP1A1 and other drug metabolising enzymes by TCDD are mediated by the AHR.

3.6.2 MC Elicits Sustained Induction of CYP1A1 by Mechanisms Other than Persistence of the Parent Compound

There is a paucity of data with respect to persistent induction of CYP1A1 by readily metabolisable, non-persistent chemicals, such as MC. Boobis *et al.*[116] showed that a single i.p. treatment of mice with MC (80 mg/kg) leads to elevated levels of total CYP content and AHH activity for up to 9 days, while Masaki *et al.*[117] reported elevated levels of total CYP content in MC-treated rats for 8 days. A single dose of MC (25 mg/kg) does not elicit a persistent effect.[108] We reported earlier that 4 daily administrations of MC[77] or dibenz[a,c]anthracene (DBA),[78] a carcinogen for which there is substantial human exposure, led to induction of ethoxyresorufin *O*-deethylase (EROD) activity that was sustained for several weeks after compound withdrawal. This persistent induction is accompanied by sustained upregulation of CYP1A1/1A2 apoproteins and their corresponding mRNA levels. On the other hand, other PAHs, such as BP and naphthacene (NAP), and the non-PAH, BNF, did not elicit this phenomenon.[78] Earlier studies have indicated that MC is metabolised rapidly in rodents *in vivo*. Takahashi dosed mice with MC by different routes (i.p., i.v., s.c.) and demonstrated that MC is instantly taken up by the liver and is eliminated from the body within 4 days.[118–120] Levine and Singer[121] reported that both BP and MC are extensively metabolised, and cleared rapidly at similar rates. MC undergoes extensive hepatic biotransformation to water-soluble metabolites, which are excreted into the bile.[122] Bresnick *et al.*[123] showed that upon i.p. administration of MC, the radioactivity in the liver at the peak time point represents only 0.5% of the administered dose, suggesting that MC is metabolised as rapidly as it enters the liver. Furthermore, *in vitro* studies have shown that MC is metabolised quickly to polar derivatives.[124]

Our observation that MC elicits a sustained induction of CYP1A1 for as many as 45 days after cessation of treatment led to the hypothesis that MC elicits this persistent induction by mechanisms other than persistence of the carcinogen. Although evidence from the literature does indicate the high susceptibility of MC to metabolism, a thorough investigation of the distribution and excretion of MC from rats had not been carried out. Therefore, we performed experiments to test the hypothesis that MC elicits sustained CYP1A1 induction by mechanisms other than retention of MC in body tissues. These studies showed rapid elimination of MC from the body, and after 15 days, when high CYP1A1 activities are expressed, the intrahepatic concentration of the parent compound was 270 pmol/g, a concentration which we have shown to be insufficient to induce CYP1A1 in liver,[125] strongly supporting our hypothesis that the phenomenon of persistent induction by MC occurs by mechanisms other than persistence of the compound, and that novel molecular mechanisms mediate this effect. We recently provided indirect evidence that some MC-DNA adducts are formed in a sequence-specific manner,[126] with AHREs being the targets. We hypothesise that covalent adduct formation in the region of the AHRE may result in sustained induction of CYP1A1 (Figure 3.5).

3.6.3 Persistent CYP1A1 Induction in Extrahepatic Tissues

The lung is a target organ for PAH-mediated carcinogenesis.[127,128] Exposure of laboratory animals to PAHs induces CYP1A1 activity in lungs.[127,128] Furthermore, cigarette smoke enhances CYP1A1 activity in lungs of humans.[129] Pulmonary CYP1A1 levels have been reported to be of prognostic significance in tobacco-related lung cancer.[130] We recently observed that MC causes sustained induction of CYP1A1 in lung. Moreover, MC elicits long-term induction of pulmonary, but not hepatic CYP1A1, in mice lacking the gene for *CYP1A2*,[130] a phenomenon that may be of mechanistic relevance to MC-mediated lung tumourigenesis. These results are of significant relevance to

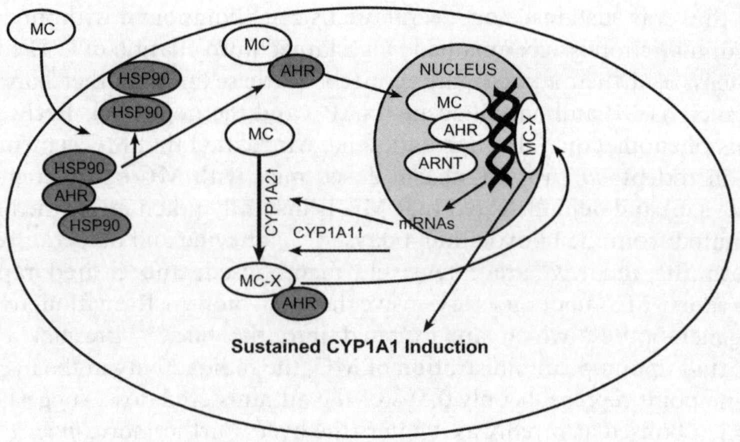

Figure 3.5 Mechanisms of sustained CYP1A1 induction by MC. We postulate that 3-methylcholanthrene (MC) is converted by CYP1A2 to a reactive metabolite MC-X. MC-X then enters the nucleus and specifically interacts covalently with AHREs or other regulatory elements on the *CYP1A1* promoter. The resulting MC-X-DNA adduct mimics the role of MC-AHR-ARNT complex in enhancing transcription of CYP1A1. HSP90, heat shock protein 90.

human health because in lung cancer patients who ceased smoking, the time required for CYP1A1 levels to return to normal is 2 to 3 months.[129,131] This is of concern because such people may be at a higher risk of toxic insult upon subsequent environmental exposures. Furthermore, Anderson *et al.*[108] and Wessner *et al.*[127] have demonstrated a positive correlation between CYP1A1 induction and lung tumour incidence. In the mammary gland, which is also a target organ for PAH-mediated carcinogenesis, a prolonged induction of CYP1A1 is observed.[125] High level of expression of CYP1A1[132] and CYP1B1,[133] and aryl hydrocarbon hydroxylase (CYP1A1/CYP1B1) activities in human mammary tissues are positively correlated with axillary lymph node status, suggesting a role for these enzymes in aggressive tumours.[134]

3.6.4 Sustained CYP1A1 Induction in Humanised Transgenic Mouse

Recent studies have utilised transgenic mice with the luciferase reporter gene to study the regulation of CYP3A[135] and 1A1 enzymes.[136] These mice offer a mechanistic model to study the mechanisms by which human promoters regulate CYP enzymes in response to xenobiotics.

3.7 Relevance of CYP1A1 Induction to Tumourigenesis

PAHs are abundant in cigarette smoke, which is one of the strongest risk factors for lung cancer. Cigarette smoking induces CYP1A1 in lungs and other organs. Several reports have suggested that lungs and lymphocytes of ex-smokers and recent smokers have high levels of CYP1A1.[137–140] Frequently, high levels of DNA adducts are noticed in these individuals.[140] Although defective DNA repair mechanisms may contribute to the persistence of adducts, it is also possible that these individuals are more susceptible to genotoxic insults by subsequent exposure to other environmental chemicals because of persistent elevation of the CYP1A1 enzyme. Studies on the mechanisms of persistent CYP1A induction could lead to significant advances in our understanding of the regulation of the xenobiotic metabolizing enzymes by PAHs, which are suspected to contribute to cancers and other disorders in humans.

3.8 CYP1A, Ki-ras Mutations, and Lung Tumourigenesis

Epidemiological studies have suggested a link between maternal exposure to carcinogenic agents and an increased incidence of childhood cancers.[141] Previous studies using these pharmacogenetic mouse models have shown that the incidence of lung tumours produced following *in utero* exposure to MC correlated positively with the responsive phenotype of the foetus.[142] In addition, the offspring from non-responsive mother had higher tumour incidences than those from responsive mothers.[142] These differences in lung tumour incidence positively correlated with the levels of expression of CYP1A1 in the responsive foetuses and mothers.[141,142] The tumours had Ki-*ras* point mutations on codons 12 and 13, suggesting activation of the Ki-*ras* oncogene. Both responsive and non-responsive mice appeared to contain the same types of mutagenic lesions, suggesting that differences in the levels of CYP1A1 expression do not alter the individual organism's susceptibility to chemically-induced cancers.[143] Miller *et al.* have suggested that the Ki-*ras* mutational activation is an early event in transplacental lung tumourigenesis, and that the type of mutation produced by exposure to chemical carcinogens can influence the carcinogenic potential of the tumour.[144–146]

Interestingly, adult mice do not display differences in tumour induction multiplicity between responsive and non-responsive phenotype.[142] The difference between CYP1A1 inducibility and tumour susceptibility in the foetal and adult mice has been attributed to the differences in the extent of inducibility of CYP1A1 by MC, the foetal mice displaying much higher induction ratios than adults.[142] Our observations[130] showing significant differences in lung CYP1A1 inducibilities between wild type and CYP1A2 (-/-) adult mice begs the question as to whether these differences would correlate with altered MC-induced tumour susceptibility and mutation spectra in these animals. Several workers have shown BP as well as MC to induce lung tumours in the susceptible A/J mice.[147–151] Thus, these studies on mechanisms of persistent induction of CYP1A enzymes by PAHs should be of immense significance in understanding the fundamental mechanisms of carcinogenesis by these compounds, and appreciating the mechanistic relationships between CYP1A expression, metabolism, DNA adduct formation, mutation spectrum and carcinogenesis.

3.9 Mechanisms of CYP1A2 Induction

CYP1A2 is predominantly a liver enzyme, which is inducible by PAHs, cigarette smoke, omeprazole, etc. Recent studies have suggested the presence of this protein in human lung.[152,153] Using CYP1A2-null animals, it has been shown that CYP1A2 plays a role in regulating some house-keeping genes other than xenobiotic metabolism.[154] The human *CYP1A1* and *1A2* genes are on chromosome 15 and are oriented head-to-head, 23.3 kb apart.[155] The spacer region might contain distinct regulatory regions for CYP1A1 and CYP1A2.[155–157] They also appear to have some common regulatory elements.[155]

3.10 CYP1A in Laboratory Animals

3.10.1 Species Differences

CYP1A1 and 1A2 are primarily regulated by the AHR. Species differences have been observed relative to the abundance of AHR. Mice exhibit strain-dependent genetic polymorphism. As described earlier, the binding of PAHs to the AHR results in dissociation of the AHR/substrate unit from a complex comprised of two molecules of heat-shock proteins (hsp90) and other proteins. The AHR interacts with ARNT to generate the heterodimeric AHR complex (AHRC). Activation of the AHRC occurs through interaction with the AHREs sequences located within the promoter regions of CYP1 genes. This interaction disrupts the chromatin structure and allows the association of other factors involved in transcriptional initiation. This regulatory process is well conserved across species and is responsible for the induction of CYP1A genes by PAHs.[158]

The AHR has been identified, cloned, and characterised in several mammalian species.[159] Species differences have been observed relative to the abundance of the AHR and the binding affinity towards TCDD, MC, BP in rat, hamster, rabbit, mouse and guinea pig.[160,161] AHR from guinea pigs is more sensitive to TCDD toxicities than hamsters, which are TCDD resistant.[161] Most of our current understanding of AHR-mediated regulation is due to studies conducted in mice.[162] Mouse and human AHR and ARNT proteins are 20% identical in amino acid sequence and are similar in structure.[163] Mice exhibit strain-dependent genetic polymorphisms in CYP1A inducibilty, with DBA/2 strain being resistant to CYP1A induction by MC, and C57BL/6 strain being highly responsive to MC.[163–167] Although some reports suggest that the lack of CYP1A induction by PAHs in the DBA/2 mouse liver is due to the rapid metabolism and elimination of these compounds, other data suggest that strain differences are due to differences in the ligand-binding affinity. The sensitive phenotype was segregated as a single autosomal dominant trait.[168] This polymorphism defines a genetic locus, which is designated as the *Ah* locus (aromatic hydrocarbon responsiveness), with Ah^b and Ah^d representing the *Ah* responsive (B6) and *Ah* non-responsive (D2) alleles, respectively.

The availability of mouse strains with high and low affinity AHRs provided a unique approach to the analysis of CYP1A induction and PAH carcinogenicity and toxicity.[168] In addition to differences in the sensitivity of the strains to both induction and toxicity by PAHs, apparent disparities are observed in the patterns of PAH-induced tumours and toxicity between Ah responsive (b/b or b/d) and non-responsive (d/d) mice; b/b and b/d mice are more prone than d/d mice to PAH-induced cancers and toxicity in target tissues in which an administered PAH is in direct contact with liver, skin, lungs, etc., whereas d/d mice are at greater risks than b/b or b/d mice of developing tumour and toxicity in organs distant to the site of administration such as bone marrow. It was hypothesised that CYP1A induction and, consequently, metabolic activation and clearance of PAHs is high in the liver, and other sites of entry, in B6 mice resulting in a significant first-pass effect of BP metabolism, leading to tumour and toxicity at proximate sites. In contrast, induction is weak in D2 mice, thus allowing more PAH to reach remote organs giving rise to increase in tumour formation and toxicity in distant sites. This observation underscores the importance of both the dosage and route of administration in the evaluation of the impact of CYP1A induction on chemical toxicity. These studies suggest a protective role for CYP1A induction in the gut and or liver against the bone marrow toxicity of high-dose oral BP. However, these studies do not distinguish whether the phenotypic difference between the responsive and non-responsive mice is due to the low activity of CYP1A, the Ah locus product or both in the D2 mice. Thus, alternatively, the activation of AHR or induction of target genes other than CYP1A may contribute to the observed differences in BP toxicity between Ah^b and Ah^d mice.[168–170]

In birds, the CYP1A orthologues are called 1A4 and 1A5.[171] In fish, there is one CYP1A enzyme,[172] CYP1A3 in trout[173] and CYP1A6 in eel.[174] CYP1A and 1A7 have also been reported in xenopus.[175]

3.10.2 Tissue-specific Expression

CYP1A1 and 1A2 are expressed in a well-defined, reproducible, and similar organ-specific manner among different animal species. Low constitutive levels of CYP1A1 mRNA can be detected by PCR in many tissues, including liver, human umbilical vein endothelial cells, lymphocytes and lung cells.[176] CYP1A1 protein is not commonly constitutively expressed, but rather is seen only after AHR activation. Recently, constitutive CYP1A1 protein was detected immunochemically in neuronal cells in rat and human brain,[177] and a novel splice variant (with an exon 6 deletion), which does not activate BP to diol epoxide adducts, has been described in human brain.[178] *In situ* hybridisation studies in mice[179] showed that TCDD induced CYP1A1 mRNA in parenchymal cells of liver, gastrointestinal tract, kidney and ovary. Most striking is the induction of CYP1A1 in vascular endothelial cells throughout the body. CYP1A2 mRNA is induced only in liver, lung and duodenum, and not in vascular endothelium. Other studies have reported induction of CYP1A1 mRNA and protein in mouse and human lung epithelial cells of various types.[180] CYP1A1 mRNA in human liver is expressed in both hepatocytes and in vascular endothelium.[181]

The induction of CYP1A in vascular endothelial cells has been reported in many mammalian species including human,[182] seal,[183] rabbit,[184] mouse,[185] porcine aorta,[186] and rat brain.[187] The organ distribution of avian TCDD-induced CYPs 1A4 and 1A5 mRNA and protein is similar to that of the mammalian orthologues, CYP1A1 and 1A2, respectively.[188] CYP1A4 mRNA and protein were induced in parenchymal cells of liver, kidney and gastrointestinal tract, and in vascular endothelium throughout the body. CYP1A5 was induced only in liver, kidney and gastrointestinal tract and not in vascular endothelium.

Tissue expression of the single CYP1A in fish, though not useful to discriminate differences in organ expression of the two CYPs, can confirm major sites for CYP1A induction across vertebrates. CYP1A is widely induced in vascular endothelium in fish, as in mammals and avians.[189,190] Fish studies also showed that the extent of CYP1A induction in endothelial cells may differ among vascular beds; i.e., CYP1A induction by TCDD was greater in eel cardiac endothelial cells than in endothelial cells from kidney or the rete mirabile, a gill organ.[191] In seabream, exposure to TCDD or BP in the diet increased CYP1A, mainly in gut epithelium, while aqueous exposures increased CYP1A mainly in the vascular endothelium.[189]

Collectively, these studies show that: (i) CYP1A1 and 1A2 are expressed in a well-defined, reproducible and similar organ-specific manner among different species. Both CYP1A1 and CYP1A2 are induced in liver and intestinal cells in all species and in lung and kidney parenchymal cells in some. CYP1A1 is induced in vascular endothelium in all species tested; (ii) CYP1A1 is induced in endothelial cells of virtually every organ, including brain, liver and heart, and by diverse AHR ligands, including PCBs, BP, and BNF[186] and in pulmonary microvascular cells also by hyperoxia.[192] CYP1A1 in vascular endothelium probably accounts for much of the CYP1A1 activity previously attributed to extrahepatic CYP1A1 expression. As vascular endothelium has major haemodynamic, inflammatory and mitogenic regulatory functions,[193] CYP1A1 induction in vascular endothelium is likely to be physiologically and toxicologically significant. In vascular smooth muscle cells of mice, CYP1A1 is induced by BP[194] and MC,[195] and this enzyme may play a role in the development of atherosclerosis induced by PAHs.[195]

3.11 The *cyp1a* Knockout Mouse Models

The construction of targeted mouse models of the *cyp1a1*, *cyp1a2*, *cyp1b1* and double knockouts of the genes made it possible to directly evaluate the role of the CYP1A subfamily in chemical toxicity and carcinogenesis in intact animals. Distinctly contradictory to the observations from *in vitro* studies that CYP1A enhances the toxicity of PAHs, the intact animal experiments, after oral dosing, revealed an essential protective role of CYP1A induction in PAH toxicity.[196] An oral dose of BP (125 mg/kg) caused death of all *cyp1a1*(−/−) mice within 30 days; whereas no mortality or signs of toxicity were observed in *cyp1a1*(+/+) mice receiving the same treatment.[197] Pathological lesions observed include striking decreases in the size of spleen and thymus, leukocytopenia, and extreme hypocellularity of the bone marrow in *cyp1a1*(−/−) mice, which may have caused lethality. Increased toxicity of BP in the *cyp1a1*(−/−) mice correlated with increased DNA adduct formation in liver, spleen and bone marrow, but not small intestine. Protection against BP-induced mortality is not observed when BP is administered at 125 mg/kg *via* i.p. injection.

A similar experiment was conducted comparing *cyp1a1*(−/−), *cyp1a2*(−/−), and *cyp1a1*(+/+) mice (oral BP of 125 mg/kg for 18 days).[198] Marked wasting, immunosuppression, and bone marrow hypocellularity were observed only in *cyp1a1*(−/−) mice. Taken together, these studies provide a conclusive proof of a protective role of CYP1A1 induction in the intestine and/or liver against the immune and bone marrow toxicity of oral BP.

The *cyp1a2*(−/−) exhibited increased toxicity from drugs that are predominantly CYP1A2 substrates. Paradoxical results were observed concerning the role of CYP1A2 in the tumourigenicity of 4-aminobiphenyl (ABP).[199] CYP1A2 enhanced the metabolic activation, ABP-DNA adduct formation and toxicity *in vitro*. However, *cyp1a2*(−/−) mice displayed increased adducts in the liver and urinary bladder by topical ABP administration, and ABP-induced hepatocellular carcinomas and preneoplastic foci, as well as ABP-induced methemoglobinaemia, were increased in these mice. Similar paradoxical roles of CYP1A2 are observed in the DNA adduct formation and malignancies induced by 2-amino-3-methylimidazo[4,5*f*]quinoline (IQ) and 2-amino-1-methyl-6-phenyl-imidazo[4,5*b*]pyridine (PhIP).[200,201] These results demonstrate alternative pathways of metabolic activation of ABP, IQ, and PhIP in the absence of CYP1A2 and protection against adduct formation of the chemicals by CYP1A2 in intact animals.

The teratogenic effects of TCDD, such as cleft palate and hydronephrosis, were assessed in *cyp1a1*(−/−) and *cyp1a2*(−/−) mice.[202] TCDD at a dose of 25 μg/kg by gavage on gestation day 10 was lethal to foetuses carried by *cyp1a2*(−/−), but not *cyp1a1*(−/−) dams. The *cyp1a2(−/−)* foetuses were more sensitive to cleft palate formation, hydronephrosis, and lethality. The effects were dependent on the maternal *cyp1a2*(−/−) genotype and correlated with more TCDD reaching the embryos from the *cyp1a2*(−/−) dams. The increased sensitivity to TCDD-induced birth defects was reverted in mice expressing human CYP1A1 and 1A2 proteins in *cyp1a2*(−/−) mice. Since TCDD binds to CYP1A2 in the liver, the results support the hypothesis that maternal mouse hepatic CYP1A2, and its induction, protects the embryos from TCDD-induced and teratogenic effects by sequestering TCDD in the maternal liver.[168]

3.12 Studies with Humanised Mice

Since species differences in the metabolism of drugs and toxicants are well known, recent studies have utilised mice expressing human *CYP1A1* and *1A2* in a *cyp1a1*(−/−) or *cyp1a2*(−/−) mouse to predict the metabolism of these compounds in humans. Differences in the metabolism of the heterocyclic amine PhIP was observed between wild type and *hCYP1A2* mice.[203] PhIP was preferentially metabolized by N^2-hydroxylation in *hCYP1A2* mice, whereas in wild type mice, 4′-hydroxylation was the predominant pathway.[203] In the case of theophylline, replacement of the mouse *cyp1a2* with human *CYP1A2* gene restored the ability to metabolise theophylline, and the metabolism changes to human profile.[204] Further studies in humanised mice would be critical for predicting the metabolism of drugs and toxicants in humans.

3.13 Role of CYP1A Enzymes in the Metabolism of Endogenous Substances

The divergence of *CYP1* and *CYP2* family precursor genes appears to have happened about 600–800 million years ago, making the *CYP1* family one of the oldest CYP families[204] (see Chapter 1). The AHR is present in all vertebrates and in C. *elegans*,[205] and TCDD induces CYP1A in every species tested.[206] The duplication of a putative *CYP1A* precursor gene into two related but distinct pairs also suggests that CYP1A underlies important protective functions.

Carcinogen activation alone cannot explain the role of CYP1A enzymes in TCDD toxicity because TCDD is neither significantly decreased nor is a substrate for either CYP. Furthermore, TCDD binds to the AHR in the parent form, and it does not require metabolic activation to produce its effects.[206] The evolutionary persistence of CYP1A1, the large magnitude and organ specificity of the induction response and new evidence that CYP1A1 and 1A2 can metabolise endogenous substrates into biologically active metabolites, all suggest that CYP1A is more than a biomarker, and the view that CYP1A induction is ‘bad’ needs reconsideration.[206]

3.13.1 Role of CYP1A Enzymes in ROS Metabolism

There are many reports of treatment with TCDD or PCBs increasing formation of ROS (reactive oxygen species): H_2O_2, superoxide, or hydroxyl radicals.[207–209] All CYPs can generate ROS during NADPH-dependent CYP catalysis[210] If the transfer of oxygen to a substrate is not tightly coupled to NADPH utilisation, some of the oxygen may be released as ROS. ROS are generated during NADPH-dependent drug metabolism by constitutive CYPs in liver microsomes and in isolated hepatocytes.[207] The rank order for ROS-generating activity by microsomes from lymphoblasts expressing human CYPs was CYP3A4>

CYP1A1 > CYP1A2 = CYP2B6[211] indicating that CYP1A enzymes, while being able to generate ROS, are not the most active enzymes in doing so, and that CYP1A1 is more active than CYP1A2.[212] Thus, one should exercise caution in attributing only adverse effects to ROS generation in response to AHR activation, and it should be appreciated that ROS have important physiological effects on gene signalling and gene activation.

3.13.2 Modulation of ROS by CYP1A Enzymes

TCDD induction of apoptosis in the yolk sac of medaka embryos was correlated with CYP1A induction, and both were attenuated by the antioxidant N-acetyl cysteine[213] supporting generation of ROS by CYP1A, and a role for ROS as a mediator of those actions of CYP1A. Varied evidence supports a role for CYP1A1 in generating ROS, and for termination or suppression of CYP1A1 action by ROS generated by CYP1A1 or other CYPs. CYP1A1 generation of H_2O_2 has been implicated in feedback regulation of CYP1A1 in HepG2 cells via the inactivation of nuclear factor 1 (NF1) activation of CYP1A1 transcription.[214] Inhibition of CYP1A1 or CYP2E1 blocked the repression of the other CYP, suggesting that ROS generated by one CYP can down-regulate another.[215] Other reports[209,212,216] also support the view that CYP1A action is curtailed by its generation of ROS.

CYP1A1, rather than CYP1A2, has been implicated in the generation of ROS. Schlezinger *et al.*[217] showed that rat liver microsomes containing baculovirus-expressed CYP1A1 increased ROS in the presence of 3,3′,4,4′ tetrachlorobiphenyl, but baculovirus-expressed CYP1A2 did not. Shertzer *et al.*[218] found lower levels of H_2O_2 in liver microsomes of *cyp1a1*(−/−)⁻ mice than the wild-type or *cyp1a2*(−/−)⁻ mice. From the additional use of selective CYP1A1, 2E1, and 1A2 inhibitors they concluded that ROS production in liver microsomes of mice treated with TCDD or other AHR activators reflected combined effects of CYP1A1 and constitutive CYPs. Generation of ROS by CYP1A1 provides a pathway by which the selective induction of CYP1A1 in endothelial cells can cause the endothelial cell and vascular damage commonly seen after TCDD or PCB exposures.[219,220] The CYP1A1 and 1A2 independence of succinate-dependent H_2O_2 production in mitochondria of TCDD-treated mice cited previously[221] contrasts with the other data showing CYP1A1 involvment in ROS production. Taken together, the role of CYP1A1 and 1A2 in the generation of ROS is not very clear and further investigations are warranted.

3.13.3 CYP-dependent Arachidonic Acid (AA) Metabolism

The membrane lipid AA is an endogenous substrate for CYP1A enzymes. As CYP-dependent AA metabolites have extensive biological actions, effects on AA metabolism have implications for CYP1A actions. CYP1A induction increases production of AA epoxides (epoxyeicosatrienoic acids [EETs]) by CYP1A2, and mono-hydroxylated products or hydroxyeicosatetraenoic acids

(HETEs) by some CYP1A1 enzymes, and it is also associated with modification of constitutive AA metabolites formed by other CYPs, particularly CYP4 family enzymes. EETs are primarily vasodilators,[222] while 20-HETE is mainly vasoconstrictive.[223,224] EETs have been proposed as mediators of nitric oxide-independent endothelial derived hyperpolarisation.[207] EETs are thought to cause vasodilation via activation of ATP dependent K^+ channels and large conductance Ca^{2+}-activated K^+ channels (BKCa channels).[207] They also have angiogenic properties in that they stimulate growth of endothelial and mesangial cells.[225]

3.13.4 AA is an Endogenous CYP1A Substrate in Humans

AA is a known substrate for at least 19 of the 57 CYPs in the human genome supporting the important role of CYP-dependent AA metabolites in homeostasis. Those CYPs include CYPs 1A1, 1A2, 1B1, 2B6, 2C8, 2C9, 2C19, 2E1, 2J2, 2U1, 3A4, 4A11, 4B1, 4F2, 4F3, 4F8, 4F12, 5A1 and 8A1.[207] EETs and 20-HETE tend to be generated by different enzyme subfamilies, and thus CYPs have redundant effects in AA metabolism. Thus, effects of AHR activation and CYP1A induction on CYP-dependent AA metabolism could be superimposed on a background of CYP AA metabolites generated by other constitutive CYPs. In general, CYPs in the same family in other species exhibit similar selectivities to the human enzymes.

CYP1A increases CYP AA metabolism in all species studied, although different products are increased to different extent. In mouse liver microsomes, TCDD-induced CYP1A increases EETs 2-fold; 19-HETE 5-fold; and 16-18 HETEs 20-fold[226] Notwithstanding the different inducers, doses, and exposure times used in the experiments in different species, the general pattern emerges that CYP1A induction increases EETs and HETEs in livers of all species investigated, except rat. In rat liver, HETEs are increased, but EETs are decreased, probably reflecting the high constitutive level of EETs in rats and CYP interactions as it will be discussed. Species differences in CYP1A-induced AA metabolism reflect differences in expression of constitutive CYPs catalysing AA metabolism, tissue distribution of CYP1A enzymes, and catalytic activities of the CYP1A enzymes. On the other hand, TCDD suppresses 20-HETE.[227,228] Thus, CYP1A enzymes can modulate the eicosanoid composition of membrane lipids *in vivo*.

3.13.5 Other Endogenous CYP1A Substrates

3.13.5.1 Other Lipids

Isoprostanes, non-enzymic peroxidation products of AA, may mediate some adverse effects of ROS. A role for CYP1A in the metabolism of these products has been suggested[229] and warrants investigation. The lipoxygenase product, lipoxinA4, has been found to activate the AHR and to competitively inhibit CYP1A1,[230] another link between eicosanoids and the AHR, but it has not yet been reported whether CYP1A enzymes are involved in lipoxin effects or metabolism. There are also links between CYP1A induction and prostaglandins (PGs). EETs can increase COX-2 (cyclooxygenase-1) and modify PG metabolism, and TCDD treatment increased cardiac PG production over the dose range in which CYP1A is increased in liver.[231]

3.13.5.2 *Uroporphyrinogen*

Uroporphyrinogen, a precursor in haem synthesis, is an endogenous substrate mainly for CYP1A2. CYP1A2 enhances its oxidation to uroporphyrin. Uroporphyrin is diverted from the normal haem biosynthetic pathway leading to uroporphyria.[206]

3.13.5.3 *Bilirubin*

Bilirubin is a substrate for CYP1A2, although CYP1A1 can also catalyse bilirubin oxidation.[232] Bilirubin and biliverdin also induce CYP1A.[233,234] Hepatic CYP1A levels are also increased in rats with congenitally high levels of bilirubin.[234]

3.13.5.4 *Oestrogens*

Human CYPs 1A1 and 1A2 metabolise 17-β oestradiol to 2- and 4-hydroxy oestradiol. Both CYPs preferentially generate 2-hydroxy-oestradiol[235,236] 4-Hydroxy-oestradiol, which is primarily generated by CYP1B1, [237] is considered the more toxic and carcinogenic metabolite, but both 2- and 4-hydroxy oestradiol can generate ROS and contribute to oxidative DNA[238] and neurologic damage.[239] On the other hand, oestradiol and catechol oestrogens also have antioxidant, cardioprotective and neuroprotective roles.[240] 17β-Oestradiol has been reported to increase CYP1A1 induction by TCDD in rat liver,[241] and in a mouse ovarian cancer cell line,[242] but to suppress CYP1A1 induction by TCDD in several human cancer cell lines[243] and in cultured human endometrial cells.[244]

3.13.5.5 *Melatonin*

Melatonin is another antioxidant and substrate for CYP1A. Again, CYP1A2, which catalyzes melatonin 6-hydroxylation and O-demethylation is the most active, but CYP1A1 (and CYP1B1) can also catalyse melatonin 6-hydroxylation.[245] The large literature on the antioxidant effects of melatonin includes reports that melatonin stimulates production of antioxidant enzymes and prevents their degradation, has anti-inflammatory activity, is a free radical scavenger and controls circadian rhythms via redox sensitive protein modulation.[246,247] Many of the affected proteins are in the Per-Arnt-Sim group, structurally related to the AHR.[248] Relationships between circadian variations in response to light, and melatonin and oestrogens in breast cancer and cancer chemotherapy have been receiving increasing attention.[249] The ability of sunlight and UV light to generate CYP1A-inducing photoproducts from the amino acid tryptophan[250] provides a link between these effects and CYP1A induction, and points to the potentially important role of light as a mediator of AHR action. Taken together, it suggests that, as for bilirubin and oestrogens, CYP1A/melatonin interactions may occur and could be related to CYP1A1 production of ROS or CYP1A-related effects on ion channels via AA products, in addition to enhanced CYP1A metabolism of melatonin.

3.13.6 Role of CYP1A Enzymes in Hyperoxic Lung Injury

Administration of supplemental oxygen is frequently encountered in the treatment of premature infants having pulmonary insufficiency[251] and in individuals suffering from acute respiratory distress syndrome (ARDS).[252] However, hyperoxia causes lung injury in experimental animals,[253–255] and contributes to chronic lung disease (bronchopulmonary dysplasia) in infants undergoing supplemental oxygen therapy.[256] The mechanisms of lung injury induced by oxygen are not completely understood, but ROS are most likely involved.[257]

The CYP1A enzymes are of particular interest to oxygen toxicity, as indicated by the increases in the hepatic and pulmonary CYP contents that are observed in *Ah-responsive* mice (C3H/HeJ), which are highly susceptible to hyperoxic lung injury, but not in *Ah-nonresponsive* (DBA/2J) mice, which are relatively resistant to hyperoxia.[258] On the other hand, Hudak *et al.*[259] demonstrated that DBA/2J mice are more susceptible to lung injury than C3H/HeJ mice, and Mansour *et al.* showed that susceptibility to hyperoxic lung injury is comparable in the *Ah-responsive* (C57BL/6J) and *non-responsive* mouse strains.[260] These observations suggest that genetic background contributes significantly to lung injury and that some of the conflicting observations could be due to differences in the parameters of lung injury that were measured. Hyperoxia induces lung CYP1A1 RNA levels and enzyme activities in lambs,[192] and exposure of adult rats to hyperoxia for 48 h leads to induction of CYP1A1/1A2 in liver,[261,262] and of CYP1A1 in lung.[180,262] Interestingly, the induction of CYP1A enzymes in liver and lung declines after continuation of hyperoxia for 60 h,[180,262] the time period that coincides with expression of overt respiratory distress in these animals, suggesting that CYP1A induction may protect against hyperoxic lung injury.[180] Mansour *et al.* observed protection against hyperoxia-induced lung injury by pretreatment of rats[263] and mice[260] with CYP1A inducers such as MC and BNF. Pineau *et al.* [264] have shown that mice lacking the *cyp1a2* gene die of what appears to be respiratory distress syndrome soon after birth, and the *cyp1a2.* (−/−) mice have developmentally immature lungs, suggesting that CYP1A2 is critical for proper lung development. However, subsequent studies have shown these mice to be viable and growing normally.[265]

Recent studies have shown that CYP1A1 can also produce ROS such as H_2O_2.[214,215] On the other hand, the protection from hyperoxic lung injury of adult rodents pretreated with BNF[229] or MC[263] has been attributed to the

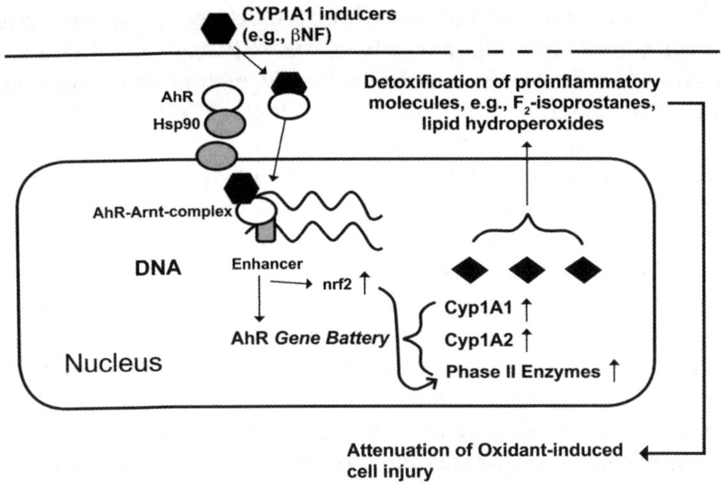

Figure 3.6 Possible mechanisms of protection by β-naphthoflavone (BNF) of hyperoxic lung injury *in vivo*. Ah receptor (AHR) ligands such as BNF induce cytochrome P450 (CYP)1A by AHR-dependent mechanisms. The induced CYP1A may ameliorate lung injury by detoxifying F_2 isoprostanes/isofurans that have been implicated in hyperoxic lung injury. HSP90, heat shock protein 90; nrf2, nuclear receptor factor-2.

induction of CYP1A1, an enzyme with high peroxidase activity.[260,263] It appears that, although CYP1A1 can generate H_2O_2, it may play a greater role as a peroxidase, leading to destruction of lipid hydroperoxides and other ROS that mediate lung injury. It is possible that CYP1A inducers such as BNF may reduce pulmonary toxicity by inducing CYP1A1/1A2 through AHR-dependent mechanisms. The high CYP1A levels may lead to destruction of compounds such as F_2 isoprostanes and isofurans, which are likely mediators of oxygen toxicity (Figure 3.6).

3.13.7 F_2 Isoprostanes, Isofurans, and Hyperoxic Lung Injury

Recently, Fessel *et al.*[266] have demonstrated increased formation of the newly discovered isofurans, which are ROS-mediated lipid peroxidation products, in lungs of mice exposed to hyperoxia. They have proposed that combined measurement of F_2-isoprostanes and isofurans provides the most accurate approach to assess free radical injury and oxidative stress *in vivo*.[266,267] The current theory surrounding these molecules, states that free radicals are specifically able to oxidise arachidonic acid within the cell membrane, and *via* a series of reactions, and transform lipids into a variety of toxic stereoisomers. The formation of a ringed structure from arachidonic acid, and its subsequent removal from the cell membrane, leads to instability and eventual cellular apoptosis.[267] These molecules, known collectively as isoprostanes and isofurans, are two groups that have been implicated in exacerbating cellular injury.[266,267] Their presence has also been correlated with the lung injury associated with bronchopulmonary dysplasia (BPD).[268–270] In fact, Cotton *et al.*[271] have recently shown that the treatment of human premature infants undergoing supplemental oxygen therapy with the CYP inhibitor cimetidine did not prevent lung injury, and the levels of F_2 isoprostanes were higher in the tracheal aspirates of these patients compared with those that did not receive cimetidine. It is our supposition that members of the *CYP1A* gene family may be intimately involved in the decrease or elimination of these compounds, either before or during the initial inflammatory event. Based on our recent results showing protection against lung injury in rats pretreated with BNF,[229] we hypothesise that CYP1A enzymes appear to play important roles in oxidative injury, and further studies on the mechanisms of regulation of CYP1A by hyperoxia, in relation to hyperoxic lung injury, may lead to the development of rational strategies, involving genetic or dietary interventions, which could in turn lead to prevention and/or treatment of chronic lung disease in preterm and term infants undergoing supplemental oxygen therapy. Furthermore, CYP1A enzymes may contribute to the amelioration of hyperoxic lung injury by metabolising lipid hydroperoxides and other ROS-mediated lipid peroxidation products (e.g., isoprostanes/isofurans) that appear to mediate lung injury (Figure 3.6). Our recent observations[272,273] showing modulation of CYP1A enzymes in lungs and livers of newborn rats suggest the utility of using animal models in the study of hyperoxic lung injury.

3.14 Protective Effects of CYP1A Induction

CYP1A induction can decrease the toxicity of xenobiotics that are metabolised by CYP1A but do not require CYP1A-mediated metabolic activation for their toxicity. For example, MC suppresses apoptosis by the heterocyclic amine derivative TRP-P-1 in cultured rat hepatocytes by accelerating the metabolic inactivation of TRP-P-1.[274] In more complex interactions, adverse effects of some carcinogens and toxins shown in cell culture experiments to be activated by inducers of CYP1A, can be circumvented or overcome by administration of the same CYP1A inducers *in vivo*, a "paradoxical" effect pointed out by Nebert *et al.*[99] Thus, although CYP1A1 increases production of BP diol-epoxides and DNA adducts in cultured cells, BP toxicity and DNA adduct formation after administration of BaP *in vivo* were greater in *cyp1a1*(−/−) mice than in *cyp1a1*(+/+) mice.[191] Similarly, production of methaemoglobin by 4-aminobiphenyl (ABP) was lower in CYP1A2(−/−) mice treated with ABP as compared to *cyp1a2*(+/+) mice.[275] Further, there was no difference in the incidence of hepatocellular carcinoma or preneoplastic foci in *cyp1a2*(+/+) and *cyp1a2* (−/−)⁻ mice treated with ABP *in vivo*.[276,277] Although CYP1A2 catalyses the carcinogenic N(2) hydroxylation of 2-amino-1-methyl-6-phenylimidazo (4,5-b)pyridine (PhIP), a heterocyclic amine in grilled meat, tumour incidence for PhIP was greater in *cyp1a2*(−/−)⁻ mice than in the wild-type.[201]

Treatment of Sprague-Dawley rats with MC decreased the formation of I-compounds (DNA modifications) from endogenous electrophiles.[77] Furthermore, CYP1A induction by MC paralleled the decrease in I-compounds over a prolonged period,[77] consistent with other cancer-related protective effects of CYP1A induction cited previously.

As mentioned earlier (*vide supra*), protection by CYP1A1 and 1A2 in hyperoxic pulmonary toxicity has been extensively documented. Pulmonary hyperoxia is a clinical problem in premature newborns and in adults with respiratory distress syndromes requiring oxygen treatment. Hyperoxia induces CYP1A1 and 1A2 in liver and 1A1 in pulmonary epithelium by an AHR-dependent mechanism.[180,278] Evidence for protection by CYP1A1 and 1A2 in this condition includes the following: (i) Treatment of mice or rats with CYP1A inducers, BNF or MC, decreases hyperoxic pulmonary toxicity, but treatment with phenobarbital, an inducer of other CYPs, is not protective;[229,260,263] (ii) CYP1A inhibitors potentiated toxicity;[279] (iii) The onset of pulmonary toxicity by hyperoxia in rats coincides with a fall in induced CYP1A levels;[180] and (iv) Hyperoxic pulmonary toxicity is increased in *cyp1a1*(−/−) and *cyp1a2*(−/−) mice.[280]

The enhanced pulmonary toxicity in *cyp1a2*(−/−) mice suggests further that although CYP1A2 is mainly hepatic, it can have extrahepatic effects, a point relevant to the discussion of extrahepatic effects of CYP1A2 induced arachidonic acid (AA) metabolites in liver (*vide supra*).

A protective effect of CYP1A2 in ROS formation has been demonstrated.[218] ROS formation was increased in *cyp1a2*(−/−) mice and suppressed by TCDD in the wild type, presumably through induction of CYP1A2, suggesting that CYP1A2 can have antioxidant activity. In addition, an increase in liver fibrosis in mice deficient in CYP1A1,[281] suggests that CYP1A1 may protect against fibrosis, and that the liver fibrosis in *ahr*(−/−) mice[282] may reflect a lack of CYP1A1.

3.14.1 Mechanisms for Beneficial Effects of CYP1A Enzymes

Nebert *et al.*[99] suggested that the protection by CYP1A *in vivo* against carcinogenesis by CYP1A inducers reflects domination of the systemic detoxification effects of induced CYP1A and phase II enzymes over local tissue-specific toxic activation effects. The phase I systemic effect seems likely to be more critical here, as toxicity was enhanced in *cyp1a1*(−/−)$^-$ mice in which phase II enzymes were unaffected.[283] Furthermore, enhanced systemic metabolism can account for the protective effects of CYP1A on metabolisable compounds, such as BP and PhIP, but not for TCDD which is not subject to significant metabolism, or for pulmonary hyperoxia, where CYP1A metabolisable inducers have not been implicated.

Enhanced antioxidant activity is another potential protective mechanism. The phase II antioxidant enzyme, NAD(P)H quinone oxidoreductase,[284] was shown to be transcriptionally enhanced after TCDD treatment by

CYP1A1-mediated ROS generation, as shown by CYP1A1 overexpression, or suppressed using CYP1A1 siRNA in HepG2 cells, supporting protective roles both for CYP1A1 and for ROS.

However, while antioxidant enzymes such as catalase, glutathione peroxidase, glutathione reductase and superoxide dismutase are increased along with CYP1A induction in some circumstances,[285] effects vary with dose of the inducer and antioxidant activity is sometimes decreased after CYP1A induction.[286] Antioxidant enzymes are not induced in wild-type mice exposed to hyperoxia[287] or in rats in association with MC protection against pulmonary injury 15 days after treatment with MC, while exhibiting sustained increase in CYP1A. Thus, enhancement of phase II or classical antioxidants may not be critical factors in the protection by CYP1A in hyperoxic pulmonary injury.

Another candidate for protection by CYP1A in pulmonary hyperoxia, proposed by Sinha *et al.*,[229] is enhanced metabolism by CYP1A of isoprostanes and isofurans. These are non-enzymic ROS-modified derivatives of AA that have been implicated in oxidative cellular injury and are considered to be the most sensitive indices of *in vivo* ROS production.[266] It has not been reported whether CYP1A enzymes metabolise isoprostanes.

Shertzer *et al.*[218] hypothesised that the protective effect of CYP1A2 against H_2O_2 production in mice reflects its action as an electron sink. They suggest that CYP1A2 accepts electrons generated during uncoupled catalysis by constitutive CYPs (CYP2E1 in particular), and in the induced state additionally by CYP1A1, thereby preventing transfer of uncoupled electrons from NADPH to oxygen. Levels of CYP1A2 in relation to other CYPs could, therefore, modulate risk of toxicity from ROS exposure. This hypothesis could help explain the increase in ROS formation by addition of planar PCBs or TCDD to incubation mixtures containing microsomes with high levels of CYP1A1,[288] and the enhanced uroporphyrinogen oxidation under similar circumstances.[289] ROS production in the combined presence of metabolism-resistant halogenated aromatic hydrocarbons and increased CYP1A1 might exceed the protective capacity of CYP1A2. Protective effects might predominate after low dose or single dose exposure to TCDD or PCBs or to PAHs, or other rapidly metabolised AHR ligands, such as photoproducts of tryptophan and dietary indole-3-carbinol and derivatives,[290] where ROS production does not exceed the protective capacity of CYP1A2. In fact, protection by CYP1A2 against ROS generated by CYP1A1 could also provide an evolutionary advantage for the two CYP1A enzymes. In addition, these studies suggest a potential interaction between CYP1A1 and 1A2.

3.15 Conclusions

The evidence presented shows that the classical activities assigned to CYP1A, metabolism and activation of exogenous toxic and carcinogenic xenobiotics, do not begin to exhaust the actions of CYP1A, nor is CYP1A induction merely an inert biomarker of AHR activation or its effects. AHR activation and CYP1A induction lead to metabolism of biologically active endogenous substrates, including AA and other eicosanoids, uroporphyrinogen, bilirubin and other bile acids, oestrogens and melatonin; the list is expected to grow. Review of the literature shows further that the role of CYP1A in the effects of TCDD and AHR activation goes beyond simply metabolising endogenous substrates. CYP1A induction may produce ROS/metabolite interactions, alterations in the balance of endogenous antioxidants, such as bilirubin, oestrogens and melatonin, and interactions with constitutive CYPs. CYP1A1 and 1A2 often have complementary or opposing effects, thereby providing protection against ROS and participating in the metabolism of endogenous substrates. Changes in K^+ and Ca^{2+} channel functions, which are common consequences of CYP1A interactions with endogenous substrates, may contribute to the effects of CYP1A induction or serve as final common pathways for diverse effects of AHR activation.

There is increasing evidence that in daily life we are exposed to many chemicals and environmental stresses which can activate the AHR (e.g., dietary indoles, tryptophan photoproducts, changes in bile acid levels, UV light, changes in oxygen tension and blood flow).[206,250,290] Frequent transient CYP1A induction responses may be part of normal, physiologic homeostatic regulation.

Further investigations to advance our understanding of the toxicological and physiological roles of CYP1A include: (i) Molecular regulation of CYP1A enzymes in relation to carcinogenesis; ii) Role of CYP1A enzymes in hyperoxic lung injury; and iii) Role of CYP1A enzymes in the metabolism of endogenous substances that may help us understand the fundamental mechanisms of cell regulation and homeostasis.

Acknowledgements

Work in the author's laboratory was supported by National Institute of Health grants (R01 ES009132, R01 H2070921 and R01 H2087174) and American Heart Association grant (Texas affiliate) (06551227).

References

1. T. Shimada, C.H. Yan, H. Yamazaki, J.C. Gautier, P.H. Beaune and F.P. Guengerich, *Mol. Pharmacol.*, 1992, **41**, 856.
2. T. Cresteil and H.J. Eisen, "Liver Cells and Drugs, Colloque INSERM", INSERM/John Libbey Erotext Ltd., London, 1988.
3. H. Schweikl, J.A. Taylor, S. Kitareewan, P. Linko, D. Nagorney and J.A. Goldstein, *Pharmacogenetics.*, 1993, **3**, 239.
4. G. Kellerman, M. Luyten-Kellermna and C.R. Shaw, *N. Engl. J. Med.*, 1973, **298**, 934.
5. B. Paigen, E. Ward, A. Reilly, L. Houten, H.L. Gurtoo, J. Minowada, K. Steenland, M.B. Havens and P. Sartori, *Cancer Res.*, 1981, **41**, 2757.
6. R. Leboeuf, M. Havens, D. Tabron and B. Paigen, *Biochim. Biophys. Acta.*, 1981, **658**, 348.

7. R.E. Kouri, C.E. McKinney, A.S. Levine, B.K. Edwards, E.S. Vesell, D.W. Nebert and T.L. McLemore, *Toxicol. Pathol.*, 1984, **12**, 44.
8. R.E. Kouri, C.E. McKinney, D.J. Slomiany, D.R. Snodgrass, N.P. Wray and T.L. McLemore, *Cancer Res.*, 1982, **42**, 5030.
9. T.L. McLemore, S. Adelberg, M.C. Liu, N.A. McMahon, S.J. Yu, W.C. Hubbard, M. Czerwinski, T.G. Wood, R. Storeng, R.A. Lubet, J.C. Eggleston, M.R. Boyd and R.N. Hinds, *J. Natl. Cancer Inst.*,1990, **82**, 1333.
10. F.P. Guengerich, "Cytochrome P450, Structure, Mechanism and Biochemistry", Plenum Press, New York, 1995, 473.
11. S.I. Hayashi, J. Watanabe, K. Nakachi and K. Kawajiri, *J. Biochem*, 1991, **110**, 407.
12. T. Tefre, D. Ryberg, A. Haugen, D.W. Nebert, V. Skaug, A. Brogger and A.L. Borresen, *Pharmacogenetics*, 1991, **1**, 20.
13. A. Hirvonen, K. Husgafvel-Pursiainen, A. Karjalainen, S. Anttila and H. Vainio, *Cancer Epidemiol. Biomarkers Prev.*, 1992, **1**, 485.
14. S.I. Hayashi, J. Watanabe, K. Nakachi and K. Kawajiri, *Nucl. Acid Res.*, 1991, **19**, 4797.
15. F. Crofts, G.N. Cosma, D. Currie, E. Taioli, P. Toniolo and S.J. Garte, *Carcinogenesis*, 1993, **14**, 1729.
16. S.J. Roberts-Thomson, M.E. McManus, R.H. Tukey, F.F. Gonzalez and G.M. Holder, *Biochem, Biophys. Res. Commun.*, 1993, **192**, 1373.
17. T. Shimada, M.V. Martin, D. Pruess-Schwarts, L.J. Marnett and F.P. Guengerich, *Cancer Res.*, 1989, **49**, 6304.
18. W. Hu, C. Sorrentino, M.S. Dension, K. Kolaja M.R. Fielden, *Mol. Pharmacol.*, 2007, **106**, in press.
19. I. Simon, F. Berhou, C. Riche, P. Beaune and D. Ratanasavanh, "Ábstracts, 5th Europenn ISSX Meeting", Tours, 1993.
20. F. Berthou, V. Carriere, D. Ratanasavanh, T. Dasduff, F. Morel, J.C. Guillouzo, A. and P. Beaune, "Abstracts, 5th Europenn ISSX Meeting", Tours, 1993.
21. G. Kellerman, M. Luyten-Kellerman and C.R. Shaw, *Am. J. Hum. Genet.*, 1973, **25**, 327.
22. Q. Ma and A.Y.H. Lu, *Chem. Res. Tox.*, 2003, **16**, 249.
23. L.S. Kaminsky and M.J. Fasco, *Crit. Rev. Toxicol.*, 1991, **21**, 407.
24. F.M. Farin and C.J. Omiecinski, *J. Toxicol. Environ. Health*, 1993, **40**, 317.
25. C. Wei, R.J. Caccavale, J.J. Kehoe, P.E. Thomas and M.M. Iba, *Cancer Lett.*, 2001, **171**, 113.
26. B. Ulrike, H.H. Barabara, T. Mario, W. Peter-Matthias and G.R. Ursula, *Toxicol. Lett.*, 2006, **164**, 278.
27. J.P. Whitlock Jr., *Chem. Res. Toxicol.*, 1993, **6**, 754.
28. L.C. Quattrochi, T. Vu and R.H. Tukey, *J. Biol. Chem.*, 1994, **269**, 6949.
29. M.A. Butler, M. Iwasaki, F.P. Guengerich and F.F. Kadlubar, *Proc. Natl. Acad. Sci. USA*, 1989, **86**, 7969.
30. M.A. Butler, N.P. Lang, J.F. Young, N.E. Caporaso, P. Vineis, R.B. Hayes, C.H. Teitel, J.P. Massengill, M.F. Lawsen and F.F. Kadlubar, *Pharmacogenetics*, 1992, **2**, 116.

31. N.P. Lang, M.A. Butler, J. Massengill, M. Lawson, R.C. Stotts, M. Hauer-Jensen and F.F. Kadlubar, *Cancer Epidemiol. Biomarkers Prev.*, 1994, **3**, 675.
32. J. Zhengwen, D. Nadine, L.F. Jorge-Nebert, M.V. Martin, F.P. Guengerich, E. Aklillu, M. Ingelman-Sundberg, G.J. Hammons, B.D. Lyn-Cook, F.F. Kadlubar, S.N. Saldana, M. Sorter, A.V. Vinke, N. Nassr, O.V. Richter and D.W. Nebert, *Pharmacogen. Genom.*, 2006, **6**, 359.
33. C.J. Patten, P.E. Thomas, R.L. Guy, M. Lee, F.J. Gonzalez, F.P. Guengerich and C.S. Yang, *Chem. Res. Toxicol.*, 1993, **6**, 511.
34. R. Dahlqvist, L. Bertilsson, D.J. Birkett, M. Eichelbaum, J. Säwe and F. Sjöqvist, *Clin. Pharmacol. Ther.*, 1984, **35**, 815.
35. G. Engle, N.G. Knebel, U. Hofman and M. Eichelbaum, in Abstract, 23rd European Workshop on Drug Metabolism, Bergamo.
36. H. Yamazaki, Z. Guo, M. Persmark, M. Mimura, K. Inoue, F.P. Guengerich and T. Shimada, *Mol. Pharmacol.*, 1994, **46**, 568.
37. L.M. Distlerath, P.E. Reilly, M.V. Martin, G.G. Davis, G.R. Wilkinson and F.P. Guengerich, *J. Biol. Chem.*, 1985, **260**, 9057.
38. T.F. Woolf, W.F. Pool, M. Kukan, S. Bezek, K. Kunze and W.F. Trager, "Abstracts, 5th North American ISSX Meeting", Tucson, 1993.
39. M.S. Faber, A. Jetter and U. Fuhr, *Basic Clin. Pharmacol. Toxicol.*, 2005, **97**, 125.
40. A.E. Rettie, K.R. Korzekwa, K.L. Kunze, R.F. Lawrence, A.C. Eddy, T. Aoyama, H. Gelboin, F.J. Gonzalez and W.F. Trager, *Chem. Res. Tox.*, 1992, **5**, 54.
41. C.H. Feldman, V.E. Hutchinson, C.E. Pippenger, T.A. Blemenfeld, B.R. Feldman and W.J. Davis, *Pediatrics*, 1980, **66**, 956.
42. D. Diaz, I. Fabre, M. Daujat, B. Saint Aubert, P. Bories, H. Michel and P. Maurel, *Gastoenterology*, 1990, **99**, 737.
43. L.C. Quattrochi and R.H. Tukey, *Mol. Pharmacol.*, 1993, **43**, 504.
44. A.G. Smith, R. Davies, T.P. Dalton, M.L. Miller, D. Judah, J. Riley, T. Gant and D.W. Nebert, *EHP Toxicogenomics*, 2003, **111**, 45.
45. K.L. Cozza, S.C. Armstrong and J. Oesterheld, "Drug Interaction Principles for Medical Practice", American Psychiatric Publishing, 2003.
46. T. Sugimura, *Science*, 1992, **258**, 603.
47. P.F. Kadlubar and G.L. Hammons, "Mammalian Cytochromes P-450", CRC Press, Boca Raton, 1987.
48. K.J. Rich and A.R. Boobis, *Micros. Res. Tech.*, 1997, **39**, 424.
49. S. Kimura, J.C. Donovan and D.W. Nebert, *J. Exp. Pathol.*, 1987, **3**, 61.
50. N. Tuteja, F.J. Gonzalez and D.W. Nebert, *Dev. Biol.*, 1985, **112**, 177.
51. K.J. Rich, J.R. Foster, R.J. Edwards, D.S. Davies and A.R. Boobis, *J. Histochem. Cytochem.*, 1993, **41**, 915.
52. D.W. Nebert, *Mol. Endocrinol.*, 1991, **5**, 1203.
53. A.R. Boobis, R.J. Edwards, J.R. Foster and K.J. Rich, *Eur. J. Pharmacol.*, 1990, **183**, 1502.
54. T. Cresteil, P. Beaune, C. Celier, J.P. Leroux and F.P. Guengerich, *J. Pharmacol. Exp. Ther.*, 1986, **236**, 269.

55. A. Dey, H. Westphal and D.W. Nebert, *Proc. Nat. Acad. Sci., USA*, 1989, **86**, 7446.
56. H.Y. Yang, M.J. Namkung and M.R. Juchau, *Biochem. Biophys. Res. Commun.*, 1991, **178**, 236.
57. M.R. Juchau, C.M. Giachelli, A.G. Fantel, J.C. Greenaway, T.H. Shepard and E.M. Faustman-Watts, *Toxicol. Appl. Pharmacol.*, 1985, **80**, 137.
58. J. Shao, P.L. Stapleton, Y.S. Lin. and E.P. Gallagher, *Drug Metab. Dispos.*, 2007, **35**, 168.
59. J.G. Slatter, I.E. Templeton, J.C. Castle, A. Kulkarni, T.H. Rushmore, K. Richards, Y. He, X. Dai, O.J. Cheng, M. Caguyong and R.G. Ulrich, *Xenobiotica*, 2006, **36**, 938.
60. M.S. Faber and U. Fuhr, *Clin. Pharmacol. Ther.*, 2004, **76**, 178.
61. L. Tomatis, "IARC: Cancer: causes, occurrence and control", IARC Scientific Publications, Lyon, 1990.
62. K. Hemminki, "Chemical Carcinogenesis and Mutagenesis I", Springer-Verlag, London, 1990.
63. C.E. Searle and O.J. Teale, "Chemical Carcinogenesis and Mutagenesis I", Springer-Verlag, London, 1990.
64. "Emerging problems and issues in environmental health. In human health and the environment some research needs", Report of the Fourth Task Force for Research and Planning in Environmental Health Science, 1992, 53.
65. M.R. Guerin, "Polycyclic Aromatic Hydrocarbons and Cancer", Academic press, New-York, 1978.
66. G. Grimmer, "Polycyclic Aromatic Hydrocarbons", CRC Press, Boca Raton, 1983.
67. M. Hall and P.L. Glover, "Chemical Carcinogenesis and Mutagenesis I", Springer-Verlag, London, 1990.
68. IARC, "Polynuclear Aromatic Compounds. Part 1, Chemical, Environmental and Experimental Data", World Health Organization, Lyon, 1983.
69. H.V. Gelboin, *Physiol. Rev.*, 1980, **60**, 1107.
70. A.D. Burdick, J.W. Davis Jr, K.J. Liu, L.G. Hudson, H. Shi, M.L. Mosnke and S.W. Burchiel, *Cancer Res.*, 2003, **63**, 7825.
71. D.W. Nebert, D.R. Nelson, M.J. Coon, R.W. Estabrook, R. Feyereisen, Y. Fujii-Kuriyama, F.J. Gonzalez, F.P. Guengerich, I.C. Gunsalus, E.F. Johnson, J.C. Loper, R. Sato, M.R. Waterman and D.J. Waxman, *DNA and Cell Biol.*, 1991, **10**, 1.
72. D.W. Nebert and F.J. Gonzalez, *Ann. Rev. Biochem.*, 1987, **56**, 945.
73. F.P. Guengerich, G.A. Dannan, S.T. Wright, M.V. Martin and L.S. Kaminsky, *Biochemistry*, 1982, **21**, 6019.
74. T.M. Guenther and D.W. Nebert, *Eur. J. Biochem.*, 1978, **91**, 4496.
75. F.P. Guengerich, *Cancer Res.*, 1988, **48**, 2946.
76. O. Pelkonen and D.W. Nebert, *Pharmacol. Rev.*, 1982, **34**, 189.
77. B. Moorthy, S. Chen, D. Li and K. Randerath, *Carcinogenesis*, 1993, **14**, 879.
78. B. Moorthy, P. Sriram and K. Randerath, *Fundam. Appl. Toxicol.*, 1994, **22**, 549.

79. D.W. Nebert, *Crit. Rev. Toxicol.*, 1989, **20**, 153.
80. S. Otto, C. Marcus, C. Pidgeon and C. Jefcoate, *Endocrinology*, 1991, **129**, 970.
81. A. Dorrenhaus, T. Muller and P.H. Roos, *Arch. Toxicol.*, 2007, **81**, 19.
82. R.C. Bowes III, A.R. Parrish, M.A. Steinberg, K.L. Willet, W. Zhao, U. Savas, C.R. Jefcoate, S.H. Safe and K.S. Ramos, *Biochem. Pharmacol.*, 1996, **52**, 587.
83. W.R. Brian, M.A. Sari, M. Iwasaki, T. Shimada, L.S. Kaminsky and F.P. Guengerich, *Biochemistry*, 1990, **29**, 11280.
84. T. Shimada, C.L. Hayes, H. Yamazaki, S. Amin, S.S. Hecht and F.P. Guengerich, *Cancer. Res.*, 1996, **56**, 2979.
85. T. Shimada, A. Sugie, T. Yamada, H. Kawazoe, M. Hashimoto, E. Azuma, T. Nakajima, K. Inoue and Y. Oda, *Xenobiotica*, 2003, **33**, 957.
86. K. Kawajiri, K. Natachi, K. Imai, S. Hayashi and J. Watanabe, *"In Xenobiotics and Cancer"*, Taylor and Francis, London, 1991.
87. F.J. Gonzalez, A.K. Jaiswal and D.W. Nebert, *Cold Spring Harbor Symp. Quart. Biol.*, 1986, **51**, 879.
88. M.L. Slattery, W. Samowitz, K. Ma, M. Murtaugh, C. Sweeney, T.R. Levin and S. Neuhausen, *Am. J. Epidemiol.*, 2004, **160**, 842.
89. K. Kawajiri, K. Nakachi, K. Imai, A. Yoshii, N. Shinoda and J. Watanabe, *FEBS Lett.*, 1990, **263**, 131.
90. K. Nakachi, K. Imai S. Hayashi and K. Kawajiri, *Cancer Res.*, 1993, **53**, 2994.
91. R.E. Kouri, C.E. McKinney, D.J. Slomiany, D.R. Snodgrass, N.P. Wray and T.L. McLemore, *Cancer Res.*, 1982, **42**, 5030.
92. E. Taioli, J. Trachman, X. Chen, P. Toniolo and S.J. Garte, *Cancer Res.*, 1995, **55**, 3757.
93. Y. Li, R.C. Millikan, D.A. Bell, L. Cui, C.K. Tse, B. Newman and K. Conway, *Breast Cancer. Res.*, 2005, **7**, R12.
94. G. Varzim, E. Monteiro, R. Silva, J. Frenandes and C. Lopes, *Eur J. Cancer Prev.*, 2003, **12**, 495.
95. B.S. Johnson, B.A. Brooks, H. Reyes, E.C. Hoffman and O. Hankinson, *Hum. Mol. Genet.*, 1992, **1**, 351.
96. H. Raunio, K. Husagavel-Pursiainen, S. Anttila, E. Hietanen, A. Hivonen and O. Pelkonen, *Gene*, 1995, **159**, 113.
97. O. Hankinson, *Arch. Biochem. Biophys.*, 2005, **15**, 379.
98. J.P. Whitlock, *Annu. Rev. Pharmacol. Toxicol.*, 1999, **39**, 103.
99. D.W. Nebert, T.P. Dalton, A.B. Okey and F.J. Gonzalez, *J. Biol. Chem.*, 2004, **4**, 23847.
100. A.B. Okey, D.S. Riddick and P.A. Harper, *Toxicol. Lett.*, 1995, **75**, 249.
101. P.D. Boucher and R.N. Hines, *Carcinogenesis*, 1995, **16**, 383.
102. F.J. Gonzalez and D.W. Nebert, *Nucleic Acids Res.*, 1985, **13**, 7269.
103. K. Sterling, T.J. Weaver, K.L. Ho, L.C. Xu and E. Bresnick, *Mol. Pharmacol.*, 1993, **44**, 560.
104. R.N. Hines, J.M. Mathis and C.S. Jacob, *Carcinogenesis*, 1988, **9**, 1599.

105. R.W. Robertson, L. Zhang, D.S. Pasco and J.B. Fagan, *Nucleic Acids Res.*, 1994, **22**, 1741.
106. L. Beebe, S.D. Fox, C.W. Riggs, S.S. Park, H.V. Gelboin, H.J. Issaq and L.M. Anderson, *Toxicol. Appl. Pharmacol.*, 1992, **114**, 16.
107. T.A. Gasiewicz, G. Rucci, E.C. Henry and R.B. Baggs, *Biochem. Pharmacol.*, 1986, **35**, 2737.
108. L.M. Anderson, D. Logsdon, S. Ruskie, S.D. Fox, H.J. Isaaq, R.M. Kovatch and C.M. Riggs, *Carcinogenesis*, 1994, **15**, 2245.
109. A. Morello, *Can.J. Biochem.*, 1965, **43**, 1289.
110. W.N. Piper, J.Q. Rose and P.G. Gehring, *Env. Health Perspect.*, 1973, **5**, 241.
111. R.D. Prokipcak and A.B. Okey, *Can. J. Physiol. Pharmacol.*, 1991, **69**, 1204.
112. H.I. Swanson and G.H. Perdew, *Arch. Biochem. Biophys.*, 1993, **302**, 167.
113. G. Rucci and T.A. Gasiewicz, *Arch. Biochem. Biophys.*, 1988, **265**, 197.
114. T. Moriguchi, H. Motohashi, T. Hosoya, O. Nakjima, S. Takahashi, S. Ohsako, Y. Aoki, N. Nishimura, C. Tohyama, Y. Fujii-Kuriyama and M. Yamamoto, *Proc. Natl. Acad. Sci. USA*, 2003, **100**, 5652.
115. A. Poland and E. Glover, *Mol. Pharmacol.*, 1974, **10**, 349–359.
116. A.R. Boobis, D.W. Nebert and J.S. Felton, *Mol. Pharmacol.*, 1977, **13**, 259.
117. R. Masaki, S. Matsuura and Y. Tashiro, *Cell Structure and Function*, 1984, **9**, 53.
118. E. Bresnick, M. Brosseau, W. Levin, L. Reik, D.E. Ryan and P.E. Thomas, *Proc. Natl. Acad. Sci. USA.*, 1981, **78**, 4083.
119. G. Takahashi, in "Polycyclic Aromatic Hydrocarbons and Cancer", Academic Press, Burlington, 1978.
120. G. Takahashi and K. Yasuhira, *Cancer Res.*, 1973, **33**, 23.
121. W.G. Levine and R.W. Singer, *J. Pharmacol. Exp. Ther.*, 1972, **183**, 411.
122. G. Takahashi, H. Shah and S. Weinhouse, *Cancer Res.*, 1977, **37**, 369.
123. E. Bresnick, R.A. Liebelt, J.G. Stevenson and J.C. Madix, *Cancer Res.*, 1967, **27**, 462.
124. P.M. Daniel and O.E. Pratt, *Nature*, 1967, **215**, 1142.
125. B. Moorthy, *J. Pharmacol. Exp. Ther.*, 2000, **294**, 313.
126. B. Moorthy, K. Muthiah, I.S. Fazili, S.R. Kondraganti, L. Wang, X.I. Couroucli and W. Jiang, *Biochem. Biophys. Res. Commun.*, 2007, **354**, 1071.
127. L.L. Wessner, M. Fan, D.O. Schaeffer, M.F. McEntee and M.S. Miller, *Carcinogenesis*, 1996, **17**, 1519.
128. H. Bartsch, E. Hietanen, S. Petruzelli, C. Giuntini, R. Saracci, A. Mussi and C.A. Angeletti, *Int. J. Cancer.*, 1990, **46**, 185.
129. S. Petruzzelli, A.M. Camus, L. Carrozzi, L. Ghelarducci, M. Rindi G. Menconni, C.A. Angeletti, M. Anotupa, E. Hietanen, A. Aitio, R. Saracci, H. Bartsch and C. Giuntini, *Cancer Res.*, 1988, **48**, 4695.
130. S.R. Kondraganti, W. Jiang and B. Moorthy, *J. Pharmacol. Exp. Ther.*, 2002, **303**, 945.

131. M. Christou, U. Savas, S. Schroeder, X. Shen, T. Thompson, M.N. Gould and C.R. Jefcoate, *Mol. Cell Endocrinol.*, 1995, **115**, 41.
132. M. Christou, U. Savas, D.C. Spink, J.F. Gierthy and C.R. Jefcoate, *Carcinogenesis*, 1994, **15**, 725.
133. G.L. Murray, C.O. Foster, T.S. Barnes, R.J. Weaver, S.W. Ewen, W.T. Melvin and M.D. Burke, *Br. J. Cancer*, 1991, **63**, 1021.
134. J.A. McKay, W.T. Melvin, A.K. Ah-See, S.W. Ewen, W.F. Greenlee, C.B. Marcus, M.D. Burke and G.I. Murray, *FEBS Lett.*, 1995, **374**, 270.
135. W. Zhang, A.F. Purchio, K. Chen, J. Wu, L. Lu, R. Coffee, P.R. Contag and D.B. West, *Drug Metab. Dispos.*, 2003, **31**, 1054.
136. S.J. Campbell, F. Carlotti, P.A. Hall, A.J. Clark and C.R. Wolf, *J. Cell Science*, 1996, **109**, 2619.
137. T.H. Rushmore, C.B. Pickett and A.Y.H. Lu, "Handbook of Experimental Pharmacology", Springer-Verlag, New York, 1994.
138. A. Puga, J. Marlowe, S. Barnes, C.Y. Chang, A. Maier, Z. Tan, J.K. Kerzee, X. Chang, M. Strockbeck and E.S. Knudsen, *Toxicology*, 2002, **181**, 171.
139. V. Vasiliou, A. Puga, C.Y. Chang, M.W. Tabor and D.W. Nebert, *Biochem Pharmacol.*, 1995, **50**, 2057.
140. O. Geneste, A.M. Camus, M. Castegnaro, S. Petruzzelli, P. Macchianini, C.A. Angeletti, C. Giuntini and H. Bartsch, *Carcinogenesis*, 1991, **12**, 1301.
141. M. Stjernfeldt, K. Berglund, J. Lindsten and J. Ludvigsson, *Lancet*, 1986, **1**, 1350.
142. M.S. Miller, A.B. Jones, S.S. Park and L.M. Anderson, *Toxicol. Appl. Pharmacol.*, 1990, **104**, 235.
143. M.S. Miller, M.R. Juchau, F.P. Guengerich, D.W. Nebert and J.L. Raucy, *Fundam. Appl. Toxicol.*, 1996, **34**, 165.
144. S. Leone-Kabler, L.L. Wessner, M.F. McEntee, R.B. D'Agostino Jr and M.S. Miller, *Carcinogenesis*, 1997, **18**, 1163.
145. M.S. Miller, *Toxicol. Appl. Pharmacol.*, 2004, **198**, 95.
146. K.M. Gressani, S. Leone-Kabler, M.G. O'Sullivan, L.D. Case, A.M. Malkinson and M.S. Miller, *Carcinogenesis*, 1999, **20**, 2159.
147. A.M. Malkinson, K.M. Koski, W.A. Evans and M.F. Festing, *Cancer Res.*, 1997, **57**, 2832.
148. X. Wang and H. Witschi, *Cancer Lett.*, 1995, **91**, 33.
149. S.S. Hecht, P.M. Kenney, M. Wang and P. Upadhyaya, *Cancer Lett.*, 2002, **187**, 87.
150. S.S. Hecht, P.M. Kenney, M. Wang and P. Upadhyaya, *Cancer Lett.*, 2001, **167**, 1.
151. S.S. Hecht, P.M. Kenney, M. Wang, N. Trushin and P. Upadhyaya, *Cancer Lett.*, 2000, **150**, 49.
152. U. Bernauer, B. Heinrich-Hirsch, M. Tonnies, W. Peter-Matthias and U. Gundert-Remy, *Toxicol. Lett.*, 2006, **164**, 278.
153. X. Ding and L.S. Kaminsky, *Annu. Rev. Pharmacol. Toxicol.*, 2003, **43**, 149.

154. D. Nebert and T. Dalton, *Nat. Rev. Cancer*, 2006, **6**, 947.
155. R. Ueda, H. Iketaki, K. Nagata, S. Kimura, F. Gonzalez, K. Kusano, T. Yoshimura and Y. Yamazoe, *Molecular Pharmacology*, 2006, **69**, 1924.
156. L.C. Quattrochi, T. Vu and R.H. Tukey, *J. Biol. Chem.*, 1994, **269**, 6949.
157. G.V. Pickwell, H. Shih and L.C. Quattrochi, *Biochem. Pharmacol.*, 2003, **65**, 1087.
158. O. Hankinson, *Ann. Rev. Pharmacol. Toxicol.*, 1995, **35**, 307.
159. M.E. Hahn, *Comp. Biochem. Phys. Part C*, 1998, **121**, 23.
160. M.S. Denison and C.F. Wilkinson, *Eur. J. Biochem.*, 1985, **147**, 429.
161. T.A. Gasiewicz and G. Rucci, *Mol. Pharmacol.*, 1984, **26**, 90.
162. P.A. Harper, J.V. Giannone, A.B. Okey and M.S. Denison, *Mol. Pharmacol.*, 1992, **42**, 603.
163. Q. Ma, *Curr. Drug Metab.*, 2001, **2**, 149.
164. P.E. Thomas, R.E. Kouri and J.J. Hutton, *Biochem. Genet.*, 1972, **6**, 157.
165. D.W. Nebert, K.F. Goujon and J.E. Gielen, *Nat. New Biol.*, 1972, **236**, 107.
166. A. Poland, E. Glover, J.R. Robinson and D.W. Nebert, *J. Biol. Chem.*, 1974, **249**, 5599.
167. C. Chang, D.R. Smith, V.S. Prasad, C.L. Sidman, D.W. Nebert and A. Puga, *Pharmacogenetics*, 1993, **3**, 312.
168. Q. Ma, A.Y. Lu, *Drug Metab. Dispos.*, 2007, in press.
169. J.R. Robinson, J.S. Felton, R.C. Levitt, S.S. Thorgeirsson and D.W. Nebert, *Mol. Pharamcol.*, 1975, **11**, 850.
170. D.W. Nebert and N.M. Jensen, *Biochem. Pharmacol.*, 1979, **28**, 149.
171. D. Gilday, M. Gannon, K. Yutzey, D. Bader and A.B. Rifkind, *J. Biol. Chem.*, 1996, **20**, 271.
172. H.G. Morrison, E.J. Weil, S.I. Karchner, M.L. Sogin and J.J. Stegeman, *Comp. Biochem. Physiol. C, Pharmacol. Toxicol. Endocrinol.*, 1998, **121**, 231.
173. A.K. Berndtson and T. Chen, *Arch Biochem. Biophys.*, 1994, **310**, 187.
174. Y. Ogino, T. Itakura, R. Mitsuo and M. Sato, *Mar. Biotechnol.*, 1999, **1**, 342.
175. Y. Fujita, H. Ohi, N. Murayama, K. Saguchi and S. Higuchi, *Arch. Biochem. Biophys.*, 1999, **371**, 24.
176. F.M. Farin, T.H. Pohlman and C.J. Omiecinski, *Toxciol. Appl. Pharmacol.*, 1994, **124**, 1.
177. S. Chinta, R. Kommaddi, C. Turman, H. Strobel and V. Ravindranath, *J. Neurochem.*, 2005, **93**, 724.
178. R.P. Kommaddi, C.M. Thurman, B. Moorthy, L. Wang, H.W. Strobel, V. Ravindranath, *J. Neurochem.*, 2007, in press.
179. A. Dey, J. Jones and D. Nebert, *Biochem. Pharmacol.*, 1999, **58**, 525.
180. X. Couroucli, S. Welty, R. Geske and B. Moorthy, *Mol. Pharmacol.*, 2002, **61**, 507.
181. R.A. McKinnon, P.D. Hall, L.D. Quattrochi, R.H. Tukey and M.E. McManus, *Hepatology*, 1991, **14**, 848.

182. B. Annas, A.L. Granberg and E.B. Brittebo, *Toxicol. Appl. Pharmacol.*, 2000, **169**, 94.
183. O.M. Hyyti, M. Nyman, M.L. Willis, H. Raunio and O. Pelkonen, *Mar. Environ. Res.*, 2001, **51**, 465.
184. J.H. Dees, B.S. Masters, U. Muller-Eberhard and E.F. Johnson, *Cancer Res.*, 1982, **42**, 1423.
185. L.M. Anderson, J.M. Ward, S.S. Park, A.B. Jones, J.L. Junker, H.V. Gelboin and J.M. Rice, *Cancer Res.*, 1987, **47**, 6079.
186. J.J. Stegeman, M.E. Hahn, R. Weisbrod, B.R. Woodin, J.S. Joy, S. Najibi and R.A. Cohen, *Mol. Pharmacol.*, 1995, **47**, 296.
187. D. Morse, A. Stein, P. Thomas and H. Lowndes, *Toxicol. Appl. Pharmacol.*, 1998, **152**, 232.
188. M. Gannon, D. Gilday and A.B. Rifkind, *Toxicol. Appl. Pharmacol.*, 2000, **164**, 24.
189. R. Smoloqitz, M. Schultz and J. Stegeman, *Carcinogenesis*, 1992, **13**, 2395.
190. J.B. Ortiz-Delgado, H. Segner and C. Sarasquete, *Aquat. Toxicol.*, 2005, **75**, 144.
191. R.A. Garrick, B.R. Woodlin and J.J. Stegeman, *In vitro Cell Dev. Biol. Anim.*, 2005, **41**, 57.
192. T.A. Hazinski, E. Noisin, I. Hamon and A. DeMatteo, *J. Clin. Invest.*, 1995, **96**, 2083.
193. J.P. Cooke, *Vasc. Med.*, 2000, **5**, 49.
194. B. Moorthy, K.P. Miller, W. Jiang, E.S. Williams, S.R. Kondraganti and K.S. Ramos, *J. Pharmacol. Exp. Ther.*, 2003, **305**, 394.
195. K.S. Ramos and B. Moorthy, *Drug Metab. Rev.*, 2005, **37**, 595.
196. D.W. Nebert, T. Dalton, A.B. Okey and F.J. Gonzales, *J. Bio. Chem.*, 2004, **279**, 23847.
197. S. Uno, T.P. Dalton, S. Derkenne, C.P. Curran, M.L. Miller, H.G. Schertzer and D.W. Nebert, *Mol. Pharmacol.*, 2004, **65**, 1225.
198. S. Uno, T.P. Dalton, N. Dragin, C.P. Curran, S. Derkenne, M. Miller, H. Shertzer, F.J. Gonzalez and D.W. Nebert, *Mol. Pharmacol.*, 2006, **69**, 1103.
199. Y. Tsuneoka, T.P. Dalton, M.L. Miller, C.D. Clay, H.G. Shertzer, G. Talaska, M. Medvedovic and D.W. Nebert, *J. Natl. Cancer Inst.*, 2003, **95**, 1227.
200. E.G. Snyderwine, M. Yu, H.A. Schut, L. Knight-Jones and S. Kimura, *Food Chem. Toxicol.*, 2002, **40**, 1529.
201. S. Kimura, M. Kawabe, A. Yu, H. Morishima, P. Fernandez-Salguero, G.J. Hammons, J.M. Ward, F.F. Kadlubar and F.J. Gonzalez, *Carcinogenesis*, 2003, **24**, 583.
202. N. Dragin, T.P. Dalton, M.L. Miller, H.G. Shertzer and D.W. Nebert, *J. Biol. Chem.*, 2006, **281**, 18591.
203. C. Cheung, X. Ma, K.W. Krausz, S. Kimura, L. Feigenbaum, T.P. Dalton, D.W. Nebert, J.R. Idle and F.J. Gonzalez, *Chem. Res. Toxicol.*, 2005, **18**, 1471.

204. S. Derkenne, C.P. Curran, H.G. Shertzer, T.P. Dalton, N. Dragin and D.W. Nebert, *Pharmacogen. Genom.*, 2005, **15**, 503.
205. D.F.V. Lewis, E. Watson and B.G. Lake, *Mutat. Res.*, 1998, **410**, 245.
206. F.J. Gonzalez and D.W. Nebert, *Trends Genet.*, 1990, **6**, 182.
207. A. Rifkind, *Drug Metabolism Reviews*, 2006, **38**, 291.
208. M. Toborek, S.W. Barger, M.P. Mattson, P. Espandiari, L.W. Robertson and B. Hennig, *J. Biochem. Toxicol.*, 1995, **10**, 219.
209. J.J. Schlezinger, J. Keller, L. Verbrugge and J.J. Stegeman, *Comp. Biochem. Physiol. C, Toxicol. Pharmacol.*, 2000, **125**, 273.
210. R.C. Zangar, D.R. Davydov and S. Verma, *Toxciol. Appl. Pharmacol.*, 2004, **199**, 316.
211. S. Puntarulo and A.I. Cederbaum, *Free. Radic. Biol. Med.*, 1989, **24**, 1324.
212. T.P. Dalton, A. Puga and H.G. Shertzer, *Chem. Biol. Interact.*, 2002, **141**, 77.
213. S.M. Cantrell, J. Joy-Schlezinger, J.J. Stegeman, D.E. Tillitt and M. Hannink, *Toxicol. Appl. Pharmacol.*, 1998, **148**, 24.
214. Y. Morel and R. Barouki, *Biochem. J.*, 1999, **342**, 481.
215. Y. Morel, I. de Waziers and R. Barouki, *Mol. Pharmacol.*, 2000, **57**, 1158.
216. J.J. Schlezinger and J.J. Stegeman, *Aqat. Toxiol.*, 2001, **52**, 101.
217. J.J. Schlezinger, R. White and J.J. Stegeman, *Mol. Pharmacol.*, 1999, **56**, 588.
218. H.G. Shertzer, C.D. Clay, M.B. Genter, S.N. Schneider, D.W. Nebert and T.P. Dalton, *Free Radic. Biol. Med.*, 2004, **36**, 605.
219. W. Dong, H. Teraoka, Y. Tsujimoto, J.J. Stegeman and T. Hiraga, *Toxicol. Sci.*, 2004, **77**, 109.
220. H. Teraoka, W. Dong, Y. Tsuijimoto, H. Iwasa, D. Endoh, N. Ueno, J.J. Stegeman, R.E. Peterson and T. Hiraga, *Biochem. Biophys. Res. Commun.*, 2003, **304**, 223.
221. A. Sneft, T. Dalton, D. Nebert, M. Genter, A. Puga, R. Hutchinson, J. Kerzee, S. Uno and H. Shertzer, *Free Radic. Biol. Med.*, 2002, **33**, 1268.
222. R.J. Roman, *Physiol. Rev.*, 2002, **82**, 131.
223. D. Sacerdoti, A. Gatta and J.C. McGiff, *Prostaglandins Other Lipid Mediat.*, 2003, **72**, 51.
224. M. Medhora, J. Daniels, K. Mundey, B. Fisslthaler, R. Busse, E.R. Jacobs and D.R. Harder, *Am. J. Physiol. Heart Circ. Physiol.*, 2003, **284**, H215.
225. M. Potente, B. Fisslthaler, R. Busse and I. Fleming, *J. Biol. Chem.*, 2003, **278**, 29619.
226. C.A. Lee, B.P. Lawrence, N.I. Kerkvliet and A.B. Rifkind, *Toxicol. Appl. Pharmacol.*, 1998, **153**, 1.
227. J.R. Falck, S. Lumin, I. Blair, E. Dishman, M.V. Martin, D.J. Waxman, F.P. Guengerich. and J.H. Capdevila, *J. Biol. Chem.*, 1990, **265**, 10244.
228. S. Huang and G.G. Gibson, *Toxicol. Appl. Pharmacol.*, 1991, **108**, 86.
229. A. Sinha, K. Muthiah, W. Jiang, X.I. Couroucli, R. Barrios and B. Moorthy, *Toxicol. Sci.*, 2005, **87**, 204.
230. C.M. Schaldach, J. Riby and L.F. Bjeldanes, *Biochemistry*, 1999, **38**, 7594.

231. C.P. Quilley and A.B. Rifkind, *Biochem. Biophys. Res. Commun.*, 1986, **136**, 582.
232. F. De Matteis, S.J. Dawson, A.R. Boobis and A. Comoglio, *Mol. Pharmacol.*, 1991, **40**, 686.
233. C.J. Sinal and J.R. Bend, *Mol. Pharmacol.*, 1997, **52**, 590.
234. J. Kapitulnik and F.J. Gonzalez, *Mol. Pharmacol.*, 1993, **43**, 722.
235. T. Aoyama, K. Korzekwa, K. Nagata, J. Gillette, H.V. Gelboin and F.J. Gonzalez, *Endocrinology*, 1990, **18**, 378.
236. Y. Tsuchiya, M. Nakajima and T. Yokoi, *Cancer Lett.*, 2005, **227**, 115.
237. C.L. Hayes, D.C. Spink, B.C. Spink, J.Q. Cao, N.J. Walker and T.R. Sutter, *Proc. Natl. Acad. Sci. USA*, 1996, **93**, 9776.
238. Z.H. Chen, Y.J. Hurh, H.K. Na, J.H. Kim, Y.J. Chun, D.H. Kim, K.S. Kang, M.H. Cho and Y.H. Surh, *Carcinogenesis*, 2004, **25**, 2005.
239. K.B. Gordon, L.M. Macrae and H.V. Carswell, *Brain Res.*, 2005, **1036**, 155.
240. J.R. Munoz-Castaneda, J. Muntane, M.C. Munoz, I. Bujalance, P. Montilla and I. Tunez, *Toxicol. Lett.*, 2006, **160**, 196.
241. S. Sarkar, N.R. Jana, J. Yonemoto, C. Tohyama and H. Sone, *Int. J. Oncol.*, 2000, **16**, 141.
242. D.S. Son, K.F. Roby, K.K. Rozman and P.F. Terranova, *Toxicology*, 2002, **176**, 229.
243. N.R. Jana, S. Sarkar, M. Ishizuka, J. Yonemoto, C. Tohyama and H. Sone, *Mol. Cell. Biol. Res. Commun.*, 2000, **4**, 174.
244. M.S. Ricci, D.G. Toscano, C.J. Mattingly and W.A. Toscano Jr, *J. Biol. Chem.*, 1999, **274**, 3430.
245. X. Ma, J.R. Idle, K.W. Krausz and F.J. Gonzalez, *Drug Metab. Dispos.*, 2005, **33**, 489.
246. S. Cuzzocrea and R.J. Reiter, *Curr. Top Med. Chem.*, 2002, **2**, 153.
247. R. Hardeland, A. Coto-Montes and B. Poeggeler, *Chronobiol. Int.*, 2003, **20**, 921.
248. Y.Z. Gu, J.B. Hogenesch and C.A. Bradfield, *Annu. Rev. Pharmacol. Toxicol.*, 2000, **40**, 519.
249. R.G. Stevens, *Epidemiology*, 2005, **16**, 254.
250. S. Diani-Moore, E. Labitzke, R. Brown, A. Garvin, L. Wong and A.B. Rifkind, *Toxicol. Sci.*, 2006, **90**, 96.
251. W.H. Northway Jr and R.C. Rosan, *Radiology*, 1968, **91**, 49.
252. T. Ozawa, S. Sugiyama, M. Hayakawa, T. Satake, F. Taki, M. Iwata and K. Taki, *Am. Rev. Respir. Dis.*, 1988, **137**, 535.
253. J.M. Clark and C.J. Lambertsen, *Pharmacol. Rev.*, 1971, **22**, 37.
254. B.A. Freeman and J.D. Crapo, *J. Biol. Chem.*, 1981, **256**, 10986.
255. C.L. Bryan and S.G. Jenkinson, *Clin. Chest. Med.*, 1988, **9**, 141.
256. S.H. Tsai, W.R. Anderson, M.B. Strickland and M. Pliego, *Radiology*, 1972, **105**, 107.
257. C.V. Smith and S.E. Welty, in: Chronic Lung Disease of Early Infancy, eds. R.D. Bland and J. Coalson, Lung Biology in Health and Disease, Marcell Dekker, Inc., N.Y., 1999, **30**, 749.

258. J.C. Gonder, R.A. Proctor and J.A. Will, *Proc. Natl. Acad. Sci USA*, 1985, **82**, 6315.
259. B.B. Hudak, L.Y. Zhang and S.R. Kleeberger, *Pharmacogenetics*, 1993, **3**, 135.
260. H. Mansour, M. Levacher, E. Azoulay-Dupis, J. Moreau, C. Marquetty and M. Gougerot-Pocidalo, *J. Appl. Physiol.*, 1988, **64**, 1376.
261. T. Okamoto, M. Mitsuhashi, I. Fujita, R.K. Sindhu and Y. Kikkawa, *Biochem. Biophys. Res. Commun.*, 1993, **197**, 878.
262. B. Moorthy, U.T. Nguyen, S. Gupta, K.D. Stewart, S.E. Welty and C.V. Smith, *Toxicol. Lett.*, 1997, **90**, 67.
263. H. Mansour, M. Brun-Pascaud, C. Marquetty, M. Gougerot-Pocidalo, J. Hakim and J.J. Pocidalo, *Am. Rev. Respir. Dis.*, 1988, **137**, 688.
264. T. Pineau, P. Fernandez-Salgeuro, S.S. Lee, T. McPhail, J.M. Ward and F.J. Gonzalez, *Proc. Natl. Acad. Sci. USA*, 1995, **92**, 5134.
265. H.C. Leonard Liang, H. Li, R.A. McKinnon, J.J. Duffy, S.S. Potter, A. Puga and D.W. Nebert, *Proc. Natl. Acad. Sci. USA*, 1996, **93**, 1671.
266. J.P. Fessel, N.A. Porter, K.P. Moore, J.R. Sheller and L.J. Roberts Jr, *Proc. Nat. Acad. Sci. USA*, 2002, **99**, 16713.
267. J.D. Morrrow and L.J. Roberts, *Am. J. Respir. Crit. Care Med.*, 2002, **166**, S25.
268. S. Goil, W.E. Trung, C. Barnes, M. Norberg, M. Rezaiekhaligh and D. Thibeault, *J. Pediatr.*, 1998, **132**, 349.
269. O.D. Saugstad, *Acta Paediatr.*, 1997, **86**, 1277.
270. M. Comporti, C. Signorini, S. Leoncini, G. Buonocore, V. Rossi and L. Ciccoli, *Free Radic. Biol. Med.*, 2004, **37**, 724.
271. R.B. Cotton, T.A. Hazinski, J.D. Morrow, L.J. Roberts, D.C. Zeldin, D.P. Lindstorm, U. Lappalainen, A.B. Law and S. Steele, *Pediatr. Res.*, 2006, **59**, 795.
272. X.I. Couroucli, Y.H. Wei, W. Jiang, K. Muthiah, L.E. Evey, R. Barrios and B. Moorthy, *Pediatr. Res.*, 2006, **59**, 401.
273. X.I. Couroucli, Y.-H. Wei, W. Jiang, R. Barrios and B. Moorthy, *J. Pharmacol. Exp. Ther.*, 2006, **317**, 946.
274. B. Shiotani, Y. Nonaka, K. Kanazawa, G. Danno and H. Ashida, *Biosci. Biotechnol. Biochem.*, 2002, **66**, 356.
275. H.G. Shertzer, T.P. Dalton, G. Talaska and D.W. Nebert, *Toxicol. Appl. Pharmacol.*, 2002, **181**, 32.
276. S. Kimura, M. Kawabe, J. Ward, H. Morishima, F.F. Kadlubar, G.J. Hammons, P. Fernandez-Salguero and F.J. Gonzalez, *Carcinogenesis*, 1999, **20**, 1825.
277. S. Kimura, M. Kawabe, A. Yu, H. Morishima, P. Fernandez-Salguero, G.J. Hammons, J.M. Ward, F.F. Kadlubar and F.J. Gonzalez, *Carcinogenesis*, 2003, **24**, 583.
278. W. Jiang, S.E. Welty, X.I. Couroucli, R. Barrios, S.R. Kondraganti, K. Muthiah, L. Yu, S.E. Avery and B. Moorthy, *J. Pharmacol. Exp. Ther.*, 2004, **310**, 512.
279. B. Moorthy, K.M. Parker, C.V. Smith, J.R. Bend and S.E. Welty, *J. Pharmacol. Exp. Ther.*, 2000, **292**, 553.

280. B. Moorthy, L. Wang, K. Muthiah, X.I. Couroucli and W. Jiang, in *Proceedings of the 14th International Conference on Cytochromes P450: Biochemistry, Biophysics and Bioinformatics Medimond S.r.l. International Proceedings, Bologna*, Italy, 2005, p. 193.
281. T.C. Peterson, P. Hodgson, P. Fernandez-Salguero, M. Neumeister and F.J. Gonzalez, *Hepatol. Res.*, 2000, **17**, 112.
282. P. Fernandez-Salguero, T. Pineau, D.M. Hilbert, T. McPhail, S.S. Lee, S. Kimura, D.W. Nebert, S. Rudikoff, J.M. Ward and F.J. Gonzalez, *Science*, 1995, **268**, 722.
283. T.P. Dalton, M.Z. Dieter, R.S. Matlib, N.L. Childs, H.G. Shertzer, M.B. Genter and D.W. Nebert, *Biochem. Biophys. Res. Commun.*, 2000, **267**, 184.
284. A. Marchand, R. Barouki and M. Garlatti, *Mol. Pharmacol.*, 2004, **65**, 1029.
285. E.Y. Park and H.M. Rho, *Mol. Cell. Biochem.*, 2002, **240**, 47.
286. P.A. Kern, R.B. Fishman, W. Song, A.D. Brown and V. Fonseca, *Toxicology*, 2002, **171**, 117.
287. L. Frank, *Free Radic. Biol. Med.*, 1991, **11**, 463.
288. H.G. Shertzer, C.D. Clay, M.B. Genter, M.C. Chames, S.N. Schneider, G.G. Oakley, D.W. Nebert and T.P. Dalton, *Free Radic. Biol. Med.*, 2004, **36**, 618.
289. M.E. Hahn and K. Chandran, *Arch. Biochem. Biophys.*, 1996, **329**, 163.
290. M.S. Denison and S.R. Nagy, *Annu. Rev. Pharmacol. Toxicol.*, 2003, **43**, 309.

CHAPTER 4

The CYP1B Subfamily

MORAG C.E. MCFADYEN[a] AND GRAEME I. MURRAY[b]

[a] School of Pharmacy, Robert Gordon University, Schoolhill, Aberdeen AB10 1FR; [b] Department of Pathology, University of Aberdeen, Medical School Buildings, Foresterhill, Aberdeen AB25 2ZD

Table of Contents

Issues in Toxicology
Cytochromes P450: Role in the Metabolism and Toxicity of Drugs and other Xenobiotics
Edited by Costas Ioannides

4.1 Introduction

Cytochrome P450 CYP1B1 is the only known member of the CYP1B subfamily. Initially identified in mouse endometrial stromal cells as a polycyclic aromatic hydrocarbon (PAH)-inducible P450[1] this enzyme was further characterised as a new P450 CYP1B subfamily.[2] In humans, CYP1B1 was isolated from a keratinocyte cell line and shown to be transcriptionally upregulated by dioxin.[3] Comparison of the CYP1B1 gene with the other two members of the CYP1 subfamily, namely CYP1A1 and CYP1A2, showed 40% homology with both CYP1A1 and CYP1A2. Despite the low degree of similarity with known members of the CYP1 family, this P450 was assigned by the P450 nomenclature committee to a new CYP1 subfamily, CYP1B1. DNA hybridisation studies indicated that there was only one member of the CYP1B1 subfamily.[3,4] The CYP1B1 gene (Genbank accession no. U03688) is contained within three exons and two introns on chromosome 2p21[4] and spans 8.5 kb of genomic DNA (Genbank accession no. U56438). The mRNA is 5.2 kb, and the open reading frame begins at the 5′ end of the second exon, in contrast to other P450s which all begin in exon 1. The open reading frame codes for a polypeptide of 543 amino acids. The long 3′-untranslated region contains multiple polyadenylation sites.[3] Characterisation of the CYP1B1 promoter has identified several regulatory elements required for maximum promoter activity.[5] Analysis of the amino acid sequence indicates that there are three potential furin cleavage sites commencing at amino acid 38, which suggests that there could be post-translational N-terminal processing. CYP1B1 is the largest known human P450, both in terms of mRNA size and the number of amino acids, but it is paradoxically the simplest in terms of intron exon structural organisation.

Unlike other P450s CYP1B1 is not expressed in human liver. Although it may be expressed at low levels in normal extrahepatic tissues, it is expressed at much higher levels in tumour cells compared with the surrounding non-tumour tissue.[6,7] Initial metabolic studies with CYP1B1, expressed in either yeast[8] or human lymphoblastoid cells[9] suggests that CYP1B1 is capable of activating, as assessed by a mutagenicity assay, putative human carcinogens including several polycyclic aromatic hydrocarbons and arylamines.[8] In addition, CYP1B1 shows high activity for the 4-hydroxylation of oestradiol[10] suggesting that CYP1B1 may also be capable of metabolising a variety of endogenous compounds involved in cell regulation. Increased 4-hydroxylation of oestradiol has been found in benign breast[11,12] and uterine tumours[13] indicating that CYP1B1 expression is increased in benign tumours.

4.2 Mechanisms of Regulation of CYP1B1

Polycyclic aromatic hydrocarbons induce a gene battery that functions to metabolise and excrete incoming xenobiotics. Potent inducers of the CYP1 family are uniformly planar and aromatic in structure. Such chemicals include natural combustion products (benzo(a)pyrenes), dietary constituents (heterocyclic amines) and manufacturing bi-products (dioxin). As such, a consensus of chemical structure would suggest these inducers invoke their gene regulation effects *via* a specific receptor protein, termed the aryl hydrocarbon (Ah) receptor. The Ah receptor is a ubiquitous cytosolic protein that responds to planar aromatic ligands by forming an activated complex with two molecules of the molecular chaperone heat shock protein 90 (hsp90), and the X-associated protein 2.[14,15] Upon ligand binding, the Ah receptor undergoes a conformational change and translocates to the nucleus. Once nuclear translocation has occurred, hsp90 is released from the Ah receptor, either at the translocation stage or upon dimerisation with another basic-helix-loop-helix partner such as ARNT (aryl hydrocarbon nuclear translocator). This heterodimeric AhR/ARNT complex binds to xenobiotic responsive elements (XRES) located upstream of the *CYP1A1*, *CYP1A2,* and *CYP1B1* genes and modulates their transcription, subsequently increasing the presence of these enzymes which are designed to oxidise and deactivate the initiating ligand.

4.2.1 Transcriptional Regulation of CYP1B1

The mechanisms controlling the expression of CYP1B1 in tumour cells remain to be completely elucidated, although the likely mechanisms are transcriptional processing and translational stability.[7] The promoter region of the CYP1B1 gene is structurally distinct from that of the two related P450s, CYP1A1 and CYP1A2 promoters, with several regulatory elements identified as required for maximal CYP1B1 activity.[5] Transcriptional regulation of CYP1B1 by dioxin or polycyclic aromatic hydrocarbons[3] is *via* the Ah receptor complex.[16] The Ah receptor is a ligand-activated receptor that on dimerisation with ARNT, is translocated from the nucleus to the cytoplasm where it binds to specific upstream sequences (XREs) of CYP1B1, and promotes its transcription. Regulation of CYP1B1 may also occur *via* non-Ah receptor mediated pathways such as cAMP, as CYP1B1 is inducible in Ah receptor-deficient mice.[17] In addition to the Ah receptor, ARNT can preferentially heterodimerise with another member of the basic-helix-loop-helix family of transcription factors, hypoxia inducible factor1α,[18,19] which is induced by hypoxia. Hif-1α is constitutively expressed in the hypoxic microenvironment of the tumour,[20] and hif-1α and its dimerisation partner ARNT bind to hypoxia responsive elements on hypoxia regulated genes. Four hypoxia responsive sequences in the upstream regulatory region of the *CYP1B1* gene have been identified, which require further characterisation. These findings highlight the possibility that hypoxia responsive elements (HREs) and XREs may regulate CYP1B1 expression either through competition or through a synergistic mechanism.

Human CYP1B1 mRNA contains multiple polyadenylation sites, and it has been suggested that there is cell type-specific alternative processing of the CYP1B1 mRNA, which may regulate the amount and /or the ability of the final transcript to be translated.[21] The use of alternative processing may also influence the stability of the mRNA. Such a mechanism has been implicated for the regulation of CYP2A5, a related mouse P450, whose enhanced expression has been suggested to play an important role in the development of liver cancer in mice.[22] The regulation of CYP2A5 has been shown to require a 44kDa protein which stabilises CYP2A5 mRNA by controlling the length of the poly A tail.[23]

4.2.1.1 Epigenetic Regulation of CYP1B1

CYP1B1 is now widely regarded as a biomarker of the neoplastic phenotype. In contrast, although CYP1B1 mRNA appears to be expressed in a wide variety of normal tissue, the protein is absent in normal liver.[24] Understanding the role of this P450 in the initiation and progression of cancer is extremely important. DNA methylation can inactivate genes and suppress gene expression directly by interfering with the binding of transcription factors or indirectly by attracting methylated DNA-binding factors that recruit histone deacetylases and precipitate an inactive chromatin structure.[25,26] CpG island methylation has been shown to be important in the transcriptional silencing of many genes. In contrast, hypomethylation is generally observed in unique sequences, including transcription control sequences that are often subject to cancer-associated hypomethylation.[26,27] Hypomethylation has been implicated in the activation of oncogenes, such as c-*myc*,[28] and may also contribute to chromosomal instability.[29]

Promoter methylation of *CYP1B1* has been associated with decreased activity of this gene.[30] Recently, increased in expression of CYP1B1 in prostate cancer was found to be regulated by promoter hypomethylation, thus confirming the importance of this regulatory mechanism in *CYP1B1* gene expression.[31] *CYP1B1* methylation takes place at multiple CpG sites within the *CYP1B1* gene, some of which are contained within key promoter elements, such as DRE (Dioxin Responsive Element) 1, DRE2, DRE3, and Sp1 binding sites at _72 and _80.[31] Methylation at these sites may decrease the accessibility of DNA-binding sites for proteins involved in AhR-mediated regulation[31] and may alter oestrogen-mediated regulation of CYP1B1. DNA methylation of *CYP1B1* has also been associated with survival in breast cancer patients treated with tamoxifen.[32] Histone methylation has also been associated with alterations in chromatin structure; therefore, gene expression of *CYP1B1* could also be regulated through chromatin remodeling. Chromatin structure is also altered by histone acetylation, and histone H3 acetylation was observed in the far upstream enhancer region and other CYP1B1 promoter elements through the interaction of histone acetyltransferase and CREB-binding protein.[33] Additionally, promoter methylation of some *CYP1B1* effectors and associated metabolic enzymes (including several steroid receptor genes and catechol-*O*-methyltransferase) have also been linked to differential gene expression in hormone-dependent cancers compared with normal tissue.[34–39] Thus, epigenetic regulation through methylation and acetylation of histones within the *CYP1B1* promoter region is a key determinant of CYP1B1 transcription, and the degree of epigenetic regulation may be tissue-specific, with those tissues relying on cAMP-mediated transcription of *CYP1B1* most likely having a different chromatin structure than other tissues.

4.2.1.2 Regulation of CYP1B1 by MicroRNAs

A recent study by Tsuchiya *et al.*,[40] demonstrated the regulation of CYP1B1 by a micro RNA known as miR-27b. Micro RNAs are small noncoding RNAs that regulate gene expression through translational repression or mRNA cleavage. In this study the authors suggest that human *CYP1B1* gene is post-transcriptionally regulated by miR-27b. A significant inverse association was observed between the expression levels of miR-27b and CYP1B1 protein in cancerous tissue. Thus, the decreased expression of miR-27b may be one of the reasons for the high levels of CYP1B1 protein expressed in tumour tissue.

4.2.1.3 Trace Metal Regulation of CYP1B1

Exposure to combinations of environmental trace metals and PAH-mixtures has been suggested to diminish the extent of CYP1 induction by PAHs. The influence of arsenite on benzo(a)pyrene (B[a]P)-induced levels of CYP1B1 in the human breast cancer cell line, T-47D, has been tested using PCR and rates of 4-methoxylation of oestrogen.[41] Arsenite at 1, 3, or 10 μM did not affect levels of CYP1B1 mRNA;[4] however, it did affect CYP1B1 activity (4-hydroxylation of oestrogen) markedly and dose responsively. These results were interpreted to indicate that the effect of arsenite was not at the level of transcription. When mouse embryo fibroblasts from C57BL/6J mice were treated with arsenite, B[a]P, or mixtures of these two agents, CYP1B1 mRNA levels were variously affected, depending on the concentrations.[42] Mixtures of 5 μM B[a]P and 1 or 5 μM arsenite led to lower levels of CYP1B1 mRNA than were obtained with 5 μM B[a]P alone. In no case did a mixture of B[a]P and arsenite synergistically increase levels of CYP1B1 mRNA above the levels induced individually by arsenite or B[a]P.

4.3 Tissue Specific Expression of CYP1B1

CYP1B1 displays distinct species- and tissue -specific patterns of expression. In humans, CYP1B1 is characteristically and selectively expressed in tumours (see next section) while in both rat and mouse CYP1B1 shows expression in a limited number of normal tissues. In rat CYP1B1 is present in adrenal gland while in murine tissues it is present in kidney and uterus.

4.3.1 CYP1B1 and Disease

4.3.1.1 Tumour Selective Expression of CYP1B1

CYP1B1 is over expressed in a wide range of tumours[6] and is widely regarded as a biomarker of the neoplastic phenotype, with enhanced expression being detected in a number of malignant tissues and tumour-derived and transformed cell lines, including prostate cancer cells,[43] lung carcinoma cells,[44] transformed B lymphocytes[45] and immortalised oral keratinocytes.[46] CYP1B1 also demonstrates a similarly high level of expression in metastatic disease.[47] There has been considerable interest in the role of CYP1B1 in tumour development and progression, and as a diagnostic marker and therapeutic target. However CYP1B1 is not expressed in normal human liver, and only selectively (in extrahepatic tissues) at a low level, under specific defined physiological circumstances.

4.3.1.2 Glaucoma

CYP1B1 also appears to be a developmentally regulated gene. Several deleterious mutations or truncating mutations in the human *CYP1B1* gene in different populations was shown to co-segregate with primary congenital glaucoma (PCG), an autosomal recessive disorder characterised by increased intraocular pressure resulting in enlargement of cornea and damage to the optic nerve that may lead to blindness.[48–51] Homozygosity for these deleterious mutations was shown to be associated with severe PCG. Several CYP1B1 mutations in various ethnic backgrounds have been implicated in its pathogenesis. *CYP1B1* is expressed in the developing and mature ciliary body of the mouse eye, and it has been speculated that mutations in this gene may contribute to the abnormal elevation in the intraocular pressure in the in the PCG patients.[51] However, CYP1B1 knockout mice did not show a glaucoma phenotype;[52] species differences in substrate specificity and tissue distribution between the human and mouse forms of CYP1B1 may account for this.

A recent study[53] indicated that the abnormalities in the development of the trabecular meshwork in PCG may result from diminished or absent metabolism of important endogenous substrates in the ciliary epithelium due to non-functional CYP1B1 enzyme. Previously, Choudhary and colleagues proposed that as the CYP1B1 orthologues were very similar in both species, both must have had a common evolutionary role in eye development; they showed that both orthologues can synthesise retinoic acid but cannot degrade the active morphogen, retinoic acid.[54] It was suggested that CYP1B1 may be involved in generating retinoic acid during a specific time point when the morphogen is necessary for trabecular meshwork development.

4.4 Role of CYP1B1 in Drug Metabolism

The major groups of compounds shown to be metabolised by CYP1B1 include polycyclic aromatic hydrocarbons, aromatic amines and several steroid hormones, most notably oestradiol (Table 4.1). This latter substrate is selectively hydroxylated at the C4 position by CYP1B1, and the 4-hydroxylation of oestradiol is probably the most specific CYP1B1 metabolic reaction. The best characterised model substrate of CYP1B1 is ethoxyresorufin which is deethylated to resorufin. This substrate is also metabolised by other P450s, most notably CYP1A1, so ethoxyresorufin deethylation is an ideal reaction in

Table 4.1 Summary of CYP1B1 substrates.

	CYP1B1 substrates
Steroid hormones	17β-oestradiol
	Testosterone
Polycyclic aromatic hydrocarbons	benzpyrene,
	benzpyrene-4,5-diol
	benzpyrene-7,8-diol,
	dibenzopyrene
	benzanthracene
	dimethylbenzanthracene
Aromatic and heterocyclic amines	2-amino-1-methy-6-phenylimidazo[4,5-b] pyridine
	2-aminoanthracene
Model substrates	Ethoxyresorufin
	Ethoxycoumarin

combination with recombinant or expressed CYP1B1, but is less than ideal when there is a mixture of P450s such as in tissue samples. With respect to therapeutic drugs, CYP1B1 interacts with a range of anticancer drugs and of most relevance is its ability to deactivate a there.

Previous studies demonstrated that CYP1B1 metabolically inactivates a range of structurally diverse anticancer drugs (docetaxel, doxorubicin, paclitaxel, mitoxantrone and tamoxifen). Recently, in a trial of 93 breast cancer patients treated with paclitaxel, those homozygous for the CYP1B1*3 variant allele (4326 c > G; L432V) were found to have a significantly longer progression-free survival than patients with at least one valine allele ($P = 0.037$). It is currently unclear whether the CYP1B1*3 polymorphism is a specific marker for outcome to paclitaxel therapy or a biological marker indicating that more aggressive therapy may be required for patients carrying this variant allele.[55]

Further studies with the semi-synthetic taxane, docetaxel, currently recommended as adjuvant therapy in early node positive breast cancer in combination with cyclophosphamide and adriamycin (TAC) demonstrated *in vitro* cytotoxic effectiveness of docetaxel in Chinese hamster ovary cells transfected with human CYP1B1.[47] Indeed, Bournique and Lemarie have confirmed our findings[47] that CYP1B1 interacts with and reduces individual anticancer drug cytotoxicity.[56] Our *in vitro* studies have previously highlighted the use of specific inhibitors to CYP1B1 in modulating the cytotoxic profile of a range of structurally diverse anticancer drugs with CYP1B1.[47,57] Chang and coworkers demonstrated that the stilbene *trans*-resveratrol inhibited the catalytic activity and suppressed the constitutive gene expression of *CYP1B1*.[58] The ability to inhibit CYP1B1, resulting in increased efficacy of chemotherapeutic intervention, may have important clinical implications.

CYP1B1 is the principal catalyst of 4-hydroxytamoxifen *trans-cis*-isomerisation, which converts the primary potent antiostrogen *trans*-4-hydroxytamoxifen to the weak oestrogen agonist *cis*-4-hydroxytamoxifen.[59]

4.4.1 CYP1B1 Metabolism of Specific Drugs

4.4.1.1 2-(4-amino-3-methylphenyl)-5-fluorobenzothiazole

The evolution of new anticancer drugs involves the combined forces of chemistry, pharmacy and pharmacology. The L-lysylamide prodrug, Phortress, a water-soluble fluorinated benzothiazole derivative,[60] is one such drug that has evolved in this manner. This planar hydrophobic aminophenyl benzothiazole analogue is a potent agonist of the aryl hydrocarbon (Ah) receptor and switches on cytochrome P450 *CYP1A1* gene expression.[61] CYP1A1 induction is regulated through the ligand activated transcription factor, Ah receptor, and its dimerisation partner, the Ah nuclear translocator; therefore, before CYP1A1 can be induced, the Ah receptor and Ah nuclear translocator need to be expressed. CYP1A1 can then be induced and convert the fluorinated prodrug Phortress into a reactive species resulting in cellular damage and subsequent apoptosis through the formation of Phortress-derived DNA adducts. Induction of CYP1A1- catalysed biotransformation of 2-(4-aminophenyl)benzothiazole within tumour cells is essential for drug activation. Although members of the CYP1 family of P450 enzymes show enhanced expression in a variety of solid tumours and antioestrogen-resistant breast cancer cell lines,[62] patients require to be screened for the presence of CYP1A1 in their tumour tissue, as not all patients may express CYP1A1 at the site of the tumour.

CYP1B1 may also be involved in the metabolic bioactivation of Phortress; human recombinant CYP1B1 supersomes metabolised the nonfluorinated benzothiazole precursor of Phortress (*i.e.* DF503).[61] Indeed, benzothiazoles have been shown to modulate the expression of CYP1B1 in benzothiazole-sensitive cell lines and to increase CYP1B1 mRNA in[61] MCF-7 cell lines.

4.4.1.2 Aryl Oxime Prodrugs

CYP1B1 is now widely acknowledged as a highly promising target for prodrug activated anticancer therapy[6] as it provides the opportunity for selective activation to occur in tumour cells without toxicity in normal cells. Indeed, CYP1B1 has undergone initial evaluation in human tumour xenografts as a target for a novel group of aryl oxime prodrugs catalysed by this enzyme. Bioactivation of these prodrugs results in the release of nitric oxide, subjecting the tumour cells to lethal doses of nitrosative stress without harming normal healthy cells.[63]

4.4.2 CYP1B1 Activated Prodrugs in Development

A range of pyrrolo-indole and pyrrolo-quinoline derivatives have been patented as prodrugs and are currently under preclinical investigation as targets for aromatic oxidation (*e.g.* hydroxylation) by CYP1B1 intra-tumourally, resulting in the release of activated DNA alkylating agents which would cause cell death or adversely effect the tumour cells. Several novel therapeutic strategies utilising natural based compounds as prodrugs, including isoflavonoids and stilbenes, targeted at CYP1B1 have now been patented and are currently in preclinical development. It is anticipated that one or more of these compounds will enter clinical trials in the near future.

4.5 Inhibitors of CYP1B1

CYP1B1 catalyses the *O*-demethylation of biochanin A (a principal isoflavonoid found in red clover) to produce genistein;[64] this widely occurring isoflavonoid, has recently been shown to inhibit several carcinogenic signalling pathways through reduced gene expression of CYP1B1, EGFR (Epidermal Growth Factor Receptor), egr (early growth response)-1, AKT2, NELL2 (Neural Epidermal growth factor-like 2) and DNA ligase III.[65] It has been demonstrated that the *trans*-resveratrol inhibited the catalytic activity and suppressed the constitutive gene expression of *CYP1B1*.[58,66,67] The ability to inhibit CYP1B1, resulting in increased efficacy of chemotherapeutic intervention, may have important clinical implications and provide a mechanism by which the efficacy of the current generation of cytotoxic anticancer drugs can be enhanced.

Use of inhibitors of CYP1B1 as co-therapy has the potential to enhance the efficacy of both current and novel anticancer drugs. Drug resistance is one of the major problems in the treatment of cancer. A range of pharmacological mechanisms combine to result in a poorer response to chemotherapy and overall decreased patient survival. A greater understanding of the factors involved in anti-cancer drug resistance would be of considerable benefit. It has previously been demonstrated in our laboratory that CYP1B1 metabolically inactivates a range of structurally diverse anticancer drugs (docetaxel, doxorubicin, paclitaxel, mitoxantrone and tamoxifen), and as a consequence of this action CYP1B1 also mediates anticancer drug resistance.[47,57] In patients with ovarian cancers receiving docetaxel there was poorer survival in those whose tumours showed a high level of CYP1B1 expression.[47] Moreover, our *in vitro* studies have previously highlighted the use of specific low-molecular weight chemical inhibitors to CYP1B1 in modulating the cytotoxic profile of a range of structurally diverse anticancer drugs.[47,57] However, there are currently no particularly suitable animal models for studying CYP1B1 activity in tumours, and the development and validation of appropriate models represents a considerable challenge.

4.6 Drug Metabolism and Pharmacogenetics of CYP1B1

There are pronounced inter-individual and inter-ethnic variations in the capacity of individuals to metabolise drugs and other xenobiotics. Some of these variations are the result of alterations in the level and specific activity of individual cytochrome P450s expressed by the individual. These can be the results of one or a number of changes in the gene structure of an individual P450, for instance changes in the promoter region can alter the amount of a particular P450 expressed. Single base changes in the coding region of a specific P450 gene may alter the activity of that P450; indeed, single nucleotide polymorphisms (SNPs) may result in alternative splicing and protein truncations resulting in the absence of a particular P450 expressed in the tissue of some people. For the human *CYP1B1* gene, variant alleles including *CYP1B1*2* to *CYP1B1*26* have been identified. In 2000, Hanna and colleagues reported that 5 allelic variants demonstrated 2.4-3.4 fold higher catalytic efficiency than the wild type *CYP1B1*, with respect to 4 hydroxylation of oestradiol.[68] Indeed, the V432L G (Val) allele has been associated with higher activity, resulting in increased oxidation of benzo(a)pyrene and formation of 4-hydroxyoestradiol than the C (Leu) allele). An increase in induction of CYP1B1 by PAHs and/or dioxins has been reported and associated with the G *versus* C allele of the V432L variant. Two polymorphic variants at codon Ala119Ser and codon Leu432- Val of CYP1B1 have been reported to have an association with breast or endometrial cancer risk.[69,70] It has been shown that the 4-hydroxylase activity of oestradiol by these variant enzymes are 2–4 fold higher than those by the wild-type enzyme.[68,71] Thus, genetic polymorphisms of CYP1B1 may lead to alterations in oestrogen metabolism and a risk for oestrogen-associated cancers. DeRoos and colleagues recently showed an increased non-Hodgkin's lymphoma risk associated with the G allele.[72] Previous reports have shown increased risk of colorectal cancer, ovarian, breast and prostate cancers associated with this polymorphism, in addition to being a putative susceptibility factor in smoking related head and neck squamous cell cancer (HNSCC).

A recent Swedish population based case-control study in women comparing normal 1st trimester pregnancies with women who experienced a miscarriage in the first trimester suggested that carriers of the *CYP1B1* 432 Val/Val genotype were at a higher risk of miscarriage (odds ratio = 1.46; 95% confidence interval, 1.02–2.08).[73] This group also noted a significant interaction between genotype and caffeine intake. These findings suggest that *CYP1B1* Val432Leu polymorphism is associated with first-trimester miscarriage, and it may also modify the risk among coffee drinkers.

According to Finnstrom *et al.*,[74] the expression of CYP1B1 is significantly higher in leukocytes of women than men, and Lin *et al.*,[75] reported that CYP1A1 and CYP1B1 levels in non-induced lymphocytes were significantly higher in female nonsmokers than in male nonsmokers, suggesting also gender differences in CYP1B1.

4.7 Conclusions

CYP1B1 is a unique P450 with distinct structural and metabolic properties. It possesses the simplest P450 exon intron gene structure and the resulting polypeptide is one of the largest P450 proteins. The protein is overexpressed in tumours and is not expressed in normal liver. CYP1B1 shows distinct substrate specificity capable of metabolising polycyclic aromatic hydrocarbons, aromatic amines and steroid hormones. Most notably and characteristically is its C4 hydroxylation of oestradiol making this reaction the most specific for assaying CYP1B1 activity.

References

1. U. Savas, M. Christou and C.R. Jefcoate, *Carcinogenesis*, 1993, **14**, 2013–2018.
2. U. Savas, K.K. Bhattacharyya, M. Christou, D.L. Alexander and C.R. Jefcoate, *J. Biol. Chem.*, 1994, **269**, 14905–14911.
3. T.R. Sutter, Y.M. Tang, C.L. Hayes, Y.Y. Wo, E.W. Jabs, X. Li, H. Yin, C.W. Cody and W.F. Greenlee, *J. Biol. Chem.*, 1994, **269**, 13092–13099.
4. Y.M. Tang, Y.Y.P. Wo, J. Stewart, A.L. Hawkins, C.A. Griffin, T.R. Sutter and W.F. Greenlee, *J. Biol. Chem.*, 1996, **271**, 28324.
5. Y.Y.P. Wo, Y.J. Stewart and J.W.F. Greenlee, *J. Biol. Chem.*, 1997, **272**, 26702–26707.
6. G.I. Murray, M.C. Taylor, M.C. McFadyen, J.A. McKay, W.F. Greenlee, M.D. Burke MD and W.T. Melvin, *Cancer Res.*, 1997, **57**, 3026–3031.
7. G.I. Murray, W.T. Melvin, W.F. Greenlee and M.D. Burke, *Annu. Rev. Pharmacol. Toxicol.*, 2001, **41**, 297–316.
8. T. Shimada, C.L. Hayes, H. Yamazaki, S. Amin, S.S. Hecht, F.P. Guengerich and T.R. Sutter, *Cancer Res.*, 1996, **56**, 2979–2984.
9. C.L. Crespi, B.W. Penman, D.T. Steimer, T. Smith, C.S. Yang and T.R. Sutter, *Carcinogenesis*, 1997, **12**, 83–89.
10. C.L. Hayes, D.C. Spink, B.C. Spink, J.Q. Cao, N.J. Walker and T.R. Sutter, *Proc. Natl. Acad. Sci. USA*, 1996, **93**, 9776–9781.
11. J.G. Liehr and M.J. Ricci, *Proc. Natl. Acad. Sci. USA*, 1996, **93**, 3294–3296.
12. J.G. Liehr, *Biochem. Soc. Trans.*, 1999, **27**, 318–323.
13. J.G. Liehr, M.J. Ricci, C.R. Jefcoate, E.V. Hannigan, J.A. Hokanson and B.T. Zhu, *Proc. Natl. Acad. Sci. USA*, 1995, **92**, 9220–9224.
14. L.A. Carver, J.J. LaPres, S. Jain, E.E. Dunham and C.A. Bradfield, *J. Biol. Chem.*, 1998, **273**, 33580–33587.
15. Q. Ma and J.P Whitlock Jr., *J. Biol. Chem.*, 1997, **272**, 8878–8884.
16. H.I. Swanson and C.A. Bradfield, *Pharmacogenetics*, 1993, **3**, 213–230.
17. D.Y. Ryu, P.A. Levie, P. Fernandez-Salguero, F.J. Gonzalez and E. Hodgson, *Mol. Pharmacol.*, 1996, **50**, 443–446.
18. G.L. Wang, B.H. Jiang, E.A. Rue and G.L. Semenza, *Proc. Natl. Acad. Sci. USA*, 1995, **92**, 5510–5514.
19. K. Gradin, J. McGuire, R.H. Wenger, I. Kvietikova, M.L. Whitelaw, R. Toftgard, L. Tora, M. Gassmann and L. Poellinger, *Mol. Cell. Biol.*, 1996, **16**, 5221–5231.
20. W.A. Denny, *Lancet Oncol.*, 2000, **1**, 25–29.
21. S.E. Shehin, R.O. Stephenson and W.F. Greenlee, *J. Biol. Chem.*, 2000, **275**, 6770–6776.

22. U.M. Wastl, W. Rossmanith, M.A. Lang, A.M. Camus-Randon, B. Grasl-Kraupp, W. Bursch and R. Schulte-Hermann, *Mol. Carcinog.*, 1998, **22**, 229–234.
23. O. Geneste, F. Raffalli and M.A. Lang, *Biochem. J.*, 1996, **313**, 1029–1037.
24. R.J. Edwards, D.A. Adams, P.S. Watts, D.S. Davies and A.R. Boobis, *Biochem. Pharmacol.*, 1998, **56**, 377–387.
25. M.R. Rountree, K.E. Bachman and S.B. Baylin, *Nat. Genet.*, 2000, **25**, 269–277.
26. R. Singal and G.D. Ginder, *Blood*, 1999, **93**, 4059–4070.
27. M. Ehrlich, *Oncogene.*, 2002, **21**, 5400–5413.
28. R.M. Sharrad, J.A. Royds, S. Rogers and A.J. Shorthouse, *Br. J. Cancer*, 1992, **65**, 667–672.
29. W.A. Schulz, J.P. Elo, A.R. Florol, S. Pennanen, S. Santourlidis, R. Engers, M. Buchardt, H.H. Seifert and T. Visakorpi, *Genes Chromosomes Cancer*, 2002, **35**, 58.
30. M. Widschwendter, K.D. Siegmund, H.M. Muller, H. Fiegl, C. Marth, E. Muller-Holzner, P.A. Jones and P.W. Laird, *Cancer Res.*, 2004, **64**, 3807–3813.
31. T. Tokizane, H. Shiina, M. Igawa M, H. Enokida, S. Urakami, T. Kawakami, T. Ogishima, S.T. Okino, L.C. Li, Y. Tanaka, N. Nonomura, A. Okuyama and R. Dahiya, *Clin. Cancer Res.*, 2005, **11**, 5793–5801.
32. W. Han, D. Kang, I.A. Park, S.W. Kim, J.Y. Bae, K.W. Chung and D.Y. Noh, *Clin. Cancer Res.*, 2004, **10**, 124–130.
33. W. Zheng and C.R. Jefcoate, *Mol. Pharmacol.*, 2005, **67**, 499–512.
34. M. Sasaki, K. Nakajima, G. Perinchery, T. Fujino, B.R. Oh, S. Fujimoto and R. Dahiya, *Oncol. Rep.*, 2000, **7**, 369–373.
35. M. Sasaki, B.R. Oh, A. Dharia, S. Fujimoto and R. Dahiya, *Mol. Carcinog.*, 2000, **29**, 59–66.
36. M. Sasaki, A. Dharia, B.R. Oh, Y. Tanaka, S. Fujimoto and R. Dahiya, *Cancer Res.*, 2001, **61**, 97–102.
37. M. Sasaki, L. Kotcherguina, A. Dharia, S. Fujimoto and R. Dahiya, *Cancer Res.*, 2001, **61**, 3262–3266.
38. M. Sasaki, Y. Tanaka, G. Perinchery, G.A. Dharia, I. Kotcherguina, S. Fujimoto and R. Dahiya, *J. Natl. Cancer Inst.*, 2002, **94**, 384–390.
39. M. Sasaki, M. Kancuchi, N. Sakuragi and R. Dahiya, *Cancer Res.*, 2003, **63**, 3101–3106.
40. Y. Tsuchiya, M. Nakajima, S. Takagi, T. Taniya and T. Yokoi, *Cancer Res.*, 2006, **66**, 9090–9098.
41. D.C. Spink, B.H. Katz, M.M. Hussain, B.C. Spink, S.J. Wu, N. Liu, R. Pause and L.S. Kaminsky, *Drug Metab. Dispos.*, 2002, **30**, 262–269.
42. S. Kann, M.Y. Huang, C. Estes, J.F. Reichard, M.A. Sartor, Y. Xia and A. Puga, *Mol. Pharmacol.*, 2005, **68**, 336–346.
43. H. Chaib, E.K. Cockrell, M.A. Rubin and J.A. Macoska, *Neoplasia*, 2001, **3**, 43–52.
44. G.M. Hellmann, W.R. Fields and D.J. Doolittle, *Toxicol. Sci.*, 2000, **61**, 154–163.

45. T.C. Islam, J. Lindvall, A. Wennborg, L.J. Branden, H. Rabbani and C.I. Smith, *Eur. J. Immunol.*, 2002, **32**, 982–993.
46. M. Vondracek, D.A. Weaver, Z. Sarang, J.J. Hedberg, J.C. Willey, L. Warngard and R.C. Grafstrom, *Int. J. Cancer.*, 2002, **99**, 776–782.
47. M.C. McFadyen, H.L. McLeod, F.C. Jackson, W.T. Melvin, J. Doehmer and G.I. Murray, *Biochem. Pharmacol.*, 2001, **62**, 207–212.
48. I. Stoilov, A.N. Akarsu and M. Sarfarazi, *Hum. Mol. Genet.*, 1997, **6**, 641–647.
49. I. Stoilov, A.N. Akarsu, I. Alozie, A. Child, M. Barsoum-Homsy, M.E. Turacli, M. Or, R.A. Lewis, N. Ozdemir, G. Brice, S.G. Aktan, L. Chevrette, M. Coca-Prados M and M. Sarfarazi, *Am. J. Hum. Genet.*, 1998, **62**, 573–584.
50. M. Plasilova, I. Stoilov, M. Sarfarazi, L. Kadasi, E. Ferakova and V. Ferak, *J. Med. Genet.*, 1999, **36**, 290–294.
51. B.A. Bejjani, D.W. Stockton, R.A. Lewis, K.F. Tomey, D.K. Dueker, M. Jabak, W.F. Astle and J.R. Lupski, *Hum. Mol. Genet.*, 2000, **9**, 367–374.
52. J.T. Buters, J. Doehmer and F.J. Gonzalez, *Drug Metab. Rev.*, 1999, **31**, 437–447.
53. M. Doshi, C. Marcus, B.A. Bejjani and D.P. Edward, *Exp. Eye. Res.*, 2006, **82**, 24–32.
54. D. Choudhary, I. Jansson, I. Stoilov, M. Sarfarazi and J.B. Schenkman, *Drug Metab. Dispos.*, 2004, **32**, 840–847.
55. S. Marsh, G. Somlo, X. Li, P. Frankel, C.R. King, W.D. Shannon, H.L. McLeod, T.W. Synold, *Pharmacogenomics J.*, advance online publication, January 16, 2007; doi:10.1038/sj.tpj.6500434.
56. B. Bournique and and A. A. Lemarie, *Drug Metab. Dispos.*, 2002, **30**, 1149–1152.
57. B. Rochat, J.M. Morsman, G.I. Murray, W.D. Figg and H.L. McLeod, *J. Pharmacol. Exp. Ther.*, 2001, **296**, 537–541.
58. T.K. Chang, W.B. Lee and H.H. Ko, *Can. J. Physiol. Pharmacol.*, 2000, **78**, 874–881.
59. H.K. Crewe, L.M. Notley, R.M. Wunsch, M.S. Lennard and E.M. Gillam, *Drug Metab. Dispos.*, 2002, **30**, 869–874.
60. V. Trapani, V. Patel, C.O. Leong, H.P. Ciolino, G.C. Yeh, C. Hose, J.B. Trepel, M.F. Stevens, E.A. Sausville and A.I. Loaiza-Perez, *Br. J. Cancer*, 2003, **88**, 599–605.
61. C.O. Leong, M. Gaskell, E.A. Martin, R.T. Heydon, P.B. Farmer, M.C. Bibby, P.A. Cooper, J.A. Double, T.D. Bradshaw and M.F. Stevens, *Br. J. Cancer*, 2003, **88**, 470–477.
62. B.L. Brockdorff, J. Skouv, B.E. Reiter and A.E. Lykkesfeldt, *Int. J. Cancer*, 2000, **88**, 90–96.
63. P. Wardman, *Curr. Pharm. Des.*, 2002, **8**, 1363–1374.
64. D.W. Roberts, D.R. Doerge, M.I. ChurchwelL, G. Gamboa da Costa, M.M. Marques and W.H. Tolleson, *J. Agric. Food Chem.*, 2004, **52**, 6623–6632.
65. J. Bai, N. Sata, H. Nagai, T. Wada, K. Yoshida, H. Mano, F. Sata and R. Kishi, *Pancreas*, 2004, **29**, 93–98.

66. Y.J. Chun, S. Kim, D. Kim, S.K. Lee and F.P. Guengerich, *Cancer Res.*, 2001, **61**, 8164–8170.
67. Y.J. Chun and S. Kim, *Med. Res. Rev.*, 2003, **23**, 657–668.
68. I.H. Hanna, S. Dawling, N. Roodi, F.P. Guengerich and F.F. Parl, *Cancer Res.*, 2000, **60**, 3440–3444.
69. N.A. Kocabas, S. Sardas, S. Cholerton, A.K. Daly and A.E. Karakaya, *Arch. Toxicol.*, 2002, **76**, 643–649.
70. M. Sasaki, Y. Tanaka, M. Kaneuchi, N. Sakuragi and R. Dahiya, *Cancer Res.*, 2003, **63**, 3913–3918.
71. L.R. Bailey, N. Roodi, W.D. Dupont and F.F. Parl, *Cancer Res.*, 1998, **58**, 5038–5041.
72. A.J. De Roos, L.S. Gold, S. Wang, P. Hartge, J.R. Cerhan, W. Cozen, M. Yeager, S. Chanock, N. Rothman and R.K. Severson, *Cancer Epidemiol. Biomarkers Prev.*, 2006, **15**, 1647–1653.
73. A.H. Karypidis, T. Soderstrom, A. Nordmark, F. Granath, S. Cnattingius and A. Rane, *Fertil. Steril.*, 2006, **86**, 1498–1503.
74. N. Finnstrom, B. Ask, M.L. Dahl, M. Gadd and A. A. Rane, *Pharmacogenomics, J.*, 2002, **2**, 111–116.
75. P. Lin, S.W. Hu and T.H. Chang, *Toxicol. Sci.*, 2003, **71**, 20–26.

CHAPTER 5
The CYP2A Subfamily

HANNU RAUNIO,[1] JUKKA HAKKOLA[2] AND OLAVI PELKONEN[2]

[1] Department of Pharmacology and Toxicology, Box 1627, 70211 Kuopio, Finland (HR); [2] Department of Pharmacology and Toxicology, University of Oulu, Box 5000, 90014 Oulu, Finland

Table of Contents

Issues in Toxicology
Cytochromes P450: Role in the Metabolism and Toxicity of Drugs and other Xenobiotics
Edited by Costas Ioannides

5.1 Introduction

Among the major CYP subfamilies participating in xenobiotic metabolism, the CYP2A subfamily has received relatively little attention. This is due to the facts that the human CYP2A enzymes have very restricted substrate specificities and they are minor hepatic forms in quantitative terms. However, interest in human CYP2A forms has increased considerably after the CYP2A6 and CYP2A13 enzymes were shown to be high-affinity metabolisers of nicotine and several carcinogenic nitrosamines. In addition, the genes encoding these enzymes are highly polymorphic with marked differences in allele distribution patterns in various populations. It is thought that several alleles of the *CYP2A6* gene are linked with smoking behaviour and increased susceptibility to tobacco smoke induced cancer. In this chapter, the basic features of the CYP2A enzymes are reviewed. Special attention will be paid to the role of experimental animals as models for human risk assessment of compounds that are metabolised by enzymes in this subfamily.

5.2 Species and Tissue Distribution

Several genes have been found in the CYP2A subfamily in different species. Table 5.1 summarises the genes found to date, the corresponding proteins, tissues in which the genes are predominantly expressed, and typical substrates. CYP2A6 was purified[1] and cloned[2,3] from human liver in the early 1990s, and was found to be a high-affinity coumarin 7-hydroxylating enzyme. CYP2A6 is mainly expressed in the liver, where it represents from less than 1% to 13% of the total liver CYP content. Minor amounts of CYP2A6 are also found in the nasal mucosa, oesophagus, and lung.[4,5] CYP enzymes orthologous to CYP2A6 have been detected at relatively high concentrations in the olfactory tissues of several animal species, indicating that these enzymes could participate in the metabolism of odorants and the processing of olfactory signals.[6] CYP2A7 is a true gene that produces mRNA in human liver. However, for reasons that are not entirely clear, no active CYP2A7 protein is produced. CYP2A13 is preferentially expressed in respiratory tissues, including nasal mucosa, trachea and lung. It is an efficient catalyst of the metabolism of tobacco-related *N*-nitrosamines.[6]

Little is known about the substrate and tissue preferences of CYP2A enzymes in species other than human, such a mouse, and rat. It is clear, however, that

Table 5.1 CYP2A genes, their products, predominant tissues of expression and examples of substrates. Pseudogenes are not listed.[4,5,6,8–13]

Gene	*Protein*	*Predominant tissue*	*Substrates*
Human			
CYP2A6	CYP2A6	Liver	Coumarin, nicotine, cotinine
CYP2A7	CYP2A7	Liver	No functional protein
CYP2A13	CYP2A13	Nasal epithelium, airways	Nicotine, *N*-nitrosamines
Mouse			
Cyp2a4	CYP2A4	Liver, kidney	Testosterone, other steroids
Cyp2a5	CYP2A5	Liver, kidney, lung, nasal mucosa	Coumarin, nicotine
Cyp2a12	CYP2A12	Liver	Testosterone
Cyp2a22	CYP2A22	Not known	Not known
Rat			
CYP2A1	CYP2A1	Liver, testis	Testosterone, aminopyrine
CYP2A2	CYP2A2	Liver	Testosterone, aminopyrine
CYP2A3	CYP2A3	Lung	Testosterone, hexamethylphosphoramide
Hamster			
CYP2A8	CYP2A8	Liver	Benzphetamine
CYP2A9	CYP2A9	Liver	Testosterone
CYP2A14	CYP2A14	Liver	7-ethoxycoumarin
CYP2A15	CYP2A15	Liver	Testosterone
CYP2A16	CYP2A16	Liver	Testosterone, coumarin
CYP2A17	CYP2A17	Not known	Not known
Rabbit			
CYP2A10	CYP2A10	Olfactory mucosa	Coumarin, hexamethylphosphoramide
CYP2A11	CYP2A11	Olfactory mucosa	Hexamethylphosphoramide
Cow			
CYP2A13	CYP2A13	Olfactory mucosa, liver	Testosterone, coumarin (?)
Pig			
CYP2A19	CYP2A19	Liver	Coumarin
Rhesus monkey			
CYP2A23	CYP2A23	Liver	Not known
CYP2A24	CYP2A24	Liver	Not known
Dog			
CYP2A13	CYP2A13	Liver	Not known
CYP2A25	CYP2A25	Liver	Not known

substantial species differences exist in the catalytic activities and regulation of CYP2A enzymes. Six CYP2A genes have been found in two hamster species (Syrian hamster and Chinese hamster), two in rabbits, two in monkeys (both cynomolgus and rhesus monkeys), and one in pig and dog (Table 5.1). The orthologous relationships of these genes are unclear, but the encoded proteins have overlapping activities with human, mouse, and rat enzymes.[7] In this chapter, attention will focus mainly on the human and mouse CYP2A enzymes.

5.3 Substrate Specificity

The CYP2A enzymes can be distinguished by their preference for two types of model substrates, coumarin and 3-ketosteroids (Figure 5.1). CYP2A6 is the classical coumarin 7-hydroxylating enzyme in human liver.[4] The human CYP2A enzymes do not possess the steroid 15α- and 7α-hydroxylase activities characteristic of rodent CYP2A enzymes (Table 5.1); similarly 17β-oestradiol and oestrone are poor substrates for CYP2A6.[14] Mouse CYP2A5 and human CYP2A6 are probably true orthologues, since they are similar in their preference for coumarin 7-hydroxylation and their poor activity in steroid metabolism. Moreover, their genes are highly similar (84%) and reside at corresponding positions in the chromosomal gene clusters containing CYP2A, CYP2B, CYP2F, CYP2G, CYP2S and CYP2T subfamilies.[15]

In mice, CYP2A4 metabolises testosterone to 15α-hydroxytestosterone, whereas CYP2A5 is a coumarin 7-hydroxylase. CYP2A12 has steroid 7α-hydroxylase activity.[10] Rat CYP2A1 is also a steroid 7α-hydroxylase, but rat CYP2A2 has predominantly testosterone 15α-hydroxylase activity and has broader specificity for xenobiotics than CYP2A1. CYP2A3 was originally cloned from rat lung, and this enzyme has testosterone 15α-activity and is relatively weak in coumarin 7-hydroxylation. Negishi and co-workers[16] showed that the *Cyp2a4* gene in *Mus musculus* is absent in the closely related *Mus spretus*, suggesting that *Cyp2a5* is ancestral to *Cyp2a4*. The detection of multiple *CYP2A* loci in many species indicates that this subfamily underwent one or more duplications early in mammalian evolution.[11]

Site-directed mutational studies have been very helpful in elucidating the roles of critical amino acids in substrate specificity of CYP2A enzymes. These studies have revealed the presence of the topologically common residues, each

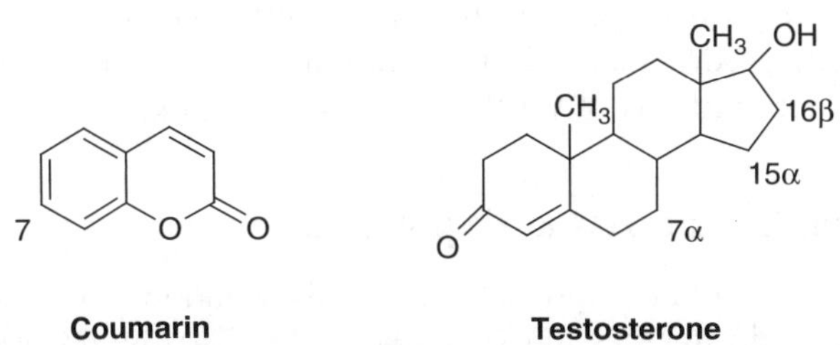

Figure 5.1 Two model substrates for CYP2A enzymes, coumarin and testosterone. The carbon atoms preferentially oxidised are shown.

of which can determine the specificities of various CYP forms. A single amino acid substitution is sufficient to alter the substrate specificity of CYP2A enzymes. Lindberg and Negishi[17] showed in their seminal paper in 1990 that the substrate specificity of mouse CYP2A5 is converted from coumarin 7- to testosterone 15α-hydroxylase activity by the substitution of Phe at position 209 with Leu. Placing Asn at this position confers a novel corticosterone 15α-hydroxylase activity to CYP2A5. Residue 209 in mouse CYP2A5 corresponds to a residue at position 206 in rat CYP2B1 that regulates its steroid hydroxylase activity.[18] These and related studies provided valuable information about the active sites of CYP2A enzymes well before the first 3-dimensional (3D) structures of mammalian CYPs were determined.

5.4 Developmental and Tissue Selective Regulation and Effect of Physiological and Pathological Factors

The regulation of most CYP2A forms is poorly characterised. Most studies on regulation have been carried out in rats and mice or cell models derived from these species. In particular, the mouse *Cyp2a5* and rat *CYP2A3* genes encoding enzymes with coumarin 7-hydroxyse activity have been studied quite extensively. The regulation of the CYP2A subfamily enzymes was recently reviewed by Su and Ding.[19]

5.4.1 Ontogenic Expression

Prenatal expression of CYP2A enzymes is low or absent in human liver and experimental animals such as mouse and rat.[20–22] Hepatic CYP2A1 expression peaks at the age of 3 weeks in rat and subsequently drops dramatically in males but is maintained at relatively high levels in females. In contrast, hepatic CYP2A2 expression in male rats reaches the highest level by the age of 14 weeks but is not detected in females at any age.[23] Shimada an co-workers[24] found low but detectable levels of CYP2A6 immunoreactive protein in a single human neonatal liver sample studied. Coumarin 7-hydroxylase activity, measured with coumarin as the *in vivo* probe drug, is similar at the ages of 6 to 13 years to the adult activity.[25] Interestingly, in human nasal mucosa CYP2A protein level is higher during the foetal period than in adulthood but, as in adults, the CYP2A13 form predominates.[26] The mechanisms governing these developmental expression patterns of CYP2A enzymes are unknown.

5.4.2 Regulation of Tissue Expression

The major sites of CYP2A expression are the liver, kidney and respiratory tract including lung and olfactory mucosa. The tissue distribution of CYP2A proteins varies greatly among species and the respective genes.[19] Tissue-selective expression of genes is regulated by the availability of a right combination of critical transcription factors and cofactors, and accessibility of the *cis*-acting binding sites in chromatin structure.[27] In liver, the hepatocyte nuclear factor 4 (HNF-4) binds to the proximal promoter of several *CYP2A* genes and contributes to, or is critical for, hepatic expression of mouse CYP2A4 and CYP2A5 and human CYP2A6.[28–31] The *CYP2A6* promoter is also activated by the octamer transcription factor-1 and CCAAT-box/enhancer binding protein α.[31] *Cyp2a5* gene expression in the liver is dependent, in addition to HNF-4, on nuclear factor 1 (NF1) interacting with a binding site at the proximal promoter.[30]

Interestingly, the NF1-binding element is conserved also in the rat *CYP2A3* gene, which is expressed at high levels in olfactory mucosa but is not in liver. Several reports by Ding and co-workers indicate that this NF1 element (also termed the nasal predominant transcriptional activating (NPTA) element) plays an important role in tissue distribution of CYP2A3.[32–34] This element was found to bind members of the NF1 family in olfactory mucosa but not in liver, which may be due to the more extensive methylation of the *CYP2A3* gene in liver than in olfactory mucosa.[32,33] Furthermore, in *CYP2A3* transgenic mice the tissue distribution of CYP2A3 mimicked that of CYP2A5, suggesting that the differential expression of the corresponding genes is not determined by simple sequence differences in their promoters.[34]

5.4.3 Regulation by Physiological Factors

Many of the *CYP2A* genes are regulated by a number of physiological conditions such as gender, circadian rhythm and nutrition. In mouse and rat liver, some CYP2A forms are expressed sex dependently. Rat CYP2A2 expression is male specific while rat CYP2A1 and mouse CYP2A4 and CYP2A5 are expressed predominantly, but not exclusively, in female liver.[35,36] Growth hormone (GH) secretion pattern is a major regulator of many sex dependent genes and male specific expression of several CYP genes is dependent on pulsatile GH excretion and consequent activation and nuclear accumulation of the transcription factor signal transducer and activator of transcription 5B (STAT5b).[35] In contrast, constant GH level increases female and decreases male-specific genes.[37] Expression of the female predominant *Cyp2a4* gene was found to be dependent on constant GH secretion and subcellular localisation of STAT5b, which in turn appear to be regulated by oestrogen receptor-α.[38] Interestingly, CYP2A2 expression is not dependent on GH but remains high in hypophysectomised rat liver.[35] In contrast, CYP2A2 is regulated by the sex-dependent expression pattern of liver enriched transcription factors.[39] Expression of the human CYP2A6 in liver has been reported to be similar in both sexes.[36,40] On the other hand, nicotine metabolism, mainly mediated by CYP2A6 (*vide infra*), is faster in women than men.[41]

The mouse *Cyp2a4* and *Cyp2a5* genes display circadian expression rhythms in liver with highest levels of the proteins detected in the evening.[42,43] The circadian rhythm of these genes is mainly determined by rhythmic expression of the liver enriched transcription factor D-site-binding protein that interacts with the regulatory regions of the *Cyp2a4* and *Cyp2a5* genes.[42]

Many CYP enzymes are regulated by nutritional status. Fasting, in particular, has been shown to induce some CYP2A forms. Rat liver CYP2A1 and the associated testosterone 7α-hydroxylase activity is induced 2 to 3-fold by 48–72 hours of fasting.[44] Similarly, mouse liver CYP2A5 mRNA is induced 2-fold after 48 hours of fasting.[45] CYP2A5 is also induced by glucagon and its second messenger cAMP suggesting that the effects of starvation on CYP2A5 may be mediated by changes in circulating glucagon levels.[43]

5.4.4 Impact of Disease on CYP2A Liver Expression

Unlike most other CYP enzymes, some CYP2As are up-regulated in certain disease conditions. Mouse CYP2A5 mRNA and enzyme activity are elevated in mouse hepatomas.[46] Furthermore, some parasitic or viral infections causing chronic inflammation in liver tissue were found to be associated with high CYP2A expression in mouse and hamster liver.[47] Similarly, bacterial infection with *Citrobacter rodentium* caused induction of hepatic CYP2A5.[48] Analogously, human CYP2A6 was found to be induced in liver by *O. viverrini* infestation and in hepatocytes expressing the hepatitis B virus (HBV) surface antigen.[49] In hepatocellular carcinomas CYP2A6 expression was heterogenous among the tumour cells and high levels were present in some cells associated with chronic inflammation and cirrhosis.[50] It is not clear whether induction of CYP2A enzymes in these disease conditions is beneficial or harmful. It has been suggested that CYP2A5 induction by *Helicobacter hepaticus* may contribute to oxidative damage in progressing disease.[51] While this may be true, studies with toxic compounds suggest that oxidative stress could actually be the cause of induction.[52,53]

5.5 Regulation by Xenobiotics

Several CYP2A enzymes are up-regulated by xenobiotics, but for most CYP2A forms our current knowledge on induction patterns is quite limited. Mouse CYP2A5 has been studied most extensively with respect to regulation by foreign compounds. The *Cyp2a5* gene sequence, including its 5′ flanking region, is very similar to *Cyp2a4*, and are probably regulated by the same factors. However, many studies have not analysed these genes separately and, therefore, in some cases it is not certain if only one or both of these genes were affected. CYP2A5 mRNA, levels or the related coumarin 7-hydroxylase activity, has been reported to be induced by numerous structurally unrelated compounds with both transcriptional and post-transcriptional mechanisms being involved (Table 5.2).

Induction of *Cyp2a5* by the classical nuclear receptor CAR (constitutive androstane receptor) activators phenobarbital and TCPOBOP suggest involvement of CAR in the regulation of *Cyp2a5*. This is confirmed by lack of

Table 5.2 Xenobiotics inducing CYP2A5 mRNA or enzyme activity.

Class	*Compound*	*Reference*
Nuclear receptor/Aryl hydrocarbon receptor ligands	phenobarbital	64
	TCPOBOP	65
	rifampicin	66
	PCN	56
	TCDD	58
	3-methylcholanthrene	58
Heavy metals	Cobalt	67
	indium	68
	cerium	69
	tin	70
	cadmium	53
Porphyrinogenic agents	thioacetamide	71
	griseofulvin	71
	aminotriazole	71
Hepatotoxins	pyrazole	72
	cocaine	73
	carbon tetrachloride	74
	hexachlorobutadiene	74

TCPOBOP, 1,4-bis[2-(3,5-dichloropyridyl]benzene; PCN, pregnelonone 16α-carbonitrile; TCDD, 2,3,7,8-tetrachlorodi-benzo-p-dioxin.

Cyp2a4/5 induction by these compounds in the CAR null mouse.[54,55] However, the CAR binding motif in the *Cyp2a4/5* sequence or the precise mechanism of regulation has not been established. It is not clear if PXR (pregnane X receptor) regulates *Cyp2a4/5*. While Cai and co-workers[56] reported *Cyp2a5* induction by the PXR ligand PCN, a similar induction was not observed in another study.[54] Interestingly, human CYP2A6 was reported to be induced by PXR only in the presence of the co-activator PGC-1α (peroxisome proliferator-activated receptor-γ coactivator 1α).[57] The AHR (aryl hydrocarbon receptor) binds to the *Cyp2a5* distal promoter and mediates induction of *Cyp2a5* by TCDD and 3-methylcholanthrene;[58] similarly, the hamster *CYP2A8* gene is regulated by the AHR.[59]

Mouse CYP2A5 is induced by several heavy metals and other hepatotoxic compounds that usually down-regulate most CYP enzymes, but the mechanism of this induction is unclear. One hypothesis is that induction is mediated by altered cellular redox status and endoplasmic reticulum stress.[52] In line with this hypothesis, CYP2A5 induction by cadmium has been reported to be dependent on the stress-related transcription factor Nrf2 (Nuclear factor (erythroid-derived 2)-like 2).[53]

Induction of CYP2A5 by pyrazole is mediated by a posttranscriptional mechanism.[60] This process has been studied in detail and shown to be mediated by cytoplasmic accumulation and binding of heterogeneous nuclear ribonucleoprotein A1 (hnRNPA1) to the 3′ untranslated region of the CYP2A5 mRNA.[61,62] hnRNPA1 interacts also with the human CYP2A6 mRNA 3′ untranslated region suggesting that similar regulatory mechanisms exist also for CYP2A6.[63]

5.6 Structure of the Human CYP2A6 Enzyme

5.6.1 Molecular Modelling

Substrates of the CYP2A6 enzyme are in general relatively small, neutral or basic molecules containing one aromatic ring.[75] Different modelling (in silico) approaches have been used to elucidate the structure and function of CYP enzymes, including members in the CYP2A subfamily.[76,77] Lewis and co-workers.[78–80] pioneered the use of comparative (homology) models based on crystal structures of bacterial or other mammalian CYP forms to study CYP2A1, CYP2A4, CYP2A5 and CYP2A6 proteins. In their recent study,[81] the findings from comparative modelling with the rabbit CYP2C5 template were in agreement with the known metabolism of coumarin by human CYP forms in the CYP1, CYP2 and CYP3 families. Other groups have used similar approaches to study the active sites of CYP2A proteins.[82–84]

Different types of quantitative structure activity relationship (QSAR) methods have been also used to elucidate CYP2A enzyme function.[80,85] These studies have shown that, in general, there is a good correlation between substrate/inhibitor binding affinities and various parameters relating to active site interactions such as hydrogen bonding, π-π stacking and compound lipophilicity. Our laboratory has used a 3D-QSAR method, the Comparative Molecular Field Analysis (CoMFA), for modelling mouse CYP2A5 and human CYP2A6 enzymes. A pilot model suggested that the CYP2A5 binding site may be larger than that of CYP2A6 due to larger steric regions in the CYP2A5 CoMFA coefficient maps.[86] Another CoMFA model predicted relatively well the interaction of naphthalene with CYP2A5 and CYP2A6 active sites.[87] The inhibitory potencies of 26 naphthalene and 16 non-naphthalene derivatives were determined for human CYP2A6 and mouse CYP2A5 enzymes, and several CoMFA models were developed to ascertain the steric and electrostatic properties required for potent inhibition. The generated CoMFA models were able to accurately predict the inhibition potencies of an external test set of chemicals.[88]

The structure of CYP2A6 protein has been recently determined by X-ray crystallography.[89] The CYP2A6 catalytic domain has the canonical fold of mammalian CYPs, consisting of 16 α-helices and 4 β-sheets. The active site volume is less than one-fourth of the volumes of CYP2C9 and CYP3A4. The hydrophobic active site contains one hydrogen bond donor (Asn297) that orients coumarin for stereoselective oxidation. The closed active site cavity complements the size, shape and hydrophobicity of coumarin and methoxsalen and favours substrates that can adopt nearly planar conformations.

Recently 3D-QSAR models of CYP2A6 were constructed in our laboratory based on homology modelling and the CYP2A6 crystal structure to identify the structural characteristics that are important for inhibition of the enzyme.[90] Specifically located hydrophobic and hydrogen donor features were found to affect inhibition potency. A virtual screening procedure was used to screen over 60 000 compounds in the Maybridge chemical database. Twenty-two candidate molecules were selected and four of these were found out to be potent and selective CYP2A6 inhibitors. They represent novel structures of CYP2A6 inhibitors, especially N1-(4-fluorophenyl)cyclopropane-1-carboxamide and 4- and 2-halogen substituted benzylamine derivatives.

5.6.2 Mutagenesis Studies

Recently, some laboratories have started using random and site-directed mutagenesis techniques to gain more insight into the function of the CYP2A6 active site, and to expand its substrate specificity. In CYP2A6, the amino acid residues at position 117 and 372 are valine and arginine, respectively. Kinetic analysis revealed that the catalytic efficiency (V_{max}/K_m) of the CYP2A6 mutants Val117Ala and Arg372His was drastically reduced. These observations demonstrate that the Val at position 117 and Arg at position 372 are critical amino acid residues for coumarin 7-hydroxylation.[91]

Assessment of coumarin 7-hydroxylating activity of CYP2A6 after random mutagenesis revealed that mutants throughout the coding region can influence activity. The major reason for decreased catalytic activity was lower binding affinities.[92] A double mutant (Asn297Gln/Ile300Val) was able to oxidise bulky derivatives of indole to coloured products. This is a step towards using mammalian CYP enzymes as biocatalysts for fine chemical production.[93]

5.7 Pharmacogenetics of Human CYP2A Genes

The human *CYP2* gene cluster on chromosome 19 is highly complex, containing six distinct subfamilies.[11] These subfamilies, *CYP2A*, *CYP2B*, *CYP2F*, *CYP2G*, *CYP2S*, and *CYP2T*, span approximately 500 kb on 19q13.2 and are mixed together rather than in distinct subfamily clusters. The *CYP2A* subfamily is overall most similar to the *CYP2G* subfamily followed by the *CYP2B* subfamily. The transcribed genes within the *CYP2A* subfamily share very high genomic sequence similarity. *CYP2A6* is 97% identical to *CYP2A7* for exonic sequence, and 85% identical to *CYP2A13*, while *CYP2A7* and *CYP2A13* are 90% identical.

The *CYP2A6* and *CYP2A13* genes are highly polymorphic. Initially, the lability of the *CYP2A* locus created problems for defining *CYP2A6* and *CYP2A13* alleles accurately.[11] These problems have largely been overcome by the detailed elucidation of the *CYP2A* locus and improved genotyping assays.

In November 2006, the Human CYP Allele Nomenclature Committee homepage[94] listed 50 alleles or predicted haplotypes of the *CYP2A6* gene. Several additional SNPs (single nucleotide polymorphisms) have been found in

Table 5.3 *CYP2A6* alleles affecting CYP2A6 function.

Gene	*Change in cDNA*	*Activity*
*CYP2A6*1A*	Wild-type	Normal
*CYP2A6*1X2*	Gene duplication	Increased activity
*CYP2A6*2*	479T>A	No activity
*CYP2A6*4*	Gene deletion (four types)	No activity
*CYP2A6*5*	1436G>T, gene conversion in the 3′ region	No activity
*CYP2A6*7*	1412T>C, gene conversion in the 3′ region	Reduced activity
*CYP2A6*9*	SNP in the TATA box	Reduced activity
*CYP2A6*10*	1412T>C, 1454G>T, gene conversion in the 3′ region	Reduced activity
*CYP2A6*11*	670T>C	Reduced activity
*CYP2A6*17*	459G>A, 1093G>A, 1224C>T	Reduced activity
*CYP2A6*19*	1175A>T, 1412T>C, gene conversion in the 3′ region	Reduced activity
*CYP2A6*20*	Frameshift SNP	No activity

the gene, but the haplotypes have not yet been determined. The most important alleles that affect the catalytic activity of CYP2A6 protein are listed in Table 5.3.

The combined frequencies of the alleles lacking or showing reduced enzymatic activity (CYP2A6*2, CYP2A6*4, CYP2A6*5, CYP2A6*7, CYP2A6*9, CYP2A6*10, CYP2A6*11, CYP2A6*17, CYP2A6*19, and CYP2A6*20) were 9.1%, 21.9%, 42.9%, and 50.5% in Caucasian, African, Korean and Japanese subjects, respectively.[95] Twenty different alleles of *CYP2A13* were identified up to November 2006. At least one allele (*CYP2A13*4*) encoded a functionally deficient protein.[96]

5.8 Role in Drug Metabolism

5.8.1 Drug Substrates

CYP2A6 participates in the metabolism of several compounds that, for practical purposes, may be categorised as pharmaceuticals and toxic compounds. An exhaustive list of CYP2A6 substrates and inhibitors is provided in the review by Rendic.[97] It should be stressed, however, that in the majority of cases CYP2A6 catalyses a minor metabolic pathway, or is a high K_m, low V_{max} enzyme for the reaction contributing negligibly to the metabolism of the particular drug at relevant substrate concentrations. There are actually very few drug substrates that are cleared solely or mainly by CYP2A6, coumarin being an exception. In the same way the known inhibitors of CYP2A6 are largely non-specific. Therefore, only a short list of the most relevant drug substrates is provided here (Table 5.4).

The relatively minor role of CYP2A6 in the metabolism of pharmaceuticals becomes apparent when compared with the CYP2D6 and especially CYP3A4,

Table 5.4 Selected CYP2A6 drug substrates.[5,11,97,103]

CYP2A6 substrate	*Therapeutic class or indication*
Coumarin	constituent in many herbal preparations
Nicotine	smoking withdrawal preparations
Methoxyflurane	anaesthetic
Halothane	anaesthetic
Losigamone	antiepileptic
Valproic acid	antiepileptic
Letrozole	aromatase inhibitor
Disulfiram	alcohol withdrawal
Tegafur	anticancer drug
Doxifluridine	anticancer drug
SM-12502	experimental antithrombotic drug

SM-12502, 3,5-dimethyl-2-(3-pyridyl)thiazolidin-4-one hydrochloride (an antagonist of platelet activating factor, not in clinical use).

which are responsible for the metabolism of several dozens or even hundreds of clinically used drugs. Of the substrates listed in Table 5.4, the most selective ones for CYP2A6 are coumarin and nicotine. Most of the other pharmaceuticals listed in this table are only partially metabolised by CYP2A6, and in most cases the kinetics of metabolism have not been studied thoroughly.

Coumarin is a naturally occurring compound which is contained in a large number of herbal medicinal products used for various disorders in some European countries. Both *in vitro* and *in vivo* studies established that coumarin is very specifically metabolised by CYP2A6 and, consequently, it is the preferred probe drug for phenotyping CYP2A6 activity. In drug development the potential for a new drug to modify CYP activities is routinely evaluated by determining the effect of the drug on *in vitro* probe reactions that reflect the activity of specific CYP enzymes. The *in vitro* findings obtained with one probe substrate are usually extrapolated to the compound's potential to affect all substrates of the same enzyme. Due to this practice, it is important to use the right probe substrate and to conduct the experimental studies under optimal conditions.[98] Surveys conducted by the Food and Drug Administration (FDA) indicated that coumarin 7-hydroxylation is among the most common *in vitro* probe reactions used in industry.[99]

Due to its very low abundance in human liver, CYP2A13 does not play a major role in the systemic clearance of any drug. CYP2A13 may, however, be a major metabolising enzyme in tissues where it is preferentially expressed, especially the nasal mucosa.

5.8.2 Inhibitors

Several compounds have been shown to inhibit CYP2A6 activity. While the majority have been assessed only *in vitro*, some agents, including some clinically used drugs, have been evaluated also *in vivo* (Table 5.5). Methoxsalen, a drug

Table 5.5 Selected CYP2A6 inhibitors.

CYP2A6 inhibitor	*Reference*
Drugs	
Methoxsalen	4
Tranylcypromine	4
Pilocarpine	4
Other compounds	
Organosulphur compounds	104
4-methylbenzaldehyde	105
Menthol	106,107
Naphthalene derivatives	88
Grapefruit juice	108
Nicotine derivatives	109

used in the treatment of psoriasis, is thus far the most potent inhibitor of CYP2A6. At low concentrations methoxsalen is also fairly selective, inhibiting preferentially CYP2A6 over other CYP forms. Tranylcypromine, a monoamine oxidase inhibitor previously used in the treatment of depression, is a competitive inhibitor of CYP2A6 with poor selectivity. Pilocarpine is a cholinergic alkaloid that competitively inhibits CYP2A6 with a high affinity (K_i 1.2 μM).[4]

Much less is known about agents than inhibit CYP2A13 activity. Methoxsalen, in addition to CYP2A6,[100] inhibits also CYP2A13 potently. Benzyl isothiocyanate and phenethyl isothiocyanate inhibit both CYP2A6 and CYP2A13 by a mechanism involving enzyme inactivation.[101] In a recent study,[102] it was shown that a metabolite of nicotine is a mechanism-based inhibitor of CYP2A13, but the reactive species responsible has not been identified yet.

5.9 Role in Xenobiotic Toxicity

5.9.1 Toxic Substrates

Whereas CYP2A6 and CYP2A13 are known to metabolise only a few clinically used drugs, a greater number of compounds of toxicological significance are either activated or deactivated by these enzymes. Examples of such compounds are listed in Table 5.6. The role of CYP2A enzymes in the metabolism of nicotine and nitrosamines will be discussed in more detail.

5.9.2 Nicotine

Tobacco smoking related diseases are a major health problem worldwide. It has been estimated that about one-third of the global population over 15 years of age smokes. Smoking is associated with a higher incidence of several types of cancer, respiratory and cardiovascular diseases, gastrointestinal disorders, as well as many other medical complications.[114] Nicotine is responsible for

Table 5.6 Examples of toxic compounds metabolised by CYP2A6 and CYP2A13.[4,5,6,110–113]

Compound	*Assay/end-point*
CYP2A6	
Nicotine	*C*-Oxidation
Cotinine	3′-Hydroxylation
NNK	α-Hydroxylation, mutagenicity
NDEA	Dealkylation, mutagenicity
AFB_1	Epoxide formation, mutagenicity
MOCA	*N*-oxidation
1,3-butadiene	Monoxide formation
Quinoline	1-Oxidation
DCBN	Protein adduct formation
MTBE	*O*-demethylation
Chloroform	Trichloromethanol formation
1,7-dimethylxanthine	8-hydroxylation
Indole	Indoxyl formation
CYP2A13	
Nicotine	*C*-Oxidation
Cotinine	3′-Hydroxylation
NNK	α-Hydroxylation
4-aminobiphenyl	*N*-Hydroxylation, mutagenicity
Aflatoxin B_1	Epoxide formation
Phenacetin	*O*-deethylation

NNK, 4-methylnitrosoamino-1-(3-pyridyl)-1-butanone; NDEA, *N*-nitrosodiethylamine; AFB_1, aflatoxin B_1; MOCA, 4,4′-methylene-bis(2-chloroaniline); DCBN, 2,6-dichlorobenzonitrile; MTBE, methyl tert-butyl ether.

establishing and maintaining tobacco dependence. It has been demonstrated that nicotine-dependent smokers smoke in a fashion that maintains peripheral and central nicotine levels. Smokers titrate their cigarette consumption to maintain steady levels of nicotine in the brain. Factors influencing nicotine plasma levels, by either intake or removal, affect smoking behaviour.[115]

Nicotine is extensively metabolised in the body – about 80% of nicotine is metabolised to the inactive cotinine. Studies with expressed CYP enzymes and human liver microsomes have revealed the key role of CYP2A6 in the catabolism of nicotine.[116,117] In this two-step reaction, nicotine is first oxidised to nicotine iminium ion, and subsequently to cotinine catalysed by the cytosolic aldehyde oxidase (Figure 5.2). The CYP2A6-mediated formation of nicotine iminium ion from nicotine is the rate-limiting step in this reaction. Further downstream of this metabolic cascade, cotinine is oxidised to several metabolites by CYP enzymes, including possibly CYP2A6. CYP2A6 appears to be the sole catalyst of nicotine catabolism at low, physiologically relevant nicotine concentrations ($<50\,\mu M$).[118] CYP2A13 is also an efficient catalyst of the conversion of nicotine to cotinine.[110]

Genetic studies have indicated that individuals with deficient *CYP2A6* alleles may have slower rates of nicotine inactivation.[119,120] Several studies have shown

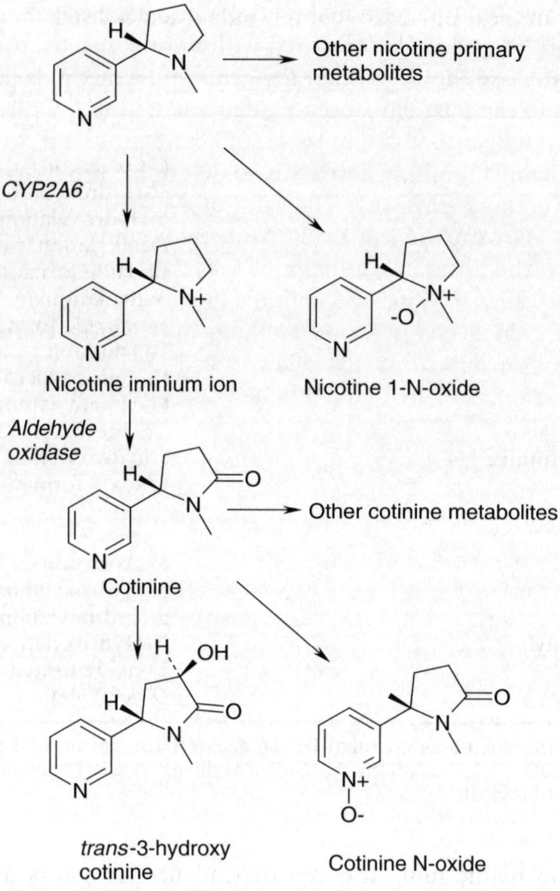

Figure 5.2 Major metabolic pathways of nicotine in humans. CYP2A6 is the major catalysing enzyme of the oxidation of nicotine to nicotine iminium ion, which is further oxidised to cotinine by aldehyde oxidase.

that slow nicotine inactivators, those with less than 50% activity *in vivo*, are approximately two times less likely to be current adult smokers, and they smoke 7–10 cigarettes less per day than people with normal metabolic rates. Fast metabolisers, with *CYP2A6* gene duplications, smoke more. Moreover, slow nicotine inactivators smoke for a shorter duration before quitting, are found more frequently among former smokers, and have increased success in quitting during clinical trials.[121] Thus, slow nicotine inactivation influences smoking behaviour and nicotine dependence. The effect of CYP2A6, however, appears to vary along the smoking continuum, and is affected by factors such as status of nicotine dependence and duration of smoking. During adolescence, having a slow inactivator genotype increases the risk for nicotine dependence while still reducing the amount smoked. From what is known about development of nicotine dependence, it is plausible that prolonged elevated nicotine levels in the brain, as a result of

CYP2A6 slow metabolism, may make adolescent poor metabolisers more vulnerable to nicotine dependence compared with normal inactivators.[121]

It should be stressed, however, that the genotyping methods used to elucidate *CYP2A6* alleles in the past have been inadequate and many rare and unknown genotypes have not been assessed in these studies. A more direct way of assessing CYP2A6 nicotine-metabolising activity is to determine the systemic clearance of nicotine *via* the cotinine pathway. The ratio of 3-hydroxycotinine to cotinine in blood has been shown to reflect CYP2A6 activity and oral clearance of nicotine.[122,123] Using this ratio as an index of CYP2A6 activity, Benowitz and co-workers[41] showed that nicotine and cotinine metabolism is faster in women than in men, and is even faster in women taking oral contraceptives. Accelerated nicotine metabolism appears to be due to oestrogen. A large epidemiological study[124] showed that the 3-hydroxycotinine to cotinine ratio was associated with sex (female > male), CYP2A6 genotype (*1B, *2, *4, *9, and *12), CYP2B6 haplotype (*4-dominant), plasma nicotine concentration, and age.

Of the current pharmacological treatments for nicotine dependence, nicotine replacement therapy is the most effective. Due to the extensive first-pass metabolism of orally ingested nicotine, it is administered *via* parenteral routes, as nasal spray, transdermal patches, lozenges or chewing gum in replacement therapy.[125] The rate of nicotine metabolism, as indicated by the 3-hydroxycotinine to cotinine ratio, could predict the effectiveness of transdermal nicotine in smoking cessation therapy.[126] In this study, individuals with higher metabolic ratios, reflecting more rapid metabolism of nicotine, were less successful in quitting smoking than individuals with low metabolic ratios.

The decreasing effect of slow CYP2A6 metabolism on smoking can be mimicked by inhibiting CYP2A6 activity *in vivo*. Proof-of-concept studies have been carried out with the CYP2A6 inhibitors methoxsalen and tranylcypromine. Tyndale, Sellers, and co-workers[127–129] have proposed that CYP2A6 inhibition as a treatment for tobacco dependence can: 1) facilitate oral nicotine as a form of nicotine replacement therapy, 2) reduce exposure to the harmful components of tobacco smoke, 3) enhance the efficacy of nicotine gum, nicotine patches and other nicotine formulations, and 4) decrease the load of carcinogens in the body. The prospect that CYP2A6 inhibitors could be used as drugs in nicotine replacement therapy has prompted efforts to devise novel inhibitor molecules. Examples of novel inhibitor structures include naphthalene[88] and nicotine derivatives.[109] To test new inhibitors *in vivo*, mouse is the preferred model since CYP2A5 catalyses nicotine to cotinine formation.[130,131]

5.9.3 *N*-Nitrosamines

There is extensive literature on the role of CYP enzymes in the metabolism of *N*-nitrosamines. Discussion will be limited to the CYP2A- mediated metabolism of the tobacco-specific *N*-nitrosamines, especially 4-(methylnitrosamino)-1-(3-pyridyl)-1-butanone (NNK), one of the most important carcinogens in tobacco smoke. NNK requires metabolic activation to form adducts with

DNA. DNA adducts are formed exclusively through CYP-catalysed hydroxylation reactions at the methylene and methyl carbons adjacent to the *N*-nitroso group (α-hydroxylation reactions). The principal metabolic pathways of NNK are shown in Figure 5.3.

It has become evident that CYP2A enzymes in different species are are excellent catalysts of NNK α-hydroxylation. Among the CYPs that have been studied, human CYP2A13 and CYP2B6, rabbit CYP2A10/11, rat CYP2A3, and mouse CYP2A5 exhibit the lowest K_m values and may thus be the most important catalysts of NNK bioactivation in the respective species (Table 5.7). CYP2A13 may be particularly important in the activation of NNK in human lung.[19,132]

In all tissues, a significant proportion of NNK is metabolised to NNAL, which is also a potent lung carcinogen.[133] NNAL is metabolically activated by CYP-mediated α-hydroxylation, with CYP2A3, CYP2A5, and CYP2A13 being efficient

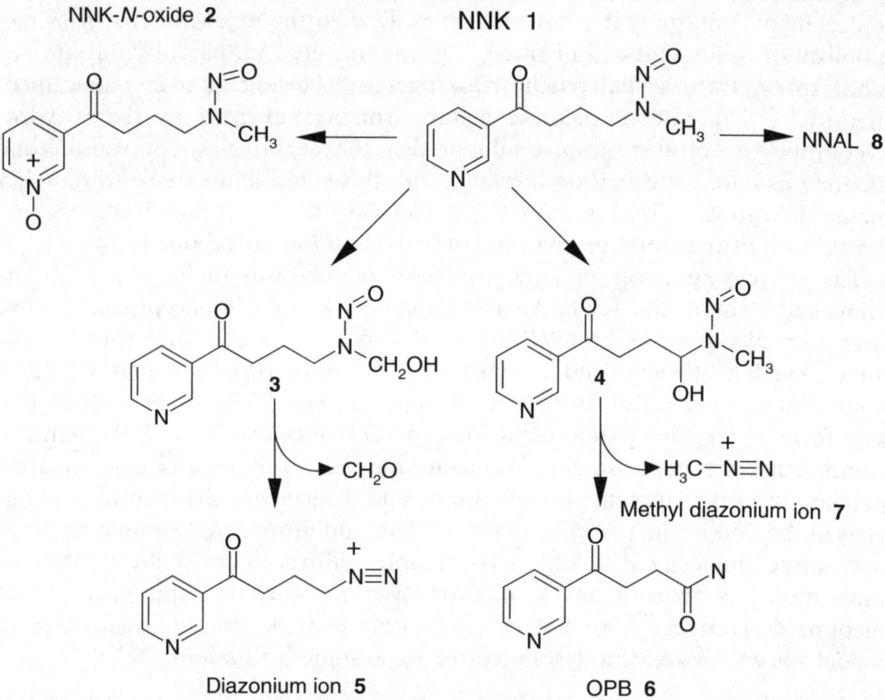

Figure 5.3 The principal metabolic pathways of NNK.[132,133] α-Hydroxylation of the methylene carbon adjacent to the *N*-nitroso group produces an unstable α-hydroxynitrosamine (**4**) that spontaneously decomposes to the methyldiazonium ion (**7**) and 4-oxo-4-(3-pyridyl)butanal (OPB **6**). α-Hydroxylation of the methyl carbon leads to an unstable α-hydroxynitrosamine (**3**) that decomposes to 4-oxo-4-(3-pyridyl)-1-butanediazonium ion (**5**) and formaldehyde. Detoxification of NNK occurs, in part, by pyridine *N*-oxidation, yielding *N*-oxides (**2**). Carbonyl reduction of NNK produces 4-(methylnitrosamino)-1-(3-pyridyl)-1-butanol (NNAL **8**).

Table 5.7 Summary of K_m values for CYP-mediated NNK metabolism.[132]

Metabolic pathway	*Low K_m (<50 μM)*	*Intermediate K_m (50-500 μM)*	*High K_m (>500 μM)*
α-methylene hydroxylation	human CYP2A13	Human CYP2A6	human CYP1A1/1A2
	rabbit CYP2A10/2A11	Rabbit CYP2G1	human CYP2D6
	rat CYP2A3	rat CYP1A1/1A2	human CYP2E1
	mouse CYP2A5	rat CYP2B1	human CYP3A4
		mouse CYP2A4	rat CYP2C6
α-methyl hydroxylation	human CYP2A13	human CYP1A1/1A2	human CYP2D6
	human CYP2B6	human CYP2A6	human CYP2E1
	rabbit CYP2A10/2A11	rat CYP1A1/1A2	human CYP3A4
	rat CYP2A3	rat CYP2B1	rat CYP2C6
	mouse CYP2A5	mouse CYP2A4	
N-oxidation		rat CYP2B1	rat CYP2C6

mediators of this reaction.[132] Kamataki and co-workers[134] studied the role of human CYP forms in the metabolic activation of tobacco-related N-nitrosamines to mutagenic products in the Salmonella mutation test. CYP2A6 was responsible for the mutagenic activation of essentially all tobacco-related N-nitrosamines studied (NNK, N-nitrosodiethylamine, N-nitrosopyrrolidine, N-nitrosopiperidine, N-nitrosonornicotine, N-nitrosoanabasine, and N-nitrosoanatabine).

5.10 CYP2A6 Polymorphism and Cancer

5.10.1 Lung Cancer

Soon after the discovery of variant alleles of the *CYP2A6* gene, it was hypothesised that CYP2A6 slow metabolisers could be protected from tobacco smoke-induced lung cancer. The first population studies indicated that this could indeed be the case.[135,136] However, subsequent studies in different populations yielded conflicting results.[137] A meta-analysis[138] of 11 studies failed to find any empirical evidence of a relationship between variant *CYP2A6* alleles and smoking status (n = 4091) or cigarette consumption (n = 1537). The meta-analysis thus suggested the *CYP2A6* gene is not associated with smoking behaviour. However, the use of broad smoking status classifications (e.g., ever- *vs.* never-smoking), which fail to account for the complex nature of gene expression (e.g., gene-gene interactions), may have obscured the relatively modest genetic influences that might have been present.

Nevertheless, an association between *CYP2A6* polymorphism and tobacco-related cancer risk is apparent in populations with a high frequency of CYP2A6 slow metabolisers, especially among the Japanese. Epidemiological studies in Japan have shown that smokers homozygous for the *CYP2A6*4C* allele have significantly lowered odds ratios for cancer risk. Other alleles reducing CYP2A6 activity also reduce the risk of lung cancer in smokers, particularly the smoking-related squamous cell carcinoma and small cell carcinoma.[139,140] CYP2A6 also activates the betel quid-specific N-nitrosamines[141] and interestingly, *CYP2A6* gene deletion was shown to reduce oral cancer risk in betel quid chewers in Sri Lanka.[142]

Growing experimental evidence supports the involvement of CYP2A enzymes in lung cancer initiation. In mouse models, methoxsalen blocks very efficiently mutagenesis of the *gpt delta* marker gene and lung tumour formation caused by NNK.[143,144]

5.10.2 Other Cancers

The human oesophagus contains CYP2A6, especially localised in the most frequent sites where oesophageal tumours arise.[145] CYP2A6 activates the potent oesophageal carcinogen *N*-nitrosobenzylmethylamine (NBzMA).[146] An excess risk of oesophageal cancer was shown related to the *CYP2A6*4* allele among non-smokers in a Chinese population.[147]

In a Japanese study,[148] the number of subjects with a homozygous *CYP2A6* gene deletion allele was significantly higher among gastric cancer patients than in healthy controls. Subdividing gastric adenocarcinoma according to tumour differentiation, patients with the well-differentiated type were 4.9-fold more likely to have the *CYP2A6* homozygote deletion genotype.

Dietary nitrosamines activated by CYP2A6 represent a potential aetiologic factor for colorectal cancer. In a case-control study[149] the distribution of CYP2A6 activity was significantly different between colorectal cancer cases and controls, with subjects in the medium and high activity groups having an increased risk. In another study,[150] inheritance of the *CYP2A6*2* allele was associated with reduced risk of colorectal cancer.

Resent studies have shown that CYP2A13 activates the carcinogens 4-aminobiphenyl[111] and aflatoxin B_1 (AFB1).[112] AFB_1 is converted by CYP2A13 to the carcinogenic AFB_1-8,9-epoxide and, in addition, to the AFM_1-8,9-epoxide. Enzyme kinetic and cytotoxicity studies indicated that CYP2A13 is more efficient than CYP2A6 in AFB_1 activation.[112]

5.11 Implications for Risk Assessment – Coumarin

Species differences in xenobiotic metabolism have a marked impact on the use of experimental animals as predictive models for humans. Risk assessment of coumarin is a good example of practical difficulties arising from species differences in the CYP2A enzymes. Coumarin is a naturally-occurring compound present in many plants, meteor including peppermint, cinnamon, sweet clover and lavender. It has been used extensively as a fixative and enhancing agent in perfumes, soaps, toothpaste, tobacco products and some alcoholic beverages.[151] In the past, coumarins have also been used to treat diverse disorders such as cancer, lymphooedema, venous insufficiency, brucellosis, and rheumatic disease. Coumarin derivatives, especially warfarin, are the most common oral anticoagulants used today. Exposure ranges from approximately 10 mg/day resulting from consumption of natural food ingredients up to 7 g/day following clinical administration.[152]

In the mid 1950s coumarin was shown to be hepatotoxic in rats and dogs. Based on these reports the use of coumarin as a food additive was banned by the FDA in 1954. Subsequent studies in the 1960s by the FDA confirmed the hepatotoxic effects in rats and dogs, but no evidence was found of coumarin-induced tumours.[151] Chronic oral bioassays carried out in the 1990s reported liver tumours in rats and mice, and lung tumours in mice, raising additional concerns about the safety of coumarin. In 1988, because of its potential hepatotoxic effects in humans, the European Commission restricted coumarin as a direct food additive to 2 mg/kg food/day. Rats were shown to be particularly susceptible to the hepatic effects, whereas lung was the major target organ in mice. Since then, extensive of research effort has focussed on understanding the aetiology of these tumours. The data support the conclusion that coumarin is not genotoxic, *i.e.* it does not react with DNA, and that cytotoxicity and regenerative hyperplasia play a major role in tumour formation.[153]

It became evident in the early 1970s that the metabolic fate of coumarin differs markedly between laboratory animals and humans, and doubts were being expressed as to the suitability of the rat as a test species in assessing the risk of coumarin to man. Indeed, it has become abundantly clear that routes of metabolism have a dramatic influence on the hepatotoxic effects of coumarin. The species-specific target organ toxicity was related to the differences in coumarin metabolism.[151] The metabolism of coumarin has been investigated in a wide range of animals species, and a simplified scheme is shown in Figure 5.4. Coumarin can be hydroxylated at all six possible carbon atoms to yield 3-, 4-, 5-, 6-, 7-, and 8-hydroxycoumarins. Opening of the lactone ring produces various metabolites including *o*-hydroxyphenylacetaldehyde (*o*-HPA), *o*-hydroxyphenylethanol (*o*-HPE), and *o*-hydroxyphenylacetic acid (*o*-HPAA). The two most important pathways are 7-hydroxylation and ring opening. The first step in the ring opening pathway is the formation of a coumarin 3,4-epoxide intermediate, which is rapidly degraded with the loss of carbon dioxide to form *o*-HPA.[153,154]

Several lines of evidence suggest that 7-hydroxylation is a detoxifying pathway whereas formation of *o*-HPA is responsible for hepatotoxicity. In most humans coumarin is extensively metabolised by hepatic CYP2A6 to 7-hydroxycoumarin which is excreted in urine as glucuronide and sulphate conjugates. While 7-hydroxylation is also the major pathway of coumarin metabolism in Old World primates, it is a minor pathway in other species. For example, 7-hydroxylation occurs poorly in the rat and most strains of mice. The high affinity of CYP2A6 for coumarin (K_m approximately 0.5 μM) implies that 7-hydroxylation is the major pathway at the low levels coumarin that we are exposed to. The occasionally observed hepatotoxicity in humans exposed to

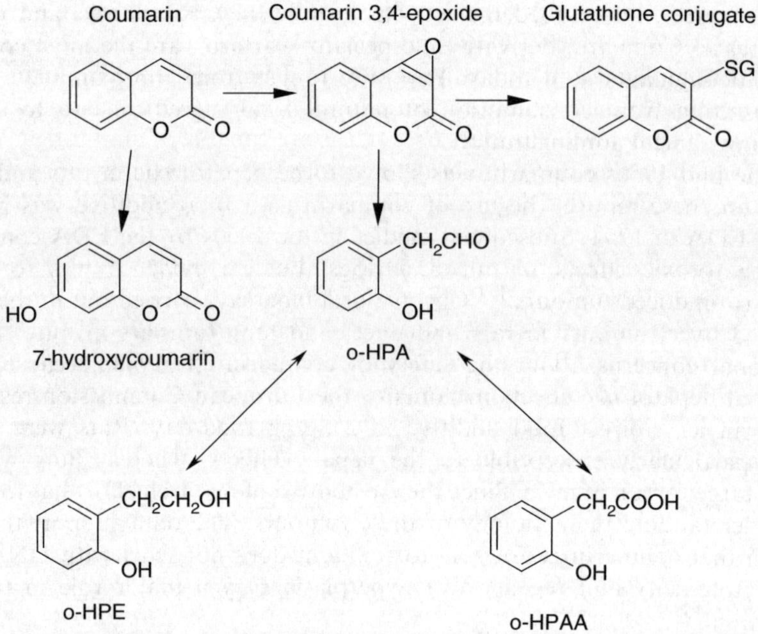

Figure 5.4 Major metabolic pathways of coumarin.[153,155] *o*-HPA, *o*-hydroxyphenylacetaldehyde; *o*-HPE, *o*-hydroxyphenylethanol; *o*-HPAA, *o*-hydroxyphenylacetic acid.

high oral doses appears to be an idiosyncratic response that is potentially influenced by multiple exogenous and endogenous factors, maybe involving also CYP2A6 poor metaboliser status. Quantitative human health risk assessment confirms the safety of coumarin exposure from natural dietary sources as well as from its use in personal care products.[155]

The lesson here is that major species differences in the detoxifying and activating metabolic pathways can have a profound impact on the outcome of toxicological tests.

5.12 Conclusions

Despite the substantial amount of information available about the members of the CYP2A subfamily, we still do not know the physiological functions of these enzymes. The CYP2A enzymes are certainly not necessary for life, since functional enzymes are lacking in many animal species. In humans, functional CYP2A6 is lacking in a substantial part, up to 50%, of some populations. CYP2A enzymes participate in the metabolism of several xenobiotics and endogenous agents (mainly steroids). There are crucial species differences in substrate specificities, tissue distribution and regulation of these enzymes. These differences must be taken into account when carrying out toxicological studies and risk assessment involving CYP2A substrates.

In marked contrast to other major xenobiotic-metabolising CYP forms, especially CYP2C9, CYP2D6 and CYP3A4, the CYP2A6 enzyme has never been implicated in clinically significant drug interactions. This is due to three reasons: 1) The systemic clearance of very few drugs is mediated by CYP2A6. 2) Only few commonly used drugs are capable of inhibiting CYP2A6 activity, methoxsalen being the prime example. 3) CYP2A6 may not be readily inducible to a clinically significant degree by any exogenous or endogenous factors.

CYP2A6 has a clear role in nicotine elimination and activation of nitrosamine procarcinogens. The suspected association between *CYP2A6* PM genotype and decreased risk of acquiring tobacco smoke-induced cancer needs further confirmation. The possible influence of CYP2A6 polymorphism on smoking habits is an intriguing area of research. Future studies should establish whether CYP2A6 status significantly affects the predisposition to nicotine addiction. It will be interesting to evaluate whether chemical inhibitors of CYP2A6 can be used as adjuvant drugs in nicotine replacement therapy.

References

1. C.H. Yun, T. Shimada and F.P. Guengerich, *Mol. Pharmacol.*, 1991, **40**, 679.
2. J.S. Miles, A.W. McLaren, L.M. Forrester, M.J. Glancey, M.A. Lang and C.R. Wolf, *Biochem. J.*, 1990, **267**, 365.
3. S. Yamano, J. Tatsuno and F.J. Gonzalez, *Biochemistry*, 1990, **29**, 1322.
4. O. Pelkonen, A. Rautio, H. Raunio and M. Pasanen, *Toxicology*, 2000, **144**, 139.
5. F.P. Guengerich, in *Cytochrome P450 Structure, Mechanism, and Biochemistry*, ed. P. Ortiz de Montellano, Kluwer Academic/Plenum Publishers, New York, 2005, p. 377.
6. X. Ding and L.S. Kaminsky, *Annu. Rev. Pharmacol. Toxicol.*, 2003, **43**, 149.
7. H. Wang, K.M. Donley, D.S. Keeney and S.M. Hoffman, *Environ. Health. Perspect.*, 2003, **111**, 1835.
8. P. Fernandez-Salguero and F.J. Gonzalez, *Pharmacogenetics*, 1995, **5**, S123.
9. T.K.H. Chang and D.J. Waxman, in *Cytochromes P450 Metabolic and Toxicological Aspects*, ed. C. Ioannides, CRC Press, Boca Raton, 1996, p. 99.
10. P. Honkakoski and M. Negishi, *Drug Metab. Rev.*, 1997, **29**, 977.
11. S.M. Hoffman, D.R. Nelson and D.S. Keeney, *Pharmacogenetics*, 2001, **11**, 687.
12. M.T. Skaanild, *Curr. Pharm. Des.*, 2006, **12**, 1421.
13. http://drnelson.utmem.edu/biblioA.html.
14. A.J. Lee, M.X. Cai, P.E. Thomas, A.H. Conney and B.T. Zhu, *Endocrinology*, 2003, **144**, 3382.

15. H. Wang, W. Tan, B. Hao, X. Miao, G. Zhou, F. He and D. Lin, *Cancer Res.*, 2003, **63**, 8057.
16. M. Negishi, T. Uno, P. Honkakoski, T. Sueyoshi, T.A. Darden and L.P. Pedersen, *Biochimie*, 1996, **78**, 685.
17. R.L. Lindberg and M. Negishi, *Nature*, 1989, **339**, 632.
18. M. Negishi, T. Uno, T.A. Darden, T. Sueyoshi and L.G. Pedersen, *FASEB J.*, 1996, **10**, 683.
19. T. Su and X. Ding, *Toxicol. Appl. Pharmacol.*, 2004, **199**, 285.
20. J. Hakkola, M. Pasanen, R. Purkunen, S. Saarikoski, O. Pelkonen, J. Maenpaa, A. Rane and H. Raunio, *Biochem. Pharmacol.*, 1994, **48**, 59.
21. R.N. Hines and D.G. McCarver, *J. Pharmacol. Exp. Ther.*, 2002, **300**, 355.
22. D. Choudhary, I. Jansson, J.B. Schenkman, M. Sarfarazi and I. Stoilov, *Arch. Biochem. Biophys.*, 2003, **414**, 91.
23. S. Imaoka, S. Fujita and Y. Funae, *Biochim. Biophys. Acta.*, 1991, **1097**, 187.
24. T. Shimada, H. Yamazaki, M. Mimura, Y. Inui and F.P. Guengerich, *J. Pharmacol. Exp. Ther.*, 1994, **270**, 414.
25. M. Pasanen, Z. Rannala, A. Tooming, E.A. Sotaniemi, O. Pelkonen and A. Rautio, *Toxicology*, 1997, **123**, 177.
26. Y. Chen, Y.Q. Liu, T. Su, X. Ren, L. Shi, D. Liu, J. Gu, Q.Y. Zhang and X. Ding, *Biochem. Pharmacol.*, 2003, **66**, 1245.
27. H. Schrem, J. Klempnauer and J. Borlak, *Pharmacol. Rev.*, 2002, **54**, 129.
28. N. Yokomori, K. Nishio, K. Aida and M. Negishi, *J. Steroid. Biochem. Mol. Biol.*, 1997, **62**, 307.
29. R. Jover, R. Bort, M.J. Gomez-Lechon and J.V. Castell, *Hepatology*, 2001, **33**, 668.
30. J. Ulvila, S. Arpiainen, O. Pelkonen, K. Aida, T. Sueyoshi, M. Negishi and J. Hakkola, *Biochem. J.*, 2004, **381**, 887.
31. M. Pitarque, C. Rodriguez-Antona, M. Oscarson and M. Ingelman-Sundberg, *J. Pharmacol. Exp. Ther.*, 2005, **313**, 814.
32. J. Zhang and X. Ding, *J. Biol. Chem.*, 1998, **273**, 23454.
33. G. Ling, C.R. Hauer, R.M. Gronostajski, B.T. Pentecost and X. Ding, *J. Biol. Chem.*, 2004, **279**, 27888.
34. T. Su, Q.Y. Zhang, J. Zhang, P. Swiatek and X. Ding, *Drug Metab. Dispos.*, 2002, **30**, 548.
35. C.A. Wiwi and D.J. Waxman, *Growth Factors*, 2004, **22**, 79.
36. A. Parkinson, D.R. Mudra, C. Johnson, A. Dwyer and K.M. Carroll, *Toxicol. Appl. Pharmacol.*, 2004, **199**, 193.
37. N.A. Pampori and B.H. Shapiro, *Endocrinology*, 1999, **140**, 1245.
38. T. Sueyoshi, N. Yokomori, K.S. Korach and M. Negishi, *Mol. Pharmacol.*, 1999, **56**, 473.
39. C.A. Wiwi and D.J. Waxman, *J. Biol. Chem.*, 2005, **280**, 3259.
40. A. Rautio, H. Kraul, A. Kojo, E. Salmela and O. Pelkonen, *Pharmacogenetics*, 1992, **2**, 227.
41. N.L. Benowitz, C.N. Lessov-Schlaggar, G.E. Swan and P. Jacob, *Clin. Pharmacol Ther.*, 2006, **79**, 480.

42. D.J. Lavery, L. Lopez-Molina, R. Margueron, F. Fleury-Olela, F. Conquet, U. Schibler and C. Bonfils, *Mol. Cell. Biol.*, 1999, **19**, 6488.
43. P. Viitala, K. Posti, A. Lindfors, O. Pelkonen and H. Raunio, *Biochem. Biophys. Res. Commun.*, 2001, **280**, 761.
44. S. Imaoka, Y. Terano and Y. Funae, *Arch. Biochem. Biophys.*, 1990, **278**, 168.
45. M. Bauer, A.C. Hamm, M. Bonaus, A. Jacob, J. Jaekel, H. Schorle, M.J. Pankratz and J.D. Katzenberger, *Physiol Genomics.*, 2004, **17**, 230.
46. V. Kobliakov, L. Kulikova, D. Samoilov and M.A. Lang, *Mol. Carcinog.*, 1993, **7**, 276.
47. G.M. Kirby, P. Pelkonen, V. Vatanasapt, A.M. Camus, C.P. Wild and M.A. Lang, *Mol. Carcinog.*, 1994, **11**, 81.
48. T.A. Richardson, M. Sherman, L. Antonovic, S.S. Kardar, H.W. Strobel, D. Kalman and E.T. Morgan, *Drug Metab. Dispos.*, 2006, **34**, 354.
49. S. Satarug, M.A. Lang, P. Yongvanit, P. Sithithaworn, E. Mairiang, P. Mairiang, P. Pelkonen, H. Bartsch and M.R. Haswell-Elkins, *Cancer Epidemiol. Biomarkers. Prev.*, 1996, **5**, 795.
50. H. Raunio, R. Juvonen, M. Pasanen, O. Pelkonen, P. Pääkkö and Y. Soini, *Hepatology*, 1998, **27**, 427.
51. M.A. Sipowicz, P. Chomarat, B.A. Diwan, M.A. Anver, Y.C. Awasthi, J.M. Ward JM, J.M. Rice, K.S. Kasprzak, C.P. Wild and L.M. Anderson, *Am. J. Pathol.*, 1997, **151**, 933.
52. W.J. Gilmore and G. M Kirby, *J. Pharmacol. Exp. Ther.*, 2004, **308**, 600.
53. A. Abu-Bakar, S. Satarug, G.C. Marks, M.A. Lang and M.R. Moore, *Toxicol. Lett.*, 2004, **148**, 199.
54. J.M. Maglich, C.M. Stoltz, B. Goodwin, D. Hawkins-Brown, J.T. Moore and S.A. Kliewer, *Mol. Pharmacol.*, 2002, **62**, 638.
55. P. Wei, J. Zhang, D.H. Dowhan, Y. Han and D.D. Moore, *Pharmacogenomics J.*, 2002, **2**, 117.
56. Y. Cai, T. Konishi, G. Han, K.H. Campwala, S.W. French and Y.J. Wan, *Eur. J. Pharm. Sci.*, 2002, **15**, 89.
57. M. Itoh, M. Nakajima, E. Higashi, R. Yoshida, K. Nagata, Y. Yamazoe and T. Yokoi, *J. Pharmacol. Exp. Ther.*, 2006, **319**, 693.
58. S. Arpiainen, F. Raffalli-Mathieu, M.A. Lang, O. Pelkonen and J. Hakkola, *Mol. Pharmacol.*, 2005, **67**, 1325.
59. K. Kurose, M. Tohkin and M. Fukuhara, *Mol Pharmacol.*, 1999, **55**, 279.
60. K. Aida and M. Negishi, *Biochemistry*, 1991, **30**, 8041.
61. O. Geneste, F. Raffalli and M.A. Lang, *Biochem. J.*, 1996, **313**, 1029.
62. F. Raffalli-Mathieu, T. Glisovic, Y. Ben-David and M.A. Lang, *Mol. Pharmacol.*, 2002, **61**, 795.
63. K. Christian, M. Lang, P. Maurel and F. Raffalli-Mathieu, *Mol Pharmacol.*, 2004, **65**, 1405.
64. A.W. Wood and A.H. Conney, *Science*, 1974, **185**, 612.
65. H. Raunio, A. Kojo, R. Juvonen, P. Honkakoski, P. Jarvinen, M.A. Lang, K. Vahakangas, H.V. Gelboin, S.S. Park and O. Pelkonen, *Biochem. Pharmacol.*, 1988, **37**, 4141.

66. M.T. Donato, P. Viitala, C. Rodriguez-Antona, A. Lindfors, J.V. Castell, H. Raunio, M.J. Gomez-Lechon and O. Pelkonen, *Drug. Metab. Dispos.*, 2000, **28**, 1321.
67. W. Legrum and K.J. Netter, *Xenobiotica*, 1980, **10**, 271.
68. S.A. Mangoura, A. Strack, W. Legrum and K.J. Netter, *Naunyn Schmiedebergs Arch. Pharmacol.*, 1989, **339**, 596.
69. P. Arvela, H. Kraul, F. Stenback and O. Pelkonen, *Toxicology*, 1991, **69**, 1.
70. B. Emde, M. Tegtmeier, B. Hahnemann and W. Legrum, *Toxicology*, 1996, **108**, 73.
71. P. Salonpaa, K. Krause, O. Pelkonen and H. Raunio, *Naunyn Schmiedebergs Arch. Pharmacol.*, 1995, **351**, 446.
72. R.O. Juvonen, S. Autio and M.A. Lang, *Biochem. Pharmacol.*, 1987, **36**, 3993.
73. P. Pellinen, F. Stenback, H. Raunio, O. Pelkonen and M. Pasanen, *Eur. J. Pharmacol.*, 1994, **292**, 57.
74. P. Pellinen, F. Stenback, A. Rautio, O. Pelkonen, M. Lang and M. Pasanen, *Naunyn Schmiedebergs Arch. Pharmacol.*, 1993, **348**, 435.
75. D.F. Lewis and M. Dickins, *Drug Discov. Today*, 2002, **7**, 918.
76. C. de Graaf, N.P. Vermeulen and K.A. Feenstra, *J. Med. Chem.*, 2005, **48**, 2725.
77. C.A. Kemp, J.D. Marechal and M.J. Sutcliffe, *Arch. Biochem. Biophys.*, 2005, **433**, 361.
78. D.F.V. Lewis and B.G. Lake, *Xenobiotica*, 1995, **25**, 585.
79. D.F. Lewis and B.G. Lake, *Xenobiotica*, 2002, **32**, 547.
80. D.F. Lewis, B.G. Lake, M. Dickins and P.S. Goldfarb, *Toxicol. In Vitro*, 2003, **17**, 179.
81. D.F. Lewis, Y. Ito and B.G. Lake, *Toxicol. In Vitro*, 2006, **20**, 256.
82. A. Gorokhov, M. Negish, E.F. Johnson, L.C. Pedersen, L. Perera, T.A. Darden and L.G. Pedersen, *Biophys. J.*, 2003, **84**, 57.
83. J.R. Jalas, M. Seetharaman, S.S. Hecht and S.E. Murphy, *Xenobiotica*, 2004, **34**, 515.
84. G.R. Stahl and H.D. Holtje, *Pharmazie*, 2005, **60**, 247.
85. D.F. Lewis, *Toxicol. In Vitro*, 2004, **18**, 89.
86. A. Poso, J. Gynther and R. Juvonen, *J. Comput. Aided Mol. Des.*, 2001, **15**, 195.
87. A. Asikainen, J. Tarhanen, A. Poso, M. Pasanen, E. Alhava and J.O. Juvonen, *Toxicol. In Vitro.*, 2003, **17**, 449.
88. M. Rahnasto, H. Raunio, A. Poso, C. Wittekindt and R.O. Juvonen, *J. Med. Chem.*, 2005, **48**, 440.
89. J. Yano, M.H. Hsu, K.J. Griffin, C.D. Stout and E.F. Johnson, *Nature Struct. Mol. Biol.*, 2005, **12**, 822.
90. M. Rahnasto, C. Wittekindt, R.O. Juvonen, M. Turpeinen, A. Petsalo, O. Pelkonen, A. Poso, G. Stahl, H.D. Höltje, H. Raunio. Pharmacogen J. 2007, Epub ahead of print.

91. X.Y. He, J. Shen, X.Y. Hu, X. Ding, A.Y. Lu and J.Y. Hong, *Arch. Biochem. Biophys.*, 2004, **427**, 143–153.
92. D. Kim D, Z.L. Wu and F.P. Guengerich, *J. Biol Chem.*, 2005, **280**, 40319.
93. Z.L. Wu, L.M. Podust and F.P. Guengerich, *J. Biol. Chem.*, 2005, **280**, 41090.
94. http://www.cypalleles.ki.se
95. M. Nakajima, T. Fukami, H. Yamanaka, E. Higashi, H. Sakai, R. Yoshida, J.T. Kwon, H.L. McLeod and T. Yokoi, *Clin. Pharmacol. Ther.*, 2006, **80**, 282.
96. S.L. Wang, X.Y. He, J. Shen, J.S. Wang and J.Y. Hong, *Toxicol. Sci.*, 2006, **94**, 38.
97. S. Rendic, *Drug Metab. Rev.*, 2002, **34**, 83.
98. R.L. Walsky and R.S. Obach, *Drug Metab. Dispos.*, 2004, **32**, 647.
99. R. Yuan, S. Madani, X.X. Wei, K. Reynolds and S.M. Huang, *Drug Metab. Dispos.*, 2002, **30**, 1311.
100. L.B. von Weymarn, Q.Y. Zhang, X. Ding and P.F. Hollenberg, *Carcinogenesis*, 2005, **26**, 621.
101. L.B. von Weymarn, J.A. Chun and P.F. Hollenberg, *Carcinogenesis*, 2006, **27**, 782.
102. L.B. von Weymarn, K.M. Brown and S.E. Murphy, *J. Pharm. Exp. Ther.*, 2006, **316**, 295.
103. H. Raunio, A. Rautio, H. Gullsten and O. Pelkonen, *Br. J. Clin. Pharmacol.*, 2001, **52**, 357.
104. K. Fujita and T. Kamataki, *Drug Metab. Disp.*, 2001, **29**, 983.
105. M. Rahnasto, H. Raunio, A. Poso and R.O. Juvonen, *Xenobiotica*, 2003, **33**, 529.
106. J.M. MacDougall, K. Fandrick, X. Zhang, S.V. Serafin and J.R. Cashman, *Chem. Res. Toxicol.*, 2003, **16**, 988.
107. N.L. Benowitz, B. Herrera and P. Jacob, *J. Pharmacol. Exp. Ther.*, 2004, **310**, 1208.
108. J. Hukkanen, P. Jacob and N.L. Benowitz, *Clin. Pharmacol. Ther.*, 2006, **80**, 522.
109. J.K. Yano, T.T. Denton, M.A. Cerny, X. Zhang, E.F. Johnson and J.R. Cashman, *J. Med. Chem.*, 2006, **49**, 6987.
110. Z. Bao, X.Y. He, X. Ding, S. Prabhu and J.Y. Hong, *Drug Metab. Dispos.*, 2005, **33**, 258.
111. M. Nakajima, M. Itoh, H. Sakai, T. Fukami, M. Katoh, H. Yamazaki, F.F. Kadlubar, S. Imaoka, Y. Funae and T. Yokoi, *Int. J. Cancer*, 2006, **19**, 2520.
112. X.Y. He, L. Tang, S.L. Wang, Q.S. Cai, J.S. Wang and J.Y. Hong, *Int. J. Cancer*, 2006, **118**, 2665.
113. T. Fukami, M. Nakajima, H. Sakai, M. Katoh, T. Yokoi, *Drug Metab. Dispos.*, 2007, **35**, 335.
114. WHO Health Report 2002 – Reducing Risks, Promoting Healthy Life. WHO, 2002.

115. N.L. Benowitz. 1996, **36**, 597.
116. M. Nakajima, T. Yamamoto, K. Nunoya, T. Yokoi, K. Nagashima, K. Inoue, Y. Funae, N. Shimada, T. Kamataki, Y. Kuroiwa, 1996, **24**, 1212.
117. E.S. Messina, R.F. Tyndale and E.M. Sellers, *J. Pharmacol. Exp. Ther.*, 1997, **282**, 1608.
118. J. Hukkanen, P. Jacob and N.L. Benowitz, *Pharmacol. Rev.*, 2005, **57**, 79.
119. R.F. Tyndale and E.M. Sellers, *Drug Metab. Dispos.*, 2001, **29**, 548.
120. C. Xu, S. Goodz, E.M. Sellers and R.F. Tyndale, *Adv. Drug Deliv. Rev.*, 2002, **54**, 1245.
121. V. Malaiyandi, E.M. Sellers and R.F. Tyndale, *Clin. Pharmacol. Ther.*, 2005, **77**, 145.
122. D. Dempsey, P. Tutka, P. Jacob, F. Allen, K. Schoedel, R.F. Tyndale and N.L. Benowitz, *Clin. Pharmacol. Ther.*, 2004, **76**, 64.
123. V. Malaiyandi, S.D. Goodz, E.M. Sellers and R.F. Tyndale, *Cancer Epidemiol. Biomarkers Prev.*, 2006, **15**, 1812.
124. E. Johnstone, N. Benowitz, A. Cargill, R. Jacob, L. Hinks, I. Day, M. Murphy and R. Walton, *Clin Pharmacol. Ther.*, 2006, **80**, 319.
125. T.P. George and S.S. O'Malley, *Trends Pharmacol. Sci.*, 2004, **25**, 42.
126. C. Lerman, R. Tyndale, F. Patterson, E.P. Wileyto, P.G. Shields, A. Pinto and N. Benowitz, *Clin. Pharmacol. Ther.*, 2006, **79**, 600.
127. E.M. Sellers, H.L. Kaplan and R.F. Tyndale, *Clin. Pharmacol. Ther.*, 2000, **68**, 35.
128. W. Zhang, T. Kilicarslan, R.F. Tyndale and E.M. Sellers, *Drug Metab. Dispos.*, 2001, **29**, 897.
129. E.M. Sellers, R.F. Tyndale and L.C. Fernandes, *Drug Discovery Today*, 2003, **8**, 487.
130. E.C.K. Siu, D.B. Wildenauer and R.F. Tyndale, *Psychopharmacology*, 2006, **184**, 401.
131. H. Raunio, N. Pokela, K. Puhakainen, M. Rahnasto, T. Mauriala, S. Auriola, R. Juvonen, *Xenobiotica*, 2007, **38**, 34.
132. J.R. Jalas, S.S. Hecht and S.E. Murphy, *Chem. Res. Toxicol.*, 2005, **18**, 95.
133. S.S. Hecht, *Nat. Rev. Cancer.*, 2003, **3**, 733.
134. T. Kamataki, K. Fujita, K. Nakayama, Y. Yamazaki, M. Miyamoto and N. Ariyoshi, *Drug Metab. Rev.*, 2002, **34**, 667.
135. M.L. Pianezza, E.M. Sellers and R.F. Tyndale, *Nature*, 1998, **393**, 750.
136. M. Miyamoto, Y. Umetsu, H. Dosaka-Akita, Y. Sawamura, J. Yokota, H. Kunitoh, N. Nemoto, K. Sato, N. Ariyoshi and T. Kamataki, *Biochem Biophys Res Commun.*, 1999, **261**, 658.
137. M. Oscarson, *Drug Metab. Dispos.*, 2001, **29**, 91.
138. B. Carter, T. Long and P. Cinciripini, *Nicotine Tob. Res.*, 2004, **6**, 221.
139. M. Nakajima, Y. Kuroiwa and T. Yokoi, *Drug Metab. Rev.*, 2002, **34**, 865.
140. T. Kamataki, M. Fujieda, K. Kiyotani, S. Iwano and H. Kunitoh, *Biochem Biophys Res Commun.*, 2005, **338**, 306.
141. M. Miyazaki, E. Sugawara, T. Yoshimura, H. Yamazaki and T. Kamataki, *Mutat. Res.*, 2005, **581**, 165.

142. Z. Topcu, I. Chiba, M. Fujieda, T. Shibata, N. Ariyoshi, H. Yamazaki, F. Sevgican, M. Muthumala, H. Kobayashi and T. Kamataki, *Carcinogenesis*, 2002, **23**, 595.
143. H. Takeuchi, K. Saoo, M. Yokohira, M. Ikeda, H. Maeta, M. Miyazaki, H. Yamazaki, T. Kamataki and K. Imaida, *Cancer Res.*, 2003, **63**, 7581.
144. M. Miyazaki, H. Yamazaki, H. Takeuchi, K. Saoo, M. Yokohira, K. Masumura, T. Nohmi, Y. Funae, K. Imaida and T. Kamataki, *Carcinogenesis*, 2005, **26**, 1947.
145. W. Godoy, R.M. Albano, E.G. Moraes, P.R. Pinho, R.A. Nunes, E.H. Saito, C. Higa, I.M. Filho, C.D. Kruel, C.C. Schirmer, R. Gurski, M.A. Lang and L.F. Pinto, *Carcinogenesis*, 2002, **23**, 611.
146. M.A. Morse, J. Lu, G.D. Stoner, S.E. Murphy and L.A. Peterson, *J. Toxicol. Environ. Health*, 1999, **58**, 397.
147. W. Tan, G.F. Chen, D.Y. Xing, C.Y. Song, F.F. Kadlubar and D.X. Lin, *Int. J. Cancer*, 2001, **95**, 96.
148. H. Tsukino, Y. Kuroda, D. Qiu, H. Nakao, H. Imai and T. Katoh, *Int. J. Cancer*, 2002, **100**, 425.
149. C. Sweeney, B.F. Coles, S. Nowell, N.P. Lang and F.F. Kadlubar, *Toxicology*, 2002, **27**, 183.
150. C. Sachse, G. Smith, M.J. Wilkie, J.H. Barrett, R. Waxman, F. Sullivan, D. Forman, D.T. Bishop and C.R. Wolf, *Carcinogenesis*, 2002, **23**, 1839.
151. A.J. Cohen, *Food Cosmet. Toxicol*, 1979, **17**, 277.
152. D. Egan, R. O'Kennedy, E. Moran, D. Cox, E. Prosser and R.D. Thornes, *Drug Metab. Rev.*, 1990, **22**, 503.
153. B.G. Lake, *Food Chem. Toxicol.*, 1999, **37**, 423.
154. O. Pelkonen, H. Raunio, M. Pasanen, A. Rautio and M.A. Lang, in *Coumarins Biology, Applications and Mode of Action*, ed. R. O'Kennedy and R.D. Thornes, Wiley, Chichester, 1997, p. 67.
155. S.P. Felter, J.D. Vassallo, B.D. Carlton and G.P. Daston, *Food Cosmet. Toxicol.*, 2006, **44**, 462.

CHAPTER 6

The CYP2B Subfamily

LAURENT CORCOS AND FRANÇOIS BERTHOU

INSERM U613/EA948, Faculty of Medicine, 22 avenue Camille Desmoulins, 29200 BREST, FRANCE

Table of Contents

Issues in Toxicology
Cytochromes P450: Role in the Metabolism and Toxicity of Drugs and other Xenobiotics
Edited by Costas Ioannides

6.1 Introduction

Originally identified in rat liver as cytochrome P450, as opposed to cytochrome P448,[1] cytochrome P4502B enzymes were later demonstrated to belong to the Phenobarbital (PB)-inducible enzyme class, whereas the cytochrome P4501A enzymes (P448) were identified as the main members of the 3-methylcholanthrene-inducible class. The P450 or P448 enzyme names were used in reference to the difference in the absorption wavelength of the haem moiety of the enzymes.[2] The isolated rat cytochrome P450 enzymes, called P-450b and P-450e, were found to be very similar, differing essentially by a short amino acid stretch (VSPAWMRE) inserted into the CYP2B2 protein.[3] The enzymes were shown to catalyse the transformation of substrates belonging either to endogenous metabolic pathways such as steroids, like testosterone, fatty acids or various chemicals. The corresponding *CYP2B* genes were proposed to have originated from a concerted evolution mechanism referred to as gene conversion following the selection of gene duplication events.[4] Originally identified in the rat, orthologous genes were rapidly discovered in mice and humans, as well as in numerous other species, from *Bacillus megaterium*[5] or *Saccharomyces cerevisiae*[6] to chicken and all mammalian species surveyed so far. Although these genes belong to the CYP2 family, they are not CYP2B members and will not be further discussed in this chapter. The CYP2B family comprises several members, which cluster to single chromosomal regions, reinforcing the proposed gene conversion hypothesis. Many studies were conducted to uncover the mechanism of their induction by phenobarbital, which was readily determined to be mainly, if not exclusively, transcriptional in nature. An important step in elucidating this mechanism in mammals was foreseen when the Fulco team demonstrated the existence of a PB-response element called the Barbie Box, which was shown to drive the induction response of P450BM3 in *Bacillus megaterium*, and proposed to play a conserved role in the rat.[7] However, this proposal did not grow to a consensus and was even further dismissed when PB-response units (PBRU) or PB-response enhancer modules (PBREM) were discovered in the gene promoter regions of rats, mice and human CYP2B family members.[8,9] An important breakthrough was made by the demonstration that the antiprogestin RU38486 could suppress *CYP2B* and *CYP2C* gene induction by PB in hepatoma cells.[10] Soon after, M. Neghishi's group made the major discovery that the Constitutive Androstane orphan nuclear Receptor (CAR) was in fact the main regulatory protein of the PB induction of the CYP2B genes.[11] A second orphan nuclear receptor, the pregnane X receptor (PXR), was then shown to mediate PB induction of the CYP3A enzyme, another typical PB response gene, which, however, is much less inducible but shows a higher basal expression level.[12] However, PXR leads the inductive response to glucocorticoids and antiglucocorticoids like RU38486. In fact RU38486 was able to counteract the activity of CAR but to activate PXR, providing a mechanistic interpretation for the original inhibition findings. Although *CYP2B* and *CYP3A* genes share common regulatory induction mechanisms, through the activation of CAR and PXR by barbiturates and/or steroid-like compounds, major differences exist in receptor sensitivity to drugs, and the substrate specificity of the CYP2B and CYP3A enzymes do not strongly overlap. More recently, the crystal structure of rabbit CYP2B4 (formerly called P450LM2) was determined and shown to be rather distinct from that of CYP2C5 (the first model of mammalian CYP active conformation), with an open structure of the enzyme that may occur as a dimer in solution.[13] Because CYP2B enzymes are intrinsically able to activate drugs, metabolism-based suicide strategies have been developed. Upon activation of the anticancer drug cyclophosphamide into phosphoramide mustard, a highly reactive and cytotoxic metabolite, cancer cell killing has been demonstrated.[14] Improved strategies using more powerful suicide systems have also been developed, including combined gene transfer with the CYP reductase, either separately or as a fusion protein, or newly engineered enzymes with lower Km for cyclophosphamide.[15,16] The development of more selective and active CYP2B suicide substrates should also help optimise the cell-killing strategies. The proofs of concept of such a suicide gene therapy have now been obtained, which should allow embarking on clinical trials in the context of selected cancer types. From their original discovery to the engineering of adapted enzymes for cancer cell suicide, CYP2Bs have retained a lot of interest from the scientific community. The goal of this review chapter is to trace back the main enzyme and gene features of the CYP2B family members as evoked in this introduction. The readers will be referred to pioneering work and to important review articles. Although not all studies can be evoked directly within this chapter, several aspects of this review will be ideally supplemented with contributions from colleagues in accompanying chapters in this book.

6.2 The CYP2B Gene Family

CYP2Bs belong to a subfamily that comprises up to 5 members in any mammalian specie, all characterised by a restricted tissue-specific expression, which is at the lower end among CYP family members in the absence of treatment with inducers. Figure 6.1 depicts a portion of the CYP phylogenetic tree that shows the evolutionary conservation among CYP2B genes, as recorded at Ensembl (http://www.ensembl.org).

Nomenclature has been homogenised through the extensive efforts of Dr. D. Nelson who has organised and regularly updated a web site dedicated to CYP-related gene family information (http://drnelson.utmem.edu/matrix.html). The various CYP2B genes from human, rat, mouse and rabbit are listed in Table 6.1.

Gene variants have been reported in rat liver, which show modified substrate specificity.[17] In addition, genetic analyses have permitted the characterisation of coumarin hydroxylase activity as being supported by CYP2A, rather than CYP2B members.[18] Importantly, as already discussed, the crystal structure of rabbit CYP2B4 was determined and shown to be rather distinct from that of CYP2C5, with an open structure of the enzyme that may occur as a dimer.[13] A summary of gene and protein features is depicted in Figure 6.2.

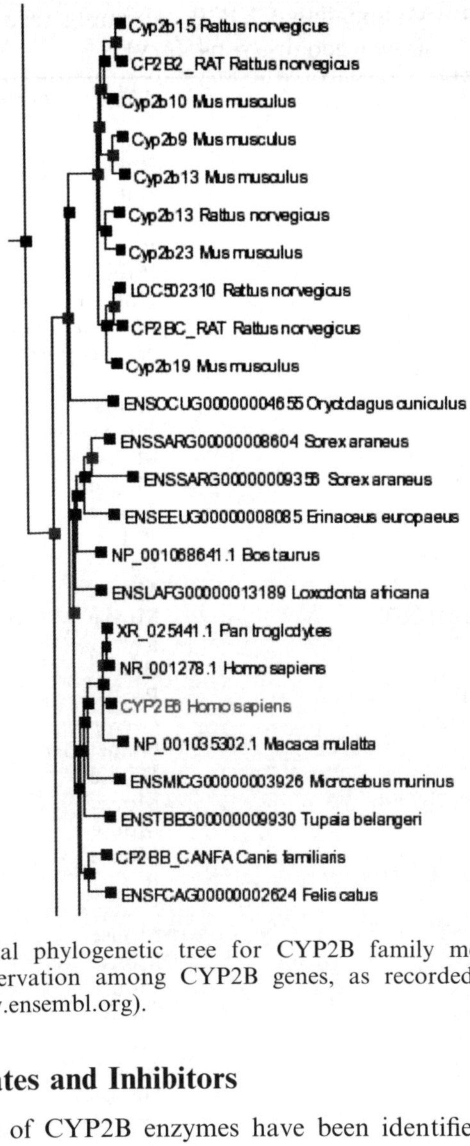

Figure 6.1 Partial phylogenetic tree for CYP2B family members. Evolutionary conservation among CYP2B genes, as recorded at Ensembl (http://www.ensembl.org).

6.2.1 Substrates and Inhibitors

Many substrates of CYP2B enzymes have been identified (Tables 6.2a and 6.2b). However, the relative metabolism efficiency strongly depends on the level and tissue distribution of the enzyme. CYP2B enzymes exhibit a relatively low degree of catalytic conservation across mammalian species, rat (CYP2B1/2B2/2B3/2B8/2B15), rabbit (CYP2B4/2B5), dog (2B11) and human (CYP2B6). For example, a striking difference in the androstenedione metabolism has been reported[19] between two closely related proteins, rabbit CYP2B4 and 2B5 differing by only 11 amino acids: CYP2B4 produces the 16-β-hydroxy metabolite while CYP2B5 produces the 16-α- and 15-α-hydroxy products. It has been suggested that the most specific catalytic probe for CYP2B activity in different mammalian species is testosterone hydroxylation in the 16β-position.

Table 6.1 Mammalian CYP2B subfamily (see http://www.icgeb.org/~p450srv).

Gene name	*Mammalian specie*
CYP2B1	Rat
CYP2B2	Rat
CYP2B3	Rat
CYP2B3-se1	Rat
CYP2B3-se2	Rat
CYP2B4	Rabbit
CYP2B6	Human
CYP2B6	*Macaca mulatta*
CYP2B6	*Macaca fasicularis*
CYP2B6	*Bos taurus*
CYP2B7P1	Human
CYP2B7P	*Bos taurus*
Cyp2b9	Mouse
Cyp2b9-de9b	Mouse
Cyp2b10	Mouse
CYP2B12	Rat
CYP2B12-de9b	Rat
Cyp2b13	Mouse
Cyp2b13-de1b2b7b	Mouse
CYP2B14P	Rat
CYP2B15	Rat
CYP2B16P	Rat
CYP2B17/2B6	*Cercopithecus aethiops*
CYP2B18	Guinea pig
Cyp2b19	Mouse
Cyp2b19-de7b8b9b	Mouse
Cyp2b20X	Mouse
Cyp2b20p1X	Mouse
CYP2B21	Rat
CYP2B22	*Sus scrofa*
Cyp2b23	Mouse
Cyp2b24pX	Mouse
Cyp2b25pX	Mouse
Cyp2b26-ps	Mouse
Cyp2b27-ps	Mouse
Cyp2b28-ps	Mouse
CYP2B29	Hamster
CYP2B30X	*Macaca mulatta*
CYP2B31	Rat
CYP2B32P	Rat

Unfortunately, testosterone 16β-hydroxlation is not suitable for human CYP2B6, due to a low turnover rate. Among the numerous substrates proposed as specific of CYP2B enzymes in a variety of species,[20] 7-ethoxy-4-trifluoromethylcoumarin can be considered as relatively specific. Accordingly, for the *in vitro* activity tests, the usual probe reactions are 7-ethoxycoumarin *O*-deethylation, benzyloxyresorufin *O*-debenzylation, pentoxyresorufin *O*-depentylation, and especially, 7-ethoxy-4-trifluoromethylcoumarin *O*-deethylation due to their easy and quick

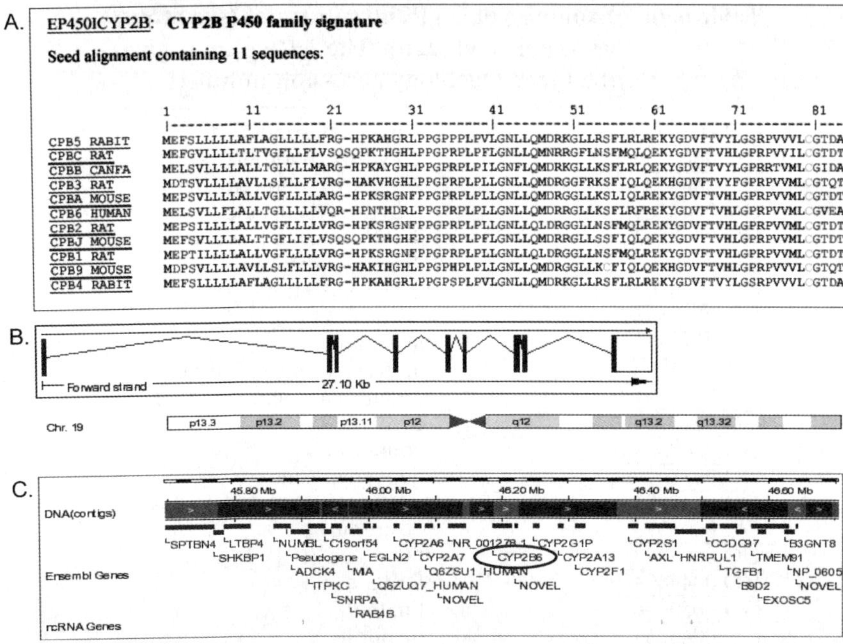

Figure 6.2 Summary of gene and protein features (http://www.ensembl.org). **A**. The peptide sequences of the main CYP2B family members are aligned for comparison. **B**. Structure of the human CYP2B6 transcript. **C**. Chromosomal context of the human CYP2B6 on chromosome 19.

Table 6.2a Specific drug substrates and inhibitors of CYP2B6.

Use	*Substrates*	*Inhibitors*
anti-tumour	Cyclophosphamide, Ifosfamide	Ticlopidine ($IC_{50} = 0.02$ μM)
analgesics	Pethidine	Clopidrogel ($IC_{50} = 0.15$ μM)
anaesthesic	Ketamine, Propofol	Clotrimazole, Itraconazole
antidepressant	Bupropion	Imidazole inhibitors (4-(4-chlorophenyl)imidazole; bifonazole). Sertaline
reverse-trancriptase inhibitor of HIV-1	Efavirenz	Raloxifene
monoamine oxidase inhibitor	Selegiline	17-α-ethinyloestradiol (17-EE)

realisation by fluorescence techniques.[21] CYP2B enzymes are very versatile catalysts with a broad range of substrates including drugs,[22] environmental pollutants[23] and endobiotics such as steroids. Numerous inhibitors[24] have been reported. However, CYP2B6 metabolises a number of drug substrates that are usually non-planar, neutral or weakly basic, fairly lipophilic with one or two hydrogen bond acceptors (Figure 6.3).

Table 6.2b Examples of CYP2B6 substrates (see http://www.ebi.ac.uk/ego). The left column gives the Gene Oncology accession numbers.

GO:0018599:	2-hydroxyisobutyrate
GO:0047546:	2-hydroxypyridine
GO:0008403:	25-hydroxycholecalciferol
GO:0047563:	3-hydroxybenzoate
GO:0018593:	4-chlorophenoxyacetate
GO:0018690:	4-methoxybenzoate
GO:0018592:	4-nitrocatechol
GO:0018595:	α-pinene
GO:0018597:	ammonia
GO:0047096:	androst-4-ene-3,17-dione
GO:0008391:	arachidonic acid
GO:0010291:	carotene
GO:0008389:	coumarin
GO:0018679:	deoxyhypusine
GO:0047882:	dibenzothiophene-5,5-dioxide
GO:0008393:	oestradiol
GO:0008393:	fatty acid (ω-1)
GO:0046424:	ferulate
GO:0050056:	linalool
GO:0045289:	luciferin
GO:0018591:	methyl tertiary butyl ether
GO:0008399:	naphthalene
GO:0018529:	nitrilotriacetate
GO:0031172:	ornithine
GO:0046409:	p-coumarate
GO:0047097:	phylloquinone
GO:0050212:	progesterone
GO:0050214:	
GO:0008401:	retinoic acid
GO:0004506:	squalene
GO:0050292:	steroid (9α)
GO:0018594:	tert-butyl alcohol
GO:0050603:	thiophene-2-carbonyl-CoA

6.3 Regulation of CYP2B Gene Expression

6.3.1 Species and Tissue Sites in Mammals

CYP2B1 and 2B2 mRNAs are both expressed at very low constitutive levels in foetal rat liver, but they are already responsive to inducers.[25,26] Human CYP2B6 is also most expressed in the liver, especially after PB treatment, and in the intestine, although basal levels may be higher in the lung where no response to the inducer can be detected.[27,28] There appear to exist important differences in abundance according to mammalian species or to ethnic groups. Hence, Caucasians express 10 times more CYP2B6 in the liver than Japanese, which translates into 0.3% versus 0.03% of total liver CYP content in Caucasians and Japanese, respectively.[29] In addition, females express higher

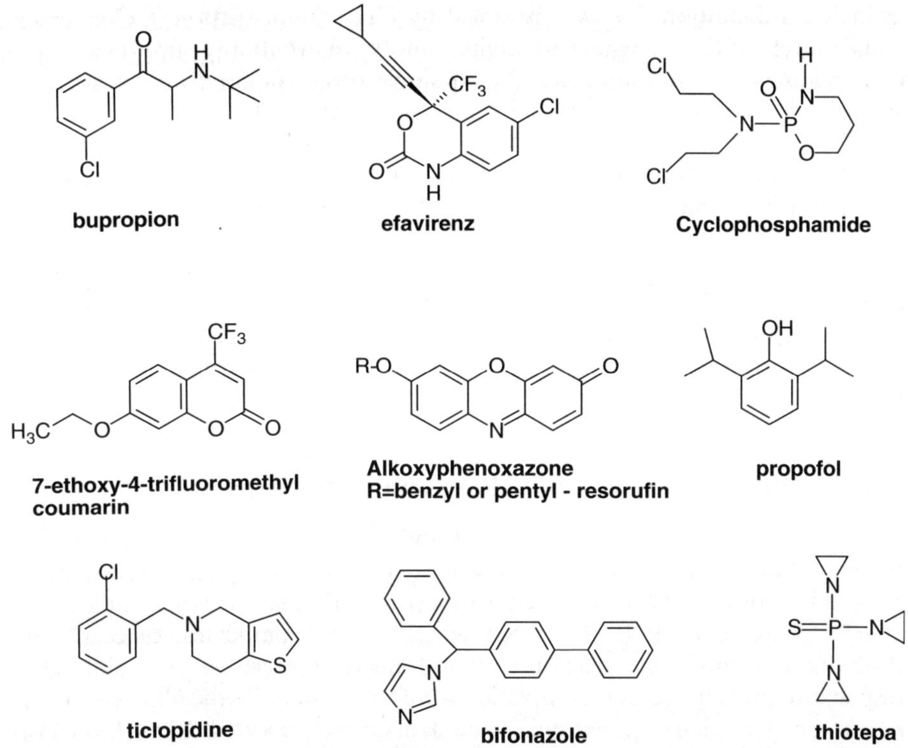

Figure 6.3 Versatile structure of 6 substances (bupropion, efavirenz, cyclophosphamide, 7-ethoxy-4-trifluoromethyl coumarin, alkoxyphenoxazone, propofol) and 3 inhibitors ticlopidine, bifonazole, thiotepa) of CYP2B6.

amounts of CYP2B6 protein (1.7-fold) than males, and 7.1% of females, but 20% of males, are poor CYP2B6 metabolisers.[30] In any case, *CYP2B6* gene expression in liver is always low and does not exceed 5–8% of total CYP content, even in inducer-treated subjects.[31]

6.3.2 Sex-dependent Regulation of CYP2B Expression

As shown by Harada and Negishi, mouse testosterone 16α-hydoxylase (CYP2B10) was expressed 10-times more in the liver of male than in female mice of the 129/J strain.[32] This sex-dependent regulation is controlled by a major gene locus with two allelic forms that can segregate between 129/J and C57BL/6J mice in which the expression level is not different between males and females.

6.3.3 Hormonal Regulation

Using the growth hormone (GH)-deficient Little mouse strain, Noshiro and Negishi showed that growth hormone suppressed liver 16α-hydroxylase activity in males, a condition that was reversed by GH administration.[33] Conversely, female levels of the enzyme were higher, and were eventually suppressed upon GH treatment when administered as a single dose. As carefully surveyed by Waxman and his group, it is rather probable that the GH-dependent pattern of CYP2B expression in liver is orchestrated by the STAT5b (Signal Transducer and Activator of Transcriptor factor 5b) transcription factor that is activated in a pulsatile fashion by GH bound to its plasma membrane receptor.[34] In addition, GH suppression is responsible for the lower responsiveness of CYP2B1 to PB in female rats.[35] However, it is apparent that the sex-dependent regulation of CYP2B expression is converse between rats and mice. If CYP2B is more expressed in male than in female rats, the reverse is true in mice.[36] No direct proof of such a dual regulation of the *CYP2B6* gene expression in humans has been provided.

6.3.4 The Influence of Pathology and Inflammation

Diabetes influences the response to PB and to GH of the *CYP2B1* gene.[37,38] Inflammation similarly has been shown to repress CYP gene expression, including that of *CYP2B*. It has been demonstrated that interleukins 1 and 6 were able to decrease CYP2B1/2 mRNA levels in rat hepatocytes, especially the PB-induced component, suggesting that transduction pathways mobilised in response to cytokine receptor activation were responsible for the observed down-regulation.[39] Similarly, Khatsenko and Kikkawa[40] reported LPS (Lipo Poly Saccharide)-mediated suppression of CYP2B1/2 induction by PB in rat liver, and Li-Masters and Morgan[41] made comparable observations in mice. Furthermore, TNF-α (Tumour Necrosis Factor α), another major pro-inflammatory mediator, has been shown to suppress mouse *CYP2B* gene expression through inhibition of CAR nuclear accumulation,[42] thereby limiting PB-induced expression.

6.3.5 Model Systems to Investigate Enzyme Activity and Gene Regulation

Various cell systems have been utilised to address the roles and regulation of CYP, including CYP2B proteins and genes. The most informative cell system was primary hepatocytes derived from humans or rodents, cultivated either as pure cultures[43,44] or as co-cultures with biliary epithelial cells.[45] These were grown as monolayers on plastic dishes or embedded in gel matrixes such as Matrigel,[46] collagen[47] or even as aggregates in suspension or spheroids.[48,49] Rat hepatoma cell lines were also used to study PB-dependent regulation[50] or were engineered to overexpress specific *CYP2B* genes for toxicological analysis.[51] In addition, tissue slices have also been used, albeit to a lesser extent, to address several toxicological endpoints.[52,53] *In vivo* studies have also been used to study gene induction by barbiturates following direct injection of plasmids into the liver.[54] Under appropriate experimental conditions, these hepatocyte-based systems are amenable to studies that can last from a few hours to several days.

The regulation of *CYP2B* gene induction by PB has been also analysed genetically in both mice and rats. Whereas the *CYP2B10* gene was systematically induced by PB in several inbred mouse strains, the CYP2B9 gene was induced, unaffected or even repressed within the same strains.[55] These differences in PB response were not related to differences in the DNA sequence encompassing the PB-response elements among the strains. More likely, they indicated the activity of modifier genes, with allelic forms being different among the strains surveyed. Similarly, in Long Evans-derived rats, a single genetic locus (with two alleles, R and r) acting dominantly was shown to control the PB response of the *CYP2B2* gene, which was suppressed in RR homozygous animals, but induced under the rr genotype. Irrespective of the genotype of the allelic form of the locus, the *CYP2B1* gene was always induced by PB.[56] Although it is presently unknown if such regulatory loci are active in humans, the use of different inbred mouse or rat strains can be developed to determine the *in vivo* contribution of the two CYP2B isoforms in drug metabolism and associated toxicity.

6.3.5.1 *Constitutive Expression in Liver Cells*

Several sequence elements have been identified to participate in the control of *CYP2B* gene regulation. In addition to the PB-response elements (PBREM in mouse *CYP2B9/2B10* and human *CYP2B6* genes and PBRU in rat *CYP2B1/2* genes), two DNase I hypersensitive sites (HS1 and HS2) have been described, together with a C/EBP (CAAT Enhancer Binding Protein) binding site in the 5′-flanking region of the rat *CYP2B1* gene.[57] These sites either encompass PBREM elements (HS2) or a CAAT box element (HS1), and thus agree with the demonstrated activity of transcription factors binding (CAR or C/EBP).[58,59]

6.3.5.2 *Induction Studies*

As mentioned earlier, xenobiotic-mediated induction of CYP2B genes has been extensively studied, especially in the case of phenobarbital (see Chapter 13). However, numerous other compounds are also CYP2B inducers, but whether they share the same inductive pathway has not always been documented. A list of some of these inducers is presented in Table 6.3.

The role of transduction pathways in the *CYP2B* gene regulation has been analysed in detail by our group and others. Using primary mouse hepatocytes, it was observed that the cAMP-dependent pathway exerted a negative role on the PB response, as treatment with cAMP-dependent protein kinase (PKA) activators dramatically inhibited PB-induced Cyp2b9/10 mRNA accumulation, whereas PKA inhibitors potentiated the PB responsiveness of this gene. Further evidence showed that the cGMP-dependent protein kinase (PKG) and Ca^{2+}/calmodulin-dependent protein kinase II played positive roles.[60] In addition, PP1 (Pyro Phosphatase 1) and/or PP2A (Pyro Phosphatase 2A) also acted positively on the PB response.[61]

Table 6.3 Short list of CYP2B inducers.

1α,25-dihydroxy vitamin D_3
2-allylisopropylacetamide
5β-pregnane-3,20-dione
Acetylaminofluorene
Allylisopropylacetamide
Bosentan
Carmabazepine
Chlordane
Cyclophosphamide
Dexamethasone
Efavirenz
Felbamate
Hyperforin
Isosafrole
Meclizine
Nevirapine
Nonplanar polychlorinated biphenyls
Oxcabazepin
Phenobarbital
Phenothiazines
Phenytoin
TCPOBOP, 1,4-bis[2-(3,5-dichloropyridyloxy)]benzene
Primidon
Rifabutin
Rifampicin
Ritonavir
RU38486
Saquinavir
Tamoxifen
TCPOBOP (mouse)
Topimarate
Trans-stilbene oxide
Troglitazone
Vigabatrin

6.4 Genetic Polymorphism and Splice Variants

Quite surprisingly, the *CYP2B6* gene displays the highest rate of genetic variants, with at least 48 members as of 2006, including several splice variants.[62] It is unclear why so many gene variations exist, but this may be related to the rather limited contribution of CYP2B6 to the metabolism of drugs in the liver in view of its low abundance (less than 8%),[63] under uninduced conditions. Genotype adaptation may thus allow a fine tuning of the enzyme under selective conditions. Alternatively, some of the genotypes may be more relevant than others in extrahepatic organs. Some of the polymorphisms may lead to increased enzymatic activity towards ethoxycoumarin.[64–66] Most of these variants may encode defective proteins. In addition, SNP polymorphisms may lead to decreased promoter activity or to variant coding sequences, accounting for the decreased metabolic rate of the HIV drug efavirenz.[67–69] Quite

Table 6.4 CYP2B6 genetic variants (http://www.cypalleles.ki.se/cyp2b6.htm): website of human cytochrome P450 alleles nomenclature committee).[82] 17-EE: 17-ethinyloestradiol.

Genetic variant of CYP2B6	*Mutations protein*	*Expression*	*Activity*
CYP2B6*1	Wild		
CYP2B6*2	R22C	decreased	
CYP2B6*3	S259R	decreased	
CYP2B6*4	K262R	decreased	Bupoprion and benzphetamine, 17-EE metabolism decreased
CYP2B6*5	R487C	decreased	
CYP2B6*6	Q172H/K262R		CPA-4 hydroxylation increased
CYP2B6*7	Q172H/K262R/R487C		
CYP2B6*8	K139E	decreased	
CYP2B6*9	Q172H		
CYP2B6*10	Q21L/R22C		
CYP2B6*11	M46V	decreased	
CYP2B6*12	G99E	decreased	
CYP2B6*13	G99E		
CYP2B6*14	K139E/Q172H/K262R		
CYP2B6*15	I391N	decreased	
CYP2B6*16	K262R/I328T	Decreased; Efavirenz impaired metabolism	Not modified
CYP2B6*17	T26S/D28G/R29T		
CYP2B6*18	I328T		
CYP2B6*19	Q172H/K262R/R236C	decreased	
CYP2B6*20	T168I/Q172H/K262R		
CYP2B6*21	P428T	decreased	
CYP2B6*22	CYP2B*1 with increased transcrption	increased	
CYP2B6*23	M459V		
CYP2B6*24	G476D		
CYP2B6*25	Q485L		

importantly, these variants may impact on the metabolism of important therapeutic drugs like artemisinin, ketamine, propofol, bupropion, cyclophosphamide and ifosfamide, thiotepa and procarbazine. However, among the most important variants (Table 6.4), none is apparently affected within the enzyme active site. In addition, because CYP2B6 also activates aflatoxin B_1,[70] which is widely produced under defective storage conditions of various cereals, into highly carcinogenic derivatives, it is highly probable that *CYP2B6* alleles may be either protective or over-active in the liver of aflatoxin B_1 exposed individuals. Similarly, cyclophosphamide metabolic activation by CYP2B6, a strategy that aims at selectively inducing cancer cell death (see below), may depend on the allelic variant selected. Correlatively, the effect of inhibitors of CYP2B6 enzyme activity, such as phenylimidazoles, may also depend on the allelic form of the gene, as shown upon comparison of mutants of the CYP2B1, CYP2B4 and CYP2B5 enzymes.[71] As of January 07, 187 SNPs have been reported (http://www.ncbi.nlm.nih.gov). Undoubtedly, additional splice or SNP variants will be uncovered in the future, which will open the path to improvements in genotype-phenotype associations, especially in the context of drug metabolism. The most important variants are listed in Table 6.4.

6.5 Suicide-based Strategies in Cancer Therapy Protocols

As mentioned, elegant CYP2B-based suicide strategies have been developped[72,73] in order to trigger cancer cell death. Based on the major contribution of CYP2B6 to the activation of cyclophosphamide (CPA), several experimental schemes have been devised. Several commonly used cancer chemotherapeutic prodrugs, including the oxazaphosphorine class, cyclophosphamide (CPA) and ifosfamide (IFA), are metabolised in the liver by a CYP-catalysed prodrug activation reaction that is required for therapeutic activity. They both undergo metabolic activation by CYP2B6 which converts them into bioactive cytotoxic metabolites. Both prodrugs are metabolised into a 4-hydroxy derivative which equilibrates with the ring-opened azophosphoramide (Figure 6.4). This aldehyde undergoes chemical decomposition to yield phosphoramide mustard and acrolein.[74] The mustard generates a highly electrophilic aziridinium species to form DNA cross-links, suggested as the key cytotoxic lesion induced in tumour cells. Although CPA and IFA are chemical isomers, each drug is metabolised by different CYP isoforms. Human CYP2B6 is the most active in CPA 4-hydroxylation while IFA 4-hydroxylation is primarily catalysed by CYP3A4. Similarly the pathway of drug inactivation by N-dechloroethylation is mainly catalysed by CYP3A4 for CPA and by CYP2B6 for CPA. As this drug inactivation yields chloracetaldehyde, a potent neurotoxic compound, the systemic distribution of metabolites formed in liver limits the use of CPA/IFA. This problem can in part be circumvented by implementation of CYP gene-directed enzyme prodrug therapy (CYP GDEPT) in tumour cells. Such a prodrug activation strategy for cancer treatment increases tumour cell exposure to cytotoxic metabolites generated locally and decreases other cell exposure to other deleterious metabolites. Preclinical studies have shown that the chemosensitivity of tumours to these prodrugs can be dramatically increased by CYP gene transfer, which confers the capability to activate the prodrug directly within the target tissue. This CYP gene-directed enzyme prodrug therapy greatly enhances the therapeutic effect of CYP-activated anticancer prodrugs without increasing host toxicity associated with systemic distribution of active drug metabolites formed by the liver.[74] Such an anticancer strategy involving the introduction of

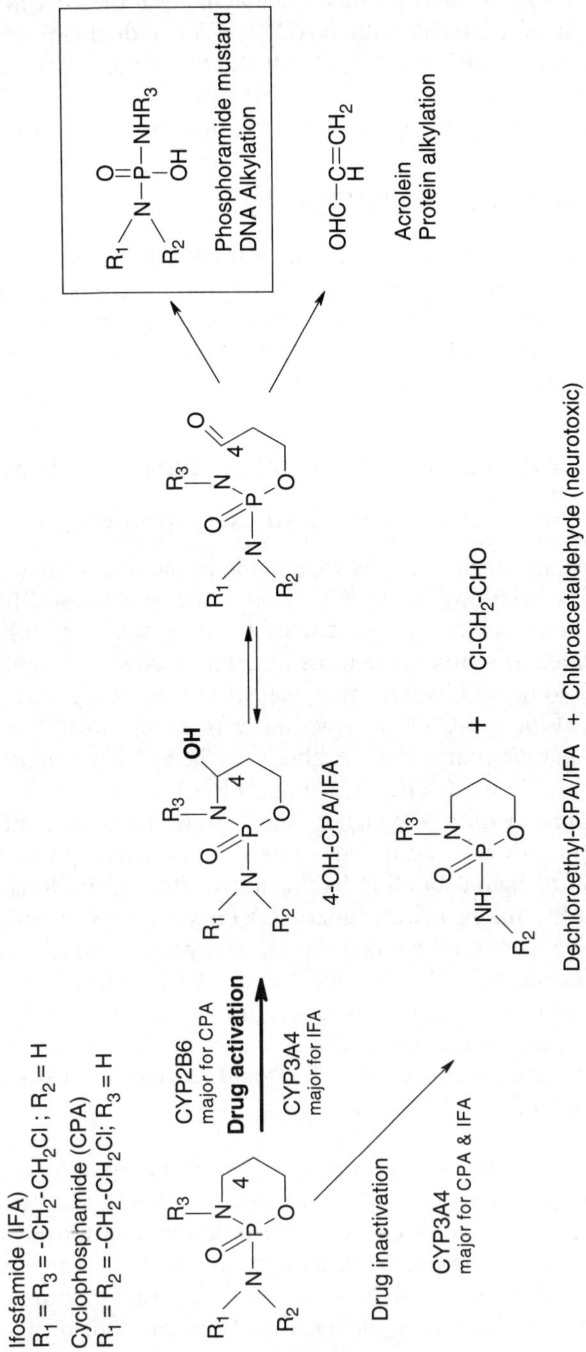

Figure 6.4 P450-catalysed pathways of CPA and IFA. R_3 = Hydrogen atom in CPA and $Cl\text{-}CH_2\text{-}CH_2$ in IFA.

CYP2B6 gene in tumour cells has been exemplified in rodent and human tumour models. Furthermore, to increase metabolic activation of CPA in tumour cells, the co-expression of CYP2B6 with NADPH-CYP reductase or with a fusion protein bearing both CYP2B and CYP reductase activities allowed higher CPA-induced cytotoxicity compared to cells expressing solely CYP2B6.[16,75] Gene directed enzyme prodrug therapy is discussed in more detail in Chapter 15.

6.6 Biochemistry of CYP2B Enzymes

Our understanding of structure-function relationships of the CYP2B enzymes has recently made remarkable progress owing to the increasing number of new CYP crystal structures.[76,77] This should allow us to get a more complete picture of the numerous mechanisms of ligand binding within the active site, and substrate specificity.

6.6.1 Structural Features of CYP2B Family Members

6.6.1.1 Structure Function by X-ray Crystallography

Recently, important breakthroughs were made in the field of mammalian CYP structure-function relationships by CYP active site modelling. Models capable of predicting the possible involvement of CYP in the metabolism of drugs or drug candidates are important tools in drug discovery. Ideally, functional information would be obtained from crystal structures. A significant step in our understanding of ligand-CYP interaction was made with the first crystal structure of a mammalian CYP, rabbit CYP2C5.[78] The structures of two additional human CYPs, CYP2C8[79] and CYP2C9 with bound warfarin[80] or flurbiprofen bound within the active site[81] were then determined. Rabbit CYP2B4 (LM2) has been crystallised[13] and has served as a prototypical model for investigation of ligand binding in the active site.[82,83] Following these first mammalian CYPs, many other human CYPs were crystallised: unliganded CYP3A4,[84] CYP3A4 bound to metyrapone and progesterone[84] or to erythromycin and ketoconazole,[85,86] CYP2D6[87] and CYP2A6 bound to coumarin.[88] These important studies raised major questions, as evoked by F.P. Guengerich: what is the nature of the active site, is it a flexible or rigid structure? what is the nature of the substrate access channel?[89,90] is the classical lock-and-key model of catalysis valuable?[90]

i. *Engineered CYP for crystallisation*: X-ray crystallography is probably the best technique to provide, at the atomic resolution, the structural information necessary to study structure-function relationships within the substrate-enzyme complex. Although the first crystal structure of a bacterial CYP was solved 20 years ago,[91] crystal structures of mammalian CYPs have been very challenging. The main difficulty was to obtain large quantities of these membrane-bound CYPs in a homogeneous state compatible with growing diffraction quality crystals. Solution ready to crystallise must contain about 0.3[84] to 0.6 mM CYP.[13] The plasmid was modified to express CYP without the trans-membrane sequence of the microsomal protein in order to improve its solubility, avoid aggregation and facilitate crystallisation. Amino acids 3–23[84] or 3–21 were truncated. However, truncation of residues 3–20 or 3–25 of CYP2B enzymes results in lower expression levels than the full-length proteins. To obviate this difficulty, several positively charged lysine residues were incorporated into the N-terminus helix to increase solubility. By this strategy, the expression levels of CYP2B enzymes increased up to 500 nM,[13] except for CYP2B6 that did not exceed 100 nM. A second modification was applied to the engineered protein, upon addition of a 4xHis tag to the C-terminus to facilitate purification by Ni^{2+} affinity chromatography. The resulting CYP was dissociated from the membrane by high salt buffer in the absence of detergent and purified to about 25 nmol/mg protein.[77]

ii. *Ligand-free CYP2B4 structure*: CYP2B6 retains the main tertiary structure of CYP, with several β-sheets and many α-helices.[88,90] The core of the protein, which is constituted by helices E, J, K and L, is highly conserved. But the original finding is that CYP2B4 reveals a large open cleft (see Protein data bank as 1P05), which extends from the protein surface directly to the haem iron between the α-helical and β-sheet domains without perturbing the overall CYP fold.[13] As a result, this open conformation easily allows the diverse range of drug substrates, even bulky ones, to access the vicinity of the haem group without the constrictions imposed by an access channel. Furthermore, the open conformation of CYP2B4 is blocked by reversible formation of a homodimer in which the open cleft is partially filled by residues of the second monomer. An alternating equilibrium between an open form and a closed form facilitates the sequestering of the substrate in the active site.

iii. *Crystal structures of CYP2B4 bound to bifonazole or imidazoles*: The binding of a substrate or inhibitor needs a large structural rearrangement towards a closed conformation compatible with the key-and-lock model. The regions surrounding the open cleft of the ligand-free structure move towards each other to sequester the inhibitor molecules.[82,83] The imidazole molecule is then tightly packed within the active site with no apparent entry or no egress channel. With another more bulky inhibitor than imidazole homologues, bifonazole, the CYP2B4 is forced to expand its active site. This finding of the Halpert laboratory was confirmed by a recent work describing the ligand binding to CYP3A4,[86] a major enzyme engaged in the metabolism of about 75% of drugs on the market. Thus, the binding of ketoconazole, a relatively large molecule, to CYP3A4 increases the size of the active site by >80%. Furthermore, two ketoconazole molecules bind in a stacked configuration. The presence of two molecules in the active site was also observed in CYP2C9 bound with warfarin.[79] Interestingly, the known site of N-demethylation of erythromycin bound to CYP3A4 is 17 Å from the haem iron. Such a structure must be considered as a non productive complex. The binding of two molecules to the active site either in CYP2C9 or CYP3A4 suggests a basis for interpretation of the atypical kinetics often displayed by these catalysts. In conclusion, of the recent findings obtained by crystallography of mammalian CYPs, it must be emphasised that the CYP enzyme has considerable flexibility, which must be considered at least as part of the basis for the broad and overlapping substrate specificity. Thus, this malleable[90] catalyst challenges any attempt to apply computational design tools with the support of relevant experimental data.

6.6.1.2 Study of Ligand-CYP Interaction by Surface Plasmon Spectroscopy (SPR)

A new detection mechanism has been recently developed[91] for low molecular weight substrate binding to haem proteins based on resonance localised surface plasmon spectroscopy. The CYP has strong electronic transitions in the visible wavelength region. Upon binding of a substrate molecule, the absorption band of CYP shifts to shorter wavelength. The event of substrate binding to a nanoparticle surface modified with CYP protein acting as receptor is monitored using UV-vis spectroscopy. It was observed, for the first time, that the binding of the substrates to the protein induces a blue-shift in the localised surface plasmon resonance (LSPR) of the nanosensors. The coupling between the molecular resonance of the substrate-free and substrate-bound CYP proteins and the nanoparticles in LSPR leads to a highly wavelength-dependent LSPR response. When the LSPR of the nanoparticles is located at a wavelength distant from the CYP resonance, an average of $\sim$19 nm red-shift is observed upon CYP binding to the nanoparticles and a $\sim$6 nm blue-shift is observed upon substrate (camphor) binding. However, this response is significantly amplified $\sim$3 to 5 times when the LSPR of the nanoparticles is located at a slightly longer wavelength than the CYP resonance, that is a 66.2 nm red-shift upon CYP binding and a 34.7 nm blue-shift upon camphor binding. This is the first example of the detection of small molecules binding to a protein modified nanoparticle surface on the basis of LSPR. This new method shows great promise as a novel means to study CYP structure-function relationships.

6.7 Conclusions and Future Prospects

Many years ago, the existence of the mammalian CYP2B subfamily has been revealed mainly through induction of a distinct haemoprotein absorbing at 450 nm, as opposed to cytochrome P448. These differences in the CO absorption spectrum indeed paved the way to uncovering a highly specialised CYP subfamily, as we know it and as we keep learning about it. The biochemistry of CYP interactions with substrates and inhibitors has been finely developed recently through major findings pertaining to CYP2B sub-family members. In addition to their contribution to several metabolic pathways, CYP2B enzymes appear to bear a specific importance depending on variations

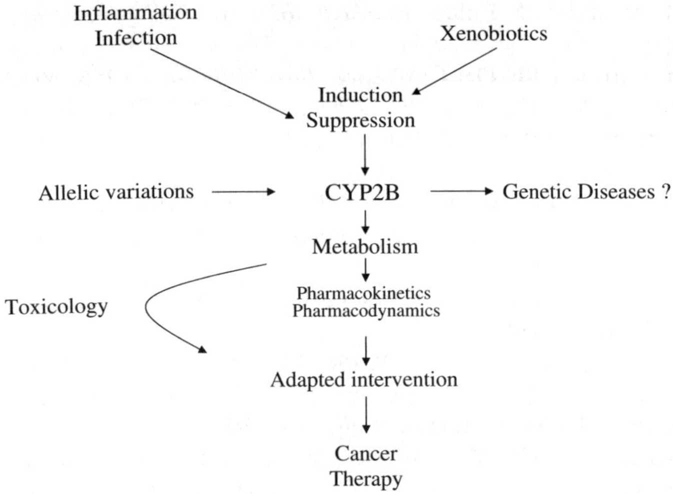

Figure 6.5 Place and prospects for CYP2B enzymes.

of their expression levels, as a result of gender specificity, inducibility and allele composition. Taking these variations into account should be part of pharmacogenetic- and pharmacogenomic-based strategies for individualised drug treatments. Already, in the field of cancer therapy, promising applications have been developed that make use of one of the hallmarks of CYPs, namely the ability to metabolically activate drugs. Apart from the cancer gene therapy approach, it may well be that specific, yet highly efficient, natural allelic forms of CYP2Bs lead to a better outcome of cancer treatment or could even be part of a cancer chemopreventive strategy in appropriate carriers. Such a prospect should, however, await that the functionality of all known allelic forms is determined, and that the best drug-CYP2B combination be identified. Figure 6.5 depicts the place and prospects for CYP2B enzymes.

Acknowledgements

This work was supported by the INSERM, the University of Bretagne Occidentale (Brest), the Faculty of Medicine (Brest), the Ligue Contre le Cancer and the region of Brittany.

References

1. T. Omura and R. Sato, *J. Biol. Chem.*, 1962, **237**, 1375–1376.
2. C. Ioannides and D.V. Parke, *Biochem. Pharmacol.*, 1987, **36**, 4197–4207.
3. M. Atchison and M. Adesnik, *J. Biol. Chem.*, 1983, **258**, 11285–11295.
4. M. Atchison and M. Adesnik, *Proc. Natl. Acad. Sci. USA*, 1986, **83**, 2300–2304.

5. B.H. Kim and A.J. Fulco, *Biochem. Biophys. Res. Commun.*, 1983, **116**, 843–850.
6. G. Bronzetti and R. Del Carratore, *Basic Life Sci.*, 1986, **39**, 343–348.
7. J.S. He and A.J. Fulco, *J. Biol. Chem.*, 1991, **266**, 7864–7869.
8. E. Trottier, A. Belzil, C. Stoltz and A. Anderson, *Gene*, 1995, **158**, 263–268.
9. P. Honkakoski and M. Negishi, *J. Biol. Chem.*, 1997, **272**, 14943–14949.
10. P.M. Shaw, M. Adesnik, M.C. Weiss and L. Corcos, *Mol. Pharmacol.*, 1993, **44**, 775–783.
11. P. Honkakoski, I. Zelko, T. Sueyoshi and M. Negishi, *Mol. Cell Biol.*, 1998, **18**, 5652–5658.
12. S.A. Kliewer, J.T. Moore, L. Wade, J.L. Staudinger, M.A. Watson, S.A. Jones, D.D. McKee, B.B. Oliver, T.M. Willson, R.H. Zetterstrom, T. Perlmann and J.M. Lehmann, *Cell*, 1998, **92**, 73–82.
13. E.E. Scott, Y.A. He, M.R. Wester, M.A. White, C.C. Chin, J.R. Halpert, E.F. Johnson and C.D. Stout, *Proc. Natl. Acad. Sci. USA*, 2003, **100**, 13196–13201.
14. L. Chen and D.J. Waxman, *Cancer Res.*, 1995, **55**, 581–589.
15. Y. Jounaidi and D.J. Waxman, *Cancer Res.*, 2000, **60**, 3761–3769.
16. M. Tychopoulos, L. Corcos, P. Genne, P. Beaune and I. de Waziers, *Cancer Gene Ther.*, 2005, **12**, 497–508.
17. M. Desrochers, M. Christou, C. Jefcoate, A. Belzil, A. Anderson, *Biochem. Pharmacol.*, 1996, Oct 25;52(8):1311–1319.
18. J.S. Miles, A.W. McLaren and C.R. Wolf, *Nucleic Acids Res.*, 1989, **17**, 8241–8255.
19. K.M. Kedzie, C.A. Balfour, G.Y. Escobar, S.W. Grimm, Y.A. He, D.J. Pepperl, J.W. Regan, J.C. Stevens and J.R. Halpert, *J. Biol. Chem.*, 1991, **266**, 22515–22521.
20. R.W. Nims and R.A. Lubet, *J. Toxicol. Environ. Health*, 1995, **46**, 271–292.
21. T.K. Chang, C.L. Crespi and D.J. Waxman, *Methods Mol. Biol.*, 2006, **320**, 97–102.
22. M. Turpeinen, H. Raunio and O. Pelkonen, *Curr. Drug Metab.*, 2006, **7**, 705–714.
23. E. Hodgson, R.L. and Rose (2006). The importance of cytochrome P450 2B6 in the human metabolism of environmental chemicals. *Pharmacol Ther.*, in press.
24. R.L. Walsky, A.V. Astuccio and R.S. Obach, *J. Clin. Pharmacol.*, 2006, **46**, 1426–1438.
25. C.M. Giachelli and C.J. Omiecinski, *J. Biol. Chem.*, 1986, **261**, 1359–1363.
26. C.J. Omiecinski, C. Hassett and P. Costa, *Mol. Pharmacol.*, 1990, **38**, 462–470.
27. M. Mimura, T. Baba, H. Yamazaki, S. Ohmori, Y. Inui, F.J. Gonzalez, F.P. Guengerich and T. Shimada, *Drug Metab. Dispos.*, 1993, **21**, 1048–1056.
28. J. Hukkanen, O. Pelkonen, J. Hakkola and H. Raunio, *Crit. Rev. Toxicol.*, 2002, **32**, 391–411.

29. T. Shimada, H. Yamazaki, M. Mimura, Y. Inui and F.P. Guengerich, *J. Pharmacol. Exp. Ther.*, 1994, **270**, 414–423.
30. V. Lamba, J. Lamba, K. Yasuda, S. Strom, J. Davila, M. L. Hancock, J.D. Fackenthal, P. K. Rogan, B. Ring, S. A. Wrighton, & E. G. Schuetz, (2003).
31. I.H. Hanna, J.R. Reed, F.P. Guengerich and P.F. Hollenberg, *Arch. Biochem. Biophys.*, 2000, **376**, 206–216.
32. N. Harada and M. Negishi, *Proc. Natl. Acad. Sci. USA*, 1985, **82**, 2024–2028.
33. M. Noshiro and M. Negishi, *J. Biol. Chem.*, 1986, **261**, 15923–15927.
34. H.W. Davey, S.H. Park, D.R. Grattan, M.J. McLachlan and D.J. Waxman, *J. Biol. Chem.*, 1999, **274**, 35331–35336.
35. B.H. Shapiro, N.A. Pampori, D.P. Lapenson and D.J. Waxman, *Arch. Biochem. Biophys.*, 1994, **312**, 234–239.
36. K. Jarukamjorn, T. Sakuma and N. Nemoto, *Biochem. Pharmacol.*, 2002, **63**, 2037–2041.
37. Y. Yamazoe, N. Murayama, M. Shimada, K. Yamauchi and R. Kato, *Arch. Biochem. Biophys.*, 1989, **268**, 567–575.
38. B.S. Donahue and E.T. Morgan, *Drug Metab. Dispos.*, 1990, **18**, 519–526.
39. Z. Abdel-Razzak, L. Corcos, A. Fautrel and A. Guillouzo, *FEBS Lett.*, 1995, **366**, 159–164.
40. O. Khatsenko and Y. Kikkawa, *J. Pharmacol. Exp. Ther.*, 1997, **280**, 1463–1470.
41. T. Li-Masters and E.T. Morgan, *Drug Metab. Dispos.*, 2001, **29**, 252–257.
42. P.J. Van Ess, S. Poloyac, M.P. Mattson and R.A. Blouin, *Pharm. Res.*, 2002, **19**, 708–712.
43. Z. Abdel-Razzak, P. Loyer, A. Fautrel, J.C. Gautier, L. Corcos, B. Turlin, P. Beaune and A. Guillouzo, *Mol. Pharmacol.*, 1993, **44**, 707–715.
44. E.L. LeCluyse, *Eur. J. Pharm. Sci.*, 2001, **13**, 343–368.
45. C. Lerche, A. Fautrel, P.M. Shaw, D. Glaise, F. Ballet, A. Guillouzo and L. Corcos, *Eur. J. Biochem.*, 1997, **244**, 98–106.
46. J. B. Ferrini, L. Pichard, J. Domergue and P. Maurel, *Chem. Biol. Interact.*, 1997, **107**, 31–45.
47. G. Tuschl and S.O. Mueller, *Toxicology*, 2006, **218**, 205–215.
48. J.Z. Tong, P. De Lagausie, V. Furlan, T. Cresteil, O. Bernard and F. Alvarez, *Exp. Cell Res.*, 1992, **200**, 326–332.
49. E.S. Tzanakakis, C.C. Hsiao, T. Matsushita, R.P. Remmel and W.S. Hu, *Cell Transplant*, 2001, **10**, 329–342.
50. L. Corcos and M.C. Weiss, *FEBS Lett.*, 1988, **233**, 37–40.
51. J. Doehmer, S. Dogra, T. Friedberg, S. Monier, M. Adesnik, H. Glatt and F. Oesch, *Proc. Natl. Acad. Sci. USA*, 1988, **85**, 5769–5773.
52. J.L. Ferrero and K. Brendel, *Adv. Pharmacol.*, 1997, **43**, 131–169.
53. X. Cui, A. Thomas, Y. Han, J. Palamanda, D. Montgomery, R.E. White, R.A. Morrison and K.C. Cheng, *J. Pharmacol. Toxicol. Methods*, 2005, **52**, 234–243.
54. Y. Park, H. Li and B. Kemper, *J. Biol. Chem.*, 1996, **271**, 23725–23728.

55. M. Damon, A. Fautrel, A. Guillouzo and L. Corcos, *Biochem. J.*, 1996, **317**(Pt 2), 481–486.
56. T.J. Dunn, A.J. Koleske, R. Lindahl and H.C. Pitot, *J. Biol. Chem.*, 1989, **264**, 13057–13065.
57. K.M. Sommer, R. Ramsden, J. Sidhu, P. Costa and C.J. Omiecinski, *Pharmacogenetics*, 1996, **6**, 369–374.
58. P.V. Luc, M. Adesnik, S. Ganguly and P.M. Shaw, *Biochem. Pharmacol.*, 1996, **51**, 345–356.
59. P. Honkakoski, R. Moore, K.A. Washburn and M. Negishi, *Mol. Pharmacol.*, 1998, **53**, 597–601.
60. N. Marc, M. Galisteo, D. Lagadic-Gossmann, A. Fautrel, F. Joannard, A. Guillouzo and L. Corcos, *Eur. J. Biochem.*, 2000, **267**, 963–970.
61. T. Kawamoto, T. Sueyoshi, I. Zelko, R. Moore, K. Washburn and M. Negishi, *Mol. Cell Biol.*, 1999, **19**, 6318–6322.
62. J.S. Miles, A.W. McLaren, F.J. Gonzalez and C.R. Wolf, *Nucleic Acids Res.*, 1990, **18**, 189.
63. C. Rodriguez-Antona and M. Ingelman-Sundberg, *Oncogene*, 2006, **25**, 1679–1691.
64. K. Nagata and Y. Yamazoe, *Drug Metab. Pharmacokinet*, 2002, **17**, 167–189.
65. T. Lang, K. Klein, J. Fischer, A.K. Nussler, P. Neuhaus, U. Hofmann, M. Eichelbaum, M. Schwab and U.M. Zanger, *Pharmacogenetics*, 2001, **11**, 399–415.
66. N. Ariyoshi, M. Miyazaki, K. Toide, Y. Sawamura and T. Kamataki, *Biochem. Biophys. Res. Commun.*, 2001, **281**, 1256–1260.
67. K. Klein, T. Lang, T. Saussele, E. Barbosa-Sicard, W.H. Schunck, M. Eichelbaum, M. Schwab and U.M. Zanger, *Pharmacogenet Genomics*, 2005, **15**, 861–873.
68. J. Wang, A. Sonnerborg, A. Rane, F. Josephson, S. Lundgren, L. Stahle and M. Ingelman-Sundberg, *Pharmaco.genet. Genomics*, 2006, **16**, 191–198.
69. D. Nolan, E. Phillips and S. Mallal, *Clin. Infect. Dis.*, 2006, **42**, 408–410.
70. M. Spatzenegger, Q. Wang, Y.Q. He, M.R. Wester, E.F. Johnson and J.R. Halpert, *Mol. Pharmacol.*, 2001, **59**, 475–484.
71. M.X. Wei, T. Tamiya, M. Chase, E.J. Boviatsis, T.K. Chang, N.W. Kowall, F.H. Hochberg, D.J. Waxman, X.O. Breakefield and E.A. Chiocca, *Hum. Gene Ther.*, 1994, **5**, 969–978.
72. L. Chen and D.J. Waxman, *Curr. Pharm. Des.*, 2002, **8**, 1405–1416.
73. P. Roy and D.J. Waxman, *Toxicol. In vitro*, 2006, **20**, 176–186.
74. L. Chen, L.J. Yu and D.J. Waxman, *Cancer Res.*, 1997, **57**, 4830–4837.
75. H. Li and T.L. Poulos, *Curr. Top. Med. Chem.*, 2004, **4**, 1789–1802.
76. Y. Zhao and J.R. Halpert, *Biochim. Biophys. Acta*, 2007, **1770**, 402–412.
77. P.A. Williams, J. Cosme, V. Sridhar, E.F. Johnson and D.E. McRee, *J. Inorg. Biochem.*, 2000, **81**, 183–190.
78. G.A. Schoch, J.K. Yano, M.R. Wester, K.J. Griffin, C.D. Stout and E.F. Johnson, *J. Biol. Chem.*, 2004, **279**, 9497–9503.
79. P.A. Williams, J. Cosme, A. Ward, H.C. Angove, D. Matak Vinkovic and H. Jhoti, *Nature*, 2003, **424**, 464–468.

80. M.R. Wester, J.K. Yano, G.A. Schoch, C. Yang, K.J. Griffin, C.D. Stout and E.F. Johnson, The structure of human cytochrome P450 2C9 complexed with flurbiprofen at 2.0- Å resolution, *J. Biol. Chem.*, 2004, **279**, 35630–35637.
81. B.K. Muralidhara, S. Negi, C.C. Chin, W. Braun and J.R. Halpert, *J. Biol. Chem.*, 2006, **281**, 8051–8061.
82. Y. Zhao, M.A. White, B.K. Muralidhara, L. Sun, J.R. Halpert and C.D. Stout, *J. Biol. Chem.*, 2006, **281**, 5973–5981.
83. J.K. Yano, M.R. Wester, G.A. Schoch, K.J. Griffin, C.D. Stout and E.F. Johnson, *J. Biol. Chem.*, 2004, **279**, 38091–38094.
84. P.A. Williams, J. Cosme, D.M. Vinkovic, A. Ward, H.C. Angove, P.J. Day, C. Vonrhein, I.J. Tickle and H. Jhoti, *Science*, 2004, **305**, 683–686.
85. M. Ekroos and T. Sjogren, *Proc. Natl. Acad. Sci. USA*, 2006, **103**, 13682–13687.
86. P. Rowland, F.E. Blaney, M.G. Smyth, J.J. Jones, V.R. Leydon, A.K. Oxbrow, C.J. Lewis, M.G. Tennant, S. Modi, D.S. Eggleston, R.J. Chenery and A.M. Bridges, Crystal structure of human cytochrome P450 2D6, *J. Biol. Chem.*, 2006, **281**, 7614–7622.
87. J.K. Yano, M.H. Hsu, K.J. Griffin, C.D. Stout and E.F. Johnson, *Nat. Struct. Mol. Biol.*, 2005, **12**, 822–823.
88. S.E. Graham and J.A. Peterson, *Arch. Biochem. Biophys.*, 1999, **369**, 24–29.
89. F.P. Guengerich, *Proc. Natl. Acad. Sci. USA*, 2006, **103**, 13565–13566.
90. T.L. Poulos, B.C. Finzel, I.C. Gunsalus, G.C. Wagner and J. Kraut, *J. Biol. Chem.*, 1985, **260**, 16122–16130.
91. J. Zhao, A. Das, X. Zhang, G.C. Schatz, S.G. Sligar and R.P. Van Duyne, Resonance surface plasmon spectroscopy: low molecular weight substrate binding to cytochrome p450, *J. Am. Chem. Soc.*, 2006a, **128**, 11004–11005.

CHAPTER 7

The CYP2C Subfamily

STEPHEN S. FERGUSON,[a] KAREN BLACK[a] AND JONATHAN P. JACKSON[b]

[a] CellzDirect, Durham, NC 27703; [b] CellzDirect, Austin TX 78754

Table of Contents

Issues in Toxicology
Cytochromes P450: Role in the Metabolism and Toxicity of Drugs and other Xenobiotics
Edited by Costas Ioannides

7.1 Introduction to the CYP2C Subfamily, Importance and Scope

The CYP2C subfamily of metalloenzymes is generally accepted as the second most pharmacologically important group of CYP450s in humans after the CYP3A subfamily due to its prominent role in human drug metabolism. The CYP2Cs are involved in the metabolism of numerous clinically-important drugs such as omeprazole, warfarin, mephenytoin and paclitaxel, to name a few.[1] In addition, the CYP2C subfamily is involved in the metabolism of several endogenous substrates important for normal physiological processes (e.g. cell signaling) such as arachidonic acid and retinoic acid.[2,3] In humans, the CYP2C subfamily is comprises four enzymes: CYP2C8, CYP2C9, CYP2C18 and CYP2C19, each of which is localised on in a single gene cluster on chromosome 10. Of these four isozymes, CYP2C9 is believed to be the most pharmacologically important due to its high expression in human liver and established role in the metabolism of numerous drugs. CYP2C8 and, to a lesser extent, CYP2C19 are known to be expressed in human liver, and have also play important role in human drug metabolism that will be described in more detail (*vide infra*). In contrast, CYP2C18, while expressed at the mRNA level in human liver, has not been shown to be expressed at the protein level and does not appear to play a significant role in drug metabolism.[4]

Although human metabolism catalysed by the CYP2C enzymes has historically been an important area of research, studies in rodent models such as rat, and, to a less frequent extent, mouse, along with studies in rabbit, dog, monkey, and pig have revealed that the CYP2C enzymes play a clear role in the hepatic metabolism of all these species. These model organisms provide essential data to better predict the effects of new chemical entities in humans. However, due to species differences on a variety of levels from enzyme structure and metabolic activity to tissue-specific expression and differential numbers of genes, and responses to inducers and inhibitors, it is often difficult to translate data from animal models to prediction of human metabolism and drug-drug interactions. Therefore, understanding the mechanisms by which these animal models (and the human system) handle xenobiotics and how these systems are altered over time by xenobiotic exposure, is necessary to effectively translate animal model data to human relevance.

In addition to the metabolism of drugs and endogenous substrates by CYP2C enzymes, advances in the areas of pharmacogenetics, selective substrates, selective inhibitors and the pathways that govern CYP2C expression (e.g. induction/constitutive regulation via nuclear receptors) have provided researchers with the tools to explore the mechanisms of pharmacokinetic and pharmacodynamic interactions. The sections below will describe recent progress made in our understanding of the CYP2C proteins on multiple levels, the theme being that our knowledge of the mechanisms by which these processes occur will lead to more rational drug design and predictive model systems.

7.2 CYP2C Chromosomal Comparisons: Human, Mouse and Rat

In humans, the CYP2C subfamily is comprised of four gene products that produce complete CYP450 gene products localised on a single locus within Chromosome 10 (q24) as depicted in Figure 7.1A (derived from both NCBI's Mapviewer utility and David Nelson's Cytochrome P450 internet site (http://drnelson.utmem.edu/CytochromeP450.html). Each gene is composed of nine exons, and the large size of the *CYP2C19* gene is the result of a large intron of approximately 25 kb.

A comparison of the genetic sequences of the coding regions of the human CYP2C subfamily is shown in the phylogenetic tree depicted in Figure 7.1B (generated by sequence alignments using VectorNTI). It is noteworthy that the CYP2C9 and CYP2C19 coding sequences are greater than 95% homologous, which distinguishes them as the most homologous members of the human CYP2C subfamily. This homology is also present in the 5′-flanking regions of these genes up to approximately 4500 bp upstream of the translation start site.

Unlike CYP450s that are conserved across mammalian species, such as CYP2E1 and CYP1A1, the CYP2C proteins do not express direct orthologues among mammal species. In fact, recent advances in the genetic sequencing of

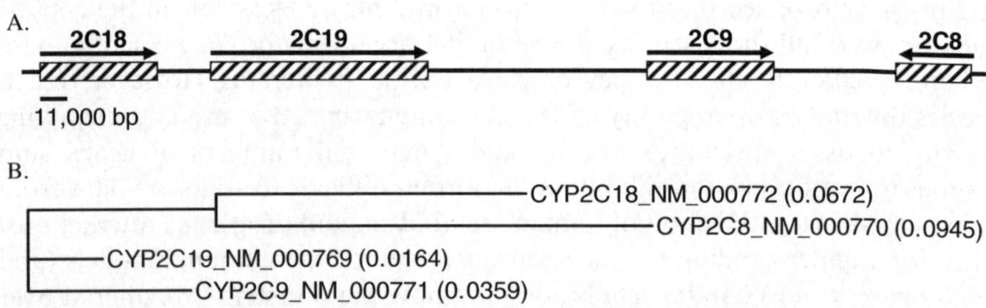

Figure 7.1 (A) Human *CYP2C* Gene Cluster (Chromosome 10q24 contig NT_030059.8). (B) Phylogenetic Tree of Human CYP2C cDNA Sequences.

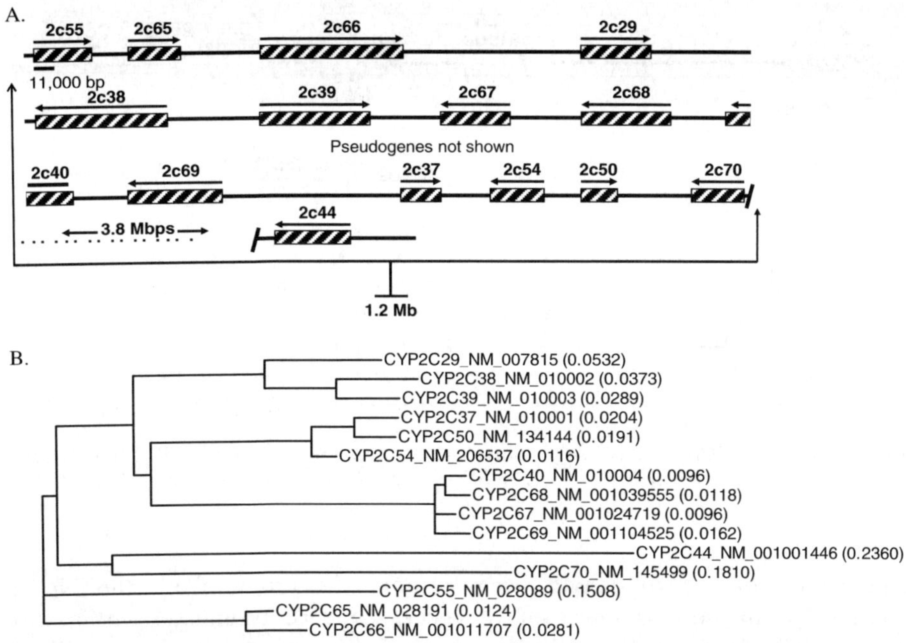

Figure 7.2 (A) Murine *CYP2C* Gene Cluster (Chromosome 19C3 contig NT_039687.7). (B) Phylogenetic Tree of Murine *CYP2C* cDNA Sequences.

the mouse and rat genomes have revealed that unlike the human CYP2C subfamily, the mouse and rat subfamilies are comprised of many more members (Figures 7.2A and 7.3A respectively).

In the mouse, all the *CYP2C* genes are localised on chromosome 19, and 14 of the 15 full length genes are located within a single cluster analogous to the human structure (Figure 7.2A).

Interestingly, one member, *CYP2C44*, is not located within this locus. This unique chromosomal location for *CYP2C44* is matched by its unique metabolic properties, expression profile and transcriptional regulation.[5,6] Comparison of the cDNA sequences of these genes reveals that *CYP2C29*, *CYP2C38* and *CYP2C39* share a high homology cluster along with *CYP2C40*, *CYP2C67*, *CYP2C68* and *CYP2C69* (Figure 7.2B), while *CYP2C44* is the most unique member of this subfamily in mice with some homologous regions with *CYP2C70*. When compared to the human CYP2C, none of the mouse genes is orthologous to any of the human genes. However, comparison of CYP2C9 with the murine *CYP2Cs* reveals that *CYP2C65* and *CYP2C66* are the most homologous in terms of protein sequence sharing 75.2% and 75.4% homology, respectively. This was also the case with CYP2C8 (72.1% and 71.5% respectively).

In the rat, all the *CYP2C* genes are localised on chromosome 1 (q53–q54) and are located within a single large cluster (Figure 7.3A). However,

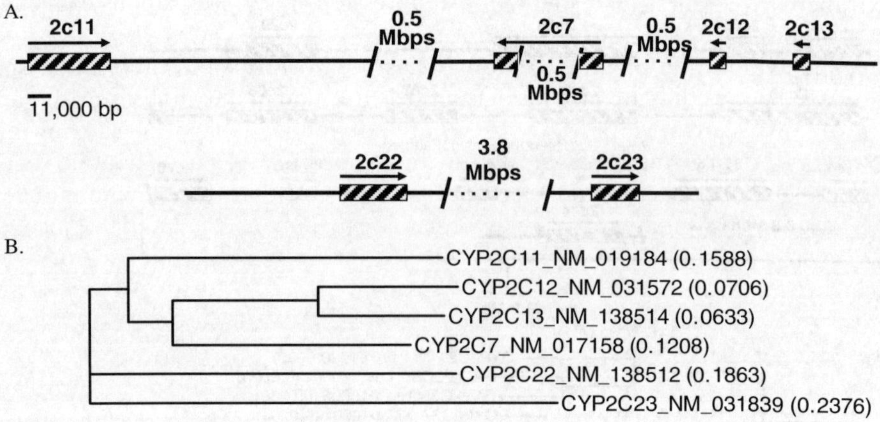

Figure 7.3 (A) Rat *CYP2C* Gene Cluster (Chromosome 1q53-q54). (B) Phylogenetic Tree of Rat *CYP2C* cDNA Sequences Currently Annoted (NCBI-2007).

comparison of the full-length genes annotated to date from the NCBI database is difficult as many of the gene sequence fragments within this region have not been localised, and several appear to be predicted sequences. Comparison of the genetic sequence of the coding regions of the well-annotated full length gene products reveals that *CYP2C12* and *CYP2C13* are relatively homologous, while better characterised genes *CYP2C7* and *CYP2C11* have relatively unique nucleotide sequences (Figure 7.3B). When compared to the human CYP2C, none of the rat genes is orthologous to any of the human genes.

Although there are similarities in the gene structure of human, mouse and rat *CYP2C*, there are marked differences in the numbers of genes and genetic sequences. The reason for this stark contrast is not clear. It has been proposed that rodent species evolved this diversity through consumption of a more metabolic-challenging diet; therefore, an enhanced ability to handle a variety of food sources (and the xenobiotic contaminants that may be included) was helpful to the rodents continued survival.[7] What is clear is that there are no direct orthologues between the human and rodent systems based on CYP450 homology or function analogous to what is known, for example, for CYP2E1. Therefore detailed knowledge indicating which CYP2C enzymes are essential for the metabolism of a drug in the mouse, for example, is not likely to be as successful a predictor of human metabolism as a translational prediction for CYP2E1 across species.[8] In addition, the CYP2C in rodents is thought to be the most highly expressed subfamily in liver, while in humans the CYP3A subfamily comprises the majority of the liver.[9] A more thorough description of specific examples species differences between human and rodent CYP2C enzymes is included in respective sections (*vide infra*), as it relates to the metabolism, endogenous expression in liver and extrahepatic tissues, and induction/suppression responses upon challenge by xenobiotics.

Some exploration of CYP2C biology has been described in rabbit, dog and monkey, species that are also used as models for human metabolism. Although the genome sequences for these species are not yet available, what is known is that CYP2C proteins are expressed in these species and participate in xenobiotic metabolism.[10] Metabolic comparisons suggest that the dog and monkey are often more similar to profiles observed in humans than rodent species; however species differences remain difficult to address due to the lack of published research in this area.

7.3 Contribution of CYP2C to Drug and Xenobiotic Metabolism

In humans the CYP2C subfamily is a major contributor to drug metabolism and appears to play an essential role in the detoxification of drugs. Numerous pharmacologically important drugs including the antiulcer omeprazole, the anti-coagulant warfarin, the anticancer drug paclitaxel, the anticonvulsant mephenytoin, and a number of non steroidal anti-inflammatory drugs (NSAIDS)[1] are metabolised by CYP2C. Table 7.1 summarises an abbreviated list of drugs, and therapeutic properties for which the CYP2Cs subfamily is thought to play a minor role in clearance. Most of the metabolism for these drugs, as it relates to clearance, occurs in the liver and/or intestine where the preponderance of CYP2C enzymes is located. However, metabolism in other tissues could also contribute to the total metabolism of these drugs. Global examination of this list of drugs and their therapeutic uses provides a perspective highlighting the diversity of therapeutic efficacy and clearance pathways, even within a specific CYP subfamily. A more detailed description of several pharmacologically important substrates for each of the important human CYP2Cs is described below.

7.3.1 Metabolism Catalysed by CYP2C8

CYP2C8 appears to be involved in the metabolism of a growing list of clinically-relevant drugs (Table 7.1). One of the most notable CYP2C8 substrates is paclitaxel (Taxol®, Figure 7.4), which was originally isolated from the tree bark of the Pacific yew tree in 1967 by Wall and Wani.[11] Paclitaxel is a mitotic inhibitor currently utilised as a leading drug for the treatment of soft tumours (sarcoma) in tissues such as breast, ovary and lung. Moreover, paclitaxel has emerged as the prototypical model substrate used to study CYP2C8-mediated metabolism via formation of a 6-α hydroxyl metabolite. Paclitaxel has also been used as an anti-proliferative agent to minimise scar tissue formation (e.g. coating shunts utilised in cardiac disorders).[12] It appears that paclitaxel is also

Table 7.1 Summary of drugs metabolised by CYP2C enzymes in humans and clinical indication.

CYP2C8	*CYP2C9*	*CYP2C19*
Diabetes	Diabetes	Ulcers
rosiglitazone	tolbutamide	omeprazole
troglitazone	glipizide	lansoprazole
pioglitazone	nateglinide	rabeprazole
repaglinide	rosiglitazone	
	glyburide	
Malaria	Inflammation/pain	Inflammation/pain
amodiaquine	diclofenac	indomethacin
chloroquine	ibuprofen	
dapsone	naproxen	
	piroxicam	
	lornoxicam	
	celecoxib	
	rofecoxib	
Cholesterol-lowering	Cholesterol-lowering	Hypertension
cerivastatin	fluvastatin	propranolol
simvastatin		
Diuretic	Diuretic	Epilepsy/Sedative
Torsemide	torsemide	diazepam
		phenobarbital
		S-mephenytoin
		phenytoin
		hexobarbital
		primidone
Cancer	Epilepsy	Muscle Relaxant
Paclitaxel	phenytoin	Carisoprodol
Diarrhoea	Clotting/Coagulation	Depression
Loperamide	(S)-warfarin	imipramine
		amitriptyline
		citalopram
	Depression	Cancer
	fluoxetine	Cyclophosphamide
	amitriptyline	Nilutamide
	Hypertension	Antiviral
	Tienilic Acid	Nelfinavir

an effective drug in the treatment of brain tumours, but issues relating to delivery to the blood-brain barrier (paclitaxel is a substrate of P-gp (ABC1B1)) have limited its efficacy in this regard.[13,14] Paclitaxel is a relatively large, multifaceted molecule that is predominantly metabolised by CYP2C8 with some contributions from CYP3A.[15] Meanwhile, structurally dissimilar drugs (Table 7.1) such as rosiglitazone (used in the treatment of type II diabetes and a peroxisome proliferator-activated receptor γ(PPARγ) agonist), troglitazone, and various cholesterol-lowering drugs like cerivastatin (3-hydroxy-3-methylglutaryl-CoA reductase (HMG-Co-A reductase inhibitor) are also metabolised

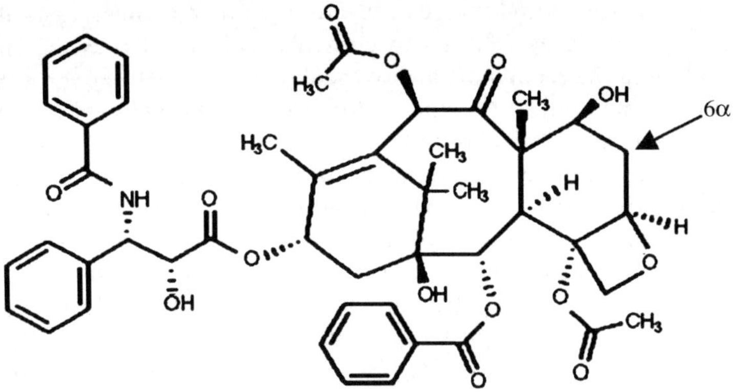

Figure 7.4 Taxol chemical structure.

to some extent by CYP2C8 in human liver.[16,17] It is not clear why these diverse substrates are metabolised by CYP2C8; however a recent crystal structure of this enzyme is allowing researchers to explore the molecular details of CYP2C8 metabolism.[18,19]

Several type II diabetes medicines are metabolised by CYP2C8. This list includes: rosiglitazone, troglitazone, pioglitazone, nateglinide and repaglinide. Repaglinide metabolism by CYP2C8 proceeds via hydroxylation of the piperidine ring system while CYP3A4 appears to also contribute by producing an aromatic amine metabolite.[20] CYP2C8 a hydroxylaties of rosiglitazone, but also catalyses the transformation to an N-demethyl metabolite.[21] Structurally-related thiazoladinediones also are metabolised to similar oxidative metabolites, catalysed by CYP2C8, with the exception of troglitazone that generates a quinine-type metabolite.[22,23]

Several antimalarial drugs are also metabolised by CYP2C8. Amodiaquine is a 4-aminoquinoline derivative that has been effectively used in the treatment of malaria over the past 50 years. These antimalarials appear to be metabolised primarily via N-deethylation transformations.[24] Metabolism of statins such as cerivastatin proceeds via both hydroxylation and oxidative demethylation.[25]

Overall, CYP2C8-mediated metabolism is emerging as an important route for the elimination of a growing number of clinically-prescribed drugs. As a result, drug-drug interactions are also being increasingly observed and are discussed further in the subsequent sections in relation to the mechanisms responsible (inhibition, induction, pharmacogenetic factors).

CYP2C8 is also involved in the metabolism of endogenous substrates in humans and rodents. Arachidonic acid (AA) is an important cell-signalling molecule in mammals, and is an endogenous substrate of CYP2C8. AA metabolites are involved in a variety of cell signalling processes including: peptide hormone secretion, modulation of electrolyte transport, vascular tone, inflammatory response, inhibition of Na/K ATPase, and serve as endothelial-derived hyperpolarising factors (EDHF).[26–35] CYP2C8 produces AA epoxide metabolites that are thought to cause, among other things, vasodilation of vasculature by serving as EDHFs to control local blood pressure in various systems, including the renal and cardiovascular systems. Moreover, CYP2C8 is involved in the metabolism of another important endogenous cell-signalling molecule, retinoic acid.[36] The potential involvement of CYP2C8 in these constitutively active processes in humans, coupled with reports demonstrating expression in a wide variety of tissues, suggests that this enzyme is important in maintaining normal human physiology.[5,37]

Multiple structure/function studies have been performed with both naturally occurring variants and diagnostic mutants to explore the molecular details of CYP2C8 enzymic activity.[38] In fact, the crystal structure of CYP2C8 has recently been published.[18] As suspected based on many of the large substrates of CYP2C8, the active site of this enzyme appears to be quite large (1438 angstroms). This structural model coupled with various structure/function studies have provided a more comprehensive understanding of CYP2C8 enzymic activity that will not be expounded upon due to the prescribed length of this chapter.

Overall CYP2C8 is an enzyme that is gaining increasing importance in drug development world due to its role in the metabolism of clinically prescribed therapeutics. In fact, the latest FDA Draft Guidance for Drug-Drug Interactions (2006) now includes CYP2C8 among the CYPs recommended for *in vitro* examination to drug developers seeking to market new drugs.[39]

7.3.2 Metabolism Catalysed by CYP2C9

CYP2C9 appears to be the most pharmacologically important of the human CYP2C enzymes, and is generally accepted as the most abundant CYP2C in human liver.[1] It is involved in the metabolism of a large number of drugs including the anticoagulants warfarin, which is one the most dangerous drugs for human health because of its narrow therapeutic index making it difficult to tailor the proper dosing strategy.

In humans, (*S*)-warfarin is metabolically deactivated by hydroxylation at the 7-position by CYP2C9.[40] Warfarin (brand names Coumadin, Jantoven, Marevan and Waran) is used in the prevention of thrombosis and embolism in many disorders. Warfarin (Figure 7.5) belongs to the coumarin class of drugs and decreases blood coagulation by interfering with vitamin K recycling metabolism through inhibition of vitamin K epoxide reductase activity for vitamin K-dependent clotting factors II, VII, IX and X, as well as the regulatory factors protein C, protein S and protein Z.[41] Warfarin is a synthetic derivative of coumarin, a chemical found naturally in many woodruff plants and at lower levels in liquorice, lavender and various other species. Interestingly, warfarin was originally developed as a rat poison. Warfarin metabolism in humans is complicated by inter-individual differences in metabolism due, in part, to variation in CYP2C9 expression and/or metabolic activity that can arise from environmental (induction/inhibition) or genetic factors described in greater

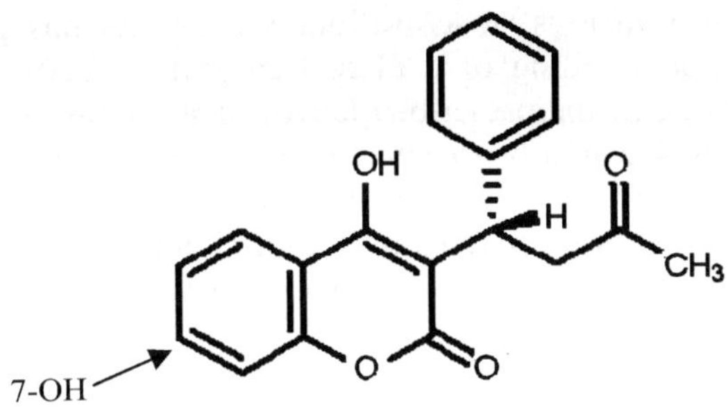

Figure 7.5 Warfarin chemical structure.

detail below. This important interaction demonstrates the need to understand the molecular details of the metabolism of therapeutic drugs.

CYP2C9 also plays a prominent role in the metabolism of numerous non-steroidal anti-inflammatory drugs (NSAIDs). For example, diclofenac, a prototypical substrate the 4′-hydroxylation of which is diagnostic for CYP2C9, is thought to be a cyclooxygenase inhibitor in the treatment of inflammation and pain relief. Tolbutamide, a type II diabetes drug, is also metabolised by CYP2C9 (hydroxylation) and tolbutamide is frequently used as an alternative diagnostic substrate for CYP2C9 activity *in vivo*.[42] The discovery that the CYP2C9 substrate sulphaphenazole is also a selective inhibitor for CYP2C9 has been an important discovery for exploring the relative contribution of CYP2C9 to the metabolism of drugs *in vivo*.[43] These and numerous other examples of pharmacologically important CYP2C9 substrates are summarised in Table 7.1.

Human CYP2C9 also catalyses the metabolism of endogenous substrates. As already discussed for CYP2C8 (*vide supra*), CYP2C9 also contributes to the metabolism of AA producing vasodilating epoxides and to a lesser degree, hydroxylated metabolites in a variety of cell signalling processes including: peptide hormone secretion, modulation of electrolyte transport, vascular tone, inflammatory response, inhibition of Na/K ATPase, and serve as endothelial-derived hyperpolarising factors (EDHF).[26–35,44] However, unlike CYP2C8, CYP2C9 is expressed at appreciable levels in the human gut, indicating that this enzyme may play a more prominent role in intestinal metabolism. It has also been suggested that CYP2C9 is the EDHF synthetase as it has often been observed to be highly expressed in cardiovascular tissues,[5] and the effects of epoxide metabolites of AA are ameliorated by inclusion of the selective CYP2C9 inhibitor sulphaphenazole.

Multiple structure/function studies have been performed with both naturally occurring variants and diagnostic mutants to explore the molecular details of CYP2C9 enzymic activity.[45] In fact, the crystal structure of CYP2C9 has recently been published.[46] As envisaged, based on the presence of acidic moieties on many CYP2C9 substrates, the active site of this enzyme contains basic amino acids that appear to be involved in substrate interactions. This structural model coupled with various structure/function studies has provided a more comprehensive understanding of CYP2C9 enzymic activity that will not be expounded upon due to the prescribed length of this chapter.

Overall CYP2C9 is generally accepted as one of the most important CYPs in human drug metabolism due to its role in the metabolism of numerous clinically prescribed drugs. Therefore, understanding the interactions (e.g. inhibition or induction) of a new chemical entity with CYP2C9 is important to effectively evaluate safety for human health.

7.3.3 Metabolism Catalysed by CYP2C19

CYP2C19 is the third member of the human CYP2C subfamily, and is important in the metabolism of a wide variety of clinically prescribed drugs in multiple therapeutic conditions. CYP2C19 is involved in the metabolism of a growing number of drugs including the anti-ulcer medication omeprazole. Omeprazole (Figure 7.6) is a proton pump inhibitor that is one of the most frequently prescribed/administered drugs in humans, that is now available over-the-counter in some countries. Omeprazole and other proton pump inhibitors are thought to act by irreversibly blocking the terminal stage of gastric acid secretion (hydrogen/potassium adenosine triphosphatase enzyme system, i.e. the H+/K+ ATPase, proton pump) in the gut. The S-enantiomer of omeprazole is the active component, and is now marketed as an enantiomeric pure form (Nexium). S-Omeprazole is metabolised by CYP2C19 to form the 5-hydroxyl metabolite whereas CYP3A4 catalyses the sulfone formation.[47] This enantiomeric selectivity, coupled with the relationship between omeprazole's susceptibility to patient variation that appears to be associated with CYP2C19 genotype, have established CYP2C19 as a major factor in metabolism.

Another well-known CYP2C19 substrate is the S-enantiomer of the anticonvulsant mephenytoin. (*S*)-mephenytoin metabolism proceeds via hydroxylation at the 4′ position catalysed by CYP2C19.[48] In fact, CYP2C19 appears to be important in the metabolism of many anticonvulsants (Table 7.1). Mephenytoin is no longer prescribed in the United States due to the production of the toxic metabolite nirvanol in humans. However mephenytoin has been used in humans in the treatment of seizures, and has been proven as a valuable tool to probe polymorphisms and inhibitors on CYP2C19 metabolism.

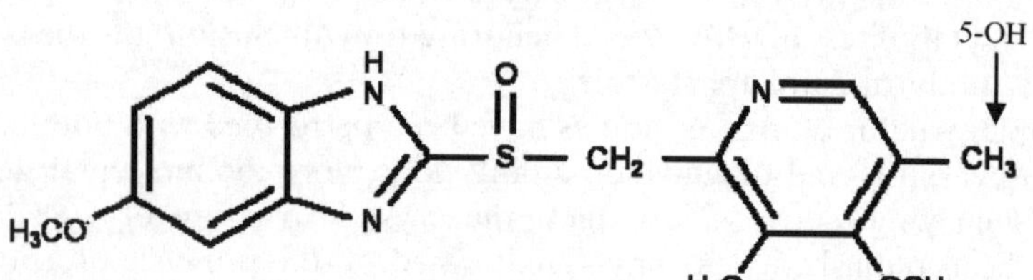

Figure 7.6 Omeprazole chemical structure.

Multiple structure/function studies have been performed with both naturally occurring variants and diagnostic mutants in order to explore the molecular details of CYP2C19 enzymic activity[49] and; the naturally occurring variants are discussed in more detail (*vide infra*).

Overall, CYP2C19 is a pharmacologically important enzyme in the metabolism of clinically-prescribed therapeutics. In addition, due to inter-individual differences in CYP2C19-mediated metabolism (*vide infra*), it is important to understand the role of CYP2C19 in the metabolism of new chemical entities to avoid potential adverse drug reactions in susceptible populations.

7.4 Pharmacogenetics of CYP2C

Due to the importance of the CYP enzymes in human drug metabolism, it is not surprising that questions relating to interindividual genetic variation may emerge. In general, humans are classified into two groups based on their ability to handle drugs; extensive metabolisers (EMs) and poor metabolisers (PMs). These phenotypic differences in population studies are often the result of a drug that is handled, in part, via a small number of metabolic pathways (e.g. CYP2D6 or CYP2C19) that are prone to genetic variability. Polymorphisms occur as a result of a number of factors such as (a) a mutation that stabilises within a population, (b) a 'founder effect', whereby an allele from a population arising from a few individuals is carried on in many of the descendants and (c) genetic drift, where an allele may increase in frequency simply through chance.

Important CYP that exhibit polymorphisms of pharmacological relevance include CYP2D6, CYP2A6 and two members of the CYP2C subfamily, CYP2C9 and CYP2C19. CYP2D6 polymorphisms are relatively frequent (12–21% in Caucasians) and can result in complete inactivation of the enzyme, which can be especially significant clinically since ~30% of prescribed drugs are metabolised by CYP2D6.[50] In order to customise patient treatment to their individual enzyme genotype, pharmacogenetics aim to identify individuals predisposed to a high risk of toxicity or poor efficacy.

Polymorphisms that produce clinically-relevant effects often result in a CYP gene product that is completely inactive. For instance, (1) if the gene is absent (in the case of *CYP2A6*4* and *CYP2D6*5*), (2) if a nonsense mutation produces a frame shift and premature stop codon that results in a truncated message and protein (*CYP2C9*3*), (3) if there is a mutation in a mRNA splice site for producing mature mRNA (*CYP2C19*2*),[51] (4) a single or group (haplotype) of mutation(s) occur that cause coding changes in CYP protein sequences that produce complete (or severe) loss of enzymic activity toward specific (or all) substrates, (5) super-metaboliser alleles due to gene duplication.[52] Interestingly, these genetic variations and the resulting differences in drug pharmacokinetics (e.g. longer half-life, larger AUC, higher Cmax, or super-rapid metabolism) can lead to either increased incidents of deleterious side effects, enhanced efficacy, or even reduced efficacy in some individuals depending on the side-effects and therapeutic window of a given drug. This can be especially important for drugs with narrow therapeutic windows, such as the anticoagulant (*S*)-warfarin, where small changes in systemic levels can lead to bleeding or poor efficacy depending to some degree on the individual's genetic CYP2C9 background.

From a mechanistic standpoint, these genetic variations can manifest into phenotypic differences on a variety of levels related to the synthesis and activity of a functional CYP enzyme in humans that are summarised in Tables 7.2–7.4.

7.4.1 CYP2C8 Pharmacogenetics

In vivo studies show clear interindividual differences in the metabolism of CYP2C8 substrates (e.g. paclitaxel, rosiglitazone, repaglinide), but there is a clear disconnect between *in vivo* and *in vitro* results. For *in vitro* studies, the *CYP2C8*3* allele has been clearly shown to be metabolically impaired.[35] However, several *in vivo* studies have failed to clearly correlate the presence of the *CYP2C8*3* allele with impaired metabolism of CYP2C8 substrates. A comprehensive list of currently known *CYP2C8* alleles is accessible at http://www.cypalleles.ki.se/cyp2c8.htm, and this current list of known CYP2C8 alleles is presented in Table 7.2. For a recent review of CYP2C8 pharmacogenetics refer to Totah and Rettie.[53]

Overall, even though many *CYP2C8* alleles have been identified, polymorphisms (to date) have not been shown to cause major clinically relevant effects on the metabolism of CYP2C8 substrates.

7.4.2 CYP2C9 Pharmacogenetics

In vivo studies show clear inter-individual differences in the metabolism of CYP2C9 substrates (e.g. warfarin). Table 7.3 summarises important alleles identified for CYP2C9. There are two main polymorphisms which are believed to contribute to the PM of (*S*)-warfarin. One is due to a *CYP2C9*11* polymorphism, possibly causing decreased enzyme stability by improper folding of the protein. The overall *in vivo* effect is a reduction in CYP2C9 (*S*)-warfarin activity.[55] A second polymorphism is in the vitamin K epoxide reductase complex subunit 1 gene (VKORC1) which codes for the vitamin K epoxide reductase enzyme. These mutations are rare but are associated with clotting factor deficiencies and warfarin resistance.[56]

In a study by Sconce *et al.* (2005),[57] the effect of age, body size, as well as CYP2C9 and VKORC1 genotype on the warfarin dose requirements were studied. *CYP2C9* homozygous wild-type patients required higher doses of (*S*)-warfarin than those with the variant *2 and *3 alleles. Likewise, wild-type VKORC1 patients required higher doses of (*S*)-warfarin than those with the variant genotype. It was concluded that age, height, and CYP2C9 genotype significantly contributed to (*S*)-warfarin and total warfarin clearance, whereas only age and body size significantly contributed to (*R*)-warfarin clearance.[57]

Table 7.2 *CYP2C8* Alleles –2007.[54]

Allele	Protein	Nucleotide changes, cDNA* *Position 1497 should be C.	Nucleotide changes, Gene (Now in the forward direction)	Effect	Enzyme activity: In vivo	Enzyme activity: In vitro	References
*CYP2C8*1A*	CYP2C8.1	None			Normal	Normal	Klose *et al.*, 1999
*CYP2C8*2*	CYP2C8.2	805A>T	1675C>T; 2010G>A; 2040_2041insT; **11054A>T**; 26562T>A; 32299C>T	**I269F**		Increased Km for paclitaxel 6a-OH	Dai *et al.*, 2001
*CYP2C8*3*	CYP2C8.3	**416G>A; 1196A>G**	1675C>T; 2010G>A; **2130G>A**; 26562T>A; **30411A>G**; 32299C>T	**R139K; K399R**		Decreased paclitaxel turnover.	Dai *et al.*, 2001
*CYP2C8*4*	CYP2C8.4	792C>G	11041C>G	I264M			Bahadur *et al.*, 2002
*CYP2C8*5*		**475delA**	**2189delA**	**Frameshift**	None		Soyama *et al.*, 2002

Table 7.2 (*Continued*).

Allele	*Protein*	*Nucleotide changes, cDNA** **Position 1497 should be C.*	*Nucleotide changes, Gene* *(Now in the forward direction)*	*Effect*	*Enzyme activity* *In vivo*	*In vitro*	*References*
*CYP2C8*6*	CYP2C8.6	511G>A	4472G>A	G171S			Hichiya *et al.*, 2005
*CYP2C8*7*		**556C>T**	**4517C>T**	**R186X**	None		Hichiya *et al.*, 2005
*CYP2C8*8*	CYP2C8.8	**556C>G**	**4517C>G**	**R186G**	Decrased		Hichiya *et al.*, 2005
*CYP2C8*9*	CYP2C8.9	740A>G	10989A>G	K247R			Hichiya *et al.*, 2005
*CYP2C8*10*	CYP2C8.10	1149G>T	26513G>T	K383N			Hichiya *et al.*, 2005
*CYP2C8*12*	CYP2C8.12	1382_1384delTTG	32184_32186delTTG	461delV			Saito *et al.*, 2007
*CYP2C8*13*	CYP2C8.13	669T>G	10918T>G	I223M			Nakajima *et al.*, 2006
*CYP2C8*14*	CYP2C8.14	712G>C	10961G>C	A238P			Nakajima *et al.*, 2006

Table 7.3 Select *CYP2C9* Alleles.[54]

		Nucleotide changes			*Enzyme activity*		
Allele	*Protein*	*cDNA*	*Gene*	*Effect*	*In vivo*	*In vitro*	*References*
*CYP2C9*1A*	CYP2C9.1	None	None		Normal	Normal	Romkes *et al.*, 1991
*CYP2C9*2A* Predicted haplotype	CYP2C9.2	**430C>T**	*−1188T>C, −1096A>G; −620G>T; −485T>A; −484C>A;* **3608C>T**	**R144C**		Decreased	Rettie *et al.*, 1994 Crespi *et al.*, 1997 Takahashi *et al.*, 2004 Several other reports
*CYP2C9*3A* Predicted haplotype	CYP2C9.3	**1075A>C**	*−1911T>C; −1885C>G; −1537G>A; −981G>A;* **42614A>C**	**I359L**	Decreased	Decreased	Sullivan-Klose *et al.*, 1996 Haining *et al.*, 1996 Several other reports
*CYP2C9*5*	CYP2C9.5	**1080C>G**	**42619C>G**	**D360E**	Decreased?	Decreased	Dickmann *et al.*, 2001 Allabi *et al.*, 2004 Allabi *et al.*, 2005
*CYP2C9*6*		**818delA**	**10601delA**	**Frame shift**	None		Kidd *et al.*, 2001

Table 7.3 (*Continued*).

		Nucleotide changes			*Enzyme activity*		
Allele	*Protein*	*cDNA*	*Gene*	*Effect*	*In vivo*	*In vitro*	*References*
*CYP2C9*8*	CYP2C9.8	**449G>A**	**3627G>A**	**R150H**	Decreased	Increased	Allabi *et al.*, 2005 Blaisdell *et al.*, 2004
*CYP2C9*11B* *Predicted haplotype*	CYP2C9.11	**1003C>T**	*−2665_−2664delTG; −1188T>C;* **42542C>T**	**R335W**		Decreased	Allabi *et al.*, 2005 Blaisdell *et al.*, 2004
*CYP2C9*12*	CYP2C9.12	**1465C>T**	**50338C>T**	**P489S**		Decreased	King *et al.*, 2004 Blaisdell *et al.*, 2004
*CYP2C9*13*	CYP2C9.13	269T>C	3276T>C	L90P	Decreased	Decreased	Si *et al.*, 2004 Guo *et al.*, 2005a Guo *et al.*, 2005b

Table 7.4 Select *CYP2C19* Alleles.[54]

Allele	Protein	Nucleotide changes: cDNA	Nucleotide changes: Gene*	Trivial name	Effect	Enzyme activity: In vivo	Enzyme activity: In vitro	References
*CYP2C19*1A*	CYP2C19.1A	None	None		None	Normal	Normal	Romkes *et al*, 1991
*CYP2C19*2A*		99C>T; **681G>A**; 990C>T; 991A>G	99C>T; **19154G>A**; 80160C>T; 80161A>G	m1; m1A	**splicing defect**; I331V	None		de Morais *et al*, 1994a
*CYP2C19*2B*		99C>T; 276G>C; **681G>A**; 990C>T; 991A>G	99C>T; 12460G>C; **19154G>A**; 80160C>T; 80161A>G	m1B	E92D; **splicing defect**; I331V	None		Ibeanu *et al*, 1998b
*CYP2C19*3A*		**636G>A**; 991A>G; 1251A>C	**17948G>A**; 80161A>G; 87313A>C	m2	**W212X**; I331V	None		de Morais *et al*, 1994b
*CYP2C19*4*		**1A>G**; 99C>T, 991A>G	**1A>G**; 99C>T; 80161A>G	m3	**GTG initiation codon**; I331V	None		Ferguson *et al*, 1998
*CYP2C19*5A*	CYP2C19.5A	**1297C>T**	**90033C>T**	m4	**R433W**	None	None	Xiao *et al*, 1997 Ibeanu *et al*, 1998a
*CYP2C19*5B*	CYP2C19.5B	99C>T; 991A>G; **1297C>T**	99C>T; 80161A>G; **90033C>T**		I331V; **R433W**	None		Ibeanu *et al*, 1998a
*CYP2C19*6*	CYP2C19.6	99C>T; **395G>A**; 991A>G	99C>T; **12748G>A**; 80161A>G	m5	**R132Q**; I331V	None	None	Ibeanu *et al*, 1998b
*CYP2C19*7*			**19294T>A**		**splicing defect**	None		Ibeanu *et al*, 1999
*CYP2C19*8*	CYP2C19.8	**358T>C**	**12711T>C**		**W120R**	None	Decreased	Ibeanu *et al*, 1999
*CYP2C19*10*	CYP2C19.10	99C>T; **680C>T**; 991A>G	99C>T; **19153C>T**; 80161A>G		**P227L**; I331V		Decreased	Blaisdell *et al*, 2002
*CYP2C19*12*	CYP2C19.12	99C>T; 991A>G; **1473A>C**	99C>T; 80161A>G; **90209A>C**		I331V; **X491C**; 26 extra aa		Unstable	Blaisdell *et al*, 2002

The main variants causing a PM phenotype in Caucasians are *CYP2C9*2* and *CYP2C9*3*. The allele frequency of CYP2C9*2 is 8–13% in Caucasians but is absent in Asians.[58] Similarly, there is a racial difference in the frequency of the *CYP2C9*3* allele, which is present in 7–9% of Caucasians and only 2.1% of Asians.[59] Depending on the substrate, *CYP2C9*2* and *CYP2C9*3* variants result in enzymes with approximately 10–40% and 5–15% of the activity of *CYP2C9*1*, respectively.[60] Recently, an allelic variant of *CYP2C9* that causes substitution of Leu359 for Ile359 has been linked with the decreased metabolic clearance of warfarin, tolbutamide and phenytoin. The frequency of this variant allele in the Japanese population is 2%, whereas in Caucasians it is 6–9%. Another variant is *CYP2C9*13*, which was identified in a Chinese poor metaboliser of lornoxicam. This variant is characterised by a mutation encoding a substitution of Leu90 for Pro90, causing a lower catalytic activity in metabolising diclofenac and lornoxicam *in vitro*.[61]

Phenytoin plasma concentrations and toxicity are known to increase in patients taking inhibitors of CYP2C19 or who have a variant allele.[62] One patient who was homozygous for the *CYP2C9*3* allele (*CYP2C9*3/*3*) and heterozygous for the *CYP2C19*2* allele (*CYP2C19*1/*2*) exhibited signs of phenytoin toxicity due to an elimination rate that was five times slower than the average (22 hours). Typically, the maximum decrease in the elimination rate (e.g. clearance) of phenytoin is 2-fold between the extremes of EMs and PMs.[62] Thus, it may be recommended that only patients such as the one described above with mutated alleles of both *CYP2C19* and *CYP2C9* should have a dosage reduction. This combination of variants is very rare among Caucasians (about 1% of the population) but more frequent in Asians (about 10%).[63]

Unlike phenytoin or phenobarbital, the newer antiepileptic drugs are not affected by genetic polymorphisms as they have been designed to have reduced susceptibility to induction and inhibition effects, and genetic polymorphisms

7.4.3 CYP2C19 Pharmacogenetics

In vivo studies show clear interindividual differences in metabolism of CYP2C19 substrates (e.g. (*S*)-mephenytoin).[64] Multiple variant alleles (*CYP2C19*2* to *CYP2C19*8*) that predict poor metabolisers have been identified (Table 7.4). The principal molecular defect in the *CYP2C19* gene responsible for the PM phenotype in mephenytoin metabolism in Caucasians is the *CYP2C19*2* allele, present in approximately 7–13% of the population.[59,65] The distribution of extensive and poor metaboliser genotypes and phenotypes shows wide interethnic differences. In EMs, about 80% of the doses of omeprazole, lansoprazole and pantoprazole seem to be cleared by CYP2C19, whereas CYP3A is more important in CYP2C19 poor metabolisers.[66] PMs of CYP2C19 experience more effective acid suppression and healing of duodenal and gastric ulcers during treatment with omeprazole and lansoprazole compared with EMs.[67] Due to the narrow therapeutic range of phenytoin, genotyping of *CYP2C19* in addition to *CYP2C9* may be needed to optimise the dosage in humans. Increased risk of toxicity of tricyclic antidepressants is likely in patients whose CYP2C19 and/or CYP2D6 activities are diminished. CYP2C19 pharmacogenetics also appear to be important in the metabolism of the antimalarial proguanil.[68]

In many cases, allelic variants result in the inactivation of the enzyme. For example, a well-known genetic polymorphism of CYP2C19 (*CYP2C19*2*) is associated with a decreased capacity to 4′-hydroxylate (*S*)-mephenytoin. It has been estimated that the defective gene causes 3% of Caucasians to be PMs of mephenytoin.[69] There is a difference in the frequency of PMs between races, whereby a higher frequency of PMs of mephenytoin has been reported in the Japanese (20%), Chinese (15–17%) and Koreans (12–16%) than in Caucasians.[70] Another defective allele is *CYP2C19*3* but this has been detected only in Asian populations.[71] There is speculation that this mutation occurred after the differentiation of the Caucasian and Asian racial groups.[60] *CYP2C19*2* and *CYP2C19*3* appear to account for more than 99% of Asian and about 87% of Caucasian poor metaboliser alleles.[72]

The proton pump inhibitor, omeprazole, is used in the short- and long-term management of acid-related intestinal disease. Omeprazole is metabolically cleared by the liver and CYP2C19 is one of the enzymes involved in its metabolism. In the Chinese Li population *CYP2C19* genotype appears to be linked to the omeprazole hydroxylation phenotype, about 90% of which can be explained by the *CYP2C19*2* and **3* alleles.[73] The differences in phenotype (healing peptic ulcers, reducing gastrooesophageal reflux disease and eradication of *Helicobacter pylori*) are reflected in the systemic drug exposure (AUC) of omeprazole, the latter being approximately 7.5- fold higher in PM than in homozygous EMs.[74] Thus, a dosing regimen based on the genotype may improve therapeutic efficacy of drugs like omeprazole.

7.4.4 Perspective on Pharmacogenetics of CYP2Cs

Polymorphisms of the human CYP2C subfamily appear to be key to our understanding of inter-individual differences in drug metabolism. It is important to assess whether the pharmacokinetics of a drug will be influenced by the levels of the enzymes responsible for their metabolism. For example, the anticancer agent indisulam, is metabolised by CYP2C9 and CYP2C19. Patients expressing the *CYP2C9*3*, *CYP2C19*2*, and *CYP2C19*3* polymorphic variants exhibited a lower elimination rate of indisulam than those carrying the normal allele (wild type or "**1*" allele), thus increasing their risk of severe haematologic toxicity.[75] In this particular case, the variant genotypes provided some predictive value for the risk of developing severe haematologic toxicity due to indisulam intake. Screening for these CYP2C polymorphisms and "pharmacogenetically guided dose adaptation" will help in the selection of the initial dose for these patients. Failure to take into account differences in the metabolic capacity of patients can lead to ineffective drug concentrations, or conversely toxic levels leading to adverse drug reactions or even death.

7.5 Drug-Drug Interactions Involving CYP2C

In drug development, one of the challenges is designing drugs that are therapeutically effective with minimal side-effects. This balance is further complicated by today's patients that are co-administered multiple drugs simultaneously (e.g. HIV, oncology, diabetic, and cardiovascular patients) to treat multiple problems. This is challenging because each individual medication alone may have been thoroughly tested in clinical trials for safety and efficacy, but co-administered drugs have often not been tested in combination. As one could imagine, testing all possible combinations of drugs that may be co-administered would be unrealistic; however recent advances in understanding the molecular mechanisms of many of these drug-drug interactions have enabled drug development scientists and clinicians to better predict the potential for interactions.

The most prominent modes of drug-drug interactions examined to date are the inhibition and induction of CYP activities. Significant progress has been made in these areas, and a more detailed description, as it relates to the CYP2C subfamily, is described below. In addition, interactions of drugs with transporters and phase II enzymes have also begun to emerge, and the interplay of these systems with the CYP system has led to a clearer understanding of idiosyncratic drug interactions that will help continue to evolve our ability to predict drug-drug interactions. Below we describe examples of drug-drug interactions involving the CYP2C subfamily and discuss the underlying mechanisms.

7.5.1 Drug-Drug Interactions - Inhibition

Drug-drug interactions involving cytochrome P450 that increase co-administered plasma drug concentrations via inhibition of metabolism are most often associated with deleterious side-effects, but may also lead to enhanced efficacy (e.g. omeprazole with CYP2C19). A drug that inhibits a specific CYP enzyme can decrease the metabolic clearance of a co-administered drug that is a substrate of the inhibited enzyme. Furthermore, the inhibited metabolic pathway could lead to decreased formation of an active metabolite, resulting in decreased efficacy. Unwanted effects are most obvious and expected when they involve drugs with a narrow therapeutic range (e.g., warfarin),[76] but is also possible with drugs having a broader therapeutic range (e.g., HMG CoA reductase inhibitors).[77] Interactions involving inhibition have been responsible for limited development potential, severe dosing restrictions and termination of development for many investigational drugs. If the potential for interactions via enzyme inhibition is not identified early during the development process, the result can be exposure to toxic drug levels in some patients that could result in a 'black box' warning on the product label or withdrawal from the market. A prominent example of a deleterious effect is the interaction between the non-sedating antihistamine, Seldane® (terfenadine) and the common antibiotic, erythromycin.[4] Patients receiving both drugs accumulated dangerously high blood levels of terfenadine due to the inhibition of CYP3A enzymes, resulting in potentially fatal arrhythmias. Based on this observation, the FDA issued a recall on Seldane® and recommended that investigational drugs be assessed, relatively early in drug development, for the potential of inhibition-related drug-drug interactions. This assessment could be made by probing the ability of new drugs to inhibit pharmacologically important drug metabolising enzymes.

In general, the mechanisms of CYP inhibition differ according to the nature of the compound involved. Some bind to the CYP reversibly, some form quasi-irreversible complexes with the prosthetic iron group, and others bind irreversibly with the prosthetic iron group or the protein moiety. Irreversible inhibitors may also cause covalent binding of the haem prosthetic group or its degradation product to the apoprotein. The most common drug-drug interaction is reversible inhibition, which is dose-dependent and can be overcome by increasing the dose of the substrate. Quasi-irreversible and irreversible inhibitors are dose-dependent and time-dependent and are referred to as 'time-dependent' or 'mechanistic' inhibitors.

Significant advances have been made in our understanding of the inhibition of the CYP2C enzymes. Table 7.5 provides a summary of drugs known to inhibit CYP2C enzyme activity in humans.

7.5.1.1 CYP2C8 Inhibition

CYP2C8, as already discussed, is emerging as a pharmacologically important enzyme due to its central role as a major human liver enzyme in the metabolism of clinically prescribed drugs. As a result, drugs that are known to inhibit CYP2C8 would be contraindicated for administration with selective CYP2C8 substrates without careful monitoring. Our understanding of the inhibition of CYP2C8 continues to emerge. One of the most notable clinically-relevant examples of a CYP2C8 inhibitor is the hyperlipidaemia drug gemfibrozil. Gemfibrozil is a cholesterol and triglyceride lowering agent which has been reported to increase the plasma concentration of a number of co-administered drugs.[78] *In vitro* studies revealed that gemfibrozil is a potent inhibitor of CYP2C9 but only a weak inhibitor of CYP2C8.[79,80] In contrast, the opposite is observed in the clinic.[78] Incubations with human liver microsomes confirmed

Table 7.5 Inhibitors of CYP2C8, CYP2C9, and CYP2C19.

CYP2C8	*CYP2C9*	*CYP2C19*
gemfibrozil	fluconazole	ticlopidine
trimethoprim	sulphaphenazole	omeprazole
quercetin	fluvoxamine	lansoprazole
rosiglitazone	amiodarone	pantoprazole
pioglitazone	fenofibrate	cimetidine
montelukast	sulfamethoxazole	indomethacin
	fluvastatin	fluoxetine

that gemfibrozil most potently inhibited CYP2C9 (with an IC_{50} of 30 μM), whereas its major metabolite, gemfibrozil glucuronide, most potently inhibited CYP2C8 (IC_{50} of 24 μM). In addition, gemfibrozil glucuronide, but not gemfibrozil, caused metabolism-dependent inhibition of CYP2C8 only.[79] The metabolism-dependent CYP2C8 inhibition was increased further by pre-incubating gemfibrozil with alamethicin-treated human liver microsomes (which increases the access to the UGTs on the inner membrane of the microsomes by causing pore-formation in the membrane) and the cofactor, UDP-glucuronic acid. Gemfibrozil also considerably increases the plasma concentrations of rosiglitazone.[81] These findings highlight the importance of assessing the time-dependent inhibitory properties of new chemical entities in both traditional liver-derived microsomal-based systems and in other metabolically competent systems capable of more liver-like handling of drugs, such as primary human hepatocyte cultures or liver slices.

Gemfibrozil is often administered in combination with other CYP2C8 substrates such as pioglitazone, rosiglitazone, repaglinide (all treatments for diabetes), cerivastatin (a lipid-lowering drug) and loperamide (a common antidiarrhoea drug). These drugs are known to be metabolized by other CYP enzymes (CYP3A4 metabolises pioglitazone, loperamide, repaglinide and cerivastatin and CYP2C9 metabolises rosiglitazone); however, the plasma concentration of these drugs in humans are affected by the co-administration of gemfibrozil.[80,82–86] The main interaction of gemfibrozil with pioglitazone, loperamide and cerivastatin is likely to be due to CYP2C8 inhibition since CYP2C8 is more important in their metabolism than CYP3A4 *in vivo*.[82,87]

In addition to gemfibrozil, trimethoprim also inhibits CYP2C8 metabolism in human microsomal systems.[88] Trimethoprim is an antibiotic often used in the treatment of urinary tract infections, and was observed to inhibit CYP2C8 with a K_i of 32 μM. The ability of trimethoprim to inhibit CYP2C8 was confirmed in a 2005 study by Hruska and coworkers where patients treated with 200 mg of trimethoprim and 8 mg of rosiglitazone displayed competitive-type inhibition of rosiglitazone metabolism resulting in a 31% increase in plasma AUC and a 27% prolongation half-life. Therefore, treatment of patients concomitantly, with trimethoprim and other CYP2C8 substrates should be done with caution, and an understanding of a new chemical's metabolism and inhibitory properties can help predict potential issues related to inhibition of CYP2C8.[89]

Pioglitazone, a thiazolidinedione antidiabetic drug, is metabolised mainly by the CYP2C8 enzyme. The leukotriene receptor antagonist montelukast has been identified as a potent inhibitor of CYP2C8 ($IC_{50} = 19.6$ nM).[90] Both montelukast and zafirlukast potently inhibit CYP2C8 activity and the metabolism of pioglitazone *in vitro*.[91] However, montelukast and zafirlukast did not increase the plasma concentrations of pioglitazone *in vivo*,[92] indicating that their inhibitory effect on CYP2C8 is negligible *in vivo*, despite their strong inhibitory effect observed *in vitro*. The results highlight the importance of *in vivo* interaction studies and of the consideration of relevant pharmacokinetic properties of drugs, including plasma protein binding data, to *in vitro-in vivo* interaction predictions.

Drug-drug interactions with CYP2C8 substrates and inhibitors highlight the importance of understanding the mechanisms by which clinically prescribed drugs interact with human liver, and emphasise the emerging role of CYP2C8 in human drug metabolism.

7.5.1.2 CYP2C9 Inhibition

CYP2C9, as described above, is the most pharmacologically important member of the CYP2C subfamily due to its pivotal role in the metabolism of drugs. Consequently, drugs that are known to inhibit CYP2C9 would be contraindicated for administration with selective CYP2C9 substrates. Table 7.5 lists some known inhibitors of CYP2C9.

The most potent CYP2C9 inhibitor to date is the antifungal drug, fluconazole. A notable drug-drug interaction, resulting from the *in vivo* inhibition of CYP2C9, is the effect of fluconazole on the metabolism of the anticonvulsant phenytoin.[93] In this instance, co-administration of 200 mg of phenytoin and 200 mg of fluconazole resulted in a 75% increase in the AUC from zero to twenty-four hours. Interactions with fluconazole and many other drugs metabolised by CYP2C9 have also been observed and include the prototypical CYP2C9 substrates tolbutamide[94] and warfarin.[95] The interaction with warfarin and other coumarin-type anticoagulants has been shown to lead to increases in prothrombin time, bleeding events (bruising, gastrointestinal bleeding, haematuria, epistaxis, and melena have been reported).[96] Clearly, these adverse effects resulting from drug interactions with a potent CYP2C9 inhibitor like fluconazole can lead to dangerous outcomes, and a clear understanding of a new chemical's ability to inhibit CYP2C9 is important when considering combination therapies.

Another CYP2C9 inhibitor that has been demonstrated to potentiate warfarin metabolism is amiodarone.[97] Amiodarone is an antiarrhythmia drug that prolongs phase 3 of the cardiac action potential which exhibits β-blocker-like and calcium channel blocker-like actions.[98] Potentiation of warfarin metabolism by amiodarone was investigated in 30 patients whose warfarin dose was adjusted to maintain the prothrombin time between 25–30% of control. Potentiation of prothrombin time was observed in 28 out of 30 patients representing a 35–65% reduction in the required dose of warfarin. In addition, the concentration of amiodarone was significantly correlated with the warfarin concentrations independent of the warfarin dose. In a study by Nolan *et al.* amiodarone has also been shown to alter the metabolism of another CYP2C9 substrate, phenytoin.[99] It was reported that treatment with 2–4 mg/kg/day phenytoin along with 200 mg daily doses of amiodarone resulted in a 71% increase in plasma AUC. Interestingly, enzyme activity assays to evaluate the kinetic parameters related to amiodarone inhibition of CYP2C9 *in vitro* showed that the parent compound only weakly affected this enzyme.[100] However, desethylamiodarone did inhibit CYP2C9 catalytic activity with a K_i of 2.3 μM along with several other CYP enzymes including CYP3A4, CYP2D6 and CYP2C19.

Sulphaphenazole is an extremely useful diagnostic inhibitor of CYP2C9 for *in vitro* studies.[101,102] In fact, sulphaphenazole is currently the preferred substrate for evaluating the ability of new chemicals to inhibit CYP2C9,[39] and is a useful *in vitro* tool for exploring the contribution of CYP2C9 in the metabolism of a given drug.

Tienilic acid (TA), a diuretic for the treatment of hypertension, is a selective time-dependent inhibitor of CYP2C9.[103,104] McGinnity and coworkers[105] determined the time-dependent kinetic parameters K_I and k_{inact} for tienilic acid with both recombinant CYP2C9 and in cultures of primary human hepatocytes with K_I values of ~2 μM and k_{inact} values of 0.05–0.19 min^{-1}, giving rise to K_{inact}/K_I ratios of ~10 to 40 min/μM, demonstrating the time-dependent nature of tienilic acid inhibition of CYP2C9. TA is known to be activated by human liver to form reactive thiophene sulfoxides that mediate the inactivation of CYP2C9.[106] Interestingly, this mechanism-based inactivation of CYP2C9 can also lead to an immune response against cytochrome P450 proteins that may lead to hepatic necrosis.[107] As a result of these deleterious interactions, and the availability of viable alternatives, TA is not frequently prescribed.

Fenofibrate is another drug that potentiates warfarin metabolism *in vivo*.[108] In this study, two patients were stabilised on long-term warfarin therapy, but when fenofibrate therapy was initiated the patients experienced 2–3 fold increases in international normalised ratio (INR) that required 30–40% reduction in their total weekly warfarin dosages. However, it has not been conclusively established that inhibition of CYP2C9 by fenofibrate is the responsible mechanism of this response.

Fluvastatin, a synthetic 3-hydroxy-3-methylglutaryl coenzyme A (HMG-CoA) reductase inhibitor, selectively and competitively inhibits CYP2C9 *in vitro*.[109] Transon *et al.*[110] showed that interactions between fluvastatin and the prototypical CYP2C9 substrate diclofenac led to ~60% increases in diclofenac Cmax levels over an 8 day period. A case report by Andrus[111] indicated that a 67-year-old man receiving a stable maintenance dosage of warfarin experienced an increased INR (without bleeding) when his atorvastatin medication was switched to fluvastatin. In addition, Kim *et al.*[112] reported that fluvastatin (40 mg twice daily) with 10 mg doses of warfarin resulted in a 26 to 42% rise in plasma AUC levels.

Fluvoxamine is an antidepressant medication that is also used in the treatment of obsessive compulsive disorder in children and social anxiety disorder. Fluvoxamine belongs to the class of serotonin re-uptake inhibitors but has minimal effects on dopamine and noradrenaline uptake.[113] Fluvoxamine is another example of a chemically-relevant CYP2C9 inhibitor that is known to modulate tolbutamide and warfarin metabolism.[114] Hemeryck *et al.*[114] demonstrated a using human liver microsomes that fluvoxamine can inhibit the metabolism of the CYP2C9 substrates tolbutamide (K_i 13.3 = μM) and (*S*)-warfarin 7-hydroxylation (K_i = 13.0). Consistent with these data, several case reports implicating fluvoxamine as the cause of drug interactions with warfarin have emerged.[115,116] Taken together, these findings suggest that fluvoxamine has the potential to provoke drug-drug interactions with CYP2C9 selective substrates. Other CYP2C9 inhibitors such as sulphamethoxazole, lovastatin, zafirlukast, sertarline, and teniposide have also been reported to participate in drug interaction.

In summary, drug-drug interactions with CYP2C9 substrates and inhibitors highlight the importance of understanding the mechanisms by which clinically prescribed drugs interact, due to the prominent role of this enzyme in the metabolism of drugs in human liver.

7.5.1.3 CYP2C19 Inhibition

As discussed above (*vide supra*), CYP2C9 and CYP2C19 share over 95% homology; therefore it is not surprising that many of the inhibitors known to inhibit CYP2C9 act similarly on CYP2C19. However, due to the significantly lower abundance of CYP2C19, relative to CYP2C9, in human liver, clear differences exist between these enzymic pathways. Table 7.5 lists several known inhibitors of CYP2C19.

Ticlopidine is an example of a CYP2C19 selective inhibitor.[117] It is a drug known to inhibit platelet aggregation by altering the function of platelet membranes. Ticlopidine is often prescribed to individuals that can not tolerate aspirin. It is oxidised by CYP2C19 to generate two major metabolites. During this oxidation CYP2C19 is inactivated in a time- and concentration-dependent manner, and is known as a mechanism-based inhibitor of CYP2C19. Ha-Duong *et al.*[118] reported a K_I of 87 μM and a k_{inact} of 0.0032 s^{-1}, and characterised ticlopidine as an efficient mechanism-based inhibitor of CYP2C19, but not as efficient as tienilic acid in the case of CYP2C9. *In vivo*, ticlopidine has been shown to impair the metabolism of the frequently prescribed antiulcer therapeutic omeprazole.[118] Ticlopidine doses of 200 mg co-administered with 20 mg doses of omeprazole resulted in an 800 to 1000-fold increase in the mean AUC ratio of omeprazole to 5-hydroxyomeprazole in extensive CYP2C19 metabolisers. Ticlopidine has also been shown to inhibit phenytoin clearance.[119] Ticlopidine is an important tool for probing the metabolism of CYP2C19 in humans and human *in vitro* model systems.

Another inhibitor that potentiates CYP2C19 metabolism is the substrate omeprazole.[120] As previously discussed in the section above, omeprazole is a valuable tool for physicians in the treatment of ulcers. Interestingly, it is also an inhibitor of CYP2C19. As Ko *et al.* reported,[120] omeprazole has a K_i of 3.1 μM for the 4-hydroxylation of (*S*)-mephenytoin while the structurally similar drug, lansoprazole, inhibited CYP2C19 with a K_i of 3.2 μM. These studies indicate that omeprazole is both a substrate and inhibitor of CYP2C19, and caution should be exercised when co-administered with other CYP2C19 substrates.

Table 7.5 lists additional examples of CYP2C19 inhibitors.[121] In summary, the drug-drug interactions involving CYP2C19 substrates and inhibitors, further emphasises the importance of appreciating the mechanisms by which drugs interact due to the prominent role CYP2C19 in the metabolism of drugs in human liver.

7.5.2 Drug-Drug Interactions - Induction and Suppression

Drug-drug interactions that decrease co-administered plasma drug concentrations via induction of metabolism are most often associated with decreased efficacy. A drug that induces a specific CYP enzyme can increase the metabolic clearance of a co-administered drug that is a substrate of the induced enzyme. Historically, cytochrome P450 induction was demonstrated in the 1960s, but the mechanisms through which this process occur, and the specific enzymes that are responsive to these inducers have only emerged over the last twenty years. Induction, as it relates to the development of a new drug, appears to be a less deleterious phenomenon in humans relative to the profound side-effects that can be observed with inhibition of CYP activity. However, one example of a drug that was denied approval by the FDA due in part to issues related to CYP induction is pilocarpine. Decreased drug efficacy due to drug-drug interactions can also lead to dangerous consequences when medical conditions go untreated. Alternatively, if the generation of a pharmacologically active metabolite is elevated as a result of induction, there is potential for increased efficacy or the appearance of deleterious side effects. Unwanted effects are most obvious and expected when they involve drugs with a narrow therapeutic range (e.g., warfarin),[76] but is also possible for drugs with a broader therapeutic range (e.g., HMG CoA reductase inhibitors).[77]

Substantial progress has been made over the last twenty years in elucidating the mechanism of CYP induction.[122–124] The most profound CYP induction responses observed in humans resulting from exposure to drugs and other xenobiotics have been observed with members of the CYP1A, CYP2B and CYP3A subfamilies.[123] However, clinically-relevant induction of other enzymes such as the CYP2C, CYP2E1, Phase II enzymes, and even transporters (e.g. P-glycoprotein) may also be pharmacologically important for understanding and predicting drug interactions in humans, and extrapolating animal and *in vitro* model data to the clinic. Induction of the CYP1A, CYP2B, and CYP3A subfamilies is known to proceed via a complex series of steps that begin with the xenobiotic inducer interacting with a specific receptor. In general, these receptors function as master regulators for manifolds of gene targets that afford an enhanced ability to eliminate xenobiotics upon challenge. For the three most sensitive CYP subfamilies, namely CYP1A, CYP2B and CYP3A, there are three receptors which, upon inducer activation, bind to the regulatory regions of these genes and upregulate their transcription.

For CYP1A induction, the aryl hydrocarbon receptor (AhR) is the ligand-binding target for chemicals such as 2,3,7,8-tetrachlorodibenzo-p-dioxin (TCDD), 3-methylcholanthrene, anthracene, polychlorinated biphenyls, among others.[123] In addition, some drugs such as omeprazole are known to induce CYP1A expression (and consequently CYP1A enzyme activity) in human (and other mammalian) liver.

For CYP2B, phenobarbital (PB) is the prototypical inducer that has been known for more than 40 years to induce the expression of liver enzymes, but until recently the mechanisms by which PB induction occurred remained a mystery. It was demonstrated by Masahiko Negishi and coworkers that

CYP2B induction by PB was controlled by the constitutive androstane receptor (CAR) in mice.[125] Interestingly, even though CAR is a nuclear receptor, it does not appear to actually bind to PB. Rather, PB elicits a translocation of CAR from the cytoplasm of the hepatocyte to the nucleus, where CAR's strong constitutive activity up-regulates transcription (possibly without a ligand, however this is not clear).[126] Although this area of research is much younger than that of AhR, it appears that PB induction in humans also functions through CAR.[127]

For CYP3A induction, rifampicin (RIF) is the prototypical inducer. In humans, CYP3A induction by RIF appears to be mediated by the pregnane X receptor (PXR). Unlike its most homologous nuclear receptor relative CAR, PXR is known to bind RIF in order to activate gene transcription. However, translocation of PXR may also play a role as studies in rodents have indicated that PXR is accumulated in the nucleus upon ligand activation. Whether the same mechanism is applicable to humans, is a controversial issue.

AhR, CAR and PXR each control CYP induction; however none of these receptors activates transcription of a single gene or gene subfamily. In fact, CAR (NR1I3) and PXR (NR1I2), which both are members of the superfamily of nuclear receptors, share a significant degree of overlap between gene targets and activators. For example, PB at high concentrations appears to activate PXR;[128] the same mechanism applies to the CYP2C substrate phenytoin. In addition, due to the high homology between CAR and PXR, they bind to similar binding sites within the regulatory regions of CYP genes. It is important to bear in mind that it is the degree of activation at a given concentration that ultimately leads to clinically-relevant responses, and the mechanisms by which these processes are observed to manifest.

A common characteristic of these three primary induction pathways that lead to clinically-relevant changes in CYP expression is that they arise from changes in *CYP* gene expression that ultimately become translated into increased enzyme expression. This is not true for all clinically-relevant induction; for example, CYP2E1 induction by ethanol is known to proceed predominantly *via* a protein stabilisation pathway[129] that does not manifest at the mRNA level or via a receptor-dependent pathway.

In discussions regarding 'induction' of *CYP* genes, it is usual to only discuss increases in expression as these have historically received the most attention. However, these pathways that control 'induction' may also play a role in regulating constitutive expression of these and other gene targets, as they are expressed in the non-induced liver. This is important because attention is focussed on 'induction' and changes in gene expression in the opposite direction that result from drug/xenobiotic exposure are not considered. These decreases in CYP expression are referred to as suppression (or repression) responses. In terms of clinical relevance, suppression has not been a point of emphasis. However, if a drug were to possess the ability to suppress the basal expression of an important CYP, say CYP2C9, then one would expect cumulative decreases in the ability of individuals to metabolise drugs that would manifest clinically as inhibition-like decreases in drug metabolism. An example of a clinically-relevant suppressor is the microtubule disrupting agent colchicine. This drug has been used in the treatment of gout and pain relief as it targets highly proliferative cells, which also improves the gout-response through decreased inflammatory action. However, colchicine has been shown to suppress multiple CYPs, including CYP2C9, by mechanisms that appear to be nuclear receptor dependent.[130] For the CYP2C subfamily, discovery of the factors governing induction of the CYP2B and CYP3A enzymes via CAR- and PXR-dependent pathways has provided molecular targets and probes to begin to explore the mechanisms controlling CYP2C induction. It is historically interesting to note that CYP2C1, CYP2C2, CYP2C3, and CYP2C4 were originally named PB-1, PB-2, PB-3, PB-4 due to the fact that these enzymes were known to be elevated following administration of phenobarbital. A more clear understanding of these formerly idiosyncratic changes in CYP2C-dependent metabolism is now known at the molecular level, and the following sections summarise our current knowledge regarding the regulation of CYP2C expression in humans, and the importance of understanding species differences when trying to translate rodent model systems to human prediction.

7.5.2.1 CYP2C8 Induction/Regulation

Of the four human CYP2C enzymes, CYP2C8 appears to be the most widely expressed in a variety of tissues.[37,131] It is highly expressed in liver, and is also expressed in various extrahepatic tissues including: kidney, adrenal gland, mammary gland, brain, lung, nasal mucosa, heart, coronary artery and aorta.[37,131–134] Therefore, multiple mechanisms may control the tissue specific expression of endogenous CYP2C8 in various tissues to govern the systemic regulation of CYP2C8-mediated metabolism of endogenous substrates such as arachidonic acid and retinoic acid.

CYP2C8, like several other prominent drug-metabolizing CYPs, has been found to be inducible by prototypical CYP inducing agents such as rifampicin and phenobarbital. For example, decreases in plasma AUC have been reported for the CYP2C8 substrate pioglitazone upon co-administration with rifampicin in patients.[135] In this study, ten healthy humans ingested 600 mg rifampicin (or placebo) once daily for 6 days and on the final day received a single 30 mg oral dose of pioglitazone. It was found that rifampicin decreased the plasma AUC by 54% and shortened the half-life from 4.9 to 2.3 hours. Rifampicin was also observed to increase pioglitazone metabolite ratios in urine by 95–98%. Moreover, CYP2C8 was involved in the rifampicin-induced clearance of rosiglitazone.[136] Bidstrup and coworkers[136] reported that rifampicin appears to act as both an inducer and inhibitor of repaglinide metabolism.[137] These data provide clear evidence that induction of CYP2C8 can lead to clinically-relevant changes in the metabolism of CYP2C8 substrates, and further demonstrate the importance of understanding a compound's ability to modulate CYP2C expression during drug development.

An important tool in probing the potential of new chemicals to induce CYP2C8 and other pharmacologically important gene products is primary cultures of human hepatocytes. This model liver system can provide an effective bridge between *in vivo* animal model studies and human responses, as they appear to model induction in the liver.[138–140] For CYP2C8, primary human hepatocyte cultures have been used to demonstrate induction by various xenobiotics including: rifampicin, phenobarbital, dexamethasone, paclitaxel, hyperforin, and 6-(4-chlorophenyl)imidazo[2,1–4] thiazole-5-carbaldehyde O-(3,4-dichlorobenzyle) *oxime* (CITCO).[140,141] These *in vitro* effects are potentially of clinical relevance provided the necessary concentrations of the inducing agents are achieved in the liver. They also imply a role for the nuclear receptors PXR, CAR and GR in the regulation of CYP2C8.

The 5′-flanking region of the *CYP2C8* gene containing the gene promoter was cloned and sequenced by several groups, however no functional exploration of this regulatory promoter regions was reported until 2004 when Bort and coworkers identified a functional HNF (hepatic nuclear factor)-3-binding site.[141] Bort and coworkers noted that the 2C8 promoter was distinctly responsive to the transcriptional activation of HNF-3 in cell lines, based on insertion of the CYP2C8 promoter reporter constructs into chromatin context versus transient transfection. In 2005 we[139] reported the cloning and functional assessment of the proximal 3000 bp of the CYP2C8 promoter along with a distal enhancer module that appears to be important in the induction of CYP2C8 transcription by PXR and its ligand activator rifampicin. In this report, functional transcription factor binding sites for CAR/PXR, HNF-4, and GR (glucocorticoid receptor) were identified. Using deletional and mutational analyses of the CYP2C8 promoter in primary human hepatocytes, no induction of the CYP2C8 promoter was evident within the proximal 3000 bp of this gene. However, a distal binding site approximately 8000 bp upstream was identified and shown to be important in the induction of CYP2C8 transcription by rifampicin. Interestingly, this effect was only observed in primary human hepatocyte cultures. This difference is also observed in the CYP2C8 regulation by HNF-4, a nuclear receptor thought to be involved in PXR-mediated CYP3A4 and CYP2C9 induction, in different cell types.[139] Therefore it is possible that chromatin context or other CYP2C8 promoter specific factors may confound results in the standard HepG2 system, analogous to that observed in the regulation of CYP2C8 by HNF3.[141]

In addition to the induction of CYP2C8, suppression has also been reported with the microtubule disruptor colchicine. This chemical is thought to modulate CYP expression by altering the basal expression of GR, which is thought to regulate CAR and PXR expression. However, GR can also directly induce the transcription of the CYP2C8 promoter via a GR-binding site within the proximal 2kb.[140] Mutational analysis demonstrated that this GR binding site, which is identical in sequence to those found in 2C9 and 2C19 promoter regions, is essential for GR-dependent dexamethasone induction.

As already discussed, CYP2C8 is expressed in a variety of tissues. However, little progress has been made in describing the factors that may be important in the regulation of CYP2C8 expression in these extrahepatic tissues. HNF-4 is known to be expressed in various tissues, as is the ubiquitously expressed GR.

In addition, analysis of the proximal promoter region of the CYP2C8 promoter reveals a perfect, putative SFRE (steroidogenic factor response element)/ERR (estrogen receptor related)α- binding site that does not appear to be present in the other human CYP2C promoter regions (unpublished observation). In preliminary experiments, this ERRα-binding site appeared to confer functional activation of CYP2C8 gene reporter constructs by ERR in HeLa cells (unpublished observations). However, due to the amorphous role of ERR in humans, it is not clear what the significance of this binding site would be to CYP2C8 cell biology. ERRα has been described to be important in the regulation of aromatase (CYP17) which, like CYP2C8, is expressed in cardiovascular tissues. Further exploration of these observations and the role of CYP2C8 in EDHF synthesis may provide important clues to enable the understanding of the role CYP2C8 may play in human cardiovascular function.

In summary, significant progress has been made in understanding the mechanisms that control CYP2C8 expression, and due to its increasingly important role in human drug metabolism, it is important to appreciate whether new drug candidates have the potential to induce CYP2C8 or are metabolised to a significant degree by this enzyme, so as to allow physicians to better anticipate possible drug-drug interactions during patient therapy.

7.5.2.2 CYP2C9 Induction/Suppression

CYP2C9 is expressed at high levels in human liver and is generally thought to be the second most abundant CYP in human liver after CYP3A4. Like CYP2C8, CYP2C9 is expressed a variety of extrahepatic tissues including: intestine, heart, coronary artery, aorta, kidney, adrenal gland and lung[134,142] Therefore, multiple mechanisms may also control the tissue-specific expression of CYP2C9 in various tissues, and govern the systemic regulation of CYP2C9-mediated metabolism of endogenous substrates such as arachidonic acid and retinoic acid.

It has been long recognised that prototypical CYP inducing agents like rifampicin could cause clinical interactions with warfarin[143,144] by increasing the metabolism of this anticoagulant. It has since been established that CYP2C9 is a major enzyme contributing to the metabolism of (*S*)-warfarin in human liver, and that CYP2C9 is inducible by prototypical inducers like rifampicin and phenobarbital.[145,146] It was also observed that patients with liver cirrhosis that were treated with rifampicin displayed marked increases in the clearance of the CYP2C9 substrate tolbutamide.[147] These clinical findings suggest that induction of CYP2C9 may be important for understanding and predicting drug-drug interactions, and recent advances in the molecular mechanisms of the factors regulating CYP2C9 expression are discussed below.

Primary cultures of human hepatocytes have been shown to be an effective model in predicting induction of human CYPs, including CYP2C9.[140,148] Induction of CYP2C9 occurs by various xenobiotics including: rifampicin, phenobarbital and dexamethasone. These *in vitro* effects are potentially of clinical relevance in humans provided the appropriate concentrations of the inducing agents are achieved in the liver. They also suggest a role for the nuclear receptors PXR, CAR and GR in the regulation of CYP2C9.

The 5′-flanking regulatory region of the *CYP2C9* gene has been functionally characterised in recent years.[128,149,150] Ibeanu and Goldstein[148] identified an HPF-1 (hepatic proliferation factor-1) site that binds to the nuclear receptor HNF-4 in the proximal promoter of the *CYP2C9* gene. Subsequently, Gerbal-Chaloin and co-workers[149] cloned approximately 2000 bp of the 5′-flanking sequence and identified putative binding sites for CAR and GR ~1800 bp upstream of the translation start site. This site was later shown by our group[127] to also be activated by ligands such as rifampicin and phenobarbital only in the presence of co-transfected PXR. We identified a more distal CAR-binding site at approximately 2800 bp upstream but this element, however, appears to only play a supportive role compared with the proximal element. Further studies of the CYP2C9 promoter revealed that the proximal HNF-4-binding site identified by Ibeanu and co-workers was essential for CYP2C9 induction by rifampicin/PXR in HepG2 cells.[151] This result is consistent with the observation that CYP3A4 induction mediated by PXR also appears to be dependent on an HNF-4-binding site[152] and was confirmed by Kamiyama and coworkers who knocked-down HNF-4 expression and observed a clear decrease in CYP2C9 expression.[153]

In addition to induction of CYP2C9, suppression has also been reported with the microtubule disruptor colchicine. This chemical is thought to modulate CYP expression by altering the basal expression of GR, which is thought to regulate CAR and PXR expression. However, GR can also directly induce the transcription of the CYP2C9 promoter via a GR-binding site within the proximal 2 kb;[149] consequently it is not clear whether the effects on CYP2C9 expression are the direct result of GR down-regulation or of secondary effect on CAR and PXR. What is clear is that CAR can play a role in the constitutive expression of *CYP2C* genes. This is revealed in the Cyp2c29 expression in CAR knock-out mice where the basal expression of Cyp2c29 is reduced by 90-fold relative to expression in wild type mice.[154] We have also shown that introduction of CAR into primary human hepatocyte cultures can up-regulate the promoter activity of CYP2C9.[155] Collectively, the reports indicate a role for HNF4 in the basal expression of CYP2C9; both CAR and HNF-4 appear to be involved in the basal regulation of *CYP2C* genes, and modulation of their activity by antagonists or factors that decrease their expression may lead to concomitant decreases in CYP2C expression analogous to those observed with colchicine. Additionally, it has been established by Pascussi and co-workers that CYP2C9[155] expression can be down-regulated by inflammatory cytokines like IL-6.[156] As a result of these known mechanisms whereby CYP2C9 expression is suppressed, coupled with the potential impact of suppression of CYP2C9 on drug pharmacokinetics (analogous to CYP2C9 inhibitors), it is important when developing new drugs to assess the potential for compounds to induce (or suppress) CYP2C9 expression.

As discussed previously, CYP2C9 is involved in the metabolism of arachidonic acid in various cell types, including cardiovascular tissues where EDHF can cause vasodilation of vasculature. Interestingly, Delozier and co-workers observed that the expression of CYP2C9 was profoundly higher in an ischaemic heart relative to multiple 'normal' hearts.[133] This observation supports previous studies linking Sudden Infant Death Syndrome and elevated CYP2C9 expression, and suggests that this enzyme may be regulated in cardiovascular tissues by hypoxia.[157] This view is further supported by the observations that the CYP2C9 inhibitor sulphaphenazole protects individuals from ischaemia/reperfusion injury. The mechanism of this protection is believed to be reduced generator of free-radicals due to inhibitor of CYP2C9 metabolism.[158] Further investigations to elucidate the mechanisms by which CYP2C9 may be regulated by hypoxia are warranted, and may also shed light on factors controlling its tissue-selective expression in humans.

In summary, significant progress has been made in understanding the mechanisms that control CYP2C9 expression, and due to its important role in human drug metabolism, it is essential to establish whether new drug candidates have the potential to induce CYP2C9 or are metabolised to a significant degree by this enzyme, so that physicians are in a better position anticipate possible drug-drug interactions during therapy.

7.5.2.3 *CYP2C19 Induction/Suppression*

CYP2C19 is expressed at relatively low levels in human liver, but appears to comprise an appreciable percentage in the intestine.[142] It is quite important in drug metabolism and undertakes the metabolism of numerous drugs including omeprazole. CYP2C19 expression is less well understood compared with CYP2C9 even though these enzymes share more than 95% homology. However, it is documented that the prototypical CYP inducer rifampicin can elicit clinical interactions with the CYP2C19 substrate (*S*)-mephenytoin.[159] In a study, 24 healthy subjects were treated with 600 mg of rifampicin for 22 days which resulted in a 3- to 8-fold increase in the 0-8 hour urinary R/S enantiomeric ration of mephenytoin following 100 mg oral administration of the racemic mixture. A concomitant 40–180% increase in the 4′-hydroxy metabolite was also observed. An important clinical consequence of CYP2C19 induction has been demonstrated with omeprazole in PMs and EMs. A 40 mg/day dose of omeprazole given to PMs over a period of 7 days caused a significant induction of CYP1A2 but, in contrast, the same dose regimen failed to induce CYP1A2 in EMs, supporting the view that EMs rapidly metabolise omeprazole to decrease the systemic exposure to this drug.[160]

Induction of CYP2C19 by various xenobiotics, such as: rifampicin, phenobarbital and dexamethasone has also been reported in cultured human hepatocytes.[161] These *in vitro* are potentially of clinical relevance provided the relevant concentrations of the inducing agents are attained in the liver.

They also raise the possibility that the nuclear receptors PXR, CAR and GR have a role in the regulation of CYP2C19.

The 5′-flanking regulatory region of the *CYP2C19* gene has recently been functionally characterised in our laboratory.[162] We cloned approximately 2000 bp of CYP2C19 5′-flanking sequence and identified putative binding sites for CAR and GR ~1800 bp upstream of the translation start site, homologous to elements shown to be functional for CYP2C9 by Gerbal-Chaloin and co-workers.[150] Interestingly, activation by CAR was quite weak in these HepG2 cell line transfected systems even though the CAR/PXR binding site was essentially identical to that of CYP2C9. The reason for this is unclear, but similar observations have been observed with the CYP2C8 promoter, and both CYP2C8 and CYP2C19 reporters also appear to be non-responsive to co-transfection with HNF-4 in HepG2 cells. Due to the established relationship between HNF-4 and PXR for both CYP2C9 and CYP3A4, a lack of functional HNF-4 response in this system might also impact responsiveness to CYP2C19, but this remains to be established.

In summary, significant progress has been made in understanding the mechanisms that control CYP2C19 expression, and due to its substantial role in human drug metabolism, it is desirable to be aware of the potential of new drugs to induce CYP2C19 or function as its substrates to allow physicians to better anticipate possible drug-drug interactions.

7.6 Conclusions and Summary

The CYP2C subfamily is clearly a critical component of the human defense system against xenobiotic challenge. Our understanding of the molecular details of these processes continues to evolve, and the use of human-like model systems, such as human hepatocyte culture systems, along with *in vivo* systems to model potential drug interactions continue to provide effective tools for the development of safer medicines.

References

1. J.A. Goldstein and S.M. de Morais, *Pharmacogenetics*, 1994, **4**, 285.
2. A.B. Rifkind, C. Lee, T.K. Chang and D.J. Waxman, *Arch. Biochem. Biophys.*, 1995, **320**, 380.
3. L.C. McSorley and A.K. Daly, *Biochem. Pharmacol.*, 2000, **60**, 517.
4. F. Läpple, O. von Richter, M.F. Fromm, T. Richter, K.P. Thon, H. Wisser, E.U. Griese, M. Eichelbaum and K.T. Kivistö, *Pharmacogenetics*, 2003, **13**, 565.
5. Tracy C. DeLozier, Cheng-Chung Tsao, Sherry J. Coulter, Julie Foley, J. Alyce Bradbury, Darryl C. Zeldin and Joyce A. Goldstein. *Journal of Pharmacology And Experimental Therapeutics*, 2004, **310**, 845.

6. Ambra Pozzi, Maria Raquel Ibanez, Arnaldo E. Gatica, Shilin Yang, Shouzuo Wei, Shaojun Mei, John R. Falck and Jorge H. Capdevila. *J. Biol. Chem.*, 2007, **282**, 17685.
7. A.G. McArthur, T. Hegelund, R.L. Cox, J.J. Stegeman, M. Litjenberg, U. Olsson, P. Sundberg and M.C. Celander, *J. Mol. Evol.*, 2003, **57**, 200–211.
8. J.F. Teiber and P.F. Hollenberg, *Carcinogenesis*, 2000, **21**, 1559.
9. N. Plant, *Biochim. Biophys. Acta.*, 2007, **1770**, 478.
10. R. Zuber, E. Anzenbacherova and P. Anzenbacher, *J. Cell. Mol. Med.*, 2002, **6**, 189.
11. M.C. Wani, H.L. Taylor, M.E. Wall, P. Coggon and A.T. McPhail, *J. Am. Chem. Soc.*, 1971, **93**, 2325.
12. C. Herdeg, M. Oberhoff, A. Baumbach, A. Blattner, D.I. Axel, S. Schröder, H. Heinle and K.R. Karsch, *J. Am. Coll. Cardiol.*, 2000, **35**, 1969–1976.
13. J.J. Heimans, J.B. Vermorken, J.G. Wolbers, C.M. Eeltink, O.W. Meijer, M.J. Taphoorn and J.H. Beijnen, *Ann. Oncol.*, 1994, **5**, 951–953.
14. S. Fellner, B. Bauer, D.S. Miller, M. Schaffrik, M. Fankhänel, T. Spruss, G. Bernhardt, C. Graeff, L. Färber, H. Gschaidmeier, A. Buschauer and G. Fricker, *J. Clin. Invest.*, 2002, **110**, 1309–1318.
15. D.S. Sonnichsen, Q. Liu, E.G. Schuetz, J.D. Schuetz, A. Pappo and M.V. Relling, *J. Pharmacol. Exp. Ther.*, 1995, **275**, 566.
16. M. Niemi, J.T. Backman, M. Granfors, J. Laitila, M. Neuvonen and P.J. Neuvonen, *Diabetologia*, 2003, **46**, 1319.
17. J.T. Backman, C. Kyrklund, M. Neuvonen and P.J. Neuvonen, *Clin. Pharmacol. Ther.*, 2002, **72**, 685.
18. Guillaume, A., Schoch, Jason, K., Yano., Michael, R., Wester, Keith J. Griffin; David Stout; Eric F. Johnson. *J. Biol. Chem.*, 2004, **279**, 9497.
19. A. Melet, C. Marques-Soares, G.A. Schoch, A.C. Macherey, M. Jaouen, P.M. Dansette, M.A. Sari, E.F. Johnson and D. Mansuy, *Biochemistry*, 2004, **43**, 15379.
20. T.B. Bidstrup, I. Bjørnsdottir, U.G. Sidelmann, M.S. Thomsen and K.T. Hansen, *Br. J. Clin. Pharmacol.*, 2003, **56**, 305.
21. S.J. Baldwin, S.E. Clarke and R.J. Chenery, *Br. J. Clin. Pharmacol.*, 1999, **48**, 424.
22. T. Jaakkola, J. Laitila, P.J. Neuvonen and J.T. Backman, *Basic Clin. Pharmacol. Toxicol.*, 2006, **99**, 44–51.
23. H. Yamazaki, A. Shibata, M. Suzuki, M. Nakajima, N. Shimada, F.P. Guengerich and T. Yokoi, *Drug. Metab. Dispos.*, 1999, **27**, 1260.
24. X.Q. Li, A. Björkman, T.B. Andersson, M. Ridderström and C.M. Masimirembwa, *J. Pharmacol. Exp. Ther.*, 2002, **300**, 399–407.
25. W. Mück, *Clin. Pharmacokinet.*, 2000, **39**, 99.
26. H.M. Falck and P. Westermark, *Clin. Exp. Immunol.*, 1983, **54**, 259.
27. J.K. Chen, J.R. Falck, K.M. Reddy, J. Capdevila and R.C. Harris, *J. Biol. Chem.*, 1998, **273**, 29254.
28. D.R. Harder, W.B. Campbell and R.J. Roman, *J. Vasc. Res.*, 1995, **32**, 79.

29. J.M. Pascual, A. McKenzie, J.R. Yankaskas, J.R. Falck and D.C. Zeldin, *J. Pharmacol. Exp. Ther.*, 1998, **286**, 772.
30. A.P. Zou, J.T. Fleming, J.R. Falck, E.R. Jacobs, D. Gebremedhin, D.R. Harder and R.J. Roman, *Am. J. Physiol.*, 1996, **270**, R228.
31. K. Node, Y. Huo, X. Ruan, B. Yang, M. Spiecker, K. Ley, D.C. Zeldin and J.K. Liao, *Science*, 1999, **285**, 1276.
32. B. Fisslthaler, R. Popp, L. Kiss, M. Potente, D.R. Harder, I. Fleming and R. Busse, *Nature*, 1999, **401**, 493.
33. J. Bauersachs, M. Christ, G. Ertl, U.R. Michaelis, B. Fisslthaler, R. Busse and I. Fleming, *Cardiovasc. Res.*, 2002, **54**, 669.
34. W.B. Campbell and D.R. Harder, *Circ. Res.*, 1999, **84**, 484.
35. D. Dai, D.C. Zeldin, J.A. Blaisdell, B. Chanas, S.J. Coulter, B.I. Ghanayem and J.A. Goldstein, *Pharmacogenetics.*, 2001, **11**, 597.
36. J. Marill, T. Cresteil, M. Lanotte and G.G. Chabot, *Mol. Pharmacol.*, 2000, **58**, 1341.
37. T.S. Klose, J.A. Blaisdell and J.A. Goldstein, *J. Biochem. Mol. Toxicol.*, 1999, **13**, 289.
38. D. Dai, D.C. Zeldin, J.A. Blaisdell, B. Chanas, S.J. Coulter, B.L. Ghanayem and J.A. Goldstein, *Pharmacogenetics*, 2001, **11**, 597.
39. Food and Drug Administration. Draft Guidance for Industry on Drug Interaction Studies—Study Design, Data Analysis, and Implications for Dosing and Labeling. http://www.fda.gov/Cder/drug/drugInteractions/guidance.htm.
40. A.E. Rettie, K.R. Korzekwa, K.L. Kunze, R.F. Lawrence, A.C. Eddy, T. Aoyama, H.V. Gelboin, F.J. Gonzalez and W.F. Trager, *Chem. Res. Toxicol.*, 1992, **5**, 54.
41. D.S. Whitlon, J.A. Sadowski and J.W. Suttie, *Biochemistry.*, 1978, **17**, 1371.
42. B.C. Jones, G. Hawksworth, V.A. Horne, A. Newlands, J. Morsman, M.S. Tute and D.A. Smith, *Drug Metab. Dispos.*, 1996, **24**, 260.
43. D.J. Elliot, B.C. Suharjono Lewis, E.M. Gillam, D.J. Birkett, A.S. Gross and J.O. Miners, *Br. J. Clin. Pharmacol.*, 2007, **64**, 450.
44. A.B. Rifkind, C. Lee, T.K. Chang and D.J. Waxman, *Arch. Biochem. Biophys.*, 1995, **320**, 380.
45. Blaisdell Joyce, Jorge-Nebert Lucia F. Coulter Sherry, Ferguson Stephen S. Lee Su-Jun, Chanas Brian, Xi Tina, Mohrenweiser Harvey, Ghanayem Burhan, Goldstein Joyce A. *Pharmacogenetics*, 2004, **14**, 527.
46. P.A. Williams, J. Cosme, A. Ward, H.C. Angove, D. Matak Vinković and H. Jhoti, *Nature*, 2003, **424**, 464.
47. W.G. Karam, J.A. Goldstein, J.M. Lasker and B.I. Ghanayem, *Drug Metab. Dispos.*, 1996, **24**, 1081.
48. S.A. Wrighton, J.C. Stevens, G.W. Becker and M. VandenBranden, *Arch. Biochem. Biophys.*, 1993, **306**, 240.
49. J. Blaisdell, H. Mohrenweiser, J. Jackson, S. Ferguson, S. Coulter, B. Chanas, T. Xi, B. Ghanayem and J.A. Goldstein, *Pharmacogenetics.*, 2002, **12**, 703.

50. R. Zuber, E. Anzenbacherová and P. Anzenbacher, *J. Cell. Mol. Med.*, 2002, **6**, 189.
51. M. Nakajima, T. Yokoi, M. Mizutani, M. Kinoshita, M. Funayama and T. Kamataki, *J. Biochem. (Tokyo)*., 1999, **125**, 803.
52. J.A. Agundez, C. Martinez, M.C. Ledesma, M.G. Ladona, J.M. Ladero and J. Benitez, *Clin. Pharmacol. Ther.*, 1994, **55**, 412–417.
53. R.A. Totah and A.E. Rettie, *Clin. Pharmacol. Ther.*, 2005, **77**, 341.
54. Web Page of the Human Cytochrome P450 (CYP) Allele Nomenclature Committee. http://www.cypalleles.ki.se/cyp2c8.htm (accessed October 2007).
55. G. Tai, F. Farin, M.J. Rieder, A.W. Dreisbach, D.L. Veenstra, C.L. Verlinde and A.E. Rettie, *Pharmacogenomics*, 2005, **15**, 475.
56. S. Rost, A. Fregin, V. Ivaskevicius, E. Conzelmann, K. Hörtnagel, H.J. Pelz, K. Lappegard, E. Seifried, I. Scharrer, E.G. Tuddenham, C.R. Müller, T.M. Strom and J. Oldenburg, *Nature*, 2004, **427**, 537.
57. E.A. Sconce, T.I. Khan, H.A. Wynne, P. Avery, L. Monkhouse, B.P. King, P. Wood, P. Kesteven, A.K. Daly and F. Kamali, *Blood*, 2005, **106**, 2329.
58. M.G. Scordo, A.P. Caputi, C. D'Arrigo, G. Fava and E. Spina, *Pharmacol. Res.*, 2004, **50**, 195–200.
59. Mizutani, T. PM frequencies of major CYPs in Asians and Caucasians. *Drug Metab. Rev.*, 2003, **35**, 99.
60. N. Zand, N. Tajik, A.S. Moghaddam and I. Milanian, *Clin. Exp. Pharmacol. Physiol.*, 2007, **34**, 102.
61. Y.H. Zhou, Q.C. Zheng, Z.S. Li, Y. Zhang, M. Sun, C.C. Sun, D. Si, L. Cai, Y. Guo and H. Zhou, *Biochimie.*, 2006, **88**, 1457.
62. M. Taguchi, K. Hongou, S. Yagi, T. Miyawaki, M. Takizawa, T. Aiba and Y. Hashimoto, *Drug Metab. Pharmacokinet.*, 2005, **20**, 107.
63. U. Klotz, *Clin. Pharmacokinet.*, 2007, **46**, 271.
64. Z. Desta, X. Zhao, J.G. Shin and D.A. Flockhart, *Clin. Pharmacokinet.*, 2002, **41**, 913.
65. S.M. De Morais, G.R. Wilkinson, J. Blaisdell, U.A. Meyer, K. Nakamura and J.A. Goldstein, *Mol. Pharmacol.*, 1994, **46**, 594.
66. A. Meyer, *Eur. J. Gastroenterol. Hepatol.*, 1996, **8**, S21.
67. U. Klotz, *Int. J. Clin. Pharmacol. Ther.*, 2006, **44**, 297.
68. J.M. Hoskins, G.M. Shenfield and A.S. Gross, *Eur. J. Clin. Pharmacol.*, 2003, **59**, 611.
69. Wilkinson, G.R., Guengerich, F.P., Branch, R.A., 0Pharmacogenetics of Drug Metabolism; Editor, Kalow W, Publisher, Pergamon Press: New York, 1992, 657.
70. K. Nakamura, F. Goto, W.A. Ray, C.B. McAllister, E. Jacqz, G.R. Wilkinson and R.A. Branch, *Clin. Pharmacol. Ther.*, 1985, **38**, 402.
71. J.L. Ruas and M.C. Lechner, *Pharmacogenetics*, 1997, **7**, 333.
72. T. Mizutani, *Drug Metab. Rev.*, 2003, **35**, 99.
73. J.H. Wang, P.Q. Li, Q.Y. Fu, Q.X. Li and W.W. Cai, *Clin. Exp. Pharmacol. Physiol.*, 2007, **34**(5-6), 421–424.

74. U. Klotz, *Int. J. Clin. Pharmacol. Ther.*, 2006, **44**, 297.
75. A.S. Zandvliet, A.D. Huitema, W. Copalu, Y. Yamada, T. Tamura, J.H. Beijnen and J.H. Schellens, *Clin. Cancer Res.*, 2007, **13**, 2970.
76. S. Lowery, K. Haley and H.I. Bussey, *Lippincotts Case Manag.*, 2005, **10**, 39.
77. G. De Angelis, *Int. J. Clin. Pract.*, 2004, **58**, 945.
78. J.T. Backman, C. Kyrklund, M. Neuvonen and P.J. Neuvonen, *Clin. Pharmacol. Ther.*, 2002, **72**, 685.
79. B.W. Ogilvie, D. Zhang, W. Li, A.D. Rodrigues, A.E. Gipson, J. Holsapple, P. Toren and A. Parkinson, *Drug Metab. Dispos.*, 2006, **34**, 191.
80. P.J. Neuvonen, M. Niemi and J.T. Backman, *Clin. Pharmacol. Ther.*, 2006, **80**, 565.
81. M. Niemi, J.T. Backman, M. Granfors, J. Laitila, M. Neuvonen and P.J. Neuvonen, *Diabetologia*, 2003, **46**, 1319.
82. T. Jaakkola, J.T. Backman, M. Neuvonen and P.J. Neuvonen, *Clin. Pharmacol. Ther.*, 2005, **77**, 404.
83. J. Kirchheiner, S. Thomas, S. Bauer, D. Tomalik-Scharte, U. Hering, O. Doroshyenko, A. Jetter, S. Stehle, M. Tsahuridu, I. Meineke, J. Brockmoller and U. Fuhr, *Clin. Pharmacol. Ther.*, 2006, **80**, 657.
84. M. Niemi, J.B. Leathart, M. Neuvonen, J.T. Backman, A.K. Daly and P.J. Neuvonen, *Clin. Pharmacol. Ther.*, 2003, **74**, 380.
85. T. Prueksaritanont, J.J. Zhao, B. Ma, B.A. Roadcap, C. Tang, Y. Qiu, L. Liu, J.H. Lin, P.G. Pearson and T.A. Baillie, *J. Pharmacol. Exp. Ther.*, 2002, **301**, 1042.
86. M. Niemi, A. Tornio, M.K. Pasanen, H. Fredrikson, P.J. Neuvonen and J.T. Backman, *Eur. J. Clin. Pharmacol.*, 2006, **62**, 463.
87. J.S. Wang, M. Neuvonen, X. Wen, J.T. Backman and P.J. Neuvonen, *Drug Metab. Dispos.*, 2002, **30**, 1352.
88. X. Wen, J.S. Wang, J.T. Backman, J. Laitila and P.J. Neuvonen, *Drug Metab. Dispos.*, 2002, **30**, 631.
89. M.W. Hruska, J.A. Amico, T.Y. Langaee, R.E. Ferrell, S.M. Fitzgerald and R.F. Frye, *Br. J. Clin. Pharmacol.*, 2005, **59**, 70.
90. R.L. Walsky, E.A. Gaman and R.S. Obach, *J. Clin. Pharmacol.*, 2005, **45**, 68.
91. R.L. Walsky, R.S. Obach, E.A. Gaman, J.P. Gleeson and W.R. Proctor, *Drug Metab. Dispos.*, 2005, **33**, 413.
92. L.I. Kajosaari, M. Niemi, J.T. Backman and P.J. Neuvonen, *Clin. Pharmacol. Ther.*, 2006, **79**, 231.
93. R.A. Blum, J.H. Wilton, D.M. Hilligoss, M.J. Gardner, E.B. Henry, N.J. Harrison and J.J. Schentag, *Clin. Pharmacol. Ther.*, 1991, **49**, 420.
94. K.L. Kunze and W.F. Trager, *Drug Metab. Dispos.*, 1996, **24**, 429.
95. T.L. Seaton, C.L. Celum and D.J. Black, *DICP*, 1990, **24**, 1177.
96. K. Laine, J. Forsstrom, P. Gronroos, K. Irjala, M. Kailajarvi and M. Scheinin, *Ther. Drug Monit.*, 2000, **22**, 503.
97. S. Almog, N. Shafran, H. Halkin, P. Weiss, Z. Farfel, U. Martinowitz and H. Bank, *Eur. J. Clin. Pharmacol.*, 1985, **28**, 257.

98. J.P. Tsikouris and C.D. Cox, *Pharmacotherapy.*, 2001, **21**, 1514.
99. P.E. Nolan, Jr., B.L. Erstad, G.L. Hoyer, M. Bliss, K. Gear and F.I. Marcus, *Am. J. Cardiol.*, 1990, **65**, 1252.
100. K. Ohyama, M. Nakajima, M. Suzuki, N. Shimada, H. Yamazaki and T. Yokoi, *Br. J. Clin. Pharmacol*, 2000, **49**, 244.
101. D.J. Back, J.F. Tjia, J. Karbwang and J. Colbert, *Br. J. Clin. Pharmacol.*, 1988, **26**, 23.
102. C. Tang, M. Shou, Q. Mei, T.H. Rushmore and A.D. Rodrigues, *J. Pharmacol. Exp. Ther.*, 2000, **293**, 453.
103. P. Jean, P. Lopez-Garcia, P. Dansette, D. Mansuy and J.L. Goldstein, *Eur. J. Biochem.*, 1996, **241**, 797.
104. A. Melet, N. Assrir, P. Jean, M. Pilar Lopez-Garcia, C. Marques-Soares, M. Jaouen, P.M. Dansette, M.A. Sari and D. Mansuy, *Arch. Biochem. Biophys.*, 2003, **409**, 80.
105. D.F. McGinnity, A.J. Berry, J.R. Kenny, K. Grime and R.J. Riley, *Drug Metab. Dispos.*, 2006, **34**, 1291.
106. M.P. López-Garcia, P.M. Dansette and D. Mansuy, *Biochemistry*, 1994, **33**, 166.
107. C. Eugène, C. Patriarche, A. Bergue and J. Quevauvilliers, *Nouv. Presse. Med.*, 1980, **9**, 1885.
108. K.Y. Kim and M.A. Mancano, *Ann. Pharmacother.*, 2003, **37**, 212.
109. D. Williams and J. Feely, *Clin. Pharmacokinet.*, 2002, **41**, 343.
110. C. Transon, T. Leemann, N. Vogt and P. Dayer, *Clin. Phamacol. Ther.*, 1995, **58**, 412.
111. M.R. Andrus, *Phamacotherapy*, 2004, **24**, 285.
112. M.J. Kim, A.N. Nafziger, A.D. Kashuba, J. Kirchheiner, S. Bauer, A. Gaedigk and J.S. Bertino, Jr., *Eur. J. Clin. Phamacol.*, 2006, **62**, 431.
113. V. Claassen, J.E. Davies, G. Hertting and P. Placheta, *Br. J. Pharmacol.*, 1977, **60**, 505.
114. A. Hemeryck, C. De Vriendt and F.M. Belpaire, *Eur. J. Clin. Phamacol.*, 1999, **54**, 947.
115. K.B. Yap and S.T. Low, *Singapore Med. J.*, 1999, **40**, 480.
116. K.K. Limke, A.R. Shelton and E.S. Elliott, *Ann. Phamacother.*, 2002, **36**, 1890.
117. N.T. Ha-Duong, S. Dijols, A.C. Macherey, J.A. Goldstein, P.M. Dansette and D. Mansuy, *Biochemistry*, 2001, **40**, 12112.
118. I. Ieiri, M. Kimura, S. Irie, A. Urae, K. Otsubo and T. Ishizaki, *Pharmacogenet Genomics*, 2005, **15**, 851.
119. S. Donahue, D.A. Flockhart and D.R. Abernethy, *Clin. Pharmacol. Ther.*, 1999, **66**, 563.
120. J.W. Ko, N. Sukhova, D. Thacker, P. Chen and D.A. Flockhart, *Drug Metab. Dispos.*, 1997, **25**, 853.
121. D.A. Flockhart, *Clin. Pharmacokinet.*, 1995, **29**, 45–52.
122. D.J. Waxman, *Arch. Biochem. Biophys.*, 1999, **369**, 11.
123. H. Wang and E.L. LeCluyse, *Clin. Pharmacokinet.*, 2003, **42**, 1331.

124. D.W. Nebert, T.P. Dalton, A.B. Okey and F.J. Gonzalez, *J. Biol. Chem.*, 2004, **279**, 23847.
125. H. Wang and M. Negishi, *Curr. Drug Metab.*, 2003, **4**, 515.
126. T. Kawamoto, T. Sueyoshi, I. Zelko, R. Moore, K. Washburn and M. Negishi, *Mol. Cell. Biol.*, 1999, **19**, 6318.
127. S. Kodama and M. Negishi, *Drug Metab. Rev.*, 2006, **38**, 75.
128. Y. Chen, S.S. Ferguson, M. Negishi and J.A. Goldstein, *J. Pharmacol. Exp. Ther.*, 2004, **308**, 495.
129. D.R. Koop and D.J. Tierney, *Bioessays.*, 1990, **12**, 429.
130. Z. Dvorak, M. Modriansky, L. Pichard-Garcia, P. Balaguer, M.J. Vilarem, J. Ulrichova, P. Maurel and J.M. Pascussi, *Mol. Pharmacol.*, 2003, **64**, 160.
131. T.S. Klose, J.A. Blaisdell and J.A. Goldstein, *J. Biochem. Mol. Toxicol.*, 1999, **13**, 289.
132. X. Ding and L.S. Kaminsky, *Ann. Rev. Pharmacol. Toxicol.*, 2003, **43**, 149.
133. T.C. Delozier, G.E. Kissling, S.J. Coulter, D. Dai, J.F. Foley, J.A. Bradbury, E. Murphy, C. Steenbergen, D.C. Zeldin and J.A. Goldstein, *Drug Metab. Dispos.*, 2007, **35**, 682.
134. A.E. Enayetallah, R.A. French, M.S. Thibodeau and D.F. Grant, *J. Histochem. Cytochem.*, 2004, **52**, 447.
135. T.B. Bidstrup, P. Damkier, A.K. Olsen, M. Ekblom, A. Karlsson and K. Brøsen, *Br. J. Clin. Pharmacol.*, 2006, **61**, 49.
136. J.Y. Park, K.A. Kim, M.H. Kang, S.L. Kim and J.G. Shin, *Clin. Pharmacol. Ther.*, 2004, **75**, 157.
137. T.B. Bidstrup, N. Stilling, P. Damkier, B. Scharling, M.S. Thomsen and K. Brosen, *Eur. J. Clin. Pharmacol.*, 2004, **60**, 109.
138. E.L. LeCluyse, *Eur. J. Pharm. Sci.*, 2001, **13**, 343.
139. J.L. Raucy, L. Mueller, K. Duan, S.W. Allen, S. Strom and J.M. Lasker, *J. Pharmacol. Exp. Ther.*, 2002, **302**, 475.
140. S.S. Ferguson, Y. Chen, E.L. LeCluyse, M. Negishi and J.A. Goldstein, *Mol. Pharmacol.*, 2005, **68**, 747.
141. R. Bort, M.J. Gómez-Lechón, J.V. Castell and R. Jover, *Arch. Biochem. Biophys.*, 2004, **426**, 63.
142. M.F. Paine, H.L. Hart, S.S. Ludington, R.L. Haining, A.E. Rettie and D.C. Zeldin, *Drug Metab. Dispos.*, 2006, **34**, 880.
143. J.A. Romankiewicz and M. Ehrman, *Ann. Intern. Med.*, 1975, **82**, 224.
144. L.D. Heimark, M. Gibaldi, W.F. Trager, R.A. O'Reilly and D.A. Goulart, *Clin. Pharmacol. Ther.*, 1987, **42**, 388.
145. J.O. Miners and D.J. Birkett, *Br. J. Clin. Pharmacol.*, 1998, **45**, 525.
146. J. Sahi, R.H. Stern, M.A. Milad, K.A. Rose, G. Gibson, X. Zheng, L. Stilgenbauer, N. Sadagopan, S. Jolley, D. Gilbert and E.L. LeCluyse, *Drug Metab. Dispos.*, 2004, **32**, 1370.
147. W. Zilly, D.D. Breimer and E. Richter, *Eur. J. Clin. Pharmacol.*, 1977, **11**, 287.

148. A. Madan, R.A. Graham, K.M. Carroll, D.R. Mudra, L.A. Burton, L.A. Krueger, A.D. Downey, M. Czerwinski, J. Forster, M.D. Ribadeneira, L.S. Gan, E.L. LeCluyse, K. Zech, P. Robertson, Jr., P. Koch, L. Antonian, G. Wagner, L. Yu and A. Parkinson, *Drug Metab. Dispos.*, 2003, **31**, 421.
149. G.C. Ibeanu and J.A. Goldstein, *Biochemistry.*, 1995, **34**, 8028.
150. S. Gerbal-Chaloin, M. Daujat, J.M. Pascussi, L. Pichard-Garcia, M.J. Vilarem and P. Maurel, *J. Biol. Chem.*, 2002, **277**, 209.
151. Y. Chen, G. Kissling, M. Negishi and J.A. Goldstein, *J. Pharmacol. Exp. Ther.*, 2005, **314**, 1125.
152. R.G. Tirona, W. Lee, B.F. Leake, L.B. Lan, C.B. Cline, V. Lamba, F. Parviz, S.A. Duncan, Y. Inoue, F.J. Gonzalez, E.G. Schuetz and R.B. Kim, *Nat. Med.*, 2003, **9**, 220.
153. Y. Kamiyama, T. Matsubara, K. Yoshinari, K. Nagata, H. Kamimura and Y. Yamazoe, *Drug Metab. Pharmacokinet.*, 2007, **22**, 287.
154. J.P. Jackson, S.S. Ferguson, R. Moore, M. Negishi and J.A. Goldstein, *Mol. Pharmacol.*, 2004, **65**, 1397.
155. S.S. Ferguson, E.L. LeCluyse, M. Negishi and J.A. Goldstein, *Mol. Pharmacol.*, 2002, **62**, 737.
156. J.M. Pascussi, S. Gerbal-Chaloin, L. Pichard-Garcia, M. Daujat, J.M. Fabre, P. Maurel and M.J. Vilarem, *Biochem. Biophys. Res. Commun.*, 2000, **274**, 707.
157. J. Prandota, *Am. J. Ther.*, 2004, **11**, 517.
158. D.J. Granville, B. Tashakkor, C. Takeuchi, A.B. Gustafsson, C. Huang, M.R. Sayen, P. Wentworth, Jr., M. Yeager and R.A. Gottlieb, *Proc. Natl. Acad. Sci. USA.*, 2004, **101**, 1321–1326.
159. H.H. Zhou, L.B. Anthony, A.J. Wood and G.R. Wilkinson, *Br. J. Clin. Pharmacol.*, 1990, **30**, 471.
160. K.L. Rost, H. Brösicke, M. Scheffler and I. Roots, *Int. J. Clin. Pharmacol. Ther. Toxicol.*, 1992, **11**, 542.
161. A. Madan, R.A. Graham, K.M. Carroll, D.R. Mudra, L.A. Burton, L.A. Krueger, A.D. Downey, M. Czerwinski, J. Forster, M.D. Ribadeneira, L.S. Gan, E.L. LeCluyse, K. Zech, P. Robertson, Jr., P. Koch, L. Antonian, G. Wagner, L. Yu and A. Parkinson, *Drug Metab. Dispos.*, 2003, **31**, 421.
162. Y. Chen, S.S. Ferguson, M. Negishi and J.A. Goldstein, *Mol. Pharmacol.*, 2003, **64**, 316.

CHAPTER 8

The CYP2D Subfamily

ULRICH M. ZANGER

Dr. Margarete Fischer-Bosch Institute of Clinical Pharmacology and the University of Tuebingen, Auerbachstr. 112, D-70376, Stuttgart, Germany

Table of Contents

Issues in Toxicology
Cytochromes P450: Role in the Metabolism and Toxicity of Drugs and other Xenobiotics
Edited by Costas Ioannides

8.1 Introduction and Molecular Genetics of the *CYP2D* Subfamily

The human cytochrome P450 (CYP) 2D6 was the first cytochrome P450 for which a classical pharmacogenetic polymorphism became known several years before the corresponding gene and protein were identified. By way of clinical observations, the antiarrhythmic and oxytocic drug sparteine and the antihypertensive debrisoquine led to the discovery of the genetic *CYP2D6* polymorphism, also known as the sparteine/debrisoquine polymorphism.[1,2] These discoveries stimulated a wealth of investigations now documented in over 3100 PubMed-listed publications on the enzyme and its catalytic properties, the corresponding gene and gene variants, and their role in drug metabolism, pharmacokinetics and pharmacodynamics, and as risk factors for numerous diseases and toxicological events. CYP2D6 metabolises over 50 clinically used drugs, most of them acting on the central nervous system or on the heart. More than 60 distinct alleles are known, some of which are associated with profound differences in expression and function, leading to four distinct phenotypes termed PM (poor metaboliser), IM (intermediate metaboliser), EM (extensive metaboliser) and UM (ultrarapid metaboliser). Both the phenotypes and the alleles occur at very different frequencies throughout different ethnic populations of the world. There is little influence on CYP2D6 expression by other factors like gender, age or typical inducing agents. Therefore, CYP2D6 represents one of the best examples of a drug metabolising enzyme under genetic control.

The *CYP2D* gene locus was first identified by using antibodies raised against purified rat debrisoquine 4-hydroxylase termed P450db1.[3] It turned out that the CYP2D subfamilies in animals have sometimes similar functions to the human enzymes, but in many aspects the animal gene subfamily structures as well as enzymic properties can also be very different. The most complete and updated compilation of cytochrome P450 sequences and other information among different species can be found on the Cytochrome P450 Homepage (http://drnelson.utmem.edu/CytochromeP450.html). *CYP2D* genes consist of 9 exons and 8 introns but the subfamily structures vary tremendeously among different animal species with respect to the number of functional genes and pseudogenes, and the function and expression of their protein products. The human *CYP2D* locus in its most common form covers about 29 kb and contains three full genes, including one functional gene, the highly polymorphic *CYP2D6*, and two closely related pseudogenes *CYP2D7P* and *CYP2D8P*.[4] The two pseudogenes are unable to produce a functional P450 protein; this is due to one single nucleotide insertion (T138 in exon 1) in the case of *CYP2D7P*, whereas numerous detrimental mutations accumulated in the coding sequence of *CYP2D8P*.[4] Nevertheless, at least *CYP2D7P* is expressed at the mRNA level in the liver and in other tissues (*vide infra*). The genomic region is highly variable between individuals with numerous gross structural variations that include full or partial gene duplications and multiduplications up to 13-fold gene amplification, deletion of the entire functional *CYP2D6* gene, and various rearrangements[5–8] as summarised by Schaeffeler.[9] For polymerase chain reaction (PCR)-based genetic diagnostic tests and for analysing mRNA transcripts it is, therefore, important to differentiate between the desired functional gene and pseudogenes using appropriately designed specific oligonucleotide primers.[10,11] Due to the large number of related genes, similar considerations of course apply to most CYP subfamilies in all species but much about this difficulty has been derived from the analysis of the human *CYP2D* locus. Remarkably, the *CYP2D* locus, initially sequenced in the course of the human genome project, was derived from an individual allele that lacked the entire *CYP2D6* gene due to a common deletion polymorphism.[6,12] The currently built 36.2 version of the human genome (available at the NCBI-server since September 2006) includes the Celera assembly and contains a complete and correctly annotated *CYP2D* locus, which however appears to be derived from a *CYP2D6*2* allele.

8.2 Catalytic Properties and Structure of CYP2D Enzymes

The substrate preference of human CYP2D6 is remarkably distinct from that of other known P450 enzymes, and this was recognised as soon as it became clear that the same enzyme catalysed the oxidative metabolism of drugs like sparteine, debrisoquine, metoprolol, amitriptyline and other compounds with alkaloidal properties. Common to these structures is a rather basic nitrogen, at some distance from the site of oxidation, and a hydrophobic domain near the site of oxidation.[13] Based on the existence of very potent inhibitors of CYP2D6, which includes the alkaloids ajmalicine and quinidine, pharmacophores for inhibitory structures were also developed.[14] Due to these early and promising successful attempts in pharmacophore modelling, CYP2D6 was often used as a paradigm for further computational approaches and refined models to predict its activity and selectivity.[15] The different modelling approaches included mechanism-based *ab-initio* calculations and 3D/4D-quantitative structure-activity relationship (QSAR) approaches, pharmacophore modelling of substrates and inhibitors, and the use of protein homology models which were based on the first crystal structures of microsomal P450s CYP2C5 and

CYP2B4.[16–18] It must be pointed out however that no single computational approach is yet able to predict reliably metabolite formation by CYP2D6, and that experimental data and model validation are often missing.

To rationalise CYP2D6 substrate selectivity, the assumption has been made that an ion pair is formed between the basic nitrogen and an acidic amino acid in the active site, which was initially believed to be Asp^{301}.[19] However, because the distance between the basic nitrogen and the site of oxidation varied quite substantially between about 5 to 10 Å, and due to other substrates not fitting the model, involvement of Glu^{216} as the principal binding residue was also proposed.[20] Site directed mutagenesis provided evidence for electrostatic interaction of amine substrates with Glu^{216} but attempts to induce new catalytic specificity profiles by substituting Asp^{301} and Glu^{216} were unsuccessful.[21,22] The two residues are also important for high affinity binding of the potent inhibitor quinidine.[23] According to the recently solved CYP2D6 crystal structure at 3.0 Å resolution, Asp^{301} forms hydrogen bonds with backbone NH-groups of Val^{119} and Phe,^{120} thus playing a role in protein stability.[24] Nevertheless, the investigators proposed that both Asp^{301} and Glu^{216} can act as binding residues for substrates and inhibitors, but that the two rotameric states (*trans* and *gauche*) of the aspartate can account for all observed distances between the basic nitrogen and the site of oxidation, and that the glutamate more likely acts to form an intermediate binding site for the substrate on the way to its final position in the active site.[24] As previous protein models were of limited value due to the low sequence identity between CYP2D6 and crystallised CYP2C5 (40%) and CYP3A4 (18%), the now available structure and future structures of CYP2D6, with various bound substrates and inhibitors, will certainly help to ultimately understand and predict the catalytic properties of this enzyme. It should also be noted that a few atypical substrates exist which lack a basic nitrogen or which are oxidised at the only basic nitrogen in the molecule, among them spirosulphonamide,[25] progesterone[26] (*vide infra*), and some N,N-dialkylated amphetamines including deprenyl.[27,28]

8.2.1 Role in Drug and Xenobiotic Metabolism

Although expressed at lower levels in the liver as compared to several other P450s, CYP2D6 makes a bigger contribution to drug metabolism because its substrate profile obviously matches that of many successfully developed drugs better than that of other enzymes. This observation is based on human P450 metabolism data base compiled and analysed by Rendic, who included data on over 2300 drugs and more than 1100 other substances.[29] The ratio of drugs to other substances interacting with CYP2D6 was 9:1, the highest of all CYPs, compared to, for example, CYP2E1 with a ratio of 1:3. The drugs interacting as substrates with CYP2D6 represented a fraction of about 15% of all drugs in the database, next to subfamilies CYP3A (36%) and CYP2C (25%). Similar analyses for inhibitors showed the reverse preference, i.e. inhibitors of CYP2D6 were more likely other xenobiotics and not drugs. Also of note, CYP2D6 showed the smallest number of interactions of the activator/inducer type and

Table 8.1 Selected major drug substrates of CYP2D6.

Therapeutic Class	*Drug*	*Pathway catalyzed by*
Analgesics/Antitussives	Codeine	O-demethylation
	Dextromethorphan	O-demethylation
	Dihydrocodeine	O-demethylation
	Ethylmorphine	O-deethylation
	Hydrocodone	N-demethylation
	Tramadol	O-demethylation
	Norcodeine	O-demethylation
	Oxycodone	O-demethylation
Anti-ADHD drug	Atomoxetine	Aromatic hydroxylation
Antiarrhythmics	Aprindine	Aromatic hydroxylation
	Encainide	O-demethylation
	Flecainide	O-dealkylation (?)
	Mexiletine	Aromatic hydroxylation
	N-Propylajmaline	Benzylic hydroxylation
	Procainamide	Arylamine N-oxidation
	Propafenone	Aromatic hydroxylation
	Sparteine	Aliphatic hydroxylation
Antidementia Drugs	Galanthamine	O-demethylation
	Nicergoline	N-demethylation
Tricyclic Antidepressants	Amitriptyline	Benzylic hydroxylation
	Clomipramine	Aromatic hydroxylation
	Desipramine	Aromatic hydroxylation
	Imipramine	Aromatic hydroxylation
	Nortriptyline	Benzylic hydroxylation
Other Antidepressants	Citalopram	N-demethylation
	Desmethylcitalopram	N-demethylation
	Fluoxetine	N-demethylation
	Fluvoxamine	unclear
	Maprotiline	unclear
	Mianserin	Aromatic hydroxylation
	Minaprine	Aromatic hydroxylation
	Mirtazapine	Aromatic hydroxylation
	Paroxetine	Demethylenation
	Venlafaxine	O-demethylation
Antidiabetic	Phenformine	Aromatic hydroxylation
Antiestrogen	Tamoxifen	Aromatic hydroxylation
Antihypertensives	Debrisoquine	Benzylic hydroxylation
	Guanoxan	Aromatic hydroxylation
	Indoramin	Aromatic hydroxylation
Antiemetics	Dolasetron	Aromatic hydroxylation
	Ondansetron	Aromatic hydroxylation
	Tropisetron	Aromatic hydroxylation
Antihistamines	Mequitazine	Aromatic Hydroxylation
	Promethazine	Aromatic hydroxylation
	Diphenhydramine	N-dealkylation
Antipsychotics	Haloperidol	N-dealkylation
	Perphenazine	N-dealkylation
	Risperidone	Aliphatic hydroxylation
	Thioridazine	Sulfoxidation
	Zuclopenthixol	N-dealkylation
Appetite suppressant	Dexfenfluramine	N-dealkylation

Table 8.1 (*Continued*).

Therapeutic Class	*Drug*	*Pathway catalyzed by*
β adrenergic blocking agents	Alprenolol	Aromatic hydroxylation
	Bufuralol	Benzylic hydroxylation
	Bunitrolol	Aromatic hydroxylation
	Bupranolol	Aromatic hydroxylation
	Carvedilol	Aromatic hydroxylation
	Metoprolol	Aliphatic hydroxylation
	Propranolol	Aromatic hydroxylation
	Timolol	O-dealkylation
Calcium antagonist	Perhexiline	Aliphatic hydroxylation
Calcimimetic	R-568*	O-demethylation
MAO-inhibitors	Amiflamine	N-demethylation
	Brofaromine	O-demethylation
Recreational drugs	Methoxyamphetamine	O-demethylation
	MDMA*	Demethylenation
	MPPP*	Aliphatic hydroxylation
Vasodilatators	Cinnarizine	Aromatic hydroxylation
	Flunarizine	Aromatic hydroxylation

ADHD: Attention Deficit/Hyperactivity Disorder; MDMA, methylenedioxymethamphethamine; MPPP, 4′-Methyl-α-pyrrolidinopropiophenone; R-568, (*R*)-*N*-(3-methoxy-phenylethyl)-3-(2′-chlorophenyl)-1-(propylamine hydrochloride); MAO: Monoamine Oxidase.

all were activators, in agreement with the well-known lack of inducibility of this enzyme. Table 8.1 contains an updated list of substrates of many different therapeutic drug classes and other xenobiotcs for which the CYP2D6-dependent pathway constitutes a major elimination route. It is noticeable that the drug profile of CYP2D6 comprises, in particular, many substances that act on the central nervous and on the cardiovascular systems. Other xenobiotic chemicals that are CYP2D6 substrates and that are of some toxicological importance include the tobacco-specific nitrosamine, 4-(methylnitrosamino)-1-(3-pyridyl)-1-butanone (NNK),[30] the heroin degradation product and Parkinson's disease-inducing agent 1-methyl 4-phenyl 1,2,3,6-tetrahydropyridine (MPTP) and several of its analogues,[31–33] as well as numerous natural plant alkaloids.[29] Furthermore, the significant contribution of CYP2D6 to the metabolism of a number of so-called designer drugs of the amphetamine type (e.g. 3,4-methylenedioxy-N-methamphetamine, MDMA, and others including MDEA and MDA), of the new benzyl- or phenyl-piperazine type (benzylpiperazine, BZP, 1-(3,4-methylenedioxybenzyl) piperazine, MDBP, and others), and of the pyrrolidinophenone type (including R,S-α-pyrrolidinopropiophenone, PPP, and 4′-methyl-α-pyrrolidinopropiophenone, MPPP; Table 8.1) must be pointed out. These drugs produce feelings of euphoria and energy, and a desire to socialise, and are very popular as rave drugs. Adverse effects of this class of drugs in humans include life-threatening serotonin syndrome, hepatotoxicity, neurotoxicity and psychopathology, and metabolites were suspected to contribute to some of the toxic effects. CYP2D6 is the major enzyme catalysing the preferred metabolic steps of several of these drugs.[34–36]

8.2.2 Potential for Drug-Drug Interactions

Drug-drug interactions are one of the most important reasons for adverse drug effects. The underlying mechanisms are most often changes in activity and expression of drug metabolising enzymes and transporters, which can be profoundly influenced by drugs and other xenobiotics. This is due to either direct inhibition at the active site (reversible/irreversible) or via modulation of gene expression (induction/repression). Numerous chemicals including drugs are known to be potent competitive or mechanism-based inhibitors of CYP2D6. Some of these drug interactions block the enzyme completely and result in the change of phenotype from EM to PM (phenocopying). This effect was demonstrated *in vivo* for the potent inhibitor quinidine,[37] the selective serotonine reuptake inhibitors (SSRIs) paroxetine,[38,39] fluoxetine and and fluvoxamine,[40] the MAO (Monoamine Oxidase)-A inhibitor moclobemide,[41] and the antiarrhythmic flecainide.[42] Drug interactions due to profound CYP2D6 inhibition *in vivo* was furthermore reported, among others, for celecoxib,[43] propoxyphene[44] and several designer drugs.[45,46]

8.2.3 Role in Endogenous Biotransformation

The fact that CYP2D6 is totally absent in a considerable fraction of humans seems to speak against an important role of this cytochrome in essential endogenous metabolic pathways. However, a possible contribution to endogenous pathways that could result in a selective advantage for the PM in development, behaviour or reproduction has been a matter of speculation since the discovery of the CYP2D6 polymorphism. First hints that pointed at potential endogenous substrates, probably among neurotransmitters or neuromodulatory amines, came from the studies of Bertilsson[47] and Llerena[48,49] who observed differences in personality between individuals with different CYP2D6 phenotype. This possibility was strongly supported by the initial findings of CYP2D6 expression in human brain[50] and of clinical association of *CYP2D6* genotypes with Parkinson's disease.[51]

Several specific and high-turnover substrates have been discovered since then, although their relevance remains to be established. An initial observation that the trace amine tryptamine was metabolised by human liver microsomes to tryptophol apparently in a CYP2D6-dependent way,[52] was not confirmed in a later study which identified MAO-A as the responsible enzyme.[53] Another trace amine, tyramine, exists in the brain but is also an exogenous compound found in cheese and wine. The conversion of tyramine to dopamine was specifically catalysed by recombinant CYP2D6 amongst 11 CYP isoforms expressed in yeast, suggesting that dopamine is in part formed from endogenous and/or exogenous tyramine by CYP2D6, with differences being noted between allelic enzyme variants.[26,54] In further studies, these researchers also found that rat CYP2D4 and human CYP2D6, which are the predominant CYP2D isoforms in the brain, possess diverse hydroxylation activities (including 6β- and 21-hydroxylase) for progesterone and 17α-hydroxyprogesterone, which are atypical CYP2D6 substrates because they lack a basic nitrogen.[26,55,56] These findings support the view that CYP2D may be involved in the regulation (metabolism and/or synthesis) of endogenous neuroactive steroids, such as progesterone and its derivatives, in brain tissues.

In a broad and systematic way Gonzalez and colleagues identified 5-methoxytryptamine, a metabolite and precursor of melatonin, as a specific and high-turnover substrate of CYP2D6 that is converted to serotonin.[57] Moreover, two related 5-methoxyindolethylamines derived from 5-hydroxytryptamine were found as additional endogenous substrates of CYP2D6, which catalyses their O-demethylation with high enzymic capacity and specificity.[53] The biological roles of these substances in the central nervous system are, however, poorly understood. Polymorphic *CYP2D6* may therefore play a role in the interconversion of these psychoactive tryptamines, including a crucial step in the serotonin-melatonin cycle.[58] Another endogenous pathway possibly involving CYP2D6 is the biosynthesis of endogenous morphine, which can apparently be synthesised at low, but physiologically significant concentrations, in various tissues and is excreted in urine. Whereas previous studies did not reveal evidence for CYP2D6 in endogenous morphine synthesis,[59] it was recently found that morphine can apparently be synthesised in human white blood cells where it may have immunomodulatory effects, from various precursors including L-3, 4-dihydroxyphenylalanine (L-DOPA), reticuline, tetrahydropapaveroline (THP), and tyramine via catalytic involvement of CYP2D6.[60]

8.3 Species Differences in *CYP2D* Molecular Genetics, Expression and Function

Mammalian *CYP2D* genes are currently available from humans, chimpanzee and other non-human primates, monkey, mouse, rat, rabbit, hamster, dog, pigs, cattle and others. This chapter presents an overview on genetic and functional aspects of the *CYP2D* subfamily enzymes in these mammalian species. Figure 8.1 depicts a phylogenetic tree composed of all currently available protein sequences together with their accession numbers.

8.3.1 CYP2D in Nonhuman Primates

The chimpanzee genome contains three *CYP2D* genes on chromosome 22. They have not yet been named or biochemically characterized but sequence comparison shows that they are highly homologous to the human genes with two of them predicted to be functional and one a pseudogene. Nonhuman primates may thus have more than one functional *CYP2D* gene which metabolize human-typical substrates like bufuralol and dextromethorphan and which are inhibited by quinidine. The first example was *CYP2D17* from the crab-eating macaque, also called cynomolgus monkey, which is extensively used in the pharmaceutical industry and in medical experimentation.[61] Two marmoset enzymes *CYP2D19* and *CYP2D30* differed in their regioselectivity for debrisoquine hydroxylation[62] and in their enantioselectivity for bufuralol

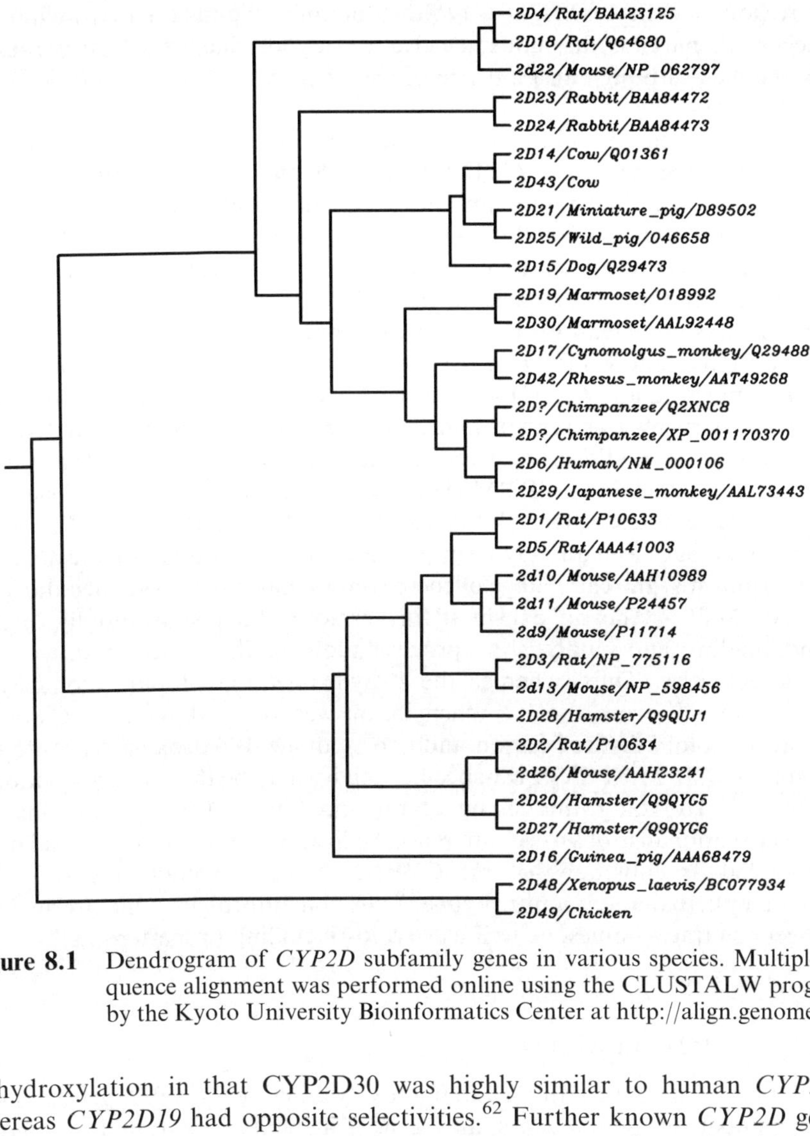

Figure 8.1 Dendrogram of *CYP2D* subfamily genes in various species. Multiple sequence alignment was performed online using the CLUSTALW program by the Kyoto University Bioinformatics Center at http://align.genome.jp/.

1′-hydroxylation in that CYP2D30 was highly similar to human *CYP2D6* whereas *CYP2D19* had opposite selectivities.[62] Further known *CYP2D* genes include *CYP2D29* from the Japanese monkey (macaca fuscata), which has catalytic properties very similar to *CYP2D6* but shows higher liver microsomal expression,[63] and the uncharacterized rhesus monkey *CYP2D42*.

8.3.2 CYP2D in Rat

The primate and rodent lineages diverged about 80 million years ago, giving sufficient time for major changes in structure and organisation of gene families. Both the rat and mouse genomes, which diverged about 20 million years ago, have greatly expanded *CYP2D* subfamilies compared to primate genomes.

The reason for the extensive *CYP2D* subfamily expansion in rodents, and especially in mice, is not understood and may be due to selective pressure from the diet, although an influence of endogenous pathways could also have played a role.

The rat genome has six functional genes but surprisingly seems to lack any *CYP2D* pseudogenes.[3,64–66] CYP2D1 and DS and CYP2D4 and CYP2D18 have over 95% similarity in amino acid sequence, and CYP2D1 is the orthologous form to human CYP2D6, sharing 82% amino acid sequence. Like other CYP2D enzymes, these six proteins are expressed in various tissues such as liver, kidney and brain, but specific tissue distribution patterns were found for their mRNAs. Whereas CYP2D1/5 was expressed in various tissues except brain, CYP2D2 and CYP2D3 mRNAs were found in liver, kidney, and small intestinal mucosa, and CYP2D4/18 mRNAs were the only forms expressed in brain, but were also present in many other tissues including adrenal glands, ovary, testis, in addition to liver, kidney and small intestinal mucosa.[67] CYP2D1, CYP2D3 and CYP2D4 were also identified in rat breast.[68] There are, however, pronounced strain differences. For example, the Dark Agouti rat strain, which had been already been proposed as a rat model for the *CYP2D6* polymorphism at the early stage of these studies, has much lower hepatic levels of CYP2D2.[69–71] The rat enzymes, furthermore, differ substantially in their ligand binding and catalytic properties, allowing the selection of enzyme-specific activities. Thus, whereas the 1′-hydroxylation of bufuralol may be catalysed by all forms, the 1′2′-ethenylation was only catalysed by CYP2D4, and propranolol 7-hydroxylation and debrisoquine 4-hydroxylation were specific for CYP2D2.[72] A CYP2D3-specific activity may be the *p*-hydroxylation of diazepam.[73] Homology modelling of rat and human CYP2D proteins was applied to rationalise observed differences in ligand binding.[74] Additional drugs investigated as substrates of rat CYP2D isoforms include methylenedioxymethamphethamine,[75] amitriptyline,[76] dextromethorphan,[77] mianserin[78] and endogenous trace amines, as well as steroids including progesterone.[56]

8.3.3 CYP2D in Mouse

At least nine different complete and putatively functional *CYP2D* genes, as well as a number of pseudogenes and gene fragments, are found on mouse chromosome 15E, a region syntenic to human 22q13.1[79] (Figure 8.1). The mouse proteins have high amino acid sequence identities (65–75%) to human CYP2D6 but they have been less well studied as compared with the rat, and functional data is incomplete. The *CYP2D22* enzyme was suggested to be the orthologue to human *CYP2D6*.[80] It is abundantly expressed in liver and at intermediate levels in the adrenal, ovary and mammary gland. However, its catalytic properties towards some typical substrates and inhibitors were unique in that it preferably catalysed N-demethylation reactions of both dextromethorphan and codeine instead of the CYP2D6-typical O-demethylation and, thus, seemed to more closely resemble CYP3A4.[81] Some mouse *CYP2D* proteins show sex-dependent expression. CYP2D9 is a male specifically expressed testosterone 16α-hydroxylase that depends on the male-typical pulsatile growth hormone pattern and signal transducer and activator of transcription (STAT) 5b signaling.[82,83] Interestingly, *Cyp2d9* transcription and activity are repressed following treatment with the Ah-receptor ligand methylcholanthrene.[84] In contrast, CYP2D4 is female dominant and apparently responsible for higher liver microsomal bufuralol 1′-hydroxylase and dextromethorphan O-demethylase activities in female mice.[85] Although various strains of mice have much lower debrisoquine 4-hydroxylase and bupropion 1′-hydroxylase activities than humans or Wistar rats, other typical CYP2D6 substrates are readily oxidised.[86] These differences are therefore due to complex species differences in CYP2D enzyme composition and substrate specificity, rendering these and other animals inadequate as valid CYP2D6 polymorphism models. More promising are the transgenic mouse lines expressing human CYP2D6. It was shown that CYP2D6-expressing mice displayed significantly higher debrisoquine hydroxylase activity than wild type mice[87] and that the CYP2D6 expression pattern in various tissues resembled that of humans, except that no expression was detected in brain.[88] If this deficiency could be overcome and endogenous *Cyp2d* genes be knocked out, these "humanised mice" could become valid models for polymorphic *CYP2D6*.

8.3.4 CYP2D in Other Species

CYP2D14 is a full length bovine CYP2D isoform, one of several highly similar clones isolated from bovine liver cDNA library but not characterised functionally.[89] CYP2D15 is the major CYP2D in dog having some enzymic activities in common with human CYP2D6 but showing some specific differences, e.g. very low debrisoquine hydroxylase activity,[90] high celecoxib hydroxylase activity[91] and inhibition sensitivity towards quinine and quinidine between that of rat CYP2D1 and human CYP2D6.[92] Dog CYP2D15, thus, appears to be quite unique in its catalytic properties. CYP2D16 is the only protein so far known from guinea pig, and it was cloned from an adrenal cDNA library because it had been shown to be expressed at high levels in microsomes of the inner zone of the gland.[93] The enzyme catalyses bufuralol 1′- and 6-hydroxylations and is inhibited by quinidine.[94] Three *CYP2D* genes have been cloned from Syrian hamster, but only CYP2D27 mRNA was detected in liver, small intestine and brain, whereas CYP2D20 and CYP2D28 were not detected. COS-7 cell expressing CYP2D27 catalysed bufuralol 1′- and debrisquine 4-hydroxylations.[95] CYP2D23 and CYP2D24 were cloned from rabbit liver cDNA library and shown to be expressed in liver, small intestine, brain (only 2D23) and stomach (only 2D24) and to catalyse bufuralol 1′-hydroxylation.[96] The porcine CYP2D25 has been recognised as a vitamin D_3 25-hydroxylase, which is unique among CYP2D enzymes, following purification from pig liver and cDNA cloning.[97]

8.4 Hepatic and Extrahepatic Expression of CYP2D Enzymes

CYP2D6 is first of all a hepatic enzyme. The fact that CYP2D6 is the only human CYP2D enzyme expressed at the protein level permitted the development of specific monoclonal and polyclonal antibodies which can be used to quantitate the protein in liver microsomes, rather accurately, by Western blot analysis.[98] Specific expression levels in liver vary dramatically from person to person from undetectable in PM (poor metaboliser) individuals, around 1 pmole per mg of microsomal protein in intermediate metabolisers and 30 or more pmoles/mg in the most active subjects.[98] Average expression levels in the liver is thus lower than those of CYP1A2, 2C8, 2C9, 2E1, and 3A4, but higher than previously believed[99] and exceeding those of 1A1, 1B1, 2A6, 2B6, 2C19, and 3A5/7.[98,100] At the RNA level, CYP2D6 expression is characterised by the occurrence of numerous splice variants.[101–103] However, with the exception of polymorphic splice sites in some *CYP2D6* poor metaboliser alleles,[104–107] and a splice variant lacking exon 6 which is associated with the intermediate metaboliser phenotype,[108] the functional significance of CYP2D6 splice variants has not been clarified.

8.4.1 Ontogeny, Induction and Regulation

With respect to ontogenic expression, CYP2D6 belongs to a group of P450 enzymes (together with CYP2E1) that are virtually undetectable in the foetal liver but whose expression surges within hours after birth, in contrast to those expressed in foetal liver (CYP3A7, CYP4A1) and those that are expressed at full level only after weeks (including CYP3A4, CYP2Cs, and CYP1A2).[109] The different developmental profiles are illustrated by the inability of human foetal liver microsomes to O-demethylate both codeine and dextromethorphan, whereas N-demethylation is present at levels comparable to adult tissue.[110] The mechanism of this ontogenic control has not been investigated for CYP2D6, but may involve post-transcriptional events as CYP2D6 mRNA has been found in human foetal liver in significant amounts before birth.[111,112] During pregnancy, oral clearance of CYP2D6 substrates, e.g. metoprolol, is substantially increased,[113] an effect that is not due to direct activation at the enzyme active site.[114] A subsequent study which involved *CYP2D6* genotyping and phenotyping using dextromethorphan revealed a decrease in the dextrorphan/dextromethorphan metabolic ratio by over 50%, demonstrating a significant effect of this physiologic condition on CYP2D6 activity.[115] A similar effect of pregnancy on activity was also found for CYP3A4 whereas CYP1A2 showed a decrease in activity.[116] The detailed results of the former study suggested an induction during the luteal phase, which is dominated by progesterone.[115] Remarkably, CYP2D6 was later identified as a progesterone 6β-, 16α-, and 21-hydroxylase.[26,117] However, no significant differences in CYP2D6 activity were found during the menstrual cycle.[118,119] Furthermore, smoking, use of oral contraceptive steroids, caffeine consumption and gender had no influence on sparteine metabolism rates.[120] CYP2D6 has also not been recognised as an inducible enzyme towards the typical inducers including phenobarbital and rifampicin,[121–123] although small genotype-dependent effects of rifampicin have been observed.[124]

Transcriptional regulation of *CYP2D6* gene expression has not been extensively studied. Nuclear factors involved in response to the typical inducers including pregnane X receptor (PXR) and constitutive androstane receptor (CAR) should not be expected to play a major role in *CYP2D6* gene transcription. An analysis of the *CYP2D6* core promoter by Cairns[125] revealed a potential role for antagonistic interactions between positively and negatively regulatory elements. They found that hepatocyte nuclear factor HNF4α could activate the *CYP2D6* promoter, whereas COUPTFI inhibited HNF4α stimulation *in vitro*. Apparently, both transcription factors competed for the same binding site, a direct repeat DR1. It was suggested that the contribution of the DR1 element in controlling the transcription of the *CYP2D6* gene depends on the balance between these positively and negatively acting transcription factors.[125] The stimulatory role of HNF4α in CYP2D6 expression was confirmed in a mouse line carrying the *CYP2D6* gene and an HNF4α conditional mutation. After deletion of HNF4α, debrisoquine 4-hydroxylase activity in *CYP2D6* humanised mice decreased more than 50%.[87] It should be noted that absence of induction in liver by typical inducers does not exclude induction by other agents, e.g. endogenous steroids, or in other tissues, as it will be discussed for brain (*vide infra*).[126]

8.4.2 Extrahepatic Expression and its Significance

Several studies have explored CYP2D6 expression in extrahepatic tissues. The preferred method is real-time reverse-transcriptase PCR but a few studies looking at the protein level have also been perfomed. Major problems with RNA studies are low expression levels (less than 5% compared with liver), possible interference with pseudogene expression and occurrence of differently spliced transcripts. Using gene-specific amplification it has been shown that CYP2D7P mRNA is expressed in liver at levels comparable to CYP2D6.[11] Thus, *CYP2D7P* is an expressed pseudogene and great care must be taken when developing RT-PCR methods to achieve specific amplification. Numerous splice variants derived from both the functional and the pseudogene(s) were described in different tissues including liver[101,102,108] lung,[127] breast[128] and brain,[103,129] but with the exception of the common *CYP2D6*4* and **41* alleles that lead to aberrant splicing (*vide infra*), the quantitative and functional relevance of splice variants has remained obscure.

8.4.2.1 Gastrointestinal Tract

Expression of cytochromes P450 and of other xenobiotic metabolising enzymes in the gastrointestinal tract occurs in the single cell layer of the mucosal enterocytes. The major CYPs expressed in the small intestine are the CYP3A, CYP2C and CYP2J2 isoforms.[130–132] The total CYP amount expressed along the entire small intestine is much lower compared with the liver but intracellular enterocytic expression levels may be even higher and not correlated to hepatic P450 content.[133] Several human studies have demonstrated that enteric CYP3A can contribute significantly to the overall first-pass metabolism of drugs.[134–136] CYP2D6 is expressed in the duodenum and jejunum but not in ileum and colon. In human and rhesus monkey intestinal microsomes, CYP2D6 was qualitatively similar to the liver enzyme[137] and interindividual variability in human intestinal mucosa was at least as extensive as that of liver.[138] Due to the low abundance and activity, the contribution to the first-pass effect of orally administered CYP2D6 substrates is probably negligible.

8.4.2.2 Central Nervous System

Expression of CYP2D6 in the human brain was one of the early landmark discoveries when a specific monoclonal antibody became available.[50] Its potential importance relates to the fact that CYP2D6 metabolises many drugs that act on the central nervous system, including antidepressants, opioids and neuroleptics, but also neurotoxins like MPTP and endogenous neuroactive amines and steroids. CYP2D6 expression in human brain samples was demonstrated initially by Western blotting,[50] and confirmed later by cDNA cloning,[139] by dextrometorphan activity determinations[140] and by immunocytochemical and *in situ* hybridisation methods.[141,142] Expression in the central nervous system is confined to certain regions and distinct neuronal cell types, including the pigmented neurons of the substantia nigra. In brain tissue samples from alcoholics and from smokers expression was found to be, on average, higher than in nonalcoholics and nonsmokers, respectively,[126,142] indicating possible tissue-specific induction in the brain and possibly explaining altered sensitivity of alcoholics to centrally acting drugs and to the mediation of neurotoxic and behavioural effects of alcohol.[143] Similar to liver and other extrahepatic tissues, numerous splice variants were detected in brain tissue samples with so far unknown functional significance. Among them, a functional revertant of CYP2D7P has been proposed to exist in brain[144] but not confirmed by others.[103]

8.4.2.3 Other Tissues

Whether CYP2D6 is expressed in lung tissue is still controversial as it was observed in one study[145] whereas others reported undetectably low levels.[146,147] Early reports on CYP2D6 transcript detection in lymphocytes are probably not reliable due to lack of specificity of the probes used. More recent studies are still controversial in that no protein or activity and only low levels or fragments of CYP2D6 transcripts were detected using specific probes in lymphocytes,[148,149] in contrast to others who found mRNA as well as expressed enzyme.[60] CYP2D6 mRNA was also detected by reverse transcriptase (RT)-PCR in human breast and breast tumour tissue.[128] If mRNA is being translated into active enzyme in breast tumours, the enzyme may contribute to the local metabolism of tamoxifen to the more potent antioestrogens, 4-hydroxytamoxifen and endoxifen.[150–152]

CYP2D6 mRNA was furthermore detected by RT-PCR in the right ventricular tissue of the heart where it may contribute to the metabolism of β-blockers.[153]

8.5 Pharmacogenetics of Human *CYP2D6*

The polymorphic expression of CYP2D6 was originally discovered in the 1970s when two research groups in England and in Germany observed that some volunteers participating in pharmacokinetic studies with debrisoquine and sparteine, respectively, developed unexpected adverse reactions. In both cases the researchers showed that the affected individuals had a substantially impaired capacity to oxidise the drug (PM phenotype), and that the metabolic defect is under monogenic control and inherited as an autosomal recessive trait.[1,2,154,155] Additional drugs were also affected by the deficient phenotype including metoprolol[156] and bufuralol.[157] Purification of the enzyme from rat and human liver[3,158] and use of specific antibodies allowed the characterisation of the molecular defect as the absence of a specific 50 kDa protein with catalytic activity towards (+)-bufuralol in liver samples of PM individuals.[159,160] Gonzalez *et al.*[101] isolated a full length human cDNA that was shown to encode a distinct, novel member of the CYP2D subfamily (P450db1) which was later renamed CYP2D6. The *CYP2D* genetic locus was sequenced[4] and defective alleles were initially identified by restriction fragment length polymorphism analysis[5] and subsequently by sequence and recombinant expression analysis.[104–106] The first allele-specific PCR-test to identify defective *CYP2D6* alleles was presented by Heim and Meyer.[10] Numerous alleles coding for nonfunctional protein products or such with increased or decreased function were subsequently identified and studied in many different populations throughout the world.[161]

8.5.1 Definition and Determination of CYP2D6 Phenotype

The most commonly used *in vivo* phenotyping procedure is by determining the ratio between the amount of unchanged drug and drug metabolite of a suitable probe drug in the urine within a certain time following administration of a single dose. As most environmental factors have not much influence on CYP2D6,[162] the urinary metabolic ratio (MR) usually represents a stable and reproducible personal parameter. The main interfering potential is enzyme inhibition caused by interaction with other CYP2D6 substrates and inhibitors having high affinity for the enzyme. Debrisoquine, dextromethorphan,[163] metoprolol,[164] sparteine and tramadol[165] are useful probe drugs although debrisoquine and sparteine are not on the market anymore. Dextrometorphan is a safe and widely available alternative to the originally used drugs. However, a disadvantage is its more extensive biotransformation in comparison to debrisoquine and sparteine, resulting in a less sensitive discrimination between EM (Extensive Metabolisers) and (Ultrarapid Metabolisers) UM.[166] More specific information on the use of probe drugs to assess the drug metabolising function of the liver *in vivo* can be found in specialised reviews.[167–169] Frequency histograms of the MR in different populations have a typical appearance which depends on the region of the world where the population studied originated from (*vide infra*).

In individuals of Caucasian origin, bimodal or trimodal distributions of the MR are typically seen with the PM subgroup (about 5 to 10% of the population) consisting of those individuals with a MR above a clearly defined antimode. The term "extensive metaboliser" (EM) was traditionally used for non-PMs, i.e. all those individuals with a MR below the antimode of a given model drug. Other populations show remarkably different phenotype profiles. Thus, Asian populations generally lack a clear polymorphic appearance but the median phenotype of the single mode has a broad distribution with a median MR that is shifted to the right (i.e. lower activity).[170,171] African populations were found to be much more diverse, and CYP2D6 phenotype patterns vary from unimodal to extensively polymorphic. PM phenotype frequencies range between 0% in Nigerians and 19% in Sun bushmen of Southern Africa, and those of UMs, if defined as carriers of gene duplications, from about 5% in African-Americans up to 29% in Ethiopians (Table 8.2). African populations also show a general "right-shift" in MR of the major part of the population and in both, Asian and African populations, it is the frequent occurrence of particular low-activity alleles (*10 and *17, respectively), which are responsible for this effect.[172,173]

Figure 8.2 shows a typical sparteine oxidation phenotype distribution in a German population.[174,175] The distribution is clearly trimodal with one larger

Table 8.2 Frequencies of major alleles with functional impact on *CYP2D6*.

Allele	*Key mutation*	*Consequence*	*f (White)*	*f (Black)*	*f (Asian)*
		Alleles with normal function			
*1	none	wild type	0.35–0.40	0.25–0.45	0.30–0.35
*2	2850C>T	R296C	0.16–0.20	0.1–0.25	0.10–0.15
		Alleles with decreased function/expression			
*10	100C>T	P34S	0.01–0.02	0.02–0.08	0.40–0.70
*17	1023C>T	T107I	<0.001	0.15–0.30	<0.001
*41	2988G>A (i6)	splice defect	0.08–0.1	<0.01 (?)	<0.02 (?)
		Alleles with increased expression			
*1×N	recombination	*dupl*CYP2D6	0.01–0.02	<0.02	<0.01
*2×N	recombination	*dupl*CYP2D6	0.01–0.10	0.01–0.30	<0.01
		Null alleles			
*3	2549delA	frame shift	0.01–0.03	<0.01	<0.01
*4	1846G>A	splice defect	0.18–0.22	0.01–0.07	<0.01
*5	recombination	*delCYP2D6*	0.03–0.05	0.03–0.06	0.03–0.05
*6	1707delT	frame shift	0.01–0.02	<0.01	<0.01

Allele frequencies (f) are shown as ranges of observed frequencies in Caucasian (White), Black African or African American (Black) or Asian individuals. Data were compiled from the literature108,173,174,176,180,187,197,203; n.d., no reliable data available.

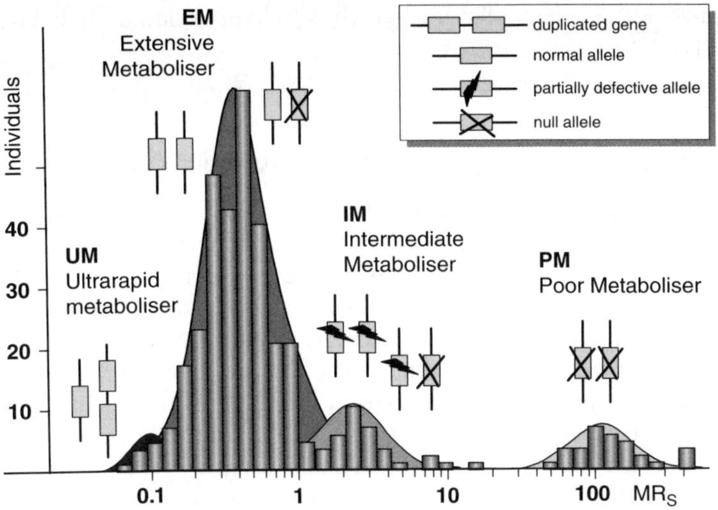

Figure 8.2 Sparteine oxidation phenotype and genotype distribution in a German population (n = 308)[175]; UM, ultrarapid metaboliser; EM, extensive metaboliser; IM, intermediate metaboliser; PM, poor metaboliser.

and one smaller subgroup having different median MR values within the EM mode (Figure 8.2). Similar observations were made with debrisoquine in other Caucasian populations.[171,176] At the high-activity end of the phenotype distribution, Bertilsson and colleagues[177] were the first to discover the phenotype termed "ultrarapid metaboliser" (UM) in a depressive woman who required an abnormally high dose of nortriptyline to attain therapeutic plasma levels. The phenomenon is now known to be due to CYP2D6 gene duplications, at least in part, but due to analytical difficulties and its relatively rare occurrence, this phenotype usually does not manifest itself as a separate subgroup in the phenotype distribution. Therefore, it is difficult to set a threshold value for this phenotype, and arbitrary values between MR = 0.15 to 0.2 have been used for debrisoquine and sparteine in various studies.[174,178–180] With dextromethorphan, the definition of UM is less reliable because this substance is almost undetectable in individuals with high activity.

CYP2D6 phenotyping is also useful for some *in vitro* applications, for example to characterise tissue samples to be used in drug metabolism studies in research or in preclinical drug development. Due to its fluorescent properties that allow highly sensitive detection, (+)-bufuralol has been used most frequently in *in vitro* investigations.[98,160,181] The stereoselective 1′-hydroxylation of (+)-bufuralol is catalysed by CYP2D6 with a K_m of approximately 5 to 10 μM. Because other CYP enzymes with higher K_m contribute to bufuralol metabolism in liver microsomes, it should either be used at a concentration of about 10 μM or, alternatively, the *in vitro* reaction can be supported by the artificial oxygen donor, cumene hydroperoxide, which mediates (+)-bufuralol 1′-hydroxylation specifically by CYP2D6.[160] Another useful *in vitro* probe drug is propafenone, which is 5′-hydroxylated almost exclusively by CYP2D6.[182]

8.5.2 Genetic Polymorphism and Interethnic Variation

A systematic nomenclature for *CYP2D6* alleles was introduced in 1996[183] and is now being maintained at the CYPallele nomenclature homepage (http://www.cypalleles.ki.se). Currently it lists 61 alleles (February 2007), which differ by certain "key mutations", i.e. sequence variations with demonstrated functional effect or such with amino acid changes, which are designated by the same allele number whereas further variants thereof are distinguished by letters.[184] For many alleles, phenotype information has been obtained from systematic *in vivo* phenotype-genotype correlation studies, studies in human liver samples, or using recombinant expression systems. For a number of the more rare variants, no phenotype information is yet available as they have been found by screening large numbers of nonphenotyped European subjects.[185] Figure 8.3 shows molecular details of the major alleles for which reliable phenotype information is available.

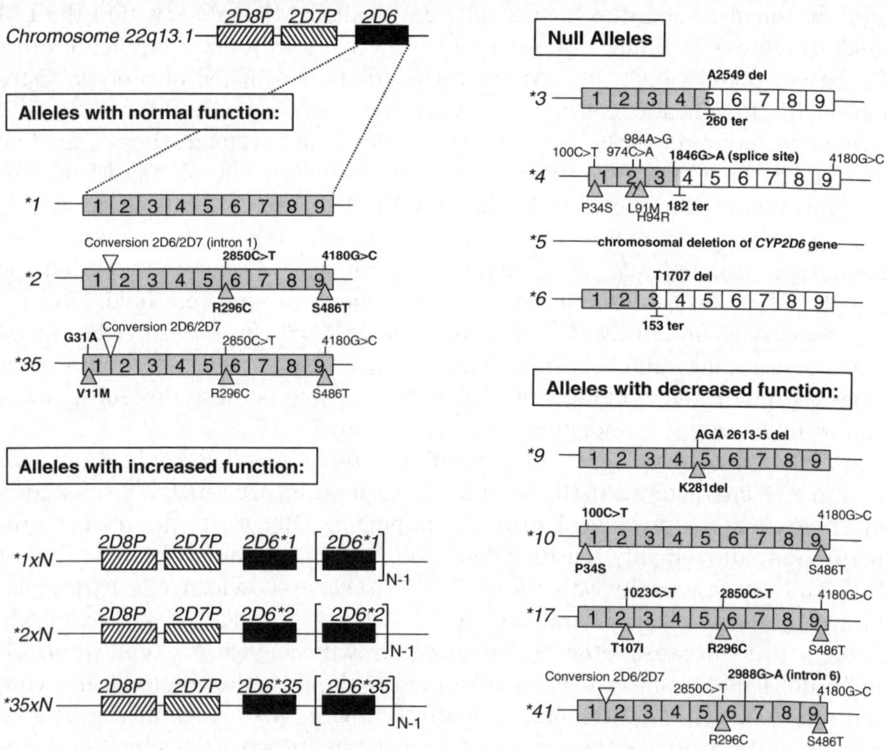

Figure 8.3 Structure of major alleles of *CYP2D6*.

8.5.2.1 Null Alleles

Molecular analysis has clearly shown that the CYP2D6 PM phenotype, as defined above, is genetically determined by the presence of two null alleles, i.e. alleles that do not encode a functional protein product. Thus, whereas for other P450 enzymes the term *poor metaboliser* is sometimes also used for individuals with a lower activity, in the case of CYP2D6 there is no detectable residual activity of the CYP2D6 enzyme. The most frequent PM allele is *CYP2D6*4*, which occurs with an allele frequency of about 20 to 25% in Caucasians and is responsible for 70 to 90 % of all PMs.[174,180] The key mutation is a G to A transition which causes a shift of the consensus acceptor splice site of the third intron by one base towards 3′-ward, thereby resulting in a spliced transcript which is one nucleotide shorter and has an interrupted reading frame.[104–106] The absence of a PM phenotype mode in Oriental populations is mainly due to the virtual absence of **4* allele. In African and in African-American populations it is present with frequencies intermediate between Asians and Caucasians.[173,186,187] In Australian Aborigines a very low frequency of **4* which was similar to that of Asians was observed.[188]

Numerous additional null alleles were found, including alleles with nonsense codons or single base pair mutations or small insertions/deletions that interrupt the reading frame or that interfere with correct splicing (*2D6*3*, **4*, **6*, **8*, **11*, **15*, **19*, **20*, **21*, **38*, **40*, **42*, **44*, **56*); lack of haem incorporation or otherwise nonfunctional full length proteins (*2D6*7*, **12*, **14*, **18*, **62*); larger chromosomal deletions of the entire *CYP2D6* gene (*2D6*5*) or *CYP2D6/2D7* hybrid genes with interrupted open reading frame (*2D6*13* and *2D6*16*; for references see CYPallele nomenclature website). *CYP2D6* gene deletion alleles are present at similar frequency in most populations.[84,173,189] Of the other null alleles, only **3* and **6* are present at an appreciable frequency in Caucasians (both about 1–2%) whereas all others appear to be very rare.

8.5.2.2 Alleles Resulting in Decreased Enzyme Activity

Several alleles are known to be associated with low but measurable CYP2D6 enzyme activity. Remarkably, a low activity phenotype is common in populations of all three major human races but the genetic basis for this is different in each case. In Caucasians *CYP2D6*9 [ΔK281]*[190] was initially detected in a family with IM (Intermediary Metaboliser) phenotype and found to express a low amount of a variant but functional CYP2D6 protein.[191] The mutant enzyme 2D6.9 (ΔK^{281}) expressed by recombinant baculovirus in insect cells had a lower specific haem content and also a decreased intrinsic clearance (CL_{int}) towards (+)-bufuralol compared with the wild-type protein, explaining its association with the IM phenotype.[98] The **9* allele has, however, a frequency of only about 1 to 2% in Caucasians and does not explain the IM phenotype in most individuals.

The C100T mutation leading to $Pro^{34}Ser$ is the key mutation in **10* and related alleles (which also includes **36* and **37*) but is present in numerous other alleles including most **4* allele variants (see cypallele nomenclature homepage). Pro[34] is part of a proline-rich region which is highly conserved among microsomal P450s and may function as a hinge between the hydrophobic membrane anchor and the globular haem-binding part of the enzyme.[192] The variant is expressed at low levels in liver[98] and has decreased but residual enzymic activity, thus explaining why individuals homozygous for **10* or with **10/*null* genotype are phenotypically IM and not PM. The *CYP2D6*10* allele has a frequency of about 2% in Caucasians but it is the dominant allele in all Oriental populations (Table 8.2) with frequencies exceeding 50 %. This explains the shift in the median MR in Asians (*vide supra*).

In Africans, a frequent allele with partially deficient function was also found, namely **17* with two nonsynonymous mutations in common with **2* and an additional T[107]I change shown to cause a decrease in activity in transfected COS cells.[193] This allele is practically absent from European Caucasians[174] but in Black populations it is present at high frequencies of up to 30%, which explains why Blacks have higher median MR values.[172,173,186,187,194]

As *9, *10, and *17 alleles were too rare to explain more than about 10 to 20% of Caucasian individuals with IM phenotype, it was to be expected that an additional low-activity allele may exist among Caucasians. Indeed, the **41* allele was found to play a comparable role to **10* and **17* in Asians and Africans, respectively. The **41* allele is genetically closely related to the functionally normal **2* allele but lacks a -1584C>G promoter polymorphism initially identified as a marker for the common low-activity allele,[195] and instead harbours a 2988G>A mutation in intron 6.[175] When the entire genomic sequence coding for *CYP2D6*41*, *CYP2D6*2* and *CYP2D6*1* alleles lacking the 5′ upstream region was recombinantly expressed in COS-1 and human Huh7 cell lines, the 2988G>A mutation present in **41* resulted in 2- to 5-fold reduced levels of CYP2D6 mRNA, apoprotein, and enzyme activity,[108] closely resembling the phenotypic properties of IM subjects in liver.[98] Thus, it was shown that the intron 6 polymorphism 2988G>A is causally responsible for markedly decreased expression and function of the **41* allele. The mechanism involves a modulation of splicing events leading to increased levels of a nonfunctional splice variant lacking exon 6 and a concurrent decrease in the functional transcript. *In silico* analysis suggested that 2988G>A modifies binding sites for the splicing factors SRp40 and SF2/ASF.[108] In Caucasians, the **41* allele has a frequency of about 8% and it is highly predictive for the IM phenotype in genotype constellations involving a null allele or another low-function allele.[175] An additional allele causing the CYP2D6 enzyme to be expressed at very low levels is **59*, which harbours the silent mutation 2939G>A in exon 6. Transient transfection of Huh7 hepatoma cells with the respective genomic *CYP2D6* sequence resulted in about 3-fold reduced levels of mRNA, immunoreactive CYP2D6 protein and propafenone hydroxylase activity of constructs carrying the 2939G>A change, demonstrating that a silent mutation can have profound effects on expression and function of *CYP2D6*.[196] There is growing evidence to support the relevance of the Caucasian IM phenotype for plasma levels of, for example, propafenone,[197] perhexiline,[198] desipramine,[199] and amitriptyline and nortriptyline.[200] Because the small residual enzyme amounts present in a patient with typical IM genotype (e.g. **41/*null*) are easily saturated, long-term treatment can lead to plasma concentrations almost as high as those observed in PM, as found in long-term treatment studies with metoprolol.[201]

8.5.2.3 Alleles Resulting in Increased Enzyme Activity

Analysis of the *CYP2D6* locus in Swedish individuals who required extremely high doses of nortriptyline uncovered alleles that have multiple copies of apparently functional *CYP2D6*.[8,202] The initial observations involved duplications and amplification of the *2 allele [R296C, S486T] but duplications are now known to have occured with various alleles including **1*, **2*, **4*, **6*, **10*, **17*, **29*, **35*, **43* and **45*.[174,180,203] The allele frequencies were ~1 to 2% in Northern European populations,[174,178,180,204] whereas much higher frequencies up to ~30% were observed in Ethiopians[205] and in Saudi Arabians.[206] It has been speculated that the remarkable preference of CYP2D6 for plant alkaloids may have been the target of natural selection in some East-African populations that depended on food containing toxic alkaloids, thereby leading to the accumulation of alleles with increased copy number and higher activity, probably also because an efficient induction mechanism was not available for this enzyme.[207] An intermediate frequency was found in Spanish populations, compatible with the extensive migrations of people from Northern Africa to the Iberian peninsula in historic times.[208,209] The mechanisms of the *CYP2D6* gene duplication and multi-duplication events most likely occurred by unequal crossover events involving a repetitive sequence which is also present in the *c-myc* gene known to be amplified in tumour cells.[210]

In human liver, significantly elevated levels of CYP2D6 protein and enzymic activity were described in a liver biopsy of an UM individual (sparteine MR 0.07) with *2D6*1/*2x2* genotype.[98] However, several subjects with similarly high expression/activity in that study did not have an increased gene copy number, and other studies also had shown that duplication alleles only predict a fraction of 15 to 30% of all individuals with UM phenotype, depending on the phenotype definition.[174,178,180,211] Searches for additional genetic variations causing increased expression or activity of CYP2D6 have remained, however, unsuccessful. In one attempt, Løvlie et al.[212] selected duplication-negative UM subjects and examined about 2 kb of *CYP2D6* promoter sequence including exon 1 and intron 1. The two most interesting variants that resulted from this screening were 31G>A ($Val^{11}Met$, **35*) in exon 1 and the above mentioned -1584C>G promoter polymorphism, both of which appeared to be over-represented among UM individuals. As already discussed (*vide supra*), the promoter polymorphism is a marker for **2* and related alleles (incl. *35) and, therefore, also marks absence of the low-activity allele **41*. In another study, the *in vivo* MR for sparteine oxidation was, however, not different between carriers of **1*, **2*, and **35*.[175] Another possibility to explain the duplication-negative UM phenotype was proposed to include functional revertants of *CYP2D7P*[144] but the genetic data of that work could not be reproduced by others.[103]

8.5.3 *CYP2D6* Genotyping

CYP2D6 genotyping strategies have to deal with two problems: one is interference with the highly homologous *2D7* and *2D8* pseudogenes and the second is the frequent occurrence of alleles with large structural alterations. The first problem is ususally circumvented by using nested PCR strategies, with a first PCR step designed to pre-amplify a *CYP2D6*-specific region.[10] Small nucleotide changes including SNPs (Single Nucleotide Polymorphisms) and insertions/deletions of one or a few bases are then detected in a second PCR step either designed as a PCR/RFLP assay or as an allele-specific PCR without subsequent digestion.[179,213] Using long-distance PCR protocols, it has become possible to amplify the entire *CYP2D6* gene within a single PCR fragment of about 5 kb, which can be used as the template for multiplex PCR to detect up to five key mutations simultaneously.[214] Similar developments have been reported by others.[215–217] By adapter ligation-mediated allele-specific amplification, even 17-plexed single-tube genotyping was reported.[218] Detection of gross structural variants was initially performed by long-PCR amplification techniques.[179,213,219,220] An easier assay based on TaqMan real-time quantification of *CYP2D6*, in relation to albumin as an internal reference gene, was shown to correctly measure the overall number of *CYP2D6* gene copies in a variety of genetic constellations.[9] It was also demonstrated that the strategy of copy number determination in combination with SNP analysis is sufficient for correct phenotype prediction. Recent applications of newer technologies used for genotyping *CYP2D6* include real-time PCR methods,[221–223] microarrays for DNA analysis,[166,224–226] pyrosequencing,[227,228] and bead arrays.[229] Although some high-density array technologies allow the analysis of practically all variants in the CYP2D6 gene, it should be pointed out that a rather limited analysis would be appropriate for most purposes, because many of the mutations are so rare that even their combined analysis does not improve phenotype prediction substantially. Genotyping should include the alleles summarised in Table 8.2, but **2* may even be omitted,[98,175] unless the study involves substrates that show a marked difference in metabolism between the wild type and the CYP2D6.2 protein variant.[230]

8.5.4 Clinical Relevance of CYP2D6 Pharmacogenetics

The clinical significance of the CYP2D6 polymorphism has been the subject of numerous earlier reviews.[231–234] Many case reports and clinical studies leave no doubt that CYP2D6-dependent polymorphic oxidation can have pharmacokinetic and pharmacodynamic consequences, but the exact effect depends strongly on the metabolic pathway and on other characteristics of the drug. For drugs with a narrow therapeutic index, which are inactivated by CYP2D6, the patient with PM phenotype may have a higher propensity to develop adverse reactions at therapeutic doses, while decreased drug effect, e.g. therapeutic failure at normal antidepressant dosages is a likely consequence the UM patient may experience. The reverse scheme must be applied to prodrugs, i.e. if the

CYP2D6-dependent metabolic step leads to the pharmacologically active substance as in the case codeine, which is O-demethylated to morphine. Thus life-threatening opioid intoxication following treatment with codeine developed in a patient who was then shown to have three or more copies of CYP2D6 consistent with the UM phenotype.[235]

In the following examples the potential of CYP2D6 genotyping for predicting pharmacokinetics, risk of adverse effects and treatment outcome has been explored in retrospective and prospective studies. In a small prospective study on desipramine, it was reported that the two patients with PM phenotype developed the highest plasma concentrations and experienced severe adverse effects, leading to dosage reduction.[236] A more than 2-fold higher frequency of null alleles was documented in the patient group that experienced adverse effects from treatment with antidepressant CYP2D6 substrates as compared with patients who tolerated the drugs well.[237] The results of the extended second study suggested greater treatment costs for patients with extremes in CYP2D6 activity (UM and PM).[238] In another prospective study with psychiatric inpatients treated with the antipsychotic drug haloperidol, which depends in part on CYP2D6 for biotransformation, significantly increased adverse effects were found in PMs, whereas a trend towards lower therapeutic effect was noted with increasing number of active *CYP2D6* genes.[239] A prospective study with cancer patients receiving chemotherapy and antiemetic treatment with CYP2D6 substrates of the 5-hydroxytryptamine type 3 receptor antagonist class, revealed that genetically defined UMs profit less from antiemetic therapy, encouraging dose adjustment based on genotype to improve treatment outcome.[240]

An important recent example of cancer therapy where CYP2D6 pharmacogenetics may have a major impact is the adjuvant treatment of breast cancer with tamoxifen. This antioestrogen is extensively metabolised in the liver to several primary and secondary metabolites. Two of these, 4-hydroxytamoxifen and the secondary metabolite endoxifen, show high affinity for the oestrogen receptor and are thought to be responsible for a large part of the antioestrogenic effect, are mainly formed by CYP2D6.[150,241] Several recent retrospective and prospective studies demonstrated a significant impact of the CYP2D6 genotype on the plasma concentrations of the active tamoxifen metabolites, treatment outcome and adverse effects.[242–246] According to these studies, patients with PM (**null/*null*) and IM (e.g. **41*/*null) genotypes, i.e. with low metabolic activation rates, have extremely low levels of the active metabolites and thus profit less from the treatment. They have a shorter time to recurrence and worse relapse-free survival relative to patients with EM genotype. Additional recent studies on the impact of CYP2D6 genotype on treatment outcome were made on metoprolol,[247–249] amitriptyline[250] and codeine.[251] However, expectations on the success of the pharmacogenetic approach may sometimes be unrealistically high. Drug toxicity and treatment outcome depend on many additional genetic and nongenetic factors, not only on a single genotype. Only in a few instances with particular drugs and treatment schemes will the patient have a clearly predictable benefit from simple genotyping. In the majority of clinical situations, multigene or multifactorial approches will be necessary to achieve an overall improvement of drug therapy, which implies minimising the risk of adverse reactions while increasing chances for effective treatment.

Acknowledgement

This work was supported by the Robert Bosch Foundation, Stuttgart, Germany.

References

1. A. Mahgoub, J.R. Idle, L.G. Dring, R. Lancaster and R.L. Smith, *Lancet*, 1977, **2**, 584.
2. M. Eichelbaum, N. Spannbrucker, B. Steincke and H.J. Dengler, *Eur. J. Clin. Pharmacol.*, 1979, **16**, 183.
3. F.J. Gonzalez, T. Matsunaga, K. Nagata, U.A. Meyer, D.W. Nebert, J. Pastewka, C.A. Kozak, J. Gillette, H.V. Gelboin and J.P. Hardwick, *DNA*, 1987, **6**, 149.
4. S. Kimura, M. Umeno, R.C. Skoda, U.A. Meyer and F.J. Gonzalez, *Am. J. Hum. Genet.*, 1989, **45**, 889.
5. R.C. Skoda, F.J. Gonzalez, A. Demierre and U.A. Meyer, *Proc. Natl. Acad. Sci. U.S.A.*, 1988, **85**, 5240.
6. A. Gaedigk, M. Blum, R. Gaedigk, M. Eichelbaum and U.A. Meyer, *Am. J. Hum. Genet.*, 1991, **48**, 943.
7. M.H. Heim and U.A. Meyer, *Genomics*, 1992, **14**, 49.
8. I. Johansson, E. Lundqvist, L. Bertilsson, M.L. Dahl, F. Sjoqvist and M. Ingelman-Sundberg, *Proc. Natl. Acad. Sci. USA*., 1993, **90**, 11825.
9. E. Schaeffeler, M. Schwab, M. Eichelbaum and U.M. Zanger, *Hum. Mutat.*, 2003, **22**, 476.
10. M. Heim and U.A. Meyer, *Lancet*, 1990, **336**, 529.
11. K. Endrizzi, J. Fischer, K. Klein, M. Schwab, A. Nussler, P. Neuhaus, M. Eichelbaum and U.M. Zanger, *Anal. Biochem.*, 2002, **300**, 121.
12. J. R. Idle, J. Corchero and F.J. Gonzalez, *Lancet*, 2000, **355**, 319.
13. T. Wolff, L.M. Distlerath, M.T. Worthington, J.D. Groopman, G.J. Hammons, F.F. Kadlubar, R.A. Prough, M.V. Martin and F.P. Guengerich, *Cancer Res.*, 1985, **45**, 2116.
14. G.R. Strobl, K.S. von, J. Stockigt, F.P. Guengerich and T. Wolff, *J. Med. Chem.*, 1993, **36**, 1136.
15. S. Ekins, M.J. de Groot and J.P. Jones, *Drug Metab. Dispos.*, 2001, **29**, 936.
16. N.P. Vermeulen, *Curr. Top. Med. Chem*, 2003, **3**, 1227.
17. D.F. Lewis, M. Dickins, B.G. Lake and P.S. Goldfarb, *Drug Metabol. Drug. Interact.*, 2003, **19**, 189.
18. C.A. Kemp, J.D. Marechal and M.J. Sutcliffe, *Arch. Biochem. Biophys.*, 2005, **433**, 361.

19. S.W. Ellis, G.P. Hayhurst, G. Smith, T. Lightfoot, M.M. Wong, A.P. Simula, M.J. Ackland, M.J. Sternberg, M.S. Lennard and G.T. Tucker, *J. Biol. Chem.*, 1995, **270**, 29055.
20. D.F. Lewis, P.J. Eddershaw, P.S. Goldfarb and M.H. Tarbit, *Xenobiotica*, 1997, **27**, 319.
21. F.P. Guengerich, G.P. Miller, I.H. Hanna, M.V. Martin, S. Leger, C. Black, N. Chauret, J.M. Silva, L. A. Trimble, J.A. Yergey and D.A. Nicoll-Griffith, *Biochemistry*, 2002, **41**, 11025.
22. F.P. Guengerich, I.H. Hanna, M.V. Martin and E.M. Gillam, *Biochemistry*, 2003, **42**, 1245.
23. L.A. McLaughlin, M.J. Paine, C.A. Kemp, J.D. Marechal, J.U. Flanagan, C.J. Ward, M.J. Sutcliffe, G.C. Roberts and C.R. Wolf, *J. Biol. Chem.*, 2005, **280**, 38617.
24. P. Rowland, F.E. Blaney, M.G. Smyth, J.J. Jones, V.R. Leydon, A.K. Oxbrow, C.J. Lewis, M.G. Tennant, S. Modi, D.S. Eggleston, R.J. Chenery and A.M. Bridges, *J. Biol. Chem.*, 2006, **281**, 7614.
25. C.A. Kemp, J.U. Flanagan, A.J. van Eldik, J.D. Marechal, C.R. Wolf, G.C. Roberts, M.J. Paine and M.J. Sutcliffe, *J. Med. Chem.*, 2004, **47**, 5340.
26. T. Niwa, T. Hiroi, D. Tsuzuki, S. Yamamoto, S. Narimatsu, T. Fukuda, J. Azuma and Y. Funae, *Brain Res. Mol. Brain Res.*, 2004, **129**, 117.
27. J.M. Grace, M.T. Kinter and T.L. Macdonald, *Chem. Res. Toxicol.*, 1994, **7**, 286.
28. M.V. Bach, R.T. Coutts and G.B. Baker, *Xenobiotica*, 2000, **30**, 297.
29. S. Rendic, *Drug. Metab. Rev.*, 2002, **34**, 83.
30. C.L. Crespi, B.W. Penman, H.V. Gelboin and F.J. Gonzalez, *Carcinogenesis*, 1991, **12**, 1197.
31. D.E. Gilham, W. Cairns, M.J. Paine, S. Modi, R. Poulsom, G.C. Roberts and C.R. Wolf, *Xenobiotica*, 1997, **27**, 111.
32. S. Modi, D.E. Gilham, M.J. Sutcliffe, L.Y. Lian, W.U. Primrose, C.R. Wolf and G.C. Roberts, *Biochemistry*, 1997, **36**, 4461.
33. T. Herraiz, H. Guillen, V.J. Aran, J.R. Idle and F.J. Gonzalez, *Toxicol. Appl. Pharmacol*, 2006, **216**, 387.
34. G.T. Tucker, M.S. Lennard, S.W. Ellis, H.F. Woods, A.K. Cho, L.Y. Lin, A. Hiratsuka, D.A. Schmitz and T.Y. Chu, *Biochem. Pharmacol.*, 1994, **47**, 1151.
35. K. Kreth, K. Kovar, M. Schwab and U.M. Zanger, *Biochem. Pharmacol.*, 2000, **59**, 1563.
36. H.H. Maurer, T. Kraemer, D. Springer and R.F. Staack, *Ther. Drug. Monit.*, 2004, **26**, 127.
37. K. Brosen, L.F. Gram, T. Haghfelt and L. Bertilsson, *Pharmacol. Toxicol.*, 1987, **60**, 312.
38. S.H. Sindrup, K. Brosen, L.F. Gram, J. Hallas, E. Skjelbo, A. Allen, G.D. Allen, S.M. Cooper, G. Mellows and T.C. Tasker, *Clin. Pharmacol. Ther.*, 1992, **51**, 278.
39. U. Jeppesen, L.F. Gram, K. Vistisen, S. Loft, H.E. Poulsen and K. Brosen, *Eur. J. Clin. Pharmacol.*, 1996, **51**, 73.

40. S. Vandel, G. Bertschy, P. Baumann, S. Bouquet, B. Bonin, T. Francois, D. Sechter and P. Bizouard, *Pharmacol. Res.*, 1995, **31**, 347.
41. L.F. Gram and K. Brosen, *Br. J. Clin. Pharmacol.*, 1993, **35**, 649.
42. W.E. Haefeli, M.J. Bargetzi, F. Follath and U.A. Meyer, *J. Cardiovasc. Pharmacol.*, 1990, **15**, 776.
43. U. Werner, D. Werner, T. Rau, M.F. Fromm, B. Hinz and K. Brune, *Clin. Pharmacol. Ther.*, 2003, **74**, 130.
44. J.M. Marraffa, L. Lang, G. Ong and D.F. Lehmann, *Clin. Pharmacol. Ther.*, 2006, **79**, 282.
45. D. Pritzker, A. Kanungo, T. Kilicarslan, R.F. Tyndale and E.M. Sellers, *J. Clin. Psychopharmacol.*, 2002, **22**, 330.
46. A. Heydari, K.R. Yeo, M.S. Lennard, S.W. Ellis, G.T. Tucker and A. Rostami-Hodjegan, *Drug. Metab. Dispos.*, 2004, **32**, 1213.
47. L. Bertilsson, C. Alm, C.C. De Las, J. Widen, G. Edman and D. Schalling, *Lancet*, 1989, **1**, 555.
48. A. Llerena, J. Cobaleda and J. Benitez, *Lancet*, 1989, **1**, 1398.
49. A. Llerena, G. Edman, J. Cobaleda, J. Benitez, D. Schalling and L. Bertilsson, *Acta. Psychiatr. Scand.*, 1993, **87**, 23.
50. R. Fonne-Pfister, M.J. Bargetzi and U.A. Meyer, *Biochem. Biophys. Res. Commun.*, 1987, **148**, 1144.
51. A. Barbeau, T. Cloutier, M. Roy, L. Plasse, S. Paris and J. Poirier, *Lancet*, 1985, **2**, 1213.
52. C. Martinez, J.A. Agundez, G. Gervasini, R. Martin and J. Benitez, *Pharmacogenetics*, 1997, **7**, 85.
53. A.M. Yu, J.R. Idle, L.G. Byrd, K.W. Krausz, A. Kupfer and F.J. Gonzalez, *Pharmacogenetics*, 2003, **13**, 173.
54. T. Hiroi, S. Imaoka and Y. Funae, *Biochem. Biophys. Res. Commun.*, 1998, **249**, 838.
55. T. Hiroi, W. Kishimoto, T. Chow, S. Imaoka, T. Igarashi and Y. Funae, *Endocrinology*, 2001, **142**, 3901.
56. W. Kishimoto, T. Hiroi, M. Shiraishi, M. Osada, S. Imaoka, S. Kominami, T. Igarashi and Y. Funae, *Endocrinology*, 2004, **145**, 699.
57. A.M. Yu, J.R. Idle, T. Herraiz, A. Kupfer and F.J. Gonzalez, *Pharmacogenetics*, 2003, **13**, 307.
58. A.M. Yu, J.R. Idle and F.J. Gonzalez, *Drug. Metab. Rev.*, 2004, **36**, 243.
59. G. Mikus, F. Bochner, M. Eichelbaum, P. Horak, A.A. Somogyi and S. Spector, *J. Pharmacol. Exp. Ther.*, 1994, **268**, 546.
60. W. Zhu, P. Cadet, G. Baggerman, K.J. Mantione and G.B. Stefano, *J. Immunol.*, 2005, **175**, 7357.
61. D.C. Mankowski, K.J. Laddison, P.A. Christopherson, S. Ekins, D.J. Tweedie and M.P. Lawton, *Arch. Biochem. Biophys.*, 1999, **372**, 189.
62. H. Hichiya, S. Kuramoto, S. Yamamoto, S. Shinoda, N. Hanioka, S. Narimatsu, K. Asaoka, A. Miyata, S. Iwata, M. Nomoto, T. Satoh, K. Kiryu, N. Ueda, S. Naito, G.T. Tucker and S.W. Ellis, *Biochem. Pharmacol.*, 2004, **68**, 165.

63. H. Hichiya, C. Takemi, D. Tsuzuki, S. Yamamoto, K. Asaoka, S. Suzuki, T. Satoh, S. Shinoda, H. Kataoka and S. Narimatsu, *Biochem. Pharmacol.*, 2002, **64**, 1101.
64. E. Matsunaga, M. Umeno and F.J. Gonzalez, *J. Mol. Evol.*, 1990, **30**, 155.
65. H. Kawashima, D.J. Sequeira, D.R. Nelson and H.W. Strobel, *J. Biol. Chem.*, 1996, **271**, 28176.
66. D.R. Nelson, L. Koymans, T. Kamataki, J.J. Stegeman, R. Feyereisen, D.J. Waxman, M.R. Waterman, O. Gotoh, M.J. Coon, R.W. Estabrook, I.C. Gunsalus and D.W. Nebert, *Pharmacogenetics*, 1996, **6**, 1.
67. T. Hiroi, S. Imaoka, T. Chow and Y. Funae, *Biochim. Biophys. Acta*, 1998, **1380**, 305.
68. H. Hellmold, J.G. Lamb, A. Wyss, J.A. Gustafsson and M. Warner, *Mol. Pharmacol.*, 1995, **48**, 630.
69. E. Matsunaga, U.M. Zanger, J.P. Hardwick, H.V. Gelboin, U.A. Meyer and F.J. Gonzalez, *Biochemistry*, 1989, **28**, 7349.
70. Y. Yamamoto, T. Tasaki, A. Nakamura, H. Iwata, A. Kazusaka, F.J. Gonzalez and S. Fujita, *Pharmacogenetics*, 1998, **8**, 73.
71. T. Schulz-Utermoehl, A.J. Bennett, S.W. Ellis, G.T. Tucker, A.R. Boobis and R.J. Edwards, *Pharmacogenetics*, 1999, **9**, 357.
72. T. Hiroi, T. Chow, S. Imaoka and Y. Funae, *Drug. Metab. Dispos.*, 2002, **30**, 970.
73. N. Sakai, K. Saito, H.S. Kim, A. Kazusaka, M. Ishizuka, Y. Funae and S. Fujita, *Drug. Metab. Dispos.*, 2005, **33**, 1657.
74. J. Venhorst, A.M. ter Laak, J.N. Commandeur, Y. Funae, T. Hiroi and N.P. Vermeulen, *J. Med. Chem.*, 2003, **46**, 74.
75. Y. Kumagai, L.Y. Lin, A. Hiratsuka, S. Narimatsu, T. Suzuki, H. Yamada, K. Oguri, H. Yoshimura and A.K. Cho, *Mol. Pharmacol.*, 1994, **45**, 359.
76. Y. Masubuchi, T. Iwasa, S. Fujita, T. Suzuki, T. Horie and S. Narimatsu, *J. Pharm. Pharmacol.*, 1996, **48**, 925.
77. R.F. Tyndale, Y. Li, N.Y. Li, E. Messina, S. Miksys and E.M. Sellers, *Drug. Metab. Dispos.*, 1999, **27**, 924.
78. T. Chow, T. Hiroi, S. Imaoka, K. Chiba and Y. Funae, *Drug. Metab. Dispos.*, 1999, **27**, 1200.
79. D.R. Nelson, D.C. Zeldin, S.M. Hoffman, L.J. Maltais, H.M. Wain and D.W. Nebert, *Pharmacogenetics*, 2004, **14**, 1.
80. N. Blume, J. Leonard, Z.J. Xu, O. Watanabe, H. Remotti and J. Fishman, *Arch. Biochem. Biophys.*, 2000, **381**, 191.
81. A.M. Yu and R.L. Haining, *Drug. Metab. Dispos.*, 2006, **34**, 1167.
82. G. Wong, K. Kawajiri and M. Negishi, *Biochemistry*, 1987, **26**, 8683.
83. H.W. Davey, R.J. Wilkins and D.J. Waxman, *Am. J. Hum. Genet.*, 1999, **65**, 959.
84. C. Lee, J.R. Hutson, V.K. Tzau and D.S. Riddick, *Drug. Metab. Dispos.*, 2006, **34**, 1530.
85. S. Lofgren, A.L. Hagbjork, S. Ekman, R. Fransson-Steen and Y. Terelius, *Xenobiotica*, 2004, **34**, 811.

86. Y. Masubuchi, T. Iwasa, S. Hosokawa, T. Suzuki, T. Horie, S. Imaoka, Y. Funae and S. Narimatsu, *J. Pharmacol. Exp. Ther.*, 1997, **282**, 1435.
87. J. Corchero, C.P. Granvil, T.E. Akiyama, G.P. Hayhurst, S. Pimprale, L. Feigenbaum, J.R. Idle and F.J. Gonzalez, *Mol. Pharmacol.*, 2001, **60**, 1260.
88. S.L. Miksys, C. Cheung, F.J. Gonzalez and R.F. Tyndale, *Drug. Metab. Dispos.*, 2005, **33**, 1495.
89. Y. Tsuneoka, Y. Matsuo, R. Higuchi and Y. Ichikawa, *Eur. J. Biochem.*, 1992, **208**, 739.
90. T. Tasaki, A. Nakamura, S. Itoh, K. Ohashi, Y. Yamamoto, M. Masuda, H. Iwata, A. Kazusaka, T. Kamataki and S. Fujita, *J. Biochem. (Tokyo)*, 1998, **123**, 162.
91. S.K. Paulson, L. Engel, B. Reitz, S. Bolten, E.G. Burton, T.J. Maziasz, B. Yan and G.L. Schoenhard, *Drug. Metab. Dispos.*, 1999, **27**, 1133.
92. F. Roussel, D.B. Duignan, M.P. Lawton, R.S. Obach, C.A. Strick and D.J. Tweedie, *Arch. Biochem. Biophys.*, 1998, **357**, 27.
93. Q. Jiang, J.M. Voigt and H.D. Colby, *Biochem. Biophys. Res. Commun.*, 1995, **209**, 1149.
94. H.D. Colby, D.M. Nowak, P.A. Longhurst, X. Zhang, J.R. Hayes and J.M. Voigt, *Pharmacology*, 2001, **62**, 229.
95. T. Oka, M. Fukuhara, F. Ushio and K. Kurose, *Comp. Biochem. Physiol. C. Toxicol. Pharmacol.*, 2000, **127**, 143.
96. Y. Yamamoto, M. Ishizuka, A. Takada and S. Fujita, *J. Biochem. (Tokyo).*, 1998, **124**, 503.
97. K. Wikvall, *Int. J. Mol. Med.*, 2001, **7**, 201.
98. U.M. Zanger, J. Fischer, S. Raimundo, T. Stuven, B.O. Evert, M. Schwab and M. Eichelbaum, *Pharmacogenetics*, 2001, **11**, 573.
99. T. Shimada, H. Yamazaki, M. Mimura, Y. Inui and F.P. Guengerich, *J. Pharmacol. Exp. Ther.*, 1994, **270**, 414.
100. D.F. Lewis, *Pharmacogenomics*, 2004, **5**, 305.
101. F.J. Gonzalez, R.C. Skoda, S. Kimura, M. Umeno, U.M. Zanger, D.W. Nebert, H.V. Gelboin, J.P. Hardwick and U.A. Meyer, *Nature*, 1988, **331**, 442.
102. J. Denson, Y. Wu, W. Yang and J. Zhang, *Biochem. Biophys. Res. Commun.*, 2005, **330**, 498.
103. A. Gaedigk, R. Gaedigk and J.S. Leeder, *Biochem. Biophys. Res. Commun.*, 2005, **336**, 1241.
104. M. Kagimoto, M. Heim, K. Kagimoto, T. Zeugin and U.A. Meyer, *J. Biol. Chem.*, 1990, **265**, 17209.
105. N. Hanioka, S. Kimura, U.A. Meyer and F.J. Gonzalez, *Am. J. Hum. Genet.*, 1990, **47**, 994.
106. A.C. Gough, J.S. Miles, N.K. Spurr, J.E. Moss, A. Gaedigk, M. Eichelbaum and C.R. Wolf, *Nature*, 1990, **347**, 773.
107. P.K. Rogan, S. Svojanovsky and J.S. Leeder, *Pharmacogenetics*, 2003, **13**, 207.

108. C. Toscano, K. Klein, J. Blievernicht, E. Schaeffeler, T. Saussele, S. Raimundo, M. Eichelbaum, M. Schwab and U.M. Zanger, *Pharmacogenet. Genomics.*, 2006, **16**, 755.
109. T. Cresteil, *Food Addit. Contam.*, 1998, **15 Suppl**, 45.
110. M.G. Ladona, B. Lindstrom, C. Thyr, P. Dun-Ren and A. Rane, *Br. J. Clin. Pharmacol.*, 1991, **32**, 295.
111. J.M. Treluyer, E. Jacqz-Aigrain, F. Alvarez and T. Cresteil, *Eur. J. Biochem.*, 1991, **202**, 583.
112. E. Jacqz-Aigrain and T. Cresteil, *Dev. Pharmacol. Ther.*, 1992, **18**, 161.
113. S. Hogstedt, B. Lindberg, D.R. Peng, C.G. Regardh and A. Rane, *Clin. Pharmacol. Ther.*, 1985, **37**, 688.
114. J.M. Pepper, M.S. Lennard, G.T. Tucker and H.F. Woods, *Pharmacogenetics*, 1991, **1**, 119.
115. M. Wadelius, E. Darj, G. Frenne and A. Rane, *Clin. Pharmacol. Ther.*, 1997, **62**, 400.
116. T.S. Tracy, R. Venkataramanan, D.D. Glover and S.N. Caritis, *Am. J. Obstet. Gynecol.*, 2005, **192**, 633.
117. T. Niwa, Y. Yabusaki, K. Honma, N. Matsuo, K. Tatsuta, F. Ishibashi and M. Katagiri, *Xenobiotica*, 1998, **28**, 539.
118. A.D. Kashuba, A.N. Nafziger, G.L. Kearns, J.S. Leeder, C.S. Shirey, R. Gotschall, A. Gaedigk and J.S. Bertino, Jr., *Pharmacogenetics*, 1998, **8**, 403.
119. L. Labbe, C. Sirois, S. Pilote, M. Arseneault, N.M. Robitaille, J. Turgeon and B.A. Hamelin, *Pharmacogenetics*, 2000, **10**, 425.
120. K.W. Bock, D. Schrenk, A. Forster, E.U. Griese, K. Morike, D. Brockmeier and M. Eichelbaum, *Pharmacogenetics*, 1994, **4**, 209.
121. K. Dilger, U. Hofmann and U. Klotz, *Clin. Pharmacol. Ther.*, 2000, **67**, 512.
122. A. Madan, R.A. Graham, K.M. Carroll, D.R. Mudra, L.A. Burton, L.A. Krueger, A.D. Downey, M. Czerwinski, J. Forster, M.D. Ribadeneira, L.S. Gan, E.L. LeCluyse, K. Zech, P. Robertson, Jr., P. Koch, L. Antonian, G. Wagner, L. Yu and A. Parkinson, *Drug. Metab. Dispos.*, 2003, **31**, 421.
123. H. Glaeser, S. Drescher, M. Eichelbaum and M.F. Fromm, *Br. J. Clin. Pharmacol.*, 2005, **59**, 199.
124. M. Eichelbaum, S. Mineshita, E.E. Ohnhaus and C. Zekorn, *Br. J. Clin. Pharmacol.*, 1986, **22**, 49.
125. W. Cairns, C.A. Smith, A.W. McLaren and C.R. Wolf, *J. Biol. Chem.*, 1996, **271**, 25269.
126. S. Miksys and R.F. Tyndale, *J. Neural. Transm. Suppl.*, 2006, **177**.
127. Z. Huang, M.J. Fasco, S. Spivack and L.S. Kaminsky, *Cancer Res.*, 1997, **57**, 2589.
128. Z. Huang, M.J. Fasco and L.S. Kaminsky, *Arch. Biochem. Biophys.*, 1997, **343**, 101.
129. S.I. Woo, L.A. Hansen, X. Yu, M. Mallory and E. Masliah, *Neurology*, 1999, **53**, 1570.

130. X. Ding and L.S. Kaminsky, *Annu. Rev. Pharmacol. Toxicol.*, 2003, **43**, 149.
131. M. Nishimura, H. Yaguti, H. Yoshitsugu, S. Naito and T. Satoh, *Yakugaku. Zasshi.*, 2003, **123**, 369.
132. M.F. Paine, H.L. Hart, S.S. Ludington, R.L. Haining, A.E. Rettie and D.C. Zeldin, *Drug. Metab. Dispos.*, 2006, **34**, 880.
133. R.O. von, O. Burk, M.F. Fromm, K.P. Thon, M. Eichelbaum and K.T. Kivisto, *Clin. Pharmacol. Ther.*, 2004, **75**, 172.
134. J.C. Kolars, W.M. Awni, R.M. Merion and P.B. Watkins, *Lancet*, 1991, **338**, 1488.
135. M.F. Paine, D.D. Shen, K.L. Kunze, J.D. Perkins, C.L. Marsh, J.P. McVicar, D.M. Barr, B.S. Gillies and K.E. Thummel, *Clin. Pharmacol. Ther.*, 1996, **60**, 14.
136. R.O. von, B. Greiner, M.F. Fromm, R. Fraser, T. Omari, M.L. Barclay, J. Dent, A.A. Somogyi and M. Eichelbaum, *Clin. Pharmacol. Ther.*, 2001, **70**, 217.
137. T. Prueksaritanont, L.M. Dwyer and A.E. Cribb, *Biochem. Pharmacol.*, 1995, **50**, 1521.
138. S. Madani, M.F. Paine, L. Lewis, K.E. Thummel and D.D. Shen, *Pharm. Res.*, 1999, **16**, 1199.
139. R.F. Tyndale, R. Sunahara, T. Inaba, W. Kalow, F.J. Gonzalez and H.B. Niznik, *Mol. Pharmacol.*, 1991, **40**, 63.
140. P. Voirol, M. Jonzier-Perey, F. Porchet, M.J. Reymond, R.C. Janzer, C. Bouras, H.W. Strobel, M. Kosel, C.B. Eap and P. Baumann, *Brain. Res.*, 2000, **855**, 235.
141. I. Siegle, P. Fritz, K. Eckhardt, U.M. Zanger and M. Eichelbaum, *Pharmacogenetics*, 2001, **11**, 237.
142. S. Miksys, Y. Rao, E. Hoffmann, D.C. Mash and R.F. Tyndale, *J. Neurochem.*, 2002, **82**, 1376.
143. S. Miksys and R.F. Tyndale, *Drug. Metab. Rev.*, 2004, **36**, 313.
144. H.V. Pai, R.P. Kommaddi, S.J. Chinta, T. Mori, M.R. Boyd and V. Ravindranath, *J. Biol. Chem.*, 2004, **279**, 27383.
145. J.M. Guidice, D. Marez, N. Sabbagh, M. Legrand-Andreoletti, C. Spire, E. Alcaide, J.J. Lafitte and F. Broly, *Biochem. Biophys. Res. Commun.*, 1997, **241**, 79.
146. K.T. Kivisto, E.U. Griese, T. Stuven, P. Fritz, G. Friedel, H.K. Kroemer and U.M. Zanger, *Pharmacogenetics*, 1997, **7**, 295.
147. H. Raunio, J. Hakkola, J. Hukkanen, O. Pelkonen, R. Edwards, A. Boobis and S. Anttila, *Arch. Toxicol. Suppl.*, 1998, **20**, 465.
148. L.A. McConnachie, B. Phillips, M. Bajpai, D.D. Shen and R.J. Ho, *Drug. Metab. Dispos.*, 2003, **31**, 1103.
149. A.B. Lind, M. Wadelius, E. Darj, N. Finnstrom, S. Lundgren and A. Rane, *Pharmacol. Toxicol.*, 2003, **92**, 295.
150. S.S. Dehal and D. Kupfer, *Cancer Res.*, 1997, **57**, 3402.
151. J.K. Coller, N. Krebsfaenger, K. Klein, K. Endrizzi, R. Wolbold, T. Lang, A. Nussler, P. Neuhaus, U. M. Zanger, M. Eichelbaum and T.E. Murdter, *Br. J. Clin. Pharmacol.*, 2002, **54**, 157.

152. Y.C. Lim, L. Li, Z. Desta, Q. Zhao, J.M. Rae, D.A. Flockhart and T.C. Skaar, *J. Pharmacol. Exp. Ther.*, 2006, **318**, 503.
153. T. Thum and J. Borlak, *Lancet*, 2000, **355**, 979.
154. L. Bertilsson, H.J. Dengler, M. Eichelbaum and H.U. Schulz, *Eur. J. Clin. Pharmacol.*, 1980, **17**, 153.
155. M. Eichelbaum, L. Bertilsson, J. Sawe and C. Zekorn, *Clin. Pharmacol. Ther.*, 1982, **31**, 184.
156. M.S. Lennard, J.H. Silas, S. Freestone and J. Trevethick, *Br. J. Clin. Pharmacol.*, 1982, **14**, 301.
157. P. Dayer, L. Balant, F. Courvoisier, A. Kupfer, A. Kubli, A. Gorgia and J. Fabre, *Eur. J. Drug. Metab. Pharmacokinet.*, 1982, **7**, 73.
158. J. Gut, T. Catin, P. Dayer, T. Kronbach, U. Zanger and U.A. Meyer, *J. Biol. Chem.*, 1986, **261**, 11734.
159. U.M. Zanger, H.P. Hauri, J. Loeper, J.C. Homberg and U.A. Meyer, *Proc. Natl. Acad. Sci. U.S.A.*, 1988, **85**, 8256.
160. U.M. Zanger, F. Vilbois, J.P. Hardwick and U.A. Meyer, *Biochemistry*, 1988, **27**, 5447.
161. U.M. Zanger, S. Raimundo and M. Eichelbaum, *Naunyn. Schmiedebergs. Arch. Pharmacol.*, 2004, **369**, 23.
162. E. Steiner, L. Iselius, G. Alvan, J. Lindsten and F. Sjoqvist, *Clin. Pharmacol. Ther.*, 1985, **38**, 394.
163. B. Schmid, J. Bircher, R. Preisig and A. Kupfer, *Clin. Pharmacol. Ther.*, 1985, **38**, 618.
164. O.O. Simooya, E. Njunju, A.R. Hodjegan, M.S. Lennard and G.T. Tucker, *Pharmacogenetics*, 1993, **3**, 205.
165. R.S. Pedersen, P. Damkier and K. Brosen, *Clin. Pharmacol. Ther.*, 2005, **77**, 458.
166. W.H. Chou, F.X. Yan, D.K. Robbins-Weilert, T.B. Ryder, W.W. Liu, C. Perbost, M. Fairchild, L.J. de, W.H. Koch and P.J. Wedlund, *Clin. Chem.*, 2003, **49**, 542.
167. K.T. Kivisto and H.K. Kroemer, *J. Clin. Pharmacol.*, 1997, **37**, 40S.
168. D.S. Streetman, J.S. Bertino, Jr. and A.N. Nafziger, *Pharmacogenetics*, 2000, **10**, 187.
169. D. Frank, U. Jaehde and U. Fuhr, *Eur. J. Clin. Pharmacol.*, 2007, **63**, 321.
170. K. Nakamura, F. Goto, W.A. Ray, C.B. McAllister, E. Jacqz, G.R. Wilkinson and R.A. Branch, *Clin. Pharmacol. Ther.*, 1985, **38**, 402.
171. L. Bertilsson, Y.Q. Lou, Y.L. Du, Y. Liu, T.Y. Kuang, X.M. Liao, K.Y. Wang, J. Reviriego, L. Iselius and F. Sjoqvist, *Clin. Pharmacol. Ther.*, 1992, **51**, 388.
172. C. Masimirembwa, I. Persson, L. Bertilsson, J. Hasler and M. Ingelman-Sundberg, *Br. J. Clin. Pharmacol.*, 1996, **42**, 713.
173. A. Gaedigk, L.D. Bradford, K.A. Marcucci and J.S. Leeder, *Clin. Pharmacol. Ther.*, 2002, **72**, 76.
174. E.U. Griese, U.M. Zanger, U. Brudermanns, A. Gaedigk, G. Mikus, K. Morike, T. Stuven and M. Eichelbaum, *Pharmacogenetics*, 1998, **8**, 15.

175. S. Raimundo, C. Toscano, K. Klein, J. Fischer, E.U. Griese, M. Eichelbaum, M. Schwab and U.M. Zanger, *Clin. Pharmacol. Ther.*, 2004, **76**, 128.
176. E. Steiner, L. Bertilsson, J. Sawe, I. Bertling and F. Sjoqvist, *Clin. Pharmacol. Ther.*, 1988, **44**, 431.
177. L. Bertilsson, A. Berg-Wistedt, L.L. Gustafsson and C. Nordin, *Ther. Drug. Monit.*, 1985, **7**, 478.
178. M.L. Dahl, I. Johansson, L. Bertilsson, M. Ingelman-Sundberg and F. Sjoqvist, *J. Pharmacol. Exp. Ther.*, 1995, **274**, 516.
179. I. Johansson, E. Lundqvist, M.L. Dahl and M. Ingelman-Sundberg, *Pharmacogenetics*, 1996, **6**, 351.
180. C. Sachse, J. Brockmoller, S. Bauer and I. Roots, *Am. J. Hum. Genet.*, 1997, **60**, 284.
181. T. Kronbach, *Methods Enzymol.*, 1991, **206**, 509.
182. H.K. Kroemer, G. Mikus, T. Kronbach, U.A. Meyer and M. Eichelbaum, *Clin. Pharmacol. Ther.*, 1989, **45**, 28.
183. A.K. Daly, J. Brockmoller, F. Broly, M. Eichelbaum, W.E. Evans, F.J. Gonzalez, J.D. Huang, J.R. Idle, M. Ingelman-Sundberg, T. Ishizaki, E. Jacqz-Aigrain, U.A. Meyer, D.W. Nebert, V.M. Steen, C.R. Wolf and U.M. Zanger, *Pharmacogenetics*, 1996, **6**, 193.
184. S.C. Sim and M. Ingelman-Sundberg, *Methods Mol. Biol.*, 2006, **320**, 183.
185. D. Marez, M. Legrand, N. Sabbagh, J.M. Guidice, C. Spire, J.J. Lafitte, U.A. Meyer and F. Broly, *Pharmacogenetics*, 1997, **7**, 193.
186. J.B. Leathart, S.J. London, A. Steward, J.D. Adams, J.R. Idle and A.K. Daly, *Pharmacogenetics*, 1998, **8**, 529.
187. E.U. Griese, S. sante-Poku, D. Ofori-Adjei, G. Mikus and M. Eichelbaum, *Pharmacogenetics*, 1999, **9**, 715.
188. E. U. Griese, K.F. Ilett, N.R. Kitteringham, M. Eichelbaum, H. Powell, R.M. Spargo, P.N. LeSouef, A.W. Musk and R.F. Minchin, *Pharmacogenetics*, 2001, **11**, 69.
189. T. Mizutani, *Drug. Metab. Rev.*, 2003, **35**, 99.
190. F. Broly and U. A. Meyer, *Pharmacogenetics*, 1993, **3**, 123.
191. R. Tyndale, T. Aoyama, F. Broly, T. Matsunaga, T. Inaba, W. Kalow, H.V. Gelboin, U.A. Meyer and F.J. Gonzalez, *Pharmacogenetics*, 1991, **1**, 26.
192. S. Yamazaki, K. Sato, K. Suhara, M. Sakaguchi, K. Mihara and T. Omura, *J. Biochem. (Tokyo)*, 1993, **114**, 652.
193. M. Oscarson, M. Hidestrand, I. Johansson and M. Ingelman-Sundberg, *Mol. Pharmacol.*, 1997, **52**, 1034.
194. M. Eichelbaum and N.M. Woolhouse, *Eur. J. Clin. Pharmacol.*, 1985 **28**, 79.
195. S. Raimundo, J. Fischer, M. Eichelbaum, E.U. Griese, M. Schwab and U.M. Zanger, *Pharmacogenetics*, 2000, **10**, 577.
196. C. Toscano, S. Raimundo, K. Klein, M. Eichelbaum, M. Schwab and U.M. Zanger, *Pharmacogenet. Genomics*, 2006, **16**, 767.
197. W.M. Cai, Y.D. Zhang, B. Chen, M.H. Cai, J.P. Luo and S.S. Ling, *Acta. Pharmacol. Sin.*, 2001, **22**, 956.

198. M.L. Barclay, S.M. Sawyers, E.J. Begg, M. Zhang, R.L. Roberts, M.A. Kennedy and J.M. Elliott, *Pharmacogenetics*, 2003, **13**, 627.
199. K.D. Furman, D.R. Grimm, T. Mueller, R.R. Holley-Shanks, R.J. Bertz, L.A. Williams, B.B. Spear and D.A. Katz, *Pharmacogenetics*, 2004, **14**, 279.
200. W. Steimer, K. Zopf, A. S. von, H. Pfeiffer, J. Bachofer, J. Popp, B. Messner, W. Kissling and S. Leucht, *Clin. Chem.*, 2004, **50**, 1623.
201. T. Rau, R. Heide, K. Bergmann, H. Wuttke, U. Werner, N. Feifel and T. Eschenhagen, *Pharmacogenetics*, 2002, **12**, 465.
202. L. Bertilsson, M.L. Dahl, F. Sjoqvist, A. berg-Wistedt, M. Humble, I. Johansson, E. Lundqvist and M. Ingelman-Sundberg, *Lancet*, 1993, **341**, 63.
203. A. Gaedigk, L. Ndjountche, K. Divakaran, B.L. Dianne, I. Zineh, T.F. Oberlander, D.C. Brousseau, D.G. McCarver, J.A. Johnson, S.W. Alander, R.K. Wayne and L.J. Steven, *Clin. Pharmacol. Ther.*, 2007 **81**, 242.
204. L. Bathum, I. Johansson, M. Ingelman-Sundberg, M. Horder and K. Brosen, *Pharmacogenetics*, 1998, **8**, 119.
205. E. Aklillu, I. Persson, L. Bertilsson, I. Johansson, F. Rodrigues and M. Ingelman-Sundberg, *J. Pharmacol. Exp. Ther.*, 1996, **278**, 441.
206. R.A. McLellan, M. Oscarson, J. Seidegard, D.A. Evans and M. Ingelman-Sundberg, *Pharmacogenetics*, 1997, **7**, 187.
207. E. Aklillu, K. Herrlin, L.L. Gustafsson, L. Bertilsson and M. Ingelman-Sundberg, *Pharmacogenetics*, 2002, **12**, 375.
208. J.A. Agundez, M.C. Ledesma, J.M. Ladero and J. Benitez, *Clin. Pharmacol. Ther.*, 1995, **57**, 265.
209. M. Ingelman-Sundberg, *Pharmacogenomics J.*, 2005, **5**, 6.
210. E. Lundqvist, I. Johansson and M. Ingelman-Sundberg, *Gene*, 1999, **226**, 327.
211. L. Bathum, I. Johansson, M. Ingelman-Sundberg, M. Horder and K. Brosen, *Pharmacogenetics*, 1998, **8**, 119.
212. R. Lovlie, A.K. Daly, G.E. Matre, A. Molven and V.M. Steen, *Pharmacogenetics*, 2001, **11**, 45.
213. A.K. Daly, K.S. Fairbrother, O.A. Andreassen, S.J. London, J.R. Idle and V.M. Steen, *Pharmacogenetics*, 1996, **6**, 319.
214. T. Stuven, E.U. Griese, H.K. Kroemer, M. Eichelbaum and U.M. Zanger, *Pharmacogenetics*, 1996, **6**, 417.
215. K. Bender, *Methods. Mol. Biol.*, 2005, **297**, 243.
216. J. Sistonen, S. Fuselli, A. Levo and A. Sajantila, *Clin. Chem.*, 2005, **51**, 1291.
217. R.L. Roberts and M.A. Kennedy, *Clin. Chim. Acta*, 2006, **366**, 348.
218. W.P. Wang, K.Y. Ni and G.H. Zhou, *Anal. Biochem.*, 2006, **355**, 240.
219. V.M. Steen, O.A. Andreassen, A.K. Daly, T. Tefre, A.L. Borresen, J.R. Idle and A.K. Gulbrandsen, *Pharmacogenetics*, 1995, **5**, 215.
220. R. Lovlie, A.K. Daly, A. Molven, J.R. Idle and V.M. Steen, *FEBS Lett.*, 1996, **392**, 30.

221. M. Hiratsuka, Y. Agatsuma, F. Omori, K. Narahara, T. Inoue, Y. Kishikawa and M. Mizugaki, *Biol. Pharm. Bull.*, 2000, **23**, 1131.
222. E. Molden, P.W. Johansen, G.H. Boe, S. Bergan, H. Christensen, H.E. Rugstad, H. Rootwelt, L. Reubsaet and G. Lehne, *Clin. Pharmacol. Ther.*, 2002, **72**, 333.
223. U.M. Stamer, B. Bayerer, S. Wolf, A. Hoeft and F. Stuber, *Clin. Chem.*, 2002, **48**, 1412.
224. G.M. Murphy, Jr., *J. Psychopharmacol.*, 2006, **20**, 72.
225. T. Heller, J. Kirchheiner, V.W. Armstrong, H. Luthe, M. Tzvetkov, J. Brockmoller and M. Oellerich, *Ther. Drug. Monit.*, 2006, **28**, 673.
226. Y. Lu, K.S. Kow-Yin, T.C. Foo, A. Hany, T.C. Quah and Y.A. Eng-Juh, *Anal. Biochem.*, 2007, **360**, 105.
227. A.L. Zackrisson and B. Lindblom, *Eur. J. Clin. Pharmacol.*, 2003, **59**, 521.
228. E. Soderback, A.L. Zackrisson, B. Lindblom and A. Alderborn, *Clin. Chem.*, 2005, **51**, 522.
229. R. Melis, E. Lyon and G.A. McMillin, *Expert. Rev. Mol. Diagn.*, 2006, **6**, 811.
230. A. Bogni, M. Monshouwer, A. Moscone, M. Hidestrand, M. Ingelman-Sundberg, T. Hartung and S. Coecke, *Toxicol. In vitro*, 2005, **19**, 621.
231. K. Brosen and L.F. Gram, *Eur. J. Clin. Pharmacol.*, 1989, **36**, 537.
232. M. Eichelbaum and A.S. Gross, *Pharmacol. Ther.*, 1990, **46**, 377.
233. P. Baumann and B. Rochat, *Int. Clin. Psychopharmacol.*, 1995, **10 Suppl 1**, 15.
234. P. Dorado, R. Berecz, E.M. Penas-Lledo, M.C. Caceres and A. Llerena, *Curr. Drug Targets.*, 2006, **7**, 1671.
235. Y. Gasche, Y. Daali, M. Fathi, A. Chiappe, S. Cottini, P. Dayer and J. Desmeules, *N. Engl. J. Med.*, 2004, **351**, 2827.
236. E. Spina, C. Gitto, A. Avenoso, G.M. Campo, A.P. Caputi and E. Perucca, *Eur. J. Clin. Pharmacol.*, 1997, **51**, 395.
237. S. Chen, W.H. Chou, R.A. Blouin, Z. Mao, L.L. Humphries, Q.C. Meek, J.R. Neill, W.L. Martin, L.R. Hays and P.J. Wedlund, *Clin. Pharmacol. Ther.*, 1996, **60**, 522.
238. W.H. Chou, F.X. Yan, L.J. de, J. Barnhill, T. Rogers, M. Cronin, M. Pho, V. Xiao, T.B. Ryder, W.W. Liu, C. Teiling and P.J. Wedlund, *J. Clin. Psychopharmacol.*, 2000, **20**, 246.
239. J. Brockmoller, J. Kirchheiner, J. Schmider, S. Walter, C. Sachse, B. Muller-Oerlinghausen and I. Roots, *Clin. Pharmacol. Ther.*, 2002, **72**, 438.
240. R. Kaiser, O. Sezer, A. Papies, S. Bauer, C. Schelenz, P.B. Tremblay, K. Possinger, I. Roots and J. Brockmoller, *J. Clin. Oncol.*, 2002, **20**, 2805.
241. M.D. Johnson, H. Zuo, K.H. Lee, J.P. Trebley, J.M. Rae, R.V. Weatherman, Z. Desta, D.A. Flockhart and T.C. Skaar, *Breast Cancer Res. Treat.*, 2004, **85**, 151.
242. M.P. Goetz, J.M. Rae, V.J. Suman, S.L. Safgren, M.M. Ames, D.W. Visscher, C. Reynolds, F.J. Couch, W.L. Lingle, D.A. Flockhart, Z. Desta, E.A. Perez and J.N. Ingle, *J. Clin. Oncol.*, 2005, **23**, 9312.

243. M.P. Goetz, S.K. Knox, V.J. Suman, J.M. Rae, S.L. Safgren, M.M. Ames, D.W. Visscher, C. Reynolds, F.J. Couch, W.L. Lingle, R.M. Weinshilboum, E.G. Fritcher, A.M. Nibbe, Z. Desta, A. Nguyen, D.A. Flockhart, E.A. Perez and J.N. Ingle, *Breast Cancer Res. Treat.*, 2007, **101**, 113.
244. J.M. Rae, M.P. Goetz, D.F. Hayes, J.N. Ingle, L. Li, A.M. Storniolo, V. Stearns and D.A. Flockhart, *Breast Cancer Res.*, 2005, **7**, E6.
245. S. Borges, Z. Desta, L. Li, T.C. Skaar, B.A. Ward, A. Nguyen, Y. Jin, A.M. Storniolo, D.M. Nikoloff, L. Wu, G. Hillman, D.F. Hayes, V. Stearns and D.A. Flockhart, *Clin. Pharmacol. Ther.*, 2006, **80**, 61.
246. W. Schroth, L. Antoniadoua, P. Fritz, M. Schwab, T. Muerdter, U.M. Zanger, W. Simon, M. Eichelbaum and H. Brauch *J. Clin. Oncol.*, submitted.
247. H. Wuttke, T. Rau, R. Heide, K. Bergmann, M. Bohm, J. Weil, D. Werner and T. Eschenhagen, *Clin. Pharmacol. Ther.*, 2002, **72**, 429.
248. I. Zineh, A.L. Beitelshees, A. Gaedigk, J.R. Walker, D.F. Pauly, K. Eberst, J.S. Leeder, M.S. Phillips, C.A. Gelfand and J.A. Johnson, *Clin. Pharmacol. Ther.*, 2004, **76**, 536.
249. R. Fux, K. Morike, A.M. Prohmer, U. Delabar, M. Schwab, E. Schaeffeler, G. Lorenz, C.H. Gleiter, M. Eichelbaum and K.T. Kivisto, *Clin. Pharmacol. Ther.*, 2005, **78**, 378.
250. W. Steimer, K. Zopf, A.S. von, H. Pfeiffer, J. Bachofer, J. Popp, B. Messner, W. Kissling and S. Leucht, *Clin. Chem.*, 2005, **51**, 376.
251. J. Kirchheiner, H. Schmidt, M. Tzvetkov, J.T. Keulen, J. Lotsch, I. Roots, J. Brockmoller, *Pharmacogenomics J.*, 2007 (in press).

CHAPTER 9

The CYP2E Subfamily

LOWELL C. OVERTON, ALICE HUDDER
AND RAYMOND F. NOVAK

Institute of Environmental Health Sciences, Wayne State University,
2727 Second Avenue, Room 4000, Detroit, MI 48201-2654

Table of Contents

Issues in Toxicology
Cytochromes P450: Role in the Metabolism and Toxicity of Drugs and other Xenobiotics
Edited by Costas Ioannides

9.1 Introduction

Cytochrome (P450) 2E1 is the only gene within the CYP2E subfamily among most mammals, with the exception of rabbits.[1] The gene is known to consist of 13 alleles that are designated alpha-numerically (e.g. CYP2E1.1 or CYP2E1.5A) largely based on when they were discovered (http://www.cypalleles.ki.se/cyp2e1.htm). None of these alleles is considered functionally polymorphic.[2] The enzyme is constitutively expressed in hepatic tissue and has been detected in lung, kidney, brain, nasal mucosa, bone marrow and peripheral blood lymphocytes. CYP2E1 expression occurs within specific regions or cell types in extrahepatic tissue (e.g. kidney). CYP2E1 has been studied extensively in the metabolism and metabolic activation of a wide variety of therapeutic agents, procarcinogens, carcinogens, as well as endogenous fatty acids and eicosanoids.

CYP2E1 expression is altered in response to a variety of xenobiotics, as well physiological and pathophysiological conditions, including nutrition (e.g. starvation, fasting, high fat diet, obesity), metabolic and endocrine disorders (e.g. obesity, diabetes), inflammation, viral infection (e.g. hepatitis C), and carcinoma (e.g. hepatocellular carcinoma). The regulation of CYP2E1 gene and protein expression is complex, and involves intracellular signalling, transcriptional, post-transcriptional, and post-translational events. Regulation also appears to be tissue specific. Recent work examining the effect of diabetic and normal insulin levels on CYP2E1 expression, exploiting contemporary knowledge of intracellular signalling pathways and components, however, has revealed that both gene transcription and mRNA translation and degradation are likely to be involved in the insulin regulated expression of CYP2E1. Additional research revealed that the ketone body acetoacetate also elevated CYP2E1 levels in primary cultured hepatocytes but only under conditions of diabetic insulin levels.

In this chapter, the distribution of CYP2E1 in animal and human tissues, gender differences in expression, the role of CYP2E1 in xenobiotic metabolism and toxicity, and its substrate specificity are reviewed. The expression of CYP2E1 is somewhat unique among the various P450s, and is regulated at the transcriptional, post-transcriptional, translational and post-translational level. Moreover, the regulation of CYP2E1 expression in response to xenobiotics, as well as endogenous cytokines and hormones, the intracellular signalling pathways and components involved in the hormonal regulation of CYP2E1, and how these components hold the key to the differential hepatic expression of CYP2E1 in disease will be discussed. The role of CYP2E1 microRNAs in the regulation of CYP2E1 mRNA translation repression and degradation is also discussed. The potential role of CYP2E1 as a biomarker for progressive and aggressive hepatic disease development is also examined. CYP2E1 ontogeny, and pharmacogenetics, including CYP2E1 polymorphisms and association with disease (i.e. gene-environment interactions) are described.

9.2 CYP2E1 Tissue and Species Distribution

CYP2E1 is a highly conserved gene. It has been cloned and sequenced from many species including human, mouse, rat, pig, dog, cow, rabbit and monkey, with orthologues found (53–56% similarity to mammalian CYP2E1) in worms (*C. elegans*) and in fungi (*M. grisea* and *N. crassa*). Expression of the CYP2E1 gene is both inducible and constitutive in mammals for many tissues.

While CYP2E1 was first identified in hepatic tissue, it has since been found in many extrahepatic tissues, including nasal mucosa, lung, kidney, breast, ovary, brain and stomach of rats,[3,4] rabbits[5–7] and humans.[4,8,9] The expression of CYP2E1 within these extrahepatic tissues, however, is generally much lower than within the liver. At the cellular level, CYP2E1 is primarily localised within the smooth endoplasmic reticulum, but has been also detected in the plasma membrane of hepatocytes.[10]

Within a given tissue, the expression of CYP2E1 is regio- and cell-specific, which may have important toxicological implications. The distribution of both CYP2E1 mRNA and protein varies within different regions of the liver. Hepatocytes exhibit the highest levels, although Kupffer cells also appear to express high levels of CYP2E1. Immunohistochemical studies have shown that CYP2E1 is localised within the perivenular region of the liver lobule (Zone 3).[4,11,12] CYP2E1 expression is elevated in this lobular region in patients that suffer from alcoholism, and alcoholic and non-alcoholic steatohepatitis.[10] *In situ* hybridisation studies have shown a three-fold increase in CYP2E1 transcripts in the acinar regions of the liver in subjects who regularly consumed ethanol;[13] CYP2E1 protein levels were substantially increased in these patients as well. Although, the ethanol-mediated increase in CYP2E1 was found in the acinus region of the liver, hepatocytes in the prevenular zone also expressed elevated levels of CYP2E1 transcripts.[13]

In other tissues, the basal expression level of CYP2E1 is lower than that described in the liver. Baseline levels of CYP2E1 in lung and kidney, for example, are 10 to 20 times lower than the baseline within liver[10] and vary by region and in response to inducers within the tissue. In the lung, CYP2E1 is primarily expressed in the bronchial mucosa and to a lesser extent, in the peripheral lung tissue.[14] However, when rats are exposed to ozone, the level of CYP2E1 is induced in the bronchioles, but not in the mucosa.[15]

CYP2E1 expression in the kidney primarily occurs in the proximal and distal tubules as demonstrated by Northern and immunoblot analysis, as well as by immunohistochemistry.[16,17] In contrast to CYP2E1 regulation in the lung, induction of CYP2E1 in the rat kidney in response to pyridine occurs in the same cells in which it is constitutively expressed. In humans, basal CYP2E1 expression has not been detected in renal cells, but it is unknown whether it is inducible in human kidney. Normal human breast tissue also expresses CYP2E1, and has been reported to be significantly increased in ductal carcinoma.[18] A low level of expression of CYP2E1 is detectable in mouse ovary and can be induced by 4-vinylcyclohexene.[19]

Induction of CYP2E1 can also vary between species. CYP2E1 is expressed within the liver of the house musk shrew (*Suncus murinus*), where the enzyme's expression is decreased approximately 24% during fasting, in contrast to rodents and humans where expression is upregulated during fasting.[20]

9.3 Age and Gender Differences in Expression

CYP2E1 expression appears to be developmentally regulated. In the rat, CYP2E1 expression begins within one day after birth[21] which has been associated with the methylation status of the gene.[22] The level of CYP2E1 remains constant throughout most of the adult life. At senescence, however, an increase in both mRNA and protein levels was reported despite a marked decrease in enzyme activity.[23] In humans CYP2E1 activity increases with age for both sexes (81% greater activity for males older than 50 years, and 87% greater activity in females older than 50 years compared with the respective 30–35 year old adults).[24]

The activity of CYPs is sexually dimorphic in the liver. Gender differences in drug and xenobiotic metabolism have been most studied in rat. Extrapolation of these data to humans must be considered carefully. The rat has additional unique CYP isoforms resulting in differential metabolism of some xenobiotics. If the compound or drug of interest is known to be metabolised by CYP2E1, however, xenobiotic metabolism by male rats closely models that of human males.[25] Male rats express CYP2E1 in liver in greater quantities than do females.[24–26] The same is true for humans, with males exhibiting higher expression of CYP2E1 than females. Consequently, clearance rates of various compounds metabolised by CYP2E1 are gender dependent.[27] In other cell types, however, this appears not to be the case as no significant gender difference in CYP2E1 activity was detected in human leukocytes.[28]

9.4 Role in Xenobiotic Metabolism and Toxicity

Ethanol oxidation and metabolism were originally considered to be the primary function of CYP2E1, lending to the idea of "ethanol specificity" for this enzyme.[29–33] Decades of research revealed that CYP2E1 catalyses the metabolism of a wide variety of xenobiotics, including therapeutic agents, environmental toxicants, procarcinogens, and carcinogens; approximately 80 in all (summarised in Table 9.1[10,34]); as well as endogenous fatty acids.[35] Despite the variety of compounds identified as CYP2E1 substrates, the majority of them are relatively low molecular weight compounds.

Several of these agents constitute protoxicants, e.g. benzene, nitrosamine, carbon tetrachloride, and acetaminophen (paracetamol) which, when

Table 9.1 Recognised substrates of cytochrome P-450 (2E1) (Modified from Lieber[10]).

Substrate	*Substrate*
Alcohols and Ketones	***Alkenes and Alkanes (Halogenated and Non-halogenated)***
Acetone	Acetol
Acetoacetate	Acetonitrile
Acetaldehyde	Acrylonitrile
Butanol	1,3-Butadiene
2-Butanone	Chloroform
Ethanol	Chloromethane
Glycerol	Dibromoethane
Isopropanol	1,1-Dichloro-2,2,2-trifluoroethane
Pentanol	2,2-Dichloro-1,1,1-trifluoroethane
	Dichloromethane
Aromatic Compounds	1,1-Dichloroethane
Acetaminophen (paracetamol)	1,1-Dichloroethylene
Aniline	1,2-Dichloropropane
Benzene	N,N-Dimethylacetamide
Bromobenzene	N,N-Dimethylformamide
Caffeine	Enflurane
Capsaicin	1,2-Epoxy-3-butene
Chlorzoxazone	Ethane
Isoniazid	Ethyl carbamate
2-Methoxyaniline	Ethylene dichloride
2-Methoxynitrobenzene	Haloethane
Phenol	Hexane
Pyridine	*β,β*-Iminodipropionitrile
p-Nitrophenol	Methoxyflurane
Pyrazole	Methyl formate
Styrene	Methylenechloride
Tamoxifen	*N*-Methylformamide
Toluene	Pentane
	Seroflurane
Ethers	1,1,1,2-Tetrafluoroethane
Diethyl Ether	1,1,2,2-Tetrafluoro-1-1(2,2,2-trifluoroethoxy) ethane
Methyl *t*-butyl ether	Thioacetamide
1,1,2,3,3,3-Hexafluoropropyl methyl ether	Tirapazamine
	1,1,1-Trichloroethane
Fatty Acids	Trichloroethane
Arachidonic Acid	Vinyl chloride
Lauric Acid	Vinyl bromide
Reducible Substrates	***Nitrosamines and azocompounds***
Carbon tetrachloride	Azoxymethane
Chromium	*N,N*-Diethylnitrosamine
Cumene hydroperoxide	*N,N*-Dimethylnitrosamine
Hydrogen Peroxide	Methylazoxymethanol
13-Hydroperoxy-9,11-octadecadienoic acid	*N*-Nitroso-2,3-dimethylmorpholine
15-Hydroperoxy-5,8,11,13-eicosatetraenoic acid	

metabolised by CYP2E1, deplete cellular glutathione, covalently bind protein, increase oxidative stress, increase immediate-early gene expression (c-fos, c-jun)[36] and cause damage to cells or tissues expressing CYP2E1. For example, metabolism of pyridine results in the production of metabolites capable of redox cycling and generating oxidative stress.[37] Inhibition of CYP2E1 activity, through the use of inhibitors, including diallyl sulphide or propyl disulphide, decreases the rate of substrate metabolism and inhibits toxicity associated with a protoxicant.[38–44] However, an exception to the protective effect of diallyl sulphide in the mortality of diabetic animals treated with thioacetamide has been reported and associated with compromised tissue repair.[44] Malotilate[45] and chlormethiazole also inhibit CYP2E1 and have been reported to be effective in preventing the development of experimental alcoholic liver disease[46] and pyrazole-induced lipopolysaccharide-induced hepatic injury in mice.[47]

In contrast, enhanced expression of CYP2E1 in cells or tissues results in significantly elevated oxidative stress and toxicity in the presence of a protoxicant, such as carbon tetrachloride or N-nitrosodimethylamine.[48–50] Consequently, enhanced expression of CYP2E1 protein, coupled with the presence of a toxicant substrate, increases target organ susceptibility to damage and potentially tumourigenesis. In addition, relatively rapid elevation of CYP2E1 protein levels can occur following exposure to xenobiotics or with altered pathophysiology.

A wide variety of inducers of CYP2E1 have been reported. These include ethanol, acetone, imidazole, pyrazole, tricholorethlyene, benzene, pyridine, dietheyether, isopropanol, thiazole, pyrazine and pryidazine.[37,51–56] One interesting and common feature associated with the inducers thiazole, pyrazine and pyridazine was a substantial decrease in hepatic CYP2E1 poly(A)+ mRNA relative to untreated animals.[57] Subsequent sections of this chapter will focus on the mechanism(s) by which xenobiotics may elevate hepatic CYP2E1 expression. It is interesting to note that mechanism(s) regulating CYP2E1 expression in renal tissue may differ from those of hepatic tissue.[58]

Multiple criteria exist for demonstrating the role of CYP2E1 in xenobiotic metabolism. These include the use of the reconstituted enzyme system consisting of purified CYP2E1, NADPH cytochrome P450 oxidoreductase and phospholipid (i.e. dilauryl phosphatidylcholine).[59] Additional approaches include the use of inducers such as ethanol, pyridine, imidazole, pyrazole, acetone, etc., isolation of subcellular fractions of smooth endoplasmic reticulum (microsomal preparations), inhibition of CYP2E1 activity with chemical inhibitors, or through use of an inhibitory antibody, the use of gene knockout animals, expression of the enzyme in either a genetically engineered bacterial or mammalian system in which CYP2E1 is the only P450 expressed,[22] or *in silico* analysis.[60,61] CYP2E1 mRNA levels are monitored using Northern blot or RT-PCR, whereas protein levels are determined using immunoblot analysis or immunohistochemistry. Metabolic activity employs substrates, inhibitors and kinetic analyses.

9.5 Substrate Specificity

CYP2E1 exhibits very broad substrate specificity.[2] Over 80 different substrates for CYP2E1 have been identified (Table 9.1[10]). Approximately 70–80% of the clinically used drugs are metabolised by cytochrome P450 1–3 family members, and CYP2E1 contributes significantly to this activity.[62,63] The vast majority of CYP2E1 substrates consist of small, hydrophobic compounds, including a variety of halogenated hydrocarbons. For a list of clinically relevant substrates of CYP2E1, please see: http://medicine.iupui.edu/flockhart/.

Benzene, styrene, nitrosamines, pyridine(s), ketones (acetone), ketone bodies, primary alcohol(s), acetaminophen (paracetamol), carbon tetrachloride, pentane, fluorocarbon refrigerants, anaesthetic agents (e.g. enflurane), isoniazid and numerous low molecular weight halogenated hydrocarbons (e.g. chloroform, dichloroethylene, trichloroethylene, vinyl chloride, acrylonitrile) have been reported to serve as substrates for both animal (rat, rabbit) and human CYP2E1.[48,64–76] The list includes many industrial solvents, halogenated hydrocarbons, and combustion by-products (phenol, pyridine, benzene, toluene, chloroform, and trichloroethylene), a variety of alcohols (e.g. ethanol, propanol, isopropanol, butanol), therapeutic agents (acetaminophen, caffeine, nicotine, chlorzoxazone), and ethers (e.g. diethyl ether, methyl *t*-butyl ether, and 1,1,2,3,3,3-hexafluoropropyl methyl ether).[10,73,77–79]

9.6 Regulation by Endogenous and Exogenous Factors

CYP2E1 regulation is very complex and is under transcriptional,[80–82] post-transcriptional,[83,84] translational,[37,85] and post-translational control.[86–90] CYP2E1 expression is regulated by development, xenobiotics, cytokines, hormones, growth factors, the ketone body acetoacetate and by disease, including inflammation, metabolic and endocrine disorders (e.g. metabolic syndrome: obesity, diabetes), hepatic cirrhosis, hepatitis C, and hepatocellular carcinoma.

9.6.1 Cytokine Regulation of CYP2E1 Expression

In 1978, Chang *et al.* noted an alteration in pharmacokinetics during inflammation and infection.[91] This is due to effects of cytokines on hepatic cytochrome P450 enzymes.[92] CYP2E1 levels are directly influenced by a variety of cytokines,[93] including interferon (IFN)-γ, tumour necrosis factor (TNF)-α, and interleukins IL-1β, IL-4 and IL-6. There have been some conflicting reports on the specific effects of individual cytokines on CYP2E1 expression, but this is largely dependent on the system (*in vivo* vs. *in vitro*) and the tissue type (liver, kidney or brain) under investigation. For the most part, the cytokines act by altering transcription of the CYP2E1 gene. Peng and Coon reported that IL-1α induces *CYP2E1* gene transcription via the binding of the transcription factor Sp1 to a 32bp response element present within the rabbit *CYP2E1* gene.[94]

Additional sites in the CYP2E1 promoter have been identified which are responsible for the regulation by IFN-γ[95] and IL-4.[96]

9.6.2 Hormonal Regulation of CYP2E1 Expression

There have been numerous reports of variable hormonal regulation of CYP2E1 in the literature, many of which seem contradictory. Endocrine control of gene expression is a complex phenomenon. Studies *in vivo* are confounded by pleiotropic effects of perturbations to the endocrine system, whereas experiments performed in transformed cell lines often yield opposing data, and caution should be used in extrapolating the results. Thus complex endocrine control of gene expression is best studied in primary cultured cells. The hormones involved in regulation of CYP2E1 include growth hormone, testosterone, thyroid hormone and insulin.

9.6.2.1 Thyroid Hormone Regulation of CYP2E1 Expression

A high dose of thyroid hormone has been shown to suppress the levels of hepatic cytochrome P450s.[97] Similarly, thyroid hormone administered to thyroidectomised rats resulted in the downregulation of CYP2E1.[98] In studies *in vitro*, however, thyroid hormone was shown to elevate expression of the rabbit *CYP2E1* gene exogenously expressed in HepG2 cells. These studies revealed the presence of a translational regulatory region in the 3′UTR of the message.[99]

9.6.2.2 Growth Hormone Effects on CYP2E1

Another hormone that appears to have an effect on the expression of CYP2E1 is growth hormone (or pituitary hormone). In hypophysectomised rats, CYP2E1 levels were elevated, but subsequently were restored to normal upon administration of growth hormone.[100] This may be a species specific phenomenon, however, because the hormone has no effect on the level of CYP2E1 expression in hypophysectomised mice.[101] When these experiments were performed in hypophysectomised diabetic rats, growth hormone had a similar restorative effect. This effect may be due to changes in glucose utilisation rather than insulin production,[102] which underscores the complexity of such studies *in vivo*.

9.6.2.3 Insulin Regulation of CYP2E1 Expression

The most interesting and best studied hormonal regulation is the effects of insulin on CYP2E1. CYP2E1 levels and activity have been monitored in obese and diabetic humans. Song *et al.* quantified CYP2E1 protein levels in type I diabetic children and found that the enzyme levels were elevated within the lymphocytes of these patients.[103] However, subsequent studies monitoring CYP2E1 levels in lymphocytes of control, Type 1 and Type 2 diabetes failed to show an elevation in CYP2E1 protein or metabolic activity.[104–106] In contrast, numerous other investigators have reported an increase in human lymphocyte CYP2E1[107–109] in diabetes as well as in animal tissues.[110–112] In addition, CYP2E1 activity in morbidly obese human subjects with non-alcoholic fatty liver disease is elevated and an association between the degree of steatosis and CYP2E1 activity and the extent of obesity and CYP2E1 activity was reported.[113]

Although it was reported that diabetes increased CYP2E1 levels and that administration of insulin (or vanadate) restored CYP2E1 levels within 48 h, it was concluded that induction of CYP2E1 was associated with elevated ketone levels.[114] This was based on a correlation ($r^2 = 0.355$) between plasma β-hydroxybutyrate and hepatic microsomal CYP2E1 (P450j). Hepatic CYP2E1 protein levels increased approximately 3-fold in four days after insulin administration to diabetic rats was terminated, although ketone bodies remained elevated. However, CYP2E1 levels returned to basal levels within 48 h of insulin administration, whereas ketone bodies remained elevated for up to four days post-insulin administration.[114] Subsequent research in primary cultured rat hepatocytes showed that 3-hydroxybutyrate failed to elevate CYP2E1 levels, while acetoacetate decreased CYP2E1 mRNA levels, activated the p70S6 kinase and increased CYP2E1 protein levels ~2.5-fold at 48 h, but only under diabetic conditions.[88,115] Insulin was reported to downregulate CYP2E1 mRNA in rat Fao hepatoma cells[116] and in primary cultured rat hepatocytes.[117] Insulin addition to Fao cells resulted in the decline of CYP2E1 and CYP2B1 mRNA levels and a decrease in CYP2E1 and 2B1 half-lives, suggesting increased mRNA degradation.[116] It is thought that the downregulation of CYP2E1 mRNA in Fao cells is achieved by mRNA destabilisation possibly through the 3′ poly-A tail.[118] Similar results of downregulation of CYP2E1 mRNA was observed when rabbit CYP2E1 mRNA was expressed in human hepatocytes.[99]

In view of these reports, experiments were designed to examine the role of insulin in the regulation of CYP2E1, 3A and 4A in primary cultured rat hepatocytes. Primary cultured rat (and human) hepatocytes offer an advantage in being able to dissect out the role of the individual hormones or metabolic products in the regulation of CYP2E1 and permit mechanistic studies in concert with hormonal, growth factor or metabolic products, to determine the components which regulate expression. Although insulin addition to primary cultured hepatocytes resulted in a decrease in CYP2E1 mRNA half-life, it actually increased the half-life of CYP2B significantly, was without significant effect on CYP3A, and slightly decreased CYP4A mRNA half-life, demonstrating a specificity of insulin effects on CYP2E1 mRNA in primary cultured rat hepatocytes.[117] CYP2E1 mRNA levels increase with decreasing insulin levels below 1 nM insulin, the normal physiologic plasma level. These data show that diabetic (or lower) levels of insulin result in the induction of CYP2E1 in primary cultured rat hepatocytes and parallel the results reported for diabetes.[86,117] The question is how does insulin, which does not function as a transcription factor, and exerts its effects externally to the cell, serve to regulate CYP2E1 expression. An additional question is, at what level, transcriptionally or post-transcriptionally, including at the level of translation, does this regulation occur. This observation resulted in subsequent research to identify the mechanism(s) by which insulin regulates CYP2E1 gene and protein expression.

9.6.2.3.1 Insulin Signalling in the Regulation of CYP2E1 Expression. In subsequent mechanistic experiments, the effects of insulin, and its physiologic antagonist glucagon, on CYP2E1 expression in primary cultured hepatocytes were examined and mechanistic experiments conducted to identify the signalling pathways and components involved.[86]

Insulin, as well as IGF-1, exerts its metabolic and mitogenic effects through binding to the insulin receptor, a member of a large class of receptors known as receptor tyrosine kinases (RTKs) (Figure 9.1). Following insulin binding to the receptor, the cytosolic tails of the insulin receptor undergo transactivation of their tyrosine kinase activity, which results in tyrosine phosphorylation. This tyrosine phosphorylation creates protein domains, such as the SH2 (i.e. Src homology 2 domain), SH3 (Src homology 3 domain) and PTB (i.e. phosphotyrosine binding domains) which recruit the Insulin Receptor Substrates (IRS). There are six members of the IRS family in hepatic tissue, however, IRS1 and IRS2 are the primary transducers of insulin's metabolic and mitogenic effects. Recruitment of IRS 1and 2 results in their undergoing tyrosine phosphorylation through tyrosine kinase activity and results in the generation of SH2, SH3, PTB and PH (Plekstrin homology) domains which serve to attract and bind additional effectors of the insulin signalling pathways, thereby further amplifying insulin signalling and downstream effects.

Two major arms of the insulin signalling pathway exist. The PI3K (phosphatidylinositol 3-Kinase), PDK1,2 (3-phosphoinositide-dependent protein kinase), serine/threonine kinase Akt/protein kinase B (PKB), mTOR (mammalian target of rapamycin), p70S6K (p70 S6 Kinase) pathway controls both gene transcription, as well as translation including the Src homologous collagen protein (Shc)/growth factor receptor binding protein 2 (Grb2)/son of sevenless (SOS), the kinases Ras and Raf, and Mitogen-activated protein kinase kinase (Mek)-Extracellular signal-regulated kinase (Erk) pathway (Figure 9.1). The principal actor in this signalling cascade is Akt, which has three family members, Akt1, 2 and 3. Akt is able to regulate gene transcription through translocation to the nucleus, where it phosphorylates and inactivates the transcription factors FoxO (Forkhead-related)1, 3a and 4. Additional stress activated protein kinases that may be activated include Jun kinase (JNK) and the stress activated protein kinase (SAPK). Mechanistic studies employing chemical inhibitors and biochemical approaches have implicated the PI3K, Akt/PKB, mTOR, p70S6k pathway in regulating CYP2E1 mRNA expression.[115,117,119]

Akt also regulates translation through activation of mTOR, which results in the phosphorylation of the eIF4E-binding protein (4EBP1) and relieves the inhibitory effect of 4EBP1 on eIF4E (eukaryotic initiation factor 4E) the rate limiting initiation factor in translation, thereby allowing translation to proceed. In addition, Akt regulates other components of the translational machinery through mTOR and the p70S6K, including S6RP (S6 Ribosomal Protein) and eIF3B (Figure 9.1). For a more detailed description of the pathways and individual components shown in Figure 9.1, see Kim and Novak (2007).[119]

Insulin concentrations greater than 0.1 nM in primary cultured hepatocytes produce a ~55% decrease (or larger) in CYP2E1 mRNA levels and the

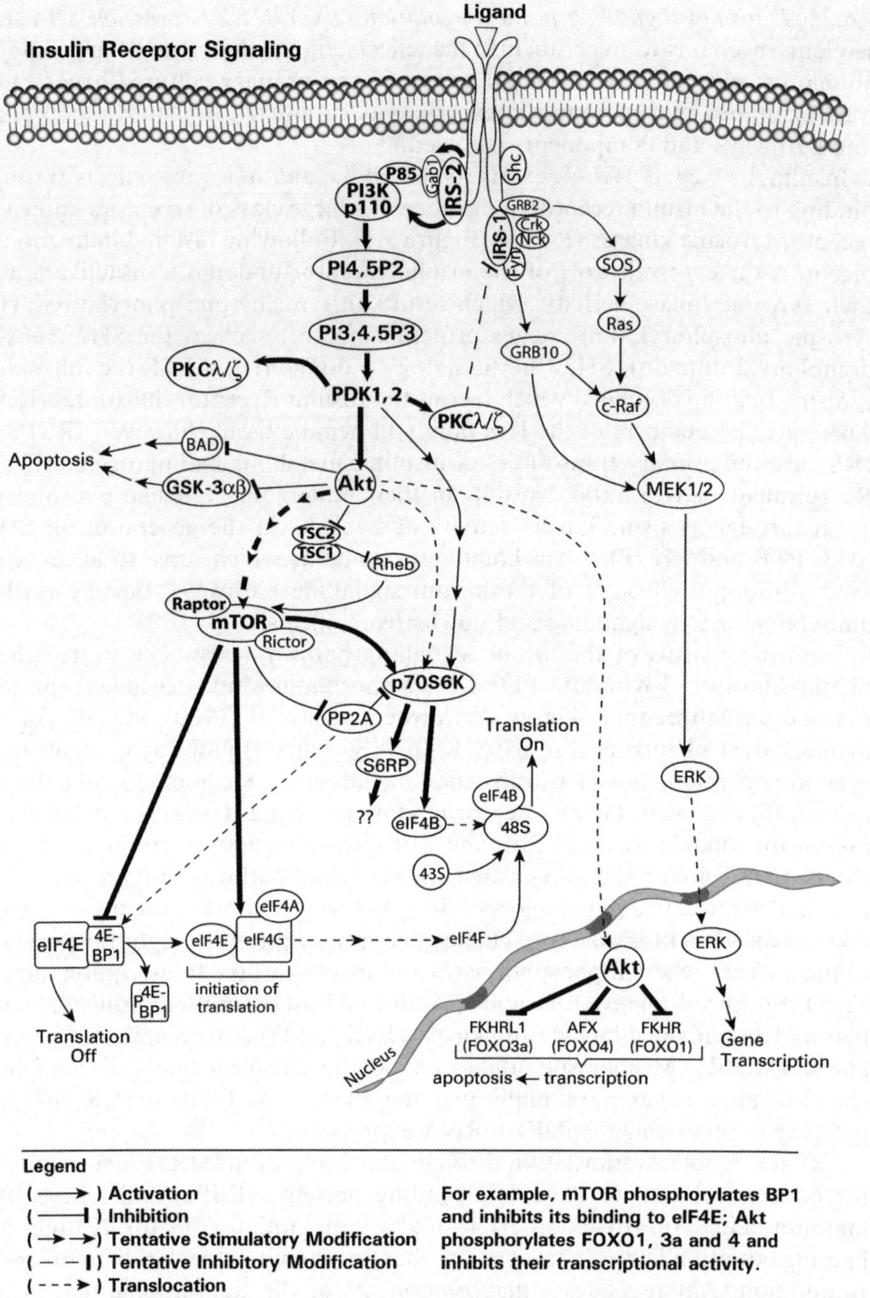

Figure 9.1 Signalling pathway for insulin. Stimulation of IRS1 and IRS 2 after the binding of the agonist insulin (or IGF1) to the insulin receptor induces activation of the Ras-Raf, Mek-Erk arm of the signaling pathway, as well as the activation of the PI3K, Akt, mTOR, p70S6k arm of this signalling cascade, the latter controlling transcriptional as well as translational events.[119] Adapted from cellsignal.com.

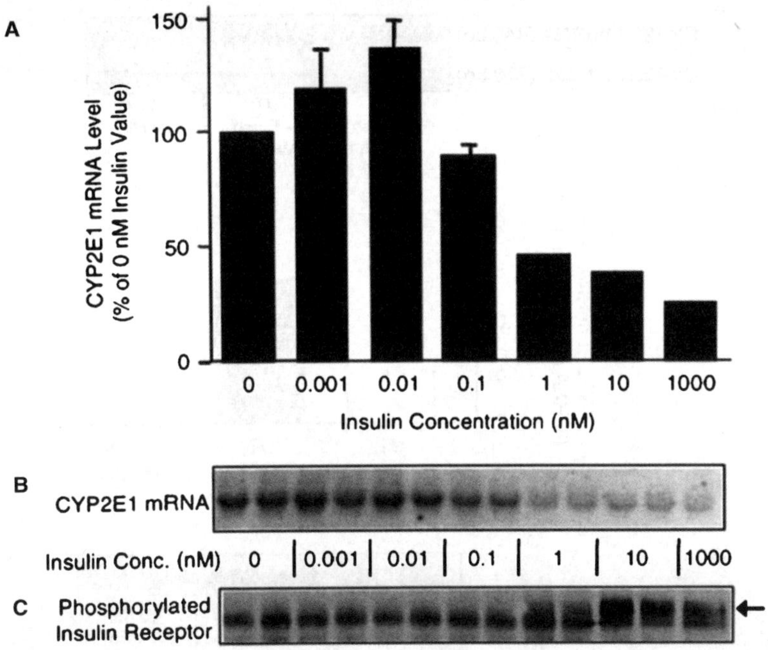

Figure 9.2 CYP2E1 mRNA levels and phosphorylation status of immunoprecipitated insulin receptor in response to insulin in primary cultured hepatocytes exposed to varying concentrations of insulin compared to cells that were not treated (UT) with insulin (Reprinted from reference 115 with permission from Wiley and Sons).

progressive decline in CYP2E1 mRNA occurs in a concentration-dependent manner[115] (Figure 9.2). CYP2E1 mRNA expression is nearly two-fold greater at diabetic insulin levels (0.1nM) than at normal physiologic (1 nM) levels, indicating that insulin has a dramatic effect on CYP2E1 expression. Our laboratory has also shown that the level of tyrosine phosphorylation of the insulin receptor, at insulin concentrations of 1, 10, and 1000 nM, corresponds with the concomitant decrease in CYP2E1 mRNA monitored at those insulin concentrations (Figure 9.2 B, C).

Using various inhibitors of PI3K (wortmannin, LY294002), mTOR (rapamycin) and Akt (Abdelmegeed *et al.*, unpublished), it was demonstrated that this cascade plays a prominent role in regulating CYP2E1 expression. Inhibitors of the Ras, Raf, Mek, ERK pathway revealed that this signalling pathway was ineffective in regulating CYP2E1 gene expression. Neither Src kinase, nor JNK or SAPK, appeared to be involved in the regulation of CYP2E1 expression under the conditions and concentrations of the inhibitors employed. Inhibitors of JNK or SAPK were also ineffective in altering insulin effects on CYP2E1 expression. Thus, PI3K, Akt, mTOR signalling appears to be the major pathway which regulates *CYP2E1* gene expression in response to insulin.

Using a heterologous nuclear RNA assay to examine CYP2E1 gene transcription, we have been able to demonstrate that insulin also decreases the transcriptional activity of the *CYP2E1* gene (Figure 9.3).

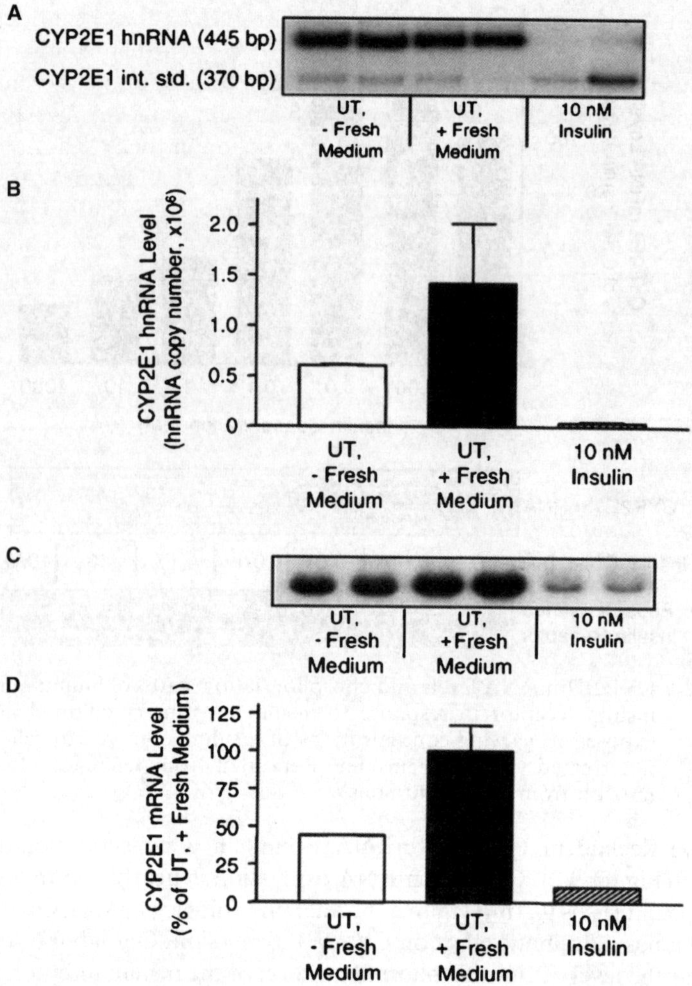

Figure 9.3 Effect of insulin on CYP2E1 gene transcription monitored by CYP2E1 hnRNA analysis. After the 4-hour plating period in the presence of 1 mmol/L insulin, hepatocytes were maintained in the absence of insulin for 48 hours. Hepatocytes were treated for 20 hours without medium change (UT, 2 Fresh Medium), with fresh medium alone (UT, 1 Fresh Medium), or with 10 nmol/L insulin in fresh medium (10 nmol/L Insulin). (A) Polymerase chain reaction products of CYP2E1 hnRNA and CYP2E1 internal standard. (B) CYP2E1 hnRNA levels plotted as hnRNA copy number, normalised to CYP2E1 internal standard band density. Data are means ± range of hnRNA band densities from 2 preparations of total RNA. (C) Northern blot of CYP2E1 mRNA levels in identical samples used for hnRNA analysis. (D) CYP2E1 mRNA levels in C plotted as a percentage of the CYP2E1 mRNA level monitored in hepatocytes treated with fresh medium (UT, 1 Fresh Medium), normalised to 7S RNA band density. (Reprinted from reference 115 with permission from Wiley and Sons.)

Although the hepatic nuclear factor 1alpha (HNF1α) has been identified as the primary transcription factor regulating *CYP2E1* gene expression, there may be other mechanism(s) controlling CYP2E1 gene transcription. Recent work[120] showed that liver specific deletion of the β-catenin gene in mice resulted in almost complete loss of CYP2E1 mRNA expression in the absence of an effect on HNF1α, suggesting that other mechanisms, including co-activators/co-regulators, may regulate CYP2E1 expression. Examination of the rat CYP2E1 DNA promoter and upstream regulatory region revealed a number of putative FoxO1, 3a and 4 binding sites, including one within the HNF1α transcription factor binding site. Whether Akt inactivation of FoxO1, 3a or 4, or whether expression levels or activity of HNF1α, or FoxO-HNF1α interactions regulate CYP2E1 gene expression is currently under investigation.

In addition to the transcriptional effects of insulin on CYP2E1 mRNA levels, it has been reported that insulin decreases the stability of the message in an FAO hepatoma cell line. Subsequent studies by Moncion *et al.* identified a 16 nucleotide sequence in the coding region (bases 115–131 following the AUG start codon) that can form a hairpin structure and negatively regulate the stability of CYP2E1 mRNA in response to insulin.[121]

9.6.2.3.2 MicroRNAs and Regulation of CYP2E1 mRNA Translation Repression and Degradation. MicroRNAs (miRNAs) constitute a class of evolutionarily conserved, endogenously expressed, noncoding RNAs which are 19–22 nucleotides in length and which bind with imperfect complementarity to the 3′-UTR (untranslated region) and negatively regulate target mRNAs.[122–125] Approximately 35% of all human genes are potential miRNA targets, and miRNAs have been associated with development,[126] tumourigenesis[127–129] and diabetes.[130] MiRNAs are associated with translating mRNAs in human cells and, specifically, have been shown to be associated with mRNAs in polysomes[131] and with actively translating mRNAs.[131,132] MiRNAs generally result in a decrease in protein expression owing to reduced mRNA translation.[125] MiRNAs regulate translation after initiation and appear to involve eIF4E. This is consistent with regulation of initiation of mRNA translation by protein-mRNA interactions at the 3′-UTR sites.[125] It is also important to note that miRNAs also promote target mRNA degradation.[125,133] MiR- 336, -352, and -333 havc bccn identified as potential regulators of CYP2E1 mRNA translation repression/degradation. Of the three miRNAs, transfection of miR-336 or -352 into hepatocytes results in suppression of CYP2E1 mRNA levels (Overton and Novak, unpublished). MicroRNAs have been reported to regulate CYP1B1 and have been shown to be decreased in breast tumour tissue.[134] MiRNAs have also been shown to regulate pancreatic islet insulin secretion.[130] Treatment of primary cultured mouse hepatocytes and diabetic mice with an anti-sense oligonucleotide directed against liver specific miR-122 resulted in decreased plasma cholesterol levels, increased hepatic fatty acid oxidation, decreased hepatic fatty acid and cholesterol synthesis, and increased activation of AMPK,[135] suggesting that miRNAs can affect other cellular processes and may have therapeutic potential.

9.6.2.3.3 Glucagon Signalling and Effects on CYP2E1 Expression. Insulin and glucagon alter CYP2E1 levels in opposing directions. Whereas insulin activates the PI3K, Akt/PKB, mTOR serine, threonine kinases and downregulates CYP2E1 mRNA expression, glucagon elevates CYP2E1 mRNA levels. Glucagon is a physiological antagonist of insulin and its administration activates several genes in an opposing manner to insulin. Glucagon functions through adenylate cyclase to elevate cAMP and activate protein kinase A (PKA) (Figure 9.4).

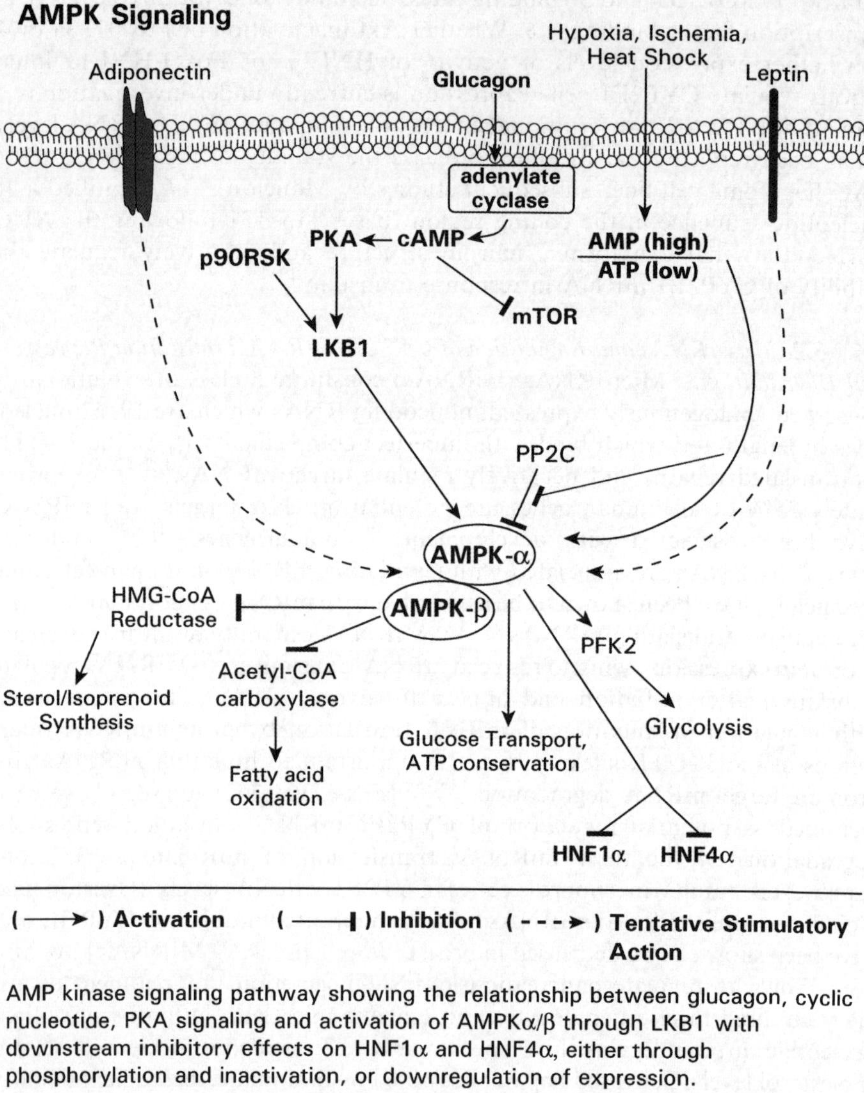

Figure 9.4 Glucagon signalling pathway involving activation of PKA, LKB1 and AMPK with inhibition of the transcription factors HNF1α and HNF4α. Adapted from cellsignal.com.

Involvement of PKA in the regulation of CYP2E1 is demonstrated by increased expression of CYP2E1 in response to glucagon, and dibutyryl-cAMP and decreased expression in response to inhibition of PKA.[86,115] Additional research examining the downstream effectors through which PKA functions, including LKB1 and AMPK (AMP kinase), is in progress. AMPK signalling inhibits HNF1α and HNF4α. HNF1α has been identified as a transcription factor which regulates CYP2E1 gene transcription.[80,82,136] It is also noteworthy that both adiponectin and leptin, which have been implicated in regulating CYP2E1 expression, are thought to stimulate AMPK (Figure 9.4).

9.6.2.3.4 Phosphatase Effects on CYP2E1 Expression. Phosphatases also counter the effects of kinase activity through removal of phosphate groups and, hence, function to elevate CYP2E1 mRNA expression by reversing phosphorylation of target proteins.[86] Therefore inhibition of phosphatase activity (Figure 9.1) by vanadate or okadaic acid will result in the suppression of CYP2E1 expression. This result is the same that would occur with increasing insulin concentration and activation of the PI3K, Akt, mTOR signalling cascade. In the streptozotocin-diabetic rat model, *in vivo* administration of vanadate has been shown to decrease CYP2E1 expression to normal levels, independent of insulin levels.[137]

9.6.3 Xenobiotic Regulation of CYP2E1 Gene Expression

9.6.3.1 Translational Regulation of CYP2E1 Expression

Evidence exists that xenobiotics (e.g. ethanol, imidazole, pyrazole, pyridine), and endogeneous agents (e.g. acetoacetate) and hormones (i.e. insulin) may regulate the expression of CYP2E1 at the level of translation.[37,51–56,88,138] Evidence has been provided that induction of CYP2E1 by pyridine occurred through increased protein synthesis in the absence of transcriptional activation.[37] Subsequent research provided evidence for increased translational efficiency in response to pyridine and acetone in rat hepatic tissue, which was employed in the examination of the polysomal distribution of CYP2E1 mRNA.[56]

Tsutsumi *et al.*, employed radiolabelling of CYP2E1 with [^{35}S] methionine and [^{3}H] aminolevulinate to label the apoprotein and heme respectively, with a view of examining the effects of ethanol on CYP2E1 expression.[139] Steady state CYP2E1 protein turnover in ethanol-fed and control animals was examined. The rates of CYP2E1 synthesis increased in ethanol-fed rats relative to controls. The authors interpreted these data to mean that induction of hepatic CYP2E1 protein by ethanol involved increased protein synthesis rather than decreased enzyme degradation. The authors suggested that the increased protein synthesis reflected the ethanol-mediated steady-state levels of CYP2E1 mRNA and/or stimulation of translation.[139] Collectively, these and subsequent data, provide a substantial case for the involvement of translation in the increased expression of CYP2E1 associated with several xenobiotics.

Several mechanisms are likely to be involved in the translational regulation of CYP2E1. These include **a)** the activation of the Akt/PKB serine, threonine kinase which regulates translation through mTOR inactivation of BP1 in the eIF4E/BP1 complex, thereby relieving the inhibitory effect of BP1 and resulting in the activation of eIF4E, the rate limiting eukaryotic factor in the initiation of translation (Figure 9.1); **b)** the secondary structure of the 5′-UTR, the coding region and/or the 3′-UTR of CYP2E1 which can regulate the rate of translation; c) specific sequences in the mRNA which target the mRNA for degradation following translation; d) proteins which bind specific regions (e.g. secondary structure) and e) microRNAs, which may regulate CYP2E1 mRNA translation repression and/or degradation.

Our laboratory has examined the polyribosomal distribution of CYP2E1 mRNA relative to the distribution of CYP2B, 3A, and 4A in rat hepatic tissue. As shown (Figure 9.5), only CYP2E1 exhibits a biphasic polysomal mRNA distribution, with CYP2E1 mRNA detected in the 40S-80S untranslated fraction, as well as in the actively translated polyribosomal fraction. In contrast CYP2B, 3A and 4A mRNAs are detected only in the actively translated polyribosomal faction.[138]

Additional research was initiated to examine the role of the 5′-UTR and 3′-UTR in translation and the role of translation in mRNA degradation. Examination of the secondary structure for the 5′UTR of CYP2E1 revealed a hairpin-like structure (−12.9 kcal/mol).

A series of recombinant RNA transcripts (rcRNA) were prepared in which the 5′-UTR, 3′-UTR and poly(A)tail were modified.[138] These transcripts were then examined for differences in polysomal distribution and for *in vitro* translational rates in a cell-free reticulocyte translation system. A brief diagrammatic

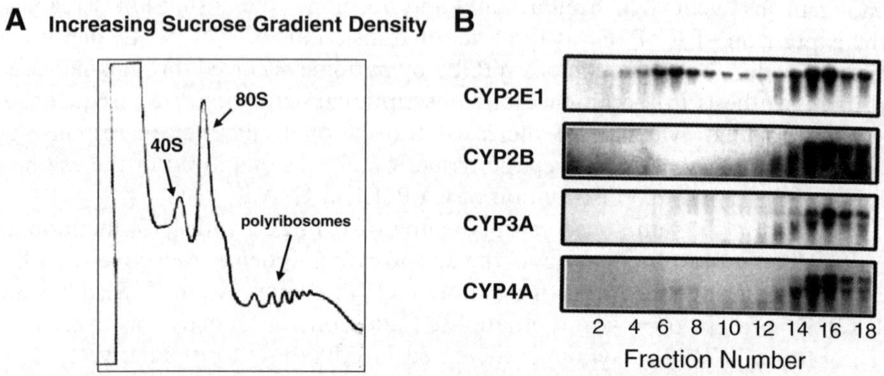

Figure 9.5 Polyribosomal distribution of rat hepatic CYP2E1, CYP2B, CYP3A and CYP4A mRNAs. Rat liver homogenates were fractionated on sucrose density gradients, and the amounts of CYP2E1, CYP2B, CYP3A and CYP4A mRNA present in the sucrose fractions were estimated by Northern blot analysis. (A) UV (260 nm) chromatogram of a sucrose gradient eluent showing a typical polysome distribution. (B) Northern blot analyses of the distribution of P450 mRNAs in the sucrose gradient fractions. (Reprinted from reference 138 with permission from Elsevier.)

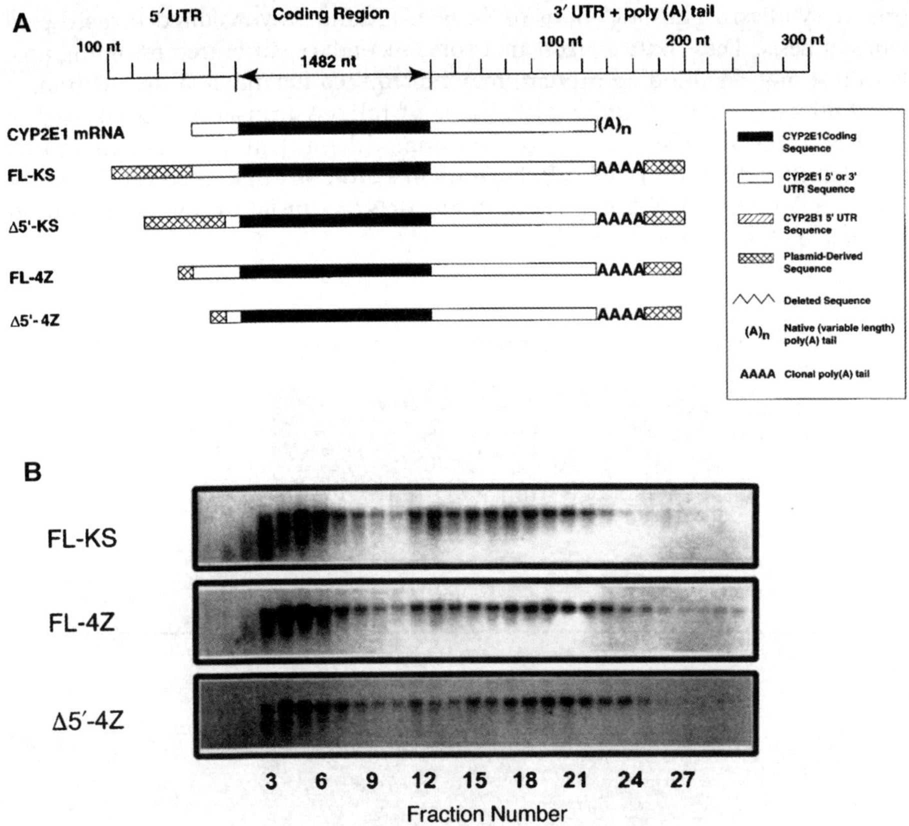

Figure 9.6 A) CYP2E1 rcRNAs used for *in vitro* translation experiments; B) Differences in polysomal distribution. (Reprinted from reference 138 with permission from Elsevier.)

representation of several of the 14 different constructs is shown in Figure 9.6A, along with the differences in polysomal distribution (9.6B).[138]

Corresponding differences in protein synthesis complement the differences in the polysomal distribution between the untranslated (60S-80S) and actively translated polyribosomal fractions of the sucrose-gradient polysomal distribution profile. The rcRNAs lacking the 28 5′UTR terminal bases of the CYP2E1 mRNA were translated into larger amounts of protein than the corresponding rcRNAs containing the entire length of the 5′UTR, suggesting that the rate of translation (i.e. translational efficiency) is governed by the 5′UTR, and perhaps the secondary structure present. Such a secondary structure might be a target for binding of protein(s) which serve to regulate the rate of translation.

In vitro translation rates of several rcRNAs, which lacked portions of the 3′UTR or rcRNAs lacking the poly(A) tail, were translated into less protein relative to the construct which yielded the largest amount of protein, whereas removal of the poly(A) tail and 3′UTR produced the most notable decrease in protein synthesis. Thus, deletions of the 3′-UTR and poly(A) tail decreased protein synthesis. These data suggest that some secondary structure or binding proteins (e.g. poly(A) binding protein) may be required for maximal translation.

Additional studies examining the rates of rcRNA translation in an *in vitro* cell-free translation system in combination with protein synthesis inhibitors revealed that inhibition of CYP2E1 protein synthesis attenuated the degradation of CYP2E1 mRNA suggesting that mRNA translation and degradation are linked (Figure 9.7, A and B).

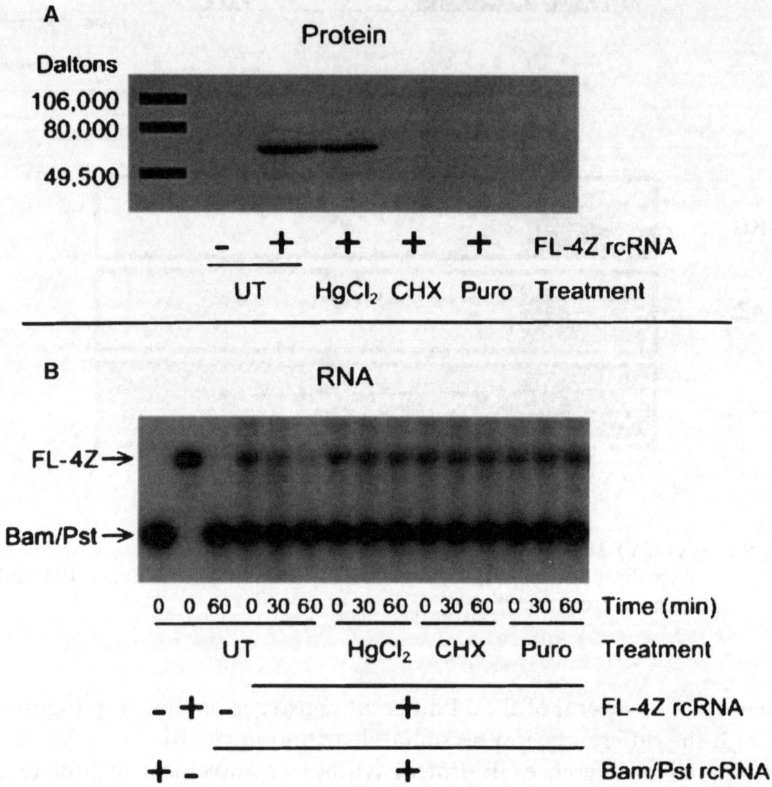

Figure 9.7 Effects of protein synthesis inhibitors on CYP2E1 rcRNA translation and stability. (A) FL-4Z rcRNA (or no RNA, second lane) was incubated with rabbit reticulocyte lysate, in either the absence (UT) or the presence of a protein synthesis inhibitor (i.e. $HgCl_2$; CHX, cycloheximide; Puro, puromycin), and incorporation of [^{35}S]methionine into the 52-kDa band was determined. (B) FL-4Z rcRNA (or no RNA, third lane) was incubated with rabbit reticulocyte lysate for 0, 30, or 60 min, in either the absence or the presence of one of the aforementioned protein synthesis inhibitors, and the amount of FL-4Z RNA remaining in the lysate was determined. The *Bam*HI–*Pst*I band (*Bam*/*Pst*) is an internal rcRNA standard. For lanes 1 and 2, 0.1 pmol FL-4Z rcRNA or *Bam*/*Pst* RNA was added directly to the RNA loading buffer for electrophoresis and blotting. (Reprinted from reference 138 with permission from Elsevier.)

9.6.3.2 *Postranslational Regulation of CYP2E1 Expression*

Regulation and inducibility of CYP2E1 by certain compounds has been reported to be controlled by the enzyme's turnover rate (or degradation) within the cell. The degradation of CYP2E1 appears to occur in two phases, one with a short half-life of 7 h and the second with a longer half-life of 37 h.[140] Some CYP2E1 substrates, such as acetone, are able to stabilise the enzyme by eliminating the rapid phase of degradation.[140] McGehee, Jr. *et al.* however, found that alcohol also had an effect on the protein half-life in FGC-4 hepatoma cells.[141] Similar results were reported in the rat, where alcohol treatment was reported to stabilise the protein by decreasing protein turnover.[142] Research by Tsutsumi *et al.* showed monophasic exponential decay curves with half-lives of 27–28 h and 17 h, respectively, which did not differ between control and ethanol-treated animals.[139] These and other studies suggested that the degradation of CYP2E1 involves the ubiquitin-proteasome pathway. Examination of CYP2E1 degradation *in vitro* revealed that ubiquitin-mediated degradation of CYP2E1 occurred and involved a predicted cytosolic domain (residues 317–340).[89] Huan *et al.*, using tetracycline-inducible HeLa cells reported that inhibitors of the proteosome significantly inhibited CYP2E1 (and CYP2B1) degradation.[143] Mutagenesis of two lysine residues, which have been implicated by Banerjee *et al.* as sites of ubiquitination, failed to cause CYP2E1 degradation in the HeLa cells.[89] CYP2E1 degradation was also reported to be slightly inhibited by inhibitors of calpain, whereas CYP2B1 was unaffected. Degradation of CYP2E1 in a Chinese hamster mutant cell line which contained a thermolabile ubiquitin-activating enzyme revealed no significant difference and suggested that CYP2E1 may be degraded by the proteosome through a ubiquitin-independent pathway.[143] It is important however, to recognise the importance of cell context in the regulation of CYP2E1 degradation.

9.6.4 Metabolic Conditions, Disease States and Oxidative Stress Alter CYP2E1 Expression

9.6.4.1 *Pathophysiologic Conditions and Disease in CYP2E1 Expression*

We have already discussed the fact that diabetes, obesity, starvation, hepatic inflammation and hormones (e.g. thyroid hormone, growth hormone) alter the expression of CYP2E1 (*vide supra*). However, several other diseases also cause alterations in CYP2E1 expression.

9.6.4.2 *Hepatocellular Carcinoma*

Diabetes is a risk factor for the development of chronic non-alcoholic liver disease (CNLD, non-alcoholic steatosis (NASH) and hepatocellular carcinoma (HCC). HCC is the fifth most common cancer in the world, and is generally fatal within one year of diagnosis of the disease. Other risk factors for development of HCC are chronic alcoholism and hepatitis infection. Cirrhosis of the liver is generally associated with chronic alcohol consumption, and either chronic alcohol consumption alone, or in combination with hepatitis, constitutes the greatest risk factor for development of HCC.

CYP2E1 levels (protein and mRNA) are decreased in HCC as compared to non-cancerous tissue. The decrease in expression of CYP2E1 has been reported to be significantly correlated with the aggressiveness of the tumour phenotype.[144] However, it is well recognised that the levels of Akt or phospho-Akt are elevated in a variety of tumour tissues and Akt plays a significant role in cell survival.[145] Examination of the phospho-Akt levels in HepG2 and Hep1c1c7 cells, which do not express readily detectable levels of CYP2E1, revealed a significant elevation of phospho-Akt levels, relative to primary cultured rat hepatocytes, which has been implicated in the downregulation of CYP2E1[115] (Zheng and Novak, unpublished data). Consequently, it is plausible that Akt or phospho-Akt levels are elevated in HCC and, as a result, CYP2E1 expression is markedly attenuated. In fact, there is a strong clinical correlation between increased levels of phospho-Akt and poor prognosis in HCC patients.[146]

9.6.4.3 Hepatitis

More than 170 million people globally are chronically infected with hepatitis C virus (HCV). Chronic HCV infection is strongly correlated with an increased risk of HCC and therefore likely contributes to over 100,000 cases of HCC annually.[147] Chronic HCV infection is characterised by a high risk of fibrosis and increasing oxidative stress.[148] In chronic alcoholics, CYP2E1 metabolism of alcohol is increased and leads to increased oxidative stress. Thus the combination of chronic alcohol consumption and chronic HCV infection may accelerate the progression of liver disease.[149] Experiments in hepatoma cells *in vitro* have shown that HCV core protein and alcohol together produce mitochondrial damage, oxidative stress, and cytotoxicity.[150]

9.6.4.4 Oxidative Stress

Oxidative stress in the cell is the result of an imbalance between prooxidants and antioxidants caused by oxidant overproduction that overwhelms the cellular antioxidant capacity.[151–153] The formation of reactive oxygen species (ROS) within the cell has been considered the main component in the formation of oxidative stress. Oxidative stress has been associated with numerous diseases, including diabetes.[154] There is a strong correlation between the onset of diabetes mellitus and increased production of free radicals, e.g. hydroxyl ($\cdot$OH) radical, and decreased antioxidant ability.[154]

CYP2E1 displays high NADPH oxidase activity because it is weakly coupled with NADPH-cytochrome P450 reductase.[155,156] Consequently, the generation of ROS by CYP2E1 and its induction by chronic ethanol exposure, have been suggested by some to be a major contributor to ethanol-induced oxidative stress and the potential for liver injury.[157] ROS production in E47 cells (i.e. HepG2 cells that constitutively express recombinant human CYP2E1) has been reported to be ~50% higher than in normal hepatocytes and associated with ~30% lower ATP levels reflecting ROS damage to Complex I.[158]

The ability to produce ROS by CYP2E1 also causes higher rates of microsomal lipid peroxidation within microsomes and liposomes enriched with CYP2E1.[159,160] Inhibition of CYP2E1 activity by antibodies against CYP2E1 suppresses the NADPH-dependent lipid peroxidation within microsomes.[160]

9.7 Pharmacogenetics and SNPs

As indicated earlier, CYP2E1 plays an important role in the bioactivation of a variety of low molecular weight toxicants, procarcinogens and carcinogens (*vide supra*). Bioactivation of these environmental agents by CYP2E1 in many cases reflects the initial event that may result in the development of cancer. CYP2E1 also increases cellular oxidative stress and the generation of ROS which lead to lipid peroxidation, altered cellular redox states, DNA damage and carcinogenesis.[161] Thus, pathophysiological conditions which result in the elevation of CYP2E1 (e.g. diabetes) may also result in increased risk of tumourigenesis. Indeed, diabetes has been identified as a risk factor for hepatocellular carcinoma.[162] Changes in enzyme level or activity as a consequence of genetic polymorphism or elevated expression in response to xenobiotics may also increase the risk of tumourigenesis. This section will briefly summarise the different polymorphisms identified in the human CYP2E1 gene. Danko and Chaschin recently published an excellent review of this topic focussing on the different polymorphisms and their role in gene-environment interactions.[163]

The human CYP2E1 gene is located on chromosome 10 and consists of 9 exons and spans more than 11 kb.[22] An illustration of the human CYP2E1 gene and location of polymorphisms is shown in Figure 9.8. McBride *et al.* identified the first human polymorphism in the CYP2E1 gene by restriction fragment length polymorphism (RFLP) analysis which revealed a loss of the restriction site *Taq* I in one allele.[164] Sequence analysis of CYP2E1 alleles from different human populations has revealed numerous polymorphisms in the promoter, 5′ and 3′ flanking regions and the introns.[165] The exons are much less polymorphic, suggesting strong selection pressure to conserve function of the gene product. The variations are most often single nucleotide polymorphisms (SNPs). There have been 128 SNPs identified in humans (EntrezSNP 2007 at: http://www.ncbi.nlm.nih.gov/entrez/query.fcgi?CMD=search&DB=snp). In addition, there have been insertions, deletions and variations in the number of repeat sequences. The majority of human polymorphisms found so far have no apparent functional consequence, while a few affect the regulation of expression or activity of the enzyme. The frequency of CYP2E1 polymorphisms has been shown to differ substantially in different ethnic populations.[166] For a complete list of alleles and allelic frequencies, please see HapMap at:

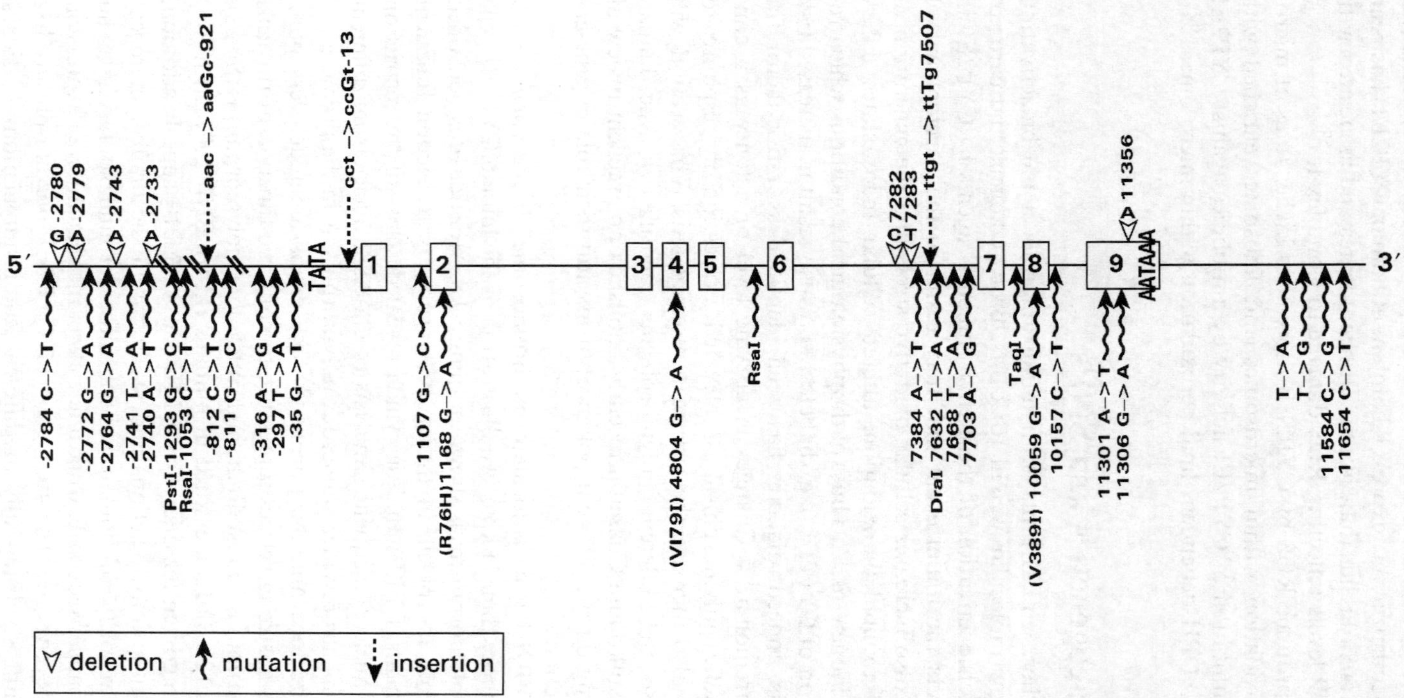

Figure 9.8 Human *CYP2E1* gene structure and location of polymorphisms. The human *CYP2E1* gene resides on chromosome 10 at 10q26.3 between 135190857 and 13520610. Exon 1 contains the 5′ untranslated region and the translation start site. Exon 9 contains the C-terminus of the protein coding sequence and the 3′ untranslated region. For reference, the TATA box is shown upstream of the transcription start site and the polyadenylation signal (AATAA) is shown in exon 9. Many of the reported human polymorphisms are shown with their approximate locations. Some of the most commonly referred to allelic variations in the literature are as follows: the c1 allele of CYP2E1 is Rsa I (+)/Pst I (–); c2 is Rsa I (–)/Pst I (+); c3 is Rsa I (+)/Pst I (+); c4 is Rsa I (–)/Pst I (–); the Dra I (D) allele contains the Dra I restriction site, while the C allele does not; and finally, the CYP2E1*2 allele is the R76H variant and the *3 allele is the V389I variant.

http://www.hapmap.org/cgi-perl/gbrowse/hapmap_B35/. To date a number (>10) of the polymorphisms of the CYP2E1 gene have been well-characterised.

Numerous polymorphisms have been identified in noncoding regions of the *CYP2E1* gene, but few of them seem to have any physiological significance. Caucasians exhibiting allelic variations in the 5′-flanking region (*Rsa* I, C-1019T), intron 6 (*Dra* I, A7766T) or intron 7 (*Taq* I) showed no significant difference in CYP2E1 activity as monitored by hepatic chlorzoxazone 6-hydroxylation.[167] Examination of additional noncoding variants A-316G, T-297A, G1107C (intron 1), and C10157Y (exon 8), singly or in combination, in a reporter system showed that they had no effect on the constitutive expression of the CYP2E1 gene.[168] The one exception was G-35T, which, in conjunction with T-297A, caused an upregulation in the expression of the reporter gene.

Coding region polymorphisms resulting in changes in the corresponding amino acid sequence have been reported. Such changes have the potential to modify secondary structure and catalytic activity of the enzyme. Two coding mutations were described and characterised by Hu *et al.*[165] In one, G10059A (exon 8) caused the amino acid change V389I, a conservative substitution. This mutation had no effect on protein function when expressed in COS-1 cells. The other, however, was a G1168A (exon 2) change resulting in an R76H amino acid substitution. This mutant protein had a reduced level of protein expression and catalytic activity in their expression system. Another coding mutation, G4804A (exon 4), caused a conservative V179I amino acid substitution with no effect seen on constitutive expression or activity of the protein.[168]

The physiological consequences of the reported polymorphisms have been difficult to determine. In some instances, an increased risk of disease was correlated with a particular polymorphism in association with other risk factors such as obesity, alcohol intake or smoking. For example, McCarver *et al.* reported elevated CYP2E1 activity in individuals carrying the CYP2E1*1D allele who were chronic alcoholics or obese.[169] It is possible that polymorphisms can affect the inducibility of the *CYPE1* gene while exerting no effect on constitutive expression. Studies examining polymorphisms in association with risk factors will help us to understand gene-environment interactions.

In view of the role of CYP2E1 in the metabolic activation of procarcinogens, the question of whether an association between CYP2E1 polymorphisms and tumourigenesis exists, has prompted numerous studies on a variety of cancers. In general, the more pronounced association of different CYP2E1 polymorphisms and cancer appeared to be those involving squamous cell oesophageal carcinoma in Chinese having the CYP2E1 C to T polymorphism and *Rsa*1 restriction site loss in the 5′-promoter region,[170,171] although for other ethnic populations no association was found, even though a trend towards increased risk was noted, especially for smokers.

In contrast, CYP2E1 polymorphisms and different genotypes in populations from an endemic region of China (Linixian County, Henan Province) appear to play a protective role in decreasing the risk of squamous cell oesophageal carcinoma.[172] A pronounced risk between the aforementioned CYP2E1 polymorphism and nasopharyngeal cancer has been reported.[173–176]

Genetic polymorphisms of CYP2E1 may also play a role in the development of hepatocellular carcinoma (HCC). A higher risk of developing HCC has been found for individuals homozygous for the c1/c1 genotype (*Pst*I or *Rsa*I fragments of CYP2E1) that smoked and consumed alcohol compared with control patients.[176] A risk of hepatocellular carcinoma has been associated with another, rare CYP2E1 allele[177] in a Spanish population and this risk increases in combination with alcoholic liver disease,[177,178] although no association was established for Japanese,[179] Korean[180] or British populations.[181]

Other cancers for which CYP2E1 polymorphisms and risk have been examined include gastric cancer, lung cancer and lung adenocarcinoma, renal/urinary tract cancer, pancreatic cancer, colorectal cancer and oral cancer. Whereas the role of CYP2E1 polymorphisms have been associated with increased risk of tumourigenesis, many of these associations were specific to ethnic populations and exacerbated by alcohol intake and diet. Given the complex nature of the initiation and promotion of cancer, the length of time for tumourigenesis and the multiple factors implicated in the tumourigenic process, additional molecular epidemiological studies employing multiple genes and/or biomarkers will be required prior to establishing a definitive role for a single gene or polymorphism.

9.8 Ontogenic Expression

Some information is available on the ontogenic and developmental expression of CYP2E1 in both human and rat. The *CYP2E1* gene in rat is transcriptionally active within one day after birth.[80] The activation of the gene is accompanied by demethylation of the cytosine residues located within the 5′-flanking region of the gene.[80] Similarly, in humans methylation of cytosines in the promoter of the *CYP2E1* gene was associated with the lack of expression in foetal liver.[182] In other tissues in humans, such as lung and kidney, expression begins prior to birth despite heavy methylation of the promoter.[183] The CYP2E1 protein appears to display some developmental fluctuation from birth through adult.[100,184] The mRNA of CYP2E1, however, appears to remain relatively constant throughout the entire lifespan.

Both rats and human possess a single CYP2E gene (CYP2E1). Rabbits, however, have two genes (CYP2E1 and CYP2E2).[185] The genes are autosomal and differentially expressed at birth.[185] Bonfils *et al.* originally suggested the possibility of two CYP2E genes within rabbits as they identified a P450 in the neonatal rabbit liver that was distinct from CYP2E1. Since the livers of foetal laboratory animals are largely devoid of a P450 monooxygenase system until after birth,[186,187] the differential expression of these two genes was determined by Peng *et al.* who showed that CYP2E2 expression begins at birth and CYP2E1 expression begins approximately at two weeks of age.[185]

References

1. D.R. Nelson, L. Koymans, T. Kamataki, J.J. Stegeman, R. Feyereisen, D.J. Waxman, M.R. Waterman, O. Gotoh, M.J. Coon, R.W. Estabrook, I.C. Gunsalus and D.W. Nebert, *Pharmacogenetics*, 1996, **6**, 1.
2. M. Ingelman-Sundberg, *Naunyn. Schmiedebergs Arch. Pharmacol.*, 2004, **369**, 89.
3. H. Wang, B. Chanas and B.I. Ghanayem, *J. Toxicol. Environ. Health A.*, 2002, **65**, 523.
4. M. Tsutsumi, J.M. Lasker, M. Shimizu, A.S. Rosman and C.S. Lieber, *Hepatology*, 1989, **10**, 437.
5. P.J. Sabourin, R.E. Tynes, R.M. Philpot, S. Winquist and A.R. Dahl, *Drug Metab. Dispos.*, 1988, **16**, 557.
6. G.G. Schnier, C.L. Laethem and D.R. Koop, *J. Pharmacol. Exp. Ther.*, 1989, **251**, 790.
7. X.X. Ding, D.R. Koop, B.L. Crump and M.J. Coon, *Mol. Pharmacol.*, 1986, **30**, 370.
8. J.L. Raucy, J.M. Lasker, C.S. Lieber and M. Black, *Arch. Biochem. Biophys.*, 1989, **271**, 270.
9. P.B. Watkins, S.A. Wrighton, P. Maurel, E.G. Schuetz, G. Mendez-Picon, G.A. Parker and P.S. Guzelian, *Proc. Natl. Acad. Sci. USA*, 1985, **82**, 6310.
10. C.S. Lieber, *Physiol. Rev.*, 1997, **77**, 517.
11. R. Buhler, K.O. Lindros, B.K. von, P. Karkkainen, J. Makinen and M. Ingelman-Sundberg, *Alcohol Alcohol Suppl.*, 1991, **1**, 311.
12. M. Ingelman-Sundberg, I. Johansson, K.E. Penttila, H. Glaumann and K.O. Lindros, *Biochem. Biophys. Res. Commun.*, 1988, **157**, 55.
13. T. Takahashi, J.M. Lasker, A.S. Rosman and C.S. Lieber, *Hepatology*, 1993, **17**, 236.
14. K. Mace, E.D. Bowman, P. Vautravers, P.G. Shields, C.C. Harris and A.M. Pfeifer, *Eur. J. Cancer*, 1998, **34**, 914.
15. K.C. Watt, C.G. Plopper, A.J. Weir, B. Tarkington and A.R. Buckpitt, *Toxicol. Appl. Pharmacol.*, 1998, **149**, 195.
16. J.A. Hotchkiss, H. Kim, F.F. Hahn, R.F. Novak and A.R. Dahl, *Toxicol. Lett.*, 1995, **78**, 1.
17. B.S. Cummings, R.C. Zangar, R.F. Novak and L.H. Lash, *Toxicology.*, 2000, **150**, 83.
18. N. Kapucuoglu, T. Coban, H. Raunio, O. Pelkonen, R.J. Edwards, A.R. Boobis and M. Iscan, *Cancer Lett.*, 2003, **196**, 153.
19. E.A. Cannady, C.A. Dyer, P.J. Christian, I.G. Sipes and P.B. Hoyer, *Toxicol. Sci.*, 2003, **73**, 423.
20. H. Nakura, S. Itoh, H. Kusano, H. Ishizone, S. Imaoka, Y. Funae, T. Yokoi and T. Kamataki, *Biochem. Pharmacol.*, 1992, **43**, 1907.
21. B.J. Song, H.V. Gelboin, S.S. Park, C.S. Yang and F.J. Gonzalez, *J. Biol. Chem.*, 1986, **261**, 16689.
22. M. Umeno, B.J. Song, C. Kozak, H.V. Gelboin and F.J. Gonzalez, *J. Biol. Chem.*, 1988, **263**, 4956.

23. V. Wauthier, V. Schenten, R.K. Verbeeck and P.B. Calderon, *Life Sci.*, 2006, **79**, 1913.
24. Z. Bebia, S.C. Buch, J.W. Wilson, R.F. Frye, M. Romkes, A. Cecchetti, D. Chaves-Gnecco and R.A. Branch, *Clin. Pharmacol. Ther.*, 2004, **76**, 618.
25. C.A. Mugford and G.L. Kedderis, *Drug Metab. Rev.*, 1998, **30**, 441.
26. E. Tanaka, *J. Clin. Pharm. Ther.*, 1999, **24**, 339.
27. R.B. Kim and D. O'shea, *Clin. Pharmacol. Ther.*, 1995, **57**, 645.
28. N. Finnstrom, B. Ask, M.L. Dahl, M. Gadd and A. Rane, *Pharmacogenomics J.*, 2002, **2**, 111.
29. C.S. Lieber and L.M. DeCarli, *Science*, 1968, **162**, 917.
30. C.S. Lieber and L.M. DeCarli, *J. Biol. Chem.*, 1970, **245**, 2505.
31. C.S. Lieber, E. Rubin and L.M. DeCarli, *Biochem. Biophys. Res. Commun.*, 1970, **40**, 858.
32. K. Comai and J.L. Gaylor, *J. Biol. Chem.*, 1973, **248**, 4947.
33. Y. Hasumura, R. Teschke and C.S. Lieber, *J. Pharmacol. Exp. Ther.*, 1975, **194**, 469.
34. J.L. Raucy, J.C. Kraner and J.M. Lasker, *Crit. Rev. Toxicol.*, 1993, **23**, 1.
35. I.A. Leclercq, G.C. Farrell, J. Field, D.R. Bell, F.J. Gonzalez and G.R. Robertson, *J. Clin. Invest.*, 2000, **105**, 1067.
36. A. Gruebele, K. Zawaski, D. Kaplan and R.F. Novak, *Drug Metab. Dispos.*, 1996, **24**, 15.
37. S.G. Kim and R.F. Novak, *Biochem. Biophys. Res. Commun.*, 1990, **166**, 1072.
38. M.J. Wargovich, *J. Nutr.*, 2006, **136**, 832S.
39. M. Shimada, L. Liu, N. Nussler, S. Jonas, J.M. Langrehr, T. Ogawa, M. Kaminishi, P. Neuhaus and A.K. Nussler, *Toxicol. Lett.*, 2006, **163**, 242.
40. D. Taubert, R. Glockner, D. Muller and E. Schomig, *Toxicol. Lett.*, 2006, **164**, 1.
41. D.M. Davenport and M.J. Wargovich, *Food Chem. Toxicol.*, 2005, **43**, 1753.
42. C.S. Yang, *Adv. Exp. Med. Biol.*, 2001, **500**, 541.
43. G.D. Loizou and J. Cocker, *Hum. Exp. Toxicol.*, 2001, **20**, 321.
44. T. Wang, K. Shankar, T.J. Bucci, A. Warbritton and H.M. Mehendale, *Toxicol. Appl. Pharmacol.*, 2001, **173**, 27.
45. S.G. Kim, J.Y. Kwak, J.W. Lee, R.F. Novak, S.S. Park and N.D. Kim, *Biochem. Biophys. Res. Commun.*, 1994, **200**, 1414.
46. T. Zima and M. Kalousova, *Alcohol. Clin. Exp. Res.*, 2005, **29**, 110S.
47. Y. Lu and A.I. Cederbaum, *Hepatology*, 2006, **44**, 263.
48. J.F. Brady, M.H. Wang, J.Y. Hong, F. Xiao, Y. Li and J.S. Yoo, *et al.*, *Toxicol. Appl. Pharmacol.*, 1991, **108**, 342.
49. C.S. Yang, J.S. Yoo, H. Ishizaki and J.Y. Hong, *Drug Metab. Rev.*, 1990, **22**, 147.
50. J.F. Brady, D.C. Li, H. Ishizaki and C.S. Yang, *Cancer Res.*, 1988, **48**, 5937.
51. K.K. Hajek and R.F. Novak, *Biochem. Biophys. Res. Commun.*, 1982, **108**, 664.

52. K.L. Kaul and R.F. Novak, *Arch. Biochem. Biophys.*, 1984, **235**, 470.
53. K.L. Kaul and R.F. Novak, *J. Pharmacol. Exp. Ther.*, 1987, **243**, 384.
54. D.R. Koop, B.L. Crump, G.D. Nordblom and M.J. Coon, *Proc. Natl. Acad. Sci. USA*, 1985, **82**, 4065.
55. S.G. Kim, D.E. Williams, E.G. Schuetz, P.S. Guzelian and R.F. Novak, *J. Pharmacol. Exp. Ther.*, 1988, **246**, 1175.
56. S.G. Kim, S.E. Shehin, J.C. States and R.F. Novak, *Biochem. Biophys. Res. Commun.*, 1990, **172**, 767.
57. S.G. Kim and R.F. Novak, *Toxicol. Appl. Pharmacol.*, 1993, **120**, 257.
58. H. Kim, S.G. Kim, M.Y. Lee and R.F. Novak, *Biochem. Biophys. Res. Commun.*, 1992, **186**, 846.
59. H.M. Peng, G.M. Raner, A.D. Vaz and M.J. Coon, *Arch. Biochem. Biophys.*, 1995, **318**, 333.
60. J.O. Miners, P.A. Smith, M.J. Sorich, R.A. McKinnon and P.I. Mackenzie, *Annu. Rev. Pharmacol. Toxicol.*, 2004, **44**, 1.
61. M. Vidali, M. Hidestrand, E. Eliasson, E. Mottaran, E. Reale and R. Rolla, *et al.*, *J. Biol. Chem.*, 2004, **279**, 50949.
62. R.J. Bertz and G.R. Granneman, *Clin. Pharmacokinet*, 1997, **32**, 210.
63. W. E. Evans and M. V. Relling, *Science*, 1999, **286**, 487.
64. F.P. Guengerich and T. Shimada, *Chem. Res. Toxicol.*, 1991, **4**, 391.
65. H. Yamazaki, Y. Inui, C.H. Yun, F.P. Guengerich and T. Shimada, *Carcinogenesis*, 1992, **13**, 1789.
66. E. Elovaara, K. Engstrom, T. Nakajima, S.S. Park, H.V. Gelboin and H. Vainio, *Xenobiotica*, 1991, **21**, 651.
67. K.E. Thummel, E.D. Kharasch, T. Podoll and K. Kunze, *Drug Metab. Dispos.*, 1993, **21**, 350.
68. M.J. Seaton, P.M. Schlosser, J.A. Bond and M.A. Medinsky, *Carcinogenesis*, 1994, **15**, 1799.
69. M.J. Olson, S.G. Kim, C.A. Reidy, J.T. Johnson and R.F. Novak, *Drug Metab. Dispos.*, 1991, **19**, 298.
70. O.S. Sohn, H. Ishizaki, C.S. Yang and E.S. Fiala, *Carcinogenesis*, 1991, **12**, 127.
71. T. Nakajima, R.S. Wang, Y. Katakura, R. Kishi, E. Elovaara, S.S. Park, H.V. Gelboin and H. Vainio, *J. Pharmacol. Exp. Ther.*, 1992, **261**, 869.
72. I. Johansson and M. Ingelman-Sundberg, *Cancer Res.*, 1988, **48**, 5387.
73. D.R. Koop, *FASEB J.*, 1992, **6**, 724.
74. F.P. Guengerich, D.H. Kim and M. Iwasaki, *Chem. Res. Toxicol.*, 1991, **4**, 168.
75. S.G. Kim and R.F. Novak, *Cancer Res.*, 1990, **50**, 5333.
76. D.R. Koop, C.L. Laethem and G.G. Schnier, *Toxicol. Appl. Pharmacol.*, 1989, **98**, 278.
77. J.L. Raucy, J.M. Lasker, J.C. Kraner, D.E. Salazar, C.S. Lieber and G.B. Corcoran, *Mol. Pharmacol.*, 1991, **39**, 275.
78. D. Lucas, J.F. Menez and F. Berthou, *Methods Enzymol.*, 1996, **272**, 115.
79. A.M. Lee, J. Yue and R.F. Tyndale, *Drug Metab. Dispos.*, 2006, **34**, 1508.
80. T. Ueno and F.J. Gonzalez, *Mol. Cell. Biol.*, 1990, **10**, 4495.

81. S.Y. Liu and F.J. Gonzalez, *DNA Cell Biol.*, 1995, **14**, 285.
82. T.E. Akiyama and F.J. Gonzalez, *Biochim. Biophys. Acta.*, 2003, **1619**, 223.
83. F.J. Gonzalez, T. Ueno, M. Umeno, B.J. Song, R.L. Veech and H.V. Gelboin, *Alcohol Alcohol Suppl.*, 1991, **1**, 97.
84. B.T. Kren, J.H. Trembley and C.J. Steer, *Am. J. Physiol.*, 1996, **270**, G763–G777.
85. R.C. Zangar, K.J. Woodcroft, T.A. Kocarek and R.F. Novak, *Drug Metab. Dispos.*, 1995, **23**, 681.
86. K.J. Woodcroft and R.F. Novak, *J. Pharmacol. Exp. Ther.*, 1999, **289**, 1121.
87. R.F. Novak and K.J. Woodcroft, *Arch. Pharm. Res.*, 2000, **23**, 267.
88. M.A. Abdelmegeed, N.J. Carruthers, K.J. Woodcroft, S.K. Kim and R.F. Novak, *J. Pharmacol. Exp. Ther.*, 2005, **315**, 203.
89. A. Banerjee, T.A. Kocarek and R.F. Novak, *Drug Metab. Dispos.*, 2000, **28**, 118.
90. B.J. Song, T. Matsunaga, J.P. Hardwick, S.S. Park, R.L. Veech, C.S. Yang, H.V. Gelboin and F.J. Gonzalez, *Mol. Endocrinol.*, 1987, **1**, 542.
91. K.C. Chang, T.D. Bell, B.A. Lauer and H. Chai, *Lancet.*, 1978, **1**, 1132.
92. K.W. Renton and L.C. Knickle, *Can. J. Physiol. Pharmacol.*, 1990, **68**, 777.
93. Z. Abdel-Razzak, P. Loyer, A. Fautrel, J.C. Gautier, L. Corcos, B. Turlin, P. Beaune and A. Guielouzo, *Mol. Pharmacol.*, 1993, **44**, 707.
94. H.M. Peng and M.J. Coon, *Arch. Biochem. Biophys.*, 2000, **382**, 129.
95. L.O. Qiu, M.W. Linder, D.M. Antonino-Green and R. Valdes, Jr., *J. Pharmacol. Exp. Ther.*, 2004, **308**, 284.
96. Z. Abdel-Razzak, M. Garlatti, M. Aggerbeck and R. Barouki, *Biochem. Pharmacol.*, 2004, **68**, 1371.
97. P. Skett and A. Weir, *Biochem. Pharmacol.*, 1983, **32**, 3115.
98. D.W. Rosenberg, G.S. Drummond and T.J. Smith, *Pharmacology*, 1995, **51**, 254.
99. H.M. Peng and M.J. Coon, *Mol. Pharmacol.*, 1999, **54**, 740.
100. D.J. Waxman, J.J. Morrissey and G.A. LeBlanc, *Endocrinology*, 1989, **124**, 2954.
101. J.Y. Hong, S.M. Ning, B.L. Ma, M.J. Lee, J.M. Pan and C.S. Yang, *Arch. Biochem. Biophys.*, 1990, **281**, 132.
102. M.H. Son, K.W. Kang, E.J. Kim, J.H. Ryu, H. Cho and S.H. Kim, *et al.*, *Chem. Biol. Interact*, 2000, **127**, 13.
103. B.J. Song, R.L. Veech and P. Saenger, *J. Clin. Endocrinol. Metab.*, 1990, **71**, 1036.
104. D. Lucas, C. Farez, L.G. Bardou, J. Vaisse, J.R. Attali and P. Valensi, *Fundam. Clin. Pharmacol.*, 1998, **12**, 553.
105. P.Y. Cheng and E.T. Morgan, *Curr. Drug Metab.*, 2001, **2**, 165.
106. L. Pucci, V. Chirulli, S. Marini, D. Lucchesi, G. Penno, P.G. Gervasi, S. Del Proto and V. Longo, *Pharmacol. Res.*, 2005, **51**, 561.
107. M.P. Hannon-Fletcher, M.J. O'Kane, K.W. Moles, Y.A. Barnett and C.R. Barnett, *Food Chem. Toxicol.*, 2001, **39**, 125.

108. Z. Wang, S.D. Hall, J.F. Maya, L. Li, A. Asghar and J.C. Gorski, *Br. J. Clin. Pharmacol.*, 2003, **55**, 77.
109. V. Haufroid, D. Ligocka, M. Buysschaert, Y. Horsmans and D. Lison, *Eur. J. Clin. Pharmacol.*, 2003, **59**, 29.
110. S. Kataoka, H. Yasui, M. Hiromura and H. Sakurai, *Life Sci.*, 2005, **77**, 2814.
111. E. Arinc, S. Arslan and O. Adali, *Arch. Toxicol.*, 2005, **79**, 427.
112. H.W. Baek, S.K. Bae, M.G. Lee and Y.T. Sohn, *J. Pharm. Sci.*, 2006, **95**, 2452.
113. M.G. Emery, J.M. Fisher, J.Y. Chien, E.D. Kharasch, E.P. Dellinger, K.V. Kowdley and K.E. Thummel, *Hepatology*, 2003, **38**, 428.
114. G.D. Bellward, T. Chang, B. Rodrigues, J.H. McNeill, S. Maines, D.E. Ryan, W. Levin and P.E. Thomas, *Mol. Pharmacol.*, 1988, **33**, 140.
115. K.J. Woodcroft, M.S. Hafner and R.F. Novak, *Hepatology*, 2002, **35**, 263.
116. I. de Waziers, M. Garlatti, J. Bouguet, P.H. Beaune and R. Barouki, *Mol. Pharmacol.*, 1995, **47**, 474.
117. K.J. Woodcroft and R.F. Novak, *Chem. Biol. Interact*, 1997, **107**, 75.
118. C. Grosset, C.Y. Chen, N. Xu, N. Sonenberg, H. Jacquemin-Sablon and A.B. Shyu, *Cell*, 2000, **103**, 29.
119. S.K. Kim and R.F. Novak, *Pharmacol. Ther.*, 2007, **113**, 88.
120. S. Sekine, B.Y. Lan, M. Bedolli, S. Feng and M. Hebrok, *Hepatology*, 2006, **43**, 817.
121. A. Moncion, N.T. Truong, A. Garrone, P. Beaune, R. Barouki and W.I. de, *J. Biol. Chem.*, 2002, **277**, 45904.
122. V. Ambros, *Nature*, 2004, **431**, 350.
123. D.P. Bartel, *Cell*, 2004, **116**, 281.
124. T. Du and P.D. Zamore, *Development*, 2005, **132**, 4645.
125. R.J. Jackson and N. Standart, *Sci. STKE.*, 2007, (**367**) rel.
126. C.Z. Chen, L. Li, H.F. Lodish and D.P. Bartel, *Science*, 2004, **303**, 83.
127. L. He and G.J. Hannon, *Nat. Rev. Genet.*, 2004, **5**, 522.
128. J. Lu, G. Getz, E.A. Miska, E. varez-Saavedra, J. Lamb, D. Peck, A. Sweet-Cordero, B.L. Ebert, R.H. Mak, A.A. Ferrando, J.R. Downing, T. Jacks, H.R. Horvitz and T.R. Golub, *Nature*, 2005, **435**, 834.
129. K.A. O'Donnell, E.A. Wentzel, K.I. Zeller, C.V. Dang and J.T. Mendell, *Nature*, 2005, **435**, 839.
130. M.N. Poy, L. Eliasson, J. Krutzfeldt, S. Kuwajima, X. Ma, P.E. Macdonald, S. Pfeffer, T. Tuschl, N. Rajewsky, P. Rorsman and M. Stoffel, *Nature*, 2004, **432**, 226.
131. P.A. Maroney, Y. Yu, J. Fisher and T.W. Nilsen, *Nat. Struct. Mol. Biol.*, 2006, **13**, 1102.
132. S. Nottrott, M.J. Simard and J.D. Richter, *Nat. Struct. Mol. Biol.*, 2006, **13**, 1108.
133. L.P. Lim, N.C. Lau, P. Garrett-Engele, A. Grimson, J.M. Schelter, J. Castle, D.P. Bartel, P.S. Linsley and J.M. Johnson, *Nature*, 2005, **433**, 769.

134. Y. Tsuchiya, M. Nakajima, S. Takagi, T. Taniya and T. Yokoi, *Cancer Res.*, 2006, **66**, 9090.
135. C. Esau, S. Davis, S.F. Murray, X.X. Yu, S.K. Pandey, M. Pear, L. Watts, S.L. Booten, M. Graham, R. Mckay, A. Subramaniam, S. Propp, B.A. Lollo, S. Freier, C.F. Bennelt, S. Bhanot and B.P. Monia, *Cell Metab.*, 2006, **3**, 87.
136. C. Cheung, T.E. Akiyama, G. Kudo and F.J. Gonzalez, *Biochem. Pharmacol.*, 2003, **66**, 2011.
137. B.S. Donahue and E.T. Morgan, *Drug Metab. Dispos.*, 1990, **18**, 519.
138. T.A. Kocarek, R.C. Zangar and R.F. Novak, *Arch. Biochem. Biophys.*, 2000, **376**, 180.
139. M. Tsutsumi, J.M. Lasker, T. Takahashi and C.S. Lieber, *Arch. Biochem. Biophys.*, 1993, **304**, 209.
140. B.J. Song, R.L. Veech, S.S. Park, H.V. Gelboin and F.J. Gonzalez, *J. Biol. Chem.*, 1989, **264**, 3568.
141. R.E. McGehee, Jr., M.J. Ronis, R.M. Cowherd, M. Ingelman-Sundberg and T.M. Badger, *Biochem. Pharmacol.*, 1994, **48**, 1823.
142. B.J. Roberts, B.J. Song, Y. Soh, S.S. Park and S.E. Shoaf, *J. Biol. Chem.*, 1995, **270**, 29632.
143. J.Y. Huan, J.M. Streicher, L.A. Bleyle and D.R. Koop, *Toxicol. Appl. Pharmacol.*, 2004, **199**, 332.
144. J.C. Ho, S.T. Cheung, K.L. Leung, I.O. Ng and S.T. Fan, *Int. J. Cancer*, 2004, **111**, 494.
145. A. Suzuki, M. Hayashida, H. Kawano, K. Sugimoto, T. Nakano and K. Shiraki, *Hepatology*, 2000, **32**, 796.
146. K. Nakanishi, M. Sakamoto, S. Yamasaki, S. Todo and S. Hirohashi, *Cancer*, 2005, **103**, 307.
147. T. Poynard, M.F. Yuen, V. Ratziu and C.L. Lai, *Lancet*, 2003, **362**, 2095.
148. M. Okuda, K. Li, M.R. Beard, L.A. Showalter, F. Scholle, S.M. Lemon and S.A. Weinman, *Gastroenterology*, 2002, **122**, 366.
149. A.S. Rosman, F. Paronetto, K. Galvin, R.J. Williams and C.S. Lieber, *Arch. Intern. Med.*, 1993, **153**, 965.
150. K. Otani, M. Korenaga, M.R. Beard, K. Li, T. Qian, L.A. Showalter, A.K. Singh, T. Wang and S.A. Weinman, *Gastroenterology*, 2005, **128**, 96.
151. H. Sies, *Am. J. Med.*, 1991, **91**, 31S.
152. J.A. Knight, *Ann. Clin. Lab. Sci.*, 1998, **28**, 331.
153. R.M. Touyz, *Hypertension*, 2004, **44**, 248.
154. M. Phillips, R.N. Cataneo, T. Cheema and J. Greenberg, *Clin. Chim. Acta.*, 2004, **344**, 189.
155. L.D. Gorsky, D.R. Koop and M.J. Coon, *J. Biol. Chem.*, 1984, **259**, 6812.
156. G. Ekstrom and M. Ingelman-Sundberg, *Biochem. Pharmacol.*, 1989, **38**, 1313.
157. I. Kessova and A.I. Cederbaum, *Curr. Mol. Med.*, 2003, **3**, 509.
158. M. Mari and A.I. Cederbaum, *J. Biol. Chem.*, 2000, **275**, 15563.
159. G. Krikun and A.I. Cederbaum, *FEBS Lett.*, 1986, **208**, 292.

160. G. Ekstrom, B.C. von and M. Ingelman-Sundberg, *Biochem. Pharmacol.*, 1989, **38**, 689.
161. M. Ingelman-Sundberg, I. Johansson, H. Yin, Y. Terelius, E. Eliasson, P. Clot and E. Albano, *Alcohol.*, 1993, **10**, 447.
162. J.A. Davila, R.O. Morgan, Y. Shaib, K.A. McGlynn and H.B. El-Serag, *Gut*, 2005, **54**, 533.
163. I.M. Danko and N.A. Chaschin, *Exp. Oncol.*, 2005, **27**, 248.
164. O.W. McBride, M. Umeno, H.V. Gelboin and F.J. Gonzalez, *Nucleic Acids Res.*, 1987, **15**, 10071.
165. Y. Hu, M. Oscarson, I. Johansson, Q.Y. Yue, M.L. Dahl, M. Tabone, S. Arinco, E. Albano and M. Ingelman-Sundberg, *Mol. Pharmacol.*, 1997, **51**, 370.
166. Y. Hu, J. Hakkola, M. Oscarson and M. Ingelman-Sundberg, *Biochem. Biophys. Res. Commun.*, 1999, **263**, 286.
167. V. Carriere, F. Berthou, S. Baird, C. Belloc, P. Beaune and W.I. de, *Pharmacogenetics*, 1996, **6**, 203.
168. K.S. Fairbrother, J. Grove, W.I. de, D.T. Steimel, C.P. Day, C.L. Crespi and A.K. Daly, *Pharmacogenetics*, 1998, **8**, 543.
169. D.G. McCarver, R. Byun, R.N. Hines, M. Hichme and W. Wegenek, *Toxicol. Appl. Pharmacol*, 1998, **152**, 276.
170. W. Tan, N. Song, G.Q. Wang, Q. Liu, H.J. Tang, F.F. Kadlubar and D.X. Lin, *Cancer Epidemiol Biomarkers Prev.*, 2000, **9**, 551.
171. X.M. Lu, Y.M. Zhang, R.Y. Lin, G. Arzi, X. Wang, Y.L. Zhang, Y. Zhang, Y. Wang and H. Wen, *World J. Gastroenterol*, 2005, **11**, 3651.
172. D.X. Lin, Y.M. Tang, Q. Peng, S.X. Lu, C.B. Ambrosone and F.F. Kadlubar, *Cancer Epidemiol Biomarkers Prev.*, 1998, **7**, 1013.
173. A. Hildesheim, C.J. Chen, N.E. Caporaso, Y.J. Cheng, R.N. Hoover, M.M. Hsu, P.H. Levine, I.H. Chen, C.S. Yang, A.K. Daly and J.R. Idle, *Cancer Epidemiol. Biomarkers Prev.*, 1995, **4**, 607.
174. A. Hildesheim, L.M. Anderson, C.J. Chen, Y.J. Cheng, L.A. Brinton, A.K. Daly, C.D. Reed, I.H. Chen, N.E. Caporaso, M.M. Hsu, J.Y. Chem, J.R. Idle, R.N. Hoover, C.S. Yang and S.K. Chhabra, *J. Natl. Cancer Inst.*, 1997, **89**, 1207.
175. N. Kongruttanachok, S. Sukdikul, S. Setavarin, V. Kerekhjanarong, P. Supiyaphun, N. Voravud, Y. Poovorawan and A. Mutirangara, *BMC Cancer*, 2001, **1**, 4.
176. M.W. Yu, A. Gladek-Yarborough, S. Chiamprasert, R.M. Santella, Y.F. Liaw and C.J. Chen, *Gastroenterology*, 1995, **109**, 1266.
177. J.M. Ladero, J.A. Agundez, A. Rodriguez-Lescure, M. az-Rubio and J. Benitez, *Gut*, 1996, **39**, 330.
178. M. Munaka, K. Kohshi, T. Kawamoto, S. Takasawa, N. Nagata, H. Itoh, S. Oda and T. Katoh, *J. Cancer Res. Clin. Oncol.*, 2003, **129**, 355.
179. S. Kato, M. Onda, N. Matsukura, A. Tokunaga, T. Tajiri, D.Y. Kim, H. Tsuruta, N. Matsuda, K. Yamoshita and Pb. Shields, *Pharmacogenetics*, 1995, **5 Spec No** S141.

180. H.S. Lee, J.H. Yoon, S. Kamimura, K. Iwata, H. Watanabe and C.Y. Kim, *Int. J. Cancer*, 1997, **71**, 737.
181. N.A. Wong, F. Rae, K.J. Simpson, G.D. Murray and D.J. Harrison, *Mol. Pathol.*, 2000, **53**, 88.
182. S.M. Jones, A.R. Boobis, G.E. Moore and P.M. Stanier, *Biochem. Pharmacol.*, 1992, **43**, 1876.
183. I. Vieira, M. Pasanen, H. Raunio and T. Cresteil, *Pharmacol. Toxicol.*, 1998, **83**, 183.
184. P.E. Thomas, S. Bandiera, S.L. Maines, D.E. Ryan and W. Levin, *Biochemistry*, 1987, **26**, 2280.
185. H.M. Peng, T.D. Porter, X.X. Ding and M.J. Coon, *Mol. Pharmacol.*, 1991, **40**, 58.
186. A.H. Neims, M. Warner, P.M. Loughnan and J.V. Aranda, *Annu. Rev. Pharmacol. Toxicol.*, 1976, **16**, 427.
187. C. Bonfils, J. Combalbert, T. Pineau, J. Angevin, C. Larroque, J. Derancourt, J.P. Capony and P. Maurel, *Eur. J. Biochem.*, 1990, **188**, 187.

CHAPTER 10

The CYP2F, CYP2G and CYP2J Subfamilies

QING-YU ZHANG AND XINXIN DING

Wadsworth Center, New York State Department of Health, and School of Public Health, State University of New York at Albany, Albany, NY 12201, USA

Table of Contents

Issues in Toxicology
Cytochromes P450: Role in the Metabolism and Toxicity of Drugs and other Xenobiotics
Edited by Costas Ioannides

Members of the *CYP2F*, *CYP2G* and *CYP2J* subfamilies were first discovered in the late 1980s and early 1990s. These genes are either preferentially (CYP2F and CYP2J, with exception of rat CYP2J3) or exclusively (CYP2G) expressed in extrahepatic tissues, where they may have important biological functions or contribute to tissue-selective xenobiotic toxicity.

10.1 CYP2F

10.1.1 Introduction

The CYP2F subfamily was first identified in humans; cDNA clones for CYP2F1 were isolated from a human lung cDNA library, which was screened using a CYP2C cDNA probe.[1] The mouse CYP2F2 cDNA was isolated soon after,[2] with the use of an antibody against a P450 protein previously purified from mouse liver.[3] It appears that the *CYP2F* subfamily consists of only a single gene in each of the mammalian species examined to date, except in humans and other primates, in which a *CYP2F* pseudogene is also present. The *CYP2F* genes are unusual members of the *CYP2* gene family because they have 10 exons; the first exon is not translated.[4] Additionally, the *CYP2F* gene intron and exon sequences from differing mammalian species seem to be more highly conserved than are the orthologous genes in other *CYP2* gene subfamilies.[5] Heterologously expressed proteins have been obtained for human CYP2F1, mouse CYP2F2, goat CYP2F3, and rat CYP2F4, for functional analysis. There are no reports of a *Cyp2f2*-null mouse. *CYP2F* genes are preferentially expressed in the respiratory tract. The best-known substrates for CYP2F enzymes include naphthalene and 3-methylindole. A comprehensive review of the metabolic mechanisms of naphthalene toxicity, including the role of CYP2F enzymes, was published recently.[6] There has been no report of any endogenous compounds that CYP2F enzymes metabolise. In addition, nothing is known about any inducers, either endogenous or xenobiotic, of the *CYP2F* genes.

10.1.2 Human CYP2F1

Human CYP2F1, initially named P450 IIF1, was first identified through cDNA cloning.[1] The cDNA clone was isolated from a human lung cDNA library.

CYP2F1 encodes a protein of 491 amino acids. The *CYP2F1* gene was localised to human chromosome 19,[1] within a *CYP2* gene cluster,[7,8] which consists of members of the *CYP2A*, *CYP2B*, *CYP2F*, *CYP2G*, *CYP2S* and *CYP2T* gene subfamilies.[9] A *CYP2F* pseudogene, *CYP2F1P*, is located at ~300 kbp from *CYP2F1*[5]. *CYP2F1P* lacks the 5′-end of the gene and the first three coding exons. It also differs from *CYP2F1* in having a 144-bp insertion in intron 5 and a 995-bp insertion in intron 6, in addition to having a number of base substitutions.[4,5]

Transcripts corresponding to CYP2F1, as well as a splice variant (corresponding to a variant cDNA clone named IIF1v), were detected at low abundance in human lung by Northern blot analysis.[1] CYP2F1 mRNA was detected by RNA-PCR in human placenta,[10,11] and in freshly isolated or primary cultures of human bronchial epithelial cells,[12,13] but not in human uterine endometrium[14] or breast tissue.[15] CYP2F1 mRNA was detected in bronchoalveolar macrophages by one group,[16,17] although it was not detected in alveolar macrophages by another.[12] In peripheral lymphocyte samples, CYP2F1 mRNA either was not detected,[16,17] or, in a study using a competitive RT-PCR method, was found to be near the detection limit.[18]

In a PCR-based quantitative study of CYP mRNA expression in various adult human organs, including adrenal gland, brain, kidney, liver, lung, placenta, prostate, small intestine, testis, and uterus, as well as in foetal liver, the highest level of CYP2F1 was found in the lung; while this CYP was also detectable (albeit at 100 times lower levels) in the testis, it was not detected, or was below the limit of quantification, in the other tissues examined.[19] The low expression or lack of expression of CYP2F1 in adult and foetal liver was also reported elsewhere.[20,21] In another study, CYP2F1 mRNA was detected by quantitative RNA-PCR in foetal human nasal mucosa (at gestational day 91) at 47-times higher levels than in the livers of the same foetuses.[22] These findings support the notion that CYP2F1 is primarily expressed in the respiratory tract. However, no quantitative analysis of CYP2F1 protein levels in various human tissues has yet been reported.

CYP2F1-immunoreactive protein was detected, with an antibody against a CYP2F1 carboxyl terminal peptide (RPFQLCLRPR),[23] on tissue microarrays of primary ovarian cancer as well as in normal ovary, and the level of immunoreactivity seen in ovarian cancer was significantly higher than in normal ovary.[24] In a similar study, CYP2F1 immunoreactivity was detected in a small fraction of colorectal cancers and lymph node metastasis, but not in normal colon.[23] CYP2F1 mRNA was not detected in breast tumour (infiltrating ductal carcinoma) or surrounding tumour-free breast tissue.[15] CYP2F1 mRNA was detected in the human bronchial epithelial cell line BEAS-2B, but not in the human alveolar epithelial cell line A549, or the hepatocellular carcinoma cell line HepG2.[4]

The *CYP2F1* gene is ~14 kbp in size; the noncoding first exon is located 1685 bp upstream of exon 2[4]. The major transcription start site was mapped to 1741 bp upstream of the ATG codon. Several putative *cis*-regulatory elements, such as SP1 (specificity protein 1), AP2 (activator protein 2), C/EBP (CCAT enhancer-binding protein), BTE (basic transcription element), were identified in the 5′-upstream region of the promoter. The minimal promoter was identified at positions –129 to +115 through reporter gene assays. A strong DNA-protein complex was detected at positions –152 to –182 by DNase I footprinting analysis using human lung nuclear extracts; the binding protein was named lung-specific factor (or LSF), because it was not detected in liver or heart. The core LSF-binding site (5′-CTCCCACGG*CACCTT*TC*CAGCTG*GCTGTGAG-3′; the underlined nucleotides are critical for maximal binding affinity) contains two E-box motifs (italicised), but the binding protein(s) in human lung nuclear extract do not seem to correspond to any known E-box factors. Although the nature of the LSF factor remains to be elucidated, it was proposed that LSF plays an important role in the tissue-selective expression of CYP2F1 in the lung.

In a subsequent study, the *CYP2F1* promoter was further characterised, with an emphasis on the role of four Sp1-like elements in the TATA-less promoter.[25] In gel-shift assays, each of the four Sp1-like sites was able to form three specific protein-DNA complexes with proteins in A549 lung cell nuclear extract. Sequential addition of each of the four elements to reporter gene constructs led to activation of promoter activity in an incremental fashion in lung-derived A549 cells, but not in liver-derived HepG2 cells. Co-transfection studies indicated that, in A549 cells, Sp1, but not Sp3, activates *CYP2F1* transcription, and that interactions between Sp1 and Sp3 proteins might further modulate CYP2F1 expression in lung cells.

Heterologously expressed CYP2F1 was active toward a number of xenobiotic compounds (Table 10.1). CYP2F1 was initially produced by heterologous expression in HepG2 cells, through the use of a vaccinia virus vector.[1]

Table 10.1 Activities of heterologously expressed CYP2F1.

Substrate	*Product(s)*	*Rate (nmol/min/ nmol P450)*	*Reference*
Ethoxycoumarin[a] (0.2 mM)	Coumarin	0.7	1
Propoxycoumarin[a] (0.2 mM)	Coumarin	0.3	1
Pentoxyresorufin[a] (0.01 mM,)	Resorufin	0.3 (25 °C)	1
Styrene[a] (5 mM)	Styrene glycol	0.1 nmol/mg cell lysate/min	26
3-methylindole[a] (0.2 mM)	3-methyleneindolenine	0.2 nmol/mg/h	28
3-methylindole[b]	3-methyleneindolenine	1.3 (Vmax; Km = 18 μM)	29
Naphthalene[b] (0.5 mM)	Naphthalene 1*R*,2*S*-oxide	0.004	29
	Naphthalene 1*S*,2*R*-oxide	0.031	
Benzene[b]	Total hydroxylated metabolites	0.01 (Vmax; Km = 3.8 μM)	32

[a]CYP2F1 was expressed in Hep G2 cells using a vaccinia virus vector.
[b]CYP2F1 was expressed in lymphoblastoid cells.

The recombinant enzyme was active in the dealkylation of ethoxycoumarin, propoxycoumarin and pentoxyresorufin.[1] However, the same enzyme preparation did not catalyse ethoxyresorufin *O*-deethylation, and it was only marginally active in benzyloxyresorufin dearylation.[1] Recombinant CYP2F1 was found to be one of the most active, among 12 human P450 isoforms tested, in the conversion of styrene, a lung toxicant, to styrene glycol;[26] however, although it was only somewhat active in the metabolic activation of another pulmonary toxin, 4-ipomeanol.[27]

Vaccinia-expressed CYP2F1 was active in the desaturation of 3-methylindole, a lung-selective toxicant, producing 3-methyleneindolenine, a toxic intermediate.[28] CYP2F1 heterologously expressed in human lymphoblastoid cells was also assayed for ability to metabolise 3-methylindole.[29] CYP2F1-containing microsomes were found to efficiently catalyse the dehydrogenation of 3-methylindole to the methylene imine,[29] consistent with an earlier finding with vaccinia-expressed CYP2F1.[28] However, CYP2F1 did not form indole-3-carbinol or 3-methyloxindole, products of hydroxylation and ring oxidation, respectively. Furthermore, at high substrate concentrations, 3-methylindole was found to inhibit CYP2F1-catalysed formation of the methylene imine, possibly as a mechanism-based inhibitor.[29] A human bronchial epithelial (BEAS-2B) cell line expressing transfected CYP2F1 was more sensitive to 3-methylindole-induced cytotoxicity than were BEAS-2B cell lines expressing transfected CYP2A6, CYP3A4 or CYP2E1; these findings further indicate the potential for CYP2F1 to mediate 3-methylindole metabolic activation in human lung cells.[30] A subsequent enzyme kinetic study of multiple P450 enzymes indicated that the 3-methylindole dehydrogenation reaction catalysed by CYP2F1 and CYP2F3 was unique, a finding suggesting that the CYP2F enzymes possess distinctive active-site structures.[31]

Recombinant CYP2F1 in lymphoblastoid cell microsomes was capable of bioactivation of naphthalene to form the pneumotoxic naphthalene epoxide.[29] CYP2F1 also metabolised benzene, an environmental and occupational toxicant, with a catalytic efficiency (Vmax/Km) of 2.6.[32] Notably, attempts have been made to produce CYP2F1 using the baculoviral expression system; however, the recombinant protein failed to form a complex with CO, apparently because of inadequate haem incorporation.[33]

Numerous single-nucleotide polymorphisms and other mutations have been reported for the *CYP2F1* gene.[5,34,35] A total of 8 *CYP2F1* alleles have been identified: *1, *2A, *2B, *3, *4, *5A, *5B, and *6 (http://www.cypalleles.ki.se).[34] The *2 alleles contain many mutations in both coding and noncoding regions, including, between positions 14 and 15 bp, at the beginning of exon 2, a C-insertion, which is a frameshift mutation, and would result in premature termination of translation at 64 bp downstream. Thus, *2A and *2B are null alleles. The *3 (Asp218Asn; Gln266His; Pro490Leu), *4 (Ser38Pro; Asp218Asn), *5 (Leu391-Pro), and *6 (Arg98Pro) alleles have not been characterised. Another coding-region variation (Gln256His) was reported,[7] but an allele containing this variation was not detected in a Caucasian population studied.[34] Interestingly, frequency of the *2A allele was 25.6% in Caucasians,[34] and frequency of (presumably) the same allele was 25% in a Chinese population examined.[35] In the Caucasian population, the *3 (7.2%) and *5A (6.1%) alleles were also common.[34] Frequency of the *2 allele was found not to be associated lung cancer risk in French Caucasians,[34] or nasopharyngeal carcinoma risk in a Chinese population.[35]

Additional coding region non-synonymous mutations and many intronic or untranslated region single-nucleotide polymorphisms are found in an Internet database (http://mutdb.org). It should be noted that the numbering of nucleotide positions differ among published studies,[7,34,35] and in Internet database; these differences cause confusion when comparing the respective findings. The presence of the pseudogene *CYP2F1P* could potentially complicate *CYP2F1* genotype analysis. It should also be noted that the originally reported CYP2F1 cDNA sequence[1] (GenBank accession number J02906) contains errors; the correct cDNA sequence (GenBank accession number NM_0007744) was determined by re-sequencing of the original cDNA clone and comparisons to the *CYP2F1* genomic sequence.[4]

10.1.3 Mouse CYP2F2

Mouse CYP2F2, also encoding a protein of 491 amino acids, was first identified through cDNA cloning from a mouse lung cDNA library; it was named P-450Nah, for naphthalene hydroxylase.[2] The library was screened with an antibody that had been shown previously to inhibit mouse lung microsomal metabolism of naphthalene.[3] The mouse *Cyp2f2* gene is located on chromosome 7,[1] within the *Cyp2a*, *Cyp2b*, *Cyp2f*, *Cyp2g*, *Cyp2s* and *Cyp2t* gene cluster.[9,36]

CYP2F2 transcript was detected through cDNA hybridisation in mouse lung.[2,37,38] CYP2F2 mRNA was also detected by Northern blot analysis in the liver, at lower levels than in the lung, but not in the kidney,[2,38] spleen, testis[39] or intestine.[2]

Pulmonary CYP2F2 mRNA in male NIH Swiss mice was not induced by intraperitoneal treatment (once a day, for 3 days) with phenobarbital (80 mg/kg), pyrazole (150 mg/kg), pregnenolone 16α-carbonitrile (75 mg/kg) or 3-methylcholanthrene (80 mg/kg); in contrast, mouse lung CYP2A5 (P-450Coh) mRNA was induced by the pyrazole treatment in the same study.[2] Treatment of non-Swiss albino and Swiss CD-1 mice with pyridine (200 mg/kg, intraperitoneal, single dose), phenobarbital (80 mg/kg, intraperitoneal, once a day, for 4 days) or β-naphthoflavone (40 mg/kg, intraperitoneal, once a day, for 3 days), known inducers of CYP2E1, CYP2B, and CYP1A, respectively, had no effect on lung microsomal styrene epoxygenase activity,[40] which, as will be described below, is primarily a CYP2F2-catalysed reaction.[41]

CYP2F2 mRNA levels were suppressed in mouse liver by diethylhexylphthalate, a peroxisome proliferator, in a dose- and time-dependent fashion.[39] An apparently alternative CYP2F2 transcript, which was smaller (by ~1 kbp) than the transcript detected in the liver, was found in the kidney through Northern blot analysis. The level of this kidney CYP2F-related transcript was also reduced by diethylhexylphthalate treatment.[39]

CYP2F2-immunoreactive protein was readily detected by immunoblot analysis in mouse lung and nasal mucosa, but it was not detected in the kidney.[38] In the liver, several immunoreactive bands were detected with the use of the polyclonal anti-2F2 antibody;[3] the identities of these proteins remain to be determined.[38] The microsomal total P450 content was found to be much higher in the airways than in other lung compartments.[42] An immunoblot analysis of CYP2F protein expression levels in various airway regions indicated that the distal airways have the highest level of CYP2F protein expression.[38] Expression of CYP2F2 protein and activity was maintained, albeit at decreased levels, in cultured explant of micro-dissected mouse lung airways.[43]

An immunohistochemical analysis detected CYP2F2 in the bronchiolar Clara cells and in alveolar type II cells of adult, male A/J mice; the level of CYP2F2 protein was not altered by a 6-month exposure of the mice to cigarette sidestream smoke, at an exposure level of 4 mg/m^3 of total suspended particulate matter in the inhalation chamber, and with an exposure schedule of 6 h/day, 5 days/week.[44] The same exposure did lead to increased expression of CYP1A1 in lung capillary endothelial cells.[44]

The developmental expression of CYP2F2 was examined in mice of ages ranging from gestational day 16 to postnatal day 63, in association with a study of the ontogenesis of Clara cells.[45] Clara cells were found to be fully differentiated by postnatal day 14 in both proximal and distal airways, as indicated by the levels of expression of Clara cell secretory protein, cilia-associated tubulin, and cytochrome P450 reductase. However, CYP2F2 protein expression, which was detected earlier in proximal than in distal airways, did not attain adult levels until after postnatal day 28.[45] Clara cell CYP2F2 mRNA expression was decreased in wild-type and $CCSP^{-/-}$ mice, both on a 129 genetic background, following exposure of the mice to 1 part per million ozone.[46] Down-regulation of CYP2F2 expression was also found in mouse hepatocellular carcinomas, as compared to normal liver.[47] In addition, CYP2F2 expression in mouse liver was decreased by a combination of transplacental exposure to arsenic (85 ppm, from gestation day 8 to 18) and postnatal treatment with the tumour promoter 12-*O*-teradecanoyl phorbol-13-acetate (TPA).[48]

Consistent with the greater susceptibility of females than males to naphthalene-induced lung injury, and the gender difference in P450-dependent naphthalene epoxygenase activity in distal airways,[49] intrapulmonary conducting airway CYP2F2 mRNA levels were higher in females than in males, although a gender difference was not found for extrapulmonary airways.[50] In female extrapulmonary airways, but not intrapulmonary airways, CYP2F2 mRNA levels were significantly lower in dioestrus than in prooestrus or oestrus stages.[50] The influence of estrus cycle on airway CYP2F2 expression, which seemed to explain the known variability in CYP2F2 expression in randomly cycling female mice,[49] should be taken into consideration when CYP2F2 function is studied in females.

The known substrates for CYP2F2 are summarised in Table 10.2. Initial studies with CYP2F2 heterologously expressed in yeast indicated that CYP2F2 was active in the metabolism of naphthalene, producing primarily the

Table 10.2 Activities of heterologously expressed CYP2F2.[a]

Substrate (concn)	Product(s)	Rate (nmol/min/nmol P450)	Reference
Naphthalene	1R,2S-naphthalene oxide	104 (Vmax; Km = 3 μM)	51
p-Nitrophenol (N.I.[b])	4-nitrocatechol	2.5	51
Ethoxyresorufin (N.I.)	Resorufin	0.0025	51
Pentoxyresorufin (N.I.)	Resorufin	0.0021	51
1-nitronaphthalene	Glutathione conjugates	17.1 (Vmax; Km = 21.5 μM)	42
2-methylnaphthalene	Glutathione conjugates	67.6 (Vmax; Km = 3.7 μM)	42
Anthracene (0.1 mM)	Total metabolites	0.14	42
Benzo(a)pyrene (0.1 mM)	Total metabolites	0.04	42
Dichloroethylene	Glutathione conjugate	2.8 (Vmax; Km = 254 μM)	68
Trichloroethylene	Chloral hydrate	13 (Vmax; Km = 114 μM)	71

[a]Recombinant CYP2F2 was produced using a baculoviral system.
[b]N.I., not indicated.

1R,2S-epoxide, but it had no detectable activity in the 3-hydroxylation of benzo(a)pyrene or the *O*-deethylation of ethoxycoumarin.[2] Functional CYP2F2 enzyme was later produced heterologously, with the use of a baculoviral expression system.[51] The recombinant protein was active towards naphthalene and 1-nitronaphthalene (the latter also a lung toxicant) as well as model P450 substrates, including *p*-nitrophenol, and at very low rates, towards ethoxyresorufin and pentoxyresorufin.[51] CYP2F2-calalysed naphthalene metabolism was highly efficient (kcat/Km; 35 $\mu M^{-1} min^{-1}$), and it produced 1R,2S- and 1S,2R-naphthalene oxides at a ratio of 66:1. The addition of cytochrome b_5 (b_5) had a slightly stimulatory effect on the reaction.[51] Recombinant CYP2F2 also metabolised 1-nitronaphthalene and 2-methylnaphthalene (another lung toxicant) with relatively high catalytic efficiencies (0.8 and 18 $\mu M^{-1} min^{-1}$, respectively), and at Km values that were considered biologically relevant.[42] The activities of CYP2F2 towards two polyaromatic hydrocarbons, anthracene and benzo(a)pyrene, were low.[42]

CYP2F2 appears to play a major role in the metabolism and respiratory-tract cytotoxicity of styrene.[41] 5-Phenyl-1-pentyne (5-P1P), an inhibitor of CYP2F2[52] and CYP2F3,[53] inhibited lung microsomal conversion of styrene to R- and S-styrene oxide in non-Swiss albino and Swiss CD-1 mice, with styrene at 2 mM and 5-P1P at 5 μM; in liver microsomes, however, 5-P1P caused essentially no inhibition.[40] 5-P1P also inhibited the formation of R- and S-styrene oxide in isolated Clara cells, which had much higher activity than did isolated type II pneumocytes.[54] Furthermore, 5-P1P prevented both the occurrence of cytotoxicity in mouse terminal bronchioles and the degeneration of the nasal olfactory mucosa (OM) that result from inhalation exposure to styrene.[41,55]

In a study of styrene metabolism in wild-type and *Cyp2e1*-knockout mice, it was found that the loss of CYP2E1 expression in the null mice led to a decrease in the rates of styrene oxidation in lung, but not in liver microsomes, relative to the wild-type mice.[56] *In vitro* studies using chemical inhibitors concurred that both CYP2E1 and CYP2F1 could be important in lung microsomal metabolism of styrene in the wild-type mice.[56] However, the *Cyp2e1*-null and wild-type mice did not differ in their susceptibility to the pulmonary toxicity of either styrene or styrene oxide.[57] Notably, diethyldithiocarbamate, the CYP2E1 inhibitor used, was found to inhibit microsomal styrene oxidase activity in *Cyp2e1*-null mice,[56] a finding that documented the limited selectivity of this inhibitor, and underscores the need for more stringent tests for the specificity of 5-P1P, the frequently used CYP2F2 inhibitor, as well as other commonly used P450 inhibitors. Indeed, 5-phenyl-1-pentyne is also an inhibitor of CYP2E1.[53,55]

The ring hydroxylation product of styrene, 4-vinylphenol, which is more potent than styrene or styrene oxide as a mouse lung toxicant,[58] is also a substrate for CYP2F.[41] Although inhibitor studies suggested that both CYP2E1 and CYP2F2 are capable of metabolising 4-vinylphenol,[59] *Cyp2e1*-null mice did not show a marked decrease in the rate of 4-vinylphenol metabolism in either liver or lung microsomes, compared to wild-type mice.[60] Yet, hepatic and pulmonary microsomal metabolism of 4-vinylphenol (0.5 mM) was significantly inhibited by 5 μM 5-P1P, by more than 50%, consistent with involvement of CYP2F2.[61] Although the metabolite(s) of 4-vinylphenol remain to be identified,[59] *in vivo* studies using P450 inhibitors, including 5-P1P, demonstrated that 4-vinylphenol requires P450-catalysed metabolism to induce hepatotoxicity and pneumotoxicity.[62]

Pretreatment of *Cyp1a1*-null mice and *Cyp1a2*-null mice with 5-P1P prevented naphthalene-induced nasal olfactory toxicity and lung haemorrhage, a finding in support of a role of CYP2F2 in naphthalene-induced nasal and lung toxicity.[63] However, indirect evidence, based on the differential effects of several known CYP inducers and inhibitors on the extent of chemically-induced lung injury and on pulmonary microsomal xenobiotic-metabolising activity, suggested that CYP2F2 is not responsible for the bioactivation or toxicity of butylated hydroxytoluene (BHT) in mouse lung.[64]

CYP2F2 has also been implicated in metabolism of benzene, a human occupational and environmental toxicant, in mouse lung. Studies using liver and lung microsomes from wild-type and *Cyp2e1*-null mice indicated that while CYP2E1 is the main benzene hydroxylase in the liver, the same isoform only accounts for 45% of microsomal benzene hydroxylase activity in the lung at a benzene concentration of 24 μM.[65] Further *in vitro* studies using chemical inhibitors confirmed that both CYP2E1 and CYP2F2 are involved in lung microsomal benzene metabolism;[65] in a study of similar design, both enzymes were also found to be significantly involved in the further metabolism of phenol (at 20 μM), the benzene metabolite, in lung microsomes.[66] Enzyme kinetic analysis of microsomal benzene metabolism in the wild-type and *Cyp2e1*-null mice suggested that CYP2F2 (or additional lung P450 isoforms other than

CYP2E1) has a higher affinity for benzene, and a lower maximal rate of phenol production, than does CYP2E1.[67]

CYP2F2 is active in the metabolic activation of 1,1-dichloroethylene (DCE), which causes toxicity in bronchiolar Clara cells.[68] Recombinant CYP2F2 was active in the formation of the reactive DCE epoxide (Km = 254 μM; Vmax/Km = 0.011 $\mu M^{-1}\,min^{-1}$), albeit with lower affinity and catalytic efficiency than were found for recombinant rat CYP2E1 (Km = 74 μM, Vmax/Km = 0.016 $\mu M^{-1}\,min^{-1}$). The epoxide was detected as 2-S-glutathionyl acetate, a conjugate with glutathione.[68] The contribution of CYP2E1 and/or CYP2F2 to DCE metabolic activation in mouse lung was further supported by the finding that *in vivo* formation of 2-S-glutathionyl acetate in the lung could be blocked by pretreatment of mice with diallyl sulphone and 5-P1P, both of which inhibit recombinant CYP2E1 and CYP2F3.[53] Pretreatment of mice with 5-P1P also blocked DCE-induced bronchiolar toxicity.[53,69] However, lung microsomal preparations from *Cyp2e1*-null mice were only 40% less active in catalysing the formation of DCE epoxide than were microsomes from wild-type (of mixed C57BL and 129/Sv genetic background) mice, indicating that other CYP enzymes, including CYP2F2, must make substantial contributions.[68] Consistent with the *in vitro* activity data was the fact that *Cyp2e1*-null mice did not show protection against DCE (75 mg/kg) induced bronchiolar toxicity.[53,69]

CYP2F2 has been shown to play a role in the metabolic activation of trichloroethylene (TCE), which induces cytotoxicity in Clara cells and damage to the bronchiolar epithelium, at intraperitoneal doses of 500 to 1000 mg/kg.[70] Recombinant CYP2F2 was capable of metabolising TCE to chloral hydrate, although the catalytic efficiency of CYP2F2 (Vmax/Km; 0.11 $\mu M^{-1}\,min^{-1}$) was lower than that of rat CYP2E1 (Vmax/Km; 0.79 $\mu M^{-1}\,min^{-1}$).[71] The involvement of CYP2E1 and/or CYP2F2 in TCE-induced bronchiolar toxicity was further suggested by the finding that pretreatment of mice with diallyl sulphone afforded protection against TCE bronchiolar toxicity and TCE-induced protein adduct formation in the lung. The likely predominant role of CYP2F2, rather than CYP2E1, in lung microsomal metabolism of TCE was supported by the findings of higher rates of TCE metabolism in *Cyp2e1*-null mice than in wild-type (of mixed C57BL and 129/Sv genetic background) mice, and the lack of protection against TCE induced bronchiolar toxicity in the *Cyp2e1*-null mice.[70]

DCE treatment led to the loss of lung microsomal immunoreactive CYP2F2, as well as CYP2E1 proteins, in CD-1 mice.[53] Treatment of mice with TCE also led to the loss of lung microsomal immunoreactive CYP2F2 and CYP2E1 proteins,[70,71] presumably as a result of *in vivo* formation of dichloroacetyl lysine adducts with CYP2E1 and CYP2F2.[70]

10.1.4 Goat CYP2F3

CYP2F-related transcripts were detected through cDNA hybridisation in the lung, but not liver, of goat.[37,72] Recombinant CYP2F3 was produced in *E. coli*, with the use of a cDNA clone isolated from goat lung.[68,72] As it has been found for recombinant CYP2F1, CYP2F3 metabolised 3-methylindole in a reconstituted system, with specific formation of 3-methyleneindolenine, the dehydrogenation product (Vmax = 0.55 nmol/min/nmol P450; Km = 0.34 mM). The reaction was stimulated by the addition of b_5, and it was inhibited by the mechanism-based P450 inhibitor 1-aminobenzotriazole. However, CYP2F3 does not appear to be the major isoform in goat lung microsomes for this reaction, given that the addition of anti-CYP2F2 led to only a 20% inhibition of the rate of 3-methyleneindolenine production with the substrate at 0.5 mM; instead, CYP4B2 seemed to be the main contributor.

Recombinant CYP2F3 metabolised naphthalene, specifically to the 1*R*,2*S*-oxide, the putative toxic intermediate, but it was not active in the *O*-deethylation of 7-ethoxycoumarin.[72] CYP2F3 was active in the hydroxylation of *p*-nitrophenol and chloroxazone,[53] known substrates for CYP2E1. The activities of recombinant CYP2F3 in the hydroxylation of these two substrates were inhibited by diallyl sulphone and 5-P1P.[53] CYP2F3 was also active in the formation of 2-S-glutathionyl acetate from DCE (Vmax = 0.35/min; Km = 0.5 mM).[68]

10.1.5 Rat CYP2F4

CYP2F4 mRNA was found to be abundant in rat lung by Northern blot analysis using a CYP2F1[1] or CYP2F4 DNA probe.[38] CYP2F4 mRNA was also detected in the liver, at lower levels than in the lung, but it was not detected in the kidney.[38] The levels of CYP2F4 mRNA in rat lung and liver were, respectively, 4 and 8 times lower than CYP2F2 mRNA levels in mouse lung and liver. An anti-CYP2F2 antibody was found to have comparable affinities for recombinant CYP2F2 and CYP2F4, in a quantitative immunoblot analysis.[38] In the rat lung, CYP2F-immunoreactive protein was detected by means of the anti-CYP2F2 antibody in various subcompartments of the airway, once again at levels substantially lower than those in various airway subcompartments of the mouse lung; the distal airways had the highest levels of expression. However, the OM had much higher levels of CYP2F-immunoreactive protein than did any of the other respiratory tissues examined. Within the OM, CYP2F4 immunoreactivity was detected intracellularly in the supporting cells and Bowman's glands, but not on the airway surface.[73] CYP2F4 protein was not detected in the kidney; in the liver, several immunoreactive bands were detected with the use of the polyclonal anti-CYP2F2 antibody; the identity of these proteins remains to be determined.[38]

Recombinant CYP2F4 was generated using a baculoviral expression system with a cDNA isolated from rat lung.[33] CYP2F4 metabolised naphthalene to 1R, 2S-naphthalene oxide, and 1-nitronaphthalene to both the 5,6- and the 7,8-epoxides. The kinetic parameters for CYP2F4 catalysed epoxygenation of naphthalene (Vmax = 107 nmol/min/nmol P450, Km = 3 μM) and 1-nitronaphthalene (Vmax = 25 nmol/min/nmol P450, Km = 18 μM) were similar to those for mouse CYP2F2.[33]

CYP2F4 was active in the metabolic activation of TCE, producing the reactive metabolite chloral hydrate (Vmax = 17 nmol/min/nmol P450;

Km = 64 μM).[71] However, CYP2F4 was not as efficient as CYP2E1 in this reaction; the catalytic efficiency (Vmax/Km) for chloral hydrate production was 0.27 for CYP2F4 and 0.79 for CYP2E1.[71] CYP2F4 was also active in the formation of the reactive DCE epoxide from DCE (Vmax = 74 nmol/min/nmol P450; Km = 5.6 mM), although the efficiency was quite low.[68]

Based on the effects of 5-P1P as a CYP2F inhibitor, CYP2F4 was implicated in the metabolic activation of 4-vinylphenol, a metabolite of styrene, in rat liver and lung.[62] CYP2F-generated metabolites are believed to mediate respiratory tract toxicity of styrene in rats as well as in mice,[41] although metabolism of styrene or its metabolites by rat CYP2F4 has not been examined directly.

10.1.6 CYP2F in Other Species

CYP2F-related transcripts were detected through cDNA hybridisation in rabbit lung, but not liver.[37] An apparently functional *CYP2F* gene (*CYP2F5*), as well as a *CYP2F* pseudogene (*CYP2F5P*), was identified in gorilla.[5] In the rhesus macaque, a CYP2F-related protein was detected by anti-CYP2F2 polyclonal antibodies in microsomes from the ethmoturbinates, which contain OM, but not in microsomes from the maxillary turbinates or the lung.[38] Presence of a CYP2F-related protein in OM microsomes was supported by the mass-spectrometric detection of two unique CYP2F-derived tryptic peptides in a gel slice that corresponded to the immunoreactive band detected by the anti-CYP2F2 antibody on an immunoblot. Notably, the signal intensity of the CYP2F-related protein detected in monkey OM microsomes was 10- and 20-fold lower than the intensity in rat and mouse OM microsomes, respectively,[38] a finding suggestive of much lower levels of CYP2F gene expression in the primates than in rodents.

10.1.7 Perspective

Much remains to be learned regarding the regulation, function and genetic polymorphisms of the *CYP2F* genes. There is clear evidence that human CYP2F1 is a functional enzyme. However, studies designed to attain a better understanding of the quantitative differences in activities towards important xenobiotic compounds, such as naphthalene, between CYP2F1 and the CYP2F enzymes in experimental animals will be critical for risk assessment. The remarkable disparity in levels of expression of *CYP2F* genes among various species – a phenomenon that likely leads to species differences in susceptibility to the toxicity of environmental or occupational chemicals – provides myriad opportunities to identify genetic mechanisms that control *CYP2F* gene expression. In this regard, the unusual presence of an untranslated first exon in the *CYP2F* genes suggests the utility of regulatory mechanisms that differ from those employed by other *CYP2* genes. On the other hand, it remains to be determined whether expression of *CYP2F* is at least partly coordinated with that of other *CYP2* genes in the *CYP2A-2B-2F-2G-2S-2T* gene cluster; most of the latter are expressed in the respiratory tract.

Genetic polymorphisms in *CYP2F1* may underlie inter-individual differences in susceptibility to respiratory-tract toxicants. Therefore, studies designed to link genetic variations with altered CYP2F1 function or expression are essential. In this regard, the recently discovered high-frequency null allele (*CYP2F1*2*)[34,35] should be very useful. Nonetheless, functional characterisation of the other alleles is necessary. Such studies are challenging, given that a robust heterologous expression system, such as baculovirus-mediated expression in insect cells, has not been established for CYP2F1; this limitation clearly hampers our ability to perform biochemical studies of CYP2F1 and its variants. Furthermore, useful tools, such as antibodies that allow specific detection and quantification of CYP2F1 protein in human lung tissues, have yet to be developed or validated. Nonetheless, recent developments in quantitative RNA-PCR technology and allelic expression analysis,[74] combined with the availability of highly sensitive, high-resolution mass spectrometers for protein identification and quantification, should facilitate efforts to identify *CYP2F1* alleles having varied expression in human tissues.

Numerous published studies on potential functions of CYP2F enzymes in animal models have relied on P450 inhibitors that are not highly specific. A *Cyp2f2*-null mouse model will thus be important for confirmation of the *in vivo* function of this enzyme in the respiratory-tract toxicity of numerous xenobiotic compounds. Additionally, a combination *CYP2F1*-transgenic and *Cyp2f2*-knockout model, which represents a "CYP2F-humanised" mouse, will be valuable (i) for determinations of the *in vivo* function of CYP2F1, particularly in relation to human risk assessment; (ii) for identification of chemical inducers or pathologic conditions that alter CYP2F1 expression; and (iii) for exploring the molecular mechanisms that regulate *CYP2F1* gene expression.

10.2 CYP2G1

10.2.1 Introduction

The *CYP2G* gene subfamily was first identified in rabbits; the CYP2G1 protein was purified to homogeneity from OM microsomes.[75] The first CYP2G1 cDNA was cloned from rat OM.[76] The *CYP2G* gene has nine exons, a typical structure among members of the *CYP2* gene family. Purified or heterologously expressed CYP2G1 proteins have been obtained from rabbits, mice and cows, for functional analysis. A *Cyp2g1*-null mouse has also been generated.

CYP2G1 is unique in that its expression shows strict tissue specificity; such expression occurs only in the olfactory chemosensory organs, i.e., the OM and the vomeronasal organ (an organ involved in pheromone detection). The CYP2G subfamily is also rare among *CYP* genes in the respect that although the gene is clearly functional in rodents, rabbits and cows, and probably also in pigs and monkeys, the human orthologues have become pseudogenes. The best-known substrates for CYP2G1 are steroid hormones.

10.2.2 Rabbit CYP2G1

Rabbit CYP2G1 was identified as a unique P450 based on a number of biochemical parameters, including amino acid composition, absorption spectra, peptide maps and amino-terminal amino acid sequences.[75] The purified protein was named P-450NMb, for nasal microsomal form b. According to the yield of purified P-450NMb from the purification protocol,[75,77] and subsequent immunochemical analysis using an anti-NMb antibody,[78] CYP2G1 was found to be a major P450 isoform in rabbit nasal OM microsomes. However, CYP2G1 protein was not detected in microsomes from the nasal respiratory mucosa, or any other tissues examined, including brain, oesophageal mucosa, heart, intestinal mucosa, kidney, liver and lung. Rabbit CYP2G1 was predominantly in the high-spin state in the ferric form.[75]

The identity of P-450NMb was confirmed when the rabbit CYP2G1 (or P450IIG1) cDNA sequence was obtained through the screening of a nasal mucosal cDNA library with antibodies to P-450NMb.[79] The full-length CYP2G1 protein was predicted to contain 494 amino acids, and the predicted amino acid sequence essentially matched the partial amino acid sequences determined for purified P-450NMb at the amino terminus.[75] Southern blot analysis of genomic DNA revealed that there is only a single *CYP2G1* gene in the rabbit genome. Northern blot analysis confirmed tissue specificity of CYP2G1 expression, and, through the use of differing cDNA probes, evidence was obtained for the occurrence of two CYP2G1 transcripts (3.6- and 2.1-kb) in the OM, presumably resulting from either alternate splicing or the use of alternate polyadenylation sites.[79] CYP2G1 protein was detected in rabbit OM microsomes as early as 2 days before birth.[80] The relatively early onset of expression of CYP2G1, as well as several other P450 isoforms, in the OM led to speculations that these enzymes have important physiological or toxicological functions in the neonatal period.[80]

As shown in Table 10.3, rabbit CYP2G1 is active toward a number of substrates. Purified CYP2G1 was active in the oxidation of ethanol and butanol, the *N*-deethylation of N-nitrosodiethylamine (NDEA), the *O*-deethylation of phenacetin, the *N*-demethylation of hexamethylphosphoramide (HMPA), and the hydroxylation of testosterone at multiple positions.[75] Antibody inhibition studies indicated that an unidentified metabolite of testosterone was uniquely formed in OM microsomes, and that the reaction was catalysed by CYP2G1.[78] Purified CYP2G1 also metabolised arachidonic acid (AA) in a reconstituted system, producing several metabolites, including compounds corresponding in retention time to ω-2 and ω-3 hydroxyeicosatetraenoic acids (HETEs) and the four epoxyeicosatrienoic acids (EETs), primarily 14,15-EET.[81] The identity of these metabolites has yet to be confirmed, but none corresponded to 20-HETE, which accounted for 57% of the total metabolites detected in rabbit nasal microsomal reactions.[81] On the other hand, CYP2G1 did not show any remarkable activity towards prostaglandin E2, retinoic acid or retinal.[82,83]

Rabbit CYP2G1 was active in the α-hydroxylation of 4-(methylnitrosamino)-1-(3-pyridyl)-1-butanone (NNK), forming the keto aldehyde, but

Table 10.3 Activities of purified, reconstituted rabbit CYP2G1.

Substrate (concn)	*Metabolite(s)*	*Rate (nmol/min/ nmol P450)*	*Reference*
HMPA (1 mM)	Formaldehyde	5.3	75
Phenacetin (2 mM)	Acetaldehyde	3.1	75
NDEA (4 mM)	Acetaldehyde	3.8	75
Ethanol (80 mM)	Acetaldehyde	4.9	75
Butanol (30 mM)	Butyl aldehyde	4.3	184
S-nicotine (0.5 mM)	N′-oxide	0.62	85
	$\Delta^{1',5'}$-iminium ion	0.31	
Aflatoxin B_1 (100 μM)	N7-guanine DNA adduct	0.04 (0.63/15 min)	87
NNK	Keto aldehyde	0.7 (Vmax)	84
Arachidonic acid (7.2 μM)	18-HETE, 17-HETE, and the four EETs	Not indicated	81
Androstenedione (5 μM)	Multiple positions, not identified	1.84	89
		1.58 (with b_5)	
5α-dihydrotestosterone (5 μM)	Multiple positions, not identified	1.04	89
		2.74 (with b_5)	
Oestradiol (5 μM)	2- Hydroxyoestradiol	0.24	89
		0.44 (with b_5)	
	4-Hydroxyoestradiol	0.10	
		0.22 (with b_5)	
Testosterone (5 μM)	11α- 15α-, 16α-, and 16β-Hydroxytestosterone, and 15-keto-testosterone	0.84	89
		1.34 (with b_5)	
Testosterone (150 μM)	15α- Hydroxytestosterone	0.89	75,82
	16α- Hydroxytestosterone	0.48	
	15 Keto-testosterone	0.66	
	Unidentified	0.20	
Progesterone (5 μM)	16α-Hydroxyprogesterone	1.20	89
		1.20 (with b_5)	
	16-Keto-progesterone	0.26	
		0.46 (with b_5)	
Progesterone	16α-hydroxyprogesterone	1.8 (Vmax)	89
19-nortestosterone	Not indicated	Not indicated	89
17-hydroxyprogesterone	Not indicated	Not indicated	89
16α-hydroxyprogesterone	Not indicated	Not indicated	89

HMPA, Hexamethylphosphoramide; NDEA, nitrosodiethylamine; HETE, hydroxyeicosatetraenoic acid; EET, epoxyeicosatrienoic acid.

not the keto alcohol, in a reconstituted system. However, with a Km of 186 μM, and a Vmax of 0.7 nmol/min/nmol P450, it was clearly not an efficient enzyme for this reaction.[84] CYP2G1 also had a low activity toward S-nicotine, producing the N-oxide and the $\Delta^{1',5'}$-iminium ion (the precursor to cotinine), but not nornicotine, at a substrate concentration of 0.5 mM.[85] However, rabbit CYP2G1 was not active towards the herbicide 2,6-dichlorobenzonitrile (DCBN), a nasal mucosa-specific toxicant.[86]

Purified rabbit CYP2G1 was active in the formation of aflatoxin (AF) B_1-N7-guanine DNA adduct in a reconstituted system, with AFB_1 at 100 μM, but it was not active in the formation of other metabolites that were formed by rabbit nasal microsomes, including AFM_1 (9α-hydroxy-AFB_1), AFP1 (4-desmethyl-AFB_1), and six unidentified polar metabolites that were not formed by liver microsomes.[87] CYP2G1 was one of the most effective P450 enzymes in the deformylation of cyclohexane carboxaldehyde to form cyclohexene,[82] but it was not active in the deformylation of citronellal.[88]

Purified rabbit CYP2G1 appears to be most efficient in the metabolism of steroid compounds.[89] A number of steroid sex hormones, including androstenedione, oestradiol, progesterone, testosterone and 5α-dihydrotestosterone, were found to be CYP2G1 substrates. At a substrate concentration of 5 μM, these compounds were metabolised at rates of 0.34–1.84 nmol/min/nmol P450 in the absence of b_5, and 0.66–2.74 nmol/min/nmol P450 in the presence of b_5, in a reconstituted system. The effects of b_5 were substrate-dependent; whereas metabolism of 5α-dihydrotestosterone was stimulated, the metabolism of androstenedione was slightly inhibited, by b_5 addition. Furthermore, addition of b_5 altered product stoichiometry for testosterone and progesterone. Among the five steroids, androstenedione had the highest turnover number in the absence of b_5, while testosterone had the highest turnover number in the presence of b_5. CYP2G1 was also active towards 19-nortestosterone, 17-hydroxyprogesterone, and 16α-hydroxyprogesterone, findings further indicative of a broad substrate specificity towards steroid compounds.

CYP2G1 metabolised androstenedione, oestradiol, testosterone and 5α-dihydrotestosterone at multiple positions but, for androstenedione and 5α-dihydrotestosterone, the metabolites formed have not been identified. Oestradiol was converted to both 2-hydroxy and 4-hydroxyoestradiol, whereas testosterone was converted to 11α- 15α-, 16α- and 16β-hydroxytestosterone, and possibly also 15-keto-testosterone.[78,89] On the other hand, progesterone was metabolised only at the 16 position, initially forming 16α-hydroxyprogesterone; that was apparently further oxidised by CYP2G1 to 16-keto-progesterone, in a reaction that was stimulated by the addition of b_5.[89]

Progesterone metabolism by rabbit CYP2G1 was efficient. The Km and Vmax values for the formation of 16α-hydroxyprogesterone in a reconstituted system, determined in the absence of b_5, were 2.0 μM and 1.8 nmol/min/nmol P450, respectively, yielding a catalytic efficiency of ~1.[89] A similar Km value was found for CYP2G1-catalysed 16α-hydroxyprogesterone formation in rabbit OM microsomes. CYP2G1 was active in progesterone hydroxylation at a physiological concentration of 15 nM. CYP2G1 activity was inhibited completely by ketoconazole and partially by α-naphthoflavone; the latter might serve both as an inhibitor and, at higher concentrations, as an activator of this enzyme.[89]

Antibody inhibition data indicated that CYP2G1 was responsible for P450-catalysed progesterone hydroxylase activities in rabbit OM microsomes at substrate concentrations of 15 nM or 2.0 μM; in addition, CYP2G1-independent metabolites were observed at substrate concentrations higher than 10 μm.[89] CYP2G1 was apparently also a major contributor to rabbit OM microsomal metabolism of testosterone and oestradiol at substrate concentrations lower than 10 μM. There was no gender difference in the expression of CYP2G1, or in OM microsomal metabolism of testosterone, oestradiol, and progesterone.[78,89] Interestingly, CYP2G1-catalysed progesterone metabolism in OM microsomes (at a substrate concentration of 2.0 μM) was inhibited by the boar pheromones 5α-androst-16-en-3-one (as a competitive inhibitor) and 5α-androst-16-en-3α-ol, at concentrations as low as 2 μM, as well as by a number of odourant compounds at higher concentrations, including octanol, 3,7-dimethyl-1-octanol, anisole and benzyl benzoate.[89] These findings led to speculations that CYP2G1 functions to regulate tissue levels of systemically derived endogenous steroid compounds in the OM; such compounds would potentially be involved in modulating pheromone signal transduction and regeneration of the olfactory epithelium.[89] It remains to be determined whether the boar pheromones are substrates for CYP2G1.

10.2.3 Rat CYP2G1

The rat CYP2G1, originally named P-450olf1 or P-450 IIG1, was initially identified by cDNA cloning, using a rat CYP2B2 cDNA probe and a rat OM cDNA library.[76] The deduced amino acid sequence of rat CYP2G1 is 83% identical to that of rabbit CYP2G1.[79] RNA blot analysis detected CYP2G1 expression in rat olfactory neuroepithelium, but not in rat brain, heart, intestine, liver, lung or tongue.[76,90] Analysis of the *Rattus norvegicus* genome led to the conclusion that there is only a single *CYP2G* gene in the rat.[90] Expression of rat CYP2G1 mRNA was detected as early as embryonic day 20, contemporaneous with the emergence of Bowman's glands[91]; expression increased continuously postnatally in both males and females, reaching a first plateau at approximately 3 weeks after birth.[90] The tissue specificity[76] and the temporal pattern of postnatal increase in gene expression[90] led to the suggestion that rat CYP2G1 has a potential role in olfactory function.

The cellular distribution of CYP2G1 protein in rat nasal cavity was examined using an antibody to rabbit CYP2G1.[92] CYP2G1 immunoreactivity was detected in the OM, the septal organ of Masera (which contains bipolar olfactory receptor neurons), and the vomeronasal organ, but it was essentially not detected in the respiratory mucosa. In the OM, immunoreactivity was detected in the sustentacular cells in the neuroepithelium and in the acinar cells of the Bowman's glands in the submucosa. In the septal organ, CYP2G1 immunoreactivity was found to be at the highest level among all of the nasal regions examined, and it was detected in the sustentacular cells in the epithelium. In the vomeronasal organ, immunoreactivity was detected in the acinar cells of the vomeronasal glands. Interestingly, CYP2G1 immunoreactivity was also detected in the mucociliary complex overlying the chemosensory epithelia in the OM, septal organ and vomeronasal organ; this pattern suggests expression of the protein in the microvillus membrane of the epithelial cells or in extracellular mucus. OM-specific expression of rat CYP2G1 protein was also demonstrated by immunohistochemical studies using an anti-peptide antibody, which detected CYP2G1 immunoreactivity in the Bowman's glands of the OM, but not in the liver, lung, kidney or nasal respiratory mucosa.[93] These findings led to the proposal that CYP2G1 and other nasal P450 enzymes play roles in perireceptor processes associated with chemosensory transduction, including xenobiotic metabolism and odourant/pheromone clearance.[92,93]

Little is known about whether rat CYP2G1 expression can be induced. CYP2G1 protein levels, along with the levels of CYP2A proteins, were significantly decreased at 48 h following a single intraperitoneal injection of coumarin, a naturally occurring fragrant compound and a poison, at 50 mg/kg.[94] It was not apparent whether the reduction in CYP2G1 expression resulted from direct inactivation of CYP2G1 protein by a reactive coumarin metabolite, or whether it was a consequence of the coumarin-induced tissue-selective cytotoxicity. CYP2G1 expression was abolished following exposure of rats to methyl bromide gas, under conditions that destroyed nearly all of the olfactory neurons and sustentacular cells.[95] Restoration of CYP2G1 expression to near-normal level occurred by 8 weeks after toxicant exposure, but only in epithelial regions that showed successful regeneration of the olfactory receptor neurons.

To date, no rat CYP2G1 protein, either purified or heterologously expressed, has been characterised. Little is known about which compounds are metabolised by rat CYP2G1, although it could be inferred from the strong sequence similarity between rat and mouse CYP2G1 that the two proteins have similar substrates.

10.2.4 Mouse CYP2G1

Mouse CYP2G1-immunoreactive protein was detected in the OM by immunoblot analysis using an anti-rabbit CYP2G1 antibody.[96] Within the OM, immunohistochemical analysis detected CYP2G1 immunoreactivity only in the supporting cells and Bowman's glands. In unilateral naris closure or olfactory bulbectomy mouse models, the expression of CYP2G1 immunoreactive protein in the non-neuronal cells was reduced in epithelial regions where olfactory receptor neurons had been lost.[96] Decreases in CYP2G1-immunoreactive protein were also observed in the OM in mice experiencing necrosis of the olfactory epithelium following exposure to DCBN[97] or coumarin.[94] The preferential expression of mouse CYP2G1 in the non-neuronal cells of the OM was confirmed in a more recent study, in which global gene expression profiles were compared between isolated olfactory sensory neurons and neighbouring cells.[98]

A mouse CYP2G1 cDNA clone was isolated from the OM of C57BL/6 mice; the deduced amino acid sequence of the mouse CYP2G1 protein is 82% identical to that of rabbit CYP2G1, and 95% identical to that of rat CYP2G1.[99] Southern blot analysis and subsequent genomic cloning studies[36,100,101] indicated that there is only one copy of the *Cyp2g1* gene in the mouse genome. The *Cyp2g1* gene is located in the Cyp2a, Cyp2b, Cyp2f, Cyp2g, Cyp2s, Cyp2t gene cluster on mouse chromosome 7.[36] RNA blot analysis detected CYP2G1 mRNA in OM, but not in brain, heart, intestine, kidney, liver, lung, spleen, testis or thymus.[99] A subsequent study detected CYP2G1 by RNA-PCR in dissected vomeronasal glands; immunoblot analysis showed that the combined levels of CYP2A5 and CYP2G1 immunoreactive proteins (the two proteins were not resolved) appeared to be at least 10 times lower in vomeronasal organ microsomes than in OM microsomes.[102] However, CYP2G1 was not detected by RNA-PCR in the lateral nasal gland (LNG), also known as Steno's gland.[101]

The mouse *Cyp2g1* gene is about 11 kbp long. A comparison of coding-region nucleotide sequences between C57BL/6 and 129/Sv strains revealed six variations, three of which were expected to result in changes in amino-acid sequence (Asp192Asn; Arg203Lys; and Val370Ile). It remains to be determined whether these sequence changes alter CYP2G1 function.[100]

The *Cyp2g1* gene appears to have a TATA-less promoter. Although consensus sequences for numerous known transcription factors are present in the 5′-regulatory region of the *Cyp2g1* gene, none has been validated experimentally.[100] However, a 3.6-kbp *Cyp2g1* 5′-flanking sequence was found to be sufficient for OM-specific expression of a β-galactosidase reporter gene in transgenic mice.[100] Nonetheless, this promoter sequence was not sufficient for *Cyp2g1*-like zonal and cell type-specific expression, a finding that suggests the involvement of additional, more distal regulatory sequences in the regulation of the *Cyp2g1* expression. Expression of the transgene, as well as expression of the endogenous mouse *Cyp2g1* gene, was detected at as early as 2 days after birth.[100]

Purified, heterologously-expressed mouse CYP2G1 differs from the purified rabbit CYP2G1[75] in being primarily in the low-spin state in the ferric form.[103] Immunoblot analysis using an antibody that recognises both CYP2A5 and CYP2G1 indicated that the two proteins combined can account for >30% of total microsomal P450 proteins in mouse OM.[103]

Recombinant mouse CYP2G1 protein, obtained through heterologous expression using a baculoviral vector in insect cells,[99] was active towards a number of substrates (as summarised in Table 10.4). In a reconstituted system, mouse CYP2G1 was active towards testosterone and progesterone, with apparent Km values in the micromolar range. CYP2G1 produced 15α-, 15β-, and 2β-hydroxytestosterone, as well as an unidentified minor product, from testosterone, with 15α-hydroxytestosterone being the most abundant. CYP2G1 produced two metabolites from progesterone; the identity of these metabolites has yet to be determined, but they did not co-migrate with a number of available standards upon HPLC analysis, including 2α-, 6α-, 6β-, 11α-, 11β-, 12α-, 16α-, 17α-, 18-, 19-, 20α-, 20β-, and 21-hydroxyprogesterone; 6-keto, 11-keto-, and 19-keto-progesterone; and 6β,21-dihydroxyprogesterone.[99] The CYP2G1-formed testosterone and progesterone metabolites were apparently the same as those formed in mouse OM microsomes, a finding that suggested a prominent role for CYP2G1 in mouse OM microsomal metabolism of these two sex steroid hormones. In this regard, OM microsomes were much more active than hepatic microsomes in catalysing these reactions.[99] Recombinant

Table 10.4 Activities of heterologously expressed mouse CYP2G1.

Substrate (concn)	*Metabolite(s)*	*Rate (nmol/min/nmol P450)*	*Reference*
Acetaminophen (paracetamol) (0.5 mM)	Protein adduct (with added BSA)	0.06	103
	3-OH-acetaminophen	1.50	
	GS-acetaminophen	10.8	
Acetaminophen	3-OH-acetaminophen	2.0 (Vmax; Km = 410 μM)	103
	GS-acetaminophen	16 (Vmax; Km = 610 μM)	
DCBN (3 μM)	Protein adduct (with added BSA)	0.008	103
		0.020 (with b_5)	
DCBN	DCBN-glutathione adduct	0.2 (Vmax; Km = 3 μM)	103
4-Nitrophenol	4-Nitrocatechol	10.5	103
Coumarin	7-Hydroxycoumarin	3.3 (Vmax; Km = 25 μM)	105
	o-HPA	4.2 (Vmax; Km = 7 μM)	
Testosterone	15α-, 15β-, 2β-Hydroxy-testosterone, and one unidentified	1.7 (Vmax; Km = 5.1 μM) (total metabolites)	99
Progesterone	Not identified (X1 and X2)	0.8 (Vmax; Km = 7.6 μM) (total metabolites)	99
Oestradiol (10 μM)	2-Hydroxyoestradiol	Minor product	102
	4- Hydroxyoestradiol	Minor product	
	Unidentified	Major products	

BSA, bovine serum albumin; DCBN, 2,6-dichlorobenzonitrile; *o*-HPA, *o*-hydroxyphenyl-acetaldehyde.

CYP2G1 also metabolised oestradiol,[102] producing 2- and 4-hydroxyoestradiol as minor products, as well as two major, unidentified metabolites having retention times shorter than those of the catechol oestrogens, when oestradiol was at 10 μM.

Purified, heterologously-expressed mouse CYP2G1 was active in the metabolic activation of DCBN and acetaminophen (paracetamol), both known to cause OM toxicity in mice.[103,104] The activity of mouse CYP2G1 towards DCBN, but not acetaminophen, was increased by the addition of b_5.[103] CYP2G1 activity toward DCBN (at 3 μM) was inhibited by 8-methoxsalen, 5-methoxsalen and metyrapone; 50% inhibition was achieved by these inhibitors at 2–10 μM, with 8-methoxsalen being the most effective. Results from antibody inhibition studies and studies with chemical inhibitors indicated that CYP2G1 and/or CYP2A5 plays a major role in metabolic activation of DCBN and acetaminophen in mouse OM microsomes, although distinction between the relative contributions of the two P450 enzymes was not possible. In this regard, CYP2A5 and CYP2G1 were indistinguishable by their responses towards a number of chemical inhibitors, including 8-methoxsalen, 5-methoxsalen, metyrapone, flavone, α-naphthoflavone, 4-methylpyrazole, 1-aminobenzotriazole and coumarin,[103] as well as diallyl sulphide,[104] findings suggestive of similar active-site structures for these two enzymes.

CYP2G1 metabolised coumarin primarily to *o*-hydroxyphenyl-acetaldehyde (*o*-HPA), and also to five other minor metabolites: 3- and 7-hydroxycoumarin plus three unidentified metabolites.[105] The same metabolites were also detected in incubations with OM microsomes, with *o*-HPA being the most abundant. Antibody inhibition studies indicated that CYP2G1 made a substantial contribution to OM microsomal *o*-HPA formation.

A *Cyp2g1*-null mouse strain was recently reported.[101] Homozygous *Cyp2g1*-null mice did not have any reproductive or developmental phenotype or any histological abnormalities in the OM or other organs examined; they also seemed to have normal olfactory ability when tested in a food-finding behaviour assay. There were no compensatory increases in the expression of CYP1A2, 2B, 2C, 2E1, 2J or 3A, NADPH-cytochrome P450 reductase, or microsomal epoxide hydrolase, in the OM; however, a substantial reduction in expression of the *Cyp2a5* gene, which is located downstream of the *Cyp2g1* gene in the mouse genome,[36] was found in the liver, the LNG, and the kidney, although not the OM, of adult *Cyp2g1*-null mice.[101] That effects of the disruption of *Cyp2g1* gene on "neighbouring" genes did not extend to other, more distally located *Cyp* genes in the *Cyp2* gene cluster on chromosome 7 was evident from the fact that the hepatic level of CYP2B proteins, and that of CYP2F2 mRNA were not different between the *Cyp2g1*-null and wild-type strains.

Relative to OM microsomes from wild-type mice, OM microsomes from *Cyp2g1*-null mice had ~60% lower rates of testosterone 15α-hydroxylation and of progesterone metabolism, and >70% lower rate of metabolic activation of coumarin; a corresponding decrease in the rate of 7-hydroxylation of coumarin was not seen.[101] These findings were consistent with results of previous *in vitro* studies that used recombinant CYP2G1. However, OM microsomes from *Cyp2g1*-null mice showed only marginal reductions in the rate of DCBN protein adduct formation or the rate of acetaminophen-glutathione adduct formation, compared to rates in wild-type mice. The minimal contribution of CYP2G1 to acetaminophen metabolic activation in the OM was confirmed by *in vivo* toxicity studies, in which similar extents of acetaminophen-induced OM toxicity were found in the *Cyp2g1*-null mice and in the two parental wild-type strains. Nevertheless, the LNG, which is also a major target organ, and in which the rate of acetaminophen metabolism was not altered even though CYP2A5 levels were drastically decreased in association with the *Cyp2g1* disruption, was protected from acetaminophen toxicity in the *Cyp2g1*-null mice. This unexpected finding led to the proposal that either CYP2G1, located in OM, and acting through a paracrine pathway, or CYP2A5, located in the LNG, and acting locally, can regulate the expression of genes important for mechanisms of chemical toxicity in the LNG via endogenous CYP2A5/CYP2G1 substrates and metabolites, including steroid hormones.[101] LNG, which surrounds the maxillary sinus, is the major site for the synthesis and secretion of odourant binding proteins,[106] and it also secretes large amounts of immunoglobulin A.[107] It is a known target for the toxic effects of a number of inhaled and systemically administered chemicals.[101]

10.2.5 Human CYP2G

Attempts were made to detect CYP2G-related proteins in human OM using the anti-rabbit CYP2G1 antibody; however, no specific CYP2G immunoreactivity was detected in autopsy nasal tissues, in which CYP2A immunoreactivity was abundantly detected.[108] An analysis of human genomic DNA provided evidence for the presence of at least two *CYP2G*-like genes in the human genome, one or both of which were transcribed in human nasal mucosa.[109] Subsequent studies indicated that there are two copies of the *CYP2G* gene in the human genome, but both have become pseudogenes.[110] These two pseudogenes were originally named *CYP2GP1* and *CYP2GP2*,[110] but have since been renamed *CYP2G1P* and *CYP2G2P*[9]. *CYP2G1P* is located between *CYP2A7* and *CYP2A18PC*, whereas *CYP2G2P* is located between *CYP2A13* and *CYP2A18PN*, in the *CYP2* gene cluster on human chromosome 19.[9,110,111] *CYP2GP1* has a single nucleotide deletion in exon 2 and a 2.4-kb deletion between exons 3 and 7, whereas *CYP2G2P* has two nonsense mutations, one each in exons 1 and 3. Studies of the genetic polymorphisms of the *CYP2G* genes in more than 200 individuals indicated that a functional *CYP2G* allele is rare, if it ever occurs, in humans.[110]

10.2.6 CYP2G in Other Species

CYP2G1-related mRNA was detected in bovine olfactory epithelium, but not in nasal respiratory epithelium.[76] A bovine P450 enzyme (P-450bov3), presumably corresponding to rabbit CYP2G1 based on amino-terminal sequence and immunoreactivity with an anti-rabbit CYP2G1 antibody, was partially purified from OM microsomes of cattle.[112] P-450bov3, which is in the high-spin state in the ferric form, appeared to account for approximately 10% of total P450 in OM microsomes. *In vitro* metabolic studies on purified P-450bov3, in a reconstituted system, indicated that P-450bov3 was similar to rabbit CYP2G1 in substrate specificity, being active in the hydroxylation of testosterone and progesterone, as well as in the metabolism of a number of xenobiotic compounds. The profiles of major testosterone and progesterone metabolites generated by P-450bov3 were identical to the profiles produced by rabbit CYP2G1, findings suggestive of conservation of the active-site structures between the two orthologous enzymes. The rates of testosterone and progesterone metabolism by partially purified bovine CYP2G1 are shown in Table 10.5. The xenobiotic substrates, including aniline, 7-ethoxycoumarin, coumarin, benzphetamine, 7-ethoxyresorufin, HMPA and NDEA, were apparently metabolised at low

Table 10.5 Testosterone and progesterone metabolism by partially purified bovine CYP2G1.[112]

Substrate (concn)	*Products*	*Rate (nmol/min/ nmol P450)*
Testosterone (500 μM)	15α-Hydroxytestosterone	0.48
	Unknown (15-keto-testosterone?)	0.31
	16α-Hydroxytestosterone	0.16
	17-Hydroxytestosterone	0.12
	6β-Hydroxytestosterone	<0.01
Progesterone (50 μM)	16α-Hydroxyprogesterone	2.9
	Unknown (16-keto-progesterone?)	0.47
	17α-Hydroxyprogesterone	0.10
	21-Hydroxyprogesterone	0.04
	11α-Hydroxyprogesterone	<0.01

rates; however, since substrate concentrations were not indicated, it is difficult to compare the reported rates with those for CYP2G1 from rabbits or mice.[112]

CYP2G1 immunoreactive protein was also detected by immunoblot analysis in the pig, in OM microsomes, but not in nasal respiratory mucosa or liver microsomes.[113] Although CYP2G1 expression has not been reported in monkeys, the rhesus monkey seems to have one functional *CYP2G1* gene and one non-functional *CYP2G* gene (corresponding to human *CYP2G2P*), as indicated by sequence analysis for regions spanning exon 1 to exon 2 of the monkey *CYP2G* genes.[110]

10.2.7 Perspective

We still know very little about the function and regulation of CYP2G1. The strict tissue-specific expression of CYP2G1 in the OM is unusual among mammalian microsomal P450 genes, and suggests the involvement of tissue-enriched transcription factors.[114] Knowledge of the molecular mechanisms that control the robust tissue-specific and cell type-specific expression of this gene in the OM will likely shed light on the regulation of numerous other genes that render the OM such a specialised sensory organ, an organ that is critical to survival in mammals. The substrates identified thus far suggest that CYP2G1 functions to regulate cellular and tissue-level homeostasis of endogenous steroid hormones in response to varied tissue burdens of xenobiotic compounds; these foreign compounds either serve as competitive (direct) CYP2G1 inhibitors, or else indirectly down-regulate CYP2G1 expression through cytotoxic mechanisms. The availability of the *Cyp2g1*-null mouse model will facilitate studies designed to test these hypotheses. While the *Cyp2g1*-null mice appear to be otherwise normal, the physiological role of steroid metabolism in the OM remains to be determined. Additional substrates also need to be identified. Finally, since humans do not have a functional CYP2G1, compounds that elicit nasal toxicity in animal models via CYP2G1-catalysed metabolic activation are unlikely to cause the same toxicity in humans. Therefore, the *Cyp2g1*-null mouse, which, as noted earlier, represents a "CYP2G-humanised" mouse model,[115] should be valuable for assessing potential human risk for drugs or environmental chemicals that are known to cause nasal toxicity in animal studies.

10.3 CYP2J

10.3.1 Introduction

The CYP2J subfamily in humans, and probably also in monkeys, contains only a single gene (*CYP2J2*), but it contains at least two genes in rabbits (*CYP2J1* and P450ib2), and multiple genes in rats (*CYP2J3*, *CYP2J4*, *CYP2J10*, *CYP2J13*, *CYP2J16*) and mice (*Cyp2j5*, *Cyp2j6*, *Cyp2j7*, *Cyp2j8*, *Cyp2j9*, *Cyp2j11*, *Cyp2j12*, and *Cyp2j13*). (For a more complete list of available CYP2J sequences, refer to http://drnelson.utmem.edu/CytochromeP450.html.) Nomenclature is largely based on the order of discovery of the genes; however, some of the names, such as monkey *CYP2J2* and rat *CYP2J13*, were given based on presumed orthology to the corresponding human and mouse gene, respectively. Rabbit P450ib2 has been sequenced only for its first 20 amino acids.[116] In general, it is difficult to identify orthologues among the *CYP2J* genes in differing species. A phylogenetic tree, showing the evolutionary distances, based on amino acid sequence identity, among characterised CYP2J proteins is shown in Figure 10.1.

Heterologously expressed or purified proteins have been obtained for CYP2J1, CYP2J2, CYP2J3, CYP2J4, CYP2J5, CYP2J6, CYP2J8 and CYP2J9, for functional analysis. A *Cyp2j*-knockout mouse has not been generated, but a *CYP2J2*-transgenic mouse has been reported. *CYP2J* genes are expressed in multiple tissues. Unlike CYP2F or CYP2G1, one or more CYP2Js have been

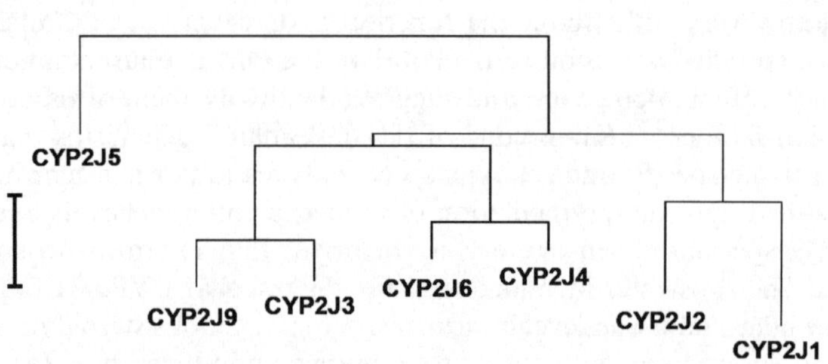

Figure 10.1 Evolutionary distances among characterized CYP2J proteins. The GenBank accession numbers for the sequences analysed are: CYP2J1: P52786; CYP2J2: NP_000766; CYP2J3: P51590; CYP2J4: NP_075414; CYP2J5: NP_034137; CYP2J6: NP_034138; CYP2J9: NP_083255. The phylogram was generated using the GrowTree program, with use of the Kimura-distance correction and the neighbour-joining method, in the Accelrys GCG SeqWeb software (San Diego, CA, USA). The vertical scale bar represents 10 substitutions per 100 residues.

found to be inducible by treatment with xenobiotic compounds, such as pyrazole, in rats and mice. The best-known substrates for CYP2J enzymes include AA and other fatty acids, although at least some of the CYP2Js are also involved in drug metabolism. The CYP2J subfamily and its biological function in relation to AA metabolism was reviewed previously;[117] remarkable progress has been made since on the function, regulation and clinical implications of the *CYP2J* genes.

10.3.2 Rabbit CYP2J

CYP2J1 was initially purified from rabbit small intestine, and was named P-450ib.[118] A full-length cDNA clone for CYP2J1 was later isolated from a cDNA library of rabbit small intestinal mucosa.[119] CYP2J1 encodes a protein of 501 amino acids. A second CYP2J isoform, named P450ib2, was later isolated from rabbit small intestine.[116] This protein was identified as a putative member of the CYP2J subfamily, because the first 20 amino acids of P450ib2 were highly similar to those of CYP2J1 (also named P450ib1), and because P450ib2 reacted with antibodies to CYP2J1, while antibodies to P450ib2 also recognised CYP2J1.

CYP2J1 was found, through RNA blot analysis, to be expressed preferentially in the small intestine. In immunoblot analysis, antibodies to either P450ib1 or P450ib2 produced positive signals in microsomes from small intestine, but not in microsomes from other tissues examined, including liver, kidney, lung, colon and stomach.[116] In immunohistochemical analysis, strong CYP2J immunoreactivity was detected in epithelial cells in various intestinal segments, including duodenum, jejunum and ileum, but not in liver, kidney, lung, stomach or colon. CYP2J-immunoreactive proteins were also detected in rabbit aortic microsomes[120] and in microsomes of pulmonary artery and peripheral lung.[121]

The two rabbit CYP2J enzymes seem to have similar substrate specificity; both were active in the *N*-demethylation of benzphetamine and aminopyrine,[116,118,119] while neither was active towards 7-ethoxycoumarin, 7-ethoxyresorufin, testosterone, laurate, or prostaglandin A1.[116] The activity of rabbit CYP2J towards endogenous compounds has not been directly demonstrated with the use of purified or expressed proteins. However, indirect evidence suggested that rabbit aortic microsomal CYP2J enzymes were involved in the epoxygenation of 15-hydroperoxyeicosatetraenoic acid, leading to the formation of vasodilating metabolites.[120]

10.3.3 Rat CYP2J

Although five genes (*CYP2J3*, *CYP2J4*, *CYP2J10*, *CYP2J13*, *CYP2J16*) and eight pseudogenes (*CYP2J3P1*, *CYP2J3P2*, *CYP2J4-de6b*, *CYP2J5P*, *CYP2J16-de2b5b9b*, *CYP2J16-de5c6c9c*, *CYP2J17P* and *CYP2J18P*) have been identified in the rat *CYP2J* subfamily (http://drnelson.utmem.edu/CytochromeP450.html), only two, *CYP2J3* and *CYP2J4*, have been characterised to date. CYP2J3 cDNA was first identified in rat liver and heart RNA; it encodes a protein of 502 amino acids.[122] Notably, a P450 protein purified earlier from rat liver, named P450 3b,[123] is believed to be the same as CYP2J3.[124] CYP2J4 cDNA was first identified in RNA from rat small intestine; it encodes a protein of 501 amino acids.[125]

Northern blot analysis indicated that CYP2J3 mRNA was abundant in rat liver, but the transcript was detected only at trace levels in other tissues examined, including heart, lung, kidney, stomach and small intestine;[122] this finding was later confirmed by quantitative RNA-PCR analysis of relative CYP2J3 levels in liver, heart and kidney.[126] The levels of CYP2J3 mRNA did not differ substantially between male and female rat liver.[126]

A CYP2J-related protein was detected using a polyclonal antibody to CYP2J2 in a number of rat tissues, including heart, liver,[122] lung,[127] pancreas,[128] oesophagus, stomach, small intestine and colon.[129] Expression of CYP2J protein does not seem to have strict cell-type specificity. In the lung,[127] CYP2J proteins were detected in ciliated and nonciliated airway epithelial cells, vascular smooth muscle, vascular endothelium, alveolar macrophages and, to a lesser extent, alveolar epithelial cells. In heart, CYP2J immunoreactivity was detected mainly in cardiomyocytes.[122] In pancreas,[128] CYP2J protein was detected primarily in multiple cell types in the islets of Langerhans, with the highest levels found in the glucagon-producing α-cells. In the intestine, CYP2J proteins were detected in intestinal epithelial cells, autonomic ganglion cells, smooth muscle cells and vascular endothelial cells.[129] The identities of the CYP2J2-immunoreactive proteins remain to be determined.

CYP2J4 was detected by RNA blot and immunoblot analysis in liver, intestine, OM, kidney, heart and lung.[125] In the small intestine, CYP2J4 was most abundant in the distal duodenum and jejunum, with the levels of CYP2J4 protein decreasing towards the ileum; within the intestinal villi, CYP2J4 levels were higher in the villus cells than in crypt cells.[130] The level of CYP2J4 mRNA was much higher than that of CYP2J3 mRNA, in the small intestine.[130]

The first evidence for significant induction of a CYP2J protein by xenobiotic treatment was reported for CYP2J4[131]; pyrazole treatment (either intraperitoneal or oral) increased CYP2J4 protein levels in enterocyte microsomes by 3- to 4-fold, and it increased CYP2J4 mRNA level in small intestinal epithelial cells by 2.5-fold. Pyrazole also induced CYP2J4 protein in the liver and OM. In the same study, enterocyte microsomal CYP2J4 protein levels were found to be unchanged by treatment with β-naphthoflavone, diminished by dexamethasone, and slightly increased by phenobarbital.[131] However, hepatic CYP2J3 protein was not induced by acetone, β-naphthoflavone, clofibrate or phenobarbital.[122]

Expression of CYP2J proteins is also regulated by a variety of pathophysiological conditions. In contrast to the induction of CYP2J4 by pyrazole treatment, CYP2J4 protein was detected at reduced levels in lung microsomes from rats with acute *Pseudomonas* pneumonia, at 44 h after exposure of rat lungs to *Pseudomonas* bacteria.[132] The pneumonia-induced reduction in lung microsomal CYP2J4 protein levels was partially prevented by treatment with an inhibitor of the inducible nitric oxide synthase at 40 h after exposure to the pathogen,[133] implicating the involvement of nitric oxide in the mechanisms of infection-related CYP2J4 suppression. In another study, the CYP2J4 mRNA level was found to be significantly decreased in the lungs at 20 h after induction of microbial sepsis by caecal ligation and puncture in adult Sprague-Dawley rats.[134] Fasting decreased the levels of CYP2J proteins in the intestine,[135] but not in the liver.[136]

In the kidney cortex and outer medulla of spontaneously hypertensive rats (SHRs), expression of a CYP2J2-immunoreactive protein was increased, relative to that in normotensive Wistar-Kyoto (WKY) rats; however, this overexpressed CYP2J-like protein appeared to differ from CYP2J3 and CYP2J4 in immunoreactivity and electrophoretic migration rates.[137] It remains to be determined whether this CYP2J2-immunoreactive protein corresponds to one of the uncharacterised rat CYP2J isoforms. The levels of CYP2J2-immunoreactive protein were also found to be significantly decreased in renal microvessels after a combined treatment of rats with angiotensin II and a high-salt diet.[138] In contrast, renal microsomal CYP2J protein (detected by anti-CYP2J2) levels were higher in pregnant than in non-pregnant rats, a finding further supported by the results of immunohistochemical analysis of CYP2J2-immunoreactivity in renal tubules.[139] Furthermore, the increase in renal microsomal CYP2J expression in pregnant rats was abolished by treatment of the animals with 6-(2-propargyloxyphenyl)hexanoic acid, an inhibitor of AA epoxygenases.[139] In male SHRs, levels of CYP2J2-immunoreactive protein in renal cortex microsomes, but not levels of CYP2J3 or CYP2J4 mRNA, were higher in rats fed a diet containing borage oil than in rats fed a diet containing sesame oil.[140] In obese Zucker rats, levels of CYP2J2-immunoreactive protein in the mesenteric artery were $\sim$40% lower than the levels in lean Zucker rats.[141] In male rats, hepatic CYP2J3 mRNA levels were not altered by hypophysectomy.[142] In the same study, CYP2J3 mRNA levels were found to be increased by $\sim$40% in isolated rat hepatocytes after 24 h in culture.[142]

Most functional studies for CYP2J3 and CYP2J4 have focussed on endogenous compounds (Table 10.6). Both CYP2J3[122] and CYP2J4[125] were expressed in insect cells by means of recombinant baculovirus. CYP2J3 was purified from rat liver in another study, and it was also heterologously expressed in *E. coli.*[126] CYP2J3 was active in metabolising AA to 8,9-, 11,12-, and 14,15-EETs, and, at much reduced abundance, 5,6-EET; 19-HETE was also a metabolite.[122] CYP2J3 also metabolises linoleic acid to form primarily linoleic acid monoepoxides.[143] Purified CYP2J3, either from rat liver or from a bacterial expression system, was highly active in the 25-hydroxylation of vitamin (Vit) D_3, the first step of Vit D activation,[144] and in the 25-hydroxylation of 1α-hydroxyvitamin (OH-Vit) D_3.[126] In contrast, the purified enzyme preparations were not active in 2α- or 16α-hydroxylation of testosterone. CYP2J3 can also 25-hydroxylate Vit D_2, although it catalyses this reaction less efficiently than it does the 25-hydroxylation of Vit D_3.[145]

Similar to CYP2J3, CYP2J4 metabolises AA, producing 19-HETE, as well as two EETs.[125] Interestingly, CYP2J4 may play a major role in the metabolism of retinal to retinoic acid in rat small intestine.[130,131] In a reconstituted system,

Table 10.6 Activities of CYP2J3, CYP2J4, CYP2J5 and CYP2J9.

Substrate (concn)	*Metabolite(s)*	*Rate (nmol/min/ nmol P450)*	*Reference (expression system)*
CYP2J3			
Arachidonic acid (50–75 μM)	19-HETE, 8,9-, 11,12-, and 14,15-EETs	0.2	122 (baculoviral)
Linoleic acid (50–100 μM)	epoxyoctadecenoic acid and hydroxyoctadecadienoic acid	0.16[a]	143 (baculoviral)
Vit D_3	25-OH-Vit D_3	3.3 (Vmax; Km = 0.79 μM)	126 (*E. coli*)
1α-OH-vit D_3	1,25-$(OH)_2$ Vit D_3	22 (Vmax; km = 0.60 μM)	126 (*E. coli*)
Vit D_3 (50 μM)	25-OH-Vit D_3	1.4	145 (*E. coli*)
Vit D_2 (50 μM)	25-OH-Vit D_2	0.86	145 (*E. coli*)
1α-OH-vit D_3 (10 μM)	1,25-$(OH)_2$ Vit D_3	22	145 (*E. coli*)
Ebastine (10 μM)	Hydroxyebastine	6.4	124 (purified from liver)
CYP2J4			
Arachidonic acid (18 μM)	19-HETE, two EETs	0.15	125 (baculoviral)
all-*trans*-retinal	All-*trans*-retinoic acid	20 (Vmax; km = 54 μM	130 (baculoviral)
9-*cis*-retinal	9-*cis*-retinoic acid	21 (Vmax; km = 49 μM)	130 (baculoviral)
Testosterone (10 μM)	Not indicated	0.005	125 (baculoviral)
Progesterone (10 μM)	Not indicated	0.019	125 (baculoviral)
CYP2J5			
Arachidonic acid (100 μM)	Mid-chain EETs (~50%), mid-chain HETEs,	4.5	148 (baculoviral)
Linoleic acid (50–100 μM)	epoxyoctadecenoic acid and hydroxyoctadecadienoic acid	0.25[a]	143 (baculoviral)
CYP2J9			
Arachidonic acid (100 μM)	19-HETE (66%), EETs (22%), other HETEs (12%)	0.37	149 (baculoviral)
Linoleic acid (50–100 μM)	epoxyoctadecenoic acid and hydroxyoctadecadienoic acid	0.08[a]	143 (baculoviral)

[a]The original paper indicated 1000-fold higher values, which apparently resulted from an error in the units.
HETE, hydroxyeicosatetraenoic acid; EET, epoxyeicosatrienoic acid.

heterologously-expressed CYP2J4 catalysed the transformation of all-*trans* and 9-*cis* retinal to the corresponding retinoic acids.[130] Enterocyte microsomal activity in the conversion of all-*trans* retinal to all-*trans* retinoic acid was increased in association with induction of CYP2J4 by pyrazole; furthermore, the majority of the microsomal activity was inhibited by an anti-CYP2J4 antibody.[131] It is not known whether CYP2J3 or other CYP2J enzyme also metabolise retinal. Additionally, CYP2J4 had low activities toward testosterone and progesterone.[125]

Few studies have examined the ability of the rat CYP2J enzymes to metabolise drugs or other xenobiotic compounds. CYP2J4 was essentially inactive toward R- or S-warfarin.[125] However, CYP2J3 is active in ebastine hydroxylation[124] and benzphetamine *N*-demethylation.[146]

10.3.4 Mouse CYP2J

The mouse *Cyp2j* gene cluster is located on chromosome 4, in close proximity to the *Cyp4a* gene cluster.[147] There are a total of eight *Cyp2j* genes, as well as a number of pseudogenes; the eight full-length genes, each evidently consisting of 9 exons, are all arranged in the same orientation, in the order *Cyp2j5*, *Cyp2j9*, *Cyp2j6*, *Cyp2j8*, *Cyp2j11*, *Cyp2j7*, *Cyp2j12*, *Cyp2j13*[9]. The size of the gene cluster is approximately 650 kbp.[9]

Only three of the mouse CYP2Js, namely, CYP2J5, CYP2J6, and CYP2J9, have been characterised. CYP2J5[148] and CYP2J6[146] cDNAs were cloned from a mouse liver cDNA library, and both proteins consist of 501 amino acid residues. Interestingly, whereas CYP2J6 is 73% identical to CYP2J5 in amino acid sequence, it is 94% identical to rat CYP2J4.[117] CYP2J9, which was cloned from a mouse brain cDNA library, encodes a protein of 502 amino acids, and it has the highest (91%) sequence identity to rat CYP2J3.[149]

CYP2J5 mRNA and protein, detected with an antibody against a CYP2J5-specific peptide, are abundant in the kidney; they are also found in the liver, but at lower levels.[148] Renal expression of CYP2J5 was detected before birth, whereas hepatic expression of CYP2J5 was not detected until 1 week after birth. Within the kidney, CYP2J5 mRNA appeared to be located in the proximal tubules of the cortex and outer medulla.[148] CYP2J5 protein and mRNA levels in kidney microsomes are higher in adult males than in adult females; the CYP2J5 expression in the kidney was upregulated by androgen and downregulated by oestrogen.[150]

CYP2J6 mRNA is abundant in epithelial cells of the small intestine, and also in other tissues at lower levels, including brain, colon, heart, kidney, lung, liver, skeletal muscle, stomach and testis.[146] However, an antibody against a CYP2J6-specific peptide did not detect a CYP2J6-sized protein in microsomes of small intestine or any other tissues examined, and recombinant CYP2J6 was rapidly degraded when mixed with intestinal microsomes, findings that led to the conclusion that CYP2J6 is an unstable protein.[146]

CYP2J9 mRNA is present at relatively high levels in the brain.[149] Although CYP2J9 mRNA appeared to be most abundant in the cerebellum, it was also detected in other brain regions.[149] In the cerebellum, CYP2J9 mRNA was abundant in the Purkinje cells, as revealed by *in situ* hybridisation with a specific riboprobe. CYP2J9 mRNA was also detected by RNA-PCR in the kidney and colon, but not in heart, liver, lung, small intestine, spleen, stomach or testis.[149]

An anti-CYP2J2 polyclonal antibody detected multiple proteins in various mouse tissues, on immunoblots;[146,149] the same antibody also detected CYP2J2-immunoreactive proteins in renal proximal tubules and collecting ducts,[148] and in cerebellar Purkinje cells.[149] However, it was not possible to distinguish among differing mouse CYP2J isoforms in these experiments. CYP2J proteins have also been detected by immunoblot analysis with an anti-CYP2J4 antibody in a variety of mouse tissues, including liver, lung, kidney, heart, eye, OM and small intestine; these proteins were most abundant in the small intestine and OM, and least abundant in the lung and eye.[151] Within the same study, a protein-specific RNA-PCR experiment was conducted: CYP2J5 mRNA was detected in the kidney and liver, but not in eye, lung, OM, heart and small intestine, whereas CYP2J6 was detected in all tissues examined.[151] Recombinant CYP2J8 protein has also been produced in insect cells.[149]

CYP2J4-related P450 proteins were induced at both the protein and mRNA levels by pyrazole, in the eye, liver, lung, kidney, OM and small intestine, but not in the heart.[151] The extent of induction of CYP2J protein increased with increasing pyrazole dose (80 to 200 mg/kg, intraperitoneal administration, once daily for 3 days), and the induction was observed in two differing mouse strains (C57BL/6 and DBA/2). The mechanism of pyrazole-induced CYP2J expression remains unclear; nevertheless, it appears to differ from the mechanism of CYP2A5 induction by pyrazole, given that CYP2J was induced by pyrazole in tissues that did not show CYP2A5 induction, such as the small intestine and OM.[151] The specific mouse CYP2J protein(s) that were induced by pyrazole have not been identified. In a study using an anti-CYP2J6-peptide antibody, CYP2J6 protein was not detected in microsomes from small intestine, liver, heart or kidney, after pyrazole treatment of mice for 3 days; however, it remains to be determined whether or not CYP2J6 is actually inducible by pyrazole, since the same antibody did not detect CYP2J6 protein in untreated animals.[146]

A CYP2J2-related protein was induced by $\sim$40% in the heart by chronic treatment of BALA/c mice with cocaine (30 mg/kg daily, intraperitoneal injection, for 14 days); the induction was suppressed by pretreatment of the animals with either of two P450 inhibitors, cimetidine and metyrapone.[152] Nose-only inhalation exposure of female mice to mercury vapour (4.0 mg/m^3, 2 h/day for 3 days) led to a $\sim$50% increase in CYP2J9 mRNA level in the brain.[149]

CYP2J5[148] and CYP2J9,[149] but neither CYP2J6[146] nor CYP2J8,[149] are active in the metabolism of AA. However, CYP2J5 and CYP2J9 generated somewhat distinct AA metabolite profiles: CYP2J5 produced 14,15-EET as the most abundant product, and also 8,9- and 11,12-EETs, as well as 11- and 15-HETEs;[148] CYP2J9, in contrast, produced 19-HETE as the main product, and the EETs as minor products.[149] Both CYP2J5 and CYP2J9 also catalysed the monoepoxygenation of linoleic acid.[143,149]

CYP2J5 was reportedly active in the oxidation of testosterone, diclofenac and bufuralol.[117] CYP2J6 was active towards benzphetamine when assayed in intact insect Sf9 cells, but not when assayed in microsomal preparations from these cells.[146]

10.3.5 Human CYP2J2

Unlike the *CYP2J* subfamilies in rodents and rabbits, the human *CYP2J* subfamily has only a single gene, *CYP2J2*. A cDNA clone for CYP2J2 was initially cloned from a human liver cDNA library; it encodes a protein of 502 amino acids.[153] The *CYP2J2* gene was mapped to human chromosome 1.[147] The *CYP2J2* gene is ~40 kbp in size, and contains nine exons.[154] Initial studies using Northern blot analysis detected abundant CYP2J2 transcript in the heart and, at lower levels in the liver, ileum, jejunum and colon.[153] However, CYP2J2 mRNA was near or below the detection limit in the brain, kidney, lung, ovary, and testis.[153] An apparent lack of correlation between levels of CYP2J2 mRNA and protein was noted for some of the tissues: an anti-CYP2J2 antibody detected CYP2J2-immunoreactive protein on Western blots in microsomes from heart, kidney, liver and skeletal muscle and, at much lower levels, in the lung and jejunum, but not in microsomal preparations from the aorta or vena cava.[153] Notably, this anti-CYP2J2 antibody detected three bands in the "P450 region" in human liver microsomes, a finding suggestive of cross reaction with other, as-yet unidentified, P450 proteins in the liver. In this regard, another phenomenon that has not yet been clarified is the fact that the endogenous microsomal CYP2J2-immunoreactive protein detected in various extrahepatic tissues migrated faster than did purified, recombinant CYP2J2 protein that has been generated using the baculoviral system, during electrophoresis.[153] In another study, a CYP2J2 protein heterologously expressed in COS-1 cells apparently co-migrated with the CYP2J protein detected in human intestine on immunoblots, a finding that implicated potential differences in protein modification, for CYP2J2 heterologously expressed in mammalian and insect cells.[155]

Subsequent studies confirmed that, in addition to abundant expression in human heart,[153] CYP2J2 mRNA and protein are relatively abundantly expressed in human lung[127] and intestine.[129] In lung, CYP2J2 protein was detected primarily in ciliated airway epithelial cells, although its expression was also detected in a variety of other lung cell types, including nonciliated epithelial cells, vascular smooth muscle and endothelial cells and, at even lower levels, alveolar macrophages.[127] In the gastrointestinal tract, CYP2J2 mRNA was detected primarily in the small intestine and colon, whereas CYP2J2 protein was detected throughout the tract.[129] Within the intestine, CYP2J2 protein and mRNA were detected mainly in epithelial cells, smooth muscle cells, vascular endothelial cells and autonomic ganglion nerve cells.[129]

CYP2J2 protein was also detected in pancreas, where CYP2J-immunoreactivity was found mainly in the islets of Langerhans.[128] Within the islets of Langerhans, CYP2J2 protein was abundant in α-, β-, δ-, and PP cells, with the highest level seen in the glucagon-producing α-cells.[128] Additionally, CYP2J2 mRNA was detected by RNA-PCR in human seminal vesicles[156] and in cultured human coronary artery endothelial cells.[157,158] In the cultured endothelial cells, the level of CYP2J2 mRNA was not altered by treatment of the cells with β-naphthoflavone or nifedipine; both treatments did induce CYP2C mRNA in the same cells.[157] Expression of CYP2J2 mRNA was detected in another study in human saphenous veins, where CYP2J2 mRNA was expressed at higher levels in varicose veins than in control veins.[159]

CYP2J2 mRNA and protein were detected in human foetal liver and nasal mucosa.[160] At gestational days 91–125, microsomal levels of CYP2J2 protein were lower in the nasal mucosae than in the livers of the same foetuses.[160] In another study,[161] CYP2J2 mRNA was detected in foetal brain, heart, intestine, kidney, liver and lung; the level of foetal liver CYP2J2 mRNA as well as that of CYP2J2 immunoreactive protein varied substantially among samples. In the same study, three aberrant CYP2J2 splicing variants were detected, with all of them expected to give rise to truncated proteins, due to premature termination of translation.[161]

The mechanisms of regulation of *CYP2J2* expression have been studied in human hepatoma-derived HepG2 cells.[162,163] *CYP2J2* expression in cultured HepG2 cells was down-regulated at a pretranslational level by hypoxia, under conditions that increased the expression of c-Jun and c-Fos.[162] Activator protein-1 (AP-1)-like regulatory elements are present in the 5′-flanking region of the *CYP2J2* gene, and results of reporter gene assays confirmed that the *CYP2J2* promoter can be regulated by c-Jun. Deletion analysis of the *CYP2J2* promoter region identified a c-Jun-responsive module (at −152/−50 bp), which contained an AP-1-like element at −56/−63 bp. In gel-shift assays, the AP-1-like element interacted with c-Jun, but not c-Fos. In a subsequent study,[163] sequence at −105/−95 bp was found to also bind c-Jun. Furthermore, mutation of either the −105/−95 bp or the −56/−63 bp c-Jun binding site led to only a partial loss of transactivation by c-Jun, with the mutation of the former, −105/−95 bp binding site, having the greater impact on transactivation. However, mutation of both elements simultaneously led to a complete loss of transactivation by c-Jun. The -105/-95 bp binding site also differs from the −56/−63 bp c-Jun binding site in being able to bind either c-Jun homodimer or c-Fos/c-Jun heterodimer. Thus, the hypoxia-induced downregulation of CYP2J2 expression appears to be mediated through the c-Jun-responsive module, which contains an uncommon AP-1 element at –105/−95 bp and a second AP-1-like element at −56/−63 bp.

Several putative Sp1 elements are present in the *CYP2J2* proximal promoter region;[162,164] one of these (at −50/−40 bp) was found to be functional in transfection assays, in which a −50G > T mutation led to a 48% decrease in reporter gene activity.[164] The −50G > T mutation corresponds to the −76G > T variation (numbered from the ATG codon) found in the *CYP2J2*7* allele. It has been proposed that Sp1 and c-Jun, which bind at neighbouring elements, interact so as to regulate *CYP2J2* expression.[163]

CYP2J2 metabolised both endogenous compounds, including AA,[153] linoleic acid[143] and Vit D,[145] and drugs, including ebastine[124,165] and astemizole.[155,166] CYP2J2 was also active towards testosterone, diclofenac and bufuralol,[117] but not benzphetamine.[153] The known substrates and metabolites for CYP2J2 are summarised in Table 10.7.

CYP2J2 metabolised AA to 5,6-, 8,9-, 11,12-, and 14,15-EETs, and to a lesser extent, 19-HETE.[153] The most abundant *in vitro* product was 14,15-EET, which

Table 10.7 Metabolc activity of CYP2J2.

Substrate (concn)	*Metabolite(s)*	*Rate (nmol/min/ nmol P450)*	*Reference (expression system)*
Arachidonic acid (100 μM)	5,6-, 8,9-, 11,12-, and 14,15-EETs	0.1 (at 30°C)	153 (baculoviral)
Linoleic acid (50–100 μM)	Epoxyoctadecenoic acid	0.16[a]	143 (baculoviral)
Vit D_3	25-OH-Vit D_3	0.087 (Vmax; Km = 7.7 μM)	145 (*E. coli*)
Vit D_2	25-OH-Vit D_2	0.16 (Vmax; km = 2.0 μM)	145 (*E. coli*)
1α-OH-Vit D_3	1,25-$(OH)_2$ Vit D_3	2.2 (Vmax; Km = 4.4 μM)	145 (*E. coli*)
Ebastine	Hydroxyebastine	8.20 (Vmax; km = 18.3 μM)	165 (baculoviral)
Hydroxyebastine	Carebestine	9.86 (Vmax; km = 0.75 μM)	165 (baculoviral)
Ebastine	Hydroxyebastine	40.6 (Vmax; km = 1.3 μM)	124 (yeast)
Ebastine (100 μM)	Hydroxyebastine	5.5	155 (COS-1)
Astemizole	Desmethylastemizole	1.1 (Vmax; km = 0.65 μM)	155 (COS-1)

[a]The original paper indicated 1000-fold higher values, which apparently resulted from an error in the units.
EET, epoxyeicosatrienoic acid.

was formed primarily as the 14R,15S-enantiomer; this feature in enantiomeric selectivity differentiates CYP2J2 from several other AA-metabolising CYPs.[117,153] Based on the observation that the EETs isolated from human heart are similar to those formed *in vitro* by CYP2J2, in terms of the chirality of the enantiomers, it was suggested that CYP2J2 contributes to EET formation *in vivo* in this organ in humans.[153] Recombinant CYP2J2 also catalyses the 25-hydroxylation of Vit D_3, Vit D_2, and 1α-OH-Vit D_3, with the respective turnover numbers increasing in that order.[145] Notably, CYP2J2 was much less active in the hydroxylation of Vit D_3 than was rat CYP2J3.

CYP2J2 is highly active in the hydroxylation of ebastine, an antihistamine.[124,165] CYP2J2 appears to be the main ebastine hydroxylase in human intestine. The addition of anti-CYP2J2 antibody to human intestinal microsomal reaction mixtures led to a ~70% inhibition of ebastine hydroxylase activity.[124] Heterologously-expressed CYP2J2, either in yeast or in insect cells, was the most active among more than 10 CYP enzymes tested for ebastine metabolism in reconstituted systems.[124,165] CYP2J2 also catalysed the sequential metabolism of hydroxyebastine to carebastine, but not the dealkylation of carebastine.[165]

CYP2J2, heterologously expressed in COS-1 cells, also catalysed the *O*-demethylation of astemizole, another antihistamine drug, with an apparent Km value of 0.65 μM.[155] CYP2J2 could be involved in human small intestinal microsomal astemizole *O*-demethylation, as suggested by the high correlation observed between CYP2J2 protein levels and astemizole *O*-demethylase activities in small intestinal microsomes,[155] and the inhibition of small intestinal astemizole *O*-demethylation activities by compounds capable of inhibiting CYP2J2.[155,166] CYP2J2 can also contribute to astemizole *O*-demethylation in human liver, although the extent of its contribution appears to be lower in liver than in small intestine.[166]

Compounds found to inhibit the astemizole *O*-demethylation activity of recombinant CYP2J2 included the known CYP2J2 substrates AA and ebastine,[155] as well as others, α-naphthoflavone, ketoconazole, tranylcypromine, troglitazone and terfenadine[166]; most of the latter have yet to be evaluated in terms of whether they are actual CYP2J2 substrates. Preliminary studies indicated that terfenadine is a substrate for CYP2J2.[161] In a more recent study, a series of compounds related to terfenadine were found to be high-affinity, selective CYP2J2 inhibitors.[167] The known P450 epoxygenase inhibitors N-methylsulphonyl-6-(2-proparglyloxyphenyl)-hexanamide and 17-octadecynoic acid also function as CYP2J2 inhibitors.[168,169]

The potential biological functions of human CYP2J2 have been examined in a series of elegant studies.[143,168–176] Most of these studies focussed more on the ability of CYP2J2 and CYP2J2 metabolites, i.e., EETs, to modulate biological processes, than on the relative contributions of CYP2J2 and other P450 enzymes capable of epoxygenation of AA.

In human endothelial cells, EETs, either added exogenously, or produced by transfected CYP2J2, were found to be antiinflammatory, a property distinct from the known vasodilatory role of these eicosanoids.[170] Also in endothelial cells, CYP2J2 and EETs showed fibrinolytic activity, apparently as a result of EET-induced increases in the expression of tissue plasminogen activator.[171] In a hypoxia-reoxygenation injury model that used cultured bovine aortic endothelial cells, transfection of CYP2J2 led to cytoprotection, possibly through a reduction in oxidative stress.[172] In vascular smooth muscle cells, CYP2J2 overexpression led to a mild inhibition of serum- and PDGF-induced cell migration.[173]

A transgenic mouse model with cardiomyocyte-specific CYP2J2 overexpression was found to be normal in terms of cardiac morphology and basal contractile function; however, these mice showed enhanced functional recovery of the heart after an episode of ischaemia, and reperfusion.[168] Subsequent mechanistic studies indicated that the cardioprotective effects associated with CYP2J2 overexpression in the heart were P450-dependent, and they involved the activation of the mitochondrial ATP-sensitive K+ channel and phospho-p42/p44 mitogen-activated protein kinase.[168] In addition, cardiomyocytes from the CYP2J2 transgenic mice were found to have significantly greater densities of L-type Ca^{2+} currents[174] and baseline K-ATP current[176] than were characteristic of wild-type cardiomyocytes.[174] The vascular protective effects of CYP2J2, and its eicosanoid metabolites, have been reviewed by others.[117,175,177]

In addition to its activities in cardiovascular and endothelial cells, CYP2J2 may also play a role in tumourigenesis. CYP2J2 was strongly expressed in human tumour tissues and carcinoma cell lines.[169] Furthermore, over-expression of CYP2J2, or addition of EETs, led to enhanced cell proliferation and impeded tumour necrosis factor alpha (TNF-alpha)-induced apoptosis in cultured carcinoma cell lines, whereas transfection of an antisense 2J2, or addition of an epoxygenase inhibitor, had effects opposite to those of CYP2J2 overexpression.[169]

A total of 10 *CYP2J2* alleles have been identified (http://www.cypalleles.ki.se). The cDNA used for the production of recombinant CYP2J2[153] corresponded to the *1 allele.[154] The *2 (Thr143Ala), *3 (Arg158Cys), *4 (Ile192Asn), *5 (Asp342Asn), and *6 (Asn404Tyr), *8 (Gly312Arg), *9 (Pro351Leu) and *10 (Pro115Leu) alleles all give rise to changes in amino-acid sequence[154,161,178], whereas the *7 (−76G >T) allele confers a decrease in transcription, as a result of the loss of an Sp1 transcription factor binding site in the CYP2J2 promoter.[164] The CYP2J2.1 through CYP2J2.6 proteins were compared for their ability to metabolise AA and linoleic acid in reconstituted systems.[154] The CYP2J2.2, CYP2J2.3 and CYP2J2.6 proteins had reduced activity towards both substrates, relative to CYP2J2.1, whereas CYP2J2.4 protein had reduced activity towards AA but normal activity towards linoleic acid, and the CYP2J2.5 protein had normal activity towards both AA and linoleic acid. The CYP2J2.8 protein, which did not bind CO in spectral assays, was essentially inactive toward astemizole and ebastine, whereas the CYP2J2.9 protein had normal astemizole *O*-demethylation and ebastine hydroxylation activities.[178] Recombinant CYP2J2.10 protein failed to incorporate haem, and was inactive towards terfenadine.[161]

CYP2J2*2-*7 alleles as well as several other single nucleotide polymorphisms were found not to be associated with hypertension, in a case-control study of an African-American population.[179] In another study, the CYP2J2*2, *3, *4, *6, and *7 alleles were found not to be associated with urinary excretion of ebastine metabolites (desalkylebastine and carebastine) in healthy Caucasians subjects.[180] On the other hand, the CYP2J2*7 allele was significantly associated with lower plasma levels of stable EET metabolites, and with an increased risk of coronary artery disease.[164,177] The CYP2J2*7 allele is relatively common in a range of ethnic groups, with allele frequencies ranging from 2.6% to 17%,[154,164,178,181] whereas the other allelic variants appear to be either rare overall, or more strongly restricted to specific ethnic groups.[154,161,178]

10.3.6 CYP2J in Other Species

Few studies have examined CYP2J expression or function in other species. In pigs, a CYP2J-related protein was detected in the ovary, and the level of this protein decreased at ovulation.[182] CYP2J mRNA was also detected in porcine coronary artery endothelial cells.[157] *CYP2J* genes have been detected in a number of other animal species, such as cynomolgus monkeys, chickens and cattle (http://drnelson.utmem.edu/CytochromeP450.html). A zebrafish CYP2J cDNA was recently identified; it presumably encodes a protein of 496 amino acids and shows ~50% amino acid sequence identity with rodent and human CYP2J proteins.[183]

10.3.7 Perspective

The well-established biological activities of CYP2J substrates and metabolites, such as AA, Vit D, retinoids, EETs and HETEs, together with the preferential expression of human CYP2J2 and most animal CYP2J proteins in diverse extrahepatic tissues, such as intestine, lung, kidney, brain and heart, suggest that the CYP2J enzymes play important and diverse biological functions. CYP2J2 also appears to be important for the disposition of selected drugs. CYP2J function can be influenced by genetic polymorphisms that affect CYP2J2 activity or expression. CYP2J expression can also be altered by exposure to xenobiotic inducers or induction of pathologic conditions. Thus, studies are needed (i) to identify additional endogenous and xenobiotic substrates for various CYP2J enzymes in humans and rodents; (ii) to identify additional CYP2J inducers and the mechanisms of induction; (iii) to determine whether human CYP2J2 is inducible by compounds known to induce CYP2J in rodents; and (iv) to achieve an understanding of the basic mechanisms of tissue-selective CYP2J expression. Studies are also needed to determine the relative roles of CYP2J and other P450 enzymes in controlling cellular, organ-specific or systemic levels of various endogenous compounds, particularly those of AA metabolites; such information is critical for assessing the functional impact of CYP2J2 genetic polymorphisms. In this regard, the continued development of effective and CYP2J-specific chemical inhibitors, as well as the generation of CYP2J knockout mouse models, are important goals. The presence in rodents of multiple *CYP2J* genes, many of which have not been characterised, makes it difficult to extrapolate findings from animal studies to humans. A "CYP2J-humanised" mouse model, in which all the *Cyp2j* genes are deleted, and are replaced with the human *CYP2J2* gene, will be valuable in these ongoing efforts.

Acknowledgments

We thank Drs. Laurence Kaminsky and Adriana Verschoor for reading the manuscript. This work was supported in part by United States Public Health Service grants CA092596 and ES07462 from the National Institutes of Health.

References

1. P.T. Nhamburo, S. Kimura, O.W. McBride, C.A. Kozak, H.V. Gelboin and F.J. Gonzalez, *Biochemistry*, 1990, **29**, 5491.
2. J.K. Ritter, I.S. Owens, M. Negishi, K. Nagata, Y.Y. Sheen, J.R. Gillette and H.A. Sasame, *Biochemistry*, 1991, **30**, 11430.

3. K. Nagata, B.M. Martin, J.R. Gillette and H.A. Sasame, *Drug Metab. Dispos.*, 1990, **18**, 557.
4. B.A. Carr, J. Wan, R.N. Hines and G.S. Yost, *Journal of Biological Chemistry*, 2003, **278**, 15473.
5. N. Chen, S.E. Whitehead, A.W. Caillat, K. Gavit, D.R. Isphording, D. Kovacevic, M.B. McCreary and S.M.G. Hoffman, *Mutation Research-Fundamental & Molecular Mechanisms of Mutagenesis*, 2002, **499**, 155.
6. A. Buckpitt, B. Boland, M. Isbell, D. Morin, M. Shultz, R. Baldwin, K. Chan, A. Karlsson, C. Lin, A. Taff, J. West, M. Fanucchi, L. Van Winkle and C. Plopper, *Drug Metabolism Reviews*, 2002, **34**, 791.
7. S.M.G. Hoffman, P. Fernandezsalguero, F.J. Gonzalez and H.W. Mohrenweiser, *J. Mol. Evol.*, 1995, **41**, 894.
8. S.M.G. Hoffman, D.R. Nelson and D.S. Keeney, *Pharmacogenetics*, 2001, **11**, 687.
9. D.R. Nelson, D.C. Zeldin, S.M. Hoffman, L.J. Maltais, H.M. Wain and D.W. Nebert, *Pharmacogenetics*, 2004, **14**, 1.
10. J. Hakkola, M. Pasanen, J. Hukkanen, O. Pelkonen, J. Maenpaa, R.J. Edwards, A.R. Boobis and H. Raunio, *Biochem. Pharmacol.*, 1996, **51**, 403.
11. J. Hakkola, H. Raunio, R. Purkunen, O. Pelkonen, S. Saarikoski, T. Cresteil and M. Pasanen, *Biochem. Pharmacol.*, 1996, **52**, 379.
12. J.C. Willey, E. Coy, C. Brolly, M.J. Utell, M.W. Frampton, E. Hammersley, W.G. Thilly, D. Olson and K. Cairns, *American Journal of Respiratory Cell & Molecular Biology*, 1996, **14**, 262.
13. D.M. Runge, T.W. Stock, T. Lehmann, C. Taege, U. Bernauer, D.B. Stolz, S. Hofmann and H. Foth, *Archives of Toxicology*, 2001, **75**, 335.
14. J. Hukkanen, M. Mantyla, L. Kangas, P. Wirta, J. Hakkola, P. Paakki, S. Evisalmi, O. Pelkonen and H. Raunio, *Pharmacology & Toxicology*, 1998, **82**, 93.
15. M. Iscan, T. Klaavuniemi, T. Coban, N. Kapucuoglu, O. Pelkonen and H. Raunio, *Breast Cancer Research & Treatment*, 2001, **70**, 47.
16. J. Hukkanen, J. Hakkola, S. Anttila, R. Piipari, A. Karjalainen, O. Pelkonen and H. Raunio, *Mol. Carcinog.*, 1997, **20**, 224.
17. H. Raunio, J. Hakkola, J. Hukkanen, O. Pelkonen, R. Edwards, A. Boobis and S. Anttila, *Arch. Toxicol. Suppl.*, 1998, **20**, 465.
18. B.C. Krovat, J.H. Tracy and C.J. Omiecinski, *Toxicological Sciences*, 2000, **55**, 352.
19. M. Nishimura, H. Yaguti, H. Yoshitsugu, S. Naito and T. Satoh, *Yakugaku Zasshi-Journal of the Pharmaceutical Society of Japan*, 2003, **123**, 369.
20. M.R. Andersen, F.M. Farin and C.J. Omiecinski, *DNA & Cell Biology*, 1998, **17**, 231.
21. M. Furukawa, M. Nishimura, D. Ogino, R. Chiba, I. Ikai, N. Ueda, S. Naito, S. Kuribayashi, M.A. Moustafa, T. Uchida, H. Sawada, T. Kamataki, Y. Funae and M. Fukumoto, *Cancer Science*, 2004, **95**, 520.

22. X.L. Zhang, Q.Y. Zhang, D.Z. Liu, T. Su, Y. Weng, G.Y. Ling, Y. Chen, J. Gu, B. Schilling and X. Ding, *Drug Metabolism & Disposition*, 2005, **33**, 1423.
23. M. Kumarakulasingham, P.H. Rooney, S.R. Dundas, C. Telfer, W.T. Melvin, S. Curran and G.I. Murray, *Clinical Cancer Research*, 2005, **11**, 3758.
24. D. Downie, M.C.E. McFadyen, P.H. Rooney, M.E. Cruickshank, D.E. Parkin, I.D. Miller, C. Telfer, W.T. Melvin and G.I. Murray, *Clinical Cancer Research*, 2005, **11**, 7369.
25. J. Wan, B.A. Carr, N.S. Cutler, D.L. Lanza, R.N. Hines and G.S. Yost, *Drug Metabolism & Disposition*, 2005, **33**, 1244.
26. T. Nakajima, E. Elovaara, F.J. Gonzalez, H.V. Gelboin, H. Raunio, O. Pelkonen, H. Vainio and T. Aoyama, *Chemical Research in Toxicology*, 1994, **7**, 891.
27. M. Czerwinski, T.L. McLemore, R.M. Philpot, P.T. Nhamburo, K. Korzekwa, H.V. Gelboin and F.J. Gonzalez, *Cancer Res.*, 1991, **51**, 4636.
28. J. Thorntonmanning, M.L. Appleton, F.J. Gonzalez and G.S. Yost, *J. Pharmacol. Expt. Ther.*, 1996, **276**, 21.
29. D.L. Lanza, E. Code, C.L. Crespi, F.J. Gonzalez and G.S. Yost, *Drug Metabolism & Disposition*, 1999, **27**, 798.
30. W.K. Nichols, R. Mehta, K. Skordos, K. Mace, A. M. A. Pfeifer, B.A. Carr, T. Minko, S.W. Burchiel and G.S. Yost, *Toxicological Sciences*, 2003, **71**, 229.
31. D.L. Lanza and G.S. Yost, *Drug Metabolism & Disposition*, 2001, **29**, 950.
32. P.L. Sheets, G.S. Yost and G.P. Carlson, *Journal of Biochemical & Molecular Toxicology*, 2004, **18**, 92.
33. R.M. Baldwin, M.A. Shultz and A.R. Buckpitt, *J. Pharmacol. Expt. Ther.*, 2005, **312**, 857.
34. G. Tournel, C. Cauffiez, I. Billaut-Laden, D. Allorge, D. Chevalier, F. Bonnifet, E. Mensier, J.J. Lafitte, M. Lhermitte, F. Broly and J.M. Lo-Guidice, *Mutat. Res.*, 2007, **617**, 79.
35. J.H. Jiang, Z. Li, G. Su, W.H. Jia, R.H. Zhang, X.J. Yu, M. Zhang, J.M. Wen and Y.X. Zeng, *Zhonghua Yi. Xue. Yi. Chuan Xue. Za Zhi.*, 2006, **23**, 383.
36. H.Y. Wang, K.M. Donley, D.S. Keeney and S. M. G. Hoffman, *Environmental Health Perspectives*, 2003, **111**, 1835.
37. S. Ramakanth, J.R. Thorntonmanning, H.F. Wang, H. Maxwell and G.S. Yost, *Toxicol. Lett.*, 1994, **71**, 77.
38. R.M. Baldwin, W.T. Jewell, M.V. Fanucchi, C.G. Plopper and A.R. Buckpitt, *J. Pharmacol. Expt. Ther.*, 2004, **309**, 127.
39. X.Q. Ye, L.M. Lu and S.S. Gill, *Biochemical & Biophysical Research Communications*, 1997, **239**, 660.
40. G.P. Carlson, *Journal of Toxicology & Environmental Health*, 1997, **51**, 477.
41. G. Cruzan, G.P. Carlson, K.A. Johnson, L.S. Andrews, M.I. Banton, C. Bevan and J.R. Cushman, *Regulatory Toxicology & Pharmacology*, 2002, **35**, 308.

42. M.A. Shultz, D. Morin, A.M. Chang and A. Buckpitt, *J. Pharmacol. Expt. Ther.*, 2001, **296**, 510.
43. L.S. Vanwinkle, A.R. Buckpitt and C.G. Plopper, *American Journal of Respiratory Cell & Molecular Biology*, 1996, **14**, 586.
44. K.E. Pinkerton, J.L. Peake, I. Espiritu, M. Goldsmith and H. Witschi, *Inhalation Toxicology*, 1996, **8**, 927.
45. M.V. Fanucchi, M.E. Murphy, A.R. Buckpitt, R.M. Philpot and C.G. Plopper, *American Journal of Respiratory Cell & Molecular Biology*, 1997, **17**, 302.
46. G.W. Mango, C.J. Johnston, S.D. Reynolds, J.N. Finkelstein, C.G. Plopper and B.R. Stripp, *American Journal of Physiology-Lung Cellular & Molecular Physiology*, 1998, **19**, L348.
47. Y. Yamada, H. Karasaki, K. Matsushima, G.H. Lee and K. Ogawa, *Laboratory Investigation*, 1999, **79**, 1059.
48. J. Liu, Y.X. Xie, B.A. Merrick, J. Shen, D.M.K. Ducharme, J. Collins, B.A. Diwan, D. Logsdon and M.P. Waalkes, *Toxicology & Applied Pharmacology*, 2006, **213**, 216.
49. L.S. Van Winkle, A.D. Gunderson, J.A. Shimizu, G.L. Baker and C.D. Brown, *American Journal of Physiology-Lung Cellular & Molecular Physiology*, 2002, **282**, L1122.
50. R.L. Stelck, G.L. Baker, K.M. Sutherland and L.S. Van Winkle, *Drug Metabolism & Disposition*, 2005, **33**, 1597.
51. M.A. Shultz, P.V. Choudary and A.R. Buckpitt, *J. Pharmacol. Expt. Ther.*, 1999, **290**, 281.
52. A. Chang, A.R. Buckpitt, C.G. Plopper and W. Alworth, *Toxicologist*, 1996, **30**, 72.
53. A.C. Simmonds, B.I. Ghanayem, A. Sharma, C.A. Reilly, B. Millen, G.S. Yost and P.G. Forkert, *J. Pharmacol. Expt. Ther.*, 2004, **310**, 855.
54. D.E. Hynes, D.B. DeNicola and G.P. Carlson, *Toxicological Sciences*, 1999, **51**, 195.
55. T. Green, R. Lee, A. Toghill, S. Meadowcroft, V. Lund and J. Foster, *Chemico-Biological Interactions*, 2001, **137**, 185.
56. G.P. Carlson, *Journal of Toxicology & Environmental Health*, 2003, **66**, 861.
57. G.P. Carlson, *Toxicol. Lett.*, 2004, **150**, 335.
58. G. Cruzan, G.P. Carlson, M. Turner and W. Mellert, *Journal of Toxicology & Environmental Health Part A*, 2005, **68**, 229.
59. G.P. Carlson, A. A. P. Rivera and N.A. Mantick, *Journal of Toxicology & Environmental Health Part A*, 2001, **63**, 541.
60. K.M. Vogie, N.A. Mantick and G.P. Carlson, *Journal of Toxicology & Environmental Health Part A*, 2004, **67**, 145.
61. G.P. Carlson, *Journal of Toxicology & Environmental Health Part A*, 2004, **67**, 905.
62. G.P. Carlson, *Toxicology*, 2002, **179**, 129.
63. M.B. Genter, J. Marlowe, J.K. Kerzee, N. Dragin, A. Puga, T.P. Dalton and D.W. Nebert, *Biochemical & Biophysical Research Communications*, 2006, **348**, 120.

64. R.D. Verschoyle, J. Martin and D. Dinsdale, *Xenobiotica*, 1997, **27**, 853.
65. M.W. Powley and G.P. Carlson, *Toxicology*, 2001, **169**, 187.
66. M.W. Powley and G.P. Carlson, *Toxicol. Lett.*, 2001, **125**, 117.
67. P.L. Sheets and G.P. Carlson, *Journal of Toxicology & Environmental Health Part A*, 2004, **67**, 421.
68. A.C. Simmonds, C.A. Reilly, R.M. Baldwin, B.I. Ghanayem, D.L. Lanza, G.S. Yost, K.S. Collins and P.G. Forkert, *Drug Metabolism & Disposition*, 2004, **32**, 1032.
69. A.C. Simmonds, B.I. Ghanayem, A. Sharma, C.A. Reilly, B. Millen, G.S. Yost and P.G. Forkert, *J. Pharmacol. Expt. Ther.*, 2004, **311**, 853.
70. P.G. Forkert, B. Millen, L.H. Lash, D.A. Putt and B.I. Ghanayem, *J. Pharmacol. Expt. Ther.*, 2006, **316**, 520.
71. P.G. Forkert, R.M. Baldwin, B. Millen, L.H. Lash, D.A. Putt, M.A. Shultz and K.S. Collins, *Drug Metabolism & Disposition*, 2005, **33**, 1429.
72. H. Wang, D.L. Lanza and G.S. Yost, *Archives of Biochemistry & Biophysics*, 1998, **349**, 329.
73. M.B. Genter, G.S. Yost and A.E. Rettie, *Journal of Biochemical & Molecular Toxicology*, 2006, **20**, 139.
74. X.L. Zhang, M. Caggana, T.L. Cutler and X. Ding, *J. Pharmacol. Expt. Ther.*, 2004, **311**, 373.
75. X. Ding and M.J. Coon, *Biochemistry*, 1988, **27**, 8330.
76. P. Nef, J. Heldman, D. Lazard, T. Margalit, M. Jaye, I. Hanukoglu and D. Lancet, *Journal of Biological Chemistry*, 1989, **264**, 6780.
77. X. Ding and M.J. Coon, *Methods in Enzymology.*, 1991, **206**, 603.
78. X. Ding and M.J. Coon, *Mol. Pharmacol.*, 1990, **37**, 489.
79. X. Ding, T.D. Porter, H.M. Peng and M.J. Coon, *Archives. of Biochemistry & Biophysics.*, 1991, **285**, 120.
80. X. Ding, H.M. Peng and M.J. Coon, *Mol. Pharmacol.*, 1992, **42**, 1027.
81. R.M. Laethem, C.L. Laethem, X. Ding and D.R. Koop, *Journal of Pharmacology & Experimental. Therapeutics.*, 1992, **262**, 433.
82. X. Ding and M.J. Coon, *Handbook Exp. Pharmacol.*, 1993, **105**, 351.
83. E.S. Roberts, A.D. Vaz and M.J. Coon, *Mol. Pharmacol.*, 1992, **41**, 427.
84. J.Y. Hong, X. Ding, T.J. Smith, M.J. Coon and C.S. Yang, *Carcinogenesis*, 1992, **13**, 2141.
85. D.E. Williams, X. Ding and M.J. Coon, *Biochemical & Biophysical. Research. Communications.*, 1990, **166**, 945.
86. X. Ding, D.C. Spink, J.K. Bhama, J.J. Sheng, A.D. Vaz and M.J. Coon, *Mol. Pharmacol.*, 1996, **49**, 1113.
87. D.A. Putt, X. Ding, M.J. Coon and P.F. Hollenberg, *Carcinogenesis*, 1995, **16**, 1411.
88. E.S. Roberts, A.D. Vaz and M.J. Coon, *Proc. Natl. Acad. Sci. U.S.A.*, 1991, **88**, 8963.
89. X. Ding and M.J. Coon, *Archives. of Biochemistry & Biophysics.*, 1994, **315**, 454.
90. P. Nef, T.M. Larabee, K. Kagimoto and U.A. Meyer, *Journal of Biological Chemistry*, 1990, **265**, 2903.

91. T. Margalit and D. Lancet, *Brain Res. Dev. Brain Res.*, 1993, **73**, 7.
92. Y. Chen, M.L. Getchell, X. Ding and T.V. Getchell, *Neuroreport.*, 1992, **3**, 749.
93. K. Zupko, Y. Poria and D. Lancet, *Eur. J. Biochem.*, 1991, **196**, 51.
94. J. Gu, V.E. Walker, T.W. Lipinskas, D.M. Walker and X. Ding, *Toxicology & Applied Pharmacology*, 1997, **146**, 134.
95. J.E. Schwob, S.L. Youngentob and R.C. Mezza, *J. Comp. Neurol.*, 1995, **359**, 15.
96. E. Walters, K. Buchheit and J.A. Maruniak, *J. Neurosci. Res.*, 1992, **33**, 103.
97. E. Walters, K. Buchheit and J.A. Maruniak, *Toxicology*, 1993, **81**, 113.
98. T.T. Yu, J.C. McIntyre, S.C. Bose, D. Hardin, M.C. Owen and T.S. McClintock, *J. Comp. Neurol.*, 2005, **483**, 251.
99. Z.C. Hua, Q.Y. Zhang, T. Su, T.W. Lipinskas and X. Ding, *Archives of Biochemistry & Biophysics*, 1997, **340**, 208.
100. X.L. Zhuo, J.E. Schwob, P.J. Swiatek and X. Ding, *Archives of Biochemistry & Biophysics*, 2001, **391**, 127.
101. X. Zhuo, J. Gu, M.J. Behr, P.J. Swiatek, H.D. Cui, Q.Y. Zhang, Y.Q. Xie, D.N. Collins and X. Ding, *J. Pharmacol. Expt. Ther.*, 2004, **308**, 719.
102. J. Gu, C. Dudley, T. Su, D.C. Spink, Q.Y. Zhang, R.L. Moss and X. Ding, *Biochemical & Biophysical Research Communications*, 1999, **266**, 262.
103. J. Gu, Q.Y. Zhang, M.B. Genter, T.W. Lipinskas, M. Negishi, D.W. Nebert and X. Ding, *J. Pharmacol. Expt. Ther.*, 1998, **285**, 1287.
104. M.B. Genter, H.C. Liang, J. Gu, X. Ding, M. Negishi, R.A. Mckinnon and D.W. Nebert, *Biochem. Pharmacol.*, 1998, **55**, 1819.
105. X.L. Zhuo, J. Gu, Q.Y. Zhang, D.C. Spink, L.S. Kaminsky and X. Ding, *J. Pharmacol. Expt. Ther.*, 1999, **288**, 463.
106. S.H. Snyder, P.B. Sklar and J. Pevsner, *Journal of Biological Chemistry*, 1988, **263**, 13971.
107. D.R. Adams, D.W. Deyoung and R. Griffith, *J. Anat.*, 1981, **132**, 29.
108. M.L. Getchell, Y. Chen, X. Ding, D.L. Sparks and T.V. Getchell, *Annals. of Otology., Rhinology. & Laryngology.*, 1993, **102**, 368.
109. J. Sheng and X. Ding, *Biochemical & Biophysical. Research. Communications.*, 1996, **218**, 570.
110. J.J. Sheng, J.C. Guo, Z.C. Hua, M. Caggana and X. Ding, *Pharmacogenetics*, 2000, **10**, 667.
111. J.J. Sheng, Z.C. Hua, J.C. Guo, M. Caggana and X. Ding, *Drug Metabolism & Disposition*, 2001, **29**, 4.
112. V. Longo, G. Amato, A. Santucci and P.G. Gervasi, *Biochem. J.*, 1997, **323**, 65.
113. S. Marini, V. Longo, A. Mazzaccaro and P.G. Gervasi, *Xenobiotica*, 1998, **28**, 923.
114. G.Y. Ling, J. Gu, M.B. Genter, X.L. Zhuo and X. Ding, *Chemico-Biological Interactions*, 2004, **147**, 247.
115. X. Ding and L.S. Kaminsky, *Annual Review of Pharmacology and Toxicology*, 2003, **43**, 149.

116. K. Koike, E. Kusunose, Y. Nishikawa, K. Ichihara, S. Inagaki, H. Takagi, Y. Kikuta and M. Kusunose, *Biochemical & Biophysical Research Communications*, 1997, **232**, 643.
117. P.E. Scarborough, J.X. Ma, W. Qu and D.C. Zeldin, *Drug Metabolism Reviews*, 1999, **31**, 205.
118. K. Ichihara, E. Kusunose, M. Kaku, S. Yamamoto and M. Kusunose, *Biochim. Biophys. Acta*, 1985, **831**, 99.
119. Y. Kikuta, K. Sogawa, M. Haniu, M. Kinosaki, E. Kusunose, Y. Nojima, S. Yamamoto, K. Ichihara, M. Kusunose and Y. Fujii-Kuriyama, *J. Biol. Chem.*, 1991, **266**, 17821.
120. S.L. Pfister, N. Spitzbarth, D.C. Zeldin, P. Lafite, D. Mansuy and W.B. Campbell, *Archives of Biochemistry & Biophysics*, 2003, **420**, 142.
121. D.L. Zhu, M. Bousamra, D.C. Zeldin, J.R. Falck, M. Townsley, D.R. Harder, R.J. Roman and E.R. Jacobs, *American Journal of Physiology-Lung Cellular & Molecular Physiology*, 2000, **278**, L335.
122. S. Wu, W.N. Chen, E. Murphy, S. Gabel, K.B. Tomer, J. Foley, C. Steenbergen, J.R. Falck, C.R. Moomaw and D.C. Zeldin, *Journal of Biological Chemistry*, 1997, **272**, 12551.
123. S. Imaoka, Y. Terano and Y. Funae, *Arch. Biochem. Biophys.*, 1990, **278**, 168.
124. T. Hashizume, S. Imaoka, M. Mise, Y. Terauchi, T. Fujii, H. Miyazaki, T. Kamataki and Y. Funae, *J. Pharmacol. Expt. Ther.*, 2002, **300**, 298.
125. Q.Y. Zhang, X. Ding and L.S. Kaminsky, *Archives of Biochemistry & Biophysics*, 1997, **340**, 270.
126. T. Yamasaki, S. Izumi, H. Ide and Y. Ohyama, *Journal of Biological Chemistry*, 2004, **279**, 22848.
127. D.C. Zeldin, J. Foley, J.X. Ma, J.E. Boyle, J. M. S. Pascual, C.R. Moomaw, K.B. Tomer, C. Steenbergen and S. Wu, *Mol. Pharmacol.*, 1996, **50**, 1111.
128. D.C. Zeldin, J. Foley, J.E. Boyle, C.R. Moomaw, K.B. Tomer, C. Parker, C. Steenbergen and S. Wu, *Endocrinology*, 1997, **138**, 1338.
129. D.C. Zeldin, J. Foley, S.M. Goldsworthy, M.E. Cook, J.E. Boyle, J.X. Ma, C.R. Moomaw, K.B. Tomer, C. Steenbergen and S. Wu, *Mol. Pharmacol.*, 1997, **51**, 931.
130. Q.Y. Zhang, G. Raner, X. Ding, D. Dunbar, M.J. Coon and L.S. Kaminsky, *Archives of Biochemistry & Biophysics* 1998, **353**, 257.
131. Q.Y. Zhang, X. Ding, D. Dunbar, L. Cao and L.S. Kaminsky, *Drug Metabolism & Disposition*, 1999, **27**, 1123.
132. A. Yaghi, J.A. Bradbury, D.C. Zeldin, S. Mehta, J.R. Bend and D.G. McCormack, *American Journal of Physiology-Lung Cellular & Molecular Physiology*, 2003, **285**, L1099.
133. A. Yaghi, J.R. Bend, C.D. Webb, D.C. Zeldin, S. Weicker, S. Mehta and D.G. McCormack, *American Journal of Physiology-Lung Cellular & Molecular Physiology*, 2004, **286**, L1260.
134. X. Cui, R.Q. Wu, M. Zhou, H.H. Simms and P. Wang, *Critical Care Medicine*, 2004, **32**, 1186.

135. M.F. Oleksiak, S. Wu, C. Parker, W. Qu, R. Cox, D.C. Zeldin and J.J. Stegeman, *Arch. Biochem. Biophys.*, 2003, **411**, 223.
136. W. Qu, R.A. Rippe, J.X. Ma, P. Scarborough, C. Biagini, F.T. Fiedorek, G.S. Travlos, C. Parker and D.C. Zeldin, *Mol. Pharmacol.*, 1998, **54**, 504.
137. Z.G. Yu, L.M. Huse, P. Adler, L. Graham, J.X. Ma, D.C. Zeldin and D.L. Krotez, *Mol. Pharmacol.*, 2000, **57**, 1011.
138. X.Y. Zhao, D.M. Pollock, E.W. Inscho, D.C. Zeldin and J.D. Imig, *Hypertension*, 2003, **41**, 709.
139. Y.Q. Zhou, H.H. Chang, J. Du, C.Y. Wang, Z. Dong and M.H. Wang, *American Journal of Physiology-Renal Physiology*, 2005, **288**, F221.
140. Z.G. Yu, V.Y. Ng, P. Su, M.M. Engler, M.B. Engler, Y. Huang, E. Lin and D.L. Kroetz, *J. Pharmacol. Expt. Ther.*, 2006, **317**, 732.
141. X.Y. Zhao, A. Dey, O.P. Romanko, D.W. Stepp, M.H. Wang, Y.Q. Zhou, L.M. Jin, J.S. Pollock, R.C. Webb and J.D. Imig, *American Journal of Physiology-Regulatory Integrative & Comparative Physiology*, 2005, **288**, R188.
142. M. Rahmaniyan, K. Patrick and N.H. Bell, *American Journal of Physiology-Endocrinology & Metabolism*, 2005, **288**, E753.
143. J.H. Moran, L.A. Mitchell, J.A. Bradbury, W. Qu, D.C. Zeldin, R.G. Schnellmann and D.F. Grant, *Toxicology & Applied Pharmacology*, 2000, **168**, 268.
144. D.E. Prosser and G. Jones, *Trends in Biochemical Sciences*, 2004, **29**, 664.
145. I. Aiba, T. Yamasaki, T. Shinki, S. Izumi, K. Yamamoto, S. Yamada, H. Terato, H. Ide and Y. Ohyama, *Steroids*, 2006, **71**, 849.
146. J. Ma, J.A. Bradbury, L. King, R. Maronpot, L.S. Davis, M.D. Breyer and D.C. Zeldin, *Biochem. Pharmacol.*, 2002, **64**, 1447.
147. J.X. Ma, S. Ramachandran, F.T. Fiedorek and D.C. Zeldin, *Genomics*, 1998, **49**, 152.
148. J.X. Ma, W. Qu, P.E. Scarborough, K.B. Tomer, C.R. Moomaw, R. Maronpot, L.S. Davis, M.D. Breyer and D.C. Zeldin, *Journal of Biological Chemistry*, 1999, **274**, 17777.
149. W. Qu, J.A. Bradbury, C.C. Tsao, R. Maronpot, G.J. Harry, C.E. Parker, L.S. Davis, M.D. Breyer, M.P. Waalkes, J.R. Falck, J.Y. Chen, R.L. Rosenberg and D.C. Zeldin, *Journal of Biological Chemistry*, 2001, **276**, 25467.
150. J. Ma, J. Graves, J.A. Bradbury, Y. Zhao, D.L. Swope, L. King, W. Qu, J. Clark, P. Myers, V. Walker, J. Lindzey, K.S. Korach and D.C. Zeldin, *Mol. Pharmacol.*, 2004, **65**, 730.
151. Q. Xie, Q.Y. Zhang, Y. Zhang, T. Su, J. Gu, L.S. Kaminsky and X. Ding, *Drug Metabolism & Disposition*, 2000, **28**, 1311.
152. J.F. Wang, Y.K. Yang, M.F. Sullivan, J.Y. Min, J.B. Cai, D.C. Zeldin, Y.F. Xiao and J.P. Morgan, *Experimental Biology & Medicine*, 2002, **227**, 182.
153. S. Wu, C.R. Moomaw, K.B. Tomer, J.R. Falck and D.C. Zeldin, *Journal of Biological Chemistry*, 1996, **271**, 3460.

154. L.M. King, J.X. Ma, S. Srettabunjong, J. Graves, J.A. Bradbury, L.P. Li, M. Spiecker, J.K. Liao, H. Mohrenweiser and D.C. Zeldin, *Mol. Pharmacol.*, 2002, **61**, 840.
155. S. Matsumoto, T. Hirama, T. Matsubara, K. Nagata and Y. Yamazoe, *Drug Metabolism & Disposition*, 2002, **30**, 1240.
156. J. Bylund, N. Finnstrom and E.H. Oliw, *Biochemical & Biophysical Research Communications*, 1999, **261**, 169.
157. B. Fisslthaler, N. Hinsch, T. Chataigneau, R. Popp, L. Kiss, R. Busse and I. Fleming, *Hypertension*, 2000, **36**, 270.
158. J. Borlak, M. Walles, K. Levsen and T. Thum, *Drug Metabolism & Disposition*, 2003, **31**, 888.
159. C. Bertrand-Thiebault, L. Ferrari, O. Boutherin-Falson, M. Kockx, S. Desquand-Billiald, J.M. Fichelle, R. Nottin, J.F. Renaud, A.M. Batt and S. Visvikis, *Clinical & Experimental Pharmacology & Physiology*, 2004, **31**, 295.
160. J. Gu, T. Su, Y. Chen, Q.Y. Zhang and X. Ding, *Toxicology & Applied Pharmacology*, 2000, **165**, 158.
161. A. Gaedigk, D.W. Baker, R.A. Totah, R. Gaedigk, R.E. Pearce, C.A. Vyhlidal, D.C. Zeldin and J.S. Leeder, *J. Pharmacol. Expt. Ther.*, 2006, **319**, 523.
162. N.Y. Marden, E. Fiala-Beer, S.H. Xiang and M. Murray, *Biochem. J.*, 2003, **373**, 669.
163. N.Y. Marden and M. Murray, *Biochem. J.*, 2005, **391**, 631.
164. M. Spiecker, H. Darius, T. Hankeln, M. Soufi, A.M. Sattler, J.R. Schaefer, K. Node, J. Borgel, A. Mugge, K. Lindpaintner, A. Huesing, B. Maisch, D.C. Zeldin and J.K. Liao, *Circulation*, 2004, **110**, 2132.
165. K.H. Liu, M.G. Kim, D.J. Lee, Y.J. Yoon, M.J. Kim, J.H. Shon, C.S. Choi, Y.K. Choi, Z. Desta and J.G. Shin, *Drug Metabolism & Disposition*, 2006, **34**, 1793.
166. S. Matsumoto, T. Hirama, H.J. Kim, K. Nagata and Y. Yamazoe, *Xenobiotica*, 2003, **33**, 615.
167. P. Lafite, S. Dijols, D. Buisson, A.C. Macherey, D.C. Zeldin, P.M. Dansette and D. Mansuy, *Bioorganic & Medicinal Chemistry Letters*, 2006, **16**, 2777.
168. J. Seubert, B.C. Yang, J.A. Bradbury, J. Graves, L.M. DeGraff, S. Gabel, R. Gooch, J. Foley, J. Newman, L. Mao, H.A. Rockman, B.D. Hammock, E. Murphy and D.C. Zeldin, *Circulation Research*, 2004, **95**, 506.
169. J.G. Jiang, C.L. Chen, J.W. Card, S.L. Yang, J.X. Chen, X.N. Fu, Y.G. Ning, X. Xiao, D.C. Zeldin and D.W. Wang, *Cancer Res.*, 2005, **65**, 4707.
170. K. Node, Y.Q. Huo, X.L. Ruan, B.C. Yang, M. Spiecker, K. Ley, D.C. Zeldin and J.K. Liao, *Science*, 1999, **285**, 1276.
171. K. Node, X.L. Ruan, J.W. Dai, S.X. Yang, L. Graham, D.C. Zeldin and J.K. Liao, *Journal of Biological Chemistry*, 2001, **276**, 15983.
172. B.C. Yang, L. Graham, S. Dikalov, R.P. Mason, J.R. Falck, J.K. Liao and D.C. Zeldin, *Mol. Pharmacol.*, 2001, **60**, 310.

173. J.X. Sun, X. Sui, J.A. Bradbury, D.C. Zeldin, M.S. Conte and J.K. Liao, *Circulation Research*, 2002, **90**, 1020.
174. Y.F. Xiao, Q.G. Ke, J.M. Seubert, J.A. Bradbury, J. Graves, L.M. DeGraff, J.R. Falck, K. Krausz, H.V. Gelboin, J.P. Morgan and D.C. Zeldin, *Mol. Pharmacol.*, 2004, **66**, 1607.
175. M. Spiecker and J.K. Liao, *Archives of Biochemistry & Biophysics*, 2005, **433**, 413.
176. T. Lu, D. Ye, X.L. Wang, J.M. Seubert, J.P. Graves, J.A. Bradbury, D.C. Zeldin and H.C. Lee, *Journal of Physiology-London*, 2006, **575**, 627.
177. M. Spiecker and J.K. Liao, *Trends in Cardiovascular Medicine*, 2006, **16**, 204.
178. S.S. Lee, H.E. Jeong, K.H. Liu, J.Y. Ryu, T. Moon, C.N. Yoon, S.J. Oh, C.H. Yun and J.G. Shin, *Pharmacogenetics & Genomics*, 2005, **15**, 105.
179. A.W. Dreisbach, S. Japa, A. Sigel, M.B. Parenti, A.E. Hess, S.L. Srinouanprachanh, A.E. Rettie, H. Kim, F.M. Farin, L.L. Hamm and J. J. L. Lertora, *American Journal of Hypertension*, 2005, **18**, 1276.
180. G. Gervasini, S. Vizcaino, J.A. Carrillo, M.J. Caballero and J. Benitez, *British Journal of Clinical Pharmacology*, 2006, **62**, 177.
181. H.J. Wang, Y. Jiang, Y. Liu, C.Z. Lin, G. Cheng, X.P. Chen, B.T. Hao, W. Tan, D.X. Lin and F.C. He, *Clinica Chimica Acta*, 2006, **365**, 125.
182. J.W. Newman, J.E. Stok, J.D. Vidal, C.J. Corbin, Q. Huang, B.D. Hammock and A.J. Conley, *Endocrinology*, 2004, **145**, 5097.
183. L. Wang, J.H. Yao, L. Chen, J.Z. Chen, J.L. Xue and W. Jia, *Biochemical & Biophysical Research Communications*, 2007, **352**, 850.
184. X. Ding and M.J. Coon, *Drug Metabolism & Disposition.*, 1990, **18**, 742.

CHAPTER 11

The CYP3 Family

DAVID J. GREENBLATT, PING HE, LISA L. VON MOLTKE AND MICHAEL H. COURT

Department of Pharmacology and Experimental Therapeutics, Tufts University School of Medicine and Tufts-New England Medical Center Boston MA, USA

Table of Contents

11.1 Introduction

The CYP3A enzymes are of major importance in human biology and clinical therapeutics.[1,2] They are the most of abundant of the CYP enzymes in the

Issues in Toxicology
Cytochromes P450: Role in the Metabolism and Toxicity of Drugs and other Xenobiotics
Edited by Costas Ioannides

liver,[3] and are the only CYPs present in substantive amounts in the enteric mucosa of the gastrointestinal tract. Substrate specificity is broad; CYP3A mediates the biotransformation of numerous endogenous substances, environmental chemicals of potential toxicological relevance, and medications used in clinical therapeutics (Table 11.1).

The significance of CYP3A in terms of human evolution and species preservation is a topic of logical speculation. The lack of a phenotypic "null" status for CYP3A – corresponding for example to CYP2D6 "poor metabolisers" – supports the notion that CYP3A is essential. Homeostasis of many endogenous steroid hormones, including those required for sexual maturation and reproduction, is dependent on CYP3A.[4,5] A number of environmental chemicals, both therapeutic and toxic, derived from plants and fungi are in fact substrates for metabolism by CYP3A enzymes. Examples include the cinchona alkaloids, the taxanes, opiates, aflatoxins, and the immunosuppressants cyclosporine, tacrolimus and sirolimus. Further clues derive from the anatomic distribution of human CYP3A, and its unique kinetic characteristics. The dual localisation of CYP3A in enteric mucosa and liver provides the species with "two shots" at protection against potentially toxic environmental chemicals before they reach the systemic circulation (Figure 11.1). The property of binding cooperativity,

Table 11.1 Representative CYP3A substrate drugs used in clinical practice.

Clearance completely or nearly completely dependent on CYP3A	*Clearance partially dependent on CYP3A*
Alfentanyl	Amitriptyline
Alprazolam	Citalopram
Atorvastatin	Clozapine
Buspirone	Dextromethorphan
Carbamazepine	Diazepam
Cyclosporine	Imipramine
Eletriptan	Methadone
Erythromycin	Omeprazole
Felodipine	Sertraline
Midazolam	Telithromycin
Nefazodone	Voriconazole
Nifedipine	Zolpidem
Quetiapine	
Quinidine	
Ritonavir	
Saquinavir	
Sildenafil	
Simvastatin	
Tacrolimus	
Tadalafil	
Trazodone	
Triazolam	
Vardenafil	
Verapamil	

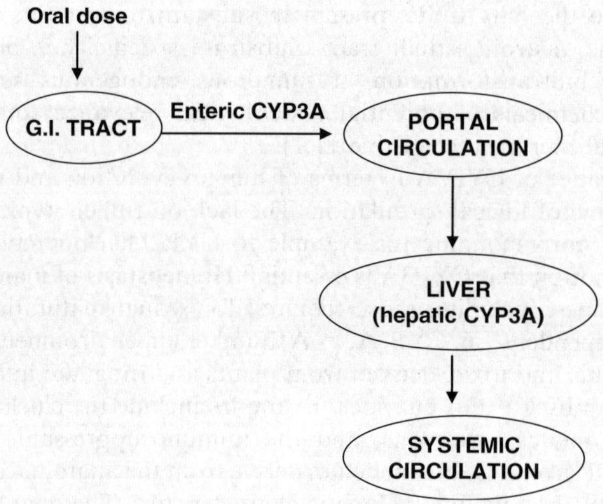

Figure 11.1 Schematic diagram of the fate of an orally-administered CYP3A substrate drug prior to reaching the systemic circulation. Metabolism by CYP3A4 and CYP3A5 is possible during passage across the enteric mucosa of the proximal small bowel as well as during passage through the liver.

yielding homotropic autoactivation, is consistent with a protective effect of increased CYP3A activity upon exposure to high and potentially toxic quantities of exogenous substrate.[6] Finally, there is extensive (but not complete) overlap in affinity of substrates for metabolism by CYP3A along with efflux transport by P-glycoprotein.[7–9] Many of the natural substances described above are substrates for both, consistent with a coincident "protective" purpose.

A notable observation not supporting the central importance of CYP3A is that no life-threatening medical consequences are known to arise from extended treatment with highly potent CYP3A inhibitors such as ketoconazole[10] and ritonavir.[11] Both of these drugs produce what amounts to a chemically-induced "poor metaboliser" phenotype. The antifungal agent ketoconazole has been available since the early 1980s, and the viral protease inhibitor ritonavir since the mid 1990s. Extended exposure to ketoconazole is associated with antiandrogenic effects, and a lipodystrophy syndrome has been linked to extended treatment with ritonavir and other protease inhibitors.[12] These sequelae are possibly attributable to CYP3A inhibition, but this is not proven. In any case, the sequelae are significant but not ominous medical consequences, raising questions as to how essential it is in adults for CYP3A phenotype to be maintained within a "normal" range.

The unique protective function of hepatic and enteric CYP3A enzymes may produce complications in drug development and clinical therapeutics. New chemical entities found to be complete or nearly complete substrates for clearance by CYP3A may actually be dropped from subsequent development. Such candidates often are seen as facing therapeutic or competitive obstacles, largely attributable to the possibility of drug interactions with other agents that are CYP3A inducers or inhibitors, and restrictive labelling that could result. Some marketed drugs, such as terfenadine, astemizole and cisapride were actually withdrawn for that reason.[13–16] Similar concerns apply to new drug candidates that themselves are found to be significant CYP3A inducers or inhibitors. Some CYP3A substrate drugs that meet a pressing medical need, such as the viral protease inhibitors saquinavir and lopinavir, have been brought through the development process and approved for clinical use despite the drawbacks and obstacles. These two drugs have the disadvantage of poor net oral bioavailability, due to some combination of hepatic/enteric presystemic extraction together with enteric efflux transport by P-glycoprotein (P-gp). So significant is this problem that saquinavir and lopinavir are, with few exceptions, combined with the CYP3A/P-gp inhibitor ritonavir for purposes of "boosting" or "augmentation" of oral bioavailability.[17–19]

The topic of this review is the translational pharmacology of CYP3A enzymes in humans. The principal focus will be a number of contemporary and/or controversial issues of current importance.

11.2 Pharmacogenomics

The four genes encoding the relevant CYP3A proteins (CYP3A4, 5, 7, and 43) are all located in a 231-kb cluster on chromosome band 7q21-q22.1. The relationship of CYP3A genomic variants to CYP3A protein expression and activity *in vitro* and *in vivo* has been extensively investigated over the last decade.[20–25] The prevailing contemporary interpretation of the existing data base is that a great deal of information has been generated, collectively demonstrating very little in the way of meaningful associations between CYP3A genotype and *in vivo* phenotypic metabolic activity. Furthermore, there is substantial disconnect between *in vitro* studies and human pharmacokinetic studies *in vivo*.

CYP3A7 is a foetal enzyme, and its expression is silenced after birth. Reports exist of persistently detectable CYP3A7 mRNA in adulthood, but there is no available evidence that CYP3A7 is of any significance in terms of *in vivo* drug-metabolising activity. CYP3A43 is an adult enzyme, and is known to be localised in prostate.[26,27] There is no known functional significance of CYP3A43 identified to date.

CYP3A4 is the dominant isoform in humans. Numerous studies have demonstrated that CYP3A4 is the most abundant of the human hepatic cytochromes P450 and also is (along with CYP3A5) the only cytochrome P450 of functional importance in the enteric mucosa of the gastrointestinal tract. Expression of hepatic and enteric CYP3A4 is not coordinately regulated; levels expressed at the two sites are not intercorrelated.[28]

Many single nucleotide polymorphisms (SNPs) have been identified in the CYP3A4 locus.[29–32] Some of these SNPs are associated with reduced functional metabolic capacity in various *in vitro* systems.[33] However, there is no substantial evidence that any CYP3A4 SNP is associated with clinically important differences in clearance of CYP3A substrates *in vivo*.[31,34–42] Consistent with this is the observation that phenotypic distribution of CYP3A metabolic activity *in vitro* and *in vivo* is unimodal rather than multimodal (bimodal or trimodal),[43–46] thereby essentially excluding the existence of common "null" alleles coding for low or zero protein expression or function. Finally, all CYP3A4 SNPs with demonstrated functional consequences are of low prevalence in the population. Nonetheless it cannot be fully excluded that one or more SNPs (such as CYP3A4*20)[33] could account for low CYP3A metabolic activity noted in some unusual "outlying" subjects in a number of human studies.[46–48]

The CYP3A4*1B polymorphism is located in the 5′-regulatory region of the *CYP3A4* gene. This SNP received considerable attention when it was first described, due to the statistical association with poor outcome in patients with prostate cancer.[49] A link of CYP3A4*1B to altered testosterone metabolism was speculated as possible mechanism, but the study did not evaluate plasma testosterone or any other index of metabolic activity. A few subsequent reports provided some evidence supporting associations of the CYP3A4*1B polymorphism with human disease, but other studies have not.[50–57] In careful clinical phenotype-genotype studies using CYP3A probe substrates such as midazolam, a detectable phenotypic importance of CYP3A4*1B was not found.[31,34–40,58] Thus the mechanism of the linkage of the CYP3A4*1B polymorphism to human disease, if a link actually exists, remains unexplained. It has been postulated that a partial linkage disequilibrium of CYP3A4*1B with CYP3A5*3 may explain the association, inasmuch as CYP3A5 is found in prostate and may play a role in local androgen metabolism.[59] However this is a speculative explanation without experimental support.

CYP3A5 shares approximately 90% sequence homology with CYP3A4. CYP3A5 is expressed in liver and gastrointestinal enteric mucosa, as well as a number of other tissues including prostate and kidney.[60] In contrast to CYP3A4, there is evidence that CYP3A5 is polymorphically expressed.[59–63] The CYP3A5*3 variant, which is the most prevalent form in many human populations, actually confers low or zero functional CYP3A phenotypic activity, whereas the less prevalent CYP3A5*1 codes for active protein. Expression of immunoactive CYP3A5 protein in human liver samples is consistent with genotype. Since CYP3A5*1 has greater prevalence in the African-American population compared with Caucasians, the possibility is raised – politically alluring to some – that a racial difference in CYP3A phenotype may exist. This has been supported by *in vitro* studies of CYP3A metabolic activity in genotyped human liver samples. However, the outcomes of clinical studies are not consistent with this scheme. Human CYP3A metabolic phenotype is, at most, weakly associated with CYP3A5 genotype, and there is little or no evidence of a race-associated difference in metabolic activity.[31,34–36,41,64–66] Again this demonstrates the *in vitro-in vivo* disconnect described previously. At present, the abundance of evidence indicates that CYP3A5, and its

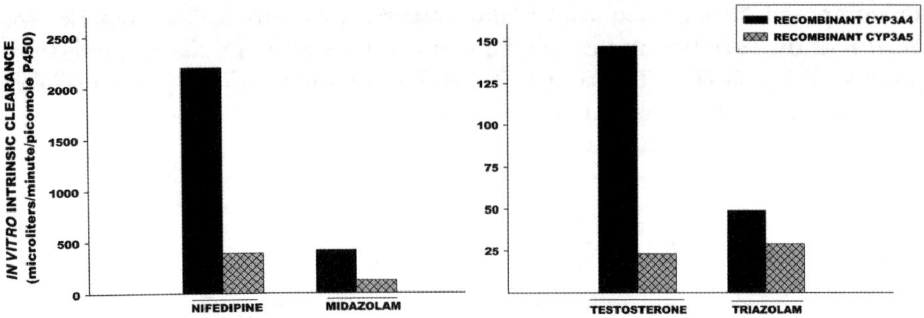

Figure 11.2 *In vitro* intrinsic clearance of four CYP3A substrates by recombinant CYP3A4 and recombinant CYP3A5 (see reference 70).

polymorphic variants, have at most a minor effect on human CYP3A phenotype *in vivo*.[61,67] However, there may be a few substrates, such as alprazolam, for which CYP3A5 genotype has a more significant role in the modulation of drug clearance.[68,69]

The value of comparative *in vitro* studies of CYP3A4 and CYP3A5 has been enhanced by the availability of relatively specific, commercially-available antibodies for separate immunoquantitation of the two proteins, as well as distinct recombinant enzymes for evaluation of enzyme function. Using these tools, it has been established, for example, that *in vitro* intrinsic metabolic clearance for a number of substrates (midazolam, triazolam, testosterone, nifedipine) *via* recombinant CYP3A4 exceeds that for CYP3A5 (Figure 11.2).[70] In addition, inhibitory potencies of an index inhibitor (ketoconazole), as well as a number of other inhibitors, are greater for CYP3A4 compared with CYP3A5.[70,71] However, a major obstacle for translation of *in vitro* results to humans is that all known CYP3A substrates are metabolised by both CYP3A4 and CYP3A5, and no substrate is specific for one or the other. Therefore, the relative *in vivo* contributions of the two enzymes to net CYP3A metabolic activity cannot be resolved by customary probe substrate methodology.

The wide individual variability in CYP3A expression and function is commonly discussed in the literature, but the metrics used to quantitate variation generally have been poorly defined. In many sources the term "fold variation" is used. A "50-fold variation in CYP3A expression" may be claimed, but the term remains undefined. If it refers to the ratio of maximum value divided by minimum value, this is not a useful metric of variation, since it is driven by the outlying values at both ends. There is a need for a consensus in the field on the meaning of "variability" and how it is best quantitated. We suggest the use of the coefficient of variation (CV), calculated as the arithmetic standard deviation divided by the arithmetic mean, expressed as percent. In any case, it is clear that "individual variability" in CYP3A expression and function observed among human liver microsomal samples *in vitro* substantially exceeds variability in CYP3A metabolic phenotype *in vivo*.[3,72] This is probably attributable to the additional variability superimposed through acquisition and processing of human liver samples for *in vitro* study: variation in the time between donor death and tissue harvesting/preservation; source of the sample (transplant donor *vs.* autopsy *vs.* surgical); mode of demise of deceased donors (acute illness or trauma *vs.* chronic disease); and the donor's medication exposure.

11.3 Enzyme Kinetics of CYP3A

Most work on the enzyme kinetic features of CYP3A enzymes has focussed on CYP3A4. This enzyme is generally characterised as low-affinity and high-capacity (high K_m, high V_{max}). There are exceptions for some substrates such as midazolam, which has a K_m value in the low micromolar range for the principal metabolic pathway (1-hydroxylation).[73,74] However for the majority of CYP3A substrates, the apparent K_m *in vitro* greatly exceeds the usual range of concentrations encountered *in vivo*, with few clinical examples of non-linear kinetics.

For many substrates, *in vitro* studies assessing reaction velocity in relation to substrate concentration yield a sigmoidal pattern, in contrast to Michaelis-Menten (hyperbolic) kinetics. The sigmoidal profile is consistent with homotropic (positive) cooperativity or autoactivation, in which binding of one substrate molecule facilitates binding of the next molecule, effectively increasing affinity to that vacant site.[75–83] This also implies multiple ligand binding sites, and changes in enzyme conformation attributable to ligand binding. Affected substrates include amitriptyline, carbamazepine, diazepam, triazolam, alprazolam, nifedipine, progesterone and testosterone.

Sigmoidal kinetics attributable to positive cooperativity complicates the interpretation of *in vitro* data, and *in vivo* extrapolation for purposes of quantitative reaction phenotyping.[84–88] With a sigmoidal kinetic pattern, the substrate concentration yielding a reaction velocity equal to 50% of V_{max} (S_{50}) is not conceptually equivalent to K_m in a Michaelis-Menten model. In addition, the concept of *in vitro* intrinsic clearance – calculated as V_{max}/K_m in the Michaelis-Menten model – is not applicable to sigmoidal kinetics. The quantity V_{max}/S_{50} has been used as a surrogate for intrinsic clearance in a number of studies, but this is applied empirically without theoretical support.

Heterotropic activation is another possible consequence of multiple substrate binding sites. With this mechanism, binding of one substrate can increase binding affinity for a different substrate. Among the first recognised "activators" was 7,8-benzoflavone (α-naphthoflavone),[89] and a number of others have been subsequently identified.

Inhibitory allosteric effects are also possible. When substrate binding inhibits subsequent binding of other substrate molecules (negative homotropy), the observed consequence is substrate inhibition.[75–83] Complex situations may arise involving concurrent heterotropic activation and inhibition. An example is the concurrent effect of testosterone, simultaneously activating triazolam α-hydroxylation while inhibiting 4-hydroxylation.[70,90]

It should be emphasised that all of these findings derive from molecular and *in vitro* models. There are no known clinical consequences or correlates of the sigmoidal kinetic profile of a number of CYP3A substrates.

11.4 Individual Variability in CYP3A Metabolic Phenotype

Despite implications to the contrary, individual variation in human CYP3A metabolic phenotype *in vivo* is in the same range as that for other CYP enzymes. Among relatively homogenous populations, such as groups of healthy, medication-free young male volunteers that typically participate in clinical pharmacokinetic studies, the between-subject CV for area under the plasma concentration curve (AUC) or clearance of a CYP3A probe such as midazolam or triazolam generally will not exceed 40–60%.[45,47,91–101] Similar variability is observed for substrates of CYP2D6,[102–105] CYP2C9[106–109] and CYP1A2.[110–113] In contrast to between-subject variation, within-subject variation in clearance of a CYP3A substrate is in the range of 10–15%.[114] This has been determined in studies in which the same dose of the same substrate (midazolam) was administered to the same subject on multiple occasions.[114]

Clearance values of different CYP3A probe substrates across individuals are not necessarily highly correlated with or "predictive" of each other.[98,115–120] This is because of the complex relationship of net CYP3A metabolic phenotype to the intrinsic clearance characteristics of the specific substrate, the route of administration, and the relative dependence of net clearance on hepatic versus enteric CYP3A activity. For low-clearance substrates, such as alprazolam and erythromycin, net clearance will be dependent primarily upon hepatic CYP3A, and essentially independent of route of administration and hepatic blood flow. For higher clearance substrates, such as buspirone, midazolam, and triazolam, intravenous clearance will depend on a combination of hepatic metabolism and hepatic blood flow, whereas clearance after oral dosage will depend on a combination of hepatic and enteric extraction, being independent of hepatic blood flow (Figure 11.3). Since each individual substrate is characterised by a unique combination of these factors, it is not surprising that phenotypes derived from different substrates are not predictive of each other. In any case, CYP3A probes that are also substrates for transport by P-gp (such as erythromycin and cyclosporine) are probably not the best choices, since their net disposition will reflect both CYP3A and P-gp effects.[121–123]

The possibility of CYP3A metabolic phenotyping without the use of an exogenous probe has received considerable attention. The ratio of urinary excretion of 6-β-hydroxycortisol divided by cortisol is the most extensively evaluated endogenous index;[124,125] the outcomes are equivocal at best. The principal drawback is that the 6-β-hydroxycortisol/cortisol excretion ratio, unlike clearance of other exogenous index substrates, can be quite variable

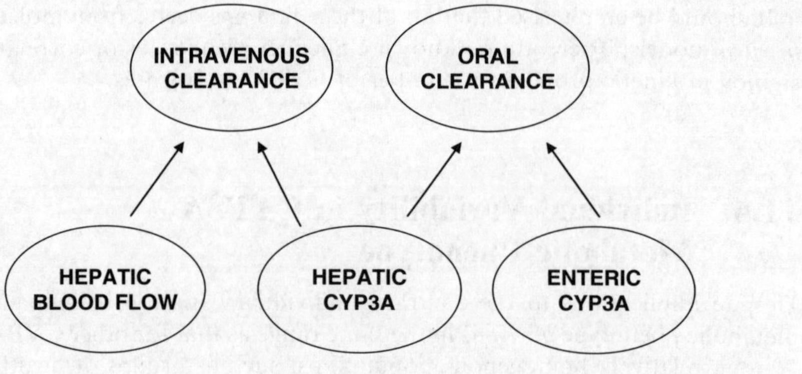

Figure 11.3 Determinants of metabolic clearance of CYP3A drug substrates when given intravenously or orally. I.V. clearance depends on hepatic CYP3A and (for high-clearance drugs) on hepatic blood flow. Oral clearance depends on a combination of enteric and hepatic CYP3A, but is not dependent on hepatic blood flow.

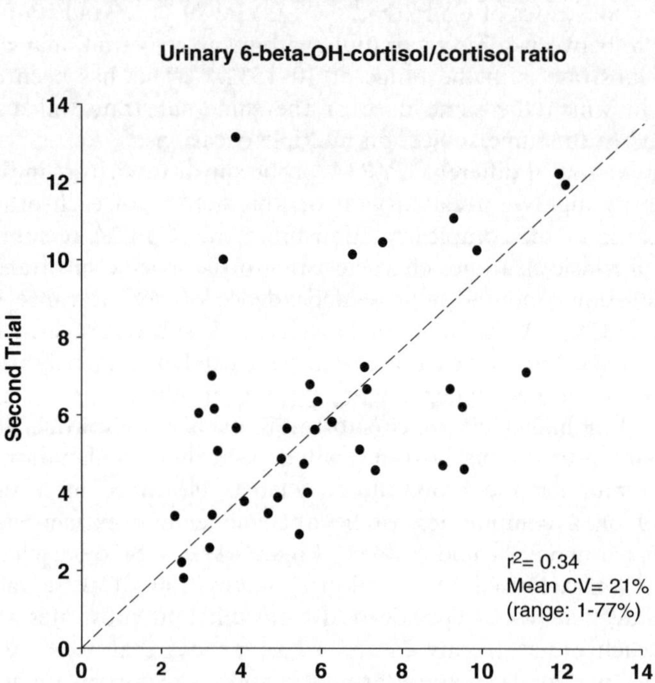

Figure 11.4 Within-subject variability in the urinary 6β-hydroxycortisol/cortisol ratio determined in the same individuals on two occasions (see reference 128).

from time to time within the same individual (Figure 11.4).[126–129] As such, the ratio may adequately reflect large differences in CYP3A phenotype as caused, for example, via extensive enzyme induction by rifampin. However the ratio appears to be relatively insensitive to more subtle individual differences in metabolic activity. Taken together, the evidence does not support the use of the 6-β-hydroxycortisol/cortisol ratio as a CYP3A phenotypic index.

11.5 Age and Gender Effects on CYP3A Phenotype

The possible modulation of CYP3A metabolic activity by age and gender is of both scientific and public health importance. With CYP3A substrate clearance values having a CV in the range of 50–60% of the mean, identifying individually appropriate dosage levels for therapeutic agents metabolised by CYP3A may necessarily involve an iterative process of trial dosage, observation and dosage revision as necessary. More precise prediction of clearance based on demographic factors such as age and gender could facilitate accuracy of initial dose selection, with reduction of the need for revision through trial and error.

11.5.1 *In vitro* and Experimental Studies

The data base from *in vitro* studies of human liver samples is large, but the findings are inconsistent and inconclusive. Some studies show reduced expression of immunoactive CYP3A and/or reduced metabolic activity in liver samples from older humans.[130–132] Other studies show little or no effect of age.[3,133,134] The intrinsic limitations of *in vitro* systems have already been discussed (*vide supra*), and are especially pertinent to assessment of advancing age as an independent determinant of protein expression or metabolic function. Demise in younger donors is more likely to result from trauma or sudden unexpected illness. Preservation of "normal" liver is more likely, and metabolic activity could actually be increased (induced) by factors such as high-dose exogenous corticosteroids used to treat head trauma. In contrast, elderly donors are more likely to have suffered from extended illness (cancer, cardiovascular disease, Alzheimer's etc.) with associated pharmacologic treatment, poor nutrition, inactivity and general debility. Studies of gender effects on CYP3A *in vitro* likewise are inconsistent and inconclusive.[131–136] It is important to note that should age and gender be independent modulators of CYP3A expression/function, outcomes of *in vitro* "population" studies could be confounded, depending on the age/gender composition of the samples available to the investigator.

In studies of Fischer-344 rats, reduced expression of immunoreactive CYP3A, and reduced CYP3A metabolic activity, has been observed in liver samples obtained from aging animals.[137] This was not explained by reduced testosterone concentrations in the older animals; enteric CYP3A, on the other hand, was not influenced by age. The rodent model may not be applicable to humans for many reasons, including that the CYP3A enzymes are different between the species. Nonetheless, the animal model could provide the opportunity to study the regulatory mechanisms through which CYP3A expression declines with age.

11.5.2 Clinical Studies

A large number of clinical pharmacokinetic studies evaluated the influence of age and gender on human CYP3A phenotype, and on age-related changes in drug disposition in general.[138–145] Numerous CYP3A substrates have been studied, and trial designs vary widely in terms of sample size, method of subject selection and approach to statistical analysis. Reviews of the topic generally are not fully comprehensive, largely because the scope of the published literature is so large. The conclusions of a review may be dependent on which specific studies are included in the evaluation and which are omitted.

Based on a literature review that is as complete as possible, there is extensive, but not completely consistent, evidence that weight-normalised metabolic clearance of CYP3A substrates declines with age (Table 11.2).[138] There is also evidence that effects of age and gender are not independent; in many studies the decrease in clearance with increasing age among male subjects is far greater than the clearance decrease in women. For this reason, studies that combine data for men and women may obscure an age effect. Enhanced pharmacodynamic effects of CYP3A substrate drugs in the elderly have also been demonstrated in many studies. This may be attributable to the higher plasma drug concentrations in elderly subjects, but also from an independent effect of age on intrinsic drug sensitivity.[146] The net clinical implication is that lower doses of most CYP3A substrate drugs are recommended for the elderly.

Table 11.2 CYP3A substrate drugs for which reduced clearance in the elderly has been demonstrated in clinical studies.[a]

Adinazolam
Alfentanyl
Alprazolam
Amlodipine
Bromazepam
Clarithromycin
Diltiazem
Erythromycin
Felodipine
Midazolam
Nefazodone
Nifedipine
Prednisolone
Quinidine
Tirilizad
Trazodone
Triazolam
Verapamil
Zolpidem
Zopiclone

[a]See Reference 138.

The data on gender effects on CYP3A phenotype are murky.[138,145,147–154] We have concluded that most studies comparing young men and young women show minimal or no difference in weight-normalised clearance. Some studies show higher clearance in women, but generally the gender difference is small in magnitude and unlikely to be of clinical importance. There is little or no evidence to suggest that use of hormonal contraceptive preparations or oestrogen supplementation has a significant effect on CYP3A phenotype.

11.6 Drug Interactions with CYP3A Substrates

Pharmacokinetic drug interactions involving drugs metabolised by CYP3A are of major contemporary scientific and public health importance.[155–157] The visibility of the topic increased substantially in the mid to late 1990s, in connection with adverse reactions to the nonsedating antihistamine terfenadine (Seldane).[13–16] Under usual circumstances, terfenadine itself is a prodrug, being nearly completely biotransformed to fexofenadine by CYP3A enzymes through presystemic extraction. However, in rare patients in whom CYP3A metabolic activity was depressed due to co-treatment with ketoconazole or erythromycin, significant concentrations of intact terfenadine did appear in the circulation, producing serious and even fatal cardiac arrhythmias due to its arrhythmogenic properties. Terfenadine was withdrawn from the market, as were astemizole and cisapride, two other drugs with similar properties. The calcium channel antagonist mibefradil was likewise withdrawn from clinical use, being a strong CYP3A inhibitor. While epidemiologic studies demonstrate that clinically important or hazardous drug interactions involving CYP3A actually are unusual relative to the prevalence of treatment with multiple drugs,[158] the sensitivity of the public and the scientific community to the drug interaction issue has nonetheless been heightened. In the process of drug discovery and development, drug candidates found to be predominant CYP3A substrates, and/or significant CYP3A inhibitors/inducers, frequently are dropped from further development due to actual or perceived liabilities or hazards attributable to drug interactions.

Drug interactions with CYP3A substrates may occur via metabolic inhibition or metabolic induction. These are mechanistically different events with different clinical consequences (Table 11.3).

11.6.1 Drug Interactions via Metabolic Inhibition

A number of drugs used in clinical practice are inhibitors of CYP3A metabolic activity (Table 11.4). Some of these, including the azole antifungal agents ketoconazole and itraconazole, and the viral protease inhibitor ritonavir are highly potent inhibitors, such that coadministration with CYP3A substrate drugs could produce important and possibly hazardous impairment of clearance and elevation of plasma levels.[10,11] New drug candidates that are CYP3A

Table 11.3 Comparison of inhibition and induction.

	Inhibition	*Induction*
Mechanism:	***Direct chemical effect on enzyme***	***Increased protein synthesis***
Time-course of onset	Rapid	Slow
Time-course of offset	Rapid	Slow
In vitro model	Cell homogenates	Cultured hepatocytes
Metric for inhibition or induction potency	K_i or IC_{50}	Not established
Clinical consequences:		
Effect on clearance of victim	Decreased	Increased
Effect on plasma levels of victim	Elevated	Reduced
Principal therapeutic concern	Toxicity	Reduced efficacy

Table 11.4 Representative CYP3A inhibitors.

Strong	*Moderate or weak*
Clarithromycin	Atazanavir
Erythromycin	Delavirdine
Itraconazole	Fluconazole
Ketoconazole	Fluvoxamine
Nefazodone	Grapefruit juice
Ritonavir	Lopinavir
Telithromycin	Nelfinavir
Voriconazole	Saquinavir

substrates generally are subjected to ketoconazole or ritonavir drug interaction studies, designed to delineate the "worst case" drug interaction scenario. A number of other inhibitors are less potent, and will produce inhibitory drug interactions that are quantitatively smaller, and may or may not be of clinical importance. Ultimately, the magnitude and clinical importance of an inhibitory drug interaction will depend on the properties of the substrate as well as the potency of and systemic exposure to the inhibitor. For substrates (such as midazolam) that ordinarily undergo high presystemic extraction after oral dosage, CYP3A inhibition may profoundly increase oral bioavailability through impaired presystemic extraction, resulting in very large increases in substrate plasma concentrations. These interactions may be clinically hazardous or – in the case of deliberate "boosting" – elevate plasma concentrations of a drug with very low bioavailability, such as saquinavir or lopinavir, into a theoretically effective range.[17–19] In contrast, CYP3A substrate drugs with low presystemic extraction and high oral bioavailability under control conditions will also have impaired clearance with coadministration of a potent CYP3A inhibitor, but the effect will be evident mainly as a prolongation of half-life rather than a change in bioavailability. Comparative studies of midazolam, triazolam, and alprazolam coadministered with ketoconazole illustrate this point (Table 11.5).[73,159–162]

Table 11.5 Comparative pharmacokinetics of oral alprazolam, triazolam, and midazolam, with and without coadministration of ketoconazole.[a]

	Alprazolam (1.0 mg)	*Triazolam (0.25 mg)*	*Midazolam (6.0 mg)*
Principal metabolites	α-OH, 4-OH	α-OH, 4-OH	α-OH, 4-OH
Ketoconazole in vitro K_i (μM)			
vs. α-OH pathway	<0.1	<0.1	<0.1
vs. 4-OH pathway	<0.1	<0.1	<0.1
Kinetic properties without inhibitor (control condition)			
C_{max} (ng/ml)	14.7	2.6	19
Elimination half-life (hr)	15.2	3.0	1.7
Oral clearance (ml/min)	86	444	2289
Kinetic properties with ketoconazole coadministration			
C_{max} (ng/ml)	16.1	5.4	81
Elimination half-life (hr)	59	18.3	9.9
Oral clearance (ml/min)	27	40	147
Ratio of total AUC (with ketoconazole divided by control)	3.98	13.7	15.1

[a]See References 73, 159–162.

Chemical inhibition may be reversible or irreversible (mechanism-based). For reversible inhibitors such as ketoconazole, the inhibitory effect will be detectable only as long as the inhibitor is in contact with the enzyme. For irreversible inhibitors such as erythromycin and atazanavir, recovery from inhibition will depend on the intrinsic rate of enzyme turnover (regeneration).[163–166] While reversible and mechanism-based inhibitors can be distinguished on theoretical and biochemical bases, the clinical distinction is less clear. It is likely that many inhibitors are mixed reversible and nonreversible. Further, CYP3A enzyme turnover is relatively rapid. Attainment of the full inhibitory effect is of relatively rapid onset after the start of exposure to the inhibitor.[167] Recovery of CYP3A metabolic function may likewise be rapid after exposure to the inhibitor is terminated.[94,168]

Grapefruit juice appears to be one of the few nutrient or natural products that produce clinically important CYP3A inhibition in humans. The furanocoumarin derivative 6′,7′-dihydroxybergamottin (DHB) appears to be the principal component of grapefruit juice accounting for the CYP3A inhibitory effect.[94,168–170] DHB is a mechanism-based (irreversible) inhibitor (Figure 11.5) that acts on enteric but not hepatic CYP3A. Drug interactions with grapefruit juice occur only with orally-administered substrates that ordinarily undergo significant presystemic extraction by enteric CYP3A.[169] Recovery of CYP3A function after exposure to grapefruit juice is consistent with the time-course of enzyme turnover.[94]

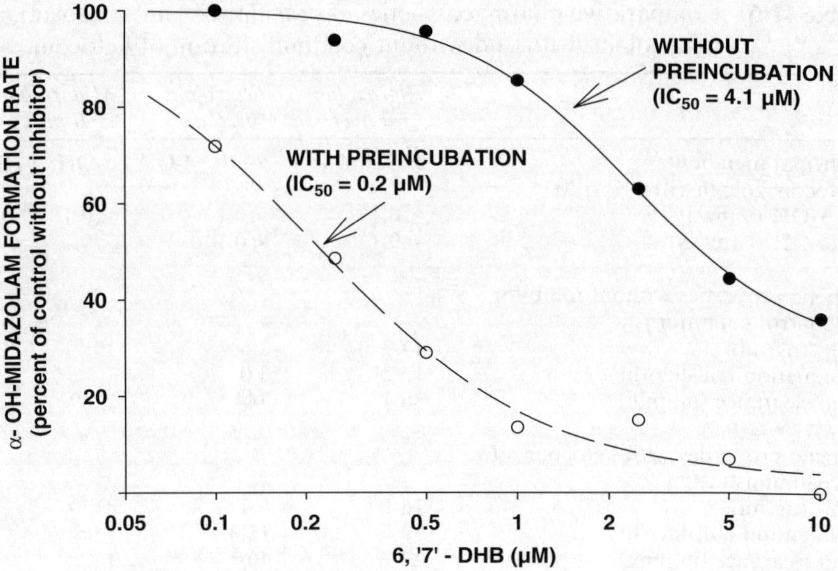

Figure 11.5 *In vitro* study of the effect of 6,′ 7′-dihydroxybergamottin (6′,7′-DHB) on α-hydroxylation of midazolam by a human liver microsomal preparation. Reaction velocities are expressed as a fraction (percentage) of the control velocity with no inhibitor present. IC_{50} represents the inhibitor concentration reducing the reaction velocity to 50% of the control value. Preincubation of the inhibitor with microsomal protein and NADPH prior to addition of the substrate (midazolam) shifts the inhibition curve leftward, reducing the IC_{50} from 4.1 micromolar to 0.2 micromolar. This is consistent with irreversible (mechanism-based) inhibition. See reference 94 for details of methodology.

The use of *in vitro* data to anticipate and predict clinical pharmacokinetic drug interactions has received extensive research attention over the last 15 years.[171–186] Customary *in vitro* systems can be used to generate quantitative estimates of an inhibitor's potency using the inhibition constant (K_i) or a 50% inhibitory concentration (IC_{50}). An *in vivo* inhibitor exposure ([I]) can be estimated or measured. These terms are combined to yield a predictive estimate of substrate clearance inhibition *in vivo*, usually designated as the ratio of substrate AUC with coadministration of the inhibitor (AUC_I) divided by AUC in the uninhibited control state (AUC_O). In its most straightforward form, the "linkage" equation going from *in vitro* to *in vivo* is:

$$1 + \frac{[I]}{K_i} = \frac{AUC_I}{AUC_O}$$

Unfortunately, the accuracy of the predictive scaling scheme to date has been bleak. There is general agreement that $[I]/K_i$ values in excess of 10.0 are associated with a high probability of a "significant" clinical interaction, whereas a value of [I]/K_i less than 0.1 makes an interaction unlikely.[187,188] Beyond these broad qualitative guidelines, reasonable quantitative prediction has been mostly unsuccessful. Reasons for poor scaling accuracy can be categorised as follows: methodologic limitations or artefact in the determination of K_i or IC_{50}; *in vivo* physiologic regulation of clearance not reflected *in vitro*; and uncertainty in identifying a value of [I] that reflects actual enzyme exposure *in vivo* (Table 11.6).

Actual enzyme-available inhibitor exposure is critically needed for accurate scaling, but is an elusive parameter. Systemic plasma inhibitor concentrations (total or unbound) can be directly measured. Intraportal "hepatic inlet" concentrations (total or unbound) may be more relevant for accurate scaling, but these are not easily measured, and must be estimated by various indirect means involving a series of assumptions.[182] In any case, data aggregations have been created in which actual AUC_I/AUC_O ratios from drug interaction studies of CYP3A substrates are compared to predicted ratios based on *in vitro* K_i values together with various options for [I].[175] Systemic total and unbound [I], as well as total and unbound hepatic inlet [I], all yield poor predictive accuracy (Figure 11.6). In the 1990s we made the empirical observation that many lipophilic drugs that are inhibitors of CYP3A (and other CYP enzymes) undergo extensive uptake into hepatic tissue, such that intrahepatic concentrations are far higher than total systemic concentrations, let alone unbound systemic concentration.[159,160,186,189] Accounting for this hepatic uptake phenomenon improved the accuracy of predictive scaling. Subsequent studies have verified "concentrative uptake" into hepatic tissue, and specifically into hepatocytes.[190–193] Our understanding of intracellular distribution of CYP

Table 11.6 Factors complicating *in vitro-in vivo* predictive scaling of pharmacokinetic drug interactions.

In vitro determination of K_i or IC_{50}

- Nonspecific *in vitro* binding of substrate or inhibitor
- Consumption (metabolism) of substrate or inhibitor
- Irreversible (mechanism-based) inhibition
- Effect of solvents on substrate kinetics or inhibitor potency

Physiological factors not reflected *in vitro*

- Oral *vs.* parenteral administration of substrate or inhibitor
- Flow-dependent intravenous clearance
- Non-CYP hepatic clearance
- Extrahepatic clearance

Determination of enzyme-available inhibitor concentration

- Plasma protein binding restricting hepatic uptake
- Concentrative hepatic uptake
- Interdose fluctuation in inhibitor exposure
- Contribution of inhibitor metabolites to net inhibition

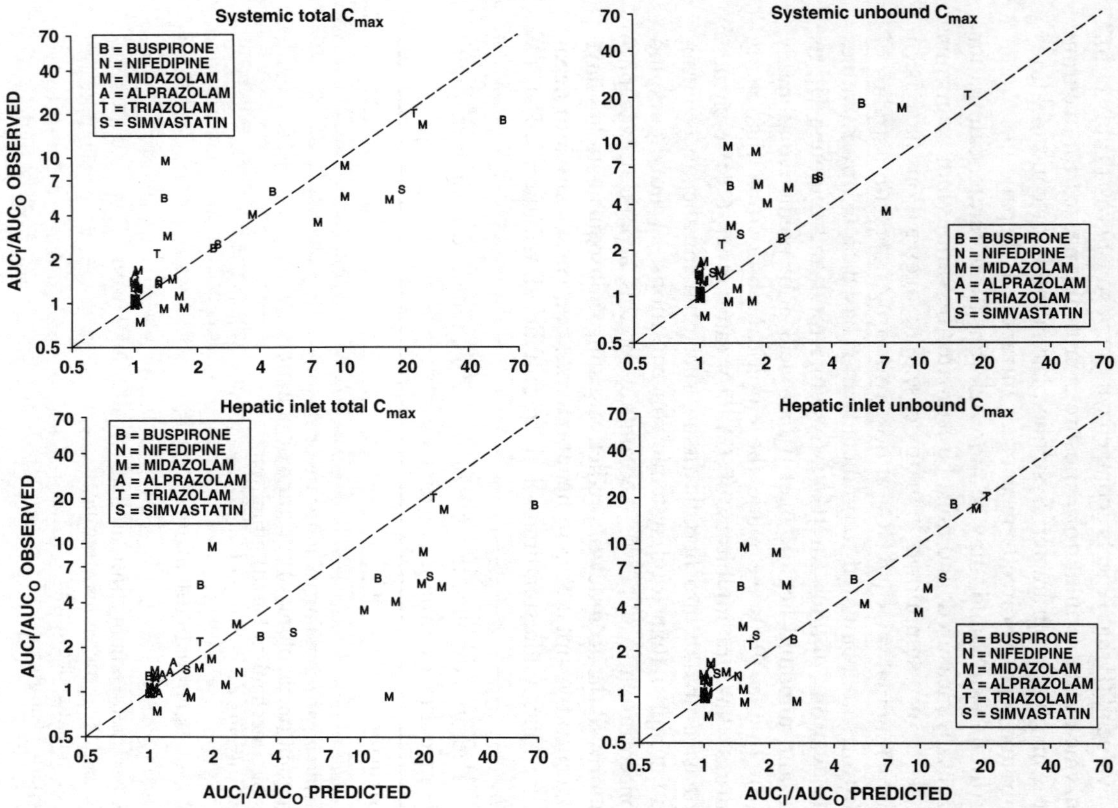

Figure 11.6 Observed (y-axis) values of AUC_I/AUC_O reported in clinical drug interaction studies in relation to predicted (x-axis) AUC_I/AUC_O values based on *in vitro* K_i values, for six different CYP3A substrates (buspirone, nifedipine, midazolam, alprazolam, triazolam, simvastatin) and a variety of inhibitors.[175] Predictions were based on four different inhibitor concentration schemes: systemic total C_{max}, systemic unbound C_{max}, hepatic inlet total C_{max}, and hepatic inlet unbound C_{max}. Note that all four inhibitor concentration schemes yield poor predictive accuracy.

inhibitors, and the factors that regulate quantitative enzyme exposure to inhibitory chemicals, is sparse. It is likely that an advance in predictive accuracy of drug interaction scaling will be dependent on an improved understanding of intracellular inhibitor distribution and the local factors regulating access of inhibitors to CYP enzymes.

11.6.2 Drug Interactions Involving Metabolic Induction

Pharmacologic induction of CYP3A metabolic activity involves an upregulation of protein synthesis, yielding higher quantities of metabolically active protein at steady-state.[194–204] Compared with inhibition, induction is a slower process. A number of days following initial exposure to the inducer may be required for CYP3A phenotype activity to reach its higher steady-state value.[205] A similar time interval is required for "de-induction" and return to baseline metabolic activity when exposure to the inducing agent is discontinued.[205] While inhibition can be studied *in vitro* using cell homogenates, assessment of induction requires cultured intact hepatocytes or immortalized cells with viable protein synthetic capability.[206–208] It is important to distinguish induction from heterotropic or homotropic activation, which involves an immediate increase in enzyme activity due to a conformational change, as opposed to an increase in the quantity of protein.

The pregnane X receptor (PXR), in the category of "orphan nuclear receptors," is a critical regulatory element mediating induction of CYP3A expression by chemical inducers.[209–217] Ligand binding to PXR is followed by heterodimerisation with retinoid X receptor (RXR), and subsequent binding to a cognate 5′-upstream response element, resulting in transcriptional activation. Some data suggests that genetic polymorphisms in PXR may underlie some component of individual variability in response to inducing agents.[218–220] An area of active investigation is the role of PXR and a number of candidate endogenous ligands as explanations for variability in CYP3A phenotypic expression.

The clinical consequence of CYP3A induction is increased clearance and reduced plasma levels of the substrate (victim) drug, with the principal concern being reduced therapeutic efficacy. Rifampin is recognised as the most powerful CYP3A inducer.[221] Examples of other clinically relevant inducers include carbamazepine, phenytoin, efavirenz, phenobarbital, corticosteroids and St. John's wort. Because regulation of CYP3A and P-gp expression share the same PXR-dependent mechanism, concurrent induction of CYP3A and P-gp is common.

In vitro K_i or IC_{50} values can be used as quantitative metrics to estimate the inhibitory potency of chemical inhibitors. No correspondingly straightforward quantitative measure is available for *in vitro* studies of the potential for induction. Typically, concentration-dependent upregulation of immunoreactive CYP3A protein expression and/or CYP3A metabolic activity is evaluated in cultured human hepatocytes. Rifampin is studied concurrently as a "positive control," and potency of the candidate inducer is expressed as a fraction of the presumed "maximum" induction effect of rifampin.[212,222] mRNA expression does not necessarily parallel protein expression and is not trustworthy as a surrogate.[223]

Compared to *in vitro* studies of inhibition using microsomal preparations, the *in vitro* evaluation of induction is expensive and complex, requiring the availability of viable human hepatocytes. No animal species has been identified as a substitute for human tissue.

11.7 Comment

Understanding and explaining individual variability in phenotypic metabolic activity of CYP3A enzymes continues to be a major challenge in molecular and clinical pharmacology. Some sources of variability can be identified and quantitated, but a large fraction of variation remains unexplained. At most, a small proportion of variability is attributed to identified genomic variants in coding or promoter regions corresponding to CYP3A4 or CYP3A5. It may be that genomic variation in transcription elements such as PXR will explain a component of phenotypic variability, along with identification of one or more endogenous ligands mediating transcriptional activation. At present, the state of knowledge does not allow confident pre-treatment selection of fully individualised doses of CYP3A substrate drugs in clinical practice. The contemporary approach to therapy continues to be initial dose estimation, followed by monitoring and iterative adjustment based on plasma levels and clinical response.

Acknowledgements

This chapter was supported by grants AG-17880, AI-58784, GM-61834, GM-74369, DK-58496, and DA-13834 from the Department of Health and Human Services.

References

1. F.P. Guengerich, *Drug Development Research*, 2000, **49**, 4–16.
2. F.P. Guengerich, *Drug Metab. Rev.*, 2004, **36**, 159–197.
3. T. Shimada, H. Yamazaki, M. Mimura, Y. Inui and F.P. Guengerich, *Journal of Pharmacology and Experimental Therapeutics*, 1994, **270**, 414–423.
4. Y. Tsuchiya, M. Nakajima and T. Yokoi, *Cancer Lett.*, 2005, **227**, 115–124.
5. A.M. Yu, K. Fukamachi, K.W. Krausz, C. Cheung and F.J. Gonzalez, *Endocrinology*, 2005, **146**, 2911–2919.
6. W.M. Atkins, W.D. Lu and D.L. Cook, *J. Biol. Chem.*, 2002, **277**, 33258–33266.
7. R.B. Kim, C. Wandel, B. Leake, M. Cvetkovic, M.F. Fromm, P.J. Dempsey, M.M. Roden, F. Belas, A.K. Chaudhary, D.M. Roden, A.J. Wood and G.R. Wilkinson, *Pharm. Res.*, 1999, **16**, 408–414.

8. C. Wandel, R.B. Kim, S. Kajiji, P. Guengerich, G.R. Wilkinson and A.J. Wood, *Cancer Res.*, 1999, **59**, 3944–3948.
9. M.D. Perloff, E. Störmer, L.L. von Moltke and D.J. Greenblatt, *Pharmaceutical Research*, 2003, **20**, 1177–1183.
10. K. Venkatakrishnan, L.L. von Moltke and D.J. Greenblatt, *Clinical Pharmacokinetics*, 2000, **38**, 111–180.
11. L.L. von Moltke, D.J. Greenblatt, J.M. Grassi, B.W. Granda, S.X. Duan, S.M. Fogelman, J.P. Daily, J.S. Harmatz and R.I. Shader, *Journal of Clinical Pharmacology*, 1998, **38**, 106–111.
12. C.A. Wanke, J.M. Falutz, A. Shevitz, J.P. Phair and D.P. Kotler, *Clin. Infect. Dis.*, 2002, **34**, 248–259.
13. P.K. Honig, D.C. Wortham, K. Zamani, D.P. Conner, J.C. Mullin and L.R. Cantilena, *Journal of American Medical Association*, 1993, **269**, 1513–1518.
14. L.L. von Moltke, D.J. Greenblatt, S.X. Duan, J.S. Harmatz and R.I. Shader, *Journal of Clinical Pharmacology*, 1994, **34**, 1222–1227.
15. B.P. Monahan, C.L. Ferguson, E.S. Killeavy, B.K. Lloyd, J. Troy and L.R. Cantilena, *JAMA*, 1990, **264**, 2788–2790.
16. R.L. Woosley, Y. Chen, J.P. Freiman and R.A. Gillis, *Jama*, 1993, **269**, 1532–1536.
17. H.L. Sham, D.J. Kempf, A. Molla, K.C. Marsh, G.N. Kumar, C.M. Chen, W. Kati, K. Stewart, R. Lal, A. Hsu, D. Betebenner, M. Korneyeva, S. Vasavanonda, E. McDonald, A. Saldivar, N. Wideburg, X. Chen, P. Niu, C. Park, V. Jayanti, B. Grabowski, G.R. Granneman, E. Sun, A.J. Japour and D.W. Norbeck, *Antimicrobial Agents and Chemotherapy*, 1998, **42**, 3218–3224.
18. A. Hsu, G.R. Granneman, G. Cao, L. Carothers, T. el-Shourbagy, P. Baroldi, K. Erdman, F. Brown, E. Sun and J.M. Leonard, *Clinical Pharmacology & Therapeutics*, 1998, **63**, 453–464.
19. D.J. Kempf, K.C. Marsh, G. Kumar, A.D. Rodrigues, D.F. Denissen, E. McDonald, M.J. Kukulka, A. Hsu, G.R. Granneman, P.A. Baroldi, E. Sun, D. Pizzuti, J.J. Plattner, D.W. Norbeck and L.M. Leonard, *Antimicrob. Agents Chemother.*, 1997, **41**, 654–660.
20. O. Burk and L. Wojnowski, *Naunyn Schmiedebergs Arch. Pharmacol.*, 2004, **369**, 105–124.
21. L. Wojnowski, *Ther. Drug Monit.*, 2004, **26**, 192–199.
22. J.K. Lamba, Y.S. Lin, E.G. Schuetz and K.E. Thummel, *Adv. Drug Deliv. Rev.*, 2002, **54**, 1271–1294.
23. M. Ingelman-Sundberg, *Naunyn Schmiedebergs Arch. Pharmacol.*, 2004, **369**, 89–104.
24. F.P. Guengerich, *Annual Review of Pharmacology and Toxicology*, 1999, **39**, 1–7.
25. E.G. Schuetz, *Mol. Pharmacol.*, 2004, **65**, 279–281.
26. K. Gellner, R. Eiselt, E. Hustert, H. Arnold, I. Koch, M. Haberl, C.J. Deglmann, O. Burk, D. Buntefuss, S. Escher, C. Bishop, H.G. Koebe,

U. Brinkmann, H.P. Klenk, K. Kleine, U.A. Meyer and L. Wojnowski, *Pharmacogenetics*, 2001, **11**, 111–121.
27. A. Westlind, S. Malmebo, I. Johansson, C. Otter, T.B. Andersson, M. Ingelman-Sundberg and M. Oscarson, *Biochemical and Biophysical Research Communications*, 2001, **281**, 1349–1355.
28. K.S. Lown, J.C. Kolars, K.E. Thummel, J.L. Barnett, K.L. Kunze, S.A. Wrighton and P.B. Watkins, *Drug Metabolism and Disposition*, 1994, **22**, 947–955.
29. S.J. Lee and J.A. Goldstein, *Pharmacogenomics*, 2005, **6**, 357–371.
30. S.J. Lee, K.A. Usmani, B. Chanas, B. Ghanayem, T. Xi, E. Hodgson, H.W. Mohrenweiser and J.A. Goldstein, *Pharmacogenetics*, 2003, **13**, 461–472.
31. P. He, M.H. Court, D.J. Greenblatt and L.L. von Moltke, *Clinical Pharmacology and Therapeutics*, 2005, **77**, 373–387.
32. P. He, M.H. Court, D.J. Greenblatt and L.L. von Moltke, *Drug Metab. Dispos.*, 2006, **34**, 1198–1207.
33. A. Westlind-Johnsson, R. Hermann, A. Huennemeyer, B. Hauns, G. Lahu, N. Nassr, K. Zech, M. Ingelman-Sundberg and O. von Richter, *Clin. Pharmacol. Ther.*, 2006, **79**, 339–349.
34. E.R. Lepper, S.D. Baker, M. Permenter, N. Ries, R.H. van Schaik, P.W. Schenk, D.K. Price, D. Ahn, N.F. Smith, G. Cusatis, R.G. Ingersoll, S.E. Bates, R.H. Mathijssen, J. Verweij, W.D. Figg and A. Sparreboom, *Clin. Cancer Res.*, 2005, **11**, 7398–7404.
35. C.B. Eap, T. Buclin, E. Hustert, G. Bleiber, K.P. Golay, A.C. Aubert, P. Baumann, A. Telenti and R. Kerb, *Eur. J. Clin. Pharmacol.*, 2004, **60**, 231–236.
36. M.D. Floyd, G. Gervasini, A.L. Masica, G. Mayo, A.L. George Jr, K. Bhat, R.B. Kim and G.R. Wilkinson, *Pharmacogenetics*, 2003, **13**, 595–606.
37. E. Garcia-Martin, C. Martinez, R.M. Pizarro, F.J. Garcia-Gamito, H. Gullsten, H. Raunio and J.A. Agundez, *Clin. Pharmacol. Ther.*, 2002, **71**, 196–204.
38. L.P. Rivory, H. Qin, S.J. Clarke, J. Eris, G. Duggin, E. Ray, R.J. Trent and J.F. Bishop, *European Journal of Clinical Pharmacology*, 2000, **56**, 395–398.
39. S.E. Ball, A. Scatina, J. Kao, G.M. Ferron, R. Fruncillo, P. Mayer, I. Weubryb, M. Guida, P.J. Hopkins, N. Warner and J. Hall, *Clinical Pharmacology and Therapeutics*, 1999, **66**, 288–294.
40. C. Wandel, J.S. Witte, J.M. Hall, C.M. Stein, A.J.J. Wood and G.R. Wilkinson, *Clinical Pharmacology & Therapeutics*, 2000, **68**, 82–91.
41. P.S. Shih and J.D. Huang, *Drug Metab. Dispos.*, 2002, **30**, 1491–1496.
42. C.B. Eap, J. Fellay, T. Buclin, G. Bleiber, K.P. Golay, M. Brocard, P. Baumann and A. Telenti, *Pharmacogenetics*, 2004, **14**, 255–260.
43. Y. Horsmans, J.P. Desager and C. Harvengt, *Pharmacol. Toxicol.*, 1992, **71**, 258–261.
44. C. Funck-Brentano, P.Y. Boelle, C. Verstuyft, C. Bornert, L. Becquemont and J.M. Poirier, *Eur. J. Clin. Pharmacol.*, 2005, **61**, 821–829.

45. H. Friedman, D.J. Greenblatt, E.S. Burstein, J.S. Harmatz and R.I. Shader, *British Journal of Clinical Pharmacology*, 1986, **22**, 639–642.
46. A. Kassai, G. Toth, M. Eichelbaum and U. Klotz, *Clinical Pharmacokinetics*, 1988, **15**, 319–325.
47. D.J. Greenblatt, M. Divoll, D.R. Abernethy, L.J. Moschitto, R.B. Smith and R.I. Shader, *British Journal of Clinical Pharmacology*, 1983, **15**, 303–309.
48. J.W. Dundee, P.S. Collier, R.J.T. Carlisle and K.W. Harper, *British Journal of Clinical Pharmacology*, 1986, **21**, 425–429.
49. T.R. Rebbeck, J.M. Jaffe, A.H. Walker, A.J. Wein and S.B. Malkowicz, *Journal of the National Cancer Institute*, 1998, **90**, 1225–1229.
50. P.L. Paris, P.A. Kupelian, J.M. Hall, T.L. Williams, H. Levin, E.A. Klein, G. Casey and J.S. Witte, *Cancer Epidemiol. Biomarkers Prev.*, 1999, **8**, 901–905.
51. C. Keshava, E.C. McCanlies and A. Weston, *Am. J. Epidemiol.*, 2004, **160**, 825–841.
52. S.J. Plummer, D.V. Conti, P.L. Paris, A.P. Curran, G. Casey and J.S. Witte, *Cancer Epidemiol. Biomarkers Prev.*, 2003, **12**, 928–932.
53. R. Aplenc, W. Glatfelter, P. Han, E. Rappaport, M. La, A. Cnaan, M.A. Blackwood, B. Lange and T. Rebbeck, *Br. J. Haematol.*, 2003, **122**, 240–244.
54. J.G. Blanco, M.J. Edick, M.L. Hancock, N.J. Winick, T. Dervieux, M.D. Amylon, R.O. Bash, F.G. Behm, B.M. Camitta, C.H. Pui, S.C. Raimondi and M.V. Relling, *Pharmacogenetics*, 2002, **12**, 605–611.
55. F.F. Kadlubar, G.S. Berkowitz, R.R. Delongchamp, C. Wang, B.L. Green, G. Tang, J. Lamba, E. Schuetz and M.S. Wolff, *Cancer Epidemiol. Biomarkers Prev.*, 2003, **12**, 327–331.
56. A.N. Tucker, K.A. Tkaczuk, L.M. Lewis, D. Tomic, C.K. Lim and J.A. Flaws, *Cancer Lett.*, 2005, **217**, 61–72.
57. R.A. Kittles, W. Chen, R.K. Panguluri, C. Ahaghotu, A. Jackson, C.A. Adebamowo, R. Griffin, T. Williams, F. Ukoli, L. Adams-Campbell, J. Kwagyan, W. Isaacs, V. Freeman and G.M. Dunston, *Hum. Genet.*, 2002, **110**, 553–560.
58. L.L. von Moltke, T.H. Tran, M.M. Cotreau and D.J. Greenblatt, *Journal of Clinical Pharmacology*, 2000, **40**, 200–204.
59. G.R. Wilkinson, *Clin. Pharmacol. Ther.*, 2004, **76**, 99–103.
60. A.K. Daly, *Clin. Pharmacokinet.*, 2006, **45**, 13–31.
61. J.A. Williams, J. Cook and S.I. Hurst, *Drug Metab. Dispos.*, 2003, **31**, 1526–1530.
62. H.G. Xie, A.J. Wood, R.B. Kim, C.M. Stein and G.R. Wilkinson, *Pharmacogenomics*, 2004, **5**, 243–272.
63. K.E. Thummel, *Pharmacogenetics*, 2003, **13**, 585–587.
64. J.Y. Park, K.A. Kim, P.W. Park, O.J. Lee, D.K. Kang, J.H. Shon, K.H. Liu and J.G. Shin, *Clin. Pharmacol. Ther.*, 2006, **79**, 590–599.
65. T. Yamamoto, T. Kubota, T. Ozeki, M. Sawada, S. Yokota, Y. Yamada, Y. Kumagai and T. Iga, *Clin. Chim. Acta.*, 2005, **362**, 147–154.

66. M. Wong, R.L. Balleine, M. Collins, C. Liddle, C.L. Clarke and H. Gurney, *Clin. Pharmacol. Ther.*, 2004, **75**, 529–538.
67. A. Westlind-Johnsson, S. Malmebo, A. Johansson, C. Otter, T.B. Andersson, I. Johansson, R.J. Edwards, A.R. Boobis and M. Ingelman-Sundberg, *Drug Metab. Dispos.*, 2003, **31**, 755–761.
68. A. Wennerholm, A. Allqvist, J.O. Svensson, L.L. Gustafsson, R.A. Mirghani and L. Bertilsson, *Eur. J. Clin. Pharmacol.*, 2005, **61**, 113–118.
69. N. Hirota, K. Ito, T. Iwatsubo, C.E. Green, C.A. Tyson, N. Shimada, H. Suzuki and Y. Sugiyama, *Biopharmaceutics & Drug Disposition*, 2001, **22**, 53–71.
70. K.C. Patki, L.L. von Moltke and D.J. Greenblatt, *Drug Metab. Dispos.*, 2003, **31**, 938–944.
71. M.G. Soars, K. Grime and R.J. Riley, *Xenobiotica*, 2006, **36**, 287–299.
72. J.L. Dorne, K. Walton and A.G. Renwick, *Food Chem. Toxicol.*, 2003, **41**, 201–224.
73. L.L. von Moltke, D.J. Greenblatt, J. Schmider, S.X. Duan, C.E. Wright, J.S. Harmatz and R.I. Shader, *Journal of Clinical Pharmacology*, 1996, **36**, 783–791.
74. M.D. Perloff, L.L. von Moltke, M.H. Court, T. Kotegawa, R.I. Shader and D.J. Greenblatt, *Journal of Pharmacology and Experimental Therapeutics*, 2000, **292**, 618–628.
75. W.M. Atkins, *Drug Discov. Today*, 2004, **9**, 478–484.
76. W.M. Atkins, *Annu. Rev. Pharmacol. Toxicol.*, 2005, **45**, 291–310.
77. J.B. Houston and A. Galetin, *Arch. Biochem. Biophys.*, 2005, **433**, 351–360.
78. S. Ekins, B.J. Ring, S.N. Binkley, S.D. Hall and S.A. Wrighton, *International Journal of Clinical Pharmacology and Therapeutics*, 1998, **36**, 642–651.
79. W. Tang and R.A. Stearns, *Curr. Drug Metab.*, 2001, **2**, 185–198.
80. E.E. Scott and J.R. Halpert, *Trends Biochem. Sci.*, 2005, **30**, 5–7.
81. M. Shou, Y. Lin, P. Lu, C. Tang, Q. Mei, D. Cui, W. Tang, J.S. Ngui, C.C. Lin, R. Singh, B.K. Wong, J.A. Yergey, J.H. Lin, P.G. Pearson, T.A. Baillie, A.D. Rodrigues and T.H. Rushmore, *Curr. Drug Metab.*, 2001, **2**, 17–36.
82. M. Shou, Q. Mei, M.W. Ettore Jr, R. Dai, T.A. Baillie and T.H. Rushmore, *Biochemical Journal*, 1999, **340**, 845–853.
83. M.L. Schrag and L.C. Wienkers, *Archives of Biochemistry and Biophysics*, 2001, **391**, 49–55.
84. L.E. Witherow and J.B. Houston, *Journal of Pharmacology and Experimental Therapeutics*, 1999, **290**, 58–65.
85. J.B. Houston and K.E. Kenworthy, *Drug Metabolism and Disposition*, 2000, **28**, 246–254.
86. K. Venkatakrishnan, L.L. von Moltke and D.J. Greenblatt, *Journal of Clinical Pharmacology*, 2001, **41**, 1149–1179.
87. K. Venkatakrishnan, L.L. von Moltke, R.S. Obach and D.J. Greenblatt, *Curr. Drug Metab.*, 2003, **4**, 423–459.
88. A.D. Rodrigues, *Biochem. Pharmacol.*, 1999, **57**, 465–480.
89. M. Shou, J. Grogan, J.A. Mancewicz, K.W. Krausz, F.J. Gonzalez, H.V. Gelboin and K.R. Korzekwa, *Biochemistry*, 1994, **33**, 6450–6455.

90. M.L. Schrag and L.C. Wienkers, *Drug Metabolism and Disposition*, 2001, **29**, 70–75.
91. D.J. Greenblatt, D.R. Abernethy, A. Locniskar, J.S. Harmatz, R.A. Limjuco and R.I. Shader, *Anesthesiology*, 1984, **61**, 27–35.
92. H.R. Ochs, D.J. Greenblatt and M. Knüchel, *Chest*, 1985, **87**, 223–226.
93. D.J. Greenblatt, J.S. Harmatz, L.L. von Moltke, C.E. Wright, A.L.B. Durol, L.M. Harrel-Joseph and R.I. Shader, *Journal of Pharmacology and Experimental Therapeutics*, 2000, **293**, 435–443.
94. D.J. Greenblatt, L.L. von Moltke, J.S. Harmatz, G. Chen, J.L. Weemhoff, C. Jen, C.J. Kelley, B.W. LeDuc and M.A. Zinny, *Clinical Pharmacology and Therapeutics*, 2003, **74**, 121–129.
95. D.J. Greenblatt, L. Gan, J.S. Harmatz and R.I. Shader, *British Journal of Clinical Pharmacology*, 2005, **60**, 244–248.
96. J.C. Gorski, D.R. Jones, B.D. Haehner-Daniels, M.A. Hamman, E.M. O'Mara and S.D. Hall, *Clinical Pharmacology and Therapeutics*, 1998, **64**, 133–143.
97. J.C. Gorski, S. Vannaprasaht, M.A. Hamman, W.T. Ambrosius, M.A. Bruce, B. Haehner-Daniels and B. Hallx, *Clin. Pharmacol. Ther.*, 2003, **74**, 275–287.
98. A.L. Masica, G. Mayo and G.R. Wilkinson, *Clin. Pharmacol. Ther.*, 2004, **76**, 341–349.
99. H.S. Lee, B.C. Goh, L. Fan, Y.M. Khoo, L. Wang, R. Lim, A.B. Ong and C. Chua, *Br. J. Clin. Pharmacol.*, 2003, **55**, 270–277.
100. J.I. Lee, D. Chaves-Gnecco, J.A. Amico, P.D. Kroboth, J.W. Wilson and R.F. Frye, *Clin. Pharmacol. Ther.*, 2002, **72**, 718–728.
101. C.B. Eap, T. Buclin, G. Cucchia, D. Zullino, E. Hustert, G. Bleiber, K.P. Golay, A.C. Aubert, P. Baumann, A. Telenti and R. Kerb, *Eur. J. Clin. Pharmacol.*, 2004, **60**, 237–246.
102. S.H. Preskorn, J. Alderman, M. Chung, W. Harrison, M. Messig and S. Harris, *Journal of Clinical Psychopharmacology*, 1994, **14**, 90–98.
103. D.R. Abernethy, D.J. Greenblatt and R.I. Shader, *Journal of Pharmacology and Experimental Therapeutics*, 1985, **232**, 183–188.
104. J. Alderman, S.H. Preskorn, D.J. Greenblatt, W. Harrison, D. Penenberg, J. Allison and M. Chung, *Journal of Clinical Psychopharmacology*, 1997, **17**, 284–291.
105. S.H. Preskorn, D.J. Greenblatt, D. Flockhart, Y. Luo, E.S. Perloff, J.S. Harmatz, B. Baker, A. Klick-Davis, Z. Desta and Z. Burtx, *J. Clin. Psychopharmacol.*, 2007, **27**, 28–34.
106. D.J. Greenblatt, D.R. Abernethy, R. Matlis, J.S. Harmatz and R.I. Shader, *Arthritis Rheum.*, 1984, **27**, 1066–1069.
107. D.J. Greenblatt, L.L. von Moltke, Y. Luo, E.S. Perloff, K.A. Horan, A. Bruce, R.C. Reynolds, J.S. Harmatz, B. Avula, I.A. Khan and P. Goldman, *Journal of Clinical Pharmacology*, 2006, **46**, 214–221.
108. D.J. Greenblatt, L.L. von Moltke, E.S. Perloff, Y. Luo, J.S. Harmatz and M.A. Zinny, *Clin. Pharmacol. Ther.*, 2006, **79**, 125–133.

109. N.M. Davies, *Clin. Pharmacokinet.*, 1995, **28**, 100–114.
110. G.B. Kaplan, D.J. Greenblatt, B.L. Ehrenberg, J.E. Goddard, M.M. Cotreau, J.S. Harmatz and R.I. Shader, *Journal of Clinical Pharmacology*, 1997, **37**, 693–703.
111. R.M. Cysneiros, D. Farkas, J.S. Harmatz, L.L. von Moltke and D.J. Greenblatt, *Clin. Pharmacol. Ther.*, 2007, **82**, 54–62.
112. K.E. Culm-Merdek, L.L. von Moltke, J.S. Harmatz and D.J. Greenblatt, *British Journal of Clinical Pharmacology*, 2005, **60**, 486–493.
113. C.P. Denaro, C.R. Brown, M. Wilson, J. Peyton and N.L. Benowitz, *Clinical Pharmacology and Therapeutics*, 1990, **48**, 277–285.
114. A.D. Kashuba, J.S. Bertino, M.L. Rocci, R.W. Kulawy, D.J. Beck and A.N. Nafziger, *Clin. Pharmacol. Ther.*, 1998, **64**, 269–277.
115. N. Yasui-Furukori, T. Kondo, T. Kubota, H. Otake, T. Ohkubo, T. Nagasaki, K. Sugawara, K. Chiba, K. Otani and S. Kaneko, *Eur. J. Clin. Pharmacol.*, 2001, **57**, 285–288.
116. M.T. Kinirons, D. O'Shea, R.B. Kim, J.D. Groopman, K.E. Thummel, A.J. Wood and G.R. Wilkinson, *Clin. Pharmacol. Ther.*, 1999, **66**, 224–231.
117. Y. Krivoruk, M.T. Kinirons, A.J.J. Wood and M. Wood, *Clinical Pharmacology and Therapeutics*, 1994, **56**, 608–614.
118. M.T. Kinirons, D. O'Shea, T.E. Downing, A.T. Fitzwilliam, L. Joellenbeck, J.D. Groopman, G.R. Wilkinson and A.J.J. Wood, *Clinical Pharmacology and Therapeutics*, 1993, **54**, 621–629.
119. P.B. Watkins, D.K. Turgeon, P. Saenger, K.S. Lown, J.C. Kolars, T. Hamilton, K. Fishman, P.S. Guzelian and J.J. Voorhees, *Clinical Pharmacology and Therapeutics*, 1992, **52**, 265–273.
120. Y.C. Chen, S.K. Gotzkowsky, A.N. Nafziger, R.W. Kulawy, M.L. Rocci Jr., J.S. Bertino Jr. and A.D. Kashuba, *Br. J. Clin. Pharmacol.*, 2006, **62**, 187–195.
121. W.L. Chiou, H.Y. Jeong, T.C. Wu and C. Ma, *Clinical Pharmacology & Therapeutics*, 2001, **70**, 305–310.
122. L.P. Rivory, K.A. Slaviero, J.M. Hoskins and S.J. Clarke, *Clinical Pharmacokinetics*, 2001, **40**, 151–158.
123. D. Kurnik, A.J. Wood and G.R. Wilkinson, *Clin. Pharmacol. Ther.*, 2006, **80**, 228–234.
124. M.M. Galteau and F. Shamsa, *Eur. J. Clin. Pharmacol.*, 2003, **59**, 713–733.
125. S.J. Kovacs, D.E. Martin, D.E. Everitt, S.D. Patterson and D.K. Jorkasky, *Clin. Pharmacol. Ther.*, 1998, **63**, 617–622.
126. O.Q. Yin, X. Shi, B. Tomlinson and M.S. Chow, *J. Clin. Pharmacol.*, 2004, **44**, 1412–1417.
127. Z. Chen, L.A. Dunning, K.E. Anderson, J.L. Holtzman and W. Zheng, *Steroids*, 2004, **69**, 67–70.
128. P. Damkier and K. Brøsen, *Clinical Pharmacology & Therapeutics*, 2000, **68**, 199–209.

129. M. Ohno, I. Yamaguchi, T. Ito, K. Saiki, I. Yamamoto and J. Azuma, *European Journal of Clinical Pharmacology*, 2000, **55**, 861–865.
130. K.C. Patki, L.L. von Moltke, J.S. Harmatz, L.M. Hesse, M.H. Court and D.J. Greenblatt, *Journal of Pharmacology and Experimental Therapeutics*, 2004, **308**, 874–879.
131. J. George, K. Byth and G.C. Farrel, *Biochemical Pharmacology*, 1995, **50**, 727–730.
132. E.A. Sotaniemi, A.J. Arranto, O. Pelkonen and M. Pasanen, *Clinical Pharmacology and Therapeutics*, 1997, **61**, 331–339.
133. C.M. Hunt, W.R. Westerkam and G.M. Stave, *Biochemical Pharmacology*, 1992, **44**, 275–283.
134. C. Transon, S. Lecoeur, T. Leeman, P. Beaune and P. Dayer, *European Journal of Clinical Pharmacology*, 1996, **51**, 79–85.
135. R. Wolbold, K. Klein, O. Burk, A.K. Nussler, P. Neuhaus, M. Eichelbaum, M. Schwab and U.M. Zanger, *Hepatology*, 2003, **38**, 978–988.
136. A. Parkinson, D.R. Mudra, C. Johnson, A. Dwyer and K.M. Carroll, *Toxicol. Appl. Pharmacol.*, 2004, **199**, 193–209.
137. J.S. Warrington, D.J. Greenblatt and L.L. von Moltke, *Journal of Pharmacology and Experimental Therapeutics*, 2004, **309**, 720–729.
138. M.M. Cotreau, L.L. von Moltke and D.J. Greenblatt, *Clinical Pharmacokinetics*, 2005, **44**, 33–60.
139. A. Hämmerlein, H. Derendorf and D.T. Lowenthal, *Clinical Pharmacokinetics*, 1998, **35**, 49–64.
140. C. Durnas, C. -M. Loi and B.J. Cusack, *Clinical Pharmacokinetics*, 1990, **19**, 359–389.
141. A.J. McLean and D.G. Le Couteur, *Pharmacol. Rev.*, 2004, **56**, 163–184.
142. A.A. Mangoni and S.H. Jackson, *Br. J. Clin. Pharmacol.*, 2004, **57**, 6–14.
143. L.L. von Moltke, D.R. Abernethy and D.J. Greenblatt, In: *Clinical Geriatric Psychopharmacology, 4th Edition*, ed. C. Salzman, Philadelphia, Lippincott Williams & Wilkins, 2005, 87–114.
144. D.J. Greenblatt, E.M. Sellers and R.I. Shader, *New England Journal of Medicine*, 1982, **306**, 1081–1088.
145. J.B. Schwartz, *Clin. Pharmacol. Ther.*, 2007, **82**, 87–96.
146. J.G. Barnhill, D.J. Greenblatt, L.G. Miller, A. Gaver, J.S. Harmatz and R.I. Shader, *Journal of Pharmacology and Experimental Therapeutics*, 1990, **253**, 1153–1161.
147. M. Gandhi, F. Aweeka, R.M. Greenblatt and T.F. Blaschke, *Annu. Rev. Pharmacol. Toxicol.*, 2004, **44**, 499–523.
148. B. Meibohm, I. Beierle and H. Derendorf, *Clin. Pharmacokinet.*, 2002, **41**, 329–342.
149. I. Beierle, B. Meibohm and H. Derendorf, *International Journal of Clinical Pharmacology and Therapeutics*, 1999, **37**, 529–547.
150. J.B. Schwartz, *Clin. Pharmacokinet.*, 2003, **42**, 107–121.

151. R.Z. Harris, L.Z. Benet and J.B. Schwartz, *Drugs*, 1995, **50**, 222–239.
152. M. Chen, L. Ma, G.L. Drusano, J.S. Bertino Jr. and A.N. Nafziger, *Clin. Pharmacol. Ther.*, 2006, **80**, 531–538.
153. B. Zhu, Z.Q. Liu, G.L. Chen, X.P. Chen, D.S. Ou-Yang, L.S. Wang, S.L. Huang, Z.R. Tan and H.H. Zhou, *Br. J. Clin. Pharmacol.*, 2003, **55**, 264–269.
154. Z. Bebia, S.C. Buch, J.W. Wilson, R.F. Frye, M. Romkes, A. Cecchetti, D. Chaves-Gnecco and R.A. Branch, *Clin. Pharmacol. Ther.*, 2004, **76**, 618–627.
155. D.J. Greenblatt and L.L. von Moltke, Pharmacokinetics and drug interactions. In: *Comprehensive Textbook of Psychiatry, 8th Edition*, ed. BJ Sadock, VA Sadock, Philadelphia. Lippincott Williams & Williams, 2005, 2699–2706.
156. D.J. Greenblatt and L.L. von Moltke, Drug-drug interactions: clinical perspective. In: *Drug-Drug Interactions.* Edited by AD Rodrigues. New York, Marcel Dekker, 2002, 565–584.
157. D.J. Greenblatt and L.L. von Moltke, Sedative-hypnotic and anxiolytic agents. In: *Metabolic Drug Interactions.* Edited by RH Levy, KE Thummel, WF Trager, PD Hansten, M Eichelbaum. Philadelphia, Lippincott, Williams and Wilkins, 2000, 259–270.
158. V. Bergk, C. Gasse, D. Rothenbacher, M. Loew, H. Brenner and W.E. Haefeli, *Clin. Pharmacol. Ther.*, 2004, **76**, 85–96.
159. L.L. von Moltke, D.J. Greenblatt, M.M. Cotreau-Bibbo, J.S. Harmatz and R.I. Shader, *British Journal of Clinical Pharmacology*, 1994, **38**, 23–31.
160. L.L. von Moltke, D.J. Greenblatt, J.S. Harmatz, S.X. Duan, L.M. Harrel, M.M. Cotreau-Bibbo, G.A. Pritchard, C.E. Wright and R.I. Shader, *Journal of Pharmacology and Experimental Therapeutics*, 1996, **276**, 370–379.
161. D.J. Greenblatt, C.E. Wright, L.L. von Moltke, J.S. Harmatz, B.L. Ehrenberg, L.M. Harrel, K. Corbett, M. Counihan, S. Tobias and R.I. Shader, *Clinical Pharmacology and Therapeutics*, 1998, **64**, 237–247.
162. S.M. Tsunoda, R.L. Velez, L.L. von Moltke and D.J. Greenblatt, *Clinical Pharmacology and Therapeutics*, 1999, **66**, 461–471.
163. S. Zhou, E. Chan, L.Y. Lim, U.A. Boelsterli, S.C. Li, J. Wang, Q. Zhang, M. Huang and A. Xu, *Curr. Drug Metab.*, 2004, **5**, 415–442.
164. S. Zhou, S. Yung Chan, B. Cher Goh, E. Chan, W. Duan, M. Huang and H.L. McLeod, *Clin. Pharmacokinet.*, 2005, **44**, 279–304.
165. R. Silverman, *Methods in Enzymology*, 1995, **249**, 240–283.
166. A.L. Blobaum, *Drug Metab. Dispos.*, 2006, **34**, 1–7.
167. T. Okudaira, T. Kotegawa, H. Imai, K. Tsutsumi, S. Nakano and K. Ohashi, *J. Clin. Pharmacol.*, 2007, **47**, 871–876.
168. K.E. Culm-Merdek, L.L. von Moltke, L. Gan, K.A. Horan, R. Reynolds, J.S. Harmatz, M.H. Court and D.J. Greenblatt, *Clin. Pharm. Ther.*, 2006, **79**, 243–254.

169. S.U. Mertens-Talcott, I. Zadezensky, W.V. De Castro, H. Derendorf and V. Butterweck, *J. Clin. Pharmacol.*, 2006, **46**, 1390–1416.
170. M.F. Paine, W.W. Widmer, H.L. Hart, S.N. Pusek, K.L. Beavers, A.B. Criss, S.S. Brown, B.F. Thomas and P.B. Watkins, *Am. J. Clin. Nutr.*, 2006, **83**, 1097–1105.
171. C. Yao and R.H. Levy, *J. Pharm. Sci.*, 2002, **91**, 1923–1935.
172. J.H. Lin, *Current Drug Metabolism*, 2000, **1**, 305–331.
173. K. Ito, T. Iwatsubo, S. Kanamitsu, K. Ueda, H. Suzuki and Y. Sugiyama, *Pharmacol. Rev.*, 1998, **50**, 387–412.
174. R.J. Bertz and G.R. Granneman, *Clinical Pharmacokinetics*, 1997, **32**, 210–258.
175. R.S. Obach, R.L. Walsky, K. Venkatakrishnan, E.A. Gaman, J.B. Houston and L.M. Tremaine, *Journal of Pharmacology and Experimental Therapeutics*, 2006, **316**, 336–348.
176. A. Galetin, H. Burt, L. Gibbons and J.B. Houston, *Drug Metab. Dispos.*, 2006, **34**, 166–175.
177. R.S. Obach, R.L. Walsky, K. Venkatakrishnan, J.B. Houston and L.M. Tremaine, *Clin. Pharmacol. Ther.*, 2005, **78**, 582–592.
178. H.S. Brown, A. Galetin, D. Hallifax and J.B. Houston, *Clin. Pharmacokinet.*, 2006, **45**, 1035–1050.
179. J.Y. Chien, A. Lucksiri, C.S. Ernest, J.C. Gorski, S.A. Wrighton and S.D. Hall, *Drug Metab. Dispos.*, 2006, **34**, 1208–1219.
180. A. Galetin, K. Ito, D. Hallifax and J.B. Houston, *Journal of Pharmacology and Experimental Therapeutics*, 2005, **314**, 180–190.
181. K. Ito, H.S. Brown and J.B. Houston, *Br. J. Clin. Pharmacol.*, 2004, **57**, 473–486.
182. S. Kanamitsu, K. Ito and Y. Sugiyama, *Pharmaceutical Research*, 2000, **17**, 336–343.
183. H.S. Brown, K. Ito, A. Galetin and J.B. Houston, *Br. J. Clin. Pharmacol.*, 2005, **60**, 508–518.
184. K.A. Bachmann and J.D. Lewis, *Ann. Pharmacother.*, 2005, **39**, 1064–1072.
185. K.A. Bachmann, *Curr. Drug Metab.*, 2006, **7**, 1–14.
186. L.L. von Moltke, D.J. Greenblatt, J. Schmider, C.E. Wright, J.S. Harmatz and R.I. Shader, *Biochemical Pharmacology*, 1998, **55**, 113–122.
187. T.D. Bjornsson, J.T. Callaghan, H.J. Einolf, V. Fischer, L. Gan, S. Grimm, J. Kao, S.P. King, G. Miwa, L. Ni, G. Kumar, J. McLeod, S.R. Obach, S. Roberts, A. Roe, A. Shah, F. Snikeris, J.T. Sullivan, D. Tweedie, J.M. Vega, J. Walsh and S.A. Wrighton, *J. Clin. Pharmacol.*, 2003, **43**, 443–469.
188. T.D. Bjornsson, J.T. Callaghan, H.J. Einolf, V. Fischer, L. Gan, S. Grimm, J. Kao, S.P. King, G. Miwa, L. Ni, G. Kumar, J. McLeod, R.S. Obach, S. Roberts, A. Roe, A. Shah, F. Snikeris, J.T. Sullivan, D. Tweedie, J.M. Vega, J. Walsh and S.A. Wrighton, *Drug Metab. Dispos.*, 2003, **31**, 815–832.

189. L.L. von Moltke, D.J. Greenblatt, M.M. Cotreau-Bibbo, S.X. Duan, J.S. Harmatz and R.I. Shader, *Journal of Pharmacology and Experimental Therapeutics*, 1994, **268**, 1278–1283.
190. K. Yamano, K. Yamamoto, H. Kotaki, S. Takedomi, H. Matsuo, Y. Sawada and T. Iga, *Drug Metab. Dispos.*, 1999, **27**, 1225–1231.
191. K. Yamano, K. Yamamoto, H. Kotaki, Y. Sawada and T. Iga, *Drug Metab. Dispos.*, 1999, **27**, 395–402.
192. K. Yamano, K. Yamamoto, H. Kotaki, S. Takedomi, H. Matsuo, Y. Sawada and T. Iga, *Journal of Pharmacology and Experimental Therapeutics*, 2000, **292**, 1118–1126.
193. K. Yamano, K. Yamamoto, M. Katashima, H. Kotaki, S. Takedomi, H. Matsuo, H. Ohtani, Y. Sawada and T. Iga, *Drug Metabolism and Disposition*, 2001, **29**, 443–452.
194. K.E. Thummel and G.R. Wilkinson, *Annual Review of Pharmacology and Toxicology*, 1998, **38**, 389–430.
195. J.H. Lin and A.Y. Lu, *Annual Review of Pharmacology and Toxicology*, 2001, **41**, 535–567.
196. G.K. Dresser, J.D. Spence and D.G. Bailey, *Clin. Pharmacokinet.*, 2000, **38**, 41–57.
197. J.H. Lin and A.Y. Lu, *Clin. Pharmacokinet.*, 1998, **35**, 361–390.
198. O. Pelkonen, J. Mäenpää, P. Taavitsainen, A. Rautio and H. Raunio, *Xenobiotica*, 1998, **28**, 1203–1253.
199. Z.Y. Zhang and Y.N. Wong, *Curr. Drug Metab.*, 2005, **6**, 241–257.
200. M. Barry and J. Feely, *Pharmacol. Ther.*, 1990, **48**, 71–94.
201. K.W. Bock, H. -P. Lipp and B.S. Bock-Hennig, *Xenobiotica*, 1990, **20**, 1101–1111.
202. D.J. Waxman and L. Azaroff, *Biochemical Journal*, 1992, **281**, 577–592.
203. B.K. Park, N.R. Kitteringham, M. Piromohamed and G.T. Tucker, *Br. J. Clin. Pharmacol.*, 1996, **41**, 477–491.
204. J.H. Lin, *Pharm. Res.*, 2006, **23**, 1089–1116.
205. E.E. Ohnhaus, *Eur. J. Clin. Pharmacol.*, 1989, **36**, 39–46.
206. E. LeCluyse, A. Madan, G. Hamilton, K. Carroll, R. DeHaan and A. Parkinson, *J. Biochem. Mol. Toxicol.*, 2000, **14**, 177–188.
207. A.P. Li, M.K. Reith, A. Rasmussen, J.C. Gorski, S.D. Hall, L. Xu, D.L. Kaminski and L.K. Cheng, *Chem. Biol. Interact.*, 1997, **107**, 17–30.
208. S.L. Ripp, J.B. Mills, O.A. Fahmi, K.A. Trevena, J.L. Liras, T.S. Maurer and S.M. de Morais, *Drug Metab. Dispos.*, 2006, **34**, 1742–1748.
209. B. Goodwin, M.R. Redinbo and S.A. Kliewer, *Annual Review of Pharmacology and Toxicology*, 2002, **42**, 1–23.
210. G.G. Gibson, N.J. Plant, K.E. Swales, A. Ayrton and W. El-Sankary, *Xenobiotica*, 2002, **32**, 165–206.
211. B.L. Urquhart, R.G. Tirona and R.B. Kim, *J. Clin. Pharmacol.*, 2007, **47**, 566–578.
212. H. Wang and E.L. LeCluyse, *Clin. Pharmacokinet.*, 2003, **42**, 1331–1357.
213. T.M. Willson and S.A. Kliewer, *Nat. Rev. Drug Discov.*, 2002, **1**, 259–266.
214. D.J. Waxman, *Archives of Biochemistry and Biophysics*, 1999, **369**, 11–23.

215. P. Honkakoski and M. Negishi, *Biochemical Journal*, 2000, **347**, 321–337.
216. U. Savas, K.J. Griffin and E.F. Johnson, *Mol. Pharmacol.*, 1999, **56**, 851–857.
217. L.A. Stanley, B.C. Horsburgh, J. Ross, N. Scheer and C.R. Wolf, *Drug Metab. Rev.*, 2006, **38**, 515–597.
218. J. Zhang, P. Kuehl and E.D. Green, *Pharmacogenetics*, 2001, **11**, 555–572.
219. J. Lamba, V. Lamba and E. Schuetz, *Curr. Drug Metab.*, 2005, **6**, 369–383.
220. P. He, M.H. Court, D.J. Greenblatt and L.L. von Moltke, *J. Clin. Pharmacol.*, 2006, **46**, 1356–1369.
221. M. Niemi, J.T. Backman, M.F. Fromm, P.J. Neuvonen and K.T. Kivistö, *Clin. Pharmacokinet.*, 2003, **42**, 819–850.
222. L.M. Hesse, Y. Sakai, D. Vishnuvardhan, A.P. Li, L.L. von Moltke and D.J. Greenblatt, *Journal of Pharmacy and Pharmacology*, 2003, **55**, 1229–1239.
223. C. Rodríguez-Antona, M.T. Donato, E. Pareja, M.J. Gómez-Lechón and J.V. Castell, *Archives of Biochemistry and Biophysics*, 2001, **393**, 308–315.

CHAPTER 12

The CYP4 Family

ALLAN E. RETTIE AND EDWARD J. KELLY

Departments of Medicinal Chemistry and Pharmaceutics, University of Washington, Seattle, WA 98195, USA

Table of Contents

Issues in Toxicology
Cytochromes P450: Role in the Metabolism and Toxicity of Drugs and other Xenobiotics
Edited by Costas Ioannides

12.1 Introduction

The CYP4 family is one of the most antique within the P450 superfamily of enzymes. As of mid-2006, some two dozen CYP4 subfamilies had been identified, the vast majority of which are found in insects where the enzyme family may have expanded, in part, to fulfil a role in insecticide resistance. In mammalian species six CYP4 subfamilies exist, comprising a total of 13 distinct CYP4 proteins in the human (Figure 12.1). Three of these subfamilies, namely CYP4A, CYP4B and CYP4F have been reasonably well studied from both a tissue distribution and substrate selectivity perspective (Table 12.1). CYP4A and 4F enzymes play important roles in the catabolic ω-oxidation of endogenous fatty acids, whereas CYP4B enzymes are more often associated with the bioactivation of certain xenobiotics. These, together with three so-called 'orphan' P450s, CYP4V, CYP4X and CYP4Z whose functions are just beginning to be explored, form the basis of this review.

Each subfamily will be considered from the perspective of enzyme multiplicity, substrate specificity, tissue distribution, gene regulation and pharmacogenetics. Within each category, emphasis will be placed wherever possible on human forms of the enzymes and their role in toxicity or disease. Several previous reviews have addressed some portion of the CYP4 landscape.[1–5] Much interesting information related to insect CYP4 enzymes (a topic not considered further here) can be found on the Web at http://p450.antibes.inra.fr and http://drnelson.utmem.edu/CytochromeP450.html.

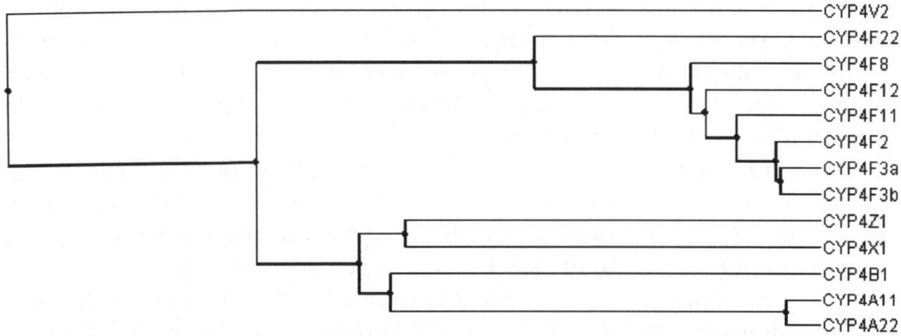

Figure 12.1 Phylogenetic tree of human CYP4 sequences. Generated using the ArboDraw software version 1.2 (Copyright © 2006, Adrian A. Canutescu and Roland L. Dunbrack Jr.).

Table 12.1 General characteristics of human CYP4 enzymes.

Enzyme	*Chromosomal localization*	*Tissue distribution*[a]	*Substrate selectivity*[a]
CYP4A11	1p33	Liver, kidney	Laurate, AA
CYP4A22	1p33	Liver, kidney (low)	?
CYP4B1	1p33	Lung, bladder	Arylamines, IPO
CYP4F2	19p13	Liver, kidney	LTB4, AA, vitamin E
CYP4F3A	19p13	Myeloid tissues	LTB4
CYP4F3B	19p13	Liver, kidney	AA, DHA
CYP4F8	19p13	Urogenital tissues	PGH1/2, AA
CYP4F11	19p13	Liver, kidney	Erythromycin (various)
CYP4F12	19p13	Liver, gut	Ebastine (various)
CYP4F22	19p13	Skin	Lipoxins?
CYP4V2	4q35	Widely distributed	?
CYP4X1	1p33	Trachea, aorta (wide)	?
CYP4Z1	1p33	Mammary tissue	?

AA – arachidonic acid
DHA – docosahexaenoic acid
IPO – 4-ipomeanol
LTB4 – leukotriene B4
[a]Entries under these categories are not intended to be exhaustive.

12.2 CYP4A

12.2.1 Multiplicity

The rat CYP4A subfamily was the first within the CYP4 gene family to be recognised to contain multiple members.[6–8] Characterisation of this new gene family was aided by its unusual functional and regulatory characteristics, exemplified by a proclivity for ω-hydroxylation of fatty acids and induction by peroxisome proliferating agents, respectively. The specificity of CYP4A enzymes for ω-oxidation facilitates the degradation of long chain fatty acids and serves as a secondary metabolic pathway for fatty acid metabolism when levels of these substrates increase during lipolysis and hepatic fatty acid uptake. The ω-hydroxylation of fatty acids and oxidation of the resulting ω-alcohols by dehydrogenases yields dicarboxylic acids which can undergo β-oxidation in peroxisomes, thus preventing accumulation of these fatty acids that could adversely affect mitochondrial function.[2]

Enzymes of the CYP4A subfamily have been identified in practically all mammalian species. In humans, two CYP4A enzymes have been identified, CYP4A11 and CYP4A22, which share 94% sequence identity within the cDNA coding region (Table 12.2), appear to have arisen as a result of a recent gene duplication on human chromosome 1. The mouse Cyp4a sub-family is the most diverse, containing eight *Cyp4a* genes (*Cyp4a10*/*12a*/*12b*/*14*/*29*/*30b*/*31*/*32*) as well as 2 pseudogenes (Cyp4a28ps, Cyp4a30as). The two *Cyp4a12* genes, *Cyp4a12a* and *Cyp4a12b*, are a result of a tandem 100 kb duplication in the Cyp4abx locus on mouse chromosome 4.[9] Rat, rabbit and pig species each have

Table 12.2 Pairwise sequence comparisons of human CYP4 enzymes.

	4A11	*4A22*	*4B1*	*4F2*	*4F3A*	*4F3B*	*4F8*	*4F11*	*4F12*	*4F22*	*4V2*	*4X1*
4A11												
4A22	94[a]											
4B1	53	52										
4F2	46	46	45									
4F3A	47	46	45	87								
4F3B	46	45	44	92	93							
4F8	45	45	44	81	77	81						
4F11	46	45	43	87	83	86	79					
4F12	46	45	43	81	78	82	78	83				
4F22	45	44	42	67	66	66	60	65	63			
4V2	33	33	31	35	36	35	35	35	36	37		
4X1	49	49	47	42	42	41	40	42	41	40	33	
4Z1	52	52	46	41	41	40	41	40	40	40	33	53

[a]Sequence identity was determined using BLASTN program with BLOSUM62 matrix (http://www.ncbi.nlm.nih.gov/blast/bl2seq/wblast2.cgi).

four *CYP4A* genes, while dogs and hamsters have 3 distinct genes. In the cow, two *CYP4A* genes have been identified while in the guinea pig, cat, minke whale and koala only a single CYP4A protein has been reported.

12.2.2 Tissue Distribution

Almost all of the CYP4A enzymes are closely associated with the metabolism of fatty acids, and together with members of the CYP4F family, are responsible for the generation of a wide variety of eicosanoid products from arachidonic acid (AA), notably the ω-hydroxylated product, 20-hydroxyeicosatetraenoic acid (20-HETE) (Figure 12.2). Eicosanoids have important biological functions throughout the body, and so the contribution of CYP4A enzymes to physiological processes that are modulated by eicosanoids will reflect, in part, their tissue distribution.

The mammalian CYP4A fatty acid hydroxylases are ubiquitously expressed with high levels found consistently in the kidney and liver. In humans, hepatic expression of CYP4A11 is substantially greater than for CYP4A22. Gene expression analysis of a panel of human liver samples revealed that, on average, the abundance of CYP4A22 transcripts was $<1\%$ of that found for CYP4A11.[10] Although CYP4A11 was initially cloned from human kidney,[11] mRNA analysis revealed a 2.5-fold higher expression in liver compared with kidney.[12] Human foetal liver (23–40 wk) also exhibited expression of CYP4A11 mRNA, but at levels less than half that seen in adult kidney. Other human tissues found to express CYP4A11 (albeit at significantly lower levels), include lung, small intestine, brain and reproductive organs.[12]

Immunohistochemical studies have confirmed expression of human CYP4A proteins in the hepatocytes of the liver lobule as well as within the kidney.[13,14]

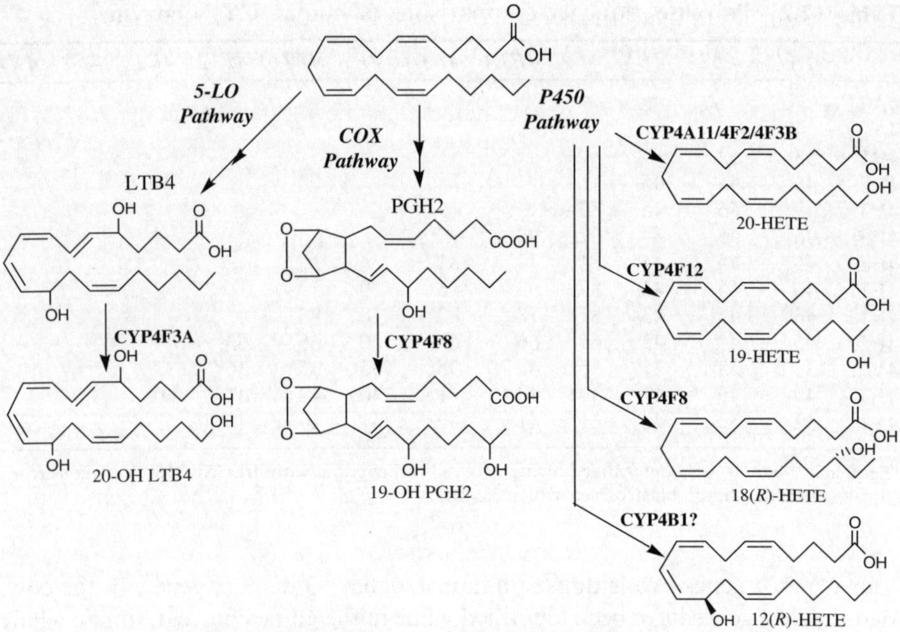

Figure 12.2 Human CYP4 enzymes involved in pathways of arachidonic acid metabolism. Human CYP4 enzymes that metabolise eicosanoid products within the 5-lipoxygenase (5-LO), cyclooxygenase (COX) and P450 pathways are shown in bold. HETE, Hydroxyeicosatetraenoic acid; LTB4 leukotriene B4; PGH2, prostaglandin H2.

In the kidney, abundant CYP4A levels were found in proximal tubules, but little immunostaining was evident in the glomeruli. In rat kidney, the proximal tubules were again identified as an important site of CYP4A expression, but the localisation in other segments of the kidney appeared to differ between rats and humans.[13] In rat liver, CYP4A1, CY4A2 and CYP4A3 are constitutively expressed.[7,8] CYP4A proteins are also detectable by Western blot analysis of rat brain, heart and lung, but levels are low compared to liver and kidney.[15]

Renal expression of CYP4A enzymes in the rat is also developmentally regulated.[16] Whereas rat CYP4A8 expression begins around weaning with no further changes, CYP4A1 and CYP4A3 peak at weaning followed by a gradual decline with increasing age. The expression of CYP4A2 in the kidney is not detectable until after weaning and then steadily increases, reaching plateau levels at $\sim$10 weeks of age. In the developing rat placenta, all four *CYP4A* genes are expressed when analysed by semi-quantitative RT-PCR with CYP4A1 exhibiting highest expression (CYP4A1 $>$ CYP4A2$>\gg$CYP4A3 approximately equal to CYP4A8).[17]

Renal expression of mouse Cyp4a enzymes has been studied in five inbred strains; (NMRI, FVB/N, 129 Sv/J, BALB/c and C57BL/6.[18] Cyp4a12a was expressed almost exclusively in male mice, whereas Cyp4a14 was selectively expressed in females. Transcripts for Cyp4a12b were almost undetectable in all of the strains examined. In the case of Cyp4a10, both sexes expressed this gene in the kidney but absolute levels were higher in females compared with males. Cyp4a12a mRNA and protein levels correlated well with ω-hydroxylation of AA and so this enzyme was concluded to be the predominant 20-HETE synthase in mouse kidney. In rabbit kidney and liver, CYP4A5 and CYP4A7 appear to be the dominant CYP4A enzymes, with the latter enzyme demonstrating high activity as a 20-HETE synthase.[19]

Pulmonary expression of rabbit CYP4A enzymes is particularly noteworthy in that prostaglandin ω-hydroxylation is induced over 100-fold in lung microsomes prepared from late gestation animals.[20] Masters and coworkers identified rabbit CYP4A4 as the specific PGE1 ω-hydroxylase induced in rabbit lung by pregnancy.[21,22] Additional interest in pulmonary CYP4A enzymes arises because, in lung, 20-HETE dilates microvessels, unlike most other tissues where this metabolite is a potent vasoconstrictor.[23] In the lung of adult male rats, CYP4A1 and CYP4A2 were expressed in pulmonary arterial endothelial and smooth muscle cells, bronchial smooth muscle and epithelial cells, type I epithelial cells and macrophages.[24] In adult male BALB/c mice, Cyp4a12 was the only Cyp4a protein expressed in the lung but absolute levels were ~100-fold less than those measured in liver.[25]

12.2.3 Substrate Specificity

Medium-chain, saturated fatty acids, exemplified by lauric acid (C12:0), are the prototypic substrates for CYP4A enzymes. These compounds are converted to the corresponding ω- (and ω-1) alcohol metabolites quite rapidly, and often with a high degree of regioselectivity. For example, rat CYP4A1 has been reported to hydroxylate lauric acid with a kcat in excess of 600 min^{-1} and an ω:ω-1 metabolite ratio of 40:1.[26] C-12 is the optimal chain length for CYP4A1-dependent fatty acid metabolism, with catalytic rates and product regioselectivity dropping as chain-length increases.[26,27] This is generally true for CYP4A enzymes across species, including the rabbit CYP4A5-7 enzymes[19] and human CYP4A11.[26]

At the time of writing, no crystal structure has yet appeared for any CYP4 protein. However, studies conducted with chimeric rat and rabbit CYP4As indicate that amino acid determinants for lauric acid binding and ω-hydroxylation reside in the first 120 N-terminal amino acids.[26,28] Moreover, detailed structure-function analysis involving a variety of lauric acid analogues suggest that the active site of rat CYP4A1 places more restraint on the nature of bulky and conformationally rigid substituents at the mid-chain region of these types of molecules, relative to the carboxylate head group or ω-terminus.[29] Recently, the size of the CYP4A1 access channel was calibrated with a series of halogen-substituted lauric acids. Failure to detect metabolism of the iodine analogue in this series prompted the conclusion that CYP4A1 possesses a highly restricted access channel having a width 3.9–4.3 Å.[30] Finally, an interesting structural peccadillo of many CYP4 enzymes, including several CYP4A proteins, is the covalent attachment of the haem prosthetic group that occurs with an appropriately positioned glutamic acid residue from the neighbouring I-helix.[31,32] The function of this post-translational modification is unknown, and CYP4 enzymes that lack this covalent haem link generally retain catalytic activity, although alterations in product regioselectivity have been reported.[33,34]

At least two CYP4A enzymes exhibit no lauric acid hydroxylase activity. Rabbit CYP4A4 instead metabolises certain prostaglandins by ω-hydroxylation (*vide infra*), and porcine CYP4A21 is an unusual chenodeoxycholic acid 6-α hydroxylase.[35] CYP4A enzymes may also play a role in the metabolism of dietary constituents as suggested by studies with phytanic acid (Figure 12.5).[36] Phytanic acid is derived from catabolism of chlorophyll, and is enriched in diets high in milk, cheese and red meat.[37] The presence of a 3-methyl substituent blocks mitochondrial β-oxidation, and so dietary phytanic acid is normally catabolised by peroxisomal α-oxidation. However, in patients with Refsum's disease, phytanic acid accumulates to high levels causing neurodegeneration as a result of mutations in the gene coding for the enzyme phytanoyl-CoA hydroxylase.[38] CYP4-mediated metabolism of this substrate is of interest because ω-hydroxylation could conceivably provide an alternative route for clearance of phytanic acid by initiating a pathway to the dicarboxylic acid which then would be a substrate for β-oxidation. Evaluation of the ω-hydroxylation activity of phytanic acid by rat and human CYP4A enzymes, as well as some rat CYP4F proteins, revealed that CYP4A1, CYP4A2, CYP4A3 and CYP4A11 displayed the highest catalytic efficiencies of the enzymes investigated.[36] However, human CYP4F enzymes were not examined in this study, and CYP4F3 catalysis may also be important.[39]

AA is perhaps the most intensively studied CYP4A substrate because 20-HETE plays an important role in the regulation of vascular tone in the coronary, pulmonary and renal circulations.[40] All four rat CYP4A enzymes can convert AA to 20-HETE with activity being greatest *in vitro* for CYP4A1, whether the enzyme is expressed in *E. coli*[26] or insect cells.[41] Metabolism of AA by CYP4A1 proceeds with turnover numbers of 6–7 min^{-1}, which are substantially smaller than those reported for the saturated fatty acids discussed above. Similar low levels of catalytic activity, relative to saturated fatty acids, were observed when rat CYP4A-dependent ω-hydroxylation of a different unsaturated fatty acid, oleic acid (C18:1), was evaluated.[26]

Turnover numbers for ω-hydroxylation of AA metabolism by purified, reconstituted CYP4A11 vary widely in the literature from zero activity[42,43] to 0.5–2 min^{-1} [11,26,44] and even as high as 49 min^{-1}.[45] Some of this variability could be due to differences in *in vitro* reconstitution conditions. Of note in this regard is the observation that cytochrome b_5 is generally required for maximal turnover by CYP4A enzymes. Indeed, rabbit CYP4A7 exhibits an absolute dependence on cytochrome b_5 for metabolism of AA.[46] To help clarify the role of CYP4 enzymes in 20-HETE formation by human liver and kidney, Lasker and coworkers employed antibodies directed against both CYP4A11 and CYP4F2. Whereas anti-CYP4A11 failed to inhibit 20-HETE formation

```
hCYP4A11     RFLAPWIGYGLLLLNGQTWFQHRRMLTPAFHYDILKPYVGLMADSVRVMLDKWEELLGQD 180
hCYP4A22     KFLAPWIGYSLLLLNGQTWFQHRRMLTPAFHYDILKPYVGLMADSVRVMLDKWEELLGQD 180
raCyp4a1     RLLAPWIGYGLLLLNGQPWFQHRRMLTPAFHYDILKPYVKNMADSIRLMLDKWEQLAGQD 179
raCyp4a2     QSLAPWIGYGLLLLNGKKWFQHRRMLTPAFHYDILKPYVKIMADSVSIMLDKWEKLDDQD 174
raCyp4a3     QFLAPWIGYGLLLLNGKKWFQHRRMLTPAFHYGILKPYVKIMADSVNIMLDKWEKLDDQD 177
raCyp4a8     RFLAPWIGYGLLLLNGQTWFQHRRMLTPAFHYDTLKPYVGIMADSVRIMLDKWEQIVGQD 178
rbCyp4a4     KLMTPWIGYGLLLLDGQTWFQHRRMLTPAFHYDILKPYVGLMVDSVQIMLDRWEQLISQD 180
rbCyp4a5     TFVAPWIGYGLLLLNGQPWFQHRRMLTPAFHYDILKPYVGLMVDSVQIMLDKWEQLVSQD 180
rbCyp4a6     RFLAPWIGYGLLLLNGQTWFQHRRMLTPAFHYDILKPYVGLMADSVQIMLDKWEQLVSQD 180
rbCyp4a7     SFLAPWIGYGLLLLEGQTWFQHRRMLTPAFHYDILKPYVGLMVDSVQVMLDKLEKLARKD 180
mCyp4a10     RLLAPWIGYGLLLLNGQPWFQHRRMLTPAFHYDILKPYVKNMADSIRLMLDKWERLADQD 179
mCyp4a12     RFLAPWIGRGLLLLDGQTWFQHRRMLTPAFHYDILKPYTEIMADSVRVMLDKWEQIVGQD 178
mCyp4a14     QFFAPWIGYGLLLLNGKKWFQHRRMLTPAFHYDILKPYVKIMADSVNIMLDKWEKLDGQD 177
               .:**** .****:*: **************. ****.  *.**: :***: *.:  :*
```

Figure 12.3 Multiple sequence alignment of the C-helix region of CYP4A isoforms. Multiple sequence alignments of human CYP4A11 and CYP4A22 with rabbit, mouse and rat CYP4A proteins using the CLUSTAL W software tool. The glycine residue at position 130 in human CYP4A11 is absolutely conserved across species but replaced by a serine residue in human CYP4A22 (highlighted in bold).

significantly in human liver microsomes,[45] metabolite production was decreased by over 30% in human kidney microsomes.[14] Therefore, CYP4A11 appears to be a greater contributor to 20-HETE formation in human kidney than in human liver, although CYP4F2 is the dominant 20-HETE synthase in both tissues.[14,45]

AA metabolism does not only involve hydroxylation to 19- and 20-HETE, but also epoxidation to 5,6-, 8,9-, 11,12- and 14,15-epoxyeicosatrienoic acids (EETs) (Figure 12.3). This activity is commonly catalysed by members of the *CYP2* gene family, with CYP2C enzymes accounting for most of the activity,[47] although some has been attributed to CYP2J enzymes.[48] With respect to CYP4A enzymes and EET metabolism, each of the four rat proteins has the capacity to hydroxylate 8,9-, 11,12- and 14,15-EET *in vitro*, at rates significantly higher than those measured for AA hydroxylation.[49] CYP4A1 was found to be the most active epoxygenase, preferring 8,9-EET as substrate and exhibiting an ω/ω-1 hydroxylation rate of 15 min^{-1}. This compares with a turnover number of 40 min^{-1} for lauric acid and 2 min^{-1} for AA. In comparison, CYP4A2, CYP4A3 and CYP4A8 preferred 11,12-EET, exhibiting turnover numbers of 7.0, 1.3 and 1.3 min^{-1}, respectively. Interestingly, in the same study, these authors found that 20-hydroxy-14,15-EET was able to effectively displace *cis*-parinaric acid from the peroxisome proliferator-activated receptor-α (PPARα) with a K_i of 3 nM, raising the possibility that these EETs (or their metabolites) could serve as endogenous ligands for this receptor.

12.2.4 Gene Regulation

The regulation of members of the *CYP4A* gene family has focused historically on the transcriptional induction mediated by chemicals and xenobiotics associated with peroxisome proliferation. This induction occurs *via* transactivation of the nuclear receptor, PPARα, following binding of an appropriate ligand.[50]

Such ligands include the prototypical peroxisome proliferator, Wyeth-14,643, as well as the fibrate class of drugs, phthalate plasticizers, chloroethylene solvents, phenoxyacetic acid herbicides, and the endogenous compounds, leukotriene B4 (LTB4) and dehydroepiandosterone sulphate.[51] In addition, a recent study that screened 200 pesticides from many different chemical classes identified several that were able to transactivate the murine PPARα receptor *in vitro*.[52] When these studies were extended *in vivo*, one of the pesticides (diclofop-methyl) induced murine hepatic Cyp4a levels to a similar degree as WY-14,643.

The intense interest in the toxicological impact of PPARα agonist stems, in large part, from the association between exposure to these xenobiotics and the development of hepatocellular carcinoma in mouse and rat model species. PPARα ligands transactivate numerous genes involved in fatty acid catabolism, including those encoding CYP4A enzymes. In cultures of primary human hepatocytes, CYP4A11 can be induced by clofibrate[53] and expression of the human gene for PPARα in "humanised" PPARα knockout mice restores PPAR ligand-induced changes in gene expression.[54] In rodents, PPAR ligands cause physiological changes beyond simple alterations in gene expression, with a proliferation of peroxisome organelles, increases in hepatocyte cell division and, ultimately, hepatocellular carcinoma.

In humans, however, PPARα agonists do not cause cancer. The reason for the species differences between rodents and humans is not clear. One early hypothesis was that levels of PPARα are approximately an order of magnitude lower in hepatocytes from humans and guinea pigs (another non-responsive species) compared with rodents.[55] In support of this, canine hepatocytes express similar levels of PPARα compared to humans,[56] and treatment of cultures of canine hepatocytes with clofibrate results in only moderate induction of CYP4A39.[57]

However, this is an overly simplistic explanation. While PPARα knockout mice are refractory to the hepatocarcinogenic effects of bezafibrate,[58] the "humanised" PPARα transgenic mice express similar levels of hepatic PPARα protein compared to normal mice, yet fail to exhibit any of the hallmarks of peroxisome proliferation and aberrant cell growth in response to Wy-14,643.[54] The latter findings suggest that there may be specific gene targets or transactivators with which the human PPARα protein is incapable of interacting. It should be noted that the induction of murine Cyp4a was similar between wild-type and PPARα "humanised" mice, and so this calls into question whether there are any physiological implications for the alterations in lipid metabolism mediated by increased Cyp4a activity. It may be the case that changes in *Cyp4a* gene expression merely serve as a biomarker of exposure to PPARα agonists, whether they be environmental toxicants or endogenous fatty acids.

In addition to changes in CYP4A expression following treatment with the aforementioned drugs and xenobiotics, certain physiological/pathological states can also affect CYP4A expression. Cyp4a mRNA levels increased 7–17 fold in the livers of rats fasted for 48 hrs, but further food deprivation to 72 hrs resulted in a return to baseline, presumably reflective of an adaptive response.[59]

The changes in *Cyp4a* gene expression could also be mimicked by inducing a diabetic state in rats by treatment with streptozotocin that was reversible with insulin administration. The alterations in hepatic fatty acid levels induced by fasting or diabetes are likely caused by inducing Cyp4a *via* PPARα, as certain fatty acids like palmitic and linoleic acid are PPARα agonists.[60] The induction of *Cyp4a* genes observed in the streptozotocin-induced rat model of diabetes is replicated in the *db/db* diabetic mutant mouse where hepatic Cyp4a protein levels were induced more than 3-fold and Cyp4a10 transcripts were induced >10-fold.[61] Associated with this increase in Cyp4a expression, the *db/db* mice were found to have fatty livers and hyperlipidaemia, as well as increases in other PPARα-responsive genes.

CYP4A gene expression is also altered in response to bacterial infection or treatment with lipopolysaccharide (LPS), but again the response is complex. Administration of LPS to F344 rats induced both hepatic and renal CYP4A,[62] but whether this was due to a direct effect of LPS or a result of hypophagia associated with LPS treatment was unclear. Subsequent studies demonstrated that the response was independent of food intake, and that there was also a sexual dimorphism in response, consistent with differences in PPARα-mediated changes in gene expression between genders.[63] The response of mice to LPS is the opposite of that observed in rats, where the majority of hepatic P450s exhibit decreased expression following treatment. In the case of mouse *Cyp4a* genes, an intraperitoneal injection of 1 mg/kg LPS resulted in 70–80% decreases in hepatic Cyp4a10 and Cyp4a14 mRNA levels.[64] Infection of mice with *Citrobacter rodentium* provided similar results.[65]

12.2.5 Pharmacogenetics

The association of genetic polymorphisms in the human *CYP4A11* and *CYP4A22* genes with physiological or pharmacological sequelae has only recently been explored. A single nucleotide polymorphism (SNP), T8590C, in *CYP4A11* has been associated with essential hypertension in Caucasians, exhibiting an adjusted odds ratio (OR) of 2.31 (95% confidence interval 1.41 to 3.78).[44] This SNP leads to a phenylalanine to serine substitution at position 434 which reduces lauric acid and AA ω-hydroxylase activity upon heterologous expression of the variant enzyme. The authors proposed that decreased 20-HETE formation in subjects with the 8590C allele resulted in hypertension due to increased sodium reabsorption in the renal tubule, in a manner similar to that observed previously in the Dahl salt-sensitive rat.[66]

A role for the CYP4A11 8590C allele in hypertension is also supported by a study that genotyped 1397 individuals from the Augsburg region of Germany.[67] Subjects that were homozygous for the CC genotype (i.e. the low activity Ser434 variant) exhibited an increased risk of hypertension compared to CT and TT genotypes, with an OR for the entire population of 3.31 (95% confidence interval, 1.38 to 7.96). While both studies are consistent with a role for CYP4A11 in hypertension, and *in vitro* data offer a mechanistic explanation, it remains to be seen whether the T8590C SNP alters 20-HETE levels and functional CYP4A-dependent activity in the renal tubule.

From data available at dbSNP (http://www.ncbi.nlm.nih.gov/projects/SNP/), and a recent study conducted in Korean subjects, it is clear that a plethora of SNPs are scattered throughout the *CYP4A11* gene, many of them occurring in intron 1. Of particular interest are the six SNPs found within coding exons, two of which (including T8590C) are non-synonymous. However, the functional significance of these polymorphisms is unknown at present.

The question of whether human CYP4A22 is a functional fatty acid hydroxylase has been addressed recently. When expressed in *E. coli*, recombinant CYP4A22 did not exhibit a typical reduced P450 CO-bound complex and had no catalytic activity towards either lauric acid or AA.[44] However, analysis of sequence alignments in the C-helical region showed that the glycine residue at position 130 of CYP4A11, which is absolutely conserved amongst CYP4A family members, is replaced by a serine residue in CYP4A22 (Figure 12.3). Site-directed mutagenesis of CYP4A11 to yield the G130S mutant enzyme resulted in a failure to incorporate haem and a loss of catalytic activity.[44] Interestingly, in the Japanese, CYP4A22 is polymorphic at the Ser130 locus, and the (functional) Gly allele is present in a substantial proportion of this population.[68]

12.3 CYP4B

12.3.1 Multiplicity

The CYP4B sub-family comprises CYP4B1 isoforms identified in rats, rabbits, mice, humans and pigs. In addition, CYP4B2 (goats and cattle) and CYP4B3 (wallaby) proteins have been described (see http://drnelson.utmem.edu/CytochromeP450.html). These species orthologues share >80% amino acid sequence homology. By far the best studied isoform is rabbit CYP4B1, which historically has been termed variously P450II, IVB1 and LM5.[4]

12.3.2 Tissue Distribution

In general, CYP4B1 is an extrahepatic P450, but tissue content varies substantially with species. In most animals CYP4B1 homologues are expressed dominantly in the lung, with low or negligible levels in uninduced liver. This is exemplified by studies conducted originally in the rabbit. These are historically interesting because CYP4B1 was the first extrahepatic P450 enzyme to be characterised in any substantive fashion, following its purification from rabbit lung microsomes.[69,70] In rabbit lung, CYP4B1 is a major P450 protein, comprising about 40% of total P450 in this organ, which contrasts with uninduced rabbit liver where the enzyme constitutes only 1–2% of total hepatic P450.[71]

Cross-species comparison of the tissue distribution of CYP4B1 is most readily accessible from studies that have quantitated mRNA levels. In the rabbit, mRNA expression is high in the lung compared to the liver and all other extrapulmonary tissues examined, including kidney, heart, and brain.[72] In constrast, mouse Cyp4b1 transcript levels were similar in the lung, kidney and brain (~20% distribution for each), and low, but detectable, in the liver and several other organs.[73] Humans transcribe CYP4B1 predominantly in the lung.[12] However, whereas catalytic activities attributable to CYP4B1 are readily measured in lung (and kidney) microsomes from a variety of experimental animals, good evidence for expression of functional human CYP4B1 is restricted largely to studies with human bladder microsomes.[74] Nonetheless, even low tissue levels of functional CYP4B1 could be important given the well recognised role of the enzyme in the bioactivation of certain furan and arylamine protoxins.

12.3.3 Substrate Specificity

Metabolic bioactivation by cytochrome P450-dependent processes is often a prerequisite for xenobiotic toxicity.[75] CYP4B1 activity is linked to metabolic bioactivation of the pneumotoxins, 4-ipomeanol (IPO) and 3-methylindole (3-MI), and a variety of mutagenic arylamine compounds such as 2-aminofluorene and 3-methoxy-4-aminoazobenzene that are associated with chemically-induced bladder and liver cancers (Figure 12.4). On the basis of these observations, CYP4B1, together with certain members of the CYP1A, CYP2E and CYP3A subfamilies, have been identified as major xenobiotic bioactivating P450 subfamilies because they catalyse formation of reactive metabolites from a wide variety of protoxins.[76]

CYP4B1 is strongly associated with bioactivation of the pneumotoxin 4-ipomeanol (IPO) (Figure 12.4). IPO, is a furan derivative formed in moldy sweet potatoes which causes lung toxicity in cattle and some other animal species, including rabbits.[77] A role for CYP4B1 in the bioactivation process was first highlighted when it was found that CYP4B1 antibodies prevented covalent binding of radiolabelled IPO to protein in pulmonary and hepatic rabbit microsomal preparations.[78] Moreover, treatment of rabbit CYP4B1-transfected human HepG2 cells with [^{14}C]-IPO resulted in the highest levels of radioactivity bound to calf thymus DNA when compared with 12 other recombinantly expressed P450s.[79] In addition, decreased covalent binding of [^{14}C]-IPO to protein was observed in rat lung microsomes following pretreatment with *p*-xylene, a CYP4B1 substrate and presumably an inhibitor of the enzyme.[80] These data demonstrate a central role for animal forms of CYP4B1 in the bioactivation of IPO to reactive intermediates that can bind to cellular protein and DNA.

The nature of the IPO reactive intermediate(s) has been the subject of several recent studies. Administration of IPO that had been labeled with deuterium in the aliphatic chain (Figure 12.4) led to the recovery of four glutathione adducts in the bile of rats all of which retained all of the original deuterium atoms.[81] This demonstrated elegantly that activation to an electrophilic intermediate occurred on the furan ring. Large-scale, *in vitro* incubations of recombinant

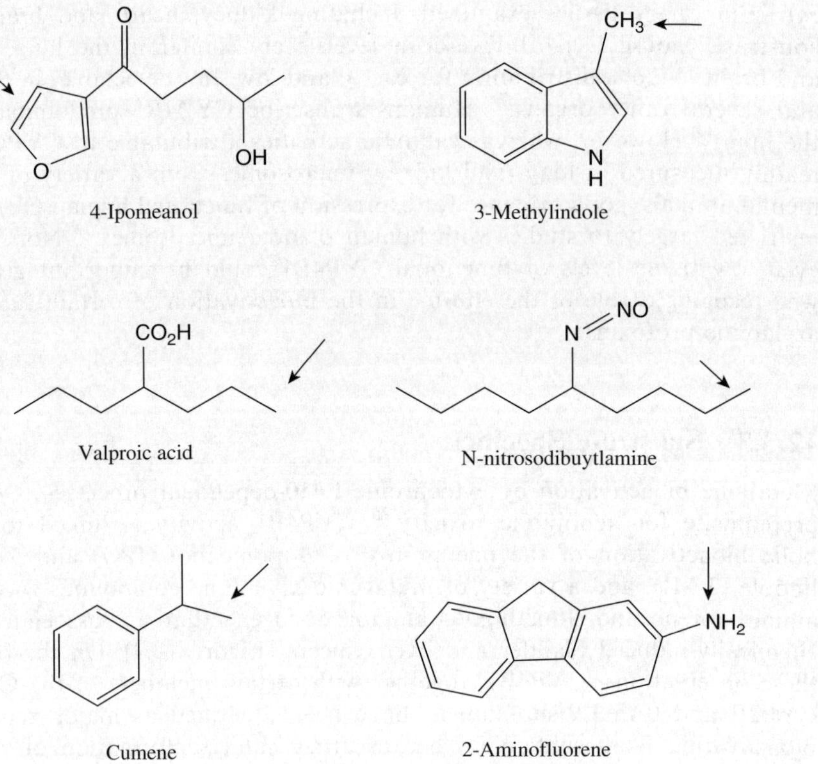

Figure 12.4 Examples of xenobiotic structures metabolised by CYP4B enzymes. Arrows designate the position of oxidative attack.

CYP4B1 with IPO in the presence of N-acetyllysine and N-acetylcysteine led to the recovery of a single major product that was identified by 2D-NMR and mass spectrometry as the N-substituted cysteinyl pyrrole derivative of IPO.[82] By analogy to model chemical studies conducted with cis-2-butene-1,4-dial,[83] the formation of this product can be rationalised by formation of an ene-dial, an electrophilic dialdehyde formed by oxidative ring-opening of the furan.

Rumenal fermentation of tryptophan produces 3-methylindole (3MI) (Figure 12.4) which causes pneumotoxicity in cattle, goats, and some laboratory animal species.[84] The reactive intermediate responsible for toxicity is likely the dehydrogenated, electrophilic 3-methyleneindolenine metabolite (3MEI).[85] This intermediate, trapped and quantitated as the N-acetylcysteine adduct, can be formed by multiple P450 enzymes including, most prominently, rabbit CYP4B1 and goat CYP4B2.[86,87]

CYP4B enzymes metabolise a variety of mutagenic and carcinogenic arylamines and azo dyes to N-hydroxy metabolites that can react with DNA *via* an electrophilic aryl nitrenium ion.[88] The arylamines, 2-aminofluorene, (2-AF) (Figure 12.4), 2-aminoanthracene (2-AA), and 2-acetylaminofluorene (2-AAF) were among the first substrates identified for rabbit CYP4B1.[71] Mouse

CYP4B1, purified initially from kidney microsomes by using *umu* gene expression as an index of mutagenic activation, bioactivates aromatic amines as well as 3-methoxy-4-aminoazobenzene and 3,3′-dichlorobenzidine, potent liver and bladder procarcinogens, respectively.[89,90] Additional xenobiotic substrates for CYP4B1 include valproic acid (VPA) and N-nitrosodibutylamine (NDBA).[91,92] These two structurally similar compounds (Figure 12.4) are associated with idiosyncratic hepatotoxicity and bladder cancer, respectively. Both VPA and NDBA undergoes CYP4B1–dependent aliphatic hydroxylation to generate ω-oxidised metabolites and, in the case of VPA, CYP4B1 forms relatively large quantities of an unusual dehydrogenated metabolite, 4-ene-VPA. Further metabolism of 4-ene-VPA by β-oxidation enzymes has been speculated to result in formation of reactive metabolites that contribute to the mitochondrial toxicity of the drug.[93,94] CYP4B1 also preferentially hydroxylates NDBA at the ω-terminus to generate N-nitrosobutyl-4-hydroxybutylamine.[92] Downstream metabolic products derived from this initial terminal hydroxylation have been implicated in bladder cancer caused by NDBA.[95]

Rabbit CYP4B1 metabolises lauric acid at both the ω and ω–1 positions,[96] in contrast to CYP4A (and CYP4F) isoforms which generally favour ω-hydroxylation of this saturated fatty acid. Interestingly, positional selectivity for CYP4B1-dependent fatty acid ω-hydroxylation increases as the chain length decreases, and this profile is recapitulated with the C7-C12 hydrocarbon analogues.[97] Therefore, the ω-hydroxylation selectivity of CYP4B1 is not governed by the presence of the carboxylic acid functionality, again in contrast to the well studied CYP4A enzymes. Indirect evidence has been presented that rabbit corneal CYP4B1 can convert AA to 12-R-HETE (Figure 12.2) and 12-hydroxyeicosatrienoic acid (12-HETrE).[98] This is of interest because of a potential role for 12-HETrE in angiogenesis.[99] However, others have failed to find metabolites of AA generated directly using the purified, reconstituted enzyme.[72]

CYP4B1 also metabolises the unfunctionalised compound, cumene (Figure 12.4), with a high degree of selectivity for ω versus ω-1 hydroxylation and significant stereoselectivity for hydroxylation on the pro-S methyl group.[100] Therefore, of all the CYP4 enzymes under discussion, CYP4B1 exhibits the broadest substrate specificity, accommodating metabolism of fatty acids, hydrocarbons, unfunctionalised aromatic compounds and arylamines. Despite this apparent diversity, common features of CYP4B1 substrates include a relatively small, compact molecular structure and the availability of a sterically unhindered position for oxidation (Figure 12.4).

Metabolism of several compounds by CYP4B1 is associated with loss of catalytic function. 11-Dodecynoic acid and 17-octadecynoic acid are acetylenic mechanism-based inhibitors of rabbit CYP4B1[96] presumably as a consequence of formation of the intermediate ketene. However, these compounds, and other acetylenic fatty acids, will also inactivate several CYP4A and CYP4F enzymes, and so are not specific CYP4B1 inhibitors. Inhalational exposure to *p*-xylene or *m*-xylene inhibits CYP4B1 activity in rats.[80,101] The mechanism is not established, but may involve formation of *p*- and *m*-tolualdehyde metabolites that require further bioactivation to a product that destroys P450.[102] Again, xylene inhibits a number of pulmonary P450s, and a specific, high affinity inhibitor of CYP4B1 that would be useful for dissecting the role of this enzyme subfamily in both fatty acid metabolism and chemical toxicity is not yet available.

12.3.4 Gene Regulation

Unlike the CYP4A enzymes which, in general, are transcriptionally regulated by the ligand-activated PPARα receptor (*vide infra*), CYP4B1 is induced by a wide variety of agents. Treatment with the prototypical P450 inducer, phenobarbital, increased the low constitutive level of hepatic CYP4B1 by a factor of ~10 in rabbits,[103] but not in other common laboratory species.[104] Aroclor 1260, but not 2,3,7,8-tetrachlorodibenzo-*p*-dioxin (TCDD), induced synthesis of CYP4B1 in the liver, but none of these compounds affected pulmonary CYP4B1 expression.[105] In the rabbit cornea, there is some evidence that clofibrate and phenobarbital may induce CYP4B1,[98] but molecular mechanisms underlying these effects have not been elucidated.

In contrast, our understanding of regulation of CYP4B1 in response to endogenous compounds and hypoxia is more advanced. The promoter region of the rabbit *CYP4B1* gene contains DNA sequences that could potentially bind heterodimers of the retinoic acid receptors including RAR/RXR, VDR/RXR, and PPAR/RXR. Indeed retinoids, but not vitamin D, caused significant transcriptional activation in HepG2 cells transiently transfected with luciferase reporter vectors containing the CYP4B1 promoter.[106] The promoter region of rabbit corneal CYP4B1 also contains DNA-binding sequences for factors known to activate gene transcription in response to hypoxia; hypoxia-inducible factor 1 (HIF-1), nuclear factor κB(NFκB), and activating protein1 (AP-1). Data obtained from electrophoretic mobility shift assays using^{32} P-labelled probes for putative DNA binding sequences of each transcription factor support a role for all three of these transactivating factors in CYP4B1 regulation under hypoxic conditions.[107]

A pronounced gender difference has been described in mouse kidney, where Cyp4b1 is expressed in a male-specific manner.[89] In male mice and rats, castration decreased CYP4B1 expression levels, which were restored partly by administration of testosterone.[108,109] Although, these data implicate androgens in the sex-specific regulation of rodent CYP4B1, promoter analysis did not reveal an androgen response element (ARE) motif,[109] and so molecular mechanisms underlying sex steroid regulation of renal CYP4B1 remain to be established.

Regulatory factors controlling the lung-selective expression of human CYP4B1 have been investigated.[110] Analysis of the promoter region of the human *CYP4B1* gene identified several regulatory elements in the first kilobase of the 5′-flanking region. The authors implicated Sp1 and/or Sp3 transcription factor binding to a proximal enhancer element acting in combination with a distal lung-selective enhancer and a liver-selective repressor motif, as critical to the pattern of expression of human CYP4B1.

12.3.5 Pharmacogenetics

Factors confounding a clear understanding of the scope of human CYP4B1 catalytic activity have been discussed in detail in a recent review.[4] Of central importance appears to be the 'mutation' of a highly conserved meander-region proline in human CYP4B1 to a serine residue not found in any other P450 sequence that may adversely affect haem binding. This situation has a parallel in the metabolic activity of CYP4A22 that was discussed above. However, unlike the CYP4A22 example, genotyping of several hundred DNA samples from Caucasians, African-Americans and Hispanic subjects did not identify a single Pro allele.[111] Six non-synonymous coding-region polymorphisms (CYP4B1**2*-**7*) have been identified in human *CYP4B1* from Caucasian and Japanese subjects.[112,113] Several of these involve amino acid changes at locations that would be expected to disrupt the activity of wild-type CYP4B1, but further evaluation been hampered by the difficulties inherent in expressing a functional form of the native protein.

12.4 CYP4F

12.4.1 Multiplicity

The mammalian CYP4F subfamily rivals the CYP4As in terms of enzyme multiplicity. First identified at the molecular level in the early 1990s by the near concurrent cloning of rat *CYP4F1* and human *CYP4F3*A,[5] four rat forms (CYP4F1, CYP4F4, CYP4F5, CYP4F6), five mouse forms, (Cyp4f13, Cyp4f14, Cyp4f16 Cyp4f16) and seven human forms (CYP4F2, CYP4F3A, CYP4F3B, CYP4F8, CYP4F11, CYP4F12 and CYP4F22) are now recognised (Table 12.1). CYP4F3A and CYP4F3B are splice variants with identical coding regions except for the interchange of exons 4 and 3, respectively.[114] The two enzymes have altered substrate specificities and tissue distributions (*vide infra*). The following discussion focusses largely on the properties of the human enzymes, genes for which are clustered together on chromosome 19. Additional information on this topic plus a detailed discussion of rodent CYP4F isoforms can be found in a recent comprehensive review.[5]

12.4.2 Tissue Distribution

In order to draw robust conclusions about the role of specific CYP4 enzymes in catalysing metabolism of a given endogenous or xenobiotic compound, particularly in an expanded subfamily like CYP4F, information about an enzyme's intrinsic catalytic efficiency (*vide infra*) is helpful, but insufficient.

Equally critical is quantitative information for the relative tissue expression levels of the P450 isoforms under investigation. However, due to a lack of human protein-specific CYP4F antibodies, only sketchy information is currently available at the protein level. Much more extensive descriptions of tissue mRNA expression have appeared. These results are summarised below and contrasted with data from peptide antibody experiments where available.

An extensive survey of the tissue distribution of CYP4F mRNA, accomplished with a cDNA probe that possessed >90% nucleotide identity to all of the human *CYP4F* genes, found strong hybridisation signals in leukocytes, bone marrow, liver, kidney and prostate.[115] Protein-specific RT-PCR then revealed CYP4F3A alone to be expressed in neutrophiles and bone marrow, and CYP4F3B and CYP4F2 to be present in human liver and kidney at 4:1 and 3:4 ratios, respectively. CYP4F3A was not detected in either human liver or kidney, and CYP4F8 was detected only in the prostate. This general profile of human CYP4 mRNA expression has been confirmed by others.[12] Northern blot analyis of human tissues localised CYP4F11 and CYP4F12 primarily in the liver.[116,117] Lower levels of expression of CYP4F11 were found in the kidney, heart and skeletal muscle, whereas modest signals for CYP4F12 expression appeared in the gastointestinal tract, heart and kidney.

C-terminal peptide antibodies have been generated towards CYP4F8 and CYP4F12,[118,119] and used to evaluate expression of the protein products in various human tissue microsomes. CYP4F12 protein was detected in human liver, kidney, gastointestinal and urogenital epithelia, thereby providing a reasonable correlation with the mRNA data. The peptide antibody against CYP4F8 also confirmed protein expression in the prostate and seminal vesicles, and further extended the tissue distribution of this protein to the epidermis, corneal epithelia, kidney and even human liver. Notably, CYP4F8 protein expression appeared to be upregulated in the epidermis of psoriatic lesions,[119] but the biological significance of this is unknown.

P450 protein distribution studies rely very heavily on the integrity of the tissues involved whether the primary endpoint is mRNA or protein analysis. When Western blotting is employed to probe enzyme localisation, cross-reactivity with other P450 isoforms can complicate interpretation unless the antibodies have been extensively characterised against all other forms of the enzyme likely to be present. These limitations not withstanding, the current picture of CYP4F expression in human tissues is consistent with important roles for these enzymes within this subfamily in both eicosanoid and xenobiotic metabolism.

12.4.3 Substrate Specificity

The prototypical substrate for the CYP4F subfamily is the endogenous eicosanoid, leukotriene B4. Indeed, the emergence of this subfamily closely followed attempts to understand how this proinflammatory molecule was inactivated in the body. LTB4 is a potent chemotactic agent primarily involved in inflammation, immune response and host defence. In 1993, Kusunose and coworkers reported the characterisation of a then novel CYP4 enzyme (CYP4F3A) from polymorphonuclear leukocytes.[120] CYP4F3A has been shown to catalyse the high affinity (Km ~ 1μM), ω-hydroxylation of LTB4 to the inactive 20-hydroxy metabolite[114,115] (Figure 12.2). Across species many, but not all, CYP4F enzymes display ω-hydroxylase activity towards LTB4. Amongst the rat CYP4F enzymes, only CYP4F1 and CYP4F4 catalyse leukotriene B4 20-hydroxylation.[121] Human CYP4F2 and CYP4F3B also metabolise LTB4, but exhibit 30-60 fold higher Km values than neutrophile CYP4F3A, whereas CYP4F8, CYP4F11 and CY4F12 have little or no activity towards this substrate.[5]

As discussed above in the section on CYP4A enzymes, much attention has been focussed on the P450-dependent metabolism of AA to ω-hydroxylated metabolites (Figure 12.2) which are implicated in the modulation of vascular tone. In human kidney, CYP4F2 has been identified as a major contributor to microsomal 20-HETE formation.[14] Studies with recombinantly expressed enzymes also highlighted a substantial 20-HETE synthase activity for CYP4F3B, but not CYP4F3A or CYP4F11.[115,122] CYP4F8 and CYP4F12 can also metabolise AA, but 18-HETE (ω-2 hydroxylation) is the preferred product.[117,123]

Metabolism of n-3 (also termed ω3) long-chain polyunsaturated fatty acids (PUFAs), such as eicosapentaenoic acid (C20:5n-3) [EPA] and docosahexaenoic acid (C22:6n-3) [DHA], is of interest because of the documented cardioprotective effects of these dietary components.[124] The structural similarity between AA, an n-6 (ω6) PUFA, and DHA and EPA make it likely, *a priori*, that CYP4 enzymes would play a role in their catabolism. While investigation of the P450-mediated metabolism of n-3 fatty acids has been neglected relative to studies involving AA, CYP4F3B has been identified as an ω-hydroxylase of EPA and DHA.[125] CYP4F8 and CYP4F12, in contrast, catalyse epoxidation of DHA.[123] Therefore, for these two enzymes, a theme of internal oxidation may be emerging, as both also catalyse ω-1/ω-2 oxidation of AA and certain PGH2 analogs[123,126] (Figure 12.2).

Lauric acid appears to be a poor substrate for members of the CYP4F family. Among rat CYP4F enzymes, only CYP4F4 was found to exhibit (minor) activity towards lauric acid.[121] Human CYP4F2-catalysed LTB4 hydroxylation was not inhibited by either lauric acid or palmitic acid (C16:0),[127] but CYP4F12 has been reported to catalyse ω-1 hydroxylation of lauric acid.[119] In contrast, much longer saturated fatty acids, *viz* C22:0, C24:0 and C26:0, are readily metabolised by CYP4F2 and CYP4F3B, but not by CYP4F3A or CYP4F12, with a high degree of positional selectivity for ω versus ω-1 hydroxylation ($>$20:1) of the C22-C26 fatty acids.[128] The metabolism of these very long-chain fatty acids is of physiological interest because they are elevated in the neurodegenerative disease, X-linked adrenoleukodystrophy.

An interesting substrate in the saturated fatty acid category is phytanic acid (Figure 12.5). As discussed above, Refsum's patients suffer from neurodegeneration as a result of a genetic deficiency in the phytanoyl-CoA 2-hydroxylase pathway.[38] Several rat and human CYP4F, as well as some CYP4A, enzymes

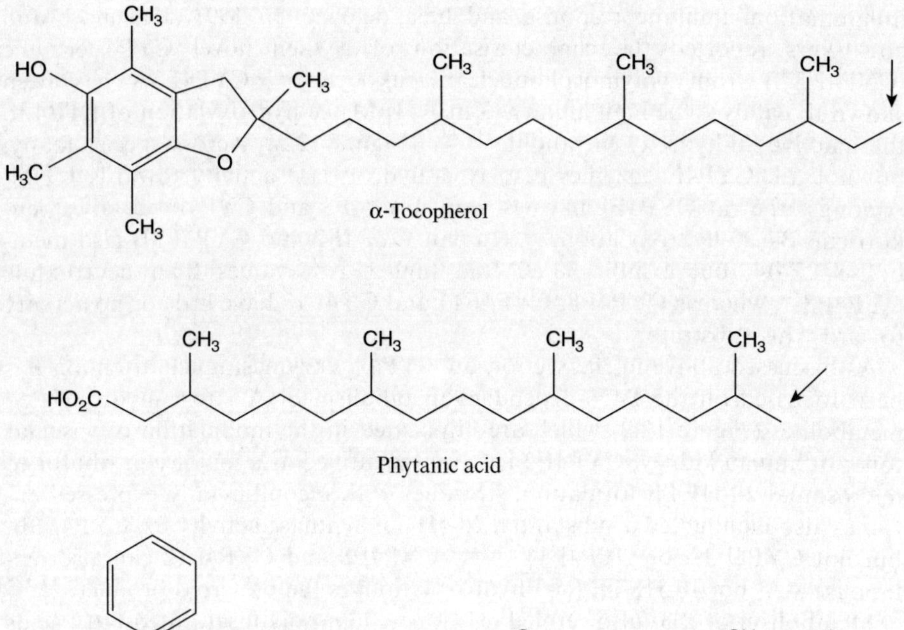

Figure 12.5 Examples of xenobiotic structures metabolised by CYP4F enzymes. Arrows designate the position of oxidative attack.

can catalyse formation of ω-hydroxyphytanic acid,[36,39] but it is not yet established which P450 enzyme is responsible for this reaction *in vivo*.

Vitamin E is a mixture of several related antioxidant compounds known as tocopherols (α, β, γ), which bear some structural similarity to phytanic acid (Figure 12.5). Metabolism of the tocopherols to urinary carboxychroman metabolites presumably relies on an initial ω-hydroxylation of the phytyl side-chain followed by β-oxidation. Tocopherol ω-hydroxylation was first thought to be due to the action of CYP3A enzymes, because ketoconazole inhibited vitamin E metabolism in mammalian cells.[129] However, subsequent studies identified CYP4F2 as the enzyme responsible for the ω-hydroxylation of γ-tocopherol.[130] In contrast, CYP4F3B and CYP4F3A exhibit little or no ω-hydroxylase activity towards γ-tocopherol which, when contrasted with metabolism of phytanic acid, suggests that the chromanol nucleus of the tocopherols may be a discriminating feature for catalysis by CYP4F2 versus CYP4F3 enzymes.

With the emphasis on fatty acid metabolism, fewer studies have examined drug and xenobiotic metabolism by CYP4F enzymes. A significant impediment here is the lack of well characterised chemical inhibitors and/or antibodies relative to those available for the common hepatic drug-metabolising P450 families. Indeed, where evidence for a significant contribution of human CYP4F enzymes to xenobiotic metabolism exists, it has been derived largely from instances where members of the CYP1-3 families failed to account for all of the microsomal turnover.

Ebastine (Figure 12.5) is an non-sedating antihistaminic drug that relies on P450-mediated hydroxylation to the active form, carebastine. In early studies with human liver microsomes and twelve recombinantly expressed P450s (but no CYP4F isoforms) only CYP3A4 catalysed formation of trace amounts of carebastine.[131] However, chemical inhibitor and antibody experiments were inconsistent with a role for CYP3A4 in this activation pathway. Subsequently, carebastine formation in monkey intestinal microsomes was shown to be inhibited by a CYP4F antibody,[132] and CYP4F12 cloned from human intestine was found to be active as an ebastine hydroxylase.[133] A head-to head steady-state kinetic comparison of CYP4F12 and CYP2J2 favoured the latter enzyme as the major human ebastine hydroxylase.[134] However, a later study reported a 10-fold higher rate of ebastine hydroxylation by CYP4F12,[135] raising the question of whether the enzyme's contribution to *in vivo* (intestinal) biaoctivation of ebastine may have been underestimated.

Recently, initial reaction phenotyping of the investigational anti-parasitic drug, DB-289 (Figure 12.5), failed to identify the human liver microsomal P450 responsible for O-demethylation of DB-289.[136] However, further screening with less commonly employed recombinant human P450s showed that CYP4F2, CYP4F3B and CYP2J2 could all catalyse this reaction. By employing CYP4F and CYP2J selective antibodies, and contrasting chemical inhibition in microsomes and recombinant enzymes with CYP4F and CYP2J substrates, the authors concluded that human CYP4F2 and/or CYP4F3B were most likely responsible for DB-289 O-demethylation in human liver microsomes.

Strobel's group has recently undertaken a semi-systematic analysis of the catalytic capabilities of CYP4F11 relative to CYP4F3A.[122] CYP4F11 was a much better catalyst (>5 times CYP4F3A's turnover) of benzphetamine N-demethylation, chlorpromazine N-demethylation and sulphoxidation, theophylline N-demethylation and eryythromycin N-demethylation, although none of the CYP4F11 turnover rates exceeded 1 min^{-1}. Erythromycin, a probe for CYP3A4, was the best substrate of those examined for CYP4F11.

In conclusion, CYP4F11, CYP4F12, CYP4F3B and CYP4F2 appear to be the most important of the human CYP4F enzymes for metabolism of drugs and xenobiotics. However, clear distinction among the various human CYP4F enzymes that may catalyse a given metabolic pathway is often complicated by the paucity of diagnostic tools available. In contrast to CYP4B, CYP4F enzymes can accommodate metabolism of larger, and sometimes bulky, structures although metabolism is often confined to the terminal end of these molecules (Figure 12.5). These observations, coupled to the fact that chemical inhibition of CYP4F activity is frequently, observed with ketoconazole, may suggest some similarities in ligand specificity between certain CYP4F enzymes and CYP3A4.

12.4.4 Gene Regulation

Within the human CYP4 family, there exists an excellent example of gene regulation by tissue-specific, alternative splicing that generates CYP4F3A in neutrophiles and CYP4F3B in liver by 'skipping' of exons 3 or 4, respectively.[114] Both enzymes contain 520 amino acids, but the selection of exon 3 versus exon 4 changes the amino acid sequence in a portion of the enzyme's active site that alters substrate specificity for LTB4 (CYP4F3A) versus arachidonic acid (CYP4F3B), as discussed above. Detailed studies of the regulatory regions of CYP4F3 reveal that alternative promoters regulate tissue-specific expression in neutrophiles versus liver.[114] Moreover, the myeloid promoter is associated with clusters of CACCT sites that can be activated by the zinc finger E-box binding homeobox 2 (ZEB-2) transcription factor to drive CYP4F3A expression.[137]

Transcriptional regulation of CYP4F gene expression differs substantially from that described for members of the CYP4A subfamily. Most strikingly, PPARα agonists like clofibrate, which induce expression of CYP4A enzymes in a PPAR-dependent manner, either downregulate or have no effect on CYP4F enzymes in the rat and mouse.[116,138] In contrast, statins upregulate expression of both CYP4A and CYP4F enzymes.[139,140] Lovastatin induced human CYP4F2, but not CYP4F3B, in HepG2 cells and human hepatocytes. The mechanism of this effect involved sterol-regulatory element binding protein-2 activation of an enhancer region of the *CYP4F2* proximal promoter.[140]

Retinoids provide a third family of ligands implicated in CYP4F regulation. These compounds are related to vitamin A and have a plethora of cellular actions due to their ability to regulate epithelial cell growth. Retinoic acid itself binds to RXR and RAR receptors which heterodimerise with a variety of nuclear receptors to mediate transcriptional regulation of numerous P450 genes. All-*trans* retinoic acid (ATRA) treatment of human leukocytes and HL60 cells induces CYP4F3A, but this response appears to be independent of the retinoid binding to RXR/RAR.[141] Interestingly, ATRA represses expression of CYP4A11 in HepaRG cells.[142]

Given the close association between CYP4F activity and LTB4 hydroxylation, it is not surprising that a substantial amount of attention has been directed towards the regulation of *CYP4F* genes in inflammation. Animal studies in this area involving LPS, cytokine and brain trauma injury models of inflammation have been reviewed recently by Strobel.[5] As with similar studies conducted on the regulation of *CYP4A* genes, the situation is complex because of the tissue- and protein-selective manner in which rat and mouse *CYP4F* genes are regulated in experimental animal models of inflammation. Some notable findings, however, have emerged from the traumatic brain injury model, wherein expression of rat CYP4F proteins increases at time-points between 24 hours and 2 weeks post-injury.[143] This temporal upregulation of CYP4F activity occurs during the resolution phase of the inflammatory process, when maximal modulation of physiological effects of leukotriene B4 and prostaglandins might be expected.

12.4.5 Pharmacogenetics

Despite ever increasing numbers of polymorphisms detected in *CYP4* genes, our knowledge of the resulting functional consequences is quite rudimentary. CYP4F12, which is capable of metabolism of a variety of xenobiotics (*vide supra*), is highly polymorphic. At least 12 SNPs have been identified in the proximal promoter region of the gene, but transfection experiments with mutated luciferase constructs did not demonstrate reproducible effects in different mammalian cell lines.[144] A further seven non-synonymous polymorphisms have been identified in the coding-region of *CYP4F12*, but catalytic efficiency towards ebastine hydroxylation was only modestly impacted by these variants *in vitro*.[135]

Three non-synonymous SNPs in *CYP4F2* have been described that result in coding-region changes in the enzyme; G12W, D269A and V433M.[145,146] Functional analysis of AA metabolism by the recombinantly-expressed G12W and V433M variants demonstrated decreased 20-HETE formation for enzymes containing the V433M mutation.[146] A functional candidate screen for coeliac disease also identified genetic variation in CYP4F2 and CYP4F3 as risk factors, but the OR of 1.3–1.8 suggests only a modest effect.[147] Therefore, in sum, there is, as yet, no strong data that link allelic variation in the CYP4F subfamily to altered xenobiotic metabolism, although a few studies hint at potential changes in endobiotic metabolism that may be related to certain inflammatory and hypertensive disease states.

12.5 'Orphan' CYP4 Enzymes

In the last few years, analysis of ever more populated genomic and transcriptomic databases has afforded new insights into the full complement of the P450 superfamily. In addition to further expansion within the CYP2 family, bioinformatic approaches have helped identify the *CYP4V2*, *CYP4X1* and *CYP4Z1* genes as additional members of the mammalian CYP4 family. Borrowing terminology from the nuclear receptor superfamily, these newly identified enzymes are currently described as 'orphan P450's because their substrate specificity/physiological role is unknown.[148]

12.5.1 CYP4X

CYP4X1 was initially identified in mouse brain[149] where the mRNA is highly and selectively expressed.[73,150] In humans, however, a much broader mRNA expression pattern is evident with the highest levels found in the aorta, trachea and skeletal muscle.[151] Like many CYP4A enzymes, human CYP4X1 is regulated by PPARα agonists and speculated to have a role in arachidonic acid metabolism.[151]

12.5.2 CYP4Z

CYP4Z1 is rather unusual because the gene, which is present in humans, is absent in the mouse.[9] Human CYP4Z1 mRNA, in contast to human CYP4X1, is selectively expressed in mammary tissue and appears to be upregulated in breast cancer tissue.[152] Also in contrast to CYP4X1, gene expression is regulated by progesterone and glucocorticoids.[151] Recent microarray analysis of ovarian cancer tissue also highlighted CYP4Z1 expression, high levels of which were associated with a poor prognosis.[153]

12.5.3 CYP4V

CYP4V genes have been identified in humans (*CYP4V2*), mice (*Cyp4v3*) and fish (*CYP4V1, CYP4V5-9*). CYP4V2 is the most distantly related of the human CYP4 family possessing only ~35% sequence identity to other family members (Figure 12.1, Table 12.2). Although no information is available yet on the substrate specificity of the enzyme, an intriguing genetic association has emerged between human CYP4V2 and a rare ocular disorder known as Bietti's corneal dystrophy (BCD).[154] This autosomal recessive disease is characterised by the presence of small yellow-white crystals and complex lipid deposits in the retina and peripheral cornea, which lead to atrophy of the retinal epithelium, constriction of the visual field and blindness in later life. This rare disease appears to be more common in East Asia, although cases have also been described recently in European patients presenting with retinal pigmentosa.[155] Potentially disruptive exonic and intronic mutations in the *CYP4V2* gene were identified originally in 23 of 25 BCD patients.[156] Subsequently, multiple investigators have corroborated this gene defect.[157,158] Interestingly, both ocular and peripheral cells obtained from BCD patients exhibit abnormal lipid metabolism.[159,160] Given the well-recognised role of other CYP4 enzymes in fatty acid metabolism, a defect in the catalytic function of the CYP4V2 enzyme in BCD patients is an attractive hypothesis, although the nature of the endogenous substrate(s) remain to be determined.

12.6 Concluding Remarks

Recent studies directed towards the CYP4F subfamily and several of the more recent 'orphan' CYP4 enzymes has expanded knowledge and interest in this group of enzymes that often seem to straddle endobiotic and xenobiotic metabolism. While a large body of data exist from exploration of the physiological roles of the CYP4A and, increasingly, the CYP4F enzymes in animal models, definitive evidence linking the catalytic activity of these enzymes to human disease states compounded by altered vascular tone and inflammation have yet to be established. 'De-orphanisation' of CYP4V2, CYP4X1 and CYP4Z1 will be an important factor in a fuller appreciation of the contributions of these enzymes to the disposition of a variety of lipid mediators that control cellular homeostasis. Future studies should also better define the role of the CYP4A, CYP4B and CYP4F enzymes in drug metabolism and toxicity. The development of specific chemical inhibitors and antibodies would facilitate this goal.

Acknowledgements

AER would like to acknowledge graduate students and postdoctoral associates (Michael B. Fisher, Kirk R. Henne, Brian R. Baer, Xianming Guan, Dieter H. Lang, Matthew J. Cheesman and Yi-Min Zheng) who have contributed substantially to some of the research from the author's laboratory that is cited in this review. Also acknowledged is grant support (RO1 GM49054) from the National Institute of General Medical Sciences.

References

1. A.E. Simpson, *Gen. Pharmacol.*, 1997, **28**, 351.
2. R.T. Okita and J.R. Okita, *Curr. Drug Metab.*, 2001, **2**, 265.
3. D.L. Kroetz and F. Xu, *Annu. Rev. Pharmacol. Toxicol.*, 2005, **45**, 413.
4. B.R. Baer and A.E. Rettie, *Drug Metab. Rev.*, 2006, **38**, 451.
5. A. Kalsotra and H.W. Strobel, *Pharmacol. Ther.*, 2006, **112**, 589.
6. J.P. Hardwick, B.J. Song, E. Huberman and F.J. Gonzalez, *J. Biol. Chem.*, 1987, **262**, 801.
7. S. Kimura, N. Hanioka, E. Matsunaga and F.J. Gonzalez, *Dna*, 1989, **8**, 503.
8. S. Kimura, J.P. Hardwick, C.A. Kozak and F.J. Gonzalez, *Dna*, 1989, **8**, 517.
9. D.R. Nelson, D.C. Zeldin, S.M. Hoffman, L.J. Maltais, H.M. Wain and D.W. Nebert, *Pharmacogenetics*, 2004, **14**, 1.
10. U. Savas, M.H. Hsu and E.F. Johnson, *Arch. Biochem. Biophys.*, 2003, **409**, 212.
11. C.N. Palmer, T.H. Richardson, K.J. Griffin, M.H. Hsu, A.S. Muerhoff, J.E. Clark and E.F. Johnson, *Biochim. Biophys. Acta.*, 1993, **1172**, 161.

12. M. Nishimura, H. Yaguti, H. Yoshitsugu, S. Naito and T. Satoh, *Yakugaku Zasshi*, 2003, **123**, 369.
13. O. Ito, Y. Nakamura, L. Tan, T. Ishizuka, Y. Sasaki, N. Minami, M. Kanazawa, S. Ito, H. Sasano and M. Kohzuki, *Mol. Cell Biochem.*, 2006, **284**, 141.
14. J.M. Lasker, W.B. Chen, I. Wolf, B.P. Bloswick, P.D. Wilson and P. K. Powell, *J. Biol. Chem.*, 2000, **275**, 4118.
15. F. Zhang, J.Q. Qian and D.W. Wang, *Acta. Pharmacol. Sin.*, 2002, **23**, 497.
16. D.L. Kroetz, L.M. Huse, A. Thuresson and M.P. Grillo, *Mol. Pharmacol.*, 1997, **52**, 362.
17. Y. Xu, G.T. Knipp and T.J. Cook, *Placenta*, 2005, **26**, 218.
18. D.N. Muller, C. Schmidt, E. Barbosa-Sicard, M. Wellner, V. Gross, H. Hercule, M. Markovic, H. Honeck, F.C. Luft and W.H. Schunck, *Biochem. J.*, 2007, **403**, 109.
19. L.J. Roman, C.N. Palmer, J.E. Clark, A.S. Muerhoff, K.J. Griffin, E.F. Johnson and B.S. Masters, *Arch. Biochem. Biophys.*, 1993, **307**, 57.
20. W.S. Powell and S. Solomon, *J. Biol. Chem.*, 1978, **253**, 4609.
21. D.E. Williams, S.E. Hale, R.T. Okita and B.S. Masters, *J. Biol. Chem.*, 1984, **259**, 14600.
22. A.S. Muerhoff, D.E. Williams, M.T. Leithauser, V.E. Jackson, M.R. Waterman and B.S. Masters, *Proc. Natl. Acad. Sci. USA*, 1987, **84**, 7911.
23. E.K. Birks, M. Bousamra, K. Presberg, J.A. Marsh, R.M. Effros and E.R. Jacobs, *Am. J. Physiol.*, 1997, **272**, L823.
24. D. Zhu, C. Zhang, M. Medhora and E.R. Jacobs, *J. Appl. Physiol.*, 2002, **93**, 330.
25. R. Le Bouquin, A. Lugnier, N. Frossard and F. Pons, *Fundam. Clin. Pharmacol.*, 2004, **18**, 181.
26. U. Hoch, Z. Zhang, D.L. Kroetz and P.R. Ortiz de Montellano, *Arch. Biochem. Biophys.*, 2000, **373**, 63.
27. M.A. Alterman, C.S. Chaurasia, P. Lu, J.P. Hardwick and R.P. Hanzlik, *Arch. Biochem. Biophys.*, 1995, **320**, 289.
28. P.A. Loughran, L.J. Roman, A.E. Aitken, R.T. Miller and B.S. Masters, *Biochemistry*, 2000, **39**, 15110.
29. R.B. Bambal and R.P. Hanzlik, *Arch. Biochem. Biophys.*, 1996, **334**, 59.
30. X. He, M.J. Cryle, J.J. De Voss and P.R. de Montellano, *J. Biol. Chem.*, 2005, **280**, 22697.
31. U. Hoch and P.R. Ortiz De Montellano, *J. Biol. Chem.*, 2001, **276**, 11339.
32. K.R. Henne, K.L. Kunze, Y.M. Zheng, P. Christmas, R.J. Soberman and A.E. Rettie, *Biochemistry*, 2001, **40**, 12925.
33. L.A. LeBrun, U. Hoch and P.R. Ortiz de Montellano, *J. Biol. Chem.*, 2002, **277**, 12755.
34. Y.M. Zheng, B.R. Baer, M.B. Kneller, K.R. Henne, K.L. Kunze and A.E. Rettie, *Biochemistry*, 2003, **42**, 4601.
35. K. Lundell, R. Hansson and K. Wikvall, *J. Biol. Chem.*, 2001, **276**, 9606.
36. F. Xu, V.Y. Ng, D.L. Kroetz and P.R. de Montellano, *J. Pharmacol. Exp. Ther.*, 2006, **318**, 835.

37. S. Patton and A.A. Benson, *Biochim. Biophys. Acta.*, 1966, **125**, 22.
38. S.J. Mihalik, J.C. Morrell, D. Kim, K.A. Sacksteder, P.A. Watkins and S.J. Gould, *Nat. Genet.*, 1997, **17**, 185.
39. J.C. Komen and R.J. Wanders, *FEBS Lett.*, 2006, **580**, 3794.
40. N. Miyata and R.J. Roman, *J. Smooth Muscle Res.*, 2005, **41**, 175.
41. X. Nguyen, M.H. Wang, K.M. Reddy, J.R. Falck and M.L. Schwartzman, *Am. J. Physiol.*, 1999, **276**, R1691.
42. H. Kawashima, T. Naganuma, E. Kusunose, T. Kono, R. Yasumoto, K. Sugimura and T. Kishimoto, *Arch. Biochem. Biophys.*, 2000, **378**, 333.
43. V. Le Quere, E. Plee-Gautier, P. Potin, S. Madec and J.P. Salaun, *J. Lipid Res.*, 2004, **45**, 1446.
44. J.V. Gainer, A. Bellamine, E.P. Dawson, K.E. Womble, S.W. Grant, Y. Wang, L.A. Cupples, C.Y. Guo, S. Demissie, C.J. O'Donnell, N.J. Brown, M.R. Waterman and J.H. Capdevila, *Circulation*, 2005, **111**, 63.
45. P.K. Powell, I. Wolf, R. Jin and J.M. Lasker, *J. Pharmacol. Exp. Ther.*, 1998, **285**, 1327.
46. P.A. Loughran, L.J. Roman, R.T. Miller and B.S. Masters, *Arch. Biochem. Biophys.*, 2001, **385**, 311.
47. J.H. Capdevila and J.R. Falck, *Biochem. Biophys. Res. Commun.*, 2001, **285**, 571.
48. Z. Yu, L.M. Huse, P. Adler, L. Graham, J. Ma, D.C. Zeldin and D.L. Kroetz, *Mol. Pharmacol.*, 2000, **57**, 1011.
49. L.A. Cowart, S. Wei, M.H. Hsu, E.F. Johnson, M.U. Krishna, J.R. Falck and J.H. Capdevila, *J. Biol. Chem.*, 2002, **277**, 35105.
50. F.J. Gonzalez, *Mol Cell Endocrinol*, 2002, **193**, 71.
51. M.A. Peraza, A.D. Burdick, H.E. Marin, F.J. Gonzalez and J.M. Peters, *Toxicol. Sci.*, 2006, **90**, 269.
52. S. Takeuchi, T. Matsuda, S. Kobayashi, T. Takahashi and H. Kojima, *Toxicol. Appl. Pharmacol.*, 2006, **217**, 235.
53. J.L. Raucy, J. Lasker, K. Ozaki and V. Zoleta, *Toxicol. Sci.*, 2004, **79**, 233.
54. K. Morimura, C. Cheung, J.M. Ward, J.K. Reddy and F.J. Gonzalez, *Carcinogenesis*, 2006, **27**, 1074.
55. J.D. Tugwood, P.R. Holden, N.H. James, R.A. Prince and R.A. Roberts, *Arch. Toxicol.*, 1998, **72**, 169.
56. M. Nagasawa, T. Ide, M. Suzuki, M. Tsunoda, Y. Akasaka, T. Okazaki, T. Mochizuki and K. Murakami, *Biochem. Pharmacol.*, 2004, **67**, 2057.
57. R.A. Graham, B. Goodwin, R.V. Merrihew, W.L. Krol and E.L. Lecluyse, *Toxicol. Sci.*, 2006, **92**, 356.
58. T. Hays, I. Rusyn, A.M. Burns, M.J. Kennett, J.M. Ward, F.J. Gonzalez and J.M. Peters, *Carcinogenesis*, 2005, **26**, 219.
59. D.L. Kroetz, P. Yook, P. Costet, P. Bianchi and T. Pineau, *J. Biol. Chem.*, 1998, **273**, 31581.
60. W. Wahli, O. Braissant and B. Desvergne, *Chem. Biol.*, 1995, **2**, 261.
61. K. Yoshinari, S. Takagi, J. Sugatani and M. Miwa, *Biol. Pharm. Bull*, 2006, **29**, 1634.

62. M.B. Sewer, D.R. Koop and E.T. Morgan, *J. Pharmacol. Exp. Ther.*, 1997, **280**, 1445.
63. S.R. Mitchell, M.B. Sewer, S.S. Kardar and E.T. Morgan, *Drug Metab. Dispos.*, 2001, **29**, 17.
64. T.A. Richardson and E.T. Morgan, *J. Pharmacol. Exp. Ther.*, 2005, **314**, 703.
65. T.A. Richardson, M. Sherman, L. Antonovic, S.S. Kardar, H.W. Strobel, D. Kalman and E.T. Morgan, *Drug Metab. Dispos.*, 2006, **34**, 354.
66. J.C. McGiff and J. Quilley, *Am. J. Physiol.*, 1999, **277**, R607.
67. B. Mayer, W. Lieb, A. Gotz, I.R. Konig, Z. Aherrahrou, A. Thiemig, S. Holmer, C. Hengstenberg, A. Doering, H. Loewel, H.W. Hense, H. Schunkert and J. Erdmann, *Hypertension*, 2005, **46**, 766.
68. M. Hiratsuka, H. Nozawa, Y. Katsumoto, T. Moteki, T. Sasaki, Y. Konno and M. Mizugaki, *Mutat. Res.*, 2006, **599**, 98.
69. E. Arinc and M. Philpot, *J. Biol. Chem.*, 1976, **251**, 3213.
70. F.P. Guengerich, *Mol. Pharmacol.*, 1977, **13**, 911.
71. L.G. Robertson, R.M. Philpot, E. Zeiger and C.R. Wolf, *Mol. Pharmacol.*, 1981, **20**, 662.
72. D.C. Zeldin, J.D. Plitman, J. Kobayashi, R.F. Miller, J.R. Snapper, J.R. Falck, J.L. Szarek, R.M. Philpot and J.H. Capdevila, *J. Clin. Invest.*, 1995, **95**, 2150.
73. D. Choudhary, I. Jansson, I. Stoilov, M. Sarfarazi and J.B. Schenkman, *Arch. Biochem. Biophys.*, 2005, **436**, 50.
74. S. Imaoka, Y. Yoneda, T. Sugimoto, T. Hiroi, K. Yamamoto, T. Nakatani and Y. Funae, *Biochem. Biophys. Res. Commun.*, 2000, **277**, 776.
75. F.P. Guengerich, *Arch. Biochem. Biophys.*, 2003, **409**, 59.
76. R.A. McKinnon and M.E. McManus, *Pathology*, 1996, **28**, 148.
77. M.R. Boyd, *Crit Rev Toxicol*, 1980, **7**, 103.
78. S.R. Slaughter, C.N. Statham, R.M. Philpot and M.R. Boyd, *J. Pharmacol. Exp. Ther.*, 1983, **224**, 252.
79. M. Czerwinski, T.L. McLemore, R.M. Philpot, P.T. Nhamburo, K. Korzekwa, H.V. Gelboin and F.J. Gonzalez, *Cancer Res.*, 1991, **51**, 4636.
80. R.D. Verschoyle, R.M. Philpot, C.R. Wolf and D. Dinsdale, *Toxicol. Appl. Pharmacol.*, 1993, **123**, 193.
81. T.M. Alvarez-Diez and J. Zheng, *Drug Metab. Dispos.*, 2004, **32**, 1345.
82. B.R. Baer, A.E. Rettie and K.R. Henne, *Chem. Res. Toxicol.*, 2005, **18**, 855.
83. L.J. Chen, S.S. Hecht and L.A. Peterson, *Chem. Res. Toxicol.*, 1997, **10**, 866.
84. G.S. Yost, *Chem. Res. Toxicol.*, 1989, **2**, 273.
85. G.S. Yost, *Adv. Exp. Med. Biol.*, 1996, **387**, 221.
86. J. Thornton-Manning, M.L. Appleton, F.J. Gonzalez and G.S. Yost, *J. Pharmacol. Exp. Ther.*, 1996, **276**, 21.
87. B.A. Carr, S. Ramakanth, G.A. Dannan and G.S. Yost, *Mol. Pharmacol.*, 2003, **63**, 1137.

88. D. Kim and F.P. Guengerich, *Annu. Rev. Pharmacol. Toxicol.*, 2005, **45**, 27.
89. S. Imaoka, T. Hiroi, Y. Tamura, H. Yamazaki, T. Shimada, M. Komori, M. Degawa and Y. Funae, *Arch. Biochem. Biophys.*, 1995, **321**, 255.
90. S. Imaoka, Y. Yoneda, T. Matsuda, M. Degawa, S. Fukushima and Y. Funae, *Biochem. Pharmacol.*, 1997, **54**, 677.
91. A.E. Rettie, P.R. Sheffels, K.R. Korzekwa, F.J. Gonzalez, R.M. Philpot and T.A. Baillie, *Biochemistry*, 1995, **34**, 7889.
92. J. Schulze, E. Richter and R.M. Philpot, *Drug Metab. Dispos.*, 1990, **18**, 398.
93. T.A. Baillie, *Chem. Res. Toxicol.*, 1988, **1**, 195.
94. W. Tang, A.G. Borel, T. Fujimiya and F.S. Abbott, *Chem. Res. Toxicol.*, 1995, **8**, 671.
95. G. Bellec, Y. Dreano, J.P. Bail, J.F. Menez and F. Berthou, *Mutat. Res.*, 1997, **377**, 199.
96. A.S. Muerhoff, D.E. Williams, N.O. Reich, C.A. CaJacob, P.R. Ortiz de Montellano and B.S. Masters, *J. Biol. Chem.*, 1989, **264**, 749.
97. M.B. Fisher, Y.M. Zheng and A.E. Rettie, *Biochem. Biophys. Res. Commun.*, 1998, **248**, 352.
98. V. Mastyugin, E. Aversa, A. Bonazzi, C. Vafaes, P. Mieyal and M.L. Schwartzman, *J. Pharmacol. Exp. Ther.*, 1999, **289**, 1611.
99. A. Mezentsev, V. Mastyugin, F. Seta, S. Ashkar, R. Kemp, D.S. Reddy, J.R. Falck, M.W. Dunn and M. Laniado-Schwartzman, *J. Pharmacol. Exp. Ther.*, 2005, **315**, 42.
100. K.R. Henne, M.B. Fisher, K.R. Iyer, D.H. Lang, W.F. Trager and A. E. Rettie, *Biochemistry*, 2001, **40**, 8597.
101. J.W. Foy, D.M. Silverman and R.A. Schatz, *J. Toxicol. Environ. Health*, 1996, **47**, 135.
102. B.R. Smith, J.L. Plummer, C.R. Wolf, R.M. Philpot and J.R. Bend, *J. Pharmacol. Exp. Ther.*, 1982, **223**, 736.
103. I.G. Robertson, C. Serabjit-Singh, J.E. Croft and R.M. Philpot, *Mol. Pharmacol.*, 1983, **24**, 156.
104. R.R. Vanderslice, B.A. Domin, G.T. Carver and R.M. Philpot, *Mol. Pharmacol.*, 1987, **31**, 320.
105. C.J. Serabjit-Singh, P.W. Albro, I.G. Robertson and R.M. Philpot, *J. Biol. Chem.*, 1983, **258**, 12827.
106. S. Ashkar, A. Mesentsev, W.X. Zhang, V. Mastyugin, M.W. Dunn and M. Laniado-Schwartzman, *J. Ocul. Pharmacol. Ther.*, 2004, **20**, 65.
107. V. Mastyugin, A. Mezentsev, W.X. Zhang, S. Ashkar, M.W. Dunn and M. Laniado-Schwartzman, *J. Cell Biochem.*, 2004, **91**, 1218.
108. S. Imaoka, Y. Yoneda, T. Sugimoto, S. Ikemoto, T. Hiroi, K. Yamamoto, T. Nakatani and Y. Funae, *Cancer Lett.*, 2001, **166**, 119.
109. J. Isern and A. Meseguer, *Biochem. Biophys. Res. Commun.*, 2003, **307**, 139.
110. M.T. Poch, N.S. Cutler, G.S. Yost and R.N. Hines, *Drug Metab. Dispos.*, 2005, **33**, 1174.

111. Y.M. Zheng, K.R. Henne, P. Charmley, R.B. Kim, D.G. McCarver, E.T. Cabacungan, R.N. Hines and A.E. Rettie, *Toxicol. Appl. Pharmacol.*, 2003, **186**, 119.
112. J.M. Lo-Guidice, D. Allorge, C. Cauffiez, D. Chevalier, J.J. Lafitte, M. Lhermitte and F. Broly, *Pharmacogenetics*, 2002, **12**, 367.
113. M. Hiratsuka, H. Nozawa, Y. Konno, T. Saito, S. Konno and M. Mizugaki, *Drug Metab. Pharmacokinet*, 2004, **19**, 114.
114. P. Christmas, S.R. Ursino, J.W. Fox and R.J. Soberman, *J. Biol. Chem.*, 1999, **274**, 21191.
115. P. Christmas, J.P. Jones, C.J. Patten, D.A. Rock, Y. Zheng, S.M. Cheng, B.M. Weber, N. Carlesso, D.T. Scadden, A.E. Rettie and R. J. Soberman, *J. Biol. Chem.*, 2001, **276**, 38166.
116. X. Cui, H. Kawashima, T.B. Barclay, J.M. Peters, F.J. Gonzalez, E.T. Morgan and H.W. Strobel, *J. Pharmacol. Exp. Ther.*, 2001, **296**, 542.
117. J. Bylund, M. Bylund and E.H. Oliw, *Biochem. Biophys. Res. Commun.*, 2001, **280**, 892.
118. K. Stark, H. Torma, M. Cristea and E.H. Oliw, *Arch. Biochem. Biophys.*, 2003, **409**, 188.
119. K. Stark, L. Schauer, G.E. Sahlen, G. Ronquist and E.H. Oliw, *Basic Clin. Pharmacol. Toxicol.*, 2004, **94**, 177.
120. Y. Kikuta, E. Kusunose, K. Endo, S. Yamamoto, K. Sogawa, Y. Fujii-Kuriyama and M. Kusunose, *J. Biol. Chem.*, 1993, **268**, 9376.
121. F. Xu, J.R. Falck, P.R. Ortiz de Montellano and D.L. Kroetz, *J. Pharmacol. Exp. Ther.*, 2004, **308**, 887.
122. A. Kalsotra, C.M. Turman, Y. Kikuta and H.W. Strobel, *Toxicol. Appl. Pharmacol.*, 2004, **199**, 295.
123. K. Stark, B. Wongsud, R. Burman and E.H. Oliw, *Arch. Biochem. Biophys.*, 2005, **441**, 174.
124. T.L. Psota, S.K. Gebauer and P. Kris-Etherton, *Am. J. Cardiol.*, 2006, **98**, 3i.
125. S.D. Harmon, X. Fang, T.L. Kaduce, S. Hu, V. Raj Gopal, J.R. Falck and A.A. Spector, *Prostaglandins Leukot. Essent Fatty Acids*, 2006, **75**, 169.
126. J. Bylund, M. Hidestrand, M. Ingelman-Sundberg and E.H. Oliw, *J. Biol. Chem.*, 2000, **275**, 21844.
127. R. Jin, D.R. Koop, J.L. Raucy and J.M. Lasker, *Arch. Biochem. Biophys.*, 1998, **359**, 89.
128. R.J. Sanders, R. Ofman, M. Duran, S. Kemp and R.J. Wanders, *J. Biol. Chem.*, 2006, **281**, 13180.
129. R.S. Parker, T.J. Sontag and J.E. Swanson, *Biochem. Biophys. Res. Commun.*, 2000, **277**, 531.
130. T.J. Sontag and R.S. Parker, *J. Biol. Chem.*, 2002, **277**, 25290.
131. T. Hashizume, M. Mise, Y. Terauchi, L. O. T. Fujii, H. Miyazaki and T. Inaba, *Drug Metab. Dispos.*, 1998, **26**, 566.
132. T. Hashizume, M. Mise, S. Matsumoto, Y. Terauchi, T. Fujii, S. Imaoka, Y. Funae, T. Kamataki and H. Miyazaki, *Drug Metab. Dispos.*, 2001, **29**, 798.

133. T. Hashizume, S. Imaoka, T. Hiroi, Y. Terauchi, T. Fujii, H. Miyazaki, T. Kamataki and Y. Funae, *Biochem. Biophys. Res. Commun.*, 2001, **280**, 1135.
134. T. Hashizume, S. Imaoka, M. Mise, Y. Terauchi, T. Fujii, H. Miyazaki, T. Kamataki and Y. Funae, *J. Pharmacol. Exp. Ther.*, 2002, **300**, 298.
135. C. Cauffiez, F. Klinzig, E. Rat, G. Tournel, D. Allorge, D. Chevalier, N. Pottier, T. Lovecchio, J.F. Colombel, M. Lhermitte, J.M. Lo-Guidice and F. Broly, *Biochem. Pharmacol.*, 2004, **68**, 2417.
136. M.Z. Wang, J.Y. Saulter, E. Usuki, Y.L. Cheung, M. Hall, A.S. Bridges, G. Loewen, O.T. Parkinson, C.E. Stephens, J.L. Allen, D.C. Zeldin, D.W. Boykin, R.R. Tidwell, A. Parkinson, M.F. Paine and J.E. Hall, *Drug Metab. Dispos.*, 2006, **34**, 1985.
137. P. Christmas, N. Carlesso, H. Shang, S.M. Cheng, B.M. Weber, F.I. Preffer, D.T. Scadden and R.J. Soberman, *J. Biol. Chem.*, 2003, **278**, 25133.
138. H. Kawashima, E. Kusunose, C.M. Thompson and H.W. Strobel, *Arch. Biochem. Biophys.*, 1997, **347**, 148.
139. T.A. Kocarek and A.B. Reddy, *Drug Metab. Dispos.*, 1996, **24**, 1197.
140. M.H. Hsu, U. Savas, K.J. Griffin and E.F. Johnson, *J. Biol. Chem.*, 2007, **282**, 5225.
141. Y. Kikuta, Y. Yamashita, S. Kashiwagi, K. Tani, K. Okada and K. Nakata, *Biochim. Biophys. Acta.*, 2004, **1683**, 7.
142. J. Antoun, Y. Amet, B. Simon, Y. Dreano, A. Corlu, L. Corcos, J.P. Salaun and E. Plee-Gautier, *FEBS Lett.*, 2006, **580**, 3361.
143. X. Cui, A. Kalsotra, A.M. Robida, D. Matzilevich, A.N. Moore, C.L. Boehme, E.T. Morgan, P.K. Dash and H.W. Strobel, *Biochim. Biophys. Acta.*, 2003, **1619**, 325.
144. C. Cauffiez, F. Klinzig, E. Rat, G. Tournel, D. Allorge, D. Chevalier, T. Lovecchio, N. Pottier, J.F. Colombel, M. Lhermitte, J.C. D'Halluin, F. Broly and J.M. Lo-Guidice, *Biochem. Pharmacol.*, 2004, **67**, 2231.
145. F. Doring, G. Rimbach and J.K. Lodge, *IUBMB Life*, 2004, **56**, 615.
146. D. E. Stec, R. J. Roman, A. Flasch and M. J. Rieder Physiol Genomics 2007 (in press).
147. C.R. Curley, A.J. Monsuur, M.C. Wapenaar, J.D. Rioux and C. Wijmenga, *Eur. J. Hum. Genet.*, 2006, **14**, 1215.
148. F.P. Guengerich, *Drug Metab. Rev.*, 2004, **36**, 159.
149. J. Bylund, C. Zhang and D.R. Harder, *Biochem. Biophys. Res. Commun.*, 2002, **296**, 677.
150. M. Al-Anizy, N.J. Horley, C.W. Kuo, L.C. Gillett, C.A. Laughton, D. Kendall, D.A. Barrett, T. Parker and D.R. Bell, *Febs. J.*, 2006, **273**, 936.
151. U. Savas, M.H. Hsu, K.J. Griffin, D.R. Bell and E.F. Johnson, *Arch. Biochem. Biophys.*, 2005, **436**, 377.
152. M.A. Rieger, R. Ebner, D.R. Bell, A. Kiessling, J. Rohayem, M. Schmitz, A. Temme, E.P. Rieber and B. Weigle, *Cancer Res.*, 2004, **64**, 2357.
153. D. Downie, M.C. McFadyen, P.H. Rooney, M.E. Cruickshank, D.E. Parkin, I.D. Miller, C. Telfer, W.T. Melvin and G.I. Murray, *Clin. Cancer Res.*, 2005, **11**, 7369.

154. G. Bietti, *Klin. Mbl. Augenheilk.*, 1937, **99**, 737.
155. A. Mataftsi, L. Zografos, E. Milla, M. Secretan and F.L. Munier, *Retina*, 2004, **24**, 416.
156. A. Li, X. Jiao, F.L. Munier, D.F. Schorderet, W. Yao, F. Iwata, M. Hayakawa, A. Kanai, M. Shy Chen, R. Alan Lewis, J. Heckenlively, R.G. Weleber, E.I. Traboulsi, Q. Zhang, X. Xiao, M. Kaiser-Kupfer, Y.V. Sergeev and J.F. Hejtmancik, *Am. J. Hum. Genet.*, 2004, **74**, 817.
157. J. Lin, K.M. Nishiguchi, M. Nakamura, T.P. Dryja, E.L. Berson and Y. Miyake, *J. Med. Genet.*, 2005, **42**, e38.
158. M. Shan, B. Dong, X. Zhao, J. Wang, G. Li, Y. Yang and Y. Li, *Mol. Vis.*, 2005, **11**, 738.
159. M.I. Kaiser-Kupfer, C.C. Chan, T.C. Markello, M.A. Crawford, R.C. Caruso, K.G. Csaky, J. Guo and W.A. Gahl, *Am. J. Ophthalmol.*, 1994, **118**, 569.
160. J. Lee, X. Jiao, J.F. Hejtmancik, M. Kaiser-Kupfer, W.A. Gahl, T.C. Markello, J. Guo and G.J. Chader, *Invest. Ophthalmol. Vis. Sci.*, 2001, **42**, 1707.

Part C

CHAPTER 13

Receptor-Mediated Regulation of Cytochromes P450

KOUICHI YOSHINARI,[a] ERIC TIEN,[c] MASAHIKO NEGISHI[c] AND PAAVO HONKAKOSKI[b]

[a] Division of Drug Metabolism and Molecular Toxicology, Graduate School of Pharmaceutical Sciences, Tohoku University, 6-3 Aramaki-Aoba, Aoba-ku, Sendai, Miyagi 980-8578, Japan; [b] Department of Pharmaceutics, University of Kuopio, P. O. Box 1627, FIN-70211, Kuopio, Finland and; [c] Pharmacogenetics Section, Laboratory of Reproductive and Developmental Toxicology, National Institute of Environmental Health Sciences, National Institutes of Health, Research Triangle Park, North Carolina 27709, USA

Table of Contents

Issues in Toxicology
Cytochromes P450: Role in the Metabolism and Toxicity of Drugs and other Xenobiotics
Edited by Costas Ioannides

13.1 Introduction

In the early 1960s, barbiturates were found to induce enzymes in the liver endoplasmic reticulum and to increase their own metabolism.[1] With this finding, the induction by drugs of drug metabolism was suggested to be the mechanism associated with drug tolerance. Subsequently, cytochrome P450 (CYP), the major enzyme that metabolises drugs, was characterised and the phenomenon of the drug induction of microsomal enzymes was conceptualised as the key regulatory system affecting pharmacological as well as toxicological consequences.[2–4] Beginning with phenobarbital (PB) and 3-methylcholanthrene (3-MC), two distinct groups of chemicals capable of inducing different forms of CYPs emerged: PB-type and 3-MC-type inducers. Many of the 3-MC types are characterised by having structures similar to aryl hydrocarbons and induce the form of CYP now called CYP1A1. Using the high-affinity and potent inducer 2,3,7,8-tetrachlorodibenzo-*p*-dioxin (TCDD), the factor now called aryl hydrocarbon receptor AhR was quickly implicated in mediating the induction by 3-MC types in the 1970s.[5] In contrast, the PB-type inducers exhibit extremely diverse structures, making it difficult to discover the factor that mediates the induction by this type. In the meantime, at least, two additional groups of inducers had been categorised: pregnane 16α-carbonitrile (PCN)-types and peroxisome proliferators that induce CYP3A and CYP4A, respectively. During the last decade, much progress has been made in identifying the constitutive active/androstane receptor (CAR) for PB-type inducers,[6] the pregnane X receptor (PXR) for PCN-type inducers[7] and the peroxisome proliferators-activating receptor α (PPARα) for peroxisome proliferators.[8] As the function of these receptors has become clear, the number of CYPs under regulation by these receptors has dramatically increased and extended from CYPs to include certain transferases (*e.g.* glutathione *S*-transferases, UDP-glucuronosyltransferases and sulphotransferases) and membrane transporters. Moreover, the drug activation of these receptors is now widely implicated in drug efficacy, drug toxicity and carcinogenicity, drug-drug interactions and drug development. This chapter primarily focuses on CAR and PXR and would update our previous reviews on the function and activation mechanism of these nuclear receptors.[9] Aspects of AhR are not covered in this review; readers should refer to recent articles.[10]

13.2 Activation Mechanisms

The nuclear receptors CAR and PXR are orphan members of the nuclear steroid/thyroid hormone receptor family (NR), and belong to the NR1I subfamily. Similar to the nuclear hormone receptors, both CAR and PXR are recognised as basically being transcription factors activated by direct binding of their agonistic drug as a ligand. Generally, this mechanism of activation is known as ligand-dependent activation. In addition, these receptors, CAR in particular, can be activated without direct binding to drugs, through a mechanism called the indirect or ligand-independent activation. Recent advances in deciphering the molecular mechanisms for both ligand- and signal-dependent activations will be described in this section.

13.3 Ligand-Dependent Mechanism

The mechanisms of ligand recognition and subsequent activation of PXR and CAR have drawn much attention for two main reasons. First, the vast number and low affinity of their ligands and significant species differences of PXR and CAR ligand-binding domains (LBDs) are unique among nuclear receptors.[11–13] This may provide some insight into how nature has accomplished binding of very diverse ligands to a single receptor. Second, because PXR and CAR regulate drug metabolism and disposition, the understanding, determination, and prediction of their ligand binding and/or activation would provide an early sign of potential drug/drug interactions due to induction of CYP enzymes by a drug candidate.

13.3.1 Ligand Binding and Activation by PXR

By December 2006, the Protein Data Bank (www.rcsb.org/pdb) contained the ligand-free human PXR LBD and four complexes with bound activators

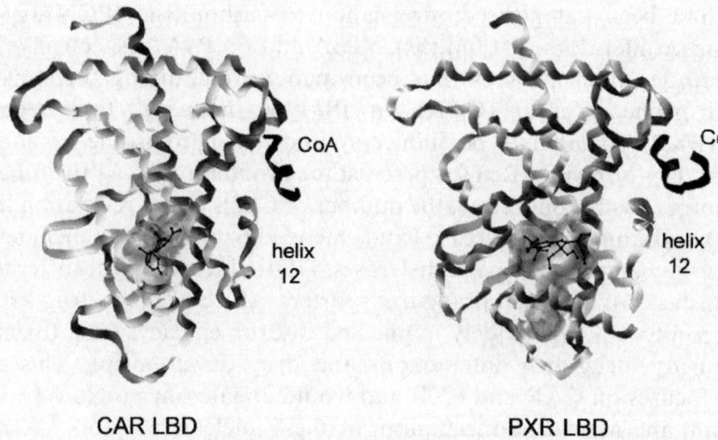

Figure 13.1 Schematic representation of agonist-bound CAR and PXR LBDs. The CAR[27] and PXR[15] LBDs are shown as grey ribbons with helix 12 indicated. The co-activator peptides (CoA) in these complexes are shown as black ribbons. The ligand-binding pockets are shown as Connolly surfaces (light grey) with agonist 5β-pregnane-3,20-dione[27] and SR12813[15] shown by black ball- and-stick rendering.

SR12813, hyperforin and rifampicin.[14–17] The overall fold of the human PXR LBD has certain differences from the three-layered α-helical "sandwich structure" of other NR LBDs: human PXR had a 45-amino acid insertion between helices 1 and 3, a five-stranded anti-parallel β-sheet and a voluminous (1,280–1,544 $Å^3$) ligand pocket (LBP) that seems to expand to accommodate diverse ligands (Figure 13.1). The flexibility of three regions near residues 178–209, 229–235 and 310–317, and movement of residues Leu-209 and His-407 upon ligand binding contribute to this accommodation. It is speculated that the first flexible region acts as a ligand entry channel.[18] The PXR LBP is formed by 20 hydrophobic, 4 polar and 4 potentially charged amino acids.[14] Distinct but overlapping sets of LBP residues interact with different PXR ligands: 13–18 contacts were made by SR12813, hyperforin and rifampicin. All three agonists share hydrogen bonding with residues Ser-247, Gln-285 and His-407 and hydrophobic interactions with residues Met-243, Trp-299 and Phe-420.[18]

Site-directed mutagenesis of PXR has been performed to elucidate species differences in PXR ligand specificity and to validate the observed LBP–ligand contacts in X-ray structures. Rifampicin and SR12813 activate the human PXR while PCN is an activator of rat and mouse PXRs. A quadruple-mutant mouse PXR gained activation by SR12813 and lost its responsiveness to PCN upon replacement of mouse residues by human PXR amino acids that were in contact with SR12813 (Leu-206, Ser-208, His-407, Arg-410).[14] Leu-308 is the key residue required for activation by ligands including rifampicin, paclitaxel and hyperforin although other regions are needed for full activity.[19,20] Leu-308 is located near one of the flexible regions and seems to contact rifampicin in the X-ray structure.[17] Gln-285 and His-407 form hydrogen bonds with human

PXR-selective rifampicin and SR12813 but their mutation to corresponding rodent residues (Gln-285→Ile and His-407→Gln) had little if any effect on the affinity of these agonists.[15,17,21] Mutation of ligand-contacting residues appears to increase the basal activity of human PXR (Ser-247→Trp, Trp-299→Ala, His-407→Ala, Arg-410→Ala)[11,14,16] and to produce only modest effects on the ligand selectivity (Phe-288→Ala, Trp-299→Met).[15] It is possible, but not yet systematically studied, that mutation of one residue forces the other LBP residues to rearrange, leading to changes in LBP contour and hydrophobicity, position of the helix 12 and subsequent co-activator recruitment.

The ligand-induced reorientation of helix 12 is a central process that NRs employ to dissociate themselves from NR co-repressors and to recruit NR co-activators in order to trigger transcription of their target genes.[22] The finding that ligand-free PXR associates with NR co-repressors which are exchanged to co-activators upon addition of an agonist[23] is consistent with this general mechanism. However, there is no clear understanding of this ligand-dependent helix 12 reorientation, because in the ligand-free PXR structure, helix 12 is already in the active position,[14] in ligand-bound complexes, only the SR12183 has a direct contact with helix 12[15] and, finally, biochemical experiments or molecular dynamics simulations addressing this problem have not yet been performed.

Activation of PXR-driven reporter genes in cell culture seems to correlate well with CYP3A4 mRNA induction.[24] Methods that determine ligand-dependent co-activator recruitment have also been useful in identifying PXR ligands.[25,26] There have been efforts to construct *in silico* models that are based on pharmacophore generation, molecular docking and quantitative structure-activity relationship (QSAR) modelling to predict PXR ligands.[18] These models have produced rather simple "rules" for PXR ligands such as one hydrogen bond acceptor and three–four hydrophobic regions. However, the usefulness of these models to accurately predict activation potential or to distinguish between PXR activators and non-activators is not clear. The observed flexibility in PXR crystal structures is another obstacle in generating reliable receptor-based prediction for PXR activators.

13.3.2 Ligand Binding and Activation by CAR

Two human CAR LBD structures with bound agonists 6-(4-chlorophenyl)-imidazo[2,1-*b*]thiazole-5-carbaldehyde *O*-(3,4-dichlorobenzyl)oxime (CITCO) and 5β-pregnanedione[27] and two mouse CAR LBD structures in complex with the inverse agonist androstenol[28] and the agonist 1,4-bis[2-(3,5-dichloropyridyloxy)]benzene (TCPOBOP)[29] were recently reported (Figure 13.1). Both CAR structures displayed two 3_{10} helices between helices 1 and 3, and a single-turn helix X (human CAR residues Leu-336 to Met-339). The helix 12 was suggested to be stabilised into active position by interactions with Lys-195 and a "barrier" of LBP residues (Phe-161, Asn-165, Phe-234, Tyr-326), and constraints imposed by helix X, thus providing explanations to the constitutive activity of CAR.[27,29] Because helix X is present in other nuclear receptors as well,[30] and the structure of the ligand-free CAR is not known, its role in ligand-dependent activation is not clear.

The CAR LBP cavity (525–675 Å^3) consists of 19 hydrophobic, 5 polar and 3 potentially charged amino acids, the last contributing to two regions for potential hydrogen bonding to ligands.[27–29] The binding of ligands is dominated by hydrophobic interactions, and direct hydrogen bonding to CAR agonists was observed for 5β-pregnanedione (His-203) and TCPOBOP (Asn-175 via a water molecule) but not for CITCO. Distinct, overlapping sets of LBP residues were in contact with human CAR agonists, as also seen with PXR,[18] and direct contacts with helix 12 were observed in the TCPOBOP–mouse CAR complex only.[27,29]

Site-directed mutagenesis and molecular dynamics simulations have provided additional insight into the basis of CAR constitutive activity, ligand selectivity and species specificity.[27–31] First, stabilising interactions of helix 12 with residues from helices 11, 5 and 4 contribute to basal activity of CAR, and mutations of key LBP residues to alanines influence the position of "barrier" residues Tyr-326 and Asn-165 and destabilise the helix 12,[30,31] a view consistent with the X-ray structures.[27–29] Second, the residue Phe-161 (Phe-171 in mouse CAR) is important for activation by agonists and residue His-203 (His-213 in mouse CAR) is essential for inhibition by androstenol[28,31] while other residues are ligand-selective including Asn-165 for tri(p-methylphenyl)phosphate and Cys-202 for ethynyloestradiol in human CAR.[31] Finally, some residues are responsible for species specificity of CAR: Substitution by the human Met-340 for Thr-350 in mouse CAR abolished TCPOBOP activation and progesterone inhibition while no change occurred with respect to ligands ethynyloestradiol and androstenol.[32,33] The presence of Met-350 would cause a steric clash with the TCPOBOP ligand.[29] A single mutation of Phe-243 → Leu in human CAR converted the inhibitory response by ethynyloestradiol to activation.[31]

Because of its high constitutive activity, cell-based CAR activation assays require modifications so as to detect weak-to-moderate inducers.[34] CAR agonists, mixed agonists and inverse agonists can be distinguished by yeast or mammalian two-hybrid systems.[31,34] Ligand-induced nuclear translocation of CAR could be an alternative way of detecting CAR-dependent inducers in primary hepatocytes.[35] Few *in silico* methods for prediction and analysis of CAR ligands have been published,[18] and QSAR studies of human CAR are in progress in our laboratory.

An expanding list of drugs, environmental pollutants and natural compounds are found to activate PXR and CAR receptors in various *in vitro* assays.[13,36] The actual effect of a chemical on drug metabolism is still very difficult to predict because: *i*) PXR and CAR may be differentially modulated by the same chemicals,[26] *ii*) the relative amounts of PXR, CAR and their co-regulators may vary,[37] *iii*) the propensity of some inducers to recruit both co-activators and co-repressors,[34] *iv*) selectivity of PXR and CAR for their common target genes may differ,[38] and *v*) the activation potency found *in vitro* is profoundly influenced by the type of assay.[24,25]

13.4 Signal-Dependent Activation

As described above, classic ligand-dependent activation of nuclear receptors such as CAR and PXR involves the physical binding of ligand to the receptor that in turn results in alteration of transcriptional activity. This, however, is not the only mechanism by which the transcriptional activity of nuclear receptors can be affected. While endogenous signals may regulate these nuclear receptors and affect xenobiotic metabolism, xenobiotics themselves can indirectly activate the nuclear receptors by regulating endogenous signals. The following section summarizes recent findings concerning cellular signals and endogenous stimuli regulating CAR and PXR (Figure 13.2).

13.4.1 CAR De-phosphorylation and PP2A

The prototypical CAR activator is the anti-seizure medication PB which triggers cytosolic CAR to translocate into the nucleus of hepatocytes.[39] Interestingly, PB does not bind directly to the receptor, but elicits its effect through a poorly defined mechanism of action. Recently, phosphorylation of CAR at serine 202 was shown to be important for regulating nuclear translocation of the receptor.[40] Mutation of serine 202 to a non-phosphorylatable alanine resulted in a form of mouse CAR that readily moved to the nucleus. Conversely, mutation of serine 202 to a phosphorylation mimicking aspartic acid residue converted mouse CAR into a receptor that was unable to move to the nucleus even in the presence of a CAR activator.

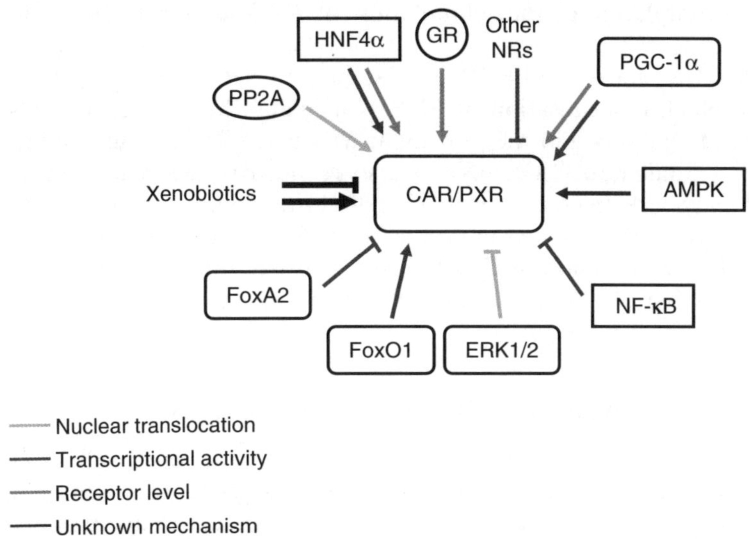

Figure 13.2 Regulation of CAR and PXR activity. In addition to cross-talk between CAR and PXR, various mechanisms and signalling pathways can regulate the activity of both receptors at multiple points along the receptor signaling pathway. Arrow colour indicates the point of regulation.

The movement of CAR into the nucleus is regulated in part by the cytoplasmic CAR retention protein (CCRP). It was shown that the sequestration of CAR in the cytosol, in the absence of activator, involves CCRP via its direct interaction with the receptor.[41] The importance of CCRP for CAR cytosolic retention raises the possibility that CCRP modification through processes such as phosphorylation, acetylation, ubiquitination and degradation play a major role in the promotion of CAR nuclear translocation.

Control of CAR nuclear translocation has been mapped to a leucine rich region near the C-terminus of the receptor. CAR does not contain a classic nuclear localisation signal (NLS), but uses this leucine rich region (LXXLXXL), termed the xenobiotic response sequence (XRS), to mediate movement to the nucleus.[42] Mutation or deletion of the XRS in CAR results in a receptor which is functionally unable to translocate into the nucleus in response to CAR activators, such as PB in mouse liver. The potential for protein-protein interactions with the XRS as well as modifications to the XRS have not been studied extensively to date, but provides another target for regulation by alternative signalling pathways in the cell.

In support of the critical role of phosphorylation/de-phosphoyrlation in the nuclear translocation of CAR, it was previously shown that treatment of mouse primary hepatocytes with the protein phosphatase inhibitor okadaic acid (OA) prevented PB-induced nuclear translocation of CAR.[39] Moreover, protein phosphatase 2A (PP2A) was recruited to the protein complex containing CAR and heat shock protein 90 (hsp90) in mouse liver after treatment with PB.[43] Collectively, these findings strongly point to the de-phosphorylation of CAR as a vital step in nuclear translocation of CAR. Endogenous signals which lead to the phosphorylation/de-phosphorylation of CAR at serine 202 remain unknown at this point.

Nuclear translocation of PXR, at least mouse PXR, is also a tightly regulated process. Nuclear localisation of PXR is mediated by a bipartite NLS in the DNA-binding domain of the protein, as well as a XRS sequence and the AF-2 domain.[44] Similar to CAR, PXR is also retained in the cytosol by CCRP.[45] Interestingly, treatment of HepG2 cells with small interfering RNA for endogenous CCRP resulted in decreased PXR activity in reporter assays suggesting that the function of CCRP in the cytosolic PXR protein complex may differ from that of CAR, and warrants further study.

13.4.2 AMP-Activated Protein Kinase (AMPK)

AMPK is an energy sensor that is regulated by various stimuli to maintain energy balance. Recent studies suggest AMPK being involved in CYP2B regulation by PB though the role of CAR is still unclear. Activation of AMPK by 5-amino-imidazole carboxamide riboside increases CYP2B expression to levels comparable to cells treated with PB in the WGA cells.[46] The WGA cell line was derived from HepG2 cells cultured in media where fructose substituted for glucose in order to promote growth of cells expressing the phenotype of adult hepatocytes such as albumin synthesis and the expression of CAR and liver metabolising enzymes. PB treatment increased cellular AMPK activity in the WGA cells, suggesting the possibility that PB signals through this kinase to affect CAR and drug metabolism. However, the mechanism underlying the effects of AMPK on CAR activity is still undefined. Mice lacking the AMPK catalytic subunits exhibit basal CYP2B expression that is elevated in the liver of untreated AMPK knockout mice, to a level comparable to PB-treated wild type mice, and treatment of AMPK knockout mice with PB does not result in any further increase.[47] In addition, CAR translocation in response to PB is unaffected in primary hepatocytes isolated from these mice suggesting that AMPK may function to activate CAR in the nucleus rather than to regulate nuclear translocation. In contrast, another study showed that AMPK activation does induce CAR nuclear translocation in mouse liver, though with much less effect than PB, while it does not result in the induction of the *Cyp2b* gene.[48] Due to the high constitutive activity of CAR, nuclear translocation of CAR is inevitably considered akin to increased transcriptional activity, but these studies with AMPK indicate otherwise. Our previous study also showed that KN-62, a Ca^{+}/calmodulin kinase inhibitor had no effect on CAR translocation in response to the CAR activator TCPOBOP, but did repress induction of CYP2B by TCPOBOP in mouse primary hepatocytes.[49]

13.4.3 PKC and PKA

It is becoming evident that kinase activity plays a major role in the regulation of CAR and PXR transcriptional activity. Activation of protein kinase C (PKC) by phorbol esters has been shown to decrease expression of PXR target genes such as CYP3A4 in mouse primary hepatocytes.[50] The proposed hypothesis is that the mechanism involves activation of PKC increasing the interaction of PXR with the known nuclear reactor co-repressor (NCoR). Conversely, protein kinase A (PKA) appears to have the opposite effect on PXR transcriptional activity. The PKA activator forskolin was shown to increase CYP3A11 expression in mouse primary hepatocytes.[51] The mechanism of this effect is suggested to be increased co-activator binding caused by the direct phosphorylation of PXR since PKA was also shown to phosphorylate mouse PXR *in vitro*. The potential role of PKC and PKA in the regulation of PXR in humans has not been examined to date. Together, these findings suggest that cell signalling cascades can significantly influence the activity of CAR and PXR, and xenobiotics that do not directly bind CAR and PXR can ultimately regulate drug metabolizing enzymes in the liver.

13.4.4 Glucocorticoid

This hormone is able to regulate the *CAR* gene at the transcriptional level. A functional glucocorticoid response element was identified in the distal promoter region of the *CAR* gene in humans.[52] This promoter fragment was regulated by the glucocorticoid dexamethasone in reporter assays, and the glucocorticoid receptor (GR) was found to bind this region using chromatin immunoprecipitation. In addition, inhibition of endogenous GR transcriptional activity through treatment with pro-inflammatory cytokines such as interleukin-6 also inhibited CAR expression, further strengthening the connection between glucocorticoids, GR activity and CAR expression.[53] Along the same lines, GR knockout mice exhibited lower expression of CYP2B, compared to wild type mice, and reduced induction of CYP2B by PB[54] which is in agreement with the hypothesis that GR is not essential for PB induction but affects CAR activity.

13.4.5 Fatty Acids

Fatty acids like arachidonic acid (AA) are precursors to a wide array of signalling molecules such as prostaglandins and leukotrienes. In primary rat hepatocytes, PB regulation of CYP2B was attenuated by AA.[55] The AA product prostaglandin E_2 (PGE_2), was found to inhibit CYP2B induction by PB, indicating that PGE_2 may be responsible for this phenomenon. PGE_2 is an upstream effector of the PKA pathway and inhibition of the down pathway attenuated the repression of CYP2B induction by PB. This finding is in contrast with the effect of PKA on PXR transcriptional activity in a previously mentioned study,[51] thus highlighting a potentially marked difference between CAR and PXR in rodents. However, as is the case with PXR, xenobiotics with the ability to regulate PKA activity could potentially impact the regulation of drug metabolising enzymes by both receptors.

13.4.6 Retinoic Acids

CAR binds as a heterodimer with retinoid X receptor (RXR) to response elements in the promoter region of target genes. RXR is a promiscuous protein and dimerises with many nuclear receptors. Previously, retinoic acids were shown to repress TCPOBOP induction of CYP2B in mouse primary hepatocytes.[56,57] Retinoic acids are ligands for the retinoic acid receptor (RAR) that is also able to heterodimerise with RXR. Thus, activation of RAR by retinoic acids promotes dimerisation with RXR which may lead to sequestration of RXR away from CAR, leading to reduced TCPOBOP induction of CYP2B. In HepG2 cells stably overexpressing CAR, the synthetic RXR ligand LG1069 lowers basal expression and blocks TCPOBOP-mediated induction of CYP2B, though the effects of LG1069 in reporter assays are unclear.[58] Regulation of the interaction of CAR or PXR with other proteins through competition or cell signalling events could have significant impact on the overall transcriptional activity of the receptors, and ultimately affect the regulation of drug metabolising enzymes.

13.4.7 Pancreatic Hormones and Starvation

It is known that insulin represses hepatic expression of certain CYP enzymes.[59] Recently, the insulin responsive forkhead transcription factor FoxO1 was characterised as the co-activator of CAR and PXR, directly binding to the receptors to increase their transcriptional activity.[60] Phosphorylation of FoxO1 results in nuclear exclusion of the protein and a decrease in the transcriptional activity. Via its membrane receptor and phosphoinositide kinase-3 (PI3K), insulin activates protein kinase B (Akt), resulting in phosphorylation and nuclear exclusion of FoxO1.[61,62] Thus, insulin can repress the CAR- and PXR-mediated transcription by removing their co-activator FoxO1 from the nucleus. In support of this function of FoxO1, CYP2B expression was upregulated in the liver of FoxO1-transgenic mice.[63] The co-activation of the nuclear receptors by FoxO1 may have much broader implications in xenobiotic metabolism, since there are many signalling pathways that can influence Akt activity, including epidermal growth factor and platelet derived growth factor.[64,65]

The PPARγ co-activator (PGC-1) is induced in the liver of fasting mice and has become an important factor in the regulation of energy metabolism as well as gluconeogenesis in the liver.[66] A key part of PGC-1α function is the regulation of PGC-1α expression. Upregulation of PGC-1α induces expression of gluconeogenic enzymes including phosphoenolpyruvate carboxykinase (PEPCK) and glucose-6-phosphatase. The transcriptional activity of Gal4-CAR in a mammalian-1-hybrid was significantly increased with PGC-1α co-expression, suggesting that PGC-1α can function as a co-activator of CAR.[67] Likewise, co-expression of PGC-1α increased PXR transcriptional activity on a synthetic DR4 responsive reporter suggesting that PGC-1α acts, in addition, as a co-activator of PXR.[68] Like many other co-activators, PGC-1α interacts directly with the ligand-binding domain of CAR and PXR. PGC-1α does not have intrinsic histone acetyltransferase activity to promote the uncoiling of DNA from histone proteins. More likely, PGC-1α recruits proteins which have this activity such as SRC-1 and the histone acetyltransferase transcriptional co-activators CBP and p300 (CBP/p300) to increase access to DNA for gene transcription.[69] PGC-1α has an established role in the regulation of the bile acid synthesis gene *CYP7A*. PGC-1α functions as a co-activator for the transcription factor hepatocyte nuclear factor-4 (HNF-4).[70] Interestingly, CAR competes with HNF-4 for binding to PGC-1α as well as binding to the promoter region *CYP7A* as shown by gel shift assay and chromatin immunoprecipitation which leads to downregulation of *CYP7A*.[71] PXR has the same effect on HNF-4 regulation of *CYP7A* but through a different mechanism. PXR binds directly to HNF-4 which displaces PGC-1α from the HNF-4 protein complex, resulting in downregulation of *CYP7A*.[72]

PGC-1α is regulated by other stimuli including insulin, dietary composition and temperature. Insulin is also a negative regulator of FoxO1 activity since insulin can activate the Akt pathway, thus leading to decreased co-activation of CAR and PXR.[59] Downregulation of PGC-1α results in insulin resistance while re-expression of PGC-1α can restore insulin sensitivity.[73,74] Other nuclear receptors such as the PPARα is known to be involved in the development of insulin resistance when PGC-1α is down-regulated,[75] and it remains to be seen whether CAR and PXR are similarly involved.

13.4.8 Growth Factors and MEK-ERK

Growth factors such as the epidermal growth factor (EGF) and the transforming growth factor (TGF) have been known for a long time to repress PB induction of CYP2B in rodents.[76] Recent reports have also linked the mitogen-activated protein kinases (MAPKs) to the suppression of drug metabolising enzymes by growth factors. Inhibition of MAP-ERK kinase (MEK) by the chemical inhibitor U0126 was shown to further increase PB induction of CYP2B in rat primary hepatocytes.[77] ERK-1 is a downstream target of MEK, and co-expression of a dominant negative form of ERK-1 significantly increased PB inducibility of a CAR-based reporter assay using the promoter of CYP2B2. Conversely, activation of MEK by either EGF or growth hormone (GH) reduced CAR transactivation of CYP2B by PB in rat primary hepatocytes and in a reporter assay using the promoter region of CYP2B1.[78] Interestingly, co-treatment of a CAR activator with EGF or GH did not affect the ability of CAR to bind to DNA, thus suggesting that the mechanism of MEK inhibition is directed elsewhere in the CAR signalling pathway, such as regulation of nuclear translocation or phosphorylation of CAR.

13.4.9 Inflammation and NF-κB

Inflammation is known to repress the expression of drug metabolising enzymes in the liver.[79] Recent discoveries allow us to make a possible link between CAR/PXR signalling and inflammation via the nuclear factor-κB (NF-κB). Activation of NF-κB by TPA in primary human hepatocytes repressed induction of *CYP3A4* gene by rifampicin suggesting that NF-κB represses PXR. The mechanism responsible for the repression of PXR transcriptional activity by NF-κB involves the direct binding of RXR to the p65 NF-κB subunit.[80] The p65 subunit was found to bind directly to the DNA-binding domain of RXR, dissociating the PXR-RXR heterodimer from its binding DNA element. Thus, activation of NF-κB by signalling molecules such as TNF-α or LPS promotes NF-κB binding to RXR, leading PXR to dissociate from the promoter region of PXR target genes. If, in fact, this is the mechanism of the repression, NF-κB should repress the function of numerous nuclear receptors that is required for their activity, including CAR.

13.5 CYP Subfamilies 2-4 Under Regulation by CAR and PXR

CAR and PXR, originally identified as orphan nuclear receptors, have been suggested to be responsible for the xenobiotic-induced expression of *CYP2B* and *CYP3A* genes, respectively. Now numerous lines of evidence demonstrate that both nuclear receptors regulate an overlapping set of genes and play vital roles in the regulation of not only CYPs but also phase II enzymes and drug

Table 13.1 Nuclear receptors and expression of *CYP* genes.

Nuclear receptors	*Ligands*	*CYP genes*[a]
CAR	xenobiotics	*CYP2As* ↑, *CYP2Bs* ↑, *CYP2Cs* ↑, *CYP3As* ↑, (*CYP4As* ↑)
PXR	xenobiotics bile acids	*CYP2As* ↑, *CYP2Bs* ↑, *CYP2Cs* ↑, *CYP3As* ↑, *CYP24* ↑, (*CYP4As* ↑)
PPARα	fatty acids	*CYP4As* ↑, *CYP7A1* ↓, (*CYP2Cs* ↓)
VDR	vitamin D3 bile acids	*CYP2Bs* ↑, *CYP2Cs* ↑, *CYP3As* ↑, *CYP24* ↑, *CYP27B1* ↓
LXR	oxysterols	*CYP7A1* ↑, (*CYP3As* ↑)
FXR	bile acids	*CYP3As* ↑, *CYP7A1* ↓, *CYP8B1* ↓,
RXR[b]	retinoic acids	*CYP2As*, *CYP2Bs*, *CYP2Cs*, *CYP3As*, *CYP4As*, *CYP7A1*, *CYP8B1*
GR	glucocorticoids	(*CYP2Bs* ↑, *CYP2Cs* ↑, *CYP3As* ↑)
HNF-4	not identified	*CYP2As* ↑, *CYP2Bs* ↑, *CYP2Cs* ↑, *CYP3As* ↑, *CYP7A1* ↑

[a]Arrows ↑ and ↓ indicate up-regulation and down-regulation, respectively. Direct binding of nuclear receptors to the promoter sequences has not been reported for the genes in parentheses.
[b]RXR is involved in the transcriptional regulation of the CYP genes by forming a heterodimer with CAR, PXR, PPARα, VDR, LXR and FXR. CAR, constitutive androstane receptor, PXR, pregnane X receptor; PPARα, peroxisome proliferator-activating receptor α; VDR, vitamin D receptor; LXR, liver X receptor; FXR, farnesoid X receptor; RXR, retinoid X receptor; GR, glucocorticoid receptor; HNF-4, hepatocyte nuclear factor-4.

transporters. This section is restricted to members of the *CYP2*, *CYP3* and *CYP4* families and the role of CAR and PXR in their regulation (Table 13.1).

13.5.1 CYP2A

CYP2A enzymes are involved in the metabolism of toxic compounds such as aflatoxin B1, nitrosamines and nicotine as well as many pharmaceutics (see Chapter 5). The exposure to xenobiotics, in turn, increases mRNA and protein levels of CYP2As in the liver of rodents and human. For example, the expression of mouse CYP2A5 is increased by treatment with several compounds including PB and TCPOBOP.[81–83] In human hepatocytes, PB and rifampicin are reported to increase the expression of CYP2A6.[83–85] Moreover, CYP2A6 mRNA levels are higher in the liver obtained from patients treated with carbamazepine than in control livers.[86] These results implicate the involvement of nuclear receptors in CYP2A induction by xenobiotics. In fact, the role of CAR and PXR in the CYP2A induction has been demonstrated by studies using receptor-null or transgenic mice. In livers of CAR knockout mice, TCPOBOP failed to increase CYP2A4 mRNA levels.[87,88] Overexpression of the constitutively active human PXR in mouse livers enhanced the mRNA levels for CYP2A4 both in wild-type and PXR-null mice.[89] More recently, cooperation of PXR and PGC-1α has been reported to be important in the transcriptional activation of *CYP2A6* gene by xenobiotics.[90]

In addition to the xenobiotic-induced and receptor-mediated transcriptional activation of CYP2A genes, post-transcriptional regulation is important for CYP2A induction by some compounds. Pyrazole is a well known compound that increases CYP2A5 mRNA levels in mouse livers.[81,91] Treatment with pyrazole increases the level of RNA-binding proteins that bind to and stabilise CYP2A5 mRNA, leading to the elevation of CYP2A5 protein content.[92–95] Nuclear receptors appear not to be involved in this mechanism.

13.5.2 CYP2B

Since PB a prototypical CYP inducer induces most preferentially CYP2B enzymes in rodent livers, its molecular mechanism has been studied by numerous researchers. In 1998, a nuclear receptor CAR was finally identified as a key transcription factor for the PB induction of *CYP2B* genes.[6]

PB-responsive enhancer module (PBREM) was first identified in mouse *Cyp2b10* gene as a regulatory *cis*-element for the transactivation of the gene by PB.[6] This 51-bp sequence is also conserved in rat *CYP2B1* and *CYP2B2* genes and human *CYP2B6* gene.[96] Mouse, rat and human PBREMs are located at −2.3 kb, −2.2 kb and −1.7 kb in the 5′-flanking region of *Cyp2b10, CYP2B1/2* and *CYP2B6* genes, respectively. These elements constitute two nuclear receptor binding motifs (DR-4) and a binding site for nuclear factor-1 (NF-1). DR-4 motifs are critical for PB-dependent transcriptional activation, while NF-1 motif seems to work as an accessory element for full induction. A heterodimer of CAR and RXR binds to both DR-4 motifs. The fundamental role of CAR in PB induction of CYP2Bs has been demonstrated with CAR-null mice, in which PB induction of *Cyp2b10* is completely lost.[97,98] In addition to PB, induction of *CYP2B* genes by various PB-type inducers, including clotrimazole, chlorpromazine, 1,1,1-trichloro-2,2-bis(4-chlorophenyl) ethane, methoxychlor and polychlorinated biphenyls, is mediated through PBREM.[6,99,100]

As already discussed, CAR is a unique nuclear receptor since it is constitutively active and its transcriptional activity is regulated by its cellular localisation. CAR is retained in the cytoplasm in untreated livers, and PB or other CYP2B inducing agents triggers its nuclear translocation.[41,43,101] PB activates both rodent and human CAR, thus inducing both the rodent and human *CYP2B* genes. This PB-induced activation of CAR does not require direct binding of the drug to CAR.[26,102] A mouse CYP2B10 inducer, TCPOBOP binds to and activates only the mouse CAR but not the human receptor[102,103] while CITCO is known as a human CAR activator increasing CYP2B6 mRNA levels in human hepatocytes.[104] Thus, ligand/activator-specificity of nuclear receptors may define species-differences in CYP2B induction by xenobiotics.

PXR was identified as a key transcriptional factor for CYP3A induction (*vide infra*) and is now known to also bind to PBREM, activating transcription of *CYP2B* genes.[38,105–107] Thus, the *Cyp2b10* gene is induced by the PXR ligand dexamethasone in the liver of CAR-null mice.[87,105] Conversely, CAR is capable of activating the *CYP3A* genes and PB induces the expression of *Cyp3a11* in the liver of PXR knockout mice[105] (*vide infra*).

In addition to PBREM, a distal enhancer module named xenobiotic-responsive enhancer module (XREM) has been identified in the promoter region of the *CYP2B6* gene.[108] It is located at −8.5 kb upstream from the transcription start site and contains several binding motifs of nuclear receptors. Cell-based reporter assays and electrophoretic mobility shift assays have demonstrated that both CAR and PXR bind to one of the motifs (DR-4) to activate the transcription, independent of PBREM. In primary cultures of human hepatocytes, maximal reporter activation was obtained with the construct containing both PBREM and XREM. Taken together, CYP2B6 induction by xenobiotics is regulated by a complex, but complementary mechanism, involving at least two closely related nuclear receptors, CAR and PXR, and two xenobiotic-responsive enhancer sequences.

13.5.3 CYP2C

The expression of human and rodent CYP2C enzymes is increased by a variety of compounds, including CAR and PXR activators such as PB and rifampicin. Transcriptional regulation of *CYP2C* genes is well characterised for human genes. The human CYP2C subfamily consists of four enzymes, namely, CYP2C8, CYP2C9, CYP2C18 and CYP2C19, and among them the expression of CYP2C8, CYP2C9 and CYP2C19 is inducible by xenobiotics in the liver.[109]

CYP2C8 is the most inducible enzyme in the liver among human CYP2C enzymes.[84,109,110] The expression of the *CYP2C8* gene is induced by both human CAR and PXR agonists in primary cultures of human hepatocytes. Consistent with these observations, CAR/RXR and PXR/RXR binding sites have been identified at −2.8 kb and −8.8 kb, respectively, in the 5′-flanking region of the *CYP2C8* gene.[111] CYP2C8 is also induced by the synthetic glucocorticoid dexamethasone, and GR-binding site has been identified at −1.9 kb in the *CYP2C8* promoter region.[111]

The expression of CYP2C9 is also activated by numerous compounds in human primary hepatocytes.[84,109,110] The fact that administration of rifampicin reduces the plasma concentrations of (*S*)-warfarin and tolbutamide, substrates of CYP2C9, implies *in vivo* induction of CYP2C9 by rifampicin.[112,113] Two nuclear receptor-binding sites have been identified at −1.8 kb and −2.9 kb of the 5′-flanking region of the *CYP2C9* gene, where both CAR and PXR can form a heterodimers by binding to RXR. More recently, HNF-4-binding sites were identified in the proximal promoter of *CYP2C9* gene, and these sites have been reported to be necessary for the maximal induction of CAR/PXR-mediated *CYP2C9* gene transcription.[114]

The constitutive expression of CYP2C19 in human livers is much lower than that of CYP2C9, and shows large interindividual variation because of genetic polymorphism. In human hepatocytes, the expression of CYP2C19 is induced by various xenobiotics similar to the other human CYP2C enzymes, CYP2C8 and CYP2C9, although it appears to be dependent on the origin of the hepatocytes.[84,110,115] Significant elevation of CYP2C19 mRNA levels has also been reported in the liver of patients treated with carbamazepine, a weak activator of human PXR.[86] Reporter and gel shift assays have identified a binding site for CAR and PXR at −1.9 kb of 5′-flanking region of the *CYP2C19* gene, and mutation of the binding motif abolishes CAR- and PXR-mediated activation of the *CYP2C19* reporter gene.[116] The same authors have also demonstrated that overexpression of human CAR in HepG2 cells increases CYP2C19 mRNA levels.

Although a number of studies have dealt with the xenobiotic induction of rodent CYP2C enzymes by xenobiotics in the mouse liver, the molecular mechanism is less characterised. Induction of murine *Cyp2c29* and *Cyp2c37* by PB and phenytoin has been reported to be mediated by CAR through its binding to the promoter regions of these genes.[117,118] In addition to mammalian CYP2C enzymes, chicken CYP2C45 is also induced by PB in LMH cells, and seems to be mediated by a nuclear receptor CXR,[119] which is a chicken homologue of mammalian PXR and CAR.[120]

13.5.4 CYP2E1

CYP2E1 metabolises various solvents such as ethanol and acetone, and is also induced by these compounds.[121] The expression of CYP2E1 is also increased during some conditions including diabetes and fasting.[121,122] Induction is regulated at the post-transcriptional level, involving the stabilisation of its mRNA and apoprotein. No nuclear receptor has been identified to be associated with the CYP2E1 induction through this mechanism.

Enzymic activity of CYP2E1 is associated with the hepatotoxicity of some chemicals such as acetaminophen (paracetamol) and carbon tetrachloride. It has been reported that CAR and PXR knockout mice are resistant to chemical-induced liver injury by these compounds, suggesting a potential role of these nuclear receptors in the hepatotoxicity of these chemicals.[123–125] However, the CAR- and/or PXR-dependent expression of CYP forms other than CYP2E1 may be the underlying mechanism for these phenomena.

13.5.5 CYP3A

In rodent livers, the synthetic glucocorticoid dexamethasone induces CYP3A enzymes. Although the dose required to induce CYP3A expression is much higher than what is needed to activate the GR, and the GR antagonist RU486 also increases CYP3A expression, the possible role of GR in CYP3A induction had been studied extensively though there are no conclusive results to date. In 1998, a mouse nuclear receptor PXR was cloned and identified as a key transcription factor for dexamethasone- and PCN-mediated induction of rodent CYP3As.[7] Subsequently, its human counterpart was identified.[126] Consistent with their importance in CYP3A induction, both human and rodents receptors are expressed mainly in the liver and small intestine, where high expression of CYP3As is observed. The principal role of PXR in CYP3A induction has been confirmed by experiments using PXR-null mice in which PCN showed no induction of CYP3A.[127,128]

Mouse PXR binds to a DR-3 motif found in the proximal region of the promoters of *CYP3A1* and *CYP3A2* genes,[7] which has been identified as a dexamethasone-responsive sequence.[129,130] In the *CYP3A4* gene, human PXR binds to an everted repeat of two half-sites separated by six nucleotides (ER-6 motif),[126] which is contained in the previously identified rifampicin-responsive element.[131] Within the sequence of ER-6, there is a DR-4 motif, which is similar to those found in *CYP2B* genes.[96]

The species-differences in the DNA sequences of PXR-binding sites and the structure of DNA binding domains of PXRs do not seem to be associated with species-differences in CYP3A induction by xenobiotics, since human and mouse PXRs show 96% amino acid identity in the DNA-binding domain and share their cognate DNA motifs: mouse and rat PXRs can activate the ER-6 element of *CYP3A4* gene and human PXR can activate DR-3 of rat *CYP3A1/2* genes.[126,131] The species-differences in CYP3A induction *in vivo*, may rather result from the ligand-selectivity of PXRs. Mouse and human PXRs share only 76% identity of amino acids in their ligand-binding domains. As mentioned above (*vide supra*) mouse PXR, but not human PXR, is activated by PCN, while human PXR, but not mouse PXR, is activated by rifampicin.[132] Thus, dexamethasone and PCN induce the expression of CYP3As in rodents but not in human cells, while rifampicin does so in humans and rabbits but not in rats and mice.[133,134]

In the human *CYP3A4* gene, a distal enhancer module named XREM is also important for the maximal activation of *CYP3A4* gene by rifampicin-activated PXR.[135] This module is located at −7.6 to −7.8 kb from the transcription start site and contains several binding motifs for nuclear receptors. Indeed, HNF-4 also binds to the module and cooperatively regulates the PXR-mediated transcription of *CYP3A4* gene.[136] As already discussed (*vide supra*), CAR is capable of activating the expression of CYP3As.[105] The transcriptional enhancement by HNF-4 is also observed for the CAR-mediated *CYP3A4* transcription through XREM.[136] Thus, both XREM and proximal ER-6 are necessary for xenobiotic induction of the *CYP3A4* gene. Using the reporter constructs containing these motifs, model mice and cell lines have been established to screen compounds that induce CYP3A4.[137,138]

CYP3A7 is the principal form of human CYP3As expressed in foetal livers. This enzyme shows 95% identity in its amino acid sequence with CYP3A4, and possesses almost identical enzymic properties. In addition to the protein-coding sequences, their promoter sequences exhibit more than 90% identity up to −8 kb from the transcription start site, which includes ER-6 and XREM. As expected from the similarity, the transcription of the *CYP3A7* gene is activated by both PXR and CAR in cultured cells.[139,140] In addition, a significant correlation between CYP3A7 mRNA levels and PXR or CAR mRNA levels has been reported in foetal livers.[141]

13.5.6 CYP4A

CYP4As catalyse the oxidation of biologically important fatty acids including arachidonic acid. The expression of *CYP4A* genes is transcriptionally activated by various xenobiotics such as hypolipidaemic fibrates and plasticiser phthalate esters.[142] Nuclear receptor PPARα mediates the chemical-induced transactivation. The role of CAR and PXR in the xenobiotic-induced expression of CYP4As has not been fully studied. A PPAR agonist, di-*n*-butyl phthalate is reported to activate rat CAR and increase the expression of CYP4A1 as well as CYP2B1 and CYP3A1 in the rat foetal liver, suggesting the involvement of CAR in the CYP4A induction.[143] The rodent PXR activator dexamethasone has been reported to increase the expression of CYP4A11 in HepG2 cells in a PPARα-independent manner.[144] On the other hand, PB induces the expression of *CYP4A10* and *CYP4A14* genes in the liver of CAR-null mice but not in wild type mice.[98] The molecular mechanism of how CAR and PXR regulate the transcription of *CYP4A* genes remains a topic for future studies.

13.6 Other Nuclear Receptors Regulate Members of the CYP2 and CYP3 Family

In addition to CAR and PXR, other nuclear receptors have been reported to play a role in the expression of drug-metabolising CYP enzymes, here which is briefly described.

13.6.1 HNF-4

HNF-4 is a nuclear receptor expressed in the liver, kidney, intestine and pancreas, and is involved in the regulation of numerous hepatic genes related to glucose, fatty acid and cholesterol metabolism, apolipoprotein synthesis and liver differentiation.[145] This receptor is also known to regulate the hepatic expression of a number of CYPs including CYP2A, CYP2B, CYP2C, CYP2D and CYP3A members.[146] The involvement of HNF-4 in the hepatic expression of these *CYP* genes has been demonstrated independently for each *CYP* gene by a combination of deoxyribonuclease I foot printing, reporter assays and gel shift assays. In contrast to these strategies, the antisense targeting technology was used to examine the role of HNF-4 in the expression of various *CYP* genes in human hepatocytes.[147] Reduction of nuclear HNF-4 protein levels significantly decreased CYP3A4, CYP3A5 and CYP2A6 mRNA levels, and modestly reduced those of CYP2B6, CYP2C9 and CYP2D6, but did not affect that of CYP2E1, although it remains to be determined whether these changes result from the reduced binding of HNF-4 to their promoter sequences.

Recent studies have indicated the existence of co-regulation by HNF-4 and CAR/PXR. HNF-4 binds to the proximal promoter of *CYP2C9* gene and its binding seems to be necessary to achieve the maximal transcriptional activation of *CYP2C9* gene by xenobiotics through CAR or PXR.[114] HNF-4 also binds to the *CYP3A4* enhancer module XREM, and modulates the PXR- and CAR-mediated transcription of this gene.[136] Thus, HNF-4 binds both proximal promoter and enhancer sequences of *CYP* genes, trans-activating their expression in the liver with or without other nuclear receptors such as CAR and PXR.

13.6.2 PPARα

Hypolipidaemic fibrates induce CYP4A expression and their cognate receptor PPARα functions as a key transcription factor for this induction. In addition to CYP4As, fibrates are also shown to induce CYP2B in rat liver and primary cultures of rat hepatocytes;[148,149] at present, however, it remains unknown whether PPARα is responsible for the CYP2B induction. Since clofibrate was shown to activate mouse CAR in cell-based reporter assays,[34] CAR may play a role in the fibrate-induced expression of CYP2Bs.

On the other hand, it has been proposed that PPARα represses the expression of CYP2Cs in rat liver. Treatment with the PPARα ligand, WY-14,643 or gemfibrozil, decreased hepatic protein levels of CYP2C11 in male and CYP2C12 in female rats,[150] and CYP2C7 mRNA and protein levels in the liver of both sexes.[151] The molecular mechanism of this suppression seems to depend on the ligand. It has been reported that nafenopin- but not dehydroepiandrosterone-mediated suppression of CYP2C11 expression is mediated by PPARα.[152] How PPARα represses the CYP2C expression is remains to be clarified.

13.6.3 VDR

Cellular action of a biologically active vitamin D_3 (1α,25-dihydroxyvitamin D_3) is mediated through the nuclear receptor vitamin D receptor (VDR). A possible role of VDR in CYP induction was first suggested by experiments demonstrating that CYP3A4 was induced by 1α,25-dihydroxyvitamin D_3 in the human colon carcinoma Caco-2 cells.[153] It was then demonstrated that VDR activates the transcription of the *CYP3A4* gene through an ER-6 motif in a human intestinal cell line.[154] Although the expression of VDR in the liver is considerably lower than that in intestines, the experiments using human primary hepatocyes have demonstrated that the expression of CYP3A4, CYP2B6 and CYP2C9 are regulated by VDR.[155] Both XREM and ER-6 are necessary for the VDR-mediated CYP3A4 induction.[155] In addition to vitamin D3s, certain bile acids are known ligands for VDR. It has been reported that VDR, rather than PXR, plays a predominant role in the induction of *CYP3A11* gene by lithocholic acid in the liver and small intestine of mice.[156]

13.6.4 FXR

Bile acid homeostasis is tightly regulated by the nuclear receptor farnesoid X receptor (FXR), which is activated by certain bile acids. CYP3As are involved in the metabolism of some bile acids, suggesting that CYP3A expression is regulated by FXR as well as PXR and CAR. In fact, it has been demonstrated that FXR activates *CYP3A4* transcription by binding to its promoter region containing XREM, and that GW4064, an FXR agonist, induces the *Cyp3a11* gene in the liver of wild-type and PXR-null mice but not FXR-null mice.[157] However, the role of FXR in CYP3A induction by bile acids may be limited because VDR and PXR also function as lithocholic acid receptors and induce CYP3A expression.[128,156,158] Furthermore, hepatic *CYP3A11* expression is increased in both wild-type and FXR-null mice after cholic acid feeding or bile duct ligation.[159–161]

13.6.5 LXR

LXR (liver X receptor) is activated by cholesterol metabolites, called oxysterols, including 22(*R*)-hydroxycholesterol and 24(*S*),25-epoxycholesterol. This nuclear receptor plays a central role in cholesterol homeostasis in conjunction with FXR. It remains unclear whether LXR directly regulates the expression of drug-metabolising CYPs. One group reported that 24(*S*),25-epoxycholesterol induced CYP3A expression in primary cultures of rat hepatocytes.[162] However, the agent was capable of increasing hepatic CYP3A mRNA levels even in LXR-null mice, and failed to do so in PXR-null mice, suggesting this oxysterol-mediated induction of CYP3As occurs through PXR not LXR.[162] Using LXR knockout mice fed standard chow or cholesterol-supplemented diet, it has been demonstrated that the expression of *CYP2B10* and *CYP3A11* is differentially modulated by LXR expression and liver cholesterol levels, although direct binding of LXR to the *Cyp2b10* or *Cyp3a11* promoter sequence has not been demonstrated.[163]

13.7 Other CYPs Influenced by CAR, PXR and Their Activators

This section gives a brief overview of how activation of CAR and PXR might influence the CYP-dependent metabolism of some endogenous compounds with ramifications on cellular and physiological homeostasis. Notable examples of this interplay include the metabolism of steroids, vitamin D and bile acids.[8]

The synthesis of steroid hormones is catalysed by CYP enzymes in families 11, 17, 19 and 21,[164] which are principally regulated by monomeric nuclear receptors within subfamilies NR5A and NR4A.[8,165,166] Steroid hormones are inactivated via several oxidising CYPs (typically CYP2B, 2C and 3A subfamilies) and conjugating enzymes[167] that are under the control of CAR and/or PXR. Early reports indicated that drug exposure can increase metabolism and clearance of steroids, the best examples being older antiepileptic drugs and rifampicin.[167,168] In extreme cases, endocrine malfunction has been observed both in experimental animals and humans,[143,169–172] although compensatory upregulation of steroid-synthesising enzymes may take place.[167,169] Another example of such hormonal imbalances is the rifampicin- and phenobarbital-induced osteomalacia. In this case, PXR agonists appear to activate the CYP24 enzyme that is responsible for inactivation of vitamin D via 24-hydroxylation.[173] Furthermore, vitamin D elimination is further enhanced by induction of CYP3A4 that catalyzes 23- and 24-hydroxylation.[174] To make matters even more complex, several CYP forms that are controlled by HNF-4 and/or PXR participate in the formation of active vitamin D.[164,175]

The catabolism of cholesterol into bile acids is mediated by CYP enzymes 7A1, 8B1, 27A1 and 3A4 along with several efflux transporters and conjugating enzymes. The disposition of drugs and xenobiotics often share the same or related enzymes, which suggests coordinated regulation and interplay between these metabolic systems.[176,177] Because elevated levels of potentially toxic bile acids can activate FXR, VDR and PXR receptors and perhaps CAR as well, there is an overall clearance of bile acids into more hydroxylated forms via activation of CYP3A4 (*vide supra*) and associated transporters and conjugating enzymes. As sterol precursors are also agonists for PXR,[178] they constitute a feed-forward pathway for enhanced elimination of bile acids.

There is an additional mechanism to further decrease levels of bile acids by blocking their synthesis via suppression of CYP7A1, 8B1 and CYP27A1. For instance, HNF4α (and in some species LXRα) positively regulate these CYP isoforms while they are suppressed by the small heterodimer partner (SHP).[177] SHP is upregulated by bile acid-activated FXR and it can block the HNF-4 and LRH1-dependent expression of CYP7A1 and of many other genes controlled by nuclear receptors.[177,179,180] The attenuation of CYP7A1 by PXR, however, is still unresolved.[161,181] It has been reported that PXR can activate CYP27A1 in intestine but not liver.[175] At present, there is very little information on the effect of CAR in bile acid-forming CYP enzymes although its role in regulating conjugating enzymes is clear.[182] However, the finding that endogenous isoprenoids may activate CAR[183] points to the possibility that CAR is intimately linked to bile acid metabolism.

13.8 Perspective

In 1998, CAR and PXR were first identified as the PB- and PCN-activated nuclear receptors mediating the drug-induced transcription of the *CYP2B* and *CYP3A* genes, respectively.[6,7] These receptors are now characterised as promiscuous transcription factors, being placed at the crossroads where both xenobiotics and endogenous stimuli co-regulate diverse liver functions including drug and energy metabolism. As the function of these receptors has broadened, the molecular mechanisms of their regulation has evolved from simple protein-DNA binding to regulation by complex protein-protein interactions. Due to the promiscuous nature of CAR and PXR, the structure-based mechanism of their ligand-elicited activation is not fully understood. It has also become clear that a simple ligand binding mechanism alone is no longer sufficient to explain the diverse functions of these receptors. Exciting future research prospects in the following areas are apparent.

13.8.1 Mechanism of Ligand-Elicited Activation

Although the X-ray crystal structures of CAR and PXR have been determined in the presence and absence of their ligands, the molecular mechanisms of how agonists stabilise helix 12 to elicit the transcriptional activity is not fully understood. In combination with further efforts to determine more X-ray crystal structures of various forms of CAR and PXR, structure-function studies using site-directed mutagenesis should continue. The ultimate goal is to predict the degree of involvement of each receptor (*i.e.* CAR, PXR, FXR, etc) in the induction of enzymes in human liver by a given drug candidate. Developing highly predictive *in silico* models and dynamic simulations is critical in our understanding of the flexible ligand binding pockets and in the development of receptor-based therapeutic drugs.

13.8.2 Direct Targets of Signal-Mediated Activation

It has long been appreciated that various signals modulate both constitutive and inducible expression of *CYP* genes. However, it is only recently that we have been able to converge some of these signals onto the functions of CAR and PXR. Even though it is known that AMPK, ERK, PKA and PKC regulate receptor-mediated transcription, the molecular basis of their regulation remains unclear. On one hand, signals may regulate direct binding of CAR and PXR to their response elements, activating receptor-mediated transcription. On the other hand, signals can be transduced to a distinct DNA element to regulate the receptor-mediated activation, making the signal-mediated regulation complex. A good example of this complexity is the synergistic activation of the *CYP2B6* gene by CAR after co-treatment with TCPOBOP and the protein phosphatase inhibitor okadaic acid (OA).[184,185] Two distinct elements within the *CYP2B6* promoter are involved in the synergism: distal PBREM, to which CAR directly binds in response to TCPOBOP, and the proximal OA response element to which OA signals CAR to associate. The direct and indirect binding of CAR result in synergism in the CYP2B6 promoter activity. This web of the complex activation mechanisms will begin to become untangled when the direct targets of these phosphorylation pathways are identified. Are they CAR and PXR themselves, and if not, how do they receive the signal of activation?

13.8.3 Finding the PB Receptor

Typified by PB, numerous CAR activators do not appear to directly bind to the receptor and are thought to activate it via a signalling mechanism. However, PB would be expected to bind to a protein that initiates PB signal to activate CAR, the PB receptor. What is the PB receptor? Where is this receptor localised within the cell? The next major breakthrough in defining the PB signal can only be made following the identification of its receptor. Recently, CAR was found to be localised at the cell membrane[186] raising the intriguing question as to whether the cell membrane bound CAR constitutes a PB receptor complex.

13.8.4 Diverse Roles in Regulating Liver Function and Disease

Insightful studies have now commenced to define the molecular basis of how CAR and PXR mediate drug-induced alteration in various liver functions such as glucose, lipid and cholesterol metabolism. CAR and PXR have been found to directly regulate the genes that encode the key enzymes involved in the metabolism of these substrates. For example, both nuclear receptors directly interact with the insulin-responsive forkhead transcription factors FoxO1 and FoxA2, repressing their ability to activate the insulin-response sequence-bearing genes: *glucose-6-phosphatase, phosphoenolpyrvate carboxykinase 1, carnitine palmitolytransferase 1 and mitochondrial HMG-CoA synthase.*[187,188] Thus, hepatic levels of gluconeogenesis, fatty acid oxidation and ketogenesis may be reduced by drugs via the receptor-Fox interactions. In fact, the *glucose-6-phosphatase and phosphoenolpyrvate carboxykinase 1* genes were down-regulated in the liver of the active PXR-bearing transgenic mice.[189] On the other hand, stearoyl-CoA desaturase 1, the key enzyme in lipogenesis, is induced by a PXR activator resulting in the hepatic accumulation of 3-hydroxybutylate, though the molecular mechanism for this process remains unexplored. It is necessary to establish whether protein-protein interactions and not the interactions with DNA are the fundamental mechanism by which CAR and PXR regulate various liver functions.

Treatment with CAR and/or PXR activators is known to cause hepatomegaly in rodents.[190,191] Moreover, chronic activation of CAR by PB and TCPOBOP led to the development of hepatocellular carcinomas in mice.[192,193] Recent studies showed that activation of PXR results in repression of inflammation by inhibiting the NF-kB target genes (*IL-2, TNF-α, IkB* and *COX-2*) in the liver of both rodents and humans.[194] CAR and PXR may directly regulate growth, proliferation and/or apoptosis signals through activation of target genes intimately involved in these cellular processes. While nuclear receptors may regulate signal molecules by acting on gene expression, their direct protein-protein interactions with signalling molecules may also be part of the molecular basis of their regulation.

References

1. H. Remmer and H.J. Merker, *Science*, 1963, **142**, 1657.
2. T. Omura and R. Sato, *J. Biol. Chem.*, 1962, **237**, 1375.
3. D.Y. Cooper, S.S. Levin, S. Narasimhulu, O. Rosenthal and R.W. Estabrook, *Science*, 1965, **147**, 400.
4. A.H. Conney, *Pharmacol. Rev.*, 1967, **19**, 317.

5. A. Poland, E. Glover and A.S. Kende, *J. Biol. Chem.*, 1976, **251**, 4936.
6. P. Honkakoski, I. Zelko, T. Sueyoshi and M. Negishi, *Mol. Cell. Biol.*, 1998, **18**, 5652.
7. S.A. Kliewer, L.T. Moore, L. Wade, J.L. Staudinger, M.A. Watson, M.A. Jones, D.D. McKee, B.B. Oliver, T.M. Willson, R.H. Zetterstrom, T. Perlman and J.M. Lehmann, *Cell*, 1998, **92**, 738.
8. I. Issemann and S. Green, *Nature*, 1990, **347**, 645.
9. P. Honkakoski and M. Negishi, *J. Biochem.*, 2000, **347**, 321.
10. Y. Fujii-Kuriyama and J. Miura, *Biochem. Biophys. Res. Commun.*, 2005, **338**, 311.
11. T.M. Willson and S.A. Kliewer, *Nat. Rev. Drug Discov.*, 2002, **1**, 259.
12. P. Honkakoski, T. Sueyoshi and M. Negishi, *Ann. Med.*, 2003, **35**, 172.
13. T.K. Chang and D.J. Waxman, *Drug Metab. Rev.*, 2006, **38**, 51.
14. R.E. Watkins, G.B. Wisely, L.B. Moore, J.L. Collins, M.H. Lambert, S.P. Williams, T.M. Willson, S.A. Kliewer and M.R. Redinbo, *Science*, 2001, **292**, 2329.
15. R.E. Watkins, P.R. Davis-Searles, M.H. Lambert and M.R. Redinbo, *J. Mol. Biol.*, 2003, **331**, 815.
16. R.E. Watkins, J.M. Maglich, L.B. Moore, G.B. Wisely, S.M. Noble, P.R. Davis-Searles, M.H. Lambert, S.A. Kliewer and M.R. Redinbo, *Biochemistry*, 2003, **42**, 1430.
17. J.E. Chrencik, J. Orans, L.B. Moore, Y. Xue, L. Peng, J.L. Collins, G.B. Wisely, M.H. Lambert, S.A. Kliewer and M.R. Redinbo, *Mol. Endocrinol.*, 2005, **19**, 1125.
18. A. Poso and P. Honkakoski, *Mini. Rev. Med. Chem.*, 2006, **6**, 937.
19. R.G. Tirona, B.F. Leake, L.M. Podust and R.B. Kim, *Mol. Pharmacol.*, 2004, **65**, 36.
20. Y. Mu, C.R. Stephenson, C. Kendall, S.P. Saini, D. Toma, S. Ren, H. Cai, S.C. Strom, B.W. Day, P. Wipf and W. Xie, *Mol. Pharmacol.*, 2005, **68**, 403.
21. T. Östberg, G. Bertilsson, L. Jendeberg, A. Berkenstam and J. Uppenberg, *Eur. J. Biochem.*, 2002, **269**, 4896.
22. H. Gronemeyer, J.-Å. Gustafsson and V. Laudet, *Nat. Rev. Drug Discov.*, 2004, **3**, 950.
23. D.R. Johnson, C.W. Li, J.C. Ghosh, L.Y. Chen and J.D. Chen, *Mol. Pharmacol.*, 2005, **69**, 99.
24. G. Luo, M. Cunningham, S. Kim, T. Burn, J. Lin, M. Sinz, G. Hamilton, C. Rizzo, S. Jolley, D. Gilbert, A. Downey, D. Mudra, R. Graham, K. Carroll, J. Xie, A. Madan, A. Parkinson, D. Christ, B. Selling, E. LeCluyse and L.S. Gan, *Drug Metab. Dispos.*, 2002, **30**, 795.
25. Z. Zhu, S. Kim, T. Chen, J.H. Lin, A. Bell, J. Bryson, Y. Dubaquie, N. Yan, J. Yanchunas, D. Xie, R. Stoffel, M. Sinz and K. Dickinson, *J. Biomol. Screen.*, 2004, **9**, 533.

26. L.B. Moore, D.J. Parks, S.A. Jones, R.K. Bledsoe, T.G. Consler, J.B. Stimmel, B. Goodwin, C. Liddle, S.G. Blanchard, T.M. Willson, J.L. Collins and S.A. Kliewer, *J. Biol. Chem.*, 2000, **275**, 15122.
27. R.X. Xu, M.H. Lambert, B.B. Wisely, E.N. Warren, E.E. Weinert, G.M. Waitt, J.D. Williams, J.L. Collins, L.B. Moore, T.M. Willson and J.T. Moore, *Mol. Cell*, 2004, **16**, 919.
28. L. Shan, J. Vincent, J.S. Brunzelle, I. Dussault, M. Lin, I. Ianculescu, M.A. Sherman, B.M. Forman and E.J. Fernandez, *Mol. Cell*, 2004, **16**, 907.
29. K. Suino, L. Peng, R. Reynolds, Y. Li, J.Y. Cha, J.J. Repa, S.A. Kliewer and H.E. Xu, *Mol. Cell*, 2004, **16**, 893.
30. B. Windshügel, J. Jyrkkärinne, J. Vanamo, A. Poso, P. Honkakoski and W. Sippl, *J. Mol. Graph. Model.*, 2007, **26**, 644.
31. J. Jyrkkärinne, B. Windshügel, J. Mäkinen, M. Ylisirniö, M. Peräkylä, A. Poso, W. Sippl and P. Honkakoski, *J. Biol. Chem.*, 2005, **280**, 5960.
32. J. Jyrkkärinne, J. Mäkinen, J. Gynther, H. Savolainen, A. Poso and P. Honkakoski, *J. Med. Chem.*, 2003, **46**, 4687.
33. A. Ueda, S. Kakizaki, M. Negishi and T. Sueyoshi, *Mol. Pharmacol.*, 2002, **61**, 1284.
34. J. Mäkinen, M. Reinisalo, K. Niemi, P. Viitala, J. Jyrkkärinne, H. Chung, O. Pelkonen and P. Honkakoski, *Biochem. J.*, 2003, **376**, 465.
35. T. Kawamoto, S. Kakizaki, K. Yoshinari and M. Negishi, *Mol. Endocrinol.*, 2000, **14**, 1897.
36. L.A. Stanley, B.C. Horsburgh, J. Ross, N. Scheer and C.R. Wolf, *Drug Metab. Rev.*, 2006, **38**, 515.
37. X. Ding and J.L. Staudinger, *J. Pharmacol. Exp. Ther.*, 2005, **314**, 120.
38. S.R. Faucette, T. Sueyoshi, C.M. Smith, M. Negishi, E.L. Lecluyse and H. Wang, *J. Pharmacol. Exp. Ther.*, 2006, **317**, 1200.
39. T. Kawamoto, T. Sueyoshi, I. Zelko, R. Moore, K. Washburn and M. Negishi, *Mol. Cell. Biol.*, 1999, **19**, 6318.
40. F. Hosseinpour, R. Moore, M. Negishi and T. Sueyoshi, *Mol. Pharmacol.*, 2006, **69**, 1095.
41. K. Kobayashi, T. Sueyoshi, K. Inoue, R. Moore and M. Negishi, *Mol. Pharmacol.*, 2003, **64**, 1069.
42. I. Zelko and M. Negishi, *Biochem. Biophys. Res. Commun.*, 2000, **277**, 1.
43. K. Yoshinari, K. Kobayashi, R. Moore, T. Kawamoto and M. Negishi, *FEBS Lett.*, 2003, **548**, 17.
44. K. Kawana, T. Ikuta, Y. Kobayashi, O. Gotoh, K. Takeda and K. Kawajiri, *Mol. Pharmacol.*, 2003, **63**, 524.
45. E.J. Squires, T. Sueyoshi and M. Negishi, *J. Biol. Chem.*, 2004, **279**, 49307.
46. F. Rencurel, A. Stenhouse, S.A. Hawley, T. Friedberg, D.G. Hardie, C. Sutherland and C.R. Wolf, *J. Biol. Chem.*, 2005, **280**, 4367.
47. F. Rencurel, M. Foretz, M.R. Kaufmann, D. Stroka, R. Looser, I. Leclerc, G. da Silva Xavier, G.A. Rutter, B. Viollet, U.A. Meyer, *Mol. Pharmacol.*, 2006, **70**, 1925.
48. S. Shindo, S. Numazawa and T. Yoshida, *Biochem. J.*, 2007, **401**, 735.

49. Y. Yamamoto, T. Kawamoto and M. Negishi, *Arch. Biochem. Biophys.*, 2003, **409**, 207.
50. X. Ding and J.L. Staudinger, *Biochem. Pharmacol.*, 2005, **69**, 867.
51. X. Ding and J.L. Staudinger, *J. Pharmacol. Exp. Ther.*, 2005, **312**, 849.
52. J.M. Pascussi, M. Busson-Le Coniat, P. Maurel and M.J. Vilarem, *Mol. Endocrinol.*, 2003, **17**, 42.
53. J.M. Pascussi, S. Gerbal-Chaloin, L. Pichard-Garcia, M. Daujat, J.M. Fabre, P. Maurel and M.J. Vilarem, *Biochem. Biophys. Res. Commun.*, 2000, **274**, 707.
54. E.G. Schuetz, W. Schmid, G. Schutz, C. Brimer, K. Yasuda, T. Kamataki, L. Bornheim, K. Myles and T.J. Cole, *Drug Metab. Dispos.*, 2000, **28**, 268.
55. C.C. Li, C.K. Lii, K.L. Liu, J.J. Yang and H.W. Chen, *J. Nutr. Biochem.*, 2006, **17**, 707.
56. S. Kakizaki, S. Karami and M. Negishi, *Drug Metab. Dispos.*, 2002, **30**, 208.
57. H. Yamada, T. Yamaguchi and K. Oguri, *Biochem. Biophys. Res. Commun.*, 2000, **277**, 66.
58. I. Tzameli, S.S. Chua, B. Cheskis and D.D. Moore, *Nucl. Recept.*, 2003, **1**, 2.
59. K.J. Woodcroft and R.F. Novak, *Chem. Biol. Interact.*, 1997, **107**, 75.
60. S. Kodama, C. Koike, M. Negishi and Y. Yamamoto, *Mol. Cell Biol.*, 2004, **24**, 7931.
61. G. Choe, S. Horvath, T.F. Cloughesy, K. Crosby, D. Seligson, A. Palotie, L. Inge, B.L. Smith, C.L. Sawyers and P.S. Mischel, *Cancer Res.*, 2003, **63**, 2742.
62. T.N. Stitt, D. Drujan, B.A. Clarke, F. Panaro, Y. Timofeyva, W.O. Kline, M. Gonzalez, G.D. Yancopoulos and D.J. Glass, *Mol. Cell*, 2004, **14**, 395.
63. W. Zhang, S. Patil, B. Chauhan, S. Guo, D.R. Powell, J. Le, A. Klotsas, R. Matika, X. Xiao, R. Franks, K.A. Heidenreich, M.P. Sajan, R.V. Farese, D.B. Stolz, P. Tso, S.H. Koo, M. Montminy and T.G. Unterman, *J. Biol. Chem.*, 2006, **281**, 10105.
64. B.M. Burgering and P.J. Coffer, *Nature*, 1995, **376**, 599.
65. P.A. Staubs, J.G. Nelson, D.R. Reichart and J.M. Olefsky, *J. Biol. Chem.*, 1998, **273**, 25139.
66. J.C. Yoon, P. Puigserver, G. Chen, J. Donovan, Z. Wu, J. Rhee, G. Adelmant, J. Stafford, C.R. Kahn, D.K. Granner, C.B. Newgard and B.M. Spiegelman, *Nature.*, 2001, **413**, 131.
67. T. Shiraki, N. Sakai, E. Kanaya and H. Jingami, *J. Biol. Chem.*, 2003, **278**, 11344.
68. S. Bhalla, C. Ozalp, S. Fang, L. Xiang and J.K. Kemper, *J. Biol. Chem.*, 2004, **279**, 45139.
69. P. Puigserver, G. Adelmant, Z. Wu, M. Fan, J. Xu, B. O'Malley and B.M. Spiegelman, *Science*, 1999, **286**, 1368.
70. J. Rhee, Y. Inoue, J.C. Yoon, P. Puigserver, M. Fan, F.J. Gonzalez and B.M. Spiegelman, *Proc. Natl. Acad. Sci. USA.*, 2003, **100**, 4012.
71. J. Miao, S. Fang, Y. Bae, J.K. Kemper, *J. Biol. Chem.,* 2006, **281**, 14537.

72. T. Li and J.Y. Chiang, *Am. J. Physiol. Gastrointest. Liver Physiol.*, 2005, **288**, G74.
73. A. Hammarstedt, P.A. Jansson, C. Wesslau, X. Yang and U. Smith, *Biochem. Biophys. Res. Commun.*, 2003, **301**, 578.
74. L.F. Michael, Z. Wu, R.B. Cheatham, P. Puigserver, G. Adelmant, J.J. Lehman, D.P. Kelly and B.M. Spiegelman, *Proc. Natl. Acad. Sci. USA.*, 2001, **98**, 3820.
75. S.H. Koo, H. Satoh, S. Herzig, C.H. Lee, S. Hedrick, R. Kulkarni, R.M. Evans, J. Olefsky and M. Montminy, *Nat. Med.*, 2004, **10**, 530.
76. J. Aubrecht, K.I. Hirsch-Ernst, V. Becker-Rabbenstein, G.F. Kahl, H. Taniguchi and M.W. Hohne, *Biochem. Pharmacol.*, 1995, **50**, 781.
77. F. Joannard, M. Rissel, D. Gilot, A. Anderson, L. Orfila-Lefeuvre, A. Guillouzo, A. Atfi and D. Lagadic-Gossmann, *Toxicol. Lett.*, 2006, **161**, 61.
78. D. Bauer, N. Wolfram, G.F. Kahl and K.I. Hirsch-Ernst, *Mol. Pharmacol.*, 2004, **65**, 172.
79. D.S. Riddick, C. Lee, A. Bhathena, Y.E. Timsit, P.Y. Cheng, E.T. Morgan, R.A. Prough, S.L. Ripp, K.K. Miller, A. Jahan and J.Y. Chiang, *Drug Metab. Dispos.*, 2004, **32**, 367.
80. X. Gu, S. Ke, D. Liu, T. Sheng, P.E. Thomas, A.B. Rabson, M.A. Gallo, W. Xie, Y. Tian, *J. Biol. Chem.,* 2006, **281**, 17882.
81. B. Hahnemann, P. Salonpaa, M. Pasanen, J. Maenpaa, P. Honkakoski, R. Juvonen, M.A. Lang, O. Pelkonen and H. Raunio, *Biochem. J.*, 1992, **286**, 289.
82. P. Honkakoski and M.A. Lang, *Arch. Biochem. Biophys.*, 1989, **273**, 42.
83. M.T. Donato, P. Viitala, C. Rodriguez-Antona, A. Lindfors, J.V. Castell, H. Raunio, M.J. Gomez-Lechon and O. Pelkonen, *Drug Metab. Dispos.*, 2000, **28**, 1321.
84. A. Madan, R.A. Graham, K.M. Carroll, D.R. Mudra, L.A. Burton, L.A. Krueger, A.D. Downey, M. Czerwinski, J. Forster, M.D. Ribadeneira, L.S. Gan, E.L. LeCluyse, K. Zech, P. Robertson Jr., P. Koch, L. Antonian, G. Wagner, L. Yu and A. Parkinson, *Drug Metab. Dispos.*, 2003, **31**, 421.
85. R.J. Edwards, R.J. Price, P.S. Watts, A.B. Renwick, J.M. Tredger, A.R. Boobis and B.G. Lake, *Drug Metab. Dispos.*, 2003, **31**, 282.
86. M. Oscarson, U.M. Zanger, O.F. Rifki, K. Klein, M. Eichelbaum and U.A. Meyer, *Clin. Pharmacol. Ther.*, 2006, **80**, 440.
87. J.M. Maglich, C.M. Stoltz, B. Goodwin, D. Hawkins-Brown, J.T. Moore and S.A. Kliewer, *Mol. Pharmacol.*, 2002, **62**, 638.
88. P. Wei, J. Zhang, D.H. Dowhan, Y. Han and D.D. Moore, *Pharmacogenomics J.*, 2002, **2**, 117.
89. J.M. Rosenfeld, R. Vargas Jr., W. Xie and R.M. Evans, *Mol. Endocrinol.*, 2003, **17**, 1268.
90. M. Itoh, M. Nakajima, E. Higashi, R. Yoshida, K. Nagata, Y. Yamazoe and T. Yokoi, *J. Pharmacol. Exp. Ther.*, 2006, **319**, 693.
91. K. Aida and M. Negishi, *Biochemistry.*, 1991, **30**, 8041.
92. O. Geneste, F. Raffalli and M.A. Lang, *Biochem. J.*, 1996, **313**, 1029.

93. M. Thulke-Gross, M. Hergenhahn, A. Tilloy-Ellul, M. Lang and H. Bartsch, *Biochem. J.*, 1998, **331**, 473.
94. A. Tilloy-Ellul, F. Raffalli-Mathieu and M.A. Lang, *Biochem. J.*, 1999, **339**, 695.
95. F. Raffalli-Mathieu, T. Glisovic, Y. Ben-David and M.A. Lang, *Mol. Pharmacol.*, 2002, **61**, 795.
96. T. Sueyoshi, T. Kawamoto, I. Zelko, P. Honkakoski and M. Negishi, *J. Biol. Chem.*, 1999, **274**, 6043.
97. P. Wei, J. Zhang, M. Egan-Hafley, S. Liang and D.D. Moore, *Nature*, 2000, **407**, 920.
98. A. Ueda, H.K. Hamadeh, H.K. Webb, Y. Yamamoto, T. Sueyoshi, C.A. Afshari, J.M. Lehmann and M. Negishi, *Mol. Pharmacol.*, 2002, **61**, 1.
99. D. Blizard, T. Sueyoshi, M. Negishi, S.S. Dehal and D. Kupfer, *Drug Metab. Dispos.*, 2001, **29**, 781.
100. X.C. Kretschmer and W.S. Baldwin, *Chem. Biol. Interact.*, 2005, **155**, 111.
101. I. Zelko, T. Sueyoshi, T. Kawamoto, R. Moore and M. Negishi, *Mol. Cell. Biol.*, 2001, **21**, 2838.
102. I. Tzameli, P. Pissios, E.G. Schuetz and D.D. Moore, *Mol. Cell. Biol.*, 2000, **20**, 2951.
103. K. Kobayashi, Y. Yamanaka, N. Iwazaki, I. Nakajo, M. Hosokawa, M. Negishi and K. Chiba, *Drug Metab. Dispos.*, 2005, **33**, 924.
104. J.M. Maglich, D.J. Parks, L.B. Moore, J.L. Collins, B. Goodwin, A.N. Billin, C.A. Stoltz, S.A. Kliewer, M.H. Lambert, T.M. Willson and J.T. Moore, *J. Biol. Chem.*, 2003, **278**, 17277.
105. W. Xie, J.L. Barwick, C.M. Simon, A.M. Pierce, S. Safe, B. Blumberg, P.S. Guzelian and R.M. Evans, *Genes Dev.*, 2000, **14**, 3014.
106. B. Goodwin, L.B. Moore, C.M. Stoltz, D.D. McKee and S.A. Kliewer, *Mol. Pharmacol.*, 2001, **60**, 427.
107. H. Wang, S.R. Faucette, D. Gilbert, S.L. Jolley, T. Sueyoshi, M. Negishi and E.L. LeCluyse, *Drug Metab. Dispos.*, 2003, **31**, 620.
108. H. Wang, S. Faucette, T. Sueyoshi, R. Moore, S. Ferguson, M. Negishi and E.L. LeCluyse, *J. Biol. Chem.*, 2003, **278**, 14146.
109. S. Gerbal-Chaloin, M. Daujat, J.M. Pascussi, L. Pichard-Garcia, M.J. Vilarem and P. Maurel, *J. Biol. Chem.*, 2002, **277**, 209.
110. J.L. Raucy, L. Mueller, K. Duan, S.W. Allen, S. Strom and J.M. Lasker, *J. Pharmacol. Exp. Ther.*, 2002, **302**, 475.
111. S.S. Ferguson, Y. Chen, E.L. LeCluyse, M. Negishi and J.A. Goldstein, *Mol. Pharmacol.*, 2005, **68**, 747.
112. J.O. Miners and D.J. Birkett, *Br. J. Clin. Pharmacol.*, 1998, **45**, 525.
113. M. Niemi, J.T. Backman, M.F. Fromm, P.J. Neuvonen and K.T. Kivisto, *Clin. Pharmacokinet.*, 2003, **42**, 819.
114. Y. Chen, G. Kissling, M. Negishi and J.A. Goldstein, *J. Pharmacol. Exp. Ther.*, 2005, **314**, 1125.
115. S. Gerbal-Chaloin, J.M. Pascussi, L. Pichard-Garcia, M. Daujat, F. Waechter, J.M. Fabre, N. Carrere and P. Maurel, *Drug Metab. Dispos.*, 2001, **29**, 242.

116. Y. Chen, S.S. Ferguson, M. Negishi and J.A. Goldstein, *Mol. Pharmacol.*, 2003, **64**, 316.
117. J.P. Jackson, S.S. Ferguson, M. Negishi and J.A. Goldstein, *Drug Metab. Dispos.*, 2006, **34**, 2003.
118. J.P. Jackson, S.S. Ferguson, R. Moore, M. Negishi and J.A. Goldstein, *Mol. Pharmacol.*, 2004, **65**, 1397.
119. M. Baader, C. Gnerre, J.J. Stegeman and U.A. Meyer, *J. Biol. Chem.*, 2002, **277**, 15647.
120. C. Handschin, M. Podvinec and U.A. Meyer, *Proc. Natl. Acad. Sci. USA.*, 2000, **97**, 10769.
121. C.S. Lieber, *Physiol. Rev.*, 1997, **77**, 517.
122. P.Y. Cheng and E.T. Morgan, *Curr. Drug Metab.*, 2001, **2**, 165.
123. J. Zhang, W. Huang, S.S. Chua, P. Wei and D.D. Moore, *Science.*, 2002, **298**, 422.
124. K.K. Wolf, S.G. Wood, J.A. Hunt, B.W. Walton-Strong, K. Yasuda, L. Lan, S.X. Duan, Q. Hao, S.A. Wrighton, E.H. Jeffery, R.M. Evans, J.G. Szakacs, L.L. von Moltke, D.J. Greenblatt, M.H. Court, E.G. Schuetz, P.R. Sinclair and J.F. Sinclair, *Drug Metab. Dispos.*, 2005, **33**, 1827.
125. Y. Yamazaki, S. Kakizaki, N. Horiguchi, H. Takagi, M. Mori and M. Negishi, *World J. Gastroenterol.*, 2005, **11**, 5966.
126. J.M. Lehmann, D.D. McKee, M.A. Watson, T.M. Willson, J.T. Moore and S.A. Kliewer, *J. Clin. Invest.*, 1998, **102**, 1016.
127. W. Xie, J.L. Barwick, M. Downes, B. Blumberg, C.M. Simon, M.C. Nelson, B.A. Neuschwander-Tetri, E.M. Brunt, P.S. Guzelian and R.M. Evans, *Nature.*, 2000, **406**, 435.
128. J.L. Staudinger, B. Goodwin, S.A. Jones, D. Hawkins-Brown, K.I. MacKenzie, A. LaTour, Y. Liu, C.D. Klaassen, K.K. Brown, J. Reinhard, T.M. Willson, B.H. Koller and S.A. Kliewer, *Proc. Natl. Acad. Sci. USA.*, 2001, **98**, 3369.
129. L.C. Quattrochi, A.S. Mills, J.L. Barwick, C.B. Yockey and P.S. Guzelian, *J. Biol. Chem.*, 1995, **270**, 28917.
130. J.M. Huss, S.I. Wang, A. Astrom, P. McQuiddy and C.B. Kasper, *Proc. Natl. Acad. Sci. USA.*, 1996, **93**, 4666.
131. J.L. Barwick, L.C. Quattrochi, A.S. Mills, C. Potenza, R.H. Tukey and P.S. Guzelian, *Mol. Pharmacol.*, 1996, **50**, 10.
132. S.A. Jones, L.B. Moore, J.L. Shenk, G.B. Wisely, G.A. Hamilton, D.D. McKee, N.C. Tomkinson, E.L. LeCluyse, M.H. Lambert, T.M. Willson, S.A. Kliewer and J.T. Moore, *Mol. Endocrinol.*, 2000, **14**, 27.
133. S.A. Wrighton, E.G. Schuetz, P.B. Watkins, P. Maurel, J. Barwick, B.S. Bailey, H.T. Hartle, B. Young and P. Guzelian, *Mol. Pharmacol.*, 1985, **28**, 312.
134. T.A. Kocarek, E.G. Schuetz, S.C. Strom, R.A. Fisher and P.S. Guzelian, *Drug Metab. Dispos.*, 1995, **23**, 415.
135. B. Goodwin, E. Hodgson and C. Liddle, *Mol. Pharmacol.*, 1999, **56**, 1329.

136. R.G. Tirona, W. Lee, B.F. Leake, L.B. Lan, C.B. Cline, V. Lamba, F. Parviz, S.A. Duncan, Y. Inoue, F.J. Gonzalez, E.G. Schuetz and R.B. Kim, *Nat. Med.*, 2003, **9**, 220.
137. G.R. Robertson, J. Field, B. Goodwin, S. Bierach, M. Tran, A. Lehnert and C. Liddle, *Mol. Pharmacol.*, 2003, **64**, 42.
138. W. Noracharttiyapot, Y. Nagai, T. Matsubara, M. Miyata, M. Shimada, K. Nagata and Y. Yamazoe, *Drug Metab. Pharmacokinet.*, 2006, **21**, 99.
139. J.M. Pascussi, Y. Jounaidi, L. Drocourt, J. Domergue, C. Balabaud, P. Maurel and M.J. Vilarem, *Biochem. Biophys. Res. Commun.*, 1999, **260**, 377.
140. G. Bertilsson, A. Berkenstam and P. Blomquist, *Biochem. Biophys. Res. Commun.*, 2001, **280**, 139.
141. C.A. Vyhlidal, R. Gaedigk and J.S. Leeder, *Drug Metab. Dispos.*, 2006, **34**, 131.
142. A.E. Simpson, *Gen. Pharmacol.*, 1997, **28**, 351.
143. M.E. Wyde, S.E. Kirwan, F. Zhang, A. Laughter, H.B. Hoffman, E. Bartolucci-Page, K.W. Gaido, B. Yan and L. You, *Toxicol. Sci.*, 2005, **86**, 281.
144. U. Savas, M.H. Hsu and E.F. Johnson, *Arch. Biochem. Biophys.*, 2003, **409**, 212.
145. F.M. Sladek, *Receptor*, 1993, **3**, 223.
146. T.E. Akiyama and F.J. Gonzalez, *Biochim. Biophys. Acta.*, 2003, **1619**, 223.
147. R. Jover, R. Bort, M.J. Gomez-Lechon and J.V. Castell, *Hepatology*, 2001, **33**, 668.
148. H.M. Wortelboer, C.A. de Kruif, A.A. van Iersel, H.E. Falke, J. Noordhoek and B.J. Blaauboer, *Biochem. Pharmacol.*, 1991, **42**, 381.
149. R.C. Zangar, K.J. Woodcroft, T.A. Kocarek and R.F. Novak, *Drug Metab. Dispos.*, 1995, **23**, 681.
150. J.C. Corton, L.Q. Fan, S. Brown, S.P. Anderson, C. Bocos, R.C. Cattley, A. Mode and J.A. Gustafsson, *Mol. Pharmacol.*, 1998, **54**, 463.
151. L.Q. Fan, H. Brown-Borg, S. Brown, S. Westin, A. Mode and J.C. Corton, *Toxicology*, 2004, **203**, 41.
152. S.L. Ripp, K.C. Falkner, M.L. Pendleton, V. Tamasi and R.A. Prough, *Mol. Pharmacol.*, 2003, **64**, 113.
153. P. Schmiedlin-Ren, K.E. Thummel, J.M. Fisher, M.F. Paine, K.S. Lown and P.B. Watkins, *Mol. Pharmacol.*, 1997, **51**, 741.
154. K.E. Thummel, C. Brimer, K. Yasuda, J. Thottassery, T. Senn, Y. Lin, H. Ishizuka, E. Kharasch, J. Schuetz and E. Schuetz, *Mol. Pharmacol.*, 2001, **60**, 1399.
155. L. Drocourt, J.C. Ourlin, J.M. Pascussi, P. Maurel and M.J. Vilarem, *J. Biol. Chem.*, 2002, **277**, 25125.
156. M. Makishima, T.T. Lu, W. Xie, G.K. Whitfield, H. Domoto, R.M. Evans, M.R. Haussler and D.J. Mangelsdorf, *Science*, 2002, **296**, 1313.
157. C. Gnerre, S. Blattler, M.R. Kaufmann, R. Looser and U.A. Meyer, *Pharmacogenetics*, 2004, **14**, 635.

158. W. Xie, A. Radominska-Pandya, Y. Shi, C.M. Simon, M.C. Nelson, E.S. Ong, D.J. Waxman and R.M. Evans, *Proc. Natl. Acad. Sci. USA.*, 2001, **98**, 3375.
159. E.G. Schuetz, S. Strom, K. Yasuda, V. Lecureur, M. Assem, C. Brimer, J. Lamba, R.B. Kim, V. Ramachandran, B.J. Komoroski, R. Venkataramanan, H. Cai, C.J. Sinal, F.J. Gonzalez and J.D. Schuetz, *J. Biol. Chem.*, 2001, **276**, 39411.
160. G.L. Guo, G. Lambert, M. Negishi, J.M. Ward, H.B. Brewer Jr, S.A. Kliewer, F.J. Gonzalez and C.J. Sinal, *J. Biol. Chem.*, 2003, **278**, 45062.
161. H.U. Marschall, M. Wagner, K. Bodin, G. Zollner, P. Fickert, J. Gumhold, D. Silbert, A. Fuchsbichler, J. Sjovall and M. Trauner, *J. Lipid. Res.*, 2006, **47**, 582.
162. S.D. Shenoy, T.A. Spencer, N.A. Mercer-Haines, M. Alipour, M.D. Gargano, M. Runge-Morris and T.A. Kocarek, *Drug Metab. Dispos.*, 2004, **32**, 66.
163. C. Gnerre, G.U. Schuster, A. Roth, C. Handschin, L. Johansson, R. Looser, P. Parini, M. Podvinec, K. Robertsson, J.A. Gustafsson and U.A. Meyer, *J. Lipid Res.*, 2005, **46**, 1633.
164. D.W. Nebert and D.W. Russell, *Lancet*, 2002, **360**, 1155.
165. R. Sirianni, J.B. Seely, G. Attia, D.M. Stocco, B.R. Carr, V. Pezzi and W.E. Rainey, *J. Endocrinol.*, 2002, **174**, R13.
166. M.H. Bassett, P.C. White and W.E. Rainey, *Endocrine Res.*, 2004, **30**, 567.
167. L. You, *Chem. Biol. Interact.*, 2004, **147**, 233.
168. G. Feuer, *Drug Metab. Rev.*, 1983, **14**, 1119.
169. Y. Zhai, H.V. Pai, J. Zhou, J.A. Amico, R.R. Vollmer and W. Xie, *Mol. Endocrinol.*, 2007, **21**, 138.
170. M.J. Morrell, *Epilepsia*, 2003, **44**(Suppl 4), 11.
171. G.D. Anderson, *Neurology*, 2004, **63**, S3.
172. Y. Kaneshiro, Y. Oda, K. Iwakiri, T. Masada, H. Iwaki, Y. Hirota, K. Kondo and K. Takaoka, *Clin. Pharmacol. Ther.*, 2006, **80**, 396.
173. J.M. Pascussi, A. Robert, M. Nguyen, O. Walrant-Debray, M. Garabedian, P. Martin, T. Pineau, J. Saric, F. Navarro, P. Maurel and M.J. Vilarem, *J. Clin. Invest.*, 2005, **115**, 177.
174. Y. Xu, T. Hashizume, M.C. Shuhart, C.L. Davis, T.F. Kalhorn, P.B. Watkins, E.G. Schuetz and K.E. Thummel, *Mol. Pharmacol.*, 2006, **69**, 56.
175. T. Li, W. Chen, J.Y. Chiang, *J. Lipid Res.* 2006, in press.
176. J.J. Eloranta and G.A. Kullak-Ublick, *Arch. Biochem. Biophys.*, 2005, **433**, 397.
177. G. Zollner, H.-U. Marschall, M. Wagner and M. Trauner, *Mol. Pharm.*, 2006, **3**, 231.
178. I. Dussault, H.D. Yoo, M. Lin, E. Wang, M. Fan, A.K. Batta, G. Salen, S.K. Erickson and B.M. Forman, *Proc. Natl. Acad. Sci. USA.*, 2003, **100**, 833.
179. B. Goodwin, S.A. Jones, R.R. Price, M.A. Watson, D.D. McKee, L.B. Moore, C. Galardi, J.G. Wilson, M.C. Lewis, M.E. Roth, P.R. Maloney, T.M. Willson and S.A. Kliewer, *Mol. Cell*, 2000, **6**, 517.

180. E. De Fabiani, N. Mitro, A.C. Anzulovich, A. Pinelli, G. Galli and M. Crestani, *J. Biol. Chem.*, 2001, **276**, 30708.
181. T. Li and J.Y. Chiang, *Am. J. Physiol.*, 2005, **88**, G74.
182. W. Xie, M.F. Yeuh, A. Radominska-Pandya, S.P. Saini, M. Negishi, B.S. Bottroff, G.Y. Cabrera, R.H. Tukey and R.M. Evans, *Proc. Natl. Acad. Sci. USA.*, 2003, **100**, 4150.
183. T.A. Kocarek and N.A. Mercer-Haines, *Mol. Pharmacol.*, 2002, **62**, 1177.
184. K. Swales, S. Kakizaki, Y. Yamamoto, K. Inoue, K. Kobayashi and M. Negishi, *J. Biol. Chem.*, 2005, **280**, 3458.
185. K. Inoue, C.H. Borchers and M. Negishi, *Biochem. J.*, 2006, **398**, 125.
186. C. Koike, R. Moore and M. Negishi, *FEBS Lett.*, 2005, **579**, 6733.
187. S. Kodama, C. Koike, M. Negishi and Y. Yamamoto, *Mol. Cell. Biol.*, 2004, **24**, 7931.
188. K. Nakamura, R. Moore, M. Negishi, T. Sueyoshi. 2007, *J. Biol. Chem.* 2007, **282**, 9768.
189. J. Zhou, Y. Zhai, Y. Mu, H. Gong, H. Uppal, D. Toma, S. Ren, R.M. Evans and W. Xie, *J. Biol. Chem.*, 2006, **281**, 15013.
190. P. Wei, J. Zhang, M. Egan-Hafley, S. Liang and D.D. Moore, *Nature*, 2000, **407**, 920.
191. J. Staudinger, Y. Liu, A. Madan, S. Habeebu and C.D. Klaassen, *Drug, Metab. Dispos.*, 2001, **29**, 1467.
192. Y. Yamamoto, R. Moore, T.L. Goldsworthy, M. Negishi and R.R. Maronpot, *Cancer Res.*, 2004, **64**, 7197.
193. W. Huang, J. Zhang, M. Washington, J. Liu, J.M. Parant, G. Lorzano and D.D. Moore, *Mol. Endocrinol.*, 2005, **19**, 1646.
194. C. Zhou, M.M. Tabb, E.L. Nelson, F. Grun, S. Verma, A. Sadatrafiei, M. Lin, S. Mallick, B.M. Forman, K.E. Thummel and B. Blumberg, *J. Clin. Invest.*, 2006, **116**, 2280.

CHAPTER 14

Modulation of Cytochromes P450 by Phytochemicals

MICHAEL MURRAY

Pharmacogenomics and Drug Development Group, Faculty of Pharmacy, University of Sydney, NSW 2006, Australia

Table of Contents

Issues in Toxicology
Cytochromes P450: Role in the Metabolism and Toxicity of Drugs and other Xenobiotics
Edited by Costas Ioannides

14.1 Introduction

14.1.1 Background to the Effects of Phytochemicals on CYP Enzymes

The effects of human exposure to phytochemicals are of increasing interest. Many herbal preparations are now being used extensively as alternative therapies even though their adverse effects and their potential for interaction with conventional drugs during concurrent administration remain under-explored. This deficiency in important information that would assist drug experts and practitioners to use herbal agents more effectively is beginning to be addressed in a systematic way. There is also some evidence that plant-derived chemicals may be potential anti-cancer therapies. Based initially on anecdotal reports, epidemiological studies have added support to this concept, which has led to more focussed clinical and experimental studies.

As indicated in previous chapters, cytochromes P450 (CYPs) are pivotal enzymes in the oxidative biotransformation of xenobiotics, including drugs and carcinogens. Drugs must be sufficiently lipophilic to be absorbed and transported to the site of their therapeutic effect. A major function of CYPs is to convert these agents to more polar products, which usually have diminished pharmacological activity and are more readily excreted. In consequence, CYPs are critical determinants of the duration of drug action. CYPs also oxidise relatively inert polycyclic aromatic hydrocarbons (PAH), food-derived heterocyclic amines, chemicals in tobacco pyrolysates and nitrosamines to highly reactive and toxic species that bind covalently to DNA bases, giving rise to genotoxicity. Because of these important roles of CYPs in drug clearance and chemical carcinogenesis there has been increasing interest in the potential of phytochemicals to modulate CYP enzyme function and gene expression. Indeed, epidemiological studies have suggested that the composition of the diet may influence tumour progression. Thus, diets high in fruits and cruciferous vegetables have been associated with decreased development of prostate and breast cancer.[1,2]

14.1.2 Mechanisms of CYP Inhibition

Factors that modulate the function of particular CYPs would be expected to influence the duration of drug action and the rate at which chemical carcinogens undergo bioactivation. Inhibition and induction of CYPs are two important processes by which phytochemicals may impair drug clearance and carcinogen activation. Inhibition can occur when two or more substrates compete for oxidation by a CYP enzyme or when a lipophilic non-substrate chemical interacts directly with the enzyme to prevent substrate oxidation.[3] These are reversible interactions in which the inhibitory agent or its metabolite may eventually diffuse out of the CYP catalytic site and restore activity.

In contrast, in irreversible inhibition the substrate is converted to a reactive intermediate that binds tightly to the CYP enzyme.[3] In mechanism-based inactivation a small group of substrates undergo CYP-dependent conversion to alkylating or arylating metabolites that are retained within the active site and irreversibly modify the CYP haem, apoprotein or both.[3,4] Autocatalytic inactivation involves the suicide processing of alkenes, alkynes, thionosulphur compounds and others to denaturing metabolites.[3,4] On the other hand, metabolite intermediate (MI)-complexation involves the generation of reactive species that bind tightly to the CYP haem without eliciting its destruction. Dissociation of the MI-complex can be achieved *in vitro* and results in the restoration of activity of the complexed CYP.[5] Thus, MI-complexation is mechanistically distinct from autocatalytic inactivation. CYP reactivation, however, is unlikely to occur *in vivo* because the MI-complex is quite stable and has been isolated intact in microsomal fractions from animals treated with MI-complex-forming chemicals.[6–8] Accordingly, restoration of MI-complexed CYP activity *in vivo* is dependent on the synthesis of a new enzyme. In principle, inhibition of CYP enzymes could be a useful strategy for decreasing the rate at which carcinogenic chemicals are activated to DNA-binding metabolites; phytochemicals that inhibit this process may be anti-carcinogens. However, it should be considered that such phytochemicals could also elicit pharmacokinetic drug-drug interactions because of the overlapping substrate specificities of certain CYPs toward carcinogens and drugs.

14.1.3 Altered Regulation of CYP Genes

CYP upregulation, or induction, occurs following the *in vivo* exposure of mammals to a wide range of lipophilic chemicals and increases CYP expression in tissues. CYP induction increases drug elimination capacity, which shortens drug half-lives and the duration of drug action, but may also mediate undesirable effects. Upregulation of CYPs involved in carcinogen activation may increase the formation of toxic products and enhance the resultant DNA damage. Phytochemicals that increase the expression of carcinogen-oxidising CYPs could similarly enhance DNA damage.

Detailed information is emerging on the complexities of mechanisms by which individual CYP genes are regulated. These mechanisms may be important in the maintenance of constitutive (basal) expression or could enable the upregulation of CYPs in response to chemical exposure (induction). In this chapter, the mechanisms involved in altered CYP regulation will be restricted to the interpretation of regulatory effects of specific phytochemicals. From such considerations it is possible to gain insight into how phytochemicals influence CYP expression in tissues.

The previous chapters in Part B of this book have documented the roles of particular CYPs in pathways of drug oxidation. It has also been shown that CYP1A, 1B and 3A are prominent in the conversion of polycyclic aromatic hydrocarbons (PAH), such as benzo[a]pyrene (BaP), to electrophilic diolepoxide metabolites.[9] CYP1A2 activates food-derived mutagens and aromatic amines to reactive *N*-hydroxy intermediates, and CYP2A enzymes activate tobacco pyrolysate products like 4-(methylnitrosamino)-1-(3-pyridyl)-1-butanone (NNK). The ethanol-inducible CYP2E1 oxidises numerous chemicals, including nitrosamines and pyrrolizidine alkaloids, to reactive intermediates that mediate toxicity.[9]

Although these CYPs are central to the activation of carcinogenic chemicals, the phase II conjugation enzymes, such as the UDP-glucuronosyltransferases, sulphotransferases and glutathione *S*-transferases, also influence the extent to which these reactive species bind to DNA. Certain dietary chemicals have been shown to activate the expression of genes that encode phase II enzymes and so may increase detoxification capacity and prevent DNA damage.[10] Dietary agents may also augment other cytoprotective factors in cells by acting as antioxidants or by promoting the synthesis of glutathione.[11] In assessing the potential anti-carcinogen benefits of phytochemicals it is essential to consider these additional factors unrelated to interactions with CYP genes.

Thousands of different chemicals have been identified in plants. Detailed information is emerging on the interactions of certain phytochemicals with CYPs. This chapter will focus on this information, with an emphasis on polyphenols, methylenedioxyphenyl compounds, psoralens, the crucifer-derived indoles and isothiocyanates and related sulphur-containing molecules.

14.2 Polyphenols

A wide variety of plants and plant products, including fruits, vegetables, grains, legumes, teas and wine contain polyphenols. Polyphenols are plant metabolites that possess one or more aromatic rings which are substituted with varying numbers of hydroxyl groups. Natural polyphenols range from simple molecules, such as phenolic acids, to polymers, such as the complex tannins. The flavonoids are the most widely studied plant polyphenols. Structurally, the flavonoids possess two aromatic rings that are linked by an oxygen-containing heterocyclic system (Table 14.1 shows the structures of commonly studied flavonoids).

14.2.1 Effects of Polyphenols on the Activation of Genotoxic Chemicals by CYPs

14.2.1.1 Modulation of CYP Activity by Flavonoids

Fruit and vegetable extracts have been shown to inhibit the genotoxic potential of food-derived mutagens and related chemicals.[12] Indeed, direct testing of citrus flavonoids has demonstrated their capacity to impair the formation of DNA adducts with BaP metabolites.[13,14] This is consistent with the inhibitory effects of flavonoids on CYP1A1-mediated BaP hydroxylation activity, which generates genotoxic products.[13,15] Indeed, the flavonoid quercetin has been shown to directly inhibit the epoxidation of (±)-*trans*-7,8-dihydro-7,8-dihydroxy-BaP to the diolepoxide 2, which ultimately binds to DNA.[16] Similarly, DNA damage induced by food-derived mutagens in cells containing over-expressed CYP1A enzymes was inhibited by chrysin, apigenin and quercetin.[17] Other CYPs with roles in carcinogen activation are also inhibited by flavonoids. Thus, naringenin and quercetin inhibited CYP2A-mediated α-hydroxylation of the tobacco carcinogen NNK.[18]

Some of the structural requirements for CYP1A inhibition by flavonoids have been described. Thus, in rat liver microsomes the more planar flavones and flavonols are generally more potent than the corresponding flavanones, which are not planar.[19] A 4-keto substituent in the flavone C-ring appears to be necessary for inhibitory potency, and a double bond between carbons 2 and 3 of the C ring and an hydroxyl group at C3 are also important (Figure 14.1). In contrast, hydroxyl substitution around the B ring, except at C4′, decreased inhibition. Further structural information for the inhibition of BaP hydroxylation in human liver by flavonoids has also been published.[20] Quercetin, kaempferol, morin and chrysin (Table 14.1) inhibited BaP hydroxylation, at least in part by impairing electron transfer to CYP via the flavoprotein NADPH-CYP-reductase. In contrast, the more hydrophobic analogues flavone, nobiletin, tangeretin and 7,8-benzoflavone (Table 14.1) enhanced the metabolism of BaP and aflatoxin B_1 to mutagenic products *in vitro*. The *in vivo* consequences of CYP activation by flavonoids that has been observed *in vitro* have not really been established, but it is conceivable that the rate of mutagen activation may be increased. However, for this to occur, appropriate concentrations of the flavonoid and carcinogen substrate would have to be present simultaneously within the enzyme active site.

Differences in the capacity of flavonoids to inhibit different CYPs that mediate carcinogen activation have been noted. Thus, Chaudhary and Willett found that CYP1B1 activity was more susceptible to inhibition than CYP1A1

Table 14.1 Structures of commonly occurring flavonoids.

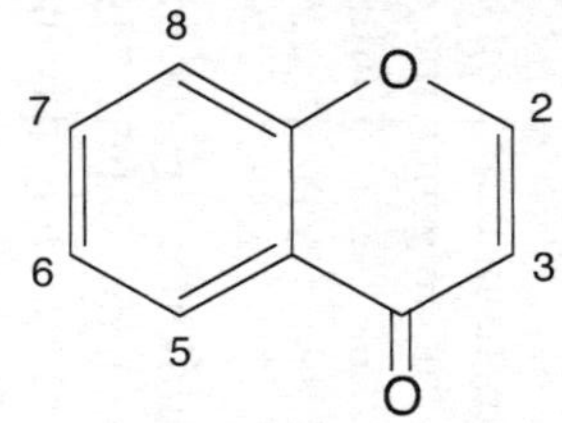

FLAVONE DERIVATIVES

Name	Substituents 2	3	5	6	7	8
Acacetin	4′-methoxyphenyl		OH		OH	
Apigenin	4′-hydroxyphenyl		OH		OH	
Baicalein	phenyl		OH	OH	OH	
Biochanin A		4′-methoxyphenyl	OH		OH	
Chrysin	phenyl		OH		OH	
Chrysoeriol	4′-hydroxy-3′-methoxyphenyl		OH		OH	
Daidzein		4′-hydroxyphenyl			OH	
Diosmetin	3′-hydroxy-4′-methoxyphenyl		OH		OH	
Fisetin	3′,4′-dihydroxyphenyl	OH			OH	
Flavanol	phenyl	OH				
Flavone	phenyl				OH	
Formononetin		4′-methoxyphenyl			OH	
Galangin	phenyl	OH	OH		OH	
Glycitein		4′-hydroxyphenyl		OCH_3	OH	
Isoxanthohumol	4′-hydroxyphenyl		OCH_3		OH	isoprenyl
Kaempferide	4′-methoxyphenyl	OH	OH		OH	
Kaempferol	4′-hydroxyphenyl	OH	OH		OH	
Luteolin	3′,4′-dihydroxyphenyl		OH		OH	

Morin	3′,5′-dihydroxyphenyl	OH	OH		OH	
Myricetin	3′,4′, 5′-trihydroxyphenyl	OH	OH		OH	
Nobiletin	3′,4′-dimethoxyphenyl		OCH_3	OCH_3	OCH_3	OCH_3
Prunetin		4′-hydroxyphenyl	OH		OCH_3	
Quercetin	3′,4′-dihydroxyphenyl	OH	OH		OH	
Scutellarein	4′-hydroxyphenyl		OH	OH	OH	
Tangeretin	4′-methoxyphenyl		OCH_3	OCH_3	OCH_3	OCH_3

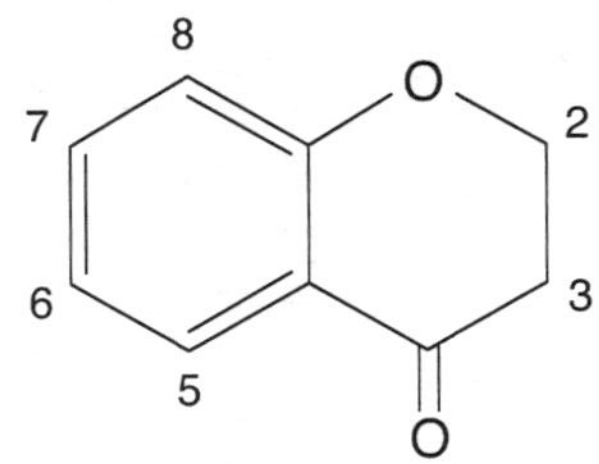

FLAVANONE DERIVATIVES

Eriodictyol	3′,4′-dihydroxyphenyl		OH			OH
Flavanone	4′-hydroxyphenyl	OH	OH			OH
Genistein		4′-hydroxyphenyl	OH			OH
Hesperitin	3′-hydroxy-4′-methoxyphenyl		OH			OH
Homoeriodictyol	4′-hydroxy-3′-methoxyphenyl		OH			OH
Naringenin	4′-hydroxyphenyl		OH			OH
Tamarixetin	3′-methoxy-4′-hydroxyphenyl	OH	OH			OH

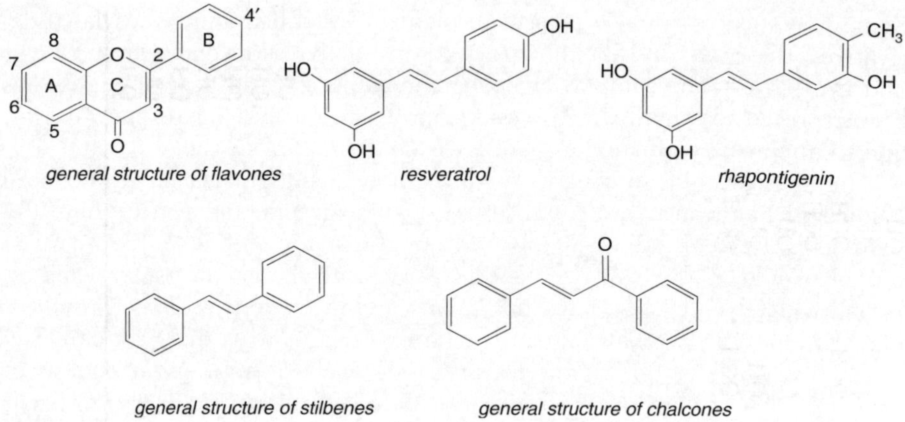

Figure 14.1 Structures of the polyphenols resveratrol and rhapontigenin and the general structures of flavones, stilbenes and chalcones.

activity.[21] Similarly, Doostdar *et al.*[22] found that morin, eriodictyol, hesperetin, acacetin, naringenin and homoeriodictyol (Table 14.1) inhibited CYP1B1 activity effectively, but only acacetin and diosmetin were significant inhibitors of CYP1A1; of the six flavonoids tested in that study only acacetin inhibited CYP1A2.

The interesting possibility has been raised that flavonoids may preferentially inhibit the oxidation of some but not all substrates of a particular CYP. Thus, α-naphthoflavone and flavone inhibited the *O*-dealkylation of two resorufin ethers by CYP1A1, but not the hydroxylation of BaP. It is feasible that flavonoids may impair the coordination of the smaller resorufin ethers within the active site of CYP1A1, but are less effective against the much larger BaP.[23] The requirements for inhibition of 2-amino-3-methylimidazo [4,5-f] quinoline (IQ) mutagenesis by flavonoids paralleled those for CYP1 inhibition and correlated with inhibition of CYPs that mediate IQ *N*-hydroxylation.[24]

14.2.1.2 Resveratrol and Structurally-related Stilbenes and Chalcones

The grape-derived phytoalexin resveratrol (Figure 14.1) is another polyphenol that has been evaluated for its antimutagenic potential. Resveratrol inhibited CYP1A activity,[25] BaP metabolite-DNA adduct formation[26] and skin tumourigenesis in 7,12-dimethylbenz(a)anthracene-treated mice *in vivo*.[27] Resveratrol may inhibit CYP1A2 by both reversible and irreversible mechanisms, with the latter likely to give rise to a more prolonged antimutagenic effect.[28]

A number of stilbenes and chalcones that resemble resveratrol were also found to be inhibitors of CYP1A-mediated ethoxyresorufin *O*-deethylation in microsomal fractions from human and rat liver.[29,30] Rhapontigenin (Figure 14.1) and other polyhydroxylated chalcones potently and selectively inhibited CYP1A1.[30]

Some polyphenols are also present in plants as esterified and glycosylated derivatives; these may also be inhibitory toward activities mediated by CYP1A1 and CYP1B1.[31] The complex chalcone curcumin, a constituent of cumin, has been reported to inhibit CYP1A and CYP2B1 in rat liver microsomes *in vitro* and to impair the induction of CYP1A by BaP *in vivo*.[32,33]

Tannins, catechins and other complex polyphenols in tea have also been studied for anticancer potential.[34–36] Epigallocatechin and epicatechin, and their gallate esters, inhibited the metabolic activation of BaP, food-derived mutagens and aflatoxin B_1 by human CYPs, with the gallate esters being especially effective.[36] Apart from decreased carcinogen activation as a result of CYP1 inhibition, the catechins also stimulated apoptosis in DNA-damaged cells. This provides further insight into the beneficial antitumour actions of catechins, since apoptosis is an important mechanism for the deletion of cells that contain damaged DNA.

It is now clear that polyphenols may directly inhibit the activities of CYPs that convert procarcinogens to genotoxic intermediates that form DNA adducts *in vitro* and in cells. Even though polyphenols are polar in nature, these polyphenols can enter cells to mediate these effects on carcinogen activation.

14.2.1.3 Inhibitory Pharmacokinetic Drug-Drug Interactions Involving Phytochemicals

Because several CYPs involved in carcinogen oxidation also participate in drug biotransformation it is not surprising that there have been reports of inhibitory drug interactions involving polyphenols. Probably the drug interaction of greatest clinical significance is that due to grapefruit juice phytochemicals. Ingestion of grapefruit juice can increase the serum concentrations of a number of coadministered CYP3A substrates. Early studies demonstrated that the citrus flavonoids naringenin and quercetin were effective inhibitors of CYP3A-dependent drug oxidation *in vitro*.[37,38] However, it now appears that these chemicals play only a minor role in the grapefruit juice interaction.[39] More recently, further *in vitro* studies indicated that flavonoids are also inhibitors of CYP1A2 and CYP2C9 in human liver microsomes.[40,41] Although the *in vivo* significance of these findings has not been fully established, it is noteworthy that CYP1A2 in particular is inhibited *in vivo* by flavonoids and other polyphenols.

There is further evidence that certain polyphenols may inhibit human CYPs that have prominent roles in drug elimination. Gallic acid is a reversible inhibitor of CYP3A, by a mechanism that appears to involve oxidation to the corresponding quinone radical.[42] The milk thistle flavolignans related to silymarin inhibited CYPs in rat liver microsomes[43] and were shown more recently to inhibit CYPs 3A in human liver.[44] The olive oil polyphenol oleuropein also inhibited CYP3A, apparently by mechanism-based inactivation.[45] Perhaps importantly, CYP1A2 and CYP2E1 were refractory to inhibition or inactivation by oleuropein. Resveratrol and the dimeric ε-viniferin are also inhibitors of drug metabolising CYPs.[46–48] Like oleuropein, resveratrol may be an inactivator of CYP3A, although this possibility remains to be substantiated.[49,50] *In vivo* studies have indicated that constituents of red wine inhibit CYP3A4 activity and increase the oral bioavailability of cisapride.[51] However, judging by the relatively low potency of resveratrol when tested directly, it is feasible that other red wine constituents may be the active inhibitory agents. Considered together, there is emerging evidence that polyphenols may inhibit drug metabolising CYPs. The oxidation of substrates of CYP1A2 and CYP-3A4 may be preferentially inhibited by these phytochemicals.

14.2.1.4 Phytochemicals and the Modulation of Biosynthetic CYPs

A number of CYPs are involved in the biosynthesis and degradation of lipophilic molecules of physiological importance, rather than xenobiotic biotransformation. CYP19 aromatase converts androgens to oestrogens by A-ring aromatisation (Figure 14.2) and was found to be inhibited by flavonoids *in vitro*.[52,53] Rotenone, a flavonoid that is present in tropical plants of the genus Lonchocorpus, potently inhibited aromatase activity.[54] Important structural

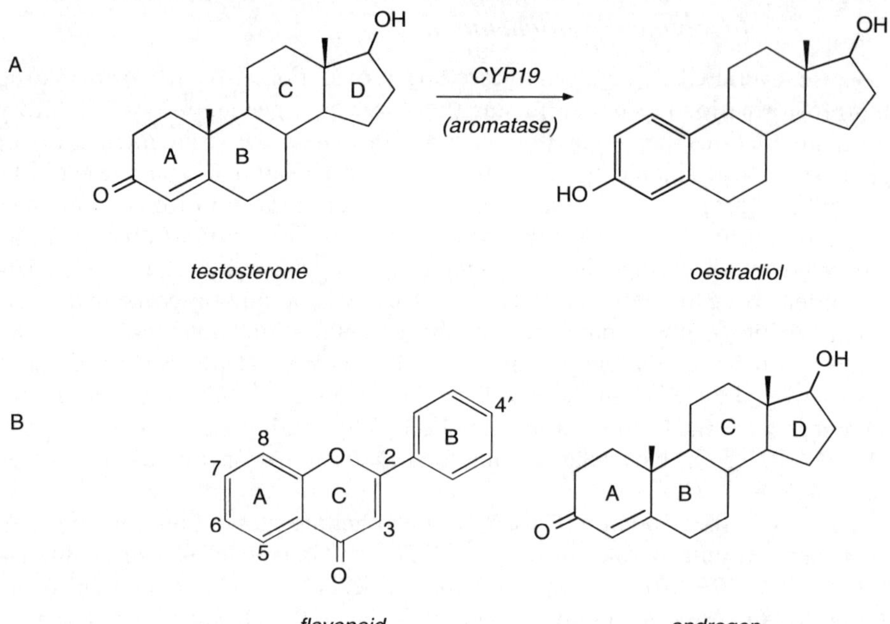

Figure 14.2 (A) CYP19 (aromatase) catalyses the aromatisation of the A-ring of androgens; shown here is the conversion of testosterone to oestradiol. (B) Structures of a typical flavonoid and the androgen testosterone. Possible overlap of the ring systems in the molecules within the CYP19 active site may mediate aromatase inhibition by flavonoids.

features of flavonoids (Figure 14.1) that contributed to CYP19 inhibition included the 4-oxo-group, which could be involved in a direct interaction with the CYP19 haem, or may inhibit electron transfer from NADPH-CYP-reductase to CYP19. The planar conformation of flavones again contributed to inhibition potency because the non-planar flavonones were less effective inhibitors. This is a similar situation to that described for the inhibitory effects of flavonoids on other CYPs.[19] The location of hydroxyl substituents around the flavonoid A-ring also influenced aromatase inhibition. By molecular modelling it was suggested that flavonoids may bind within the active site of CYP19 aromatase in an orientation that would enable the overlap of their aromatic ring systems with androgen substrates (Figure 14.2).[55]

CYP24 catalyses the 24-hydroxylation of 25-hydroxy- and 1α,25-dihydroxy-vitamin D_3, in order to deactivate vitamin D. CYP27B1 catalyses the 1α-hydroxylation of 25-hydroxy-vitamin D_3, which is important in vitamin D activation *in vivo*. Thus, the interplay of these two CYPs regulates the intracellular availability of active vitamin D metabolites, which influences cell growth and differentiation. Treatment of prostate-derived DU-145 tumour cells with the flavonoid genistein inhibited the activities of both CYPs and also altered their expression levels in cells.[56,57] Overexpression of these CYPs in tumours increases the capacity for vitamin D_3 depletion. The findings with genistein are therefore of considerable interest because it may be that flavonoids may be developed as antiproliferative agents in tumour cells that are dependent on vitamin D_3.

14.2.1.5 Modulatory Effects of Polyphenols on CYP Gene Expression

Polyphenols also modulate CYP expression in cells, which is another potential mechanism that may contribute to altered biotransformation of lipophilic substrates. Induction of CYP genes would be expected to increase rates of drug and carcinogen biotransformation, whereas CYP downregulation would decrease the rate of oxidation of these substrates.

There has been a focus on the regulatory effects of flavonoids on CYPs involved in carcinogen activation. Some flavonoids have been shown to upregulate CYP1A1 and CYP1A2 in animals, although they are considerably less potent than prototypic inducers, such as 3-methylcholanthrene.[58] Quercetin, diosmetin and biochanin A (Table 14.1) induced CYP1A1 mRNA in MCF-7 cells, thus supporting the assertion that these molecules may increase CYP gene transcription.[59,60] Interestingly, however, quercetin also modulated the induction of CYP1A1 by the potent inducer 2,3,7,8-tetrachlorodibenzo-*p*-dioxin (TCDD). TCDD is a ligand of the aryl hydrocarbon (Ah)-receptor, which interacts with planar PAH, enters the cell nucleus and binds to xenobiotic-response elements that are located in the 5′-flanks of target genes, such as CYP1A1, to activate transcription. In cells that had been transfected with CYP1A1 regulatory region reporter constructs, flavonoids were found to activate reporter activity, although to a much lesser extent than TCDD, and also to abrogate the strong activation elicited by TCDD. Thus, flavonoids may exert bimodal effects on CYP1A gene expression consisting of low potency induction and antagonism of induction mediated by AhR ligands like TCDD. The mechanism by which flavonoids stimulate CYP1A1 induction remains to be clarified, but there is great interest in their modulatory actions on CYP1A regulation. Thus, flavonoid treatments could be tailored to try to minimise the activity of protoxicant CYPs, such as CYP1A1, following exposure to TCDD and related environmental pollutants.

Similar modulatory effects on CYP1A1 induction have been observed with other polyphenols, including resveratrol and theaflavin.[61,62] CYP1A induction by the proton pump inhibitor omeprazole was also modulated by these polyphenols. This could be particularly significant because omeprazole is not a ligand for the Ah-receptor. Indeed, it is noteworthy that the receptor undergoes post-translational modification for optimal transactivation potential. Thus, omeprazole and related benzimidazoles induce CYP1A1 by activating tyrosine kinases that phosphorylate the Ah-receptor;[63] these kinases may be inhibited by flavonoids, and possibly other polyphenols.[64] It is interesting to speculate that polyphenols could impair CYP1A1 induction by decreasing the phosphorylation of the Ah-receptor.

Induction of other CYPs by flavonoids has also been reported. In rat liver, flavone and tangeretin induced CYP2B proteins, but again their potencies were considerably lower than that of phenobarbital.[58] The CYP2B subfamily is inducible by ligands of the constitutive androstane receptor (CAR) and pregnane X receptor (PXR),[65] which are members of the steroid hormone receptor superfamily. Quercetin increased CYP2B mRNA in human hepatocytes, but was inactive in a cell-based assay using a PXR-driven reporter.[66] This suggests that induction does not involve the PXR, but a role for the alternative regulator CAR has not yet been evaluated. Similar findings have been made using grapeseed extract, silymarin and kava, except that kava activated the PXR-reporter system in transfected cells. Like the Ah-receptor, steroid hormone receptors undergo complex phosphorylation/dephosphorylation steps for optimal transactivation capacity, so that it is conceivable that polyphenols modulate receptor phosphorylation to influence the extent of CYP induction.

In addition to CYP induction, there is emerging evidence that polyphenols may also downregulate certain CYPs in liver. Low concentrations of resveratrol and flavonoids downregulated CYP19 aromatase mRNA and protein in SK-BR-3 cells, which is consistent with repression of transcription.[54] The complex polyphenol tannic acid also decreased the hepatic and renal expression of CYP2E1 protein and its associated activity in mice, but the level of the regulatory impairment was unclear.[67] It is feasible that CYP downregulation may also be associated with altered phosphorylation, but to this point an evaluation of the effects of flavonoids on pathways of constitutive CYP regulation has not emerged.

14.2.1.6 Biotransformation of Polyphenols

Flavonoids have been shown to be competitive inhibitors of CYP-mediated substrate oxidations, which is consistent with reports that they are alternative substrates. Major pathways of flavonoid biotransformation include aromatic hydroxylation and *O*-demethylation of methoxy substituents.[68] Hydroxylation by CYP1A appears to be favoured when the B-ring is minimally hydroxylated,[69–71] and CYP1B1 is an important *O*-demethylase that converts the red clover isoflavonoids biochanin A to genistein and formononetin to daidzein. In human liver CYP 2C19 and CYP2D6 may also mediate the *O*-demethylation pathway.[72]

Most studies of the effects of flavonoids on CYP activity and expression have been undertaken *in vitro*. However, bioavailability is a critical factor that influences the efficacy of flavonoids *in vivo*. If administered in the diet flavonoids must be absorbed from the gut and enter the systemic circulation in order to penetrate tissues to reach the sites of action and biotransformation. In general, the bioavailability of flavonoids is limited by low absorption and rapid elimination. Many flavonoids are readily conjugated by phase II enzymes to glucuronides and sulphates, which restricts their systemic availability and decreases the concentrations that can be attained in tissues.[73–75] Resveratrol is partially absorbed from the gut, but is detected in many tissues soon after ingestion. Clinical studies have shown that only low concentrations of free resveratrol are achieved in human plasma and that the polyphenol is extensively glucuronidated.[76] Thus, the activity of phase II enzymes is a critical factor that influences the available concentrations of flavonoids *in vivo*.

14.3 Methylenedioxyphenyl Compounds

Methylenedioxyphenyl compounds (MDPs) are another important group of phytochemicals found in many plants and oils, including parsnips, carrots, parsley, nutmeg, sesame seeds, pepper and sassafras. Naturally occurring MDPs elicit a wide range of interactions with CYPs that have been intensively studied. In addition, synthetic MDP derivatives, typified by piperonyl butoxide, have been used commercially as synergists that increase the toxicity of coadministered pesticides to insects. The mechanism of this synergistic effect involves the inhibition by MDPs of pesticide-oxidising CYPs in insects.

14.3.1 Role of CYPs in the Formation of Inhibitory Metabolites of MDP Compounds

CYPs oxidise MDPs to the corresponding catechols, which are stable products, and reactive metabolites, which bind tightly to CYP haem to inhibit substrate oxidation (Figure 14.3). The precise inhibitory actions of particular MDP compounds are strongly related to their structure. Reactive metabolites of MDPs form metabolite intermediate (MI) complexes that are especially stable

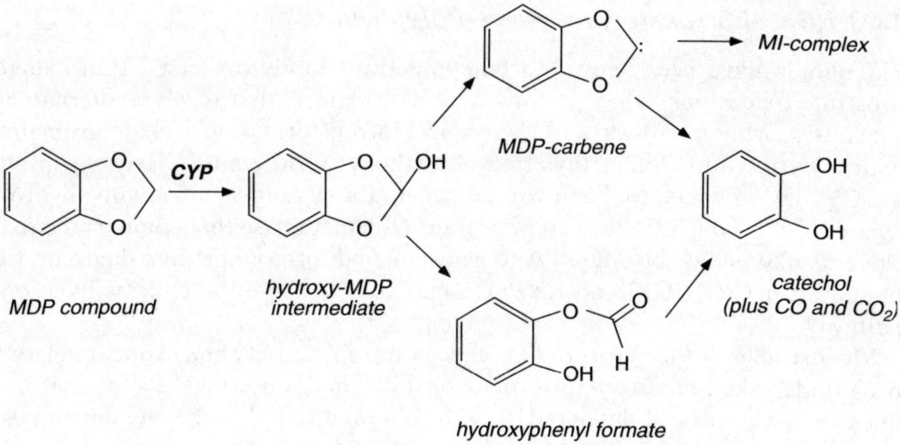

Figure 14.3 Oxidation of MDP compounds at the methylenedioxy carbon to form the hydroxy-MDP intermediate that can generate the MDP-carbene resulting in the formation of the inhibitory MI-complex. Alternately, hydroxy-MDP may be hydrolysed to the hydroxyphenyl formate derivative that decomposes to the catechol, releasing in the process carbon monoxide and carbon dioxide.

when the CYP haem is in the ferrous form but, in the oxidised state, the MI-complex can be displaced by lipophilic substrates, restoring the activity of the previously sequestered CYP.[77,78] The hydrophobic and electronic character of MDP substituent groups influences the extent to which the analogues form MI-complexes.[79–81] MDPs carrying electron donating substituents and aliphatic side chains tend to generate stable MI-complexes, whereas electron withdrawing substituents promote the formation of carbon monoxide from the methylenedioxy carbon.[79]

Naturally occurring MDPs also elicit CYP inhibition. Extracts of goldenseal (*Hydrastis canadensis*) contain the MDP-based alkaloids berberine and hydrastine.[82] Hydrastine inhibited human CYP3A4 but not CYP2C9 or CYP-2D6, whereas berberine was more effective against CYP2D6. The shrub kava (*Piper methysticum*) is indigenous to islands of the south pacific, and a kava aqueous extract is used by inhabitants of these islands to promote a sense of well being. Kava extracts and kavalactones, especially methysticin and dihydromethysticin, inhibited several human drug-metabolising CYPs *in vitro*. Inhibition was potentiated by preincubation with NADPH-fortified microsomes, which is consistent with MI-complex formation. Inhibition and possible MI-complexation of CYP2C9 and CYP2C19 activities and, to a lesser extent, CYP3A4 activity in human liver microsomes was noted.[83] *In vivo* administration of goldenseal and kava extracts to human volunteers altered the metabolic ratios of certain CYP substrates. Thus, goldenseal inhibited the clearance of CYP2D6 and CYP3A4/5 substrates while kava inhibited CYP2E1.[84]

More recently, several MDP-based lignans were isolated from the herbal medicinal plant *Piper cubeba*. These agents, especially (-)-clusin and

(-)-dihydroclusin, were potent and selective mechanism-based inhibitors of CYP3A4.[85] Piperine, an MDP compound present in pepper, and several related molecules containing intact MDP moieties inhibited the oxidation of CYP1A substrates in microsomal fractions from control, phenobarbital- and 3-methylcholanthrene-induced rat liver.[86]

In hepatic microsomes from rats that had been pretreated with certain MDP compounds a proportion of the CYP is present as a stable MI-complex. *In vitro* dissociation of the microsomal complex with lipophilic substrates selectively reactivates the inhibited CYP. Aminopyrine *N*-demethylase activity was reactivated after displacement of the MDP metabolite from its MI-complex, which suggests that CYP2B and CYP3A may be sequestered in these microsomes.[6,78] This was substantiated by the increased 6ß- and 16ß-hydroxylation of the steroid substrate androstenedione that occurred after *in vitro* displacement of the isosafrole metabolite from its MI-complex with CYP.[8] In contrast, displacement of the MDP MI-complex did not increase activities mediated by CYP1A1 and CYP2C11, suggesting that neither of these CYPs is present as an inhibited complex in liver.

Safrole is a weak carcinogen that undergoes biotransformation to a reactive genotoxic intermediate (Figure 14.4). As is the case with other carcinogenic chemicals, CYPs play a central role in this activation step. The nature of the aromatic substituents is an important determinant of carcinogenic potential. Thus, allyl-substituted MDP analogues, like safrole and estragole, but not the propenyl-substituted isosafrole, undergo CYP-mediated 1′-hydroxylation in hepatic microsomes. This is the initial activation step and is followed by sulphonation, which is catalysed by sulphotransferases, and then DNA adduction.[87,88] In human liver CYP2C9*1, CYP2A6, CYP2D6*1, and CYP2E1 have been identified as the principal catalysts in safrole 1′-hydroxylation.[89]

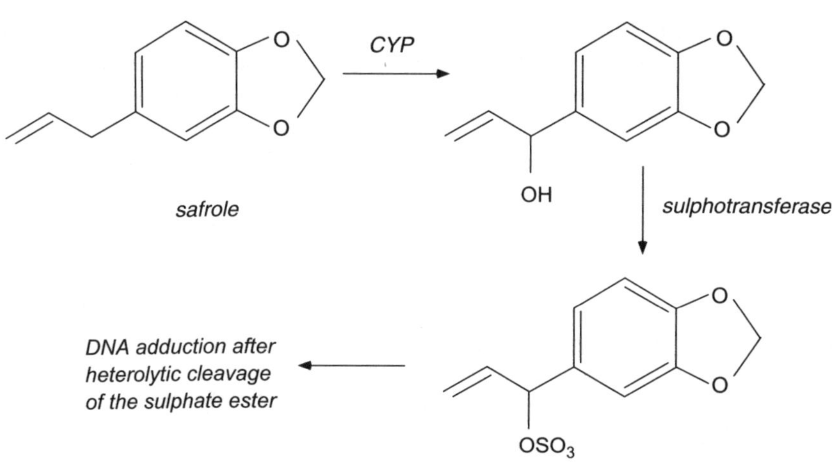

Figure 14.4 CYP-dependent oxidation of safrole to the benzylic alcohol, followed by sulphotransferase-mediated formation of the sulphate ester that can undergo heterolytic cleavage to form DNA adducts.

14.3.2 Induction of CYPs by MDP Compounds

The relationships between the structure of MDP compounds and their capacity to induce CYP gene expression have, to some extent, been investigated. The naturally-occurring MDP compounds safrole, isosafrole and myristicine, are effective inducers of several CYPs.[90] These agents all contain aliphatic substituent groups within the aromatic system. Structure-activity studies with a series of synthetic 5-*n*-alkoxy-MDPs led to similar findings. Thus, induction of CYP1A activity was well correlated in parabolic relationships with the hydrophobicity of the alkoxy substituent, and reached optimal potency with chain lengths containing five or six carbon atoms. *In vivo* formation of MI-complexes was also significant with the C5-C6 compounds, with those formed by the *n*-butoxy to *n*-hexoxy-substituted derivatives being particularly stable.[91] Administration of synthetic and naturally-occurring MDP compounds that possessed electron-withdrawing substituents did not generate MI-complexes of significant magnitude and did not significantly induce CYP activities.[6,79,80,90]

Administration of piperine, an MDP compound that is a constituent of black pepper, to rats upregulated CYP1A and CYP2B, but decreased CYP2E1 expression and its associated activities by $\sim 30\%$.[92] Hexobarbital sleeping time was prolonged after short-term (up to 8-hr) *in vivo* treatment of rats with piperine.[93] These latter studies were undertaken with single doses of piperine and it remains to be established whether longer term treatment might promote MI-complexation, as well as CYP induction, as was reported previously.[92]

One of the most intensively studied MDP compounds is isosafrole, which alters the expression of a number of CYPs, inducing CYP3A, CYP2B1, CYP1A1 and CYP1A2, and suppressing CYP2C11.[94] The mechanisms by which the expression of these CYPs are altered are likely to be multi-factorial, possibly involving the Ah-receptor, PXR and CAR, but remain largely undefined at this stage. CYP1A2 was induced in both wild-type and Ah-receptor-defective mice, which suggests that this process is Ah-receptor independent. In contrast, CYP1A1 was not induced in Ah-receptor defective mice that were treated with isosafrole.[95–97] Thus, the Ah-receptor is essential for CYP1A1 induction despite the structural dissimilarities between isosafrole and more established ligands for the receptor. It is possible that isosafrole may modulate a different step in activation of the receptor, such as a post-translational step involving phosphorylation, or the MDP-metabolite may stabilise CYP1A against proteolysis.[98]

14.4 Furanocoumarins (Psoralens)

14.4.1 Anticarcinogen and Drug-Interaction Effects Mediated by CYP Inhibition

Apiaceous vegetables, such as carrots, parsnips, celery and parsley, contain furanocoumarins (psoralens), including 8-methoxypsoralen (8-MOP) and apigenin, which inhibit CYP1A2 and other CYPs.[99] These vegetables may have anticarcinogenic properties due to the inhibitory effects of psoralens on CYP-dependent carcinogen activation. Thus, psoralens decreased the mutagenicity of aflatoxin B_1 in cell systems and inhibited DNA-adduct formation with 7,12-dimethylbenz(a)anthracene in mouse mammary glands.[99,100] Much of the initial interest focussed on CYP2A inhibition by psoralens, possibly because of the structural similarity of these molecules to coumarin, which is an established substrate for CYP2A in mammalian liver.[101,102] 8-MOP inactivated human CYP2A6 during CYP-dependent oxidation in the furan ring to reactive intermediates, most likely the epoxide and ketoenal (Figure 14.5).[102] Apart from CYP2A6, CYP2C19 and CYP2E1 were also weakly inactivated by 8-MOP.[103] When ^{14}C-8-MOP was utilised in mechanistic studies, most of the label was bound to CYP2A6 and the flavoprotein NADPH-CYP-reductase. These findings indicate that a portion of the reactive 8-MOP metabolite can leave the site of formation (CYP2A6) to arylate the nearby reductase; this is unusual behaviour, because destructive metabolites of most mechanism-based inactivators are retained within the CYPs that catalyse their formation. It may be that the reactive intermediates formed from psoralens are somewhat more stable than the radicals formed during activation of olefins, acetylenic steroids and other established suicide substrates. The 8-methoxy substituent in 8-MOP was found to be important for inactivation because the 8-hydroxy analogue elicited only competitive inhibition.[103] The hydroxyl group may be involved in hydrogen-bond formation in the active site of CYP2A6, which may reorient the molecule so that inactivation is disfavoured.

Recent experimental and clinical studies have evaluated the role of bergamottin and other constituent grapefruit juice psoralens in pharmacokinetic drug-drug interactions. Intestinal CYP3A may be involved in the presystemic oxidation of some drug substrates, such as felodipine, and inactivation of CYP3A by grapefruit juice psoralens enhances the absorption, bioavailability, serum levels and toxicity of these drugs. Bergamottin derivatives are oxidised by CYP3A enzymes in enterocytes to metabolites that degrade and inhibit intestinal CYPs.[104] Since this mechanism was established, bergamottin derivatives have also been found to inactivate, under certain conditions, other CYPs.[105,106] However, because bergamottin derivatives are not absorbed, their clinical significance would be restricted to effects on intestinal CYPs and with drugs that are major substrates for those CYPs.[104,107] It is noteworthy that the 8-geranoxy substituent in bergamottin seems to markedly alter the CYP inactivation profile of the 8-methoxy derivative 8-MOP away from CYP2A6 in favour of CYP3A4/5 and CYP2B6.[104–108] It will be interesting to evaluate the structural basis for this change in inhibition profile.

14.5 Carotene and Related Plant Pigments

14.5.1 Anticarcinogenic Effects Mediated by Altered CYP Regulation

Astaxanthin, canthaxanthin and related carotenoids (Figure 14.6) consist of polyene chains that frequently terminate in ring systems, such as the β-ionone

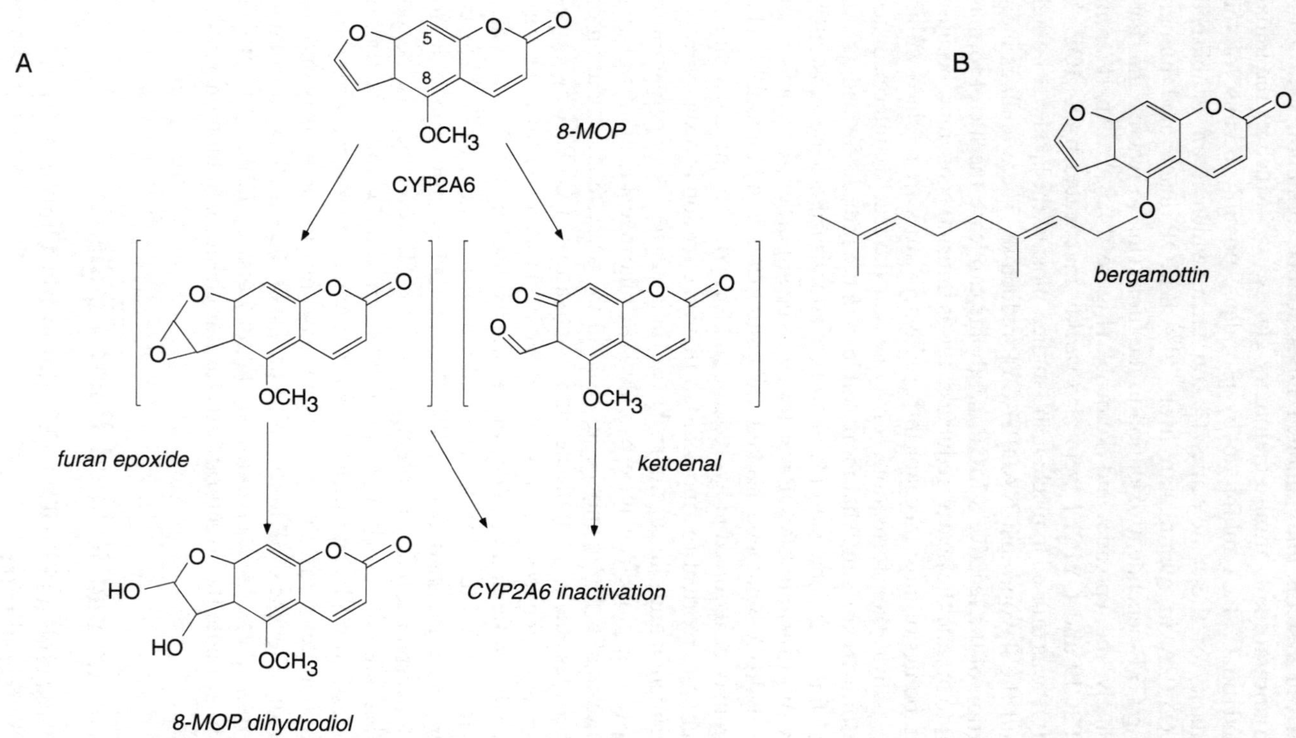

Figure 14.5 (A) Oxidation of 8-MOP by CYP2A6 to the postulated reactive epoxide and ketoenal intermediates, that generate the dihydrodiol metabolite and, inactivate CYP2A6. (B) Structure of bergamottin, an important inactivator of intestinal CYP3A4 present in grapefuit juice.

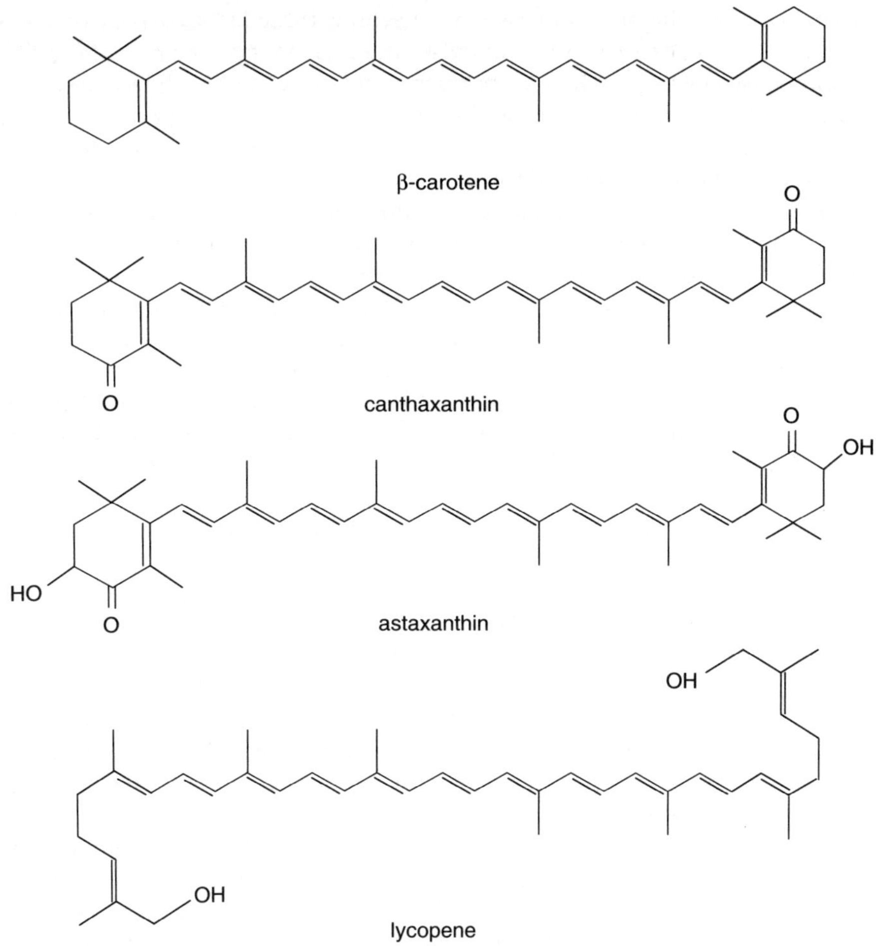

Figure 14.6 Structural similarity of β-carotene, canthaxanthin, astaxanthin and lycopene.

ring in β-carotene. β-Carotene itself is the dietary precursor of retinoids, or vitamin A derivatives and so has an important role in nutrition. These agents have generated interest as potential dietary inhibitors of tumourigenesis. Thus, it has been suggested from epidemiological and experimental studies that increased intake of fruits and vegetables containing these molecules may be beneficial against certain cancers, possibly involving altered CYP function and expression. Direct testing of the efficiency of high-dose β-carotene supplementation in intervention trials, however, appeared to increase the progression of lung cancer in smokers or in those who had been exposed to asbestos in the workplace.[109] Although these are very high risk groups, it appears that the use of carotenoid supplementation may not necessarily impart health benefits against all tumour types and further research is required to properly assess the value of these supplements.

There is contradictory information regarding the effects of β-carotene on CYP expression in experimental animals, ranging from no apparent effect[110,111] to appreciable effects.[112] These disparities could be related to differences in the timing and dosage of β-carotene, and to the particular animal species and strain involved. Indeed, some of these studies have been conducted at extremely high levels of β-carotene supplementation.

Extracts of fruits and vegetables containing carotenoids have been tested for their inhibitory effects on the mutagenicity of chemicals that require activation by CYPs.[113] In rats, a complex protocol consisting of exposure to several toxic chemicals and partial hepatectomy was used to stimulate tumourigenesis. Administration of diets that were supplemented with β-carotene, canthaxanthin, astaxanthin and, especially, lycopene during the early phase of the conditioning protocol decreased the number of positive foci induced by this regimen.[112] In another report, administration of lycopene for several days decreased CYP2E1 activity.[114] This may shed some light on the potential benefits due to dietary lycopene, because CYP2E1 is important in the activation of nitrosamines,[9] and could contribute to the apparent antitumourigenic properties of carotenoids.[112]

The capacity of carotenoids to modulate CYP expression has been tested in experimental animals. Canthaxanthin and, to a lesser extent, the somewhat more polar agent astaxanthin induced CYP1A and CYP2B, but with considerably lower potency than conventional inducers.[112,114,115] In contrast, CYP3A activity was unchanged, and CYP2E1 activity was actually decreased to 60% of control. Thus, procarcinogen activation mediated by CYP2E1 could be suppressed by such treatments. Similarly, activities mediated by the CYP-1A, and possibly also CYP2B, subfamilies were increased in the liver of rats treated with certain carotenoids.[111] High-level β-carotene supplementation effected 2–15-fold increases in activities mediated by hepatic CYP1A1/2, CYP-3A1/2, CYP2E1, CYP2B1/2 and CYP2C11 in rats.[116] These increases occurred in dose-, tissue- and gender-related fashion, but reports of CYP induction by high dose β-carotene are essentially in accord with the observations of increased carcinogenic potential from intervention studies in human subjects, and indicate caution in the use of these agents. In contrast, however, there have been reports that β-carotene and other palm carotenoids decreased hepatic BaP metabolism *in vivo* and *in vitro* in rat.[117] The regulation of CYP1A genes by carotenoids is of interest because these molecules do not fulfil the structural requirements for binding to the Ah-receptor.[118] As is the case with other types of phytochemicals that regulate CYP1A, it will now be pertinent to determine whether the chemicals elicit post-translational modification of the Ah-receptor.

In primary human hepatocytes, astaxanthin induced CYP3A4 and CYP2B6, but not CYP1A or CYP2C.[119] The involvement of the PXR and CAR steroid hormone receptors in astaxanthin-mediated induction of CYP3A4 and 2B6 should now be explored. Considered together, the inhibitory and inductive effects of carotenoids and related molecules have only been observed to date at relatively high concentrations that may be difficult to achieve using normal dietary intake.

14.6 Sulphur and Indole-Containing Compounds in Cruciferous Vegetables

The cancer chemopreventive potential of glucosinolates and their metabolites, present in broccoli and other cruciferous vegetables, has generated considerable recent interest.[120] During digestion, indoles and isothiocyanates are liberated from glucosinolate precursors and these molecules appear to offer cellular protection against DNA damage and malignant transformation. Among the most intensively studied glucosinolates are glucobrassicin, which is present in broccoli and brussels sprouts, and glucoraphin (glucoraphanin) and gluconasturtin which are present at high level in cress and turnips. Glucobrassicin is hydrolysed to liberate indoles whereas glucoraphin and gluconasturtin give rise to isothiocyanates.

14.6.1 Inhibitory Effects on CYP Enzymes

Benzyl and phenethyl isothiocyanates (BITC and PEITC, respectively) occur naturally in cruciferous vegetables and are selective mechanism-based inactivators of certain CYPs.[121] Thus, PEITC inactivated CYP2E1, but CYP1A2, CYP2A6, CYP2B6, CYP2D6, and CYP3A4 were relatively refractory to inactivation.[122] PEITC also reversibly inhibited CYP1A2, CYP2B6 and CYP2C9, but was relatively non-potent against CYP2C8, CYP2D6 and CYP3A4 activities.[122] The potential for long-term inhibition of CYP2E1 activity by PEITC makes this an interesting agent in terms of its potential chemopreventive action against nitrosamines and other toxins that are activated by this CYP.

The glucosinolate product sulforaphane inhibited nitrosamine mutagenicity, which is consistent with the capacity to inhibit CYP2E1; however, sulforaphane is considerably less potent than other isothiocyanates in this regard.[123] Limited structure-activity studies have been conducted in microsomal fractions from 3-methylcholanthrene- and phenobarbital-induced rat liver. Increasing chain length along the alkyl isothiocyanate series to the hexyl derivative increased inhibition potency against CYP1A and CYP2B, with derivatives containing longer alkyl chains being less potent.[124] It will now be useful to compare the potencies of these synthetic isothiocyanates against CYP2E1, as well as other human CYPs.

The available information suggests that the clearance of substrates for CYP2E1 may be impaired by concurrent exposure to cruciferous vegetables, although the actual quantities of vegetables required to be ingested could be quite high. Thus, in human volunteers, isothiocyanate-containing watercress extracts increased the elimination half-life and bioavailability of the CYP2E1 substrate chlorzoxazone to around 1.5-fold of pre-dose levels.[125] In contrast, no inhibition of CYP2D6-dependent debrisoquine 4-hydroxylation occurred *in vivo* despite some evidence of *in vitro* efficacy.[126] At present there is a deficiency of information as to whether cress or other cruciferous vegetables inhibit other drug metabolising CYPs *in vivo*.

Inhibition of CYP2A-mediated α-hydroxylation of the tobacco carcinogen NNK may be useful against pulmonary cancer induced by cigarette smoke.[127] A number of synthetic isothiocyanates inhibited NNK-induced lung tumourigenicity and DNA adduction.[128] It is now appropriate that these observations should now be expanded further in *in vivo* studies.

Diallyl sulphide is another sulphur-containing compound that is found in garlic and modulates CYP expression. Diallyl sulphide inhibited the biotransformation of nitrosamines[129] and the mutagenicity of vinyl chloride and *N,N*-dimethylnitrosamine,[130] consistent with CYP2E1 inhibition. Inhibitory potency was increased by preincubation of diallyl sulphide or diallyl disulphide with NADPH-supplemented microsomes, which is suggestive of mechanism-based inactivation by reactive metabolites.[129]

When assessed in healthy volunteers, garlic extracts decreased CYP2E1 activity as reflected by the serum ratio of 6-hydroxychlorzoxazone to chlorzoxazone.[131] In contrast, minimal effects were observed on the excretion of metabolites of dextromethorphan and alprazolam, indicating that garlic-derived chemicals do not inhibit CYP2D6 and CYP3A4.[132] Similarly, Gurley *et al.*[131] found little evidence that these CYPs or CYP1A2 were inhibited by garlic oil (500 mg administered three times daily for 28 days). However, there have been reports that garlic supplements altered the pharmacokinetics of the CYP3A4 substrates sequinavir and ritonavir, in volunteers.[133,134] It may be important to consider the nature of the garlic extract and its means of preparation, since the nature of the organic sulphur derivatives present may be influenced by preparation and storage.

14.6.2 CYP Induction by Indoles and Isothiocyanates

Dietary indole-3-carbinol ingestion has been shown to induce CYP1A2[135] and to accelerate the oxidation of the food-derived mutagen 2-amino-1-methyl-6-phenylimidazo[4,5-*b*] pyridine (PhIP).[136] However, this treatment also decreased PhIP-DNA adduct formation. This is due most likely to enhanced detoxification of reactive genotoxic PhIP metabolites by phase II enzymes, which are strongly induced by indole-3-carbinol and related chemicals, and which outweighs the deleterious effects of enhanced PhIP activation.[10]

Indole-3-carbinol is not itself the active inducer of CYP1A. Instead it is converted to indolo[3,2-*b*]carbazole and diindyl methane at the acidic pH of the stomach;[137] these molecules strongly activate the Ah receptor.[118,138] Interestingly, diindolyl methane was also found to induce CYP19 aromatase in human adrenocortical carcinoma cells.[139] The mechanism underlying this observation has not been established although it is unlikely to be mediated by the Ah-receptor. However, this finding could be relevant for the dietary management of oestrogen-dependent tumours in man.

Administration of diallyl sulphide to rats increased CYP2B1/2 mRNA, protein and activity in the liver and other tissues.[140,141] CYP1A1 and CYP3A1 were also induced by garlic oil and by the constituent allyl sulphides, with diallyl sulphide being the most potent.[142] These findings illustrate the complexities associated with the use of these agents as potential chemoprotective agents. Even though they appear to be useful inhibitors of CYP2E1 in particular, the observation that other CYPs may be upregulated by allyl sulphides complicates their use.

There is also evidence that treatment of animals with glucosinolates and their hydrolytic products may downregulate the expression of certain CYPs. Thus, oral administration of gluconasturtin decreased the activities of several CYPs in hepatic, renal and pulmonary preparations.[143] Although PEITC induced CYP2B1, activities mediated by CYP1A and CYP3A decreased.[144] Again, the optimal use of glucosinolates in man will be dependent on a thorough understanding of their *in vivo* action.

14.7 Other Phytochemicals

14.7.1 Induction of CYPs by Complex Terpenoids and Related Agents

The use of herbal medicines is increasing. An agent in widespread use is ginkgo biloba, which contains numerous chemicals including the diterpene ginkgolides A, B, C and J, and sesquiterpenes such as bilobalide. CYP2B1 mRNA expression in rat hepatocytes was increased strongly by bilobalide and CYP3A23 mRNA was upregulated by ginkgolide A.[145] To this point, the underlying mechanism has not been elaborated, but differential selectivity of the phytochemicals for PXR or CAR, which regulate these CYPs, is possible. In this regard, the herbal agent hyperforin, which is the major constituent in St John's wort, is a potent activator of PXR and strongly induces CYP3A4 and CYP2B6 in human tissue and in cell systems.[146,147]

β-Myrcene is an acyclic monoterpene found in the essential oils of lemongrass, hop, verbena, bay and others. Such monoterpene-containing oils are widely used as flavouring food additives, in cosmetics and in household products. The capacities of camphor, menthol, pinene, limonene and myrcene (Figure 14.7) to induce rat CYP2B mRNA and protein have been established.[148,149] Thus, *in vivo* administration of terpenoids increases CYP activity[150,151] and decreases barbiturate sleeping time.[150,152] CYP2B was strongly induced at a pretranslational level by camphor, but less so by menthol and pinene; limonene and myrcene were ineffective. In male Wistar rats, the terpenoid citral induced CYP4A and other genes consistent with stimulation of the peroxisome proliferator-activated receptor-α isoform.[153] These findings suggest that different terpenoids may activate different members of the steroid hormone receptor superfamily in rodents. An assessment of the relevance to human CYP gene regulation may result from initial studies in human hepatocytes.

14.7.2 Inhibition of CYPs by Terpenoids and Role of CYPs in Terpenoid Oxidation

The terpenoids β-myrcene, (-)-α-pinene, (+)-α-pinene, D-limonene, α-terpinene, citral and citronellal, were all found to be competitive inhibitors of CYP2B1 activity in microsomes from phenobarbital-treated rats, but CYP1A activity was relatively refractory to inhibition.[154] D-Limonene and perillyl alcohol inhibited

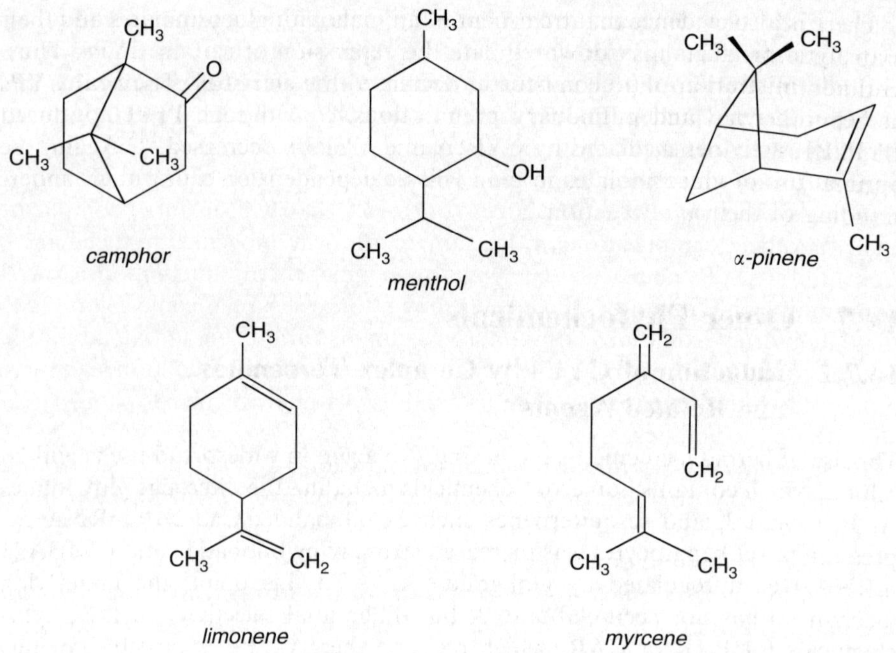

Figure 14.7 Terpenoid structures.

pulmonary tumorigenesis in mice *in vivo* by inhibiting the metabolic activation of NNK.[155] These findings suggest that certain terpenoids may also inhibit CYP2E1 and CYP2A, which are involved in NNK and nitrosamine activation. The terpenoid structure may therefore be a useful starting point for the development of cytoprotective agents. Certainly, it is an important point that these agents are active *in vivo* when administered to experimental animals.

To develop terpenoids, because of their CYP modulatory capacity, it will be important that information is available on their biotransformation pathways and interactions with individual CYPs. In this respect, limonene was oxidised by human CYP2C9 in human liver microsomes to the 6- and 7-hydroxylated analogues, carveols and perillyl alcohols, respectively.[156] Furthermore, the biotransformation of the isomeric 1,4- and 1,8-cineoles, which are constituent oils in *Citrus medica* and *Eucalyptus polybractea* respectively, is mediated by CYP3A in human liver. 1,8-Cineole formed the 2α- and 3α-hydroxy products,[157] whereas 1,4-cineole was oxidised to 2-exo-hydroxy-1,4-cineole.[158] These findings are of particular interest in that several different human CYPs may contribute to terpenoid biotransformation.

14.8 Summary and Future Issues

The present chapter has summarised the available details on the interaction of phytochemicals with CYP genes and enzymes. The consequences of these interactions may include altered rates of xenobiotic and endobiotic biotransformation, and altered CYP gene expression. Much of the focus of experimental effort to date has been to evaluate the effects of polyphenols and certain other phytochemicals on the CYPs involved in chemical carcinogenesis. In this respect some promising molecules have been identified that could be used in future anti-cancer strategies. However, it has also emerged that these chemicals exert quite complex effects on CYPs that will, in turn, complicate their therapeutic applications. For example, anti-carcinogenic phytochemicals that inhibit CYPs could also elicit pharmacokinetic drug interactions. Moreover, many CYP inhibitors have been shown to exert biphasic effects on CYP-mediated drug oxidations after *in vivo* administration; an initial inhibitory phase is often followed by a more prolonged phase of CYP induction leading to enhanced oxidation capacity. The optimal use of phytochemicals should consider these possibilities.

Future studies will continue to define the individual CYPs that are susceptible to inhibition by phytochemicals, the roles of human CYPs in phytochemical oxidation pathways and the regulatory effects of phytochemicals on CYP gene expression. These studies will focus on human CYPs, not only because the tools required to undertake such studies are becoming increasingly available, but also because these approaches will generate the most useful information.

Acknowledgements

Research in this laboratory is supported by the Australian National Health and Medical Research Council.

References

1. J. Ahn, M.D. Gammon, R.M. Santella, M.M. Gaudet, J.A. Britton, S.L. Teitelbaum, M.B. Terry, S. Nowell, W. Davis, C. Garza, A.I. Neugut and C.B. Ambrosone, *Am. J. Epidemiol.*, 2005, **162**, 943.
2. J.M. Chan, P.H. Gann and E.L. Giovannucci, *J. Clin. Oncol.*, 2005, **23**, 8152.
3. M. Murray and G.F. Reidy, *Pharmacol. Rev.*, 1990, **42**, 85.
4. M.A. Correia and P.R. Ortiz de Montellano, in *"Cytochrome P450: Structure, Mechanism, and Biochemistry"*, ed. P.R. Ortiz de Montellano, Kluwer/Plenum, New York, 2005, (3rd edition), p. 247.
5. M. Murray, *Clin. Exp. Pharmacol. Physiol.*, 1997, **24**, 465.
6. T.R. Fennell, B.C. Sweatman and J.W. Bridges, *Chem.-Biol. Interact.*, 1980, **31**, 189.
7. D. Pessayre, D. Larrey, J. Vitaux, P. Brei, J. Belghiti and J.P. Benhamou, *Biochem. Pharmacol.*, 1982, **31**, 1699.
8. M. Murray, L. Zaluzny and G.C. Farrell, *Arch. Biochem. Biophys.*, 1986, **251**, 471.

9. F.P. Guengerich, in: *"Cytochrome P450: Structure, Mechanism and Biochemistry"*, ed. P.R. Ortiz de Montellano, Kluwer/Plenum, New York, 2005, (3rd edition), p. 377.
10. C. Bonnesen, I.M. Eggleston and J.D. Hayes, *Canc. Res.*, 2001, **61**, 6120.
11. I. Rahman, S.K. Biswas and P.A. Kirkham, *Biochem. Pharmacol.*, 2006, **72**, 1439.
12. R. Edenharder, C. Leopold and M. Kries, *Mutat. Res.*, 1995, **341**, 303.
13. Z.C. Kang, S.J. Tsai and H. Lee, *Nutr. Canc.*, 1999, **35**, 175.
14. X. Wen, U.K. Walle and T. Walle, *Carcinogenesis*, 2005, **26**, 803.
15. M. Das, H. Mukhtar, D.P. Bik and D.R. Bickers, *Canc. Res.*, 1987, **47**, 760.
16. D. Schwarz, P. Kisselev and I. Roots, *Canc. Res.*, 2003, **63**, 8062.
17. S. Lautraite, A.C. Musonda, J. Doehmer, G.O. Edwards and J.K. Chipman, *Mutagenesis*, 2002, **17**, 45.
18. W.L. Bear and R.W. Teel, *Anticanc. Res.*, 2000, **20**, 3323.
19. J.Y. Moon, D.W. Lee and K.H. Park, *Xenobiotica*, 1998, **28**, 117.
20. M.K. Buening, R.L. Chang, M.T. Huang, J.G. Fortner, A.W. Wood and A.H. Conney, *Canc. Res.*, 1981, **41**, 67.
21. A. Chaudhary and K.L. Willett, *Toxicol.*, 2006, **217**, 194.
22. H. Doostdar, M.D. Burke and R.T. Mayer, *Toxicol.*, 2000, **144**, 31.
23. I.B. Tsyrlov, V.M. Mikhailenko and H.V. Gelboin, *Biochim. Biophys. Acta.*, 1994, **1205**, 325.
24. H. Lee, H.W. Wang, H.Y. Su and N.J. Hao, *Mutagenesis*, 1994, **9**, 101.
25. H.P. Ciolino and G.C. Yeh, *Mol. Pharmacol.*, 1999, **56**, 760.
26. G. Berge, S. Ovrebo, I.V. Botnen, A. Hewer, D.H. Phillips, A. Haugen and S. Mollerup, *Brit. J. Canc.*, 2004, **91**, 333.
27. M. Jang, L. Cai, G.O. Udeani, K.V. Slowing, C.F. Thomas, C.W. Beecher, H.H. Fong, N.R. Farnsworth, A.D. Kinghorn, R.G. Mehta, R.C. Moon and J.M. Pezzuto, *Science*, 1997, **275**, 218.
28. T.K. Chang, J. Chen and W.B. Lee, *J. Pharmacol. Exp. Ther.*, 2001, **299**, 874.
29. Y.J. Chun, S.Y. Ryu, T.C. Jeong and M.Y. Kim, *Drug Metab. Dispos.*, 2001, **29**, 389.
30. H. Forejtnikova, K. Lunerova, R. Kubinova, D. Jankovska, R. Marek, R. Kares, V. Suchy, J. Vondracek and M. Machala, *Toxicol.*, 2005, **208**, 81.
31. M.C. Henderson, C.L. Miranda, J.F. Stevens, M.L. Deinzer and D.R. Buhler, *Xenobiotica*, 2000, **30**, 235.
32. S. Oetari, M. Sudibyo, J.N. Commandeur, R. Samhoedi and N.P. Vermeulen, *Biochem. Pharmacol.*, 1996, **51**, 39.
33. R. Thapliyal and G.B. Maru, *Fd. Chem. Toxicol.*, 2001, **39**, 541.
34. S.-S. Chen and K.-T. Chung, *Fd. Chem. Toxicol.*, 2000, **38**, 1.
35. H. Mukhtar, M. Das, W.A. Khan, Z.Y. Wang, D.P. Bik and D.R. Bickers, *Canc. Res.*, 1988, **48**, 2361.
36. S. Muto, K. Fujita, Y. Yamazaki and T. Kamataki, *Mutat. Res.*, 2001, **479**, 197.

37. H.R. Ha, J. Chen, P.M. Leuenberger, A.U. Freiburghaus and F. Follath, *Eur. J. Clin. Pharmacol.*, 1995, **48**, 367.
38. F.P. Guengerich and D.H. Kim, *Carcinogenesis*, 1990, **11**, 2275.
39. D.G. Bailey, G.K. Dresser, J.H. Kreeft, C. Munoz, D.J. Freeman and J.R. Bend, *Clin. Pharmacol. Ther.*, 2000, **68**, 468.
40. Y. Li, E. Wang, C.J. Patten, L. Chen and C.S. Yang, *Drug Metab. Dispos.*, 1994, **22**, 566.
41. L.L. von Moltke, J.L. Weemhoff, E. Bedir, I.A. Khan, J.S. Harmatz, P. Goldman and D.J. Greenblatt, *J. Pharm. Pharmacol.*, 2004, **56**, 1039.
42. I. Stupans, H.W. Tan, A. Kirlich, K. Tuck, P. Hayball and M. Murray, *J. Pharm. Pharmacol.*, 2002, **54**, 269.
43. P. Letteron, G. Labbe, C. Degott, A. Berson, B. Fromenty, M. Delaforge, D. Larrey and D. Pessayre, *Biochem. Pharmacol.*, 1990, **39**, 2027.
44. R. Venkataramanan, V. Ramachandran, B.J. Komoroski, S. Zhang, P.L. Schiff and S.C. Strom, *Drug Metab. Dispos.*, 2000, **28**, 1270.
45. I. Stupans, M. Murray, A. Kirlich, K.L. Tuck and P.J. Hayball, *Fd. Chem. Toxicol.*, 2001, **39**, 1119.
46. B. Piver, F. Berthou, Y. Dreano and D. Lucas, *Life Sci.*, 2003, **73**, 1199.
47. G. Regev-Shoshani, O. Shoseyov and Z. Kerem, *Biochem. Biophys. Res. Commun.*, 2004, **323**, 668.
48. R. Mikstacka, A.M. Rimando, K. Szalaty, K. Stasik and W. Baer-Dubowska, *Xenobiotica*, 2006, **36**, 269.
49. W.K. Chan and A.B. Delucchi, *Life Sci.*, 2000, **67**, 3103.
50. T.K. Chang and R.K. Yeung, *Can. J. Physiol. Pharmacol.*, 2001, **79**, 220.
51. E.M. Offman, D.J. Freeman, G.K. Dresser, C. Munoz, J.R. Bend and D.G. Bailey, *Clin. Pharmacol. Ther.*, 2001, **70**, 17.
52. J.Y. Moon, X. Wang and M.E. Morris, *Toxicol. in vitro*, 2006, **20**, 187.
53. A.R. Ibrahim and Y.J. Abul-Hajj, *J. Steroid Biochem. Mol. Biol.*, 1990, **37**, 257.
54. J.T. Sanderson, J. Hordijk, M.S. Denison, M.F. Springsteel, M.H. Nantz and M. van den Berg, *Toxicol. Sci.*, 2004, **82**, 70.
55. Y.C. Kao, C. Zhou, M. Sherman, C.A. Laughton and S. Chen, *Environm. Health Perspect.*, 1998, **106**, 85.
56. H. Farhan and H.S. Cross, *Ann. N. Y. Acad. Sci.*, 2002, **973**, 459.
57. H. Farhan, K. Wahala and H.S. Cross, *J. Steroid Biochem. Mol. Biol.*, 2003, **84**, 423.
58. M.C. Canivenc-Lavier, M. Bentejac, M.L. Miller, J. Leclerc, M.H. Siess, N. Latruffe and M. Suschetet, *Toxicol. Appl. Pharmacol.*, 1996, **136**, 348.
59. H.P. Ciolino, P.J. Daschner and G.C. Yeh, *Biochem. J.*, 1999, **340**, 715.
60. E.H. Han, J.Y. Kim and H.G. Jeong, *Arch. Pharmacal Res.*, 2006, **29**, 570.
61. J.E. Lee and S. Safe, *Biochem. Pharmacol.*, 2001, **62**, 1113.
62. Q. Feng, Y. Torii, K. Uchida, Y. Nakamura, Y. Hara and T. Osawa, *J. Agric. Fd. Chem.*, 2002, **50**, 213.
63. M. Backlund and M. Ingelman-Sundberg, *Cell. Signal.*, 2005, **17**, 39.
64. G. Agullo, L. Gamet-Payrastre, S. Manenti, C. Viala, C. Remesy, H. Chap and B. Payrastre, *Biochem. Pharmacol.*, 1997, **53**, 1649.

65. S.R. Faucette, T.-C. Zhang, R. Moore, T. Sueyoshi, C.J. Omiecinski, E.L. LeCluyse, M. Negishi and H. Wang, *J. Pharmacol. Exp. Ther.*, 2007, **320**, 72.
66. J.L. Raucy, *Drug Metab. Dispos.*, 2003, **31**, 533.
67. V. Krajka-Kuzniak and W. Baer-Dubowska, *Toxicol. Lett.*, 2003, **143**, 209.
68. S.E. Nielsen, V. Breinholt, U. Justesen, C. Cornett and L.O. Dragsted, *Xenobiotica*, 1998, **28**, 389.
69. D.W. Roberts, D.R. Doerge, M.I. Churchwell, G. Gamboa da Costa, M.M. Marques and W.H. Tolleson, *J. Agric. Fd. Chem.*, 2004, **52**, 6623.
70. K.M. Atherton, E. Mutch and D. Ford, *Biochem. Pharmacol.*, 2006, **72**, 624.
71. V.M. Breinholt, S.E. Rasmussen, K. Brosen and T.H. Friedberg, *Pharmacol. Toxicol.*, 2003, **93**, 14.
72. W.H. Tolleson, D.R. Doerge, M.I. Churchwell, M.M. Marques and D.W. Roberts, *J. Agric. Fd. Chem.*, 2002, **50**, 4783.
73. J.A. Boutin, F. Meunier, P.H. Lambert, P. Hennig, D. Bertin, B. Serkiz and J.P. Volland, *Drug Metab. Dispos.*, 1993, **21**, 1157.
74. A.J. Day, F. Mellon, D. Barron, G. Sarrazin, M.R. Morgan and G. Williamson, *Free Radic. Res.*, 2001, **35**, 941.
75. D.M. Goldberg, J. Yan and G.J. Soleas, *Clin. Biochem.*, 2003, **36**, 79.
76. C. Yu, Y.G. Shin, A. Chow, Y. Li, J.W. Kosmeder, Y.S. Lee, W.H. Hirschelman, J.M. Pezzuto, R.G. Mehta and R.B. van Breemen, *Pharm. Res.*, 2002, **19**, 1907.
77. M. Dickins, C.R. Elcombe, S.J. Moloney, K.J. Netter and J.W. Bridges, *Biochem. Pharmacol.*, 1979, **28**, 231.
78. M. Murray, C.F. Wilkinson and C.E. Dube, *Toxicol. Appl. Pharmacol.*, 1983, **68**, 66.
79. L.S. Yu, C.F. Wilkinson and M.W. Anders, *Biochem. Pharmacol.*, 1980, **29**, 1113.
80. C.B. Marcus, M. Murray and C.F. Wilkinson, *Xenobiotica*, 1985, **15**, 351.
81. J.C. Cook and E. Hodgson, *Chem.-Biol. Interact.*, 1985, **54**, 299.
82. P. Chatterjee and M.R. Franklin, *Drug Metab. Dispos.*, 2003, **31**, 1391.
83. J.M. Mathews, A.S. Etheridge and S.R. Black, *Drug Metab. Dispos.*, 2002, **30**, 1153.
84. B.J. Gurley, S.F. Gardner, M.A. Hubbard, D.K. Williams, W.B. Gentry, I.A. Khan and A. Shah, *Clin. Pharmacol. Ther.*, 2005, **77**, 415.
85. T. Usia, T. Watabe, S. Kadota and Y. Tezuka, *Life Sci.*, 2005, **76**, 2381.
86. S. Koul, J.L. Koul, S.C. Taneja, K.L. Dhar, D.S. Jamwal, K. Singh, R.K. Reen and J. Singh, *Bioorg. Med. Chem.*, 2000, **8**, 251.
87. A.B. Swanson, E.C. Miller and J.A. Miller, *Biochim. Biophys. Acta*, 1981, **673**, 504.
88. Y.F. Ueng, C.H. Hsieh and M.J. Don, *Fd. Chem. Toxicol.*, 2005, **43**, 707.
89. S.M. Jeurissen, J.J. Bogaards, H.M. Awad, M.G. Boersma, W. Brand, Y.C. Fiamegos, T.A. van Beek, G.M. Alink, E.J. Sudholter, N.H. Cnubben and I.M. Rietjens, *Chem. Res. Toxicol.*, 2004, **17**, 1245.

90. C. Ioannides, M. Delaforge and D.V. Parke, *Chem.-Biol. Interact.*, 1985, **53**, 303.
91. M. Murray, C.F. Wilkinson, C. Marcus and C.E. Dube, *Mol. Pharmacol.*, 1983, **24**, 129.
92. M.H. Kang, S.M. Won, S.S. Park, S.G. Kim, R.F. Novak and N.D. Kim, *Xenobiotica*, 1994, **24**, 1195.
93. C.K. Atal, R.K. Dubey and J. Singh, *J. Pharmacol. Exp. Ther.*, 1985, **232**, 258.
94. D.J. Waxman, G.A. Dannan and F.P. Guengerich, *Biochemistry*, 1985, **24**, 4409.
95. J.C. Cook and E. Hodgson, *Chem.-Biol. Interact.*, 1986, **58**, 233.
96. N.H. Adams, P.E. Levi and E. Hodgson, *Chem.-Biol. Interact.*, 1993, **86**, 255.
97. D.Y. Ryu, P.E. Levi, P. Fernandez-Salguero, F.J. Gonzalez and E. Hodgson, *Mol. Pharmacol.*, 1996, **50**, 443.
98. R. Voorman and S.D. Aust, *Arch. Biochem. Biophys.*, 1988, **262**, 76.
99. S. Peterson, J.W. Lampe, T.K. Bammler, K. Gross-Steinmeyer and D.L. Eaton, *Fd. Chem. Toxicol.*, 2006, **44**, 1474.
100. M. Prince, C.T. Campbell, T.A. Robertson, A.J. Wellsm and H.E.. Kleiner, *Carcinogenesis*, 2006, **27**, 1204.
101. J. Maenpaa, H. Sigusch, H. Raunio, T. Syngelma, P. Vuorela, H. Vuorela and O. Pelkonen, *Biochem. Pharmacol.*, 1993, **45**, 1035.
102. L.L. Koenigs and W.F. Trager, *Biochemistry*, 1998, **37**, 10047.
103. L.L. Koenigs, R.M. Peter, S.J. Thompson, A.E. Rettie and W.F. Trager, *Drug Metab. Dispos.*, 1997, **25**, 1407.
104. K.S. Lown, D.G. Bailey, R.J. Fontana, S.K. Janardan, C.H. Adair, L.A. Fortlage, M.B. Brown and W. Guo, *J. Clin. Invest.*, 1997, **99**, 2545.
105. W. Tassaneeyakul, L.Q. Guo, K. Fukuda, T. Ohta and Y. Yamazoe, *Arch. Biochem. Biophys.*, 2000, **378**, 356.
106. H.L. Lin, U.M. Kent and P.F. Hollenberg, *J. Pharmacol. Exp. Ther.*, 2005, **313**, 154.
107. M.F. Paine, A.B. Criss and P.B. Watkins, *Drug Metab. Dispos.*, 2004, **32**, 1146.
108. M.F. Paine, A.B. Criss and P.B. Watkins, *J. Pharmacol. Exp. Ther.*, 2005, **312**, 1151.
109. W.J. Blot, J.Y. Li, P.R. Taylor, W. Guo, S. Dawsey, G.Q. Wang, C.S. Yang, S.F. Zheng, M. Gail, G.Y. Li, Y. Yu, B. Liu, J. Tangrea, Y. Sun, F. Liu, J.F. Fraumeni, Y.H. Zhang and B. Li, *J. Nat. Canc. Inst.*, 1993, **85**, 1483.
110. S. Gradelet, J. Leclerc, M.H. Siess and P.O. Astorg, *Xenobiotica*, 1996, **26**, 909.
111. C. Jewell and N.M. O'Brien, *Brit. J. Nutr.*, 1999, **81**, 235.
112. P. Astorg, S. Gradelet, R. Berges and M. Suschetet, *Nutr. Canc.*, 1997, **29**, 60.
113. R. Rauscher, R. Edenharder and K.L. Platt, *Mutat. Res.*, 1998, **413**, 129.

114. S. Gradelet, P. Astorg, J. Leclerc, J. Chevalier, M.F. Vernevaut and M.H. Siess, *Xenobiotica*, 1996, **26**, 49.
115. P. Astorg, S. Gradelet, J. Leclerc, M.C. Canivenc and M.H. Siess, *Fd. Chem. Toxicol.*, 1994, **32**, 735.
116. M. Paolini, A. Antelli, L. Pozzetti, D. Spetlova, P. Perocco, L. Valgimigli, G.F. Pedulli and G. Cantelli-Forti, *Carcinogenesis*, 2001, **22**, 1483.
117. B. Tan and F.L. Chu, *Am. J. Clin. Nutr.*, 1991, **53**, 1071S.
118. M. Gillner, J. Bergman, C. Cambillau, B. Fernstrom and J.-Å. Gustafsson, *Mol. Pharmacol.*, 1985, **28**, 357.
119. A. Kistler, H. Liechti, L. Pichard, E. Wolz, G. Oesterhelt, A. Hayes and P. Maurel, *Arch. Toxicol.*, 2002, **75**, 665.
120. M. Paolini, P. Perocco, D. Canistro, L. Valgimigli, G.F. Pedulli, R. Iori, C.D. Croce, G. Cantelli-Forti, M.S. Legator and S.Z. Abdel-Rahman, *Carcinogenesis*, 2004, **25**, 61.
121. T.C. Goosen, U.M. Kent, L. Brand and P.F. Hollenberg, *Chem. Res. Toxicol.*, 2000, **13**, 1349.
122. M. Nakajima, R. Yoshida, N. Shimada, H. Yamazaki and T. Yokoi, *Drug Metab. Dispos.*, 2001, **29**, 1110.
123. S. Barcelo, J.M. Gardiner, A. Gescher and J.K. Chipman, *Carcinogenesis*, 1996, **17**, 277.
124. C.C. Conaway, D. Jiao and F.L. Chung, *Carcinogenesis*, 1996, **17**, 2423.
125. I. Leclercq, J.P. Desager and Y. Horsmans, *Clin. Pharmacol. Ther.*, 1998, **64**, 144.
126. N. Caporaso, J. Whitehouse, S. Monkman, C. Boustead, H. Issaq, S. Fox, M.A. Morse, J.R. Idle and F.L. Chung, *Pharmacogenetics*, 1994, **4**, 275.
127. L.B. von Weymarn, J.A. Chun and P.F. Hollenberg, *Carcinogenesis*, 2006, **27**, 782.
128. M.A. Morse, *Exp. Lung Res.*, 1998, **24**, 595.
129. C.R. Morris, S.C. Chen, L. Zhou and L.M. Schopfer, *Nutr. Canc.*, 2004, **48**, 54.
130. Y.J. Surh, R.C. Lee, K.K. Park, S.T. Mayne, A. Liem and J.A. Miller, *Carcinogenesis*, 1995, **16**, 2467.
131. B.J. Gurley, S.F. Gardner, M.A. Hubbard, D.K. Williams, W.B. Gentry, Y. Cui and C.Y. Ang, *Clin. Pharmacol. Ther.*, 2002, **72**, 276.
132. J.S. Markowitz, C.L. Devane, K.D. Chavin, R.M. Taylor, Y. Ruan and J.L. Donovan, *Clin. Pharmacol. Ther.*, 2003, **74**, 170.
133. S.C. Piscitelli, A.H. Burstein, N. Welden, K.D. Gallicano and J. Falloon, *Clin. Inf. Dis.*, 2002, **34**, 234.
134. K. Gallicano, B. Foster and S. Choudhri, *Brit. J. Clin. Pharmacol.*, 2003, **55**, 199.
135. M. Xu, A.C. Bailey, J.F. Hernaez, C.R. Taoka, H.A. Schut and R.H. Dashwood, *Carcinogenesis*, 1996, **17**, 1429.
136. Y.H. He, M.D. Friesen, R.J. Ruch and H.A. Schut, *Fd. Chem. Toxicol.*, 2000, **38**, 15.
137. O. Vang, H. Frandsen, K.T. Hansen, J.B. Nielsen and O. Andersen, *Pharmacol. Toxicol.*, 1999, **84**, 59.

138. H. Liu, M. Wormke, S.H. Safe and L.F. Bjeldanes, *J. Nat. Canc. Inst.*, 1994, **86**, 1758.
139. J.T. Sanderson, L. Slobbe, G.W. Lansbergen, S. Safe and M. van den Berg, *Toxicol. Sci.*, 2001, **61**, 40.
140. J. Pan, J.Y. Hong, D. Li, E.G. Schuetz, P.S. Guzelian, W. Huang and C.S. Yang, *Biochem. Pharmacol.*, 1993, **45**, 2323.
141. C.K. Lii, C.W. Tsai and C.C. Wu, *J. Agric. Fd. Chem.*, 2006, **54**, 5191.
142. C.C. Wu, L.Y. Sheen, H.W. Chen, W.W. Kuo, S.J. Tsai and C.K. Lii, *J. Agric. Fd. Chem.*, 2002, **50**, 378.
143. D. Canistro, C.D. Croce, R. Iori, J. Barillari, G. Bronzetti, G. Poi, M. Cini, L. Caltavuturo, P. Perocco and M. Paolini, *Mutat. Res.*, 2004, **545**, 23.
144. J. Guo, D. Nikolic, L.R. Chadwick, G.F. Pauli and R.B. van Breemen, *Drug Metab. Dispos.*, 2006, **34**, 1152.
145. T.K. Chang, J. Chen and X.W. Teng, *Drug Metab. Dispos.*, 2006, **34**, 234.
146. L.B. Moore, B. Goodwin, S.A. Jones, G.B. Wisely, C.J. Serabjit-Singh, T.M. Willson, J.L. Collins and S.A. Kliewer, *Proc. Natl. Acad. Sci. USA.*, 2000, **97**, 7500.
147. B. Goodwin, L.B. Moore, C.M. Stoltz, D.D. McKee and S.A. Kliewer, *Mol. Pharmacol.*, 2001, **60**, 427.
148. C.A. Austin, E.A. Shephard, S.F. Pike, B.R. Rabin and I.R. Phillips, *Biochem. Pharmacol.*, 1988, **37**, 2223.
149. A.C. De-Oliveira, L.F. Ribeiro-Pinto, S.S. Otto, A. Goncalves and F.J. Paumgartten, *Toxicol.*, 1997, **124**, 135.
150. A.E. Wade, J.E. Holl, C.C. Hilliard, E. Molton and F.E. Greene, *Pharmacol.*, 1968, **1**, 317.
151. A.M. Malkinson, *Toxicol. Appl. Pharmacol.*, 1979, **49**, 551.
152. A.M. Malkinson and W.C. Shere, *Res. Commun. Chem. Pathol. Pharmacol.*, 1979, **25**, 607.
153. S.J. Roffey, R. Walker and G.G. Gibson, *Fd. Chem. Toxicol.*, 1990, **28**, 403.
154. A.C. De-Oliveira, L.F. Ribeiro-Pinto and J.R. Paumgartten, *Toxicol. Lett.*, 1997, **92**, 39.
155. M.A. Morse and A.L. Toburen, *Canc. Lett.*, 1996, **104**, 211.
156. M. Miyazawa, M. Shindo and T. Shimada, *Drug Metab. Dispos.*, 2002, **30**, 602.
157. M. Duisken, F. Sandner, B. Blomeke and J. Hollender, *Biochim. Biophys. Acta*, 2005, **1722**, 304.
158. M. Miyazawa, M. Shindo and T. Shimada, *Xenobiotica*, 2001, **31**, 713.

CHAPTER 15

Cytochromes P450 in Cancer Therapeutics

THOMAS K.H. CHANG

Faculty of Pharmaceutical Sciences, The University of British Columbia
Vancouver, British Columbia, V6T 1Z3, Canada

Table of Contents

Issues in Toxicology
Cytochromes P450: Role in the Metabolism and Toxicity of Drugs and other Xenobiotics
Edited by Costas Ioannides

15.1 Introduction

The standard treatment modalities available to cancer patients are surgery, radiation, and chemotherapy. Each of these modalities has its advantages and disadvantages. Considerable effort has been spent on the discovery and development of novel cytotoxic pharmaceuticals. However, the anticancer drug of first choice in the clinical management of most types of cancer, especially solid tumours, has remained more or less the same over the years. In general, lack of tumour selectivity is a major limitation of the anticancer drugs prescribed in current clinical practice. These drugs usually have a narrow therapeutic index, so that cancer chemotherapy is accompanied by adverse symptoms, many of which are severe and dose-limiting. Given that successful outcomes in anticancer drug development programs are few and far between, it is important to devise novel and innovative strategies to improve the therapeutic outcome of the anticancer drugs currently in use in clinical oncology. It is also important to enhance our understanding of tumour biology and the pharmacokinetics and pharmacodynamics of existing anticancer drugs, which may lead to the identification of previously unrecognised therapeutic targets and new therapeutic approaches.

Many of the currently available anticancer drugs are substrates for the various enzymes involved in drug biotransformation. Some of these are prodrugs, which require enzymic activation to produce the pharmacologically active metabolite(s), whereas others undergo enzymic inactivation. Members of the cytochrome P450 superfamily of enzymes play an important role in the NADPH-dependent biotransformation of a broad range of chemicals, including many anticancer drugs. Innovations and important discoveries made over the years have led to the identification of cytochromes P450 as potential therapeutic targets for cancer therapy. The overall aim of this chapter is to inform the readers of the exciting developments in cytochrome P450-based cancer therapeutics, particularly in the area of gene-directed enzyme prodrug therapy (GDEPT). This chapter will begin with an overview of 1) the role of individual enzymes in the CYP1, CYP2, and CYP3 families as catalysts of anticancer drug biotransformation and 2) tumoural expression of cytochromes P450.

15.2 Human Cytochromes P450 as Catalysts of Anticancer Drug Metabolism

The human cytochrome P450 superfamily of enzymes is comprised of 18 families, 42 subfamilies, and 57 individual genes.[1] Enzymes in the CYP1, CYP2, and CYP3 families are particularly important in the biotransformation of drugs. Several complementary experimental approaches are used to identify the individual cytochrome P450 catalysts of a drug activation or detoxification reaction.[2] These include: a) catalysis by purified or recombinant cytochrome P450 enzyme; b) correlational analysis; c) cytochrome P450 enzyme-selective (or subfamily-selective) chemical inhibitors; and d) inhibitory monospecific (or subfamily-specific) anti-cytochrome P450 antibodies. In the case of a cytotoxic drug, another approach is to determine the effect of alteration in biotransformation on drug toxicity in cell-based assays, such as those involving cells transfected with a cytochrome P450 cDNA,[3] or other cell-based assays conducted in the presence of a microsomal activation system (e.g. hepatic microsomes or cDNA-expressed cytochrome P450 enzymes).[4]

Individual human cytochrome P450 enzymes have been identified as catalysts in the biotransformation of various drugs used in clinical oncology and of experimental anticancer drugs (Table 15.1). These include: 1) prodrugs in which cytochrome P450-mediated metabolic activation is required to produce pharmacological activity (i.e., drug bioactivation); 2) parent drugs with pharmacological activity, but the overall effect may be enhanced by cytochrome P450-mediated metabolism; and 3) parent drugs with pharmacological activity, but the effect is abolished by cytochrome P450-mediated metabolism (i.e., drug inactivation). Of particular relevance to this chapter are the anticancer prodrugs.

15.2.1 Anticancer Prodrugs Bioactivated by Cytochromes P450

15.2.1.1 Cyclophosphamide

Cyclophosphamide is a commonly used anticancer drug. It is effective against a variety of solid tumours and haematological malignancies.[5] This oxazaphosphorine prodrug requires bioactivation in order to exert its bifunctional DNA-alkylating action and antitumour activity.[6] CYP2B6 is the major

Table 15.1 Anticancer drugs as substrates for individual human cytochrome P450 enzymes.

Drug	*Metabolic reaction*	*Human cytochrome P450 enzyme*	*References*
Alkylating Agents			
Cyclophosphamide	4-Hydroxylation[a]	**CYP2B6**, CYP2A6, CYP2C8, CYP2C9, CYP2C19	7,11,11–14
	N-Dechloroethylation	**CYP3A**	10,15,16,187
Dacarbazine	*N*-Demethylation[a]	**CYP1A2**, CYP1A1, CYP2E1	23
(R)-Ifosfamide	4-Hydroxylation[a]	CYP3A4 > CYP3A5	20
	N-Dechloroethylation	**CYP3A**	20,21
(S)-Ifosfamide	4-Hydroxylation[a]	CYP3A4 > CYP2B6 > CYP3A5	20
	N-Dechloroethylation	**CYP2B6**	20,21
Procarbazine	Apoprocarbazine Formation[a]	not yet identified	25
Thio-TEPA	Oxidative Desulphuration[a]	CYP3A4, CYP2B6	157
Taxanes			
Docetaxel (taxotere)	Hydroxylation[b]	**CYP3A**	115–117
Paclitaxel	6-Hydroxylation	**CYP2C8**	188,189
	3′-*p*-Hydroxylation	**CYP3A**	189,190
Epipodophyllotoxins			
Etoposide (VP-16)	*O*-demethylation[a]	**CYP3A**	191
Teniposide (VM-26)	*O*-demethylation[a]	**CYP3A**	191
Vinca Alkaloids			
Vinblastine	Unidentified	CYP3A	108,192
Vincristine	M1 metabolite formation	**CYP3A5**, CYP3A4	193
Vindesine	Unidentified	CYP3A	194
Vinorelbine	Unidentified	CYP3A	195

Table 15.1 (*Continued*).

Drug	*Metabolic reaction*	*Human cytochrome P450 enzyme*	*References*
Camptothecins			
Irinotecan (CPT-11)	Oxidation	CYP3A	196–198
Selective Oestrogen-Receptor Modulators			
Tamoxifen	*N*-demethylation	**CYP2D6**, **CYP3A4**, CYP1A1, CYP1A2, CYP3A4	199,200
	4-Hydroxylation[a]	**CYP2D6**, CYP2B6, CYP2C9, CYP3A4, CYP3A5	199,201
		CYP1A1, CYP1A2	
	α-Hydroxylation	**CYP2D6**, CYP3A4, CYP1A1, CYP1A2, CYP1B1	200
		CYP2C9	
Androgen Receptor Antagonists			
Flutamide	2-Hydroxylation[a]	**CYP1A2**, CYP1A1, CYP1B1	167,169
Other Anticancer Drugs			
All-*trans*-retinoic acid	4-Hydroxylation	CYP26A1, CYP3A7, CYP3A5 CYP2C8, CYP3A4	202,203
	18-Hydroxylation	CYP26A1, CYP3A7, CYP4A11, CYP2C8	202,203
	4-Oxidation	CYP26A1, CYP3A7, CYP4A11	202,203
	5,6-Epoxidation	CYP2C8, CYP2C9, CYP1A1, CYP3A4, CYP3A5	203
Gefitinib	M523595 formation	**CYP3A**	204
Medroxyprogesterone	Unidentified	**CYP3A**	205
Experimental Anticancer Drugs			
AQ4N	Reduction	CYP3A	42
Methoxymorpholinyl doxorubicin	Unidentified[a]	**CYP3A**	4
Morpholinyl doxorubicin	Unidentified[a]	**CYP3A**	206
Tegafur	5′-Hydroxylation[a]	**CYP2A6**, CYP1A2, CYP2C8	27,28,30
Thalidomide	5-Hydroxylation	**CYP2C19**	207
	5′-Hydroxylation	**CYP2C19**, CYP2B6	207

Prodrugs are in italics.
Enzyme shown in bold denotes major catalyst.
[a]Leads to formation of pharmacologically active metabolite.

human cytochrome P450 enzyme that catalyses the bioactivation of cyclophosphamide to form 4-hydroxycyclophosphamide.[7] This metabolite is not cytotoxic, but undergoes spontaneous ring opening to form aldophosphamide, which is chemically decomposed to the DNA-alkylating phosphoramide mustard[8], and a urotoxic byproduct, acrolein.[9] Other human hepatic microsomal cytochrome P450 enzymes that contribute to cyclophosphamide 4-hydroxylation include CYP2C9[10] and, to a lesser extent, CYP3A,[7,10] and possibly CYP2A6.[7] Pharmacogenetic analysis indicates that the CYP2B6*6 allele (Gln to His substitution at amino acid 172) is associated with higher cyclophosphamide 4-hydroxylation activity, when compared with the wild-type CYP2B6*1.[11] CYP2C enzymes, such as CYP2C8,[12] CYP2C9,[12,13] and CYP2C19,[12,13] are catalytically active in cyclophosphamide 4-hydroxylation, but they are less efficient than CYP2B6*1.[11,12] Among the various CYP2C9 allelic variants, CYP2C9*3 (Ile to Leu substitution at amino acid 359) is less active than the wild-type CYP2C9*1 in cyclophosphamide 4-hydroxylation.[12] A human pharmacogenetic study reported an association between the CYP2C9*2/*2 genotype and decreased elimination rate constant for cyclophosphamide.[14]

An inactivation pathway of cyclophosphamide involves cytochrome P450-catalysed side-chain oxidation.[8] The *N*-dechloroethylation of cyclophosphamide yields the pharmacologically inactive 2-dechloroethylcyclophosphamide and a neurotoxic and nephrotoxic byproduct, chloroacetaldehyde, in a reaction catalysed almost exclusively by CYP3A in human hepatic microsomes.[10,15,16] Chemical inhibition of CYP3A-catalysed inactivation of cyclophosphamide appears to be beneficial in that it reduces the systemic exposure to chloroacetaldehyde, but without a loss of antitumour activity, as demonstrated in an *in vivo* rat tumour model.[17]

15.2.1.2 Ifosfamide

Ifosfamide, an isomer analogue of cyclophosphamide, is also a DNA-alkylating agent. This prodrug is a chiral molecule, that is administered clinically as a racemic mixture. Ifosfamide undergoes bioactivation to form 4-hydroxyifosfamide. Similar to cyclophosphamide, the end-products of the bioactivation of isofamide are a mustard derivative (ifosfamide mustard) and the urotoxic acrolein. The side-chain inactivation products of ifosfamide are the pharmacologically inactive 2-dechloroethylifosfamide and 3-dechloroethylifosfamide and a neurotoxic and nephrotoxic byproduct, chloroacetaldehyde.[18] In contrast to cyclophosphamide 4-hydroxylation, ifosfamide 4-hydroxylation is catalysed not by CYP2B6, but by CYP3A, in human hepatic microsomes, as determined in experiments performed with racemic ifosfamide.[7] CYP2A6 appears to contribute minimally to ifosfamide 4-hydroxylation in human hepatic microsomes.[7] Recombinant CYP2C8, CYP2C9, and CYP2C19 are also catalytically active in ifosfamide 4-hydroxylation, with the CYP2C9*3 allelic variant being less efficient than the wild-type CYP2C9*1.[12] Whereas CYP3A appears to be the sole contributor to cyclophosphamide

N-dechloroethylation in human hepatic microsomes,[10,15,16] both CYP3A (~70%) and CYP2B6 (~25%) contribute to ifosfamide *N*-dechloroethylation.[16] Consistent with the *in vitro* data on a role for CYP3A in ifosfamide biotransformation, a randomised crossover human study reported decreased ifosfamide activation by an inhibitor of CYP3A, ketoconazole, whereas rifampin, an inducer of CYP3A, increased both the activation and side-chain inactivation of ifosfamide.[19]

Ifosfamide undergoes stereoselective 4-hydroxylation[20] and *N*-dechloroethylation.[20,21] CYP3A is more active in (*R*)-ifosfamide 4-hydroxylation, whereas CYP2B6 is more active in (*S*)-ifosfamide 4-hydroxylation.[20] With respect to side-chain oxidation, CYP3A preferentially catalyses the formation of (*R*)-2-dechloroethylifosfamide and (*R*)-3-dechloroethylifosfamide.[20,21] In contrast, CYP2B6 preferentially forms (*S*)-2-dechloroethylifosfamide and (*S*)-3-dechloroethylifosfamide.[20,21]

15.2.1.3 Dacarbazine

Dacarbazine is active against human melanoma, Hodgkins's disease, and soft tissue sarcoma.[22] It is a prodrug that requires oxidative metabolism to produce DNA methylation activity. CYP1A2 is the major human hepatic cytochrome P450 enzyme that catalyses the *N*-demethylation of dacarbazine to form 5-[3-methyl-triazen-1-yl]-imidazole-4-carboxamide,[23] which yields 5-[3-hydroxy-methyl-3-methyl-triazen-1-yl]-imidazole-4-carboxamide. This metabolite decomposes to amino-imidazole carboxamide and methane diazohydroxide. The ultimate DNA methylation species is thought to be a methyl carbonium ion, formed as a byproduct of the breakdown of methane diazohydroxide.[24] Other catalysts of dacarbazine *N*-demethylation include CYP1A1 and CYP2E1.[23]

15.2.1.4 Procarbazine

Procarbazine is an oral alkylating prodrug used as part of a combination chemotherapy for the management of Hodgkin's disease, non-Hodgkin's lymphoma, and brain tumours.[22] The human cytochrome P450 enzymes that catalyse the bioactivation of procarbazine have not been identified, although an earlier study indicated a role for rat CYP1A and CYP2B.[25]

15.2.1.5 Tegafur

Tegafur is a prodrug of 5-fluorouracil (5-FU),[26] which is a pyrimidine antagonist. CYP2A6 is the major catalyst of the hepatic microsomal bioactivation of tegafur to form 5′-hydroxytegafur,[27] which decomposes spontaneously to form 5-FU. Other catalysts of tegafur 5′-hydroxylation are CYP1A2 and CYP2C8.[28] Consistent with a role of CYP2A6 in tegafur bioactivation, treatment of CYP2A6-transfected cells with tegafur confers increased drug sensitivity.[29] Pharmacogenetic analysis indicates that the CYP2A6*4C and CYP2A6*11 mutant alleles are associated with increased plasma levels of tegafur as a result of decreased conversion of the parent drug to 5-FU.[30]

15.2.1.6 Methoxymorpholinyl Doxorubicin

Methoxymorpholinyl doxorubicin is an analogue of doxorubicin.[31] This experimental agent has a better toxicity profile, particularly with respect to cardiotoxcity, which is a dose-limiting toxicity of doxorubicin.[32] Methoxymorpholinyl doxorubicin is a prodrug that requires bioactivation to produce its DNA strand-breaking activity *via* topoisomerase I cleavage. Although the identity of the active metabolite(s) is still not known, it appears to be an *O*-demethylated product formed in a reaction catalysed by CYP3A4, but not CYP3A5 or CYP3A7, as assessed in a microsome prodrug co-culture assay.[4]

15.2.1.7 Thalidomide

Thalidomide was used originally in the management of morning sickness in pregnant women. However, the drug was withdrawn from the market by 1962 because of severe teratogenicity and dysmyelia.[33] Thalidomide has attracted interest in oncology over the past decade because of its inhibitory effect on angiogenesis.[34] This experimental anticancer drug has therapeutic activity against multiple myeloma[35] and may also be useful in other forms of haematological malignancies[36] and in solid tumours, such as androgen-independent prostate tumour.[37] The antiangiogenic action of this drug requires metabolic activation.[38] Among the two hydroxylated metabolites of thalidomide identified to date, 5′-hydroxythalidomide, but not 5-hydroxythalidomide, is a moderately effective inhibitor of angiogenesis.[39] CYP2C19 and, to a lesser extent, CYP2B6, catalyse thalidomide 5′-hydroxylation, whereas CYP2C19, but not CYP2B6, catalyses thalidomide 5-hydroxylation.[40]

15.2.1.8 1,4-Bis[2-(dimethylamino-N-oxide)ethyl]amino-5, 8-di-hydroxyanthracene-9,10-dione (AQ4N)

AQ4N is a prodrug that becomes activated under hypoxic conditions, and as a result, it exhibits preferential selectivity toward hypoxic tumour cells.[41] It is bioreduced by human hepatic microsomal cytochrome P450 enzymes,[42] but the identity of the individual cytochrome P450 enzymes capable of catalysing AQ4N bioreduction has not been characterised in detail. Initial evidence suggests a role for CYP3A, based on results from experiments with cytochrome P450 enzyme-selective chemical inhibitors.[42]

15.3 Drug-Metabolising Cytochrome P450 Enzyme Expression in Human Tumours

Studies have been conducted over the past decade with the aim of characterising the tumoural expression of specific drug-metabolising human cytochrome P450 enzymes,[43–45] namely those in the CYP1, CYP2, and CYP3 families.[1] Most effort has been directed toward solid malignant tumours.[43] Among these, breast tumours are the most extensively characterised (Table 15.2). Other solid tumours analysed for the expression of specific CYP1, CYP2, and CYP3 enzymes include those of the colon,[46–57] lung,[49,50,54,56–64] ovary,[49,50,56,65,66] bladder,[49,50,56,67–69] oesophagus,[49,50,54,70–72] kidney,[50,56,73–75] liver,[49,50,54,56,76–80] prostate,[56,69,81–83] stomach,[49,50,54,56,84,85] brain,[49,50,86] and testis.[49,56] A detailed examination of the published data in the literature reveals that conflicting results exist regarding tumoural expression of cytochromes P450. For example, as shown in Table 15.2, CYP1A protein was detected in 21 of 54 samples in one study[87] and 81 of 393 samples in another study,[88] whereas it was not detected in any of the 12 samples analysed in a third study.[89] Similarly, CYP2B6 protein was present in 25 of 33 samples in one study, but absent in each of the 10 samples in another study.[90] The conflicting data may be explained by a variety of reasons, including: 1) the use of different methodologies; for example, immunohistochemistry *vs.* immunoblotting; 2) the use of antibodies that were prepared against a cytochrome P450 protein from a species other than human (e.g. rat or mouse); 3) quality/age of the tissue specimens; 4) differences in the pathological nature of the tumoural tissues; and 5) ethnic diversity in the subject populations; for example, as assessed by immunohistochemistry, CYP3A protein was not detected in a panel of 34 breast tumour specimens from Japanese patients.[91]

In contrast to the findings in some of the earlier studies which reported a lack of cytochrome P450 expression in human tumour samples,[47,89] a general conclusion from the subsequent studies is that cytochrome P450 enzymes are expressed in the various solid malignant tumours. Based on the published information to date, the following human CYP1, CYP2 and CYP3 proteins have been detected in solid malignant tumours: 1) CYP1B1, CYP2C8, CYP2C9 and CYP2J2 in bladder tumours;[49,56] 2) CYP1B1 in brain tumours;[49] 3) CYP1A1, CYP1A2, CYP1B1, CYP2B6, CYP2C8, CYP2C9, CYP2D6, CYP2E1 and CYP3A4 in breast tumours;[49,56,87,88,90,92–95] 4) CYP1B1, CYP2C8, CYP2C9, CYP2D6, CYP2E1, CYP3A4, CYP3A5 and CYP3A7 in colon tumours;[49,52,55,56] 5) CYP1A, CYP1B1 and CYP3A in oesophageal tumours;[49,70] 6) CYP1B1, CYP2C8, CYP2C9 and CYP3A in renal tumours;[49,56,75] 7) CYP1A, CYP2A6, CYP2C9 and CYP3A in hepatic tumours;[50,56,76–78] 8) CYP1A, CYP1B1, CYP2C8, CYP2C9, CYP2E1 and CYP3A in pulmonary tumours;[49,56,58,60,64,96] 9) CYP1B1, CYP2C8, CYP2C9, CYP2D6, CYP2E1, CYP3A4 and CYP3A5 in ovarian tumours;[49,56,65,66] 10) CYP1A, CYP1B1, CYP2C8, CYP2C9 and CYP3A in prostate tumours;[56,69,81] 11) CYP1A, CYP1B1, CYP2C9 and CYP3A in stomach tumours;[49,56,84,85] and 12) CYP1B1, CYP2C8 and CYP2C9 in testicular tumours.[49,56] Collectively, these findings indicate that CYP1B1 expression is a common feature of these tumours.

Table 15.2 Detection of individual CYP1, CYP2, and CYP3 enzymes and of the pregnane X receptor in human breast tumours.

	Level of expression	*Methodology*	*Occurrence*[a]	*Reference*
CYP1A	protein	immunohistochemistry[b]	21 of 54	87
CYP1A	protein	immunoblotting[c]	0 of 12	89
CYP1A	protein	immunohistochemistry	81 of 393	88
CYP1A1	mRNA	RT-PCR[d]	6 of 13	93
CYP1A1	mRNA	RT-PCR[e]	1 of 20	208
CYP1A1	protein	immunoblotting	29 of 33	94
CYP1A2	mRNA	RT-PCR[d]	2 of 13	93
CYP1A2	protein	immunohistochemistry	51 of 393	88
CYP1B1	mRNA	RT-PCR[d]	12 of 13	93
CYP1B1	mRNA	RT-PCR[e]	20 of 20	208
CYP1B1	protein	immunohistochemistry	12 of 12	49
CYP1B1	protein	immunohistochemistry and immunoblotting	2 of 2	181
CYP1B1	protein	immunohistochemistry	28 of 34	91
CYP1B1	protein	immunohistochemistry	65 of 393	88
CYP2A6	mRNA	RT-PCR[e]	1 of 14	208
CYP2A6	protein	immunohistochemistry	1 of 34	91
CYP2A7	mRNA	RT-PCR[e]	0 of 7	208
CYP2A13	mRNA	RT-PCR[e]	0 of 7	208
CYP2B	protein	immunohistochemistry[c]	0 of 10	209
CYP2B	protein	immunoblotting[c]	0 of 12	89
CYP2B6	mRNA	RT-PCR[e]	12 of 14	208
CYP2B6	protein	immunoblotting	25 of 33	94
CYP2B6	protein	immunoblotting	0 of 10	90
CYP2C	protein	immunohistochemistry[c]	7 of 18	209
CYP2C	protein	immunoblotting[c]	23 of 23	209
CYP2C	protein	immunoblotting	0 of 12	89
CYP2C	mRNA	RT-PCR[d]	13 of 13	93
CYP2C	mRNA	RT-PCR[e]	5 of 6	208
CYP2C8	mRNA	RT-PCR[e]	10 of 10	92
CYP2C9	mRNA	RT-PCR[e]	10 of 10	92
CYP2C8	protein	immunoblotting	1 of 12	56
CYP2C9	protein	immunoblotting	9 of 9	92
CYP2C9	protein	immunohistochemistry	7 of 12	56
CYP2C9	protein	immunoblotting	9 of 9	90
CYP2C19	mRNA	RT-PCR[e]	0 of 10	92
CYP2D6	mRNA	RT-PCR[e]	5 of 6	208
CYP2D6	mRNA	RT-PCR[d]	10 of 13	93
CYP2E1	mRNA	RT-PCR[e]	5 of 5	208
CYP2E1	protein	immunoblotting[c]	0 of 12	89
CYP2E1	protein	immunoblotting	25 of 33	94
CYP2E1	protein	immunohistochemistry	25 of 25	95
CYP2E1	protein	immunohistochemistry	175 of 393	88
CYP2F1	mRNA	RT-PCR[e]	0 of 6	208
CYP2J2	protein	immunohistochemistry	7 of 12	56
CYP3A	protein	immunohistochemistry[c]	18 of 18	209
CYP3A	protein	immunohistochemistry[b]	12 of 54	87
CYP3A	protein	immunoblotting	0 of 12	89
CYP3A	protein	immunohistochemistry[b]	27 of 32	210

Table 15.2 (*Continued*).

	Level of expression	*Methodology*	*Occurrence*[a]	*Reference*
CYP3A	protein	immunohistochemistry	97 of 393	88
CYP3A	protein	immunohistochemistry	0 of 34	91
CYP3A4	mRNA	RT-PCR[d]	2 of 13	93
CYP3A4	mRNA	RT-PCR[e]	0 of 8	208
CYP3A4	protein	immunohistochemistry	25 of 25	95
CYP3A4	protein	immunoblotting	29 of 33	94
CYP3A4	protein	immunoblotting	11 of 11	90
CYP3A5	mRNA	RT-PCR[d]	2 of 13	93
CYP3A5	mRNA	RT-PCR[e]	0 of 4	208
CYP3A5	protein	immunoblotting	0 of 10	90
CYP3A7	mRNA	RT-PCR[e]	0 of 4	208
PXR	protein	immunohistochemistry[b]	47 of 60	105

[a]Number of positive cases out of the total number of individual samples analysed.
[b]Used anti-mouse CYP antibody.
[c]Used anti-rat CYP antibody.
In all other cases, anti-human CYP antibody was used.
[d]Block cycler PCR followed by southern blotting and chemiluminescence detection.
[e]Block cycler PCR followed by agarose gel electrophoresis and ethidium bromide staining.

CYP1B1 is an extrahepatic cytochrome P450 (see Chapter 4),[97,98] in contrast to most of the members of the CYP1, CYP2, and CYP3 families.[99] CYP1B1 is thought of as a tumour-specific cytochrome P450, based on immunohistochemical analysis of malignant tissues from 14 different types of tumour and the adjacent normal tissue sample from the same individual.[49] In this study, CYP1B1 protein was present in virtually all of the specimens analysed (sample size ranged from 6 to 12 for each tumour type), but undetectable in the normal tissue. However, as shown in subsequent investigations, CYP1B1 protein is expressed in normal tissues, such as colon[51,55] and lung.[64] It could be that the preferential expression of CYP1B1 in tumours relative to the corresponding normal tissue is dependent on the tumour type. CYP1B1 is present in the majority of the ovarian tumours analysed, but it has not been detected in any of the paired normal tissues analysed to date.[49,65,66] In breast carcinoma, oestrogen receptor status is a determinant of CYP1B1 expression. A lack of oestrogen receptor α is associated with a substantial increase in tumoural CYP1B1 expression.[100]

CYP2J2 is expressed in a variety of tissues, including the heart.[101] It catalyses the metabolic conversion of arachidonic acid to epoxyeicosatrienoic acids.[102] The tumour types in which CYP2J2 is overexpressed include squamous cell carcinoma of the oesophagus, adenocarcinoma of the lung and colon, small cell lung carcinoma, and carcinoma of the breast, stomach, and liver.[54] CYP2W1 also catalyses the metabolism of arachidonic acid and this enzyme appears to be selectively expressed in tumoral tissues of colon and lung.[57] In contrast, other cytochrome P450 enzymes are expressed to a lesser extent in tumours relative to the corresponding normal tissue. CYP1A1, CYP2B6, CYP2E1 and CYP3A4 protein levels were less in breast tumours than in the adjacent normal tissues, in an analysis of paired samples from the same individual.[94]

Several of the enzymes in the CYP2 and CYP3 families, such as CYP2B6 and CYP3A4, are regulated by a nuclear receptor known as pregnane X receptor (PXR).[103] Both PXR mRNA[104] and protein[105] have been detected in breast tumours. PXR gene expression is comparable between the tumoural and normal breast tissues.[104] Breast tumours that express the oestrogen receptor have a decreased level of PXR mRNA when compared with oestrogen receptor-negative tumours.[104]

15.3.1 Tumoural Drug Metabolism

The presence of cytochrome P450 enzymes in tumoural tissues has implications with respect to tumoural metabolism of anticancer drugs. It may provide for the local bioactivation of anticancer prodrugs to yield the pharmacologically active metabolite(s). It is also the basis for the development of novel anticancer prodrugs (see Section 15.5). In other cases, it may lead to the inactivation of pharmacologically active parent drugs; for example, the taxanes, such as paclitaxel[106] and docetaxel,[107] and the vinca alkaloids, such as vinblastine.[108] Although the levels of cytochrome P450 expression in tumours are generally much less than those in liver,[100] there is sufficient catalytic capacity in tumour microsomes to catalyse cytochrome P450-mediated anticancer drug metabolism. As shown by *in vitro* investigations, human breast tumour microsomes are capable of catalysing ifosfamide *N*-dechloroethylation,[90] which leads to the formation of pharmacologically inactive metabolites.[5] Similarly, human colon tumour microsomes metabolise paclitaxel to a pharmacologically inactive metabolite, 3′-*p*-hydroxypaclitaxel.[109]

15.3.2 Cytochrome P450 and Resistance to Cancer Chemotherapy

Tumoural cytochrome P450 expression may influence not only anticancer drug metabolism, but potentially also therapeutic efficacy. The therapeutic efficacy of many anticancer drugs is severely limited by drug resistance in tumour cells. Such resistance to chemotherapy may be intrinsic or acquired.[110] Intrinsic resistance refers to the situation in which the tumours are not responsive at all to the cancer chemotherapeutic drugs. In contrast, in acquired drug resistance, the tumours do respond initially to the cytotoxic agent, but they eventually become refractory to subsequent courses of chemotherapy. Various mechanisms are responsible for the development of resistance to a cancer chemotherapeutic drug, including overexpression of drug efflux transporters, such as P-glycoprotein.[111] In the case where the pharmacologically active parent drug is metabolised to inactive metabolites, overexpression of drug-inactivating enzymes is a potential mechanism of acquired resistance to a cancer chemotherapeutic drug. Relatively little is known about the potential role of cytochrome P450 in drug resistance. However, evidence indicates that cell lines experimentally manipulated to express high levels of CYP3A4 have a decreased sensitivity towards vinblastine.[108] Similar findings have been reported for paclitaxel in CYP2C8-expressing cells.[112] and docetaxel in CYP1B1-expressing cells.[107] Another line of evidence indicates that high CYP3A4 mRNA or protein levels in human breast tumours is associated with a poor treatment response to docetaxel, whereas low or a lack of CYP3A4 expression is associated with a greater response rate.[113,114] This may relate to the ability of docetaxel to induce CYP3A enzyme expression[63] because this drug is also a substrate for CYP3A.[115–117] Future studies will be required to determine whether cytochrome P450 enzymes play a role in clinical resistance to select anticancer drugs.

15.4 Cytochrome P450 Gene-Directed Enzyme Prodrug Therapy (GDEPT)

The administration to patients of conventional cancer chemotherapeutic drugs, at the usual therapeutic doses, often results in the development of various dose-limiting adverse effects. This primarily reflects the narrow therapeutic index of the drugs and their lack of specificity toward tumoural tissue. To overcome this major limitation, novel therapeutic strategies are needed with the overall goal of improving drug efficacy, while minimising systemic or host toxicity. An important area of cancer chemotherapeutic research over the past two decades has been the development and experimental application of gene therapy (or gene transfer) technologies as a means to specifically target the therapy toward tumour cells and increase tumour sensitivity.[118,119] Among the various candidate genes suitable for cancer gene therapy development are those that encode prodrug-activation enzymes.[120,121] This form of cancer gene therapy is referred to as gene-directed enzyme prodrug therapy (GDEPT).[122] In contrast to conventional cancer chemotherapy in which prodrug bioactivation occurs in cells (e.g. hepatocytes) that may be at some distance from the tumour cells, an advantage of GDEPT is that the bioactivation of the pharmacologically inactive prodrug occurs in tumour cells transduced with the prodrug-activation enzyme/suicide gene. The combination of the prodrug ganciclovir and the herpes simplex virus thymidine kinase (HSV-TK) enzyme represents the prototypic GDEPT.[123] As shown in Table 15.3, additional experimental GDEPT strategies involve anticancer prodrugs and the specific cytochromes P450 that catalyse their bioactivation.[124]

15.4.1 Oxazaphosphorine Anticancer Prodrug in Combination with a Cytochrome P450 Gene

The initial investigations in cytochrome P450-based GDEPT were conducted with the combination of an oxazaphosphorine prodrug, such as cyclophosphamide or ifosfamide, and the rat CYP2B1 gene.[125,126] CYP2B1 catalyses the activation of both of these prodrugs.[127,128] Although CYP2B1 is highly inducible by various drugs (e.g. phenobarbital) and other chemicals,[129] its

Table 15.3 Anticancer prodrug and gene combinations in cytochrome P450 gene-directed enzyme prodrug therapy.

Prodrug	*Gene*	*References*
Alkylating Agent		
Cyclophosphamide	CYP2B1	125,126,130,136,142–144, 211–214
	CYP2B1/NADPH-cytochrome P450 reductase	139,215
	CYP2B6	131,132,140,215,216
	CYP2B6/NADPH-cytochrome P450 reductase	131,217
	CYP2B11	135
	CYP2C8-MET-385	131
	CYP2C8-MET-385/ NADPH-cytochrome P450 reductase	131
	CYP2C9	131,218
	CYP2C19	131
	CYP3A4	131
Ifosfamide	CYP2B1	126,130,219–222
	CYP2B1/NADPH-cytochrome P450 reductase	141
Bioreductive Agent		
AQ4N	CYP1A1	151
	CYP2B6	150
	CYP3A4	149

constitutive expression is low or undetectable in normal cells. The local bioactivation of the anticancer prodrug should, therefore, occur mainly in transduced tumour cells, with little or no bioactivation in non-transduced normal cells. Consequently, this approach would be expected to provide tumour cell selectivity. Results of the initial set of proof-of-principle studies conducted *in vitro* in cultured 9L rat gliosarcoma cells[126] and MCF-7 human breast carcinoma cells,[130] and *in vivo* in solid tumour-bearing rodents[125,126] indicated that CYP2B1 gene transfer was a viable experimental cancer chemotherapeutic strategy. It sensitised the tumour cells to the cell-killing activity of cyclophosphamide and ifosfamide.

Cell culture experiments with transduced 9L rat gliosarcoma cells identified CYP2B6, CYP2C8, CYP2C9, CYP2C18-Met385, CYP2C18-Thr385, CYP2C19, and CYP3A4 as suitable genes in oxazaphosphorine-based GDEPT.[131–133] Among these genes, CYP2B6 and CYP3A4 are the most effective in increasing tumour cell sensitivity to the cytotoxic action of cyclophosphamide and ifosfamide, respectively,[131] consistent with the selective, efficient activation of these two prodrugs by CYP2B6 and CYP3A4.[7] Canine CYP2B11 is also able to catalyse cyclophosphamide and ifosfamide activation.[134] In fact, it is a lower Km and catalytically more efficient enzyme than rat CYP2B1 or human CY2B6 in activating these two produgs, as determined by enzyme kinetic analysis in cultured 9L gliosarcoma cells expressing

CYP2B1, CYP2B6 or CYP2B11.[135] A lower Km prodrug activating enzyme such as CYP2B11 provides an advantage in a GDEPT approach because lower dosages/concentrations are sufficient to elicit therapeutic activity. Cyclophosphamide concentrations of less than 100 μM are able to product cytotoxic action in cultured tumour cells transduced with CYP2B11.[135] The therapeutic outcome, as assessed by the length of the tumour-free period, is better with CYP2B11 gene therapy than with CYP2B6 gene therapy, as demonstrated in a *scid* mouse tumour xenograft model that used a metronomic 6-day repeating treatment schedule with cyclophosphamide.[135]

Oxazaphosphorine prodrug activation in cytochrome P450-expressing tumour cells is accompanied by a phenomenon known as a "bystander" cytotoxic effect,[126,136] which refers to the diffusion of the activated metabolite(s) from the transduced tumour cells to nearby non-transduced cells in the solid tumour mass in a process involving phagocytosis of the apoptotic vesicles.[137] In the case of cyclophosphamide and ifosfamide, neither phosphoramide mustard nor isophosphoramide mustard is likely to be the metabolite that diffuses to the nearby cells because they do not readily cross cell membranes. Candidate metabolites that might be responsible for the "bystander" effect include 4-hydroxycyclophosphamide, 4-hydroxyifosfamide, aldophosphamide, and the metabolic cytotoxic byproducts chloroacetaldehyde and acrolein.[124]

NADPH-cytochrome P450 reductase catalyses the obligatory two-electron transfer from NADPH to the cytochrome P450 haemoprotein, a rate-limiting step in cytochrome P450-mediated catalysis.[138] The co-expression of NADPH-cytochrome P450 reductase potentiates the *in vitro* cytotoxic action of cyclophosphamide in 9L rat gliosarcoma cells expressing CYP2B1,[139] CYP2B6,[140] CYP2B11[135] or CYP2C18.[131] Improvement in anti-tumour activity is demonstrated *in vivo* in rodents inoculated with 9L cells expressing NADPH-cytochrome P450 reductase in combination with each of these cytochrome P450 enzymes.[131,135,139] However, the potentiation by NADPH-cytochrome P450 reductase may be tumour cell type-specific. As shown in a cell culture study, the co-transfection of NADPH-cytochrome P450 reductase and CYP2B1 decreases CYP2B1 catalytic activity and attenuates cell viability in human embryonic kidney 293 cells.[141] It also decreases CYP2B1 catalytic activity in the Panc-1 and Pan02 human pancreatic cells and the T47D human breast cancer cells, whereas it increases the catalytic activity in the Hs578T human breast cancer cells.[141]

Animal studies on GDEPT with cyclophosphamide are performed by administering the prodrug mainly *via* intraperitoneal injection,[126] but also by intracarotid,[142] intracerebral,[125] or intratumoural injection.[143] The last route of administration would be preferable because it would result in a higher drug concentration at the site of cytotoxic action. A biodegradable polymer designed for the slow release of cyclophosphamide has been developed[144] as a way to achieve a better drug concentration-time profile following intratumoural injection. Pharmacokinetic analysis of the tumour tissues indicated that the intratumoural, controlled release of cyclophosphamide resulted in prolonged drug levels, a higher maximal drug concentration, and a larger area under the tumour 4-hydroxycyclophosphamide/aldophosphamide concentration-time curve (AUC) in mice injected with an oncolytic viral vector named rRp450, which consisted of a functionally altered HSV1 virus and the CYP2B1 gene. The AUC was almost 270-fold larger with the cyclophosphamide released into the tumour from the polymer when compared with the AUC following intraperitoneal drug administration.[144] These preclinical data provide an impetus for the development of innovative intratumoural prodrug delivery systems and their applications into GDEPT.

15.4.2 Clinical Trials of Oxazaphosphorine Anticancer Prodrugs and Cytochrome P450-Based GDEPT

A non-randomized, open-label, prospective phase I/II clinical trial was conducted to assess the safety and efficacy of CYP2B1 gene therapy.[145,146] Microencapsulated L293 cells expressing CYP2B1 were administered by intra-arterial instillation to 14 patients with advanced-stage pancreatic carcinoma who were not candidates for curative surgery. Low dose intravenous ifosfamide therapy was initiated two days later. There was no occurrence of allergic complications or pancreatitis in any of the subjects. The therapeutic outcome appears to be promising; the mean survival time in this group of subjects was 44 weeks, whereas it was 22 weeks in a historical control group at the same clinical centre.

MetXia-P450 is a commercial replication-deficient retroviral vector designed to encode the human CYP2B6 gene.[147] A phase I clinical study investigated the safety of the retroviral vector and the efficiency of gene transfer in a total of 12 subjects who were either advanced breast cancer patients or melanoma patients not eligible for systemic chemotherapy.[148] The vector was injected intratumourally and this was followed by low-dose oral cyclophosphamide therapy. Gene transfer, as assessed by immunochemical detection of β-galactosidase, was evident in the majority of the biopsy samples, although the extent of transduction was apparent in $<1\%$ of cells. There were no serious adverse effects attributed to the retroviral vector. An intriguing possibility arising from this study is that this therapeutic approach may elicit an antitumour immune response. The results of these phase I/II studies provide a basis for additional clinical studies of cytochrome P450-based GDEPT.

15.4.3 Experimental Anticancer Prodrug in Combination with a Cytochrome P450 Gene

15.4.3.1 AQ4N in Combination with the Human CYP3A4, CYP2B6 or CYP1A1 Gene

Tumour hypoxia is one of the factors that limit the effectiveness of conventional cancer chemotherapy. As a result, some of the drug development effort has been directed toward a class of agents known as hypoxia-activated prodrugs, such as AQ4N. It is bioreduced by human hepatic microsomal cytochrome P450 enzymes, and CYP3A appears to contribute to AQ4N bioreduction.[42] Treatment of CYP3A4-transfected RIF-1 murine fibrosarcoma cells under anoxic conditions increases cell kill, as assessed by an *in vitro* clonogenic assay.[149] However, the co-transfection of CYP3A4 and NADPH-cytochrome P450 reductase does not further enhance the cytotoxic action of AQ4N. *In vivo* experiments indicate that CYP3A4 gene therapy confers greater antitumour activity in tumour-bearing mice exposed to AQ4N and irradiation.[149] The effect of AQ4N is similarly enhanced by GDEPT strategies involving CYP2B6[150] and CYP1A1.[151]

15.4.4 Other Potential Drug/Gene Combinations in Cytochrome P450-Based GDEPT

The cytochrome P450-based GDEPT studies performed to date have focussed primarily on the oxazaphosphorine alkylating anticancer prodrugs, cyclophosphamide and ifosfamide and, to a much lesser extent, on the experimental agent, AQ4N. GDEPT investigations involving cytochrome P450 genes have not been conducted using other anticancer prodrugs, such as the alkylating agents procarbazine and dacarbazine, or the experimental agents methoxymorpholinyl doxorubicin, tegafur and thalidomide.

Cytochrome P450-based GDEPT may confer therapeutic advantage to another group of anticancer drugs in which one or more of the metabolites have greater pharmacological activity or potency than the parent drug. This group of drugs includes *N,N′,N″*-triethylenethiophosphoramide (thio-TEPA), tamoxifen and flutamide.

15.4.4.1 Thio-TEPA

Thio-TEPA is a DNA-alkylating agent with pharmacological activity against various solid tumours (e.g. breast and ovary).[152] Its clinical utility is primarily as high-dose chemotherapy in conjunction with autologous bone marrow transplantation.[153] Thio-TEPA produces interstrand DNA cross links[154] and cytotoxic activity.[155] It undergoes cytochrome P450-catalysed oxidative desulphuration to form *N,N′,N″*-triethylenephosphoramide (TEPA),[156] which produces DNA lesions,[154] and a byproduct, aziridine, which forms DNA adducts.[152] Experiments with human recombinant cytochrome P450 enzymes indicate that CYP2B6 and CYP3A4, but not CYP2A6 or CYP2E1, catalyse the metabolic conversion of thio-TEPA to TEPA.[157] Consistent with the finding that thio-TEPA is a substrate for CYP2B6, this drug is a mechanism-based inactivator of this enzyme.[158–160] CYP2B6 and CYP3A4 are therefore candidate human cytochrome P450 genes in a GDEPT strategy involving thio-TEPA. Another candidate gene is the rat CYP2B1. Among the various rat CYP2A, CYP2B and CYP2C enzymes, CYP2B1 is the most active catalyst in the biotransformation of thio-TEPA to form TEPA and aziridine.[156]

15.4.4.2 *Tamoxifen*

Tamoxifen is a nonsteroidal triphenylethylene derivative that acts as a selective oestrogen receptor modulator and used therapeutically in the management of breast cancer.[161] The biotransformation of tamoxifen produces various metabolites, including *Z*-4-hydroxytamoxifen,[162,163] which is 30 to 100-fold more potent than the parent drug for the oestrogen receptor.[164,165] CYP2D6 is a major catalyst of tamoxifen 4-hydroxylation.[162] It is relatively more efficient in catalysing the enzymic formation of this metabolite, when compared to CYP2B6, CYP2C9 or CYP2C19.[163] Thus, CYP2D6 is a candidate human cytochrome P450 gene in a GDEPT strategy involving tamoxifen.

15.4.4.3 *Flutamide*

Flutamide is a nonsteroidal antiandrogen that provides a therapeutic option in the management of prostate cancer.[166] CYP1 enzymes metabolise flutamide to form 2-hydroxyflutamide,[167] which has a greater affinity than the parent drug for the androgen receptor.[168] CYP1A2 is 21-fold more efficient than CYP1B1, and 7.5-fold more efficient than CYP1A1 in the 2-hydroxylation of flutamide.[169] Thus, CYP1A2 is a candidate human cytochrome P450 gene in a GDEPT strategy involving flutamide.

15.5 Other Cytochrome P450-Based Experimental Approaches to Cancer Chemotherapy

15.5.1 Novel Cytochrome P450-Activated Prodrugs

The preferential expression of CYP1B1 in tumour tissues relative to normal tissues,[49] at least in certain tumour types,[65] makes this an attractive target for the development of novel CYP1B1-activated anticancer prodrugs. The rationale is that the CYP1B1-catalysed bioactivation of the prodrug occurs predominantly in tumour cells, thereby mitigating the normal cells from the adverse effects that are often associated with conventional cancer chemotherapy. CYP1B1-activated anticancer prodrugs under development include: 1) *trans*-resveratrol, which is metabolised by CYP1B1 to form the cytotoxic piceatannol;[170] 2) aryl oxime prodrugs activated by CYP1B1 to yield nitric oxide as a byproduct, leading to a state of nitrosative stress;[171] 3) prodrug analogues of duocarmycin;[172] and 4) a chalcone derivative, 3,4-methylenedioxy-3′,4′,5′-trimethoxychalcone, which is activated by CYP1B1 to yield a tyrosine kinase inhibitor.[173]

The benzothiazole derivatives represent another series of compounds undergoing development as cytochrome P450-activated prodrugs. Phortress (NSC 710305) is a fluorinated benzothiazole prodrug activated by CYP1A1 (and perhaps also CYP1B1). This series of benzothiazole derivatives are potent agonists of the aryl hydrocarbon receptor and inducers of CYP1 enzymes, such as CYP1A1 and CYP1B1,[174,175] which in turn catalyse the formation of reactive metabolites, resulting in DNA adduct formation and cell cycle arrest.[176]

15.5.2 Cytochrome P450 and Cancer Immunotherapy

The preferential expression of CYP1B1 in tumour cells has led to investigations of this cytochrome P450 protein as a target for cancer immunotherapy.[177–181] CYP1B1 has been shown to be a tumour antigen recognized by cytotoxic T lymphocytes.[178] *In vitro* experiments indicate that CYP1B1-specific cytotoxic T lymphocytes are able to lyse cells from various tumour cell lines, such as the U266 multiple myeloma cells, the 36M ovarian carcinoma cells, the K029 melanoma cells and the IM-9 Epstein-Barr virus-transformed lymphoid cell lines. By comparison, they are not toxic to normal cells, including monocytes and immortalised fibroblasts.[178] HLA-A2/K^b transgenic mice vaccinated with a CYP1B1 plasmid encapsulated in biodegradable microparticles are able to elicit immune responses;[178,179] however, multiple immunisations are required.[179] Results from preclinical studies have led to clinical studies of CYP1B1 as tumour antigen in cancer immunotherapy. In an open-label, fixed-dose, single-arm phase I trial,[180] 17 subjects with cancer (including 14 with metastatic cancer) were vaccinated with ZYC300, a CYP1B1 plasmid DNA encapsulated in biodegradable poly-DL-lactide-coglycolide microparticles. Of all the subjects, six developed immune response to the CYP1B1 vaccine and none of them developed autoimmune toxicity. Development of immune response was associated with a better response to the next therapy and overall survival outcome.

15.5.3 Cytochrome P450-Based Antisense Oligomers

Several classes of anticancer drugs are inactivated by cytochrome P450 enzymes, such as CYP3A. These drugs include the taxanes, such as paclitaxel[106] and docetaxel,[107] and the vinca alkaloids, such as vinblastine.[108] Antisense technology has been employed as an experimental molecular therapeutic strategy to mitigate or prevent the expression of a cytochrome P450 gene, with the aim of delaying anticancer drug inactivation and prolonging drug action. A phosphorodiamidate morpholino antisense oligomer referred to as AVI-4557 targets the CYP3A4 gene. This antisense oligomer has been shown to inhibit CYP3A4 expression in Caco-2 cells transfected with CYP3A4 cDNA and enhance the cytotoxicity of paclitaxel.[182]

15.5.4 Small Molecule Inhibitors of Cytochrome P450

Considerable effort has been directed towards the development of small molecules to inhibit the catalytic activity of various cytochrome P450 enzymes, mainly those involved in the biosynthesis and degradation of endogenous substances. The experimental therapeutic strategies involving small molecule inhibitors include: 1) inhibition of CYP17 to suppress androgen biosynthesis in the treatment of prostate cancer;[183] 2) inhibition of CYP19 to suppress oestrogen biosynthesis in the treatment of breast cancer;[184] 3) inhibition of CYP26A to attenuate the metabolism of retinoic acid in the management of various types of cancer, including acute promyelocytic leukaemia;[185] and 4) inhibition of CYP24 and CYP27B1 to prevent the inactivation of vitamin D, which has therapeutic activity against certain types of tumours (e.g. prostate tumours).[186]

15.6 Conclusions

Cytochrome P450 enzymes, particularly those in the CYP1, CYP2, and CYP3 families, play an important role in the bioactivation and inactivation of numerous anticancer drugs. Many of these enzymes have been detected in the various types of tumour. Tumoural cytochrome P450 expression may have therapeutic consequences. It may compromise the efficacy of those anticancer drugs (e.g. docetaxel) that undergo cytochrome P450-catalysed inactivation. There is evidence indicating tumour-selective expression of cytochrome P450 (e.g. CYP1B1), at least in some tumour types. This represents an opportunity for the development of novel anticancer prodrugs designed specifically to undergo CYP1B1-catalysed bioactivation. With the application of gene transfer technologies, data from preclinical studies and initial phase I/II clinical investigations indicate that cytochrome P450-based GDEPT is a viable strategy to improve the efficacy and mitigate the host toxicity of anticancer prodrugs, including those (e.g. cyclophosphamide and ifosfamide) that are used clinically as first-line cancer chemotherapy. Other therapeutic strategies, such as cytochrome P450-based cancer immunotherapy and antisense oligomers, are still in their infancy. Overall, there is optimism for major advances in cytochrome P450-based experimental cancer therapeutics that will lead to innovative pharmacotherapeutic paradigms in clinical oncology.

Acknowledgements

This work was supported by the Canadian Institutes of Health Research and the Michael Smith Foundation for Health Research.

References

1. D.W. Nebert and D.W. Russell, *Lancet*, 2002, **360**, 1155.
2. A.Y.H. Lu, R.W. Wang and J.H. Lin, *Drug Metab. Dispos.*, 2003, **31**, 345.
3. P.A. Philip, S. Ali-Sadat, J. Doehmer, T.A. Kocarek, A. Akhtar, H. Lu and K.K. Chan, *Cancer Chemother. Pharmacol.*, 1999, **43**, 59.
4. H. Lu and D.J. Waxman, *Mol. Pharmacol.*, 2005, **67**, 212.

5. J. Zhang, Q. Tian, S.Y. Chan, S.C. Li, S. Zhou, W. Duan and Y.Z. Zhu, *Drug Metab. Rev.*, 2005, **37**, 611.
6. G.A. LeBlanc and D.J. Waxman, *Drug Metab. Rev.*, 1989, **20**, 395.
7. T.K.H. Chang, G.F. Weber, C.L. Crespi and D.J. Waxman, *Cancer Res.*, 1993, **53**, 5629.
8. M.E. de Jonge, A.D.R. Huitema, S. Rodenhuis and J.H. Beijnen, *Clin. Pharmacokinet.*, 2005, **44**, 1135.
9. L.H. Fraiser, S. Kanekal and J.P. Kehrer, *Drugs*, 1991, **42**, 781.
10. S. Ren, J.S. Yang, T.F. Kalhorn and J.T. Slattery, *Cancer Res.*, 1997, **57**, 4229.
11. H.J. Xie, U. Yasar, S. Lundgren, L. Griskevicius, Y. Terelius, M. Hassan and A. Rane, *Pharmacogenomics J.*, 2003, **3**, 53.
12. T.K.H. Chang, L. Yu, J.A. Goldstein and D.J. Waxman, *Pharmacogenetics*, 1997, **7**, 211.
13. L. Griskevicius, U. Yasar, M. Sandberg, M. Hidestrand, E. Eliasson, G. Tybring, M. Hassan and M.L. Dahl, *Eur. J. Clin. Pharmacol.*, 2003, **59**, 103.
14. R. Timm, R. Kaiser, J. Lotsch, U. Heider, O. Sezer, K. Weisz, M. Montemurro, I. Roots and I. Cascorbi, *Pharmacogenomics J.*, 2005, **5**, 365.
15. F. Bohnenstengel, U. Hofmann, M. Eichelbaum and H.K. Kroemer, *Eur. J. Clin. Pharmacol.*, 1996, **51**, 297.
16. Z. Huang, P. Roy and D.J. Waxman, *Biochem. Pharmacol.*, 2000, **59**, 961.
17. L.J. Yu, P. Drewes, K. Gustafsson, E.G.C. Brain, J.E.D. Hecht and D.J. Waxman, *J. Pharmacol. Exp. Ther.*, 1999, **288**, 928.
18. M. Furlanut and L. Franceschi, *Oncology*, 2003, **65 (Suppl. 2)**, 2.
19. T. Kerbusch, R.L. Jansen, R.A. Mathot, A.D. Huitema, M. Jansen, R.E.N. van Rijswijk and J.H. Beijnen, *Clin. Pharmacol. Ther.*, 2001, **70**, 132.
20. P. Roy, O. Tretyakov, J. Wright and D.J. Waxman, *Drug Metab. Dispos.*, 1999, **27**, 1309.
21. C.P. Granvil, A. Madan, M. Sharkawi, A. Parkinson and I.W. Wainer, *Drug Metab. Dispos.*, 1999, **27**, 533.
22. E. Chu and A.C. Sartorelli, in: "Basic and Clinical Pharmacology", ed., B.G. Katzung, The McGraw-Hill Companies, Inc., New York, 2007, 878.
23. J.M. Reid, M.J. Kuffel, J.K. Miller, R. Rios and M.M. Ames, *Clin. Cancer Res.*, 1999, **5**, 2192.
24. L. Meer, R.C. Janzer, P. Kleihues and G.F. Kolar, *Biochem. Pharmacol.*, 1986, **35**, 3243.
25. L. Goria-Gatti, A. Iannone, A. Tomasi, G. Poli and E. Albano, *Carcinogenesis*, 1992, **13**, 799.
26. H. Toide, H. Akiyoshi, Y. Minato, H. Okuda and S. Fujii, *Gann*, 1977, **68**, 553.
27. K. Ikeda, K. Yoshisue, E. Matsushima, S. Nagayama, K. Kobayashi, C.A. Tyson, K. Chiba and Y. Kawaguchi, *Clin. Cancer Res.*, 2000, **6**, 4409.

28. T. Komatsu, H. Yamazaki, N. Shimada, M. Nakajima and T. Yokoi, *Drug Metab. Dispos.*, 2000, **28**, 1457.
29. N. Murayama, K. Sai, Y. Nakajima, N. Kaniwa, S. Ozawa, Y. Ohno and J. Sawada, *Jpn. J. Cancer Res.*, 2001, **92**, 524.
30. S. Daigo, Y. Takahashi, M. Fujieda, N. Ariyoshi, H. Yamazaki, W. Koizumi, S. Tanabe, K. Saigenji, S. Nagayama, K. Ikeda and Y. Nishioka, *Pharmacogenetics*, 2002, **12**, 299.
31. E.M. Acton, G.L. Tong, C.W. Mosher and R.L. Wolgemuth, *J. Med. Chem.*, 1984, **27**, 638.
32. K.J.M. Schimmel, D.J. Richel, R.B.A. van den Brink and H.J. Guchelaar, *Cancer Treat. Rev.*, 2004, **30**, 181.
33. W. Lenz, *Am. J. Dis. Child*, 1996, **2**, 99.
34. R.J. D'Amato, M.S. Loughnan, E. Flynn and J. Folkman, *Proc. Natl. Acad. Sci. USA.*, 1994, **91**, 4082.
35. S. Singhal, J. Mehta, R. Desikan, D. Ayers, P. Robertson, P. Eddlemon, N. Munshi, E. Anaissie, C. Wilson, M. Dhodapkar, J. Zeldis and B. Barloqie, *N. Eng. J. Med.*, 1999, **341**, 1565.
36. P. Richardson, T. Hideshima and K. Anderson, *Annu. Rev. Med.*, 2002, **53**, 629.
37. M.C. Cox, W.L. Dahut and W.D. Figg, *Urol. Oncol.*, 2006, **24**, 246.
38. K.S. Bauer, S.C. Dixon and W.D. Figg, *Biochem. Pharmacol.*, 1998, **55**, 1827.
39. D.K. Price, Y. Ando, E.A. Kruger, M. Weiss and W.D. Figg, *Ther. Drug Monit.*, 2002, **24**, 104.
40. Y. Ando, E. Fuse and W.D. Figg, *Clin. Cancer Res.*, 2002, **8**, 1964.
41. M.V. Hejmadi, S.R. McKeown, O.P. Friery, I.A. McIntyre, L.H. Patterson and D.G. Hirst, *Br. J. Cancer*, 1996, **73**, 499.
42. S.M. Raleigh, E. Wanogho, M.D. Burke, S.R. McKeown and J.H. Patterson, *Int. J. Radiat. Oncol. Biol. Phys.*, 1998, **42**, 763.
43. J.H. Patterson and G.I. Murray, *Curr. Pharm. Design*, 2002, **8**, 1335.
44. K. Fujita, *Curr. Drug Metab.*, 2006, **7**, 23.
45. C. Rodriguez-Antona and M. Ingelman-Sundberg, *Oncogene*, 2006, **25**, 1679.
46. I. de Waziers, P.H. Cugnenc, A. Berger, J.P. Leroux and P.H. Beaune, *Carcinogenesis*, 1991, **12**, 905.
47. L. Massaad, I. de Waziers, V. Ribrag, F. Janot, P.H. Beaune, J. Morizet, A. Gouyette and G.G. Chabot, *Cancer Res.*, 1992, **52**, 6567.
48. J.A. McKay, G.I. Murray, R.J. Weaver, S.W.B. Ewen, W.T. Melvin and M.D. Burke, *Gut*, 1993, **34**, 1234.
49. G.I. Murray, M.C. Taylor, M.C.E. McFadyen, J.A. McKay, W.F. Greenlee, M.D. Burke and W.T. Melvin, *Cancer Res.*, 1997, **57**, 3026.
50. T. Yokose, M. Doy, T. Taniguchi, T. Shimada, M. Kakiki, T. Horie, Y. Matsuzaki and K. Mukai, *Virchows. Arch.*, 1999, **434**, 401.
51. P. Gibson, J.H. Gill, P.A. Khan, J.M. Seargent, S.W. Martin, P.A. Batman, J. Griffith, C. Bradley, J.A. Double, M.C. Bibby and P.M. Loadman, *Mol. Cancer Ther.*, 2003, **2**, 527.

52. I. Bergheim, C. Bode and A. Parlesak, *BMC Gastroenterology*, 2005, **5**, 34.
53. H. Chang, J.M. Su, C.C. Huang, L.C. Liu, C.H. Tsai, M.C. Chou and P. Lin, *Cancer Detect. Prev.*, 2005, **29**, 562.
54. J.G. Jiang, C.L. Chen, J.W. Card, S. Yang, J.X. Chen, X.N. Fu, Y.G. Ning, X. Xiao, D.C. Zeldin and D.W. Wang, *Cancer Res.*, 2005, **65**, 4707.
55. M. Kumarakulasingham, P.H. Rooney, S.R. Dundas, C. Telfer, W.T. Melvin, S. Curran and G.I. Murray, *Clin. Cancer Res.*, 2005, **11**, 3758.
56. A.E. Enayetallah, R.A. French and D.F. Grant, *J. Mol. Hist.*, 2006, **37**, 133.
57. M. Karlgren, A. Gomez, K. Stark, J. Svard, C. Rodriguez-Antona, E. Oliw, M.L. Bernal, S.R. Cajal, I. Johansson and M. Ingelman-Sundberg, *Biochem. Biophys. Res. Commun.*, 2006, **341**, 451.
58. C. Toussaint, N. Albin, L. Massaad, D. Grunenwald, G. Parise Jr., J. Morizet, A. Gouyette and G.G. Chabot, *Cancer Res.*, 1993, **53**, 4608.
59. M. Czerwinski, T.L. McLemore, H.V. Gelboin and F.J. Gonzalez, *Cancer Res.*, 1994, **54**, 1085.
60. K.T. Kivisto, A. Linder, G. Friedel, P. Beaune, C. Belloc, H.K. Kroemer and P. Fritz, *Virchows. Arch.*, 1995, **426**, 243.
61. K.T. Kivisto, E.U. Griese, P. Fritz, A. Linder, J. Hakkola, H. Raunio, P. Beaune and H.K. Kroemer, *Arch. Pharmacol.*, 1996, **353**, 207.
62. Z. Huang, M.J. Fasco, S. Spivack and L.S. Kaminsky, *Cancer Res.*, 1997, **57**, 2589.
63. K. Fujitaka, T. Oguri, T. Isobe, Y. Fujiwara and N. Kohno, *Cancer Chemother. Pharmacol.*, 2001, **48**, 42.
64. S.D. Spivack, G.J. Hurteau, M.J. Fasco and L.S. Kaminsky, *Clin. Cancer Res.*, 2003, **9**, 6002.
65. M.C.E. McFadyen, M.E. Cruickshank, I.D. Miller, H.L. McLeod, W.T. Melvin, N.E. Haites, D. Parkin and G.I. Murray, *Br. J. Cancer*, 2001, **85**, 242.
66. D. Downie, M.C.E. McFadyen, P.H. Rooney, M.E. Cruickshank, D.E. Parkin, I.D. Miller, C. Telfer, W.T. Melvin and G.I. Murray, *Clin. Cancer Res.*, 2005, **11**, 7369.
67. G.I. Murray, V.E. Taylor, J.A. McKay, R.J. Weaver, S.W.B. Ewen, W.T. Melvin and M.D. Burke, *Int. J. Exp. Path.*, 1995, **76**, 271.
68. A. Brauers, E. Manegold, R. Buettner, J.M. Baron, H.F. Merk and G. Jakse, *Cancer Detect. Prev.*, 2000, **24**, 356.
69. D.M. Carnell, R.E. Smith, F.M. Daley, P.R. Barber, P.J. Hoskin, G.D. Wilson, G.I. Murray and S.A. Everett, *Int. J. Radiat. Oncol. Biol. Phys.*, 2004, **58**, 500.
70. G.I. Murray, D. Shaw, R.J. Weaver, J.A. McKay, S.W.B. Ewen, W.T. Melvin and M.D. Burke, *Gut*, 1994, **35**, 599.
71. T. Nakajima, R.S. Wang, Y. Nimura, Y.M. Pin, M. He, H. Vainio, N. Murayama, T. Aoyama and F. Iida, *Carcinogenesis*, 1996, **17**, 1477.
72. S.J. Hughes, M.A. Morse, C.M. Weghorst, H. Kim, P.B. Watkins, F.P. Guengerich, M.B. Orringer and D.G. Beer, *Neoplasia*, 1999, **1**, 145.
73. S. Rintala, T.L.J. Tammela and R. Tuimala, *Urol. Res.*, 1998, **26**, 117.

74. Y.L. Cheung, A.C. Kerr, M.C.E. McFadyen, W.T. Melvin and G.I. Murray, *Cancer Lett.*, 1999, **139**, 199.
75. G.I. Murray, M.C.E. McFadyen, R.T. Mitchell, Y.L. Cheung, A.C. Kerr and W.T. Melvin, *Br. J. Cancer*, 1999, **79**, 1836.
76. P. Fritz, E. Behrle, P. Beaune, M. Eichelbaum and H.K. Kroemer, *Histochemistry*, 1993, **99**, 443.
77. G.I. Murray, J.A. McKay, R.J. Weaver, S.W.B. Ewen, W.T. Melvin and M.D. Burke, *J. Pathol.*, 1993, **171**, 49.
78. H. Raunio, R. Juvonen, M. Pasanen, O. Pelkonen, P. Paakko and Y. Soini, *Hepatology*, 1998, **27**, 427.
79. X.B. Man, L. Tang, X.H. Qiu, L.Q. Yang, H.F. Cao, M.C. Wu and H.Y. Wang, *World J. Gastroenterol.*, 2004, **10**, 1565.
80. M. Wu, S. Chen and X. Wu, *Med. Sci. Monit.*, 2006, **12**, BR174–BR178.
81. G.I. Murray, V.E. Taylor, J.A. McKay, R.J. Weaver, S.W.B. Ewen, W.T. Melvin and M.D. Burke, *J. Pathol.*, 1995, **177**, 147.
82. N. Finnstrom, M. Thorn, L. Loof and A. Rane, *Eur. J. Clin. Pharmacol.*, 2001, **57**, 403.
83. T. Tokizane, H. Shiina, M. Igawa, H. Enokida, S. Urakami, T. Kawakami, T. Ogishima, S.T. Okino, L.C. Li, Y. Tanaka, N. Nonomura, A. Okuyama and R. Dahiya, *Clin. Cancer Res.*, 2005, **11**, 5793.
84. G.I. Murray, M.C. Taylor, M.D. Burke and W.T. Melvin, *Br. J. Cancer*, 1998, **77**, 1040.
85. K.L. Zhang, J.X. Ma, X.Y. Chen, Y. Sun, Q.Y. Kong, J. Liu and H. Li, *Oncol. Rep.*, 2004, **12**, 1335.
86. H. Knupfer, M.M. Knupfer, M. Hotfilder and R. Prei, *Oncol. Res.*, 1999, **11**, 523.
87. G.I. Murray, R.J. Weaver, P.J. Paterson, S.W.B. Ewen, W.T. Melvin and M.D. Burke, *J. Pathol.*, 1993, **169**, 347.
88. S. Haas, C. Pierl, V. Harth, B. Pesch, S. Rabstein, T. Bruning, Y. Ko, U. Hamann, J. Justenhoven, H. Brauch and H.P. Fischer, *Int. J. Cancer*, 2006, **119**, 1785.
89. N. Albin, L. Massaad, C. Toussaint, M.C. Mathieu, J. Morizet, O. Parise, A. Gouyette and G.G. Chabot, *Cancer Res.*, 1993, **53**, 3541.
90. R. Schmidt, F. Baumann, H. Knupfer, M. Brauckhoff, L.C. Horn, M. Schonfelder, U. Kohler and R. Preiss, *Br. J. Cancer*, 2004, **90**, 911.
91. T. Oyama, M. Morita, T. Isse, N. Kagawa, S. Nakata, T. So, M. Mizukami, Y. Ichiki, K. Ono, M. Sugaya, H. Uramoto, T. Yoshimatsu, T. Hanagiri, K. Sugio, T. Kawamoto and K. Yasumoto, *Front. Biosci.*, 2005, **10**, 1156.
92. H. Knupfer, R. Schmidt, D. Stanitz, M. Brauckhoff, M. Schonfelder and R. Preiss, *Breast*, 2004, **13**, 28.
93. Z. Huang, M.J. Fasco, H.L. Figge, K. Keyomarsi and L.S. Kaminsky, *Drug Metab. Dispos.*, 1996, **24**, 899.
94. B.F. El-Rayes, S. Ali, L.K. Heilbrun, S. Lababidi, D. Bouwman, D. Visscher and P.A. Philip, *Clin. Cancer Res.*, 2003, **9**, 1705.
95. N. Kapucuoglu, T. Coban, H. Raunio, O. Pelkonen, R.J. Edwards, A.R. Boobis and M. Iscan, *Cancer Lett.*, 2003, **202**, 17.

96. K.T. Kivisto, P. Fritz, A. Linder, G. Friedel, P. Beaune and H.K. Kroemer, *Histochemistry*, 1995, **103**, 25.
97. T.K.H. Chang, J. Chen, V. Pillay, J.Y. Ho and S.M. Bandiera, *Toxicol. Sci.*, 2003, **71**, 11.
98. X. Ding and L.S. Kaminsky, *Annu. Rev. Pharmacol. Toxicol.*, 2003, **43**, 149.
99. M.A. Gibbs and N.A. Hosea, *Clin. Pharmacokinet.*, 2003, **42**, 969.
100. I. Bieche, I. Girault, E. Urbain, S. Tozlu and R. Lidereau, *Breast Cancer Res.*, 2004, **6**, R252–R263.
101. S. Wu, C.R. Moomaw, K.B. Tomer, J.R. Falck and D.C. Zeldin, *J. Biol. Chem.*, 1996, **271**, 3460.
102. K.J. Chen, J. Capdevila and R.C. Harris, *Mol. Cell Biol.*, 2001, **21**, 6322.
103. T.K.H. Chang and D.J. Waxman, *Drug Metab. Rev.*, 2006, **38**, 51.
104. H. Dotzlaw, E. Leygue, P. Watson and L.C. Murphy, *Clin. Cancer Res.*, 1999, **5**, 2103.
105. Y. Miki, T. Suzuki, K. Kitada, N. Yabuki, R. Shibuya, T. Moriya, T. Ishida, N. Ohuchi, B. Blumberg and H. Sasano, *Cancer Res.*, 2006, **66**, 535.
106. G. Kumar, S. Ray, T. Walle, Y. Huang, M. Willingham, S. Self and K. Bhalla, *Cancer Chemother. Pharmacol.*, 1995, **36**, 129.
107. M.C.E. McFadyen, H.L. McLeod, F.C. Jackson, W.T. Melvin, J. Doehmer and G.I. Murray, *Biochem. Pharmacol.*, 2001, **62**, 207.
108. D. Yao, S. Ding, B. Burchell, C.R. Wolf and T. Friedberg, *J. Pharmacol. Exp. Ther.*, 2000, **294**, 387.
109. C. Martinez, E. Garcia-Martin, R.M. Pizarro, F.J. Garcia-Gamito and J.A.G. Agundez, *Br. J. Cancer*, 2002, **87**, 681.
110. D.B. Longley and P.G. Johnston, *J. Pathol.*, 2005, **205**, 275.
111. M.M. Gottesman, T. Fojo and S.E. Bates, *Nat. Rev. Cancer*, 2002, **2**, 48.
112. E. Garcia-Martin, R.M. Pizarro, C. Martinez, Y. Gutierrez-Martin, G. Perez, R. Jover and J.A.G. Agundez, *Pharmacogenomics*, 2006, **7**, 575.
113. Y. Miyoshi, A. Ando, Y. Takamura, T. Taguchi, Y. Tamaki and S. Noguchi, *Int. J. Cancer*, 2002, **97**, 129.
114. Y. Miyoshi, T. Taguchi, S.J. Kim, Y. Tamaki and S. Noguchi, *Breast Cancer*, 2005, **12**, 11.
115. F. Marre, G.J. Sanderink, G. de Sousa, C. Gaillard, M. Martinet and R. Rahmani, *Cancer Res.*, 1996, **56**, 1296.
116. I. Royer, B. Monsarrat, M. Sonnier, M. Wright and T. Cresteil, *Cancer Res.*, 1996, **56**, 58.
117. M. Shou, M. Martinet, K.R. Korzekwa, K.W. Krausz, F.J. Gonzalez and H.V. Gelboin, *Pharmacogenetics*, 1998, **8**, 391.
118. F. McCormick, *Nat. Rev. Cancer*, 2001, **1**, 130.
119. M. Ogris, *Anti-Cancer Agents Med. Chem.*, 2006, **6**, 563.
120. S.M. Freeman, K.A. Whartenby, J.L. Freeman, C.N. Abboud and A.J. Marrogi, *Semin. Oncol.*, 1996, **23**, 31.
121. M.J. Mastrangelo, D. Berd, F.E. Nathan and E.C. Lattime, *Semin. Oncol.*, 1996, **23**, 4.

122. T.A. Connors, *Gene Ther.*, 1995, **2**, 702.
123. F.L. Moolten, *Cancer Res.*, 1986, **46**, 5276.
124. P. Roy and D.J. Waxman, *Toxicol. In vitro*, 2006, **20**, 176.
125. M.X. Wei, T. Tamiya, M. Chase, E.J. Bovistsis, T.K.H. Chang, N.W. Kowall, F.H. Hochberg, D.J. Waxman, X.O. Breakefield and E.A. Chiocca, *Hum. Gene Ther.*, 1994, **5**, 969.
126. L. Chen and D.J. Waxman, *Cancer Res.*, 1995, **55**, 581.
127. L. Clarke and D.J. Waxman, *Cancer Res.*, 1989, **49**, 2344.
128. C.S. Chen, Y. Jounaidi and D.J. Waxman, *Drug Metab. Dispos.*, 2005, **33**, 1261.
129. D.J. Waxman and L. Azaroff, *Biochem. J.*, 1992, **281**, 577.
130. L. Chen, D.J. Waxman, D. Chen and D.W. Kufe, *Cancer Res.*, 1996, **56**, 1331.
131. Y. Jounaidi, J.E.D. Hecht and D.J. Waxman, *Cancer Res.*, 1998, **58**, 4391.
132. O. Kan, L. Griffiths, D. Baban, S. Iqball, M. Uden, H. Spearman, J. Slingsby, T. Price, M. Esapa, S. Kingsman, A. Kingsman, A. Slade and S. Naylor, *Cancer Gene Ther.*, 2001, **8**, 473.
133. L.J. Zhou and X.Z. Zhu, *J. Pharmacol. Exp. Ther.*, 2000, **293**, 982.
134. C.S. Chen, J.T. Lin, K.A. Goss, Y. He, J.R. Halpert and D.J. Waxman, *Mol. Pharmacol.*, 2004, **65**, 1278.
135. Y. Jounaidi, C.S. Chen, G.J. Veal and D.J. Waxman, *Mol. Cancer Ther.*, 2006, **5**, 541.
136. M.X. Wei, T. Tamiya, R.J. Rhee, X.O. Breakefield and E.A. Chiocca, *Clin. Cancer Res.*, 1995, **1**, 1171.
137. S.M. Freeman, C.N. Abboud, K.A. Whartenby, C.H. Packman, D.S. Koeplin, F.L. Moolten and G.N. Abraham, *Cancer Res.*, 1993, **53**, 5274.
138. B.S.S. Masters and C.C. Marohnic, *Drug Metab. Rev.*, 2006, **38**, 209.
139. L. Chen, L.J. Yu and D.J. Waxman, *Cancer Res.*, 1997, **57**, 4830.
140. Y. Jounaidi and D.J. Waxman, *Cancer Res.*, 2000, **60**, 3761.
141. J. Lengler, M. Omann, D. Duvier, H. Holzmuller, W. Gregor, B. Salmons, W.H. Gunzberg and M. Renner, *Biochem. Pharmacol.*, 2006, **72**, 893.
142. Y. Manome, P.Y. Wen, L. Chen, T. Tanaka, Y. Dong, M. Yamazoe, A. Hirshowitz, D.W. Kufe and H.A. Fine, *Gene Ther.*, 1996, **3**, 513.
143. M. Aghi, T.C. Chou, K. Suling, X.O. Breakefield and E.A. Chiocca, *Cancer Res.*, 1999, **59**, 3861.
144. T. Ichikawa, W.P. Petros, S.M. Ludeman, J. Fangmeier, F.H. Hochberg, O.M. Colvin and E.A. Chiocca, *Cancer Res.*, 2001, **61**, 864.
145. M. Lohr, A. Hoffmeyer, J. Kroger, M. Freund, J. Hain, A. Holle, P. Karle, W.T. Knofel, S. Liebe, P. Muller, H. Nizze, M. Renner, R.M. Saller, T. Wagner, K. Hauenstein, W.H. Gunzberg and B. Salmons, *Lancet*, 2001, **357**, 1591.
146. M. Lohr, F. Hummel, G. Faulmann, J. Ringel, R. Saller, J. Hain, W.H. Gunzburg and B. Salmons, *Cancer Chemother. Pharmacol.*, 2002, **49(Suppl 1)** S21–S24.
147. S. Hunt, *Curr. Opin. Mol. Ther.*, 2001, **3**, 595.

148. J.P. Braybrooke, A. Slade, G. Deplanque, R. Harrop, S. Madhusudan, M.D. Forster, R. Gibson, A. Makris, D.C. Talbot, J. Steiner, L. White, O. Kan, S. Naylor, M.W. Carroll, S.M. Kingsman and A.L. Harris, *Clin. Cancer Res.*, 2005, **11**, 1512.
149. H.O. McCarthy, A. Yakkundi, V. McErlane, C.M. Hughes, G. Keilty, M. Murray, L.H. Patterson, D.G. Hirst, S.R. McKeown and T. Robson, *Cancer Gene Ther.*, 2003, **10**, 40.
150. V. McErlane, A. Yakkundi, H.O. McCarthy, C.M. Hughes, L.H. Patterson, D.G. Hirst, T. Robson and S.R. McKeown, *J. Gene Med.*, 2005, **7**, 851.
151. A. Yakkundi, V. McErlane, M. Murray, H.O. McCarthy, C. Ward, C.M. Hughes, L.H. Patterson, D.G. Hirst, S.R. McKeown and T. Robson, *Cancer Gene Ther.*, 2006, **13**, 598.
152. M.J. van Maanen, C.J.M. Smeets and J.H. Beijnen, *Cancer Treat. Rev.*, 2000, **26**, 257.
153. S. Kurian, M. Qazilbash, J. Fay, S. Wolff, R. Herzig, G. Hobbs, P. Bunner, R. Weisenborn, M. Aya-Ay, J. Lynch and S. Ericson, *Breast J.*, 2006, **12**, 531.
154. N.A. Cohen, M.J. Egorin, S.W. Snyder, B. Ashar, B.E. Wietharn, S.S. Pan, D.D. Ross and J. Hilton, *Cancer Res.*, 1991, **51**, 4360.
155. B. Miller, T. Tenenholz, M.J. Egorin, G. Sosnovsky, N.U.M. Rao and P. L. Gutierrez, *Cancer Lett.*, 1988, **41**, 157.
156. S. Ng and D.J. Waxman, *Cancer Res.*, 1991, **51**, 2340.
157. P.A. Jacobson, K. Green, A. Birnbaum and R.P. Remmel, *Cancer Chemother. Pharmacol.*, 2002, **49**, 461.
158. J.M. Rae, N.V. Soukhova, D.A. Flockhart and Z. Desta, *Drug Metab. Dispos.*, 2002, **30**, 525.
159. E. Harleton, M. Webster, N.N. Bumpus, U.M. Kent, J.M. Rae and P.F. Hollenberg, *J. Pharmacol. Exp. Ther.*, 2004, **310**, 1011.
160. T. Richter, M. Schwab, M. Eichelbaum and U.M. Zanger, *Biochem. Pharmacol.*, 2005, **69**, 517.
161. V.C. Jordan, *Nat. Rev. Cancer*, 2007, **7**, 46.
162. S.S. Dehal and D. Kupfer, *Cancer Res.*, 1997, **57**, 3402.
163. J.K. Coller, N. Krebsfaenger, K. Klein, K. Endrizzi, R. Wolbold, T. Lang, A. Nussler, P. Neuhaus, U.M. Zanger, M. Eichelbaum and T.E. Murdter, *Br. J. Clin. Pharmacol.*, 2002, **54**, 157.
164. V.C. Jordan, M.M. Colms, L. Rowsby and G. Prestwick, *J. Endocrinol.*, 1977, **75**, 305.
165. C. Fabian, L. Tilzer and L. Sternson, *Biopharm. Drug Dispos.*, 1981, **2**, 381.
166. S.M. Singh, S. Gauthier and F. Labrie, *Curr. Med. Chem.*, 2000, **7**, 211.
167. M.S. Shet, M. McPhaul, C.W. Fisher, N.R. Stallings and R.W. Estabrook, *Drug Metab. Dispos.*, 1997, **25**, 1298.
168. J.J. Morris, L.R. Hughes, A.T. Glen and P.J. Taylor, *J. Med. Chem.*, 1991, **34**, 447.
169. B. Rochat, J.M. Morsman, G.I. Murray, W.D. Figg and H.L. McLeod, *J. Pharmacol. Exp. Ther.*, 2001, **296**, 537.

170. G.A. Potter, L.H. Patterson, E. Wanogho, P.J. Perry, P.C. Butler, T. Ijaz, K.C. Ruparelia, J.H. Lamb, P.B. Farmer, L.A. Stanley and M.D. Burke, *Br. J. Cancer*, 2002, **86**, 774.
171. M.C.E. McFadyen, W.T. Melvin and G.I. Murray, *Mol. Cancer Ther.*, 2004, **3**, 363.
172. M.C.E. McFadyen and G.I. Murray, *Future Oncol.*, 2005, **1**, 259.
173. S. Sale, R.G. Tunstall, K.C. Ruparelia, P.C. Butler, G.A. Potter, W.P. Steward and A. J. Gescher, *Invest. New Drugs*, 2006, **24**, 459.
174. C.D. Hose, M. Hollingshead, E.A. Sausville and A. Monks, *Mol. Cancer Ther.*, 2003, **2**, 1265.
175. E. Brantley, V. Trapani, M.C. Alley, C.D. Hose, T.D. Bradshaw, M.F.G. Stevens, E.A. Sausville and S.F. Stinson, *Drug Metab. Dispos.*, 2004, **32**, 1392.
176. V. Trapani, V. Patel, C.O. Leong, H.P. Ciolino, G.C. Yeh, C. Hose, J.B. Trepel, M.F.G. Stevens, E.A. Sausville and A.I. Loaiza-Perez, *Br. J. Cancer*, 2003, **88**, 599.
177. J.D. Gordan and R.H. Vonderheide, *Cytotherapy*, 2002, **4**, 317.
178. B. Maecker, D.H. Sherr, R.H. Vonderheide, M.S. von Bergwelt-Baildon, N. Hirano, K.S. Anderson, Z. Xia, M.O. Butler, K.W. Wucherpfennig, C. O'sHara, G. Cole, S.S. Kwak, U. Ramstedt, A.J. Tomlinson, R.M. Chicz, L.M. Nadler and J.L. Schultze, *Blood*, 2003, **102**, 3287.
179. T.M. Luby, G. Cole, L. Baker, J.S. Kornher, U. Ramstedt and M.L. Hedley, *Clin. Immunol.*, 2004, **112**, 45.
180. J.G. Gribben, D.P. Ryan, R. Boyajian, R.G. Urban, M.L. Hedley, K. Beach, P. Nealon, U. Matulonis, S. Campos, T.D. Gilligan, P.G. Richardson, B. Marshall, D. Neuberg and L.M. Nadler, *Clin. Cancer Res.*, 2005, **11**, 4430.
181. B. Maecker, M.S. von Bergwelt-Baildon, K.S. Anderson, R.H. Vonderheide, K.C. Anderson, L.M. Nadler and J.L. Schultze, *Clin. Exp. Immunol.*, 2005, **141**, 558.
182. V. Arora, M.L. Cate, C. Ghosh and P.L. Iversen, *Drug Metab. Dispos.*, 2002, **30**, 757.
183. F. Leroux, *Curr. Med. Chem.*, 2005, **12**, 1623.
184. R.W. Brueggemeier, *Expert Opin. Pharmacother.*, 2006, **7**, 1919.
185. V.C.O. Njar, L. Gediya, P. Purushottamachar, P. Chopra, T.S. Vasaitis, A. Khandelwal, J. Mehta, C. Huynh, A. Belosay and J. Patel, *Bioorg. Med. Chem.*, 2006, **14**, 4323.
186. Y.R. Lou, S. Qiao, R. Talonpoika, H. Syvala and P. Tuohimaa, *J. Steroid Biochem. Mol. Biol.*, 2004, **92**, 317.
187. D. Walker, J.P. Flinois, S.C. Monkman, C. Beloc, A.V. Boddy, S. Cholerton, A.K. Daly, M.J. Lind, A.D.J. Pearson, P.H. Beaune and J.R. Idle, *Biochem. Pharmacol.*, 1994, **47**, 1157.
188. A. Rahman, K.R. Korzekwa, J. Grogan, F.J. Gonzalez and J.W. Harris, *Cancer Res.*, 1994, **54**, 5543.
189. T. Cresteil, B. Monsarrat, J. Dubois, M. Sonnier, P. Alvinerie and F. Gueritte, *Drug Metab. Dispos.*, 2002, **30**, 438.

190. T. Cresteil, B. Monsarrat, P. Alvinerie, J.M. Treluyer, I. Vieira and M. Wright, *Cancer Res.*, 1994, **54**, 386.
191. M.V. Relling, J. Nemec, E.G. Schuetz, J.D. Schuetz, F.J. Gonzalez and K.R. Korzekwa, *Mol. Pharmacol.*, 1994, **45**, 352.
192. X.R. Zhou-Pan, E. Seree, X.J. Zhou, M. Placidi, P. Maurel, Y. Barra and R. Rahmani, *Cancer Res.*, 1993, **53**, 5121.
193. J.B. Dennison, P. Kulanthaivel, R.J. Barbuch, J.L. Renbarger, W.J. Ehlhardt and S.D. Hall, *Drug Metab. Dispos.*, 2006, **34**, 1317.
194. X.J. Zhou, X.R. Zhou-Pan, T. Gauthier, M. Placidi, P. Maurel and R. Rahmani, *Biochem. Pharmacol.*, 1993, **45**, 853.
195. D. Beulz-Riche, P. Grude, C. Puozzo, F. Sautel, C. Filaquier, C. Riche and D. Ratanasavanh, *Fundam. Clin. Pharmacol.*, 2005, **19**, 545.
196. M.C. Haaz, L. Rivory, C. Riche, L. Vernillet and J. Rober, *Cancer Res.*, 1998, **58**, 468.
197. A. Santos, S. Zanetta, T. Cresteil, A. Deroussent, F. Pein, E. Raymond, L. Vernillet, M.L. Risse, V. Boige, A. Gouyette and G. Vassal, *Clin. Cancer Res.*, 2000, **6**, 2012.
198. K. Sai, N. Kaniwa, S. Ozawa and J. Sawada, *Drug Metab. Dispos.*, 2001, **29**, 1505.
199. H.K. Crewe, L.M. Notley, R.M. Wunsch, M.S. Lennard and E.M.J. Gillam, *Drug Metab. Dispos.*, 2002, **30**, 869.
200. M. Sharma, D.E. Shubert, M. Sharma, J. Lewis, B.P. McGarrigle, D.P. Bofinger and J.R. Olson, *Chem. Biol. Interact.*, 2003, **146**, 237.
201. H.K. Crewe, S.W. Ellis, M.S. Lennard and G.T. Tucker, *Biochem. Pharmacol.*, 1997, **53**, 171.
202. J.A. White, B. Beckett-Jones, Y.D. Guo, F.J. Dilworth, J. Bonasoro, G. Jones and M. Petkovich, *J. Biol. Chem.*, 1997, **272**, 18538.
203. J. Marill, T. Cresteil, M. Lanotte and G.G. Chabot, *Mol. Pharmacol.*, 2000, **58**, 1341.
204. D. McKillop, A.D. McCormick, A. Millar, G.S. Miles, P.J. Phillips and M. Hutchison, *Xenobiotica*, 2005, **35**, 39.
205. K. Kobayashi, N. Mimura, H. Fujii, H. Minami, Y. Sasaki, N. Shimada and K. Chiba, *Clin. Cancer Res.*, 2000, **6**, 3297.
206. A.D. Lewis, D.H.M. Lau, G.E. Duran, C.R. Wolf and B.I. Sikic, *Cancer Res.*, 1992, **52**, 4379.
207. Y. Ando, D.Y. Price, W.L. Dahut, M.C. Cox, E. Reed and W.D. Figg, *Cancer Biol. Ther.*, 2002, **1**, 669.
208. M. Iscan, T. Klaavuniemi, T. Coban, N. Kapucuoglu, O. Pelkonen and H. Raunio, *Breast Cancer Res. Treat.*, 2001, **70**, 47.
209. L.M. Forrester, G.M. Neal, D.J. Judah, M.J. Glancey and C.R. Wolf, *Proc. Natl. Acad. Sci. USA.*, 1990, **87**, 8306.
210. C. Galant, J.L. Gala, V. Van Den Berge, M. Berliere, E. Haumont and Y. Horsmans, *Pharmacol. Toxicol.*, 2001, **88**, 142.
211. M. Chase, R.Y. Chung and E.A. Chiocca, *Nature Biotech.*, 1998, **16**, 444.

212. P. Karle, P. Muller, R. Renz, R. Jesnowski, R. Saller, K. von Rombs, H. Nizze, S. Liebe, W.H. Gunzburg, B. Salmons and M. Lohr, *Adv. Exp. Med. Biol.*, 1998, **451**, 97.
213. T.M. Pawlik, H. Nakamura, J.T. Mullen, H. Kasuya, S.S. Yoon, S. Chandrasekhar, E.A. Chiocca and K.K. Tanabe, *Cancer*, 2002, **95**, 1171.
214. Y. Jounaidi and D.J. Waxman, *Cancer Res.*, 2004, **64**, 292.
215. Y. Jounaidi and D.J. Waxman, *Cancer Res.*, 2001, **61**, 4437.
216. P.S. Schwartz, C.S. Chen and D.J. Waxman, *Cancer Res.*, 2002, **62**, 6928.
217. M. Tychopoulos, L. Corcos, P. Genne, P. Beaune and I. de Waziers, *Cancer Gene Ther.*, 2005, **12**, 497.
218. D. Zhou, Y. Lu, M.S. Steiner and J.T. Dalton, *Antimicrob. Agents Chemother.*, 2000, **44**, 2659.
219. M. Lohr, P. Muller, P. Karle, J. Stange, S. Mitzner, R. Jesnowski, H. Nizze, B. Nebe, S. Liebe, B. Salmons and W.H. Gunzburg, *Gene Ther.*, 1998, **5**, 1070.
220. P. Muller, R. Jesnowski, P. Karle, R. Renz, R. Saller, H. Stein, K. Puschel, K. von Rombs, H. Nizze, S. Liebe, T. Wagner, W.H. Gunzberg, B. Salmons and M. Lohr, *Ann. N. Y. Acad. Sci.*, 1999, **880**, 337.
221. P. Karle, M. Renner, B. Salmons and W.H. Gunzberg, *Cancer Gene Ther.*, 2001, **8**, 220.
222. S. Samel, M. Keese, A. Lux, R. Jesnowski, R. Prosst, R. Saller, M. Hafner, J. Sturm, S. Post and M. Lohr, *Cancer Gene Ther.*, 2006, **13**, 65.

Subject Index